T0229008

ase note that the previous printing included a CD-ROM attached to the back of the book.

The material is now only available on the companion website:

http://www.elsevierdirect.com/companion.jsp?ISBN= 9780122035906

NUMERICAL METHODS FOR LINEAR CONTROL SYSTEMS

Design and Analysis

BISWA NATH DATTA

Department of Mathematical Sciences
Northern Illinois University
DeKalb, IL 60115

Amsterdam • Boston • Heidelberg • London • New York • Oxford
Paris • San Diego • San Francisco • Singapore • Sydney • Tokyo

ELSEVIER
ACADEMIC
PRESS

Elsevier Academic Press
525 B Street, Suite 1900, San Diego, California 92101-4495, USA
84 Theobald's Road, London WC1X 8RR, UK

This book is printed on acid-free paper. ∞

Library of Congress Cataloging-in-Publication Data
Datta, Biswa Nath
   Numerical methods for linear control systems design and analysis/B.N. Data.
      p. cm.
   Included bibliographical references and index.

      1. Control theory. 2. System analysis. 3. Linear control systems. I. Title.
   QA402.3D368 2003
   629.8'32—dc22
      ISBN: 978-0-12-203590-6   2003058331

British Library Cataloguing in Publication Data
A catalogue record for this book is available from the British Library

   ISBN: 978-0-12-203590-6

For all information on all Academic Press publications
visit our website at www.academicpress.com

Transferred to Digital Printing 2012.

To
> *my wife*
> *Karabi*
> *and*
> *my son, daughter and daughter-in law*
> *Rajarshi, Rakhi, and Swati*

CONTENTS

PREFACE

Remarkable progress has been made in both theory and applications of all important areas of control theory. Theory is rich and sophisticated. Some beautiful applications of control theory are presently being made in aerospace, biomedical engineering, industrial engineering, robotics, economics, power systems, etc. Unfortunately, the same assessment of progress does not hold in general for computations in control theory.

Many of the methods described in earlier control and systems theory text books were developed before the computer era and were based on approaches that are not numerically sound. Most of these methods, for example, require reduction of the system matrices to some condensed forms, such as a companion form or the Jordan canonical form, and it is well-known that these forms cannot, in general, be achieved in a numerically stable way.

The situation is, however, changing quite fast. In the last 20 years or so, *numerically viable algorithms have been developed for many of the common linear control problems*. Softwares based on these methods have been developed and are still being built.

Unfortunately, these methods and softwares do not seem to be widely known and easily accessible to broad groups of applied mathematicians, control theorists, and practicing control engineers. They are still largely confined in reprints and preprints (in this context it is noted that a reprint book on *"Numerical Linear Algebra Techniques for Systems and Control"* edited by R.V. Patel, A. Laub, and P. Vandooren containing a large number of important published papers in this area has recently been published by IEEE/CRC Press). The primary reason for the inaccessibility of these algorithms and the softwares, in my opinion, is that an understanding, efficient implementations, and making possible modifications of these methods needed for some applications of special interests, require an *interdisciplinary knowledge of linear algebra, numerical linear algebra, control theory, and computer science*; and such a combined expertise is hard to find.

What is, therefore, needed is a book that makes these algorithms accessible to a wide variety of users, researchers, and students.

For practicing users, it is important that the algorithms are described in a manner that is suitable for easy implementation on a wide range of computers, that important aspects of implementations are discussed, and a clear comparative study of one algorithm over the other for a given problem with respect to efficiency, storage, numerical stability, etc., is presented. The latter will help the users to choose the one most suitable for his or her applications. Furthermore, for the students and researchers, it is important that the mechanism of the development of the algorithms is clearly explained and aspects of perturbation analysis of the problems and round-off error analyses and convergence properties of the algorithms, whenever available, are included in some details.

Of course, all these need to be accomplished requiring a minimal amount of background in the areas mentioned above. This is certainly a difficult and an ambitious task. *But the present book aspires to do that and aims at reaching out to a broad spectrum of audience in a number of disciplines including mathematics, control and systems engineering, and other applications areas such as vibrations, aerospace, space-sciences, and structural and manufacturing engineering.*

The recent book on "Computational Methods for Linear Control Systems" by P. H. Petkov, N.D. Christov, and M. M. Konstantinov also aims to fulfill that need to some extent. The scope of this book is, however, much more limited than that of the present book.

The current book is an outgrowth of lecture notes compiled by the author over several years for a graduate course in *numerical methods in control theory* taught at Northern Illinois University (almost all students of this course have been mathematics students with no prior background in control theory). The book has also been used in several short courses given by the author including the SIAM short course on *Numerical Methods in Control, Signal, and Image Processing*, Seattle, August 15, 1993 and, the short course on *Numerical Methods for Linear Control and Systems* at the *International Conference on Mathematical Theory of Networks and Systems*, St. Louis, 1996. The audience of these short courses had varying backgrounds.

The book covers most important and relevant problems arising in control system design and analysis with a special emphasis on computational aspects. These include:

- Numerical solutions of state equations and frequency response computations
- Controllability, observability, and distance to controllability
- Stability, inertia, robust stability, and distance to instability
- Numerical solutions and conditioning of Lyapunov, Sylvester, and algebraic Riccati equations
- Numerical algorithms for feedback stabilization, eigenvalue and robust eigenvalue assignment and conditioning of the eigenvalue assignment problem

- Numerical algorithms for full-order and reduced-order observer design and Kalman filtering
- Realization and subspace algorithms for model identification
- Algorithms for balanced realization and model reduction
- Large-scale solutions of control problems
- H_2 and H_∞ control

The numerical algorithms described in the book have the following desirable features:

- **Efficiency.** Algorithms are of order $O(n^3)$.
- **Numerical Stability.** Algorithms are either numerically stable or composed of numerically stable computations.
- **State-of-the-art Algorithms.** The state-of-the-art algorithms for all problems have been included.
- **Comparative Study and Recommendations.** Whenever possible, a comparison of various algorithms for the same problem with respect to efficiency, numerical stability, and accuracy has been given and based on this comparative study, recommendation for practicing engineers has been made.
- **Step by Step Explanation.** All algorithms have been explained step by step with illustrative examples illustrating each step of the algorithm.
- **Software and Implementations.** Important selected software for each topic has been included.
- **MATLAB Toolkit.** There exists a MATLAB toolkit called **MATCONTROL**, implementing major algorithms in the book.
- **Algorithms for both Continuous-time and Discrete-time systems.** Algorithms are described both for continuous-time and discrete-time systems.

The discussions on theoretical aspects of control theory have been kept to a minimum, only the relevant facts have been mentioned. However, the importance and applications of the problems have been discussed to an extent to motivate the readers in mathematics and other areas of science and engineering who are not familiar with control problems. *Numerical Linear Algebra techniques needed to understand and implement the algorithms have been developed in the book itself in a concise manner without going into too much details and attempts have been made to make the techniques understandable to the readers who do not have a prior background in numerical linear algebra and numerical analysis.* Of course, people having a background in numerical analysis or numerical algebra and/or control theory will have a definite advantage.

   A *special emphasis has been given to the clear understanding of the distinction between a "bad" algorithm and a "numerically effective" algorithm.*

Some discussions on *large-scale computing in control* have been included too. The research in this area is still in its infancy, but some aspects of current research have been included to give the readers a flavor. There is an urgent need for an expanded research in this area as outlined in the 1988 NSF panel report: **"Future Directions in Control Theory: A Mathematical Perspective."** It is hoped our short coverage in this area will provide enough incentive and motivation to beginning researchers, both from control theory and applied and computational mathematics, to work in the area.

The MATLAB toolkit *MATCONTROL* will help the students and the users understand the merits and drawbacks of one algorithm over the others and possibly help a user to make a right decision in choosing an ideal algorithm for a particular application.

Organization of the Book:
The book has **fifteen** chapters. These fifteen chapters have been organized into **four parts**; each part consisting of several chapters, grouped together (roughly) with a common theme.

Part I. REVIEW OF LINEAR AND NUMERICAL LINEAR ALGEBRA

Part II. CONTROL SYSTEM ANALYSIS

Part III. CONTROL SYSTEMS DESIGN

Part IV. SPECIAL TOPICS

Chapter 15.   Large-scale Matrix Computations in Control: Krylov Subspace
               Methods
Heading: **Intended Audience**

The book can be used as a textbook for an advanced graduate course in control engineering such as *Computational Methods for Control Systems Design and Analysis* and *Computer-aided Control System Design* or for an advanced graduate topic course on *Numerical Linear Algebra Techniques in Control and Systems* in applied mathematics and scientific computing. Far more material than can be covered in one semester has been included, so professors can tailor material to particular courses and develop their own course syllabi out of the book. *Above all, the book is intended to serve as a reference book for practicing engineers and applied scientists, researchers, and graduate students.* The book is also very suitable for **self-study**.

ACKNOWLEDGMENTS

This book would not exist in its present form without the help, guidance, suggestions, and encouragement of many friends and colleagues around the world.

My deepest gratitude is to the following people, who not only read and re-read several Chapters carefully and diligently but actually made valuable contributions. They are: Professor *Kanti Datta* of Indian Institute of Technology, Kharagpur, India; Dr. *Vasile Sima* of Research Institute of Informatics, Bucharest, Romania, and two of my former Ph.D. students, *João Carvalho* (now at Universidade Federal de Rio Grande do Sol, Porto Alegre, Brazil), and *Daniil Sarkissian* (now at Mississippi State University). Drs. Carvalho and Sarkissian also wrote the M-files for the MATLAB-based software **MATCONTROL** that accompanies the book.

The following people read certain chapters carefully and thoroughly, pointed out errors, and gave comments and suggestions for improvements: Professors *Thanos Antoulas* of Rice University; *Mark Arnold* of University of Arkansas; *Daniel Boley* of University of Minnesota; *Ralph Byers* of University of Kansas; *Peter Benner* of Technical University of Berlin, Germany; *James Bunch* of University of California; *Raghu Balakrishnan* of Purdue University; *Steve Campbell* of North Carolina State University; *William Ferng* of the Boeing Company; *Roland Freund* of Lucent Technology; *Floyd Hanson* of University of Illinois, Chicago; *Nicholas Higham* of University of Manchester; *Y.P. Hong* of Northern Illinois University; *C. He* Formerly of University of Kansas; *Imad Jaimoukha* of Imperial College of Science; *James Lam* of University of Hong Kong; *Peter Lancaster* of University of Calgary; *W.-W. Lin* of National Tsing Hua University, Taiwan; *Volker Mehrmann* of Technical University of Berlin, Germany; *Vasilios Papakos* of Imperial College of Science, UK; *Daniel Pierce* of the Boeing Company; *Yitshak Ram* of Louisiana State University; *Lothar Reichel* of Kent State University; *Miloud Sadkane* of France; *Ali Sayed* of University of California at Los Angeles; *J.-G. Sun* of Umeå University, Sweden; *Francoise Tisseuer* of University of Manchester, UK; *Andras Varga* of DLR, Germany and *H. Xu* of University of Kansas.

My appreciation goes to Professors *Amit Bhaya* of Federal University of Rio de Janeiro, Brazil; *Shankar Bhattacharyya* of Texas A&M University; and *Rajnikant*

Patel of Concordia University, Canada. They carefully checked the outline of the book for the appropriateness of its contents and made some constructive suggestions.

My other graduate students, *W. Peng, C. Qi,* and *C. Hetti, and Ruyi Yu* helped me with proofreading of several chapters and writing codes for some MATCONTROL functions.

The present and the past office crew of Mathematical Sciences Department at Northern Illinois University contributed a great deal by typing the many different versions of the book and providing timely and valuable assistance with other aspects of manuscript preparation.

Thank you, *Lisa Allison, Elizabeth Buck, Joy Zaccaria, Liz Mehren,* and *Erika Tavizo.* Liz almost typed the first draft by herself. However, the present version would not exist without the continuous hard work of *Joy and Lisa.* Thanks are also due to *Dr. Eric Behr,* our computer systems manager for his continuous help with several aspects of this book. In particular, the cover design of the book was Eric's idea.

The book would have probably never been completed without the moral support, understanding, and continuous encouragement of my mathematician wife *Karabi.* She would never let me forget that the project had to be finished, no matter how hard it was and how long it took. The other members of my family, who were around me during this project: *Rajarshi,* my son and his wife *Swati*; *Rakhi,* my daughter; *Paula* my niece and her husband *Sidharth,* and *Sudeep,* my nephew; were also very supportive, encouraging, and cheerfully put up with my occasional frustrations.

It is a pleasure to acknowledge the encouragement of my colleagues Professors *Abhijit Gupta* and *Pradip Majumdar* of Northern Illinois University and my friend *Bhaskar Bandyopadhaya* of Rockford.

Thanks are also due to *Tiffany Gasbarrini* and *Michael Foster* of Elsevier Press for their interests in this project. It was a joy to work with them.

Notwithstanding my sincere efforts to minimize the errors in the book, I am sure quite a few of them will still be there.

I shall remain grateful to any reader kind enough to send his or her comments and suggestions or to point out errors in the book. I can be reached by e-mail: *dattab@math.niu.edu*

ABOUT THE AUTHOR

Biswa Nath Datta is a Professor of Mathematical Sciences and a Presidential Research Professor at *Northern Illinois University*. Professor Datta held visiting professorships at *University of Illinois* at Urbana-Champaign, *Pennsylvania State University, University of California*, San Diego, *State University of Campinas*, Brazil, as well as at many other universities and research laboratories around the world, including the *Boeing Company*. He also held short-term distinguished visiting professorships at several universities around the world.

His research interests are interdisciplinary, blending linear and numerical linear algebra with control and systems theory. He was elected to a *Fellow* of IEEE in 2000 for his interdisciplinary contributions. He was elected as an *"Academician"* by the Academy of Nonlinear Sciences in 2002.

Professor Datta is the author of more than 90 interdisciplinary papers and a book entitled *Numerical Linear Algebra and Applications*, published in 1995. He has served in the past or is presently serving on the editorial board of premier journals such as *SIAM J. Matrix Analysis and Applications, Linear Algebra and its Applications (Special Editor), Numerical Linear Algebra with Applications, the Journal of Mathematical Systems, Estimation, and Control, etc.* He is the Founding Editor and the Editor-in-Chief of the annual series—*Applied and Computational Control, Signals, and Circuits*. He has also edited four interdisciplinary books. He is also the co-author of the control engineering software package, entitled *"Control System Professional—Advanced Numerical Methods,"* Wolfram Research Inc., 2003.

He has delivered many invited talks at international conferences and numerous colloquium talks at universities and research laboratories around the world.

Professor Datta served as the *vice-Chair* of the *SIAM Linear Algebra Activity Group*, as the *Chairman* of the committee of the SIAM Prize for the Best Applied Linear Algebra Paper and a member of the Hans Schneider Prize Committee in Linear Algebra.

He also organized and chaired or co-chaired the *AMS—IMS-SIAM Joint Summer Research Conference on Linear Algebra and its Role in Systems Theory* 1984; the *SIAM Conference on Linear Algebra in Signals, Systems, and Control*, 1986, 1988, 1993, 2001; *Mathematical Theory of Networks and Systems (MTNS)*, 1996, and numerous interdisciplinary invited special sessions on control, systems, and signal processing at several linear algebra and control conferences.

LIST OF ALGORITHMS

NOTATIONS AND SYMBOLS

$\mathbb{R}$	field of real numbers
$\mathbb{R}^{m \times n}$	set of all real matrices of order $m \times n$
$\mathbb{C}^{m \times n}$	set of complex matrices of order $m \times n$
$\mathbb{C}$	field of complex numbers
ϵ	belongs to
$> \ (\geq)$	positive definiteness (semi-definiteness)
$Re(\alpha)$	real part of $\alpha \in \mathbb{C}$
$Im(\alpha)$	imaginary part of $\alpha \in \mathbb{C}$
$\delta(t)$	unit impulse
$\Omega(M)$	spectrum of the matrix M
C_M	controllability matrix
O_M	observability matrix
C_G	controllability Grammian
O_G	observability Grammian
$Ker(A), N(A)$	Kernel and nullspace of A
$S^{\perp}$	orthogonal subspace of S
■	end of proof
I_s	Idenity matrix of order s (Default for an $n \times n$ identity matrix is I)
A^T	transpose of A
A^*	complex conjugate transpose of A
A^{-1}	inverse of A
$In(A)$	inertia of A
$\beta(A)$	distance of A to a set of unstable matrices
$\mu(A, B)$	distance of (A, B) to uncontrollability
$\|G\|_{\infty}$	H_{∞} - norm of the stable transfer function $G(s)$
$diag(d_1, .., d_n)$	an $n \times n$ diagonal matrix with $d_1, ..., d_n$ on the diagonal
SVD	singular value decomposition
QR	QR factorization
trace (A)	trace of the matrix A

$\begin{bmatrix} A & B \\ C & D \end{bmatrix}$ state space realization: $C(SI - A)^{-1}B + D$

$G(s)$ transfer function matrix

$\|A\|$ 2-norm of A

$\|A\|_F$ Frobenius norm of A

$\sigma_{\min}(A)$ smallest singular value of A

$\sigma_{\max}(A)$ largest singular value of A

$\simeq$ approximately equal to

$R(A)$ range of A

$\bar{\sigma}(A)$ largest singular value of A

$\sigma_i(A)$ ith singular value of A

$\sum$ diagonal matrix containing singular values

CARE   continuous-time algebraic Riccati equation

DARE   discrete-time algebraic Riccati equation

CHAPTER 1

INTRODUCTION AND OVERVIEW

A linear time-invariant **continuous-time dynamical system in state-space** is described by the matrix differential equations of the form:

$$\dot{x}(t) = Ax(t) + Bu(t); \qquad x(t_0) = x_0, \quad t \geq t_0 \qquad (1.0.1)$$

$$y(t) = Cx(t) + Du(t), \qquad (1.0.2)$$

where A, B, C, and D are real time-invariant $n \times n$ state matrix, $n \times m (m \leq n)$ input matrix, $r \times n (r \leq n)$ output matrix, and $r \times m$ direct transmission matrix, respectively. The vectors u, x, and y are time-dependent vectors referred to as *input, state,* and *output,* respectively. The dot, "$\dot{x}$," denotes ordinary differentiation with respect to t. If $m = 1$, then the matrix B is an $n \times 1$ column vector, and is denoted by b. The control problem dealing with such an input vector is referred to as the *single-input problem* (because u is a scalar in this case). The *single-output* problem is analogously defined.

Similarly, **a linear time-invariant discrete-time dynamical system in state-space** is represented by the vector-matrix difference equations of the form:

$$x(k+1) = Ax(k) + Bu(k); \qquad x(0) = x_0, \quad k \geq 0 \qquad (1.0.3)$$

$$y(k) = Cx(k) + Du(k). \qquad (1.0.4)$$

For notational convenience, the system (1.0.1)–(1.0.2) or its discrete counterpart (1.0.3)–(1.0.4) is sometimes denoted simply by (A, B, C, D). **The matrix D will be assumed to be a zero matrix for most problems in this book.**

The transfer function matrix from u to y for the system (1.0.1)–(1.0.2) is defined as

$$\hat{y}(s) = G(s)\hat{u}(s),$$

where $\hat{u}(s)$ and $\hat{y}(s)$ are the Laplace transforms of $u(t)$ and $y(t)$ with $x(0) = 0$. Thus,

$$G(s) = C(sI - A)^{-1}B + D.$$

Sometimes, the notation

$$\left[\begin{array}{c|c} A & B \\ \hline C & D \end{array} \right] = C(sI - A)^{-1}B + D$$

will be used for simplicity.

The transfer function matrix for a discrete-time system is similarly defined.

This book deals with computational methods for control problems modeled by the systems of the above types; and with numerical analysis aspects associated with these computational methods, such as conditioning of problems, numerical stability of algorithms, accuracy of solutions, etc.

The following major topics, associated with the design and analysis of linear control system have been addressed in the book: (i) Linear and Numerical Linear Algebra, (ii) System Responses, (iii) Controllability and Observability, (iv) Stability and Inertia, (v) Lyapunov, Sylvester and Riccati Equations, (vi) Realization and Identification, (vii) Feedback Stabilization and Eigenvalue Assignment, (viii) State Estimation, (ix) Internal Balancing and Model Reduction, (x) Nearness to Uncontrollability and Instability, (xi) Sensitivity and Conditioning for Eigenvalue Assignment; Lyapunov, Sylvester, and Riccati equations, (xii) H_2 and H_∞ Control, and (xiii) Selected Control Software.

In what follows, we give an overview of each of these topics with references to the Chapter(s) and Sections(s) in which it is dealt with. *For references of the papers cited in these sections, please consult the reference sections of the associated chapters.*

1.1 LINEAR AND NUMERICAL LINEAR ALGEBRA (CHAPTER 2 AND CHAPTERS 3 AND 4)

The linear and numerical linear algebra background needed to understand the computational methods has been done in the book itself in **Chapters 2–4**.

All major aspects of numerical matrix computations including solutions and least-squares solutions of algebraic linear systems, eigenvalue and singular value computations, computations of generalized eigenvalues and eigenvectors, along with the **conditioning** of these problems and **numerically stability** of the algorithms have been covered.

Canonical Forms

A common strategy for numerically solving control problems can be described in the following steps taken in the sequence:

Step 1. The problem is transformed by reducing the matrices A, B, and C to some convenient "*condensed*" forms using transformations that preserve the desirable properties of the problem at hand.

Step 2. The transformed problem is solved by exploiting the structure of the condensed forms of the matrices A, B, and C obtained in Step 1.

Step 3. The solution of the original problem is recovered from the solution of the transformed problem.

Two condensed forms that have been used often in the past in control literature are: the *Jordan Canonical Form* (JCF) and the *Frobenius* (or *Block Companion*) Form (a variation of this is known as the *Luenberger Canonical Form*). Exploitation of rich structures of these forms often makes it much easier to solve a problem and these forms are very convenient for textbook illustrations.

Unfortunately, determination of both these forms might require very ill-conditioned similarity transformations.

Suggestions. Avoid the use of the JCF and companion canonical forms in numerical computations, and use only canonical forms that can be obtained using well-conditioned transforming matrices, such as orthogonal transformations. The Hessenberg form, the controller-Hessenberg and the observer-Hessenberg forms, the real Schur and the generalized real Schur forms, the Hessenberg-triangular form are examples of such canonical forms. These forms can be obtained via orthogonal transformations. The errors in numerical computations involving orthogonal matrix multiplications are not magnified by the process and the sensitivity of a computational problem remains unaffected by the use of orthogonal transformations.

1.2 SYSTEM RESPONSES (CHAPTER 5)

For the continuous-time system $(1.0.1)$–$(1.0.2)$, the dynamical system responses $x(t)$ and $y(t)$ for $t \geq t_0$ can be determined from the following formulas:

$$x(t) = e^{A(t-t_0)} x(t_0) + \int_{t_0}^{t} e^{A(t-s)} Bu(s)\, ds, \qquad (1.2.1)$$

$$y(t) = Cx(t) + Du(t). \qquad (1.2.2)$$

In order to study the behavior of a dynamical system, it is customary to determine the responses of the system due to different inputs. Two most common inputs are the **unit step** function and the **unit impulse**.

Thus, the **unit step response** of a system is the output that occurs when the input is the unit step function (it is assumed that $x(0) = 0$). Similarly, the **unit impulse response** is the output that occurs when the input is the unit impulse.

The **impulse response matrix** of the system $(1.0.1)$ and $(1.0.2)$ is defined by

$$H(t) = Ce^{At}B + D\delta(t),$$

where $\delta(t)$ is the Dirac delta function. The impulse response is the response of the system to a Dirac input $\delta(t)$.

Thus, to obtain different responses, one needs to compute the matrix exponential $e^{At} = I + At + (A^2 t^2/2) + \cdots$ and the integrals involving this matrix. *The computational challenge here is how to determine e^{At} without explicitly computing the matrix powers.* Finding higher powers of a matrix is computationally intensive and is a source of instability for the algorithm that requires such computations.

An obvious way to compute e^A is to use some simple canonical forms of A such as the *JCF* or a *companion form of A*. *It is shown in Chapter 5 by simple examples how such computations can lead to inaccurate results.* Computations using truncated Taylor series might also give erroneous result (see **Example 5.3.3**).

The method of choice here is either *Padé approximation with scaling and squaring* (**Algorithm 5.3.1**) or the *method based on reduction of A to real Schur form* (**Algorithm 5.3.2**).

A method (**Algorithm 5.3.3**) due to Van Loan (1978) for computing an integral involving an matrix exponentials is also described in Section **5.3.5**.

Frequency Response Computations

The frequency response plot for many different values of the frequency ω is important in the study of various important properties of linear systems. The frequency response curves indicate how the magnitude and angle of the sinusoidal steady-state response change as the frequency of the input is changed. For this, the frequency-response matrix $G(j\omega) = C(j\omega I - A)^{-1}B + D(\omega \geq 0)$ needs to be computed. Computing $G(j\omega)$ using the LU decomposition of A would require $O(n^3)$ operations per ω and is, therefore, not practical when this computation has to be done for a large number of values of ω. An efficient and practical method due to Laub (1981), based on reduction of A to a Hessenberg matrix, is presented in **Algorithm 5.5.1**, and short discussions on some other recent methods for efficient computations of the frequency-response matrix is included in **Section 5.5.2**.

1.3 CONTROLLABILITY AND OBSERVABILITY PROBLEMS (CHAPTER 6)

The system (1.0.1) is controllable or, equivalently, the pair (A, B) is controllable, if for any initial state $x(0) = x_0$ and the final state x_f, there exists an input $u(t)$ such that the solution satisfies $x(t_f) = x_f$. Several mathematically equivalent criteria of controllability are stated and proved in **Theorem 6.2.1**. The most well-known of them being Criterion (ii). Unfortunately, *this criterion does not yield to a numerically viable test for controllability (see **Example 6.6.1**).*

Similar remarks hold for other criteria. See **Example 6.6.2** in Chapter 6 which demonstrates the pitfall of the eigenvalue criterion (popularly known as the **Hautus–Popov–Belevich** criterion).

*A numerically viable test of controllability, based on the reduction to (A, B) to the **controller-Hessenberg** form, is given in Section 6.7 (**Staircase Algorithm**).*

Observability is a dual concept to controllability. Thus, all that we have said above about controllability applies equally to observability.

1.4 STABILITY AND INERTIA (CHAPTER 7)

It is well known that the uncontrolled system

$$\dot{x} = Ax(t) \tag{1.4.1}$$

is asymptotically stable if and only if all the eigenvalues of A have negative real parts.

Similarly, the discrete system

$$x(k + 1) = Ax(k) \tag{1.4.2}$$

is asymptotically stable if and only if the eigenvalues of A have moduli less than 1.

The common approaches for determining the stability of a system include (i) finding the characteristic polynomial of A followed by application of the Routh–Hurwitz test in case of continuous-time stability or the Schur–Cohn criterion in case of discrete-time stability (ii) solving and testing the positive definiteness of the solution matrix X of the associated Lyapunov equations:

$$XA + A^{T}X = -M \text{ (for continuous-time stability)} \tag{1.4.3}$$

or

$$X - A^{T}XA = M \text{ (for discrete-time stability).} \tag{1.4.4}$$

Finding the characteristic polynomial of a matrix is potentially a numerically unstable process and, furthermore, the coefficients of the characteristic polynomial can be extremely sensitive to small perturbations (**see Chapter 4**). The Lyapunov equation approach is counterproductive in the sense that the most numerically viable method for solving a Lyapunov equation, namely, the **Schur** method, is based on the reduction of A to a real Schur form, and the latter either explicitly displays the eigenvalues of A or they can be trivially computed.

It is, therefore, commonly believed that the most viable way to *test the stability of a dense system is to compute the eigenvalues of A using the universally used method, called the QR iteration with double shift* (**see Chapter 4 (Section 4.3.3)**).

Having said this, let's note that with explicit computation of eigenvalues, one gets much more than what is needed for determining the stability, and moreover, as just said, the eigenvalues can be extremely ill-conditioned. *An indirect method that neither explicitly solves a Lyapunov equation nor computes the eigenvalues, is stated in Algorithm 7.5.1.* This method was later modified by Datta and Datta (1981). **According to theoretical operations-count, both these methods are about 3–4 times faster than the eigenvalue method and several times faster than the Lyapunov equation method.**

Two important inertia theorems (**Theorem 7.4.1** and **Theorem 7.4.2**) are stated in **Section 7.4**.

1.5   LYAPUNOV, SYLVESTER, AND ALGEBRAIC RICCATI EQUATIONS (CHAPTERS 8 AND 13)

The Lyapunov equations (1.4.3) and (1.4.4) arise in (i) Stability and Robust Stability Analyses (**Chapter 7**), (ii) Model Reduction (**Chapter 14**), (iii) Internal Balancing (**Chapter 14**), and (iv) Determining H_2-norm (**Chapter 7**).

A variation of the Sylvester equation $XA + BX = M$ called the **Sylvester-observer equation**, arises in the design of observer (**Chapter 12**), and it can also be used to solve feedback stabilization and pole-placement problems (**Chapters 10 and 11**).

Solving these equations via reduction of A and/or B to a companion form or the JCF is numerically unreliable, because, as said before, these forms cannot be, in general, obtained in a numerically stable way.

Experience with numerical experiments reveal that *solving Lyapunov equations of order higher than 20 using a companion form of A yields solutions with errors as large as the solutions themselves.* **Example 8.5.1** in Chapter 8 illustrates the danger of solving a Lyapunov equation using the JCFs of A. With A and C as chosen in **Example 8.5.1**, the *solution of* (1.4.3) *via diagonalization of A (available in MATLAB function **lyap2**) is very different from the exact solution.*

The methods of choice are: (i) *The Schur method* (**Section 8.5.2**) *for the Lyapunov equation* and (ii) *The Hessenberg–Schur method* (**Algorithm 8.5.1**) *for the Sylvester equation.*

In several applications, all that is needed is the Cholesky factor L of the symmetric positive definite solution X of a Lyapunov equation, for example, the controllability and observability Grammians of a stable system needed for **balanced realization** in the context of **model reduction (Chapter 14)**.

It is numerically desirable that L is found without explicitly computing the matrix X and without forming the matrix $C^T C$ or BB^T. A method for obtaining such an L due to Hammarling both for the continuous-time and the discrete-time systems are described in **Chapter 8 (Algorithms 8.6.1 and 8.6.2)**.

The continuous-time algebraic Riccati equation (CARE):

$$XA + A^T X - XBR^{-1}B^T X + Q = 0 \qquad (1.5.1)$$

and the discrete-time algebraic Riccati equation (DARE).

$$A^T X A - X - A^T X B (R + B^T X B)^{-1} B^T X A + Q = 0 \qquad (1.5.2)$$

and their variations arise in (i) LQR and LQG Design (**Chapters 10 and 12**), (ii) Kalman Filtering (**Chapter 12**), and (iii) H_∞ Control (**Chapter 10**).

An algebraic Riccati equation may have many solutions. Of special interests, from applications viewpoints, is the unique stabilizing solution. **Numerical methods for computing such a solution are described in Chapter 13.** The stabilizing solution of the CARE may be obtained by constructing a basis of the invariant subspace corresponding to the eigenvalues with negative real parts (stable invariant subspace) of the associated Hamiltonian matrix

$$H = \begin{pmatrix} A & -S \\ -Q & -A^T \end{pmatrix}, \qquad \text{where } S = B R^{-1} B^T.$$

It is natural to construct such a basis by finding the eigendecomposition of H. *However, the eigenvector matrix can be highly ill-conditioned if H has multiple or near multiple eigenvalues.* The difficulty can be overcome by using an ordered real Schur decomposition of H. This gives rise to the **Schur method for the CARE** (Laub 1979). The Schur method for the CARE is described in **Algorithm 13.5.1**. Section 13.5.1 also contains some discussions on the Schur method for the DARE. The Schur method for the DARE is based on finding a stable invariant subspace of the associated symplectic matrix

$$M = \begin{pmatrix} A + S(A^{-1})^T Q & -S(A^{-1})^T \\ (-A^{-1})^T Q & (A^{-1})^T \end{pmatrix}.$$

The Schur methods, however, may not give an accurate solution in case R is nearly singular. This difficulty can be overcome by using an extended matrix pencil. The stabilizing solution of the CARE maybe computed by finding the *ordered generalized Schur decomposition* of this pencil using the QZ iteration algorithm. Such a method is called an *inverse-free generalized Schur method* and is described in **Algorithm 13.5.3** and **Algorithm 13.5.4**, respectively, for the CARE and DARE.

The **matrix sign-function methods** has been developed in **Section 13.5.3**. The matrix sign-function method for the CARE is based on computing the matrix sign-function of the Hamiltonian matrix H (see **Algorithm 13.5.6**). For the DARE (**Algorithm 13.5.7**), the matrix $H' = (P+N)^{-1}(P-N)$, where $P = \begin{pmatrix} A & 0 \\ -Q & I \end{pmatrix}$, and $N = \begin{pmatrix} I & S \\ 0 & A^T \end{pmatrix}$ is computed first and then the matrix sign-function method for the CARE is applied.

The matrix sign-function methods are not stable in general, unless an iterative refinement technique is used.

Any Riccati equation solver should be followed by an iterative refinement method, such as Newton's method. For detailed descriptions of **Newton's methods**, see Chapter 13 (**Section 13.5.4**). Newton's methods for the CARE and DARE are described, respectively, in **Algorithms 13.5.8 and 13.5.10**, while **Algorithms 13.5.9 and 13.5.11**, described **Newton's methods with line search** for the CARE and the DARE, respectively.

In summary, *in case R is robustly nonsingular, the Schur method or the matrix sign function method, followed by Newton's method, is recommended for the CARE. In case R is nearly singular, the inverse-free generalized Schur method* (**Algorithm 13.5.3**) *should be used.*

For the DARE, *the inverse-free generalized Schur method* (**Algorithm 13.5.4**) *is the most general purpose method and is recommended to be used in practice.* Again, the *method should be followed by Newton's iterative refinement technique.*

1.6   REALIZATION AND IDENTIFICATION (CHAPTER 9)

Given a set of a large number of Markov parameters, the problem of determining the system matrices A, B, C, and D from this set, is called a **state-space realization problem**.

There are many realizations corresponding to a given set of Markov parameters and the one of the least possible dimension of A, called a **Minimal Realization** (MR), is of practical interest. *A realization is an MR if and only if it is both controllable and observable* (**Theorem 9.2.1**).

The two MRs are related via a nonsingular transforming matrix (**Theorem 9.2.2**) and the degree of an MR is called the **McMillan degree.**

The existing algorithms for finding an MR are all based on factoring an associated block Hankel matrix of Markov parameters:

$$
M_k = \begin{pmatrix} H_1 & H_2 & \cdots & H_k \\ H_2 & H_3 & \cdots & H_{k+1} \\ \vdots & & & \\ H_k & H_{k+1} & \cdots & H_{2k-1} \end{pmatrix}
$$

(k has to be greater than or equal to the McMillan degree), where $H_i = CA^{i-1}B$ is the ith Markov parameter.

The block Hankel matrix M_k can be highly ill-conditioned and, therefore, care should be taken in obtaining its factorization. The singular value decomposition (SVD) is certainly a numerically viable procedure for such a factorization. *Two SVD-based algorithms* (**Algorithms 9.3.1 and 9.3.2**) are presented in Chapter 9.

The Markov parameters are easily generated from a transfer function matrix in case they are of a discrete-time system; indeed in this case they are just the impulse responses. However, they are not readily available for a continuous-time system.

Thus, it is more of a practical importance to identify the system matrices directly from the input–output sequence.

Two subspace identification algorithms (**Algorithms 9.4.1 and 9.4.2**) *from DeMoor et al. (1999), that do not require explicit computations of Markov parameters, is presented in Section 9.4.*

Also stated in this chapter is a subspace algorithm (**Algorithm 9.4.3**) for *frequency-domain identification.* The frequency-domain identification problem concerns finding the system matrices A, B, C, and D from a given set of measured frequency responses at a set of frequencies (not necessarily distinct).

The subspace methods are numerically stable practical methods for systems identification.

1.7   FEEDBACK STABILIZATION AND EIGENVALUE ASSIGNMENT (CHAPTERS 10 AND 11)

Suppose that the uncontrolled system (1.4.1) is not stable, then it is desirable to make it stable. If the state vector $x(t)$ is measurable, then choosing

$$u(t) = -Kx(t),$$

we obtain the **closed-loop system:**

$$\dot{x}(t) = (A - BK)x(t).$$

Mathematically, the problem is then to find matrix K such that $A - BK$ is stable.

The system (1.0.1) is said to be **stabilizable** if such a K exists.

In many practical instances, just stabilizing a system is not enough. Certain design constraints require that all the eigenvalues be placed in certain specified regions of the complex plane.

This gives rise to the well-known **Eigenvalue Assignment** (EVA) problem or the so-called **pole-placement** problem.

Computing the feedback vector f via controller canonical form or using the well-known Ackermann formula for single-input problem does not yield to a numerically viable algorithm (see **Example 11.2.1**).

Several numerically viable algorithms, based on the reduction of the pair (A, B) to the *controller-Hessenberg form* rather than the *controller-Canonical form*, or to *the real Schur form of A* have been developed in recent years and a few selected algorithms are presented in **Chapter 11**. These include (i) Recursive algorithms (**Algorithms 11.2.1 and 11.3.1**), based on evaluations of some simple recursive relations, (ii) QR and RQ type algorithms (**Algorithms 11.2.2, 11.2.3** and the one described in **Section 11.3.2**), (iii) The Schur algorithm (**Algorithm 11.3.3**) based on reduction of A to a real Schur form, and (iv) The robust EVA algorithm (**Algorithm 11.6.1**).

A parametric algorithm (**Algorithm 11.3.4**) for *Partial eigenvalue assignment (PEVA)* is described in **Section 11.3.4**. Lyapunov and Sylvester equations can also be used for feedback stabilization and EVA. Two Lyapunov based methods for feedback stabilization; one for the continuous-time system and the other for the discrete-time system, have been described in Chapter 10 (**Section 10.2**). A comparative study in tabular forms with respect to the efficiency and numerical stability of different algorithms for EVA is given in Chapter 11 (**Sections 11.7 and 11.8**). *Based on factors, such as ease of implementation, efficiency, and practical aspect of numerical stability, the author's favorites are:* **Algorithm 11.2.1** *for the single-input problem and* **Algorithm 11.3.3** *for the multi-input problem. Also, Algorithm 11.3.1 is extremely easy to use.*

1.8   STATE ESTIMATION (CHAPTER 12)

In many practical situations, the states are not fully accessible, but the designer knows the input $u(t)$ and the output $y(t)$. However, for stabilization and EVA by state feedback, for LQR and LQG design, for Kalman filters, to solve H_∞ state-feedback control problems, and others, the knowledge of the complete state vector $x(t)$ is required. Thus, the unavailable states, somehow, need to be estimated accurately from the knowledge of the matrices A, B, and C and the input and output vectors $u(t)$ and $y(t)$. Mathematically, the **state estimation problem is the problem of finding an estimate $\hat{x}(t)$ of $x(t)$ such that the error vector** $e(t) = x(t) - \hat{x}(t)$ **approaches zero as fast as possible.**

It is shown (**Theorem 12.2.1**) that if the states $x(t)$ of the system (1.0.1)–(1.0.2) are estimated by

$$\dot{\hat{x}}(t) = (A - KC)\,\hat{x}(t) + Ky(t) + Bu(t), \tag{1.8.1}$$

where the matrix K is constructed such that $A - KC$ is a stable matrix, then the error vector $e(t)$ has the property that $e(t) \to 0$ as $t \to \infty$. The observability of the pair (A, C) ensures the existence of such a matrix K.

It is clear from the above result that the state estimation problem can be solved by solving the feedback stabilization or the EVA problem for the pair (A^T, C^T).

An alternative approach for state estimation is via solution of the Sylvester equation $XA - FX = GC$ (see **Theorem 12.3.1**).

Two numerically reliable algorithms (**Algorithms 12.7.1** and **12.7.2**) for the Sylvester-observer equation, both based on the reduction of the pair (A, C) to *controller-Hessenberg forms*, have been described in Chapter 12. Furthermore, necessary and sufficient conditions for the nonsingularity of the solution X of the Sylvester-observer equation have been given in **Theorems 12.6.1** and **12.6.2**.

Optimal State Estimation: The Kalman Filter

The problem of finding the optimal steady-state estimation of the states of a stochastic system is considered in **Section 12.9**. An algorithm (**Algorithm 12.9.1**) for the state estimating using Kalman filter is described and the duality between Kalman filter and the LQR design is discussed.

The Linear Quadratic Gaussian Problem

The linear quadratic Gaussian problem (LQG) deals with optimization of a performance measure for a stochastic system. An algorithm (**Algorithm 12.10.1**) for LQG design is described in **Section 12.10.1**.

1.9   INTERNAL BALANCING AND MODEL REDUCTION (CHAPTER 14)

The model reduction is a procedure for obtaining a reduced-order model that preserves some important properties such as the stability, and is close to the original model, in some sense. One way to obtain such a model is via **internally balanced realization**. A continuous-time stable system given by (A, B, C) is internally balanced if there exists a nonsingular transforming matrix T such that $T^{-1}C_G T^{-T} = T^T O_G T = \Sigma = \mathrm{diag}(\sigma_1, \sigma_2, \cdots, \sigma_d, \sigma_{d+1}, \cdots, \sigma_n)$, where C_G and O_G are, respectively, controllability and observability Grammians. The diagonal entries $\sigma_1, \cdots, \sigma_n$ are called the **Hankel singular values**. Once the system is internally balanced, the reduced-order model can be obtained by deleting the states corresponding to the negligible Hankel singular values. Let $G(s)$ and $G_R(s)$ denote the transfer function matrices, respectively, of the original and the reduced-order models. Then a bound for the error $E = \|G(s) - G_R(s)\|_\infty$ is given in **Theorem 14.4.1**.

An algorithm (**Algorithm 14.2.1**) for constructing a balanced realization, based on the SVD of the matrix $L_o^T L_C$, where L_o and L_C are, respectively, the Cholesky factors of observability and controllability Grammians, is given in Section 14.2. *The difficulty with this method is that the transforming matrix T may be ill-conditioned* (see **Section 14.2.2**). An algorithm, based on the Schur decomposition of the matrix $C_G O_G$, that overcomes this difficulty is the Schur algorithm, **Algorithm 14.4.2**. The Schur algorithm was developed by Safonov and Chiang (1989). *It produces a reduced-order model which has the same error property as the one obtained via internal balancing.*

This chapter also contains several other algorithms for balanced realization and model reduction, including the **Square-root algorithm (Algorithm 14.2.2)** for balanced realization and **Hankel-norm approximation algorithm** for model reduction (**Algorithm 14.5.1**).

1.10   NEARNESS TO UNCONTROLLABILITY AND INSTABILITY (CHAPTERS 6 AND 7) AND ROBUST STABILITY AND STABILITY RADIUS (CHAPTERS 7 AND 10)

1.10.1   Nearness to Uncontrollability and Instability

There are systems which are theoretically perfectly controllable, but may be very close to uncontrollable systems (see the Example in **Section 6.9**).

Thus, what is important in practice is to know when a system is close to an uncontrollable system rather than asking if it is controllable or not.

A measure of distance to uncontrollability, denoted by $\mu(A, B)$, is defined (Paige 1980) as follows:

$$\mu(A, B) = \min\{\| \triangle A, \triangle B \|_2 \text{ such that the system } (A + \triangle A, B + \triangle B)$$
$$\text{is uncontrollable}\}$$

It can be shown (**Theorem 6.9.1**) (Miminis 1981; Eising 1984; (Kenney and Laub 1998) that

$$\mu(A, B) = \min \sigma_n(sI - A, B),$$

where σ_n denotes the smallest singular value. Several algorithms (Miminis 1981; Wicks and DeCarlo 1991; Elsner and He 1991) for computing $\mu(A, B)$ have been developed in the last several years. A Newton-type algorithm (**Algorithm 6.9.1**) due to Elsner and He (1991) and an SVD algorithm due to Wicks and DeCarlo (**Algorithm 6.9.2**) are described in Chapter 6.

Similar remarks hold for the stability of a system. *There are systems which are clearly stable theoretically, but in reality are very close to unstable systems.* A well-known example of such a system is the system with a 20×20 upper bidiagonal matrix A having 10s along the subdiagonal and -1 along the main diagonal. Since the eigenvalues of A are all -1, it is perfectly stable. However, if the (20, 1)th entry is perturbed to $\epsilon = 10^{-18}$ from zero, then one of the eigenvalues becomes positive, making the matrix A unstable.

A measure of the distance to instability is

$$\beta(A) = \min\{\| \triangle A \| \text{ such that } A + \triangle A \text{ is unstable}\}.$$

Again, it can be shown (Van Loan 1985) that

$$\beta(A) = \min_{\omega \in \mathbb{R}} \sigma_{\min}(A - j\omega I).$$

A bisection algorithm (**Algorithm 7.6.1**) due to Byers (1988) for estimating $\beta(A)$ is described in **Chapter 7**.

A bisection algorithm (**Algorithm 7.6.2**) for estimating the distance to a discrete unstable system is also described in this chapter.

1.10.2 Robust Stability and Stability Radius (Chapters 7 and 10)

The robust stability concerns the stability of the perturbed system:

$$\dot{x}(t) = (A + E) x(t),$$

where A is a stable matrix and E is an $n \times n$ perturbation matrix. Two robust stability results **(Theorems 7.7.1 and 7.7.2)** using Lyapunov equations are given in Chapter 7.

The stability radius of the matrix triple (A, B, C) is defined as:

$$r_{\mathbb{F}}(A, B, C) = \inf\{\bar{\sigma}(\triangle) : \triangle \in \mathbb{F}^{m \times r} \text{ and } A + B\triangle C \text{ is unstable}\},$$

where $\bar{\sigma}(M)$, following the notation of Qiu *et al.* (1995), **denotes the largest singular value of** M (i.e., $\bar{\sigma}(M) = \sigma_{\max}(M)$). For real matrices (A, B, C), $r_{\mathbb{R}}(A, B, C)$ is called the **real stability radius** and, for complex matrices (A, B, C), $r_{\mathbb{C}}(A, B, C)$ is called the **complex stability radius.**

The stability radius, thus, determines the magnitude of the smallest perturbation needed to destroy the stability of the system.

"Stability" here is referred to as either continuous-stability (with respect to the left half-plane) or discrete-stability (with respect to the unit circle).

Let $\partial\mathbb{C}_g$ denote the boundary of either the half plane or the unit circle. Let A be stable or discrete-stable.

Formulas for complex and real stability radii are given, respectively, in **Theorems 7.8.1** and **7.8.2**.

Section 10.7 of Chapter 10 deals with the relationship between the complex stability radius and Riccati equation. A **characterization of the complex stability radius** is given in **Theorem 10.7.2** using a parametric Hamiltonian matrix and the connection between complex stability radius and an algebraic Riccati equation is established in **Theorem 10.7.3**.

A simple bisection algorithm (**Algorithm 10.7.1**) for computing the complex stability radius, based on **Theorem 10.7.2**, is then described at the conclusion of this section.

1.11 SENSITIVITY AND CONDITION NUMBERS OF CONTROL PROBLEMS

The sensitivity of a computational problem is determined by its condition number. If the condition number is too large, then the solution is too sensitive to small perturbations and the problem is called an **ill-conditioned** problem.

The ill-conditioning has a direct effect on the accuracy of the solution. *If a problem is ill-conditioned, then even with a numerically stable algorithm, the accuracy of the solution cannot be guaranteed.* Thus, it is important to know if a computational problem is ill- or well-conditioned.

While the condition numbers for major problems in numerical linear algebra have been identified (**Chapter 3**), only a few studies on the sensitivities of computational problems in control have been made so far. The sensitivity study is done by theoretical perturbation analysis.

In this book, we have included perturbation analysis of the **matrix exponential problem, (Section 5.3.2)**, of the **Lyapunov and Sylvester equations (Section 8.3)**, of the **algebraic Riccati equations (Section 13.4)** and of **the state feedback and EVA problems (Sections 11.4 and 11.5)**.

1.12 H_∞-CONTROL (CHAPTER 10)

H_∞-control problems concern stabilizing perturbed versions of the original system with certain constraints on the size of the perturbations. Both state feedback and output feedback versions of H_∞-control have been considered in the literature and are stated in **Chapter 10** of this book. A simplified version of the output feedback H_∞-control problem and a result on the existence of a solution have been stated in **Section 10.6.3** of the chapter.

Solution of H_∞ Control Problems Requires Computation of H_∞-Norm.

Two numerical algorithms for computing H_∞-norm of a stable transfer function matrix: the *bisection algorithm* (**Algorithm 10.6.1**) due to Boyd *et al.* (1989), and the *two-step algorithm* (**Algorithm 10.6.2**) due to Bruinsma *et al.* (1990) are described in **Chapter 10**. Both these algorithms are based on the following well-known result (**Theorem 10.6.1**):

Let $G(s)$ be the transfer function matrix of the system (1.0.1)–(1.0.2) *and let* $\gamma > 0$ *be given, then* $\| G \|_\infty < \gamma$ *if and only if* $\sigma_{max}(D) < \gamma$ *and the matrix M_γ defined by*

$$M_\gamma = \begin{pmatrix} A + BR^{-1}D^\mathsf{T}C & BR^{-1}B^\mathsf{T} \\ -C^\mathsf{T}(I + DR^{-1}D^\mathsf{T})C & -(A + BR^{-1}D^\mathsf{T}C)^\mathsf{T} \end{pmatrix},$$

where $R = \gamma^2 I - D^\mathsf{T}D$, *has no imaginary eigenvalues.*

The implementation of the algorithms require a lower and an upper bound for the H_∞-norm. These bounds can be computed using the **Enns–Glover formula**:

$$\gamma_{lb} = \max\{\sigma_{max}(D), \sigma H_1)\}$$

$$\gamma_{ub} = \sigma_{max}(D) + 2\sum_{i=1}^{n} \sigma H_i,$$

where σ_{H_i} is the ith **Hankel singular value**. The Hankel singular value are the square-roots of the eigenvalues of the matrix $C_G O_G$, where C_G and O_G are, respectively, the controllability and observability Grammian.

1.13   SOFTWARE FOR CONTROL PROBLEMS

There now exist several high-quality numerically reliable softwares for control systems design and analysis. These include, among others:

- MATLAB-based Control Systems Tool Box
- MATHEMATICA-based Control System Professional—Advanced Numerical Methods (**CSP-ANM**)
- Fortran-based SLICOT (A Subroutine Library in Systems and Control Theory)
- MATRIX_X
- The System Identification Toolbox
- MATLAB-based Robust Control Toolbox
- μ-Analysis and Synthesis Toolbox

A MATLAB-based tool-kit, called **MATCONTROL**, is provided with this book.

A feature that distinguishes MATCONTROL and CSP-ANM from the other software is that both these software have implemented more than one (typically several) numerically viable algorithms for any given problem. This feature is specially attractive for **control education** in the classrooms, because, students, researchers, and teachers will have an opportunity to compare one algorithm over the others with respect to efficiency, accuracy, easiness for implementation, etc., without writing routines for each algorithm by themselves.

There also exist some specialized software developed by individuals for special problems. These include **polepack** developed by George Miminis (1991), **robpole** developed by Tits and Yang (1996), **Sylvplace** developed by Varga (2000) for pole placement; **ricpack** developed by Arnold and Laub (1984). for Riccati equations, HTOOLS for H_∞ and H_2 synthesis problems developed by Varga and Ionescu (1999), etc.

A brief description of some of these tool boxes appear in Appendix A. Some of the software packages developed by individuals may be obtained from the authors themselves. An internet search might also be helpful in locating these softwares.

References

For references of the papers cited in this chapter, the readers are referred to the **References** section of each chapter.

REVIEW OF LINEAR AND NUMERICAL LINEAR ALGEBRA

CHAPTER 2

A REVIEW OF SOME BASIC CONCEPTS AND RESULTS FROM THEORETICAL LINEAR ALGEBRA

Topics covered

● Basic Concepts for Theoretical Linear Algebra

2.1 INTRODUCTION

Although a first course in linear algebra is a prerequisite for this book, for the sake of completeness, we establish some notations and quickly review the basic definitions and concepts on matrices and vectors in this chapter. Fundamental results on **vector and matrix norms** are described in some details. These results will be used frequently in the later chapters of the book. *The students can review material of this chapter, as needed.*

2.2 ORTHOGONALITY OF VECTORS AND SUBSPACES

Let $u = (u_1, u_2, \ldots, u_n)^{\mathrm{T}}$ and $v = (v_1, v_2, \ldots, v_n)^{\mathrm{T}}$ be two n-dimensional column vectors. The angle θ between two nonzero vectors u and v is given by

$$\cos(\theta) = \frac{u^* v}{\|u\| \|v\|},$$

where $u^* v = \sum_{i=1}^{n} \bar{u}_i v_i$, is the *inner product* of the vectors u and v. The vectors u and v are **orthogonal** if $\theta = 90°$, that is, if $u^* v = 0$. The symbol $\perp$ is used to denote orthogonality. The set of vectors $\{x_1, x_2, \ldots, x_k\}$ in $\mathbb{C}^n$ are **mutually**

orthogonal if $x_i^* x_j = 0$ for $i \neq j$, and **orthonormal** if $x_i^* x_j = \delta_{ij}$, where δ_{ij} is the Kronecker delta function; that is, $\delta_{ii} = 1$ and $\delta_{ij} = 0$ for $i \neq j$, and "$*$" denotes complex conjugate transpose.

Let S be a nonempty subset of $\mathbb{C}^n$. Then S is called a **subspace** of $\mathbb{C}^n$ if $s_1, s_2 \in S$ implies $c_1 s_1 + c_2 s_2 \in S$, where c_1 and c_2 are arbitrary scalars. That is, S is a subspace if any linear combination of two vectors in S is also in S.

For every subspace there is a unique smallest positive integer r such that every vector in the subspace can be expressed as a linear combination of at most r vectors in the subspace; r is called the **dimension** of the subspace and is denoted by dim[S].

Any set of r linearly independent vectors from S of dim[S] $= r$ forms a **basis** of the subspace.

The **orthogonal complement** of a subspace S is defined by $S^\perp = \{y \in \mathbb{C}^n \mid y^* x = 0$ for all $x \in S\}$.

The set of vectors $\{v_1, v_2, \ldots, v_n\}$ form an **orthonormal basis** of a subspace S if these vectors form a basis of S and are orthonormal.

Two subspaces S_1 and S_2 of $\mathbb{C}^n$ are said to be **orthogonal** if $s_1^* s_2 = 0$ for every $s_1 \in S_1$ and every $s_2 \in S_2$. Two orthogonal subspaces S_1 and S_2 will be denoted by $S_1 \perp S_2$.

2.3 MATRICES

In this section, we state some fundamental concepts and results involving the *eigenvalues and eigenvectors: rank, range, nulspaces*, and the *inverse of a matrix*.

2.3.1 The Characteristic Polynomial, the Eigenvalues, and the Eigenvectors of a Matrix

Let A be an $n \times n$ matrix. Then the polynomial $p_A(\lambda) = \det(\lambda I - A)$ is called the **characteristic polynomial**. The zeros of the characteristic polynomial are called the **eigenvalues** of A. This is equivalent to the following: $\lambda \in \mathbb{C}$ is an eigenvalue of A if and only if there exists a nonzero vector x such that $Ax = \lambda x$.

The vector x is called a **right eigenvector** (or just an **eigenvector**) of A. A nonzero vector y is called a **left eigenvector** if $y^* A = \lambda y^*$ for some $\lambda \in \mathbb{C}$.

If an eigenvalue of A is repeated s times, then it is called a multiple eigenvalue of multiplicity s. If $s = 1$, then the eigenvalue is a **simple eigenvalue**.

Definition 2.3.1. *If* $\lambda_1, \lambda_2, \ldots, \lambda_n$ *are the n eigenvalues of A, then* $\max |\lambda_i|$, $i = 1, \ldots, n$ *is called the* **spectral radius** *of A. It is denoted by* $\rho(A)$.

Invariant Subspaces

A subspace S of $\mathbb{C}^n$ is called the invariant subspace or A-**invariant** if $Ax \in S$ for every $x \in S$.

Clearly, an eigenvector x of A defines a one-dimensional invariant subspace.

An A-invariant subspace $S \subseteq \mathbb{C}^n$ is called a **stable invariant subspace** if the eigenvectors in S correspond to the eigenvalues of A with negative real parts.

The Cayley–Hamilton Theorem

The **Cayley–Hamilton theorem** states that the characteristic polynomial of A is an annihilating polynomial of A. That is, if $p_A(\lambda) = \lambda^n + a_1\lambda^{n-1} + \cdots + a_n I$, then $p_A(A) = A^n + a_1 A^{n-1} + \cdots + a_n I = 0$.

Definition 2.3.2. *An $n \times n$ matrix A having fewer than n linearly independent eigenvectors is called a **defective matrix**.*

Example 2.3.1. The matrix

$$A = \begin{pmatrix} 1 & 2 \\ 0 & 1 \end{pmatrix}$$

is defective. It has only one eigenvector $\begin{pmatrix} 1 \\ 0 \end{pmatrix}$.

2.3.2 Range and Nullspaces

For every $m \times n$ matrix A, there are two important associated subspaces: the **range** of A, denoted by $R(A)$, and the **null space** of A, denoted by $N(A)$, defined as follows:

$$R(A) = \{b \mid b = Ax \text{ for some } x\},$$
$$N(A) = \{x \mid Ax = 0\}.$$

The dimension of $N(A)$ is called the **nullity** of A and is denoted by **null**(A).

2.3.3 Rank of a Matrix

Let A be an $m \times n$ matrix. Then the subspace spanned by the row vectors of A is called the **row space** of A. The subspace spanned by the columns of A is called the **column space** of A. The range of A, $R(A)$, is the same as the column space of A.

The **rank** of a matrix A is the dimension of the column space of A. It is denoted by rank(A).

An $m \times n$ matrix is said to have **full column rank if its columns are linearly independent**. The **full row rank** is similarly defined. A matrix A is said to have **full rank** if it has either full row rank or full column rank. If A does not have full rank, it is called **rank deficient**.

The best way to find the rank of a matrix in a computational setting is via the *singular value decomposition* (SVD) of a matrix (see **Chapter 4**).

2.3.4   The Inverse of a Matrix

An $n \times n$ matrix A is said to be invertible if there exists an $n \times n$ matrix B such that $AB = BA = I$. The inverse of A is denoted by A^{-1}. An invertible matrix A is often called **nonsingular**.

An interesting property of the inverse of the product of two invertible matrices is: $(AB)^{-1} = B^{-1}A^{-1}$.

Theorem 2.3.1.   *For an $n \times n$ matrix A, the following are equivalent:*

- *A is nonsingular.*
- $\det(A)$ *is nonzero.*
- $\operatorname{rank}(A) = n.$
- $N(A) = \{0\}.$
- A^{-1} *exists.*
- *A has linearly independent rows and columns.*
- *The eigenvalues of A are nonzero.*
- *For all x, $Ax = 0$ implies that $x = 0$.*
- *The system $Ax = b$ has a unique solution.*

2.3.5   The Generalized Inverse of a Matrix

Let A^* be the complex conjugate transpose of A; that is, $A^* = \left(\bar{A} \right)^{\mathrm{T}}$.

The (Moore–Penrose) **generalized inverse** of a matrix A, denoted by $A^{\dagger}$, is a unique matrix satisfying the following properties: (i) $AA^{\dagger}A = A$, (ii) $A^{\dagger}AA^{\dagger} = A^{\dagger}$, (iii) $(AA^{\dagger})^* = AA^{\dagger}$, and (iv) $(A^{\dagger}A)^* = A^{\dagger}A$.

Note: If A is square and invertible, then $A^{\dagger} = A^{-1}$.

2.3.6   Similar Matrices

Two matrices A and B are called **similar** if there exists a nonsingular matrix T such that

$$T^{-1}AT = B.$$

An important property of similar matrices: **Two similar matrices have the same eigenvalues**. However, two matrices having the same eigenvalues need not be similar.

2.3.7   Orthogonal Projection

Let S be a subspace of $\mathbb{C}^n$. Then an $n \times n$ matrix P having the properties: (i) $R(P) = S$, (ii) $P^* = P$ (P is **Hermitian**), (iii) $P^2 = P$ (P is **idempotent**) is called the **orthogonal projection** onto S or simply the **projection matrix**. We denote the orthogonal projection P onto S by P_S. **The orthogonal projection onto a subspace is unique.**

Let $V = (v_1, \ldots, v_k)$, where $\{v_1, \ldots, v_k\}$ is an orthonormal basis for a subspace S. Then,

$$P_S = V V^*$$

is the unique orthogonal projection onto S. **Note that V is not unique, but P_S is.**

A Relationship Between P_S and $P_{S\perp}$

If P_S is the orthogonal projection onto S, then $I - P_S$, where I is the identity matrix of the same order as P_S, is the orthogonal projection onto $S^\perp$. It is denoted by $P_S^\perp$.

The Orthogonal Projection onto $R(A)$

It can be shown that if A is $m \times n$ $(m \geq n)$ and has **full rank**, then the orthogonal projection P_A onto $R(A)$ is given by:

$$P_A = A(A^*A)^{-1}A^*.$$

2.4   SOME SPECIAL MATRICES

2.4.1   Diagonal and Triangular Matrices

An $m \times n$ matrix $A = (a_{ij})$ is a **diagonal matrix** if $a_{ij} = 0$ for $i \neq j$. We write $A = \text{diag}(a_{11}, \ldots, a_{ss})$, where $s = \min(m, n)$. An $n \times n$ matrix A is a **block diagonal matrix** if it is a diagonal matrix whose each diagonal entry is a square matrix. It is written as:

$$A = \text{diag}(A_{11}, \ldots, A_{kk}),$$

where each A_{ii} is a square matrix. The sum of the orders of $A_{ii}, i = 1, \ldots, k$ is n.
   An $m \times n$ matrix $A = (a_{ij})$ is an **upper triangular** matrix if $a_{ij} = 0$ for $i > j$.
   The transpose of an upper triangular matrix is **lower triangular**; that is, $A = (a_{ij})$ is lower triangular if $a_{ij} = 0$ for $i < j$.

2.4.2   Unitary (Orthogonal) Matrix

A complex square matrix U is **unitary** if $UU^* = U^*U = I$, where $U^* = \left(\overline{U}\right)^{\mathrm{T}}$. A real square matrix O is **orthogonal** if $OO^{\mathrm{T}} = O^{\mathrm{T}}O = I$. If U is an $n \times k$ matrix such that $U^*U = I_k$, then U is said to be **orthonormal**.
   Orthogonal matrices play a very important role in numerical matrix computations.

The following important properties of orthogonal (unitary) matrices are attractive for numerical computations: (i) The inverse of an orthogonal (unitary) matrix O is just its transpose (conjugate transpose), (ii) The product of two orthogonal (unitary) matrices is an orthogonal (unitary) matrix, (iii) The 2-norm and the Frobenius norm are invariant under multiplication by an orthogonal (unitary) matrix **(See Section 2.6)**, and (iv) The error in multiplying a matrix by an orthogonal matrix is not magnified by the process of numerical matrix multiplication **(See Chapter 3)**.

2.4.3   Permutation Matrix

A nonzero square matrix P is called a **permutation matrix** if there is exactly one nonzero entry in each row and column which is 1 and the rest are all zero.

Effects of Premultiplication and Postmultiplication by a permutation matrix

When a matrix A is premultiplied by a permutation matrix P, the effect is a permutation of the rows of A. Similarly, if A is postmultiplied by a permutation matrix, the effect is a permutation of the columns of A.

Some Important Properties of Permutation Matrices

- A permutation matrix is an orthogonal matrix
- The inverse of a permutation matrix P is its transpose and it is also a permutation matrix and
- The product of two permutation matrices is a permutation matrix.

2.4.4   Hessenberg (Almost Triangular) Matrix

A square matrix A is **upper Hessenberg** if $a_{ij} = 0$ for $i > j + 1$. The transpose of an upper Hessenberg matrix is a lower Hessenberg matrix, that is, a square matrix $A = (a_{ij})$ is a **lower Hessenberg matrix** if $a_{ij} = 0$ for $j > i + 1$. A square matrix A that is both upper and lower Hessenberg is **tridiagonal**.

$$\begin{pmatrix} * & * & & 0 \\ \vdots & & \ddots & \\ * & \vdots & & * \\ * & * & \cdots & * \end{pmatrix} \qquad \begin{pmatrix} * & \cdots & * & * \\ * & \cdots & * & * \\ & \ddots & \vdots & \vdots \\ 0 & & * & * \end{pmatrix}$$

$$\text{Lower Hessenberg} \qquad \text{Upper Hessenberg}$$

An upper Hessenberg matrix $A = (a_{ij})$ is **unreduced** if $a_{i,i-1} \neq 0$ for $i = 2, 3, \ldots, n$.

Similarly, a lower Hessenberg matrix $A = (a_{ij})$ is **unreduced** if $a_{i,i+1} \neq 0$ for $i = 1, 2, \ldots, n - 1$.

2.4.5 Companion Matrix

An unreduced upper Hessenberg matrix of the form

$$C = \begin{pmatrix} 0 & 0 & \cdots & \cdots & c_1 \\ 1 & 0 & \cdots & \cdots & c_2 \\ 0 & 1 & \cdots & \cdots & \\ \vdots & \ddots & \ddots & \ddots & \vdots \\ 0 & 0 & 0 & 1 & c_n \end{pmatrix}$$

is called an **upper companion matrix**. The transpose of an upper companion matrix is a **lower companion matrix**.

The **characteristic polynomial** of the companion matrix C is:

$$\det(\lambda I - C) = \det(\lambda I - C^{\mathrm{T}}) = \lambda^n - c_n \lambda^{n-1} - c_{n-1}\lambda^{n-2} - \cdots - c_2\lambda - c_1.$$

2.4.6 Nonderogatory Matrix

A matrix A is **nonderogatory** if and only if it is similar to a companion matrix of its characteristic polynomial. That is, A is a nonderogatory matrix if and only if there exists a nonsingular matrix T such that $T^{-1}AT$ is a companion matrix.

Remark

- An unreduced Hessenberg matrix is nonderogatory, but the converse is not true.

2.4.7 The Jordan Canonical Form of a Matrix

For an $n \times n$ complex matrix A, there exists a nonsingular matrix T such that

$$T^{-1}AT = J = \mathrm{diag}(J_1, \ldots, J_k),$$

where

$$J_i = \begin{pmatrix} \lambda_i & 1 & & & 0 \\ & \lambda_i & 1 & & \\ & & \ddots & \ddots & \\ 0 & & & \ddots & 1 \\ & & & & \lambda_i \end{pmatrix}$$

is $m_i \times m_i$ and $m_1 + \cdots + m_k = n$.

The matrices J_i are called **Jordan matrices** or **Jordan blocks** and J is called the **Jordan Canonical Form (JCF) of** A. For each $j = 1, 2, \ldots, k$, λ_j is the eigenvalue of A with multiplicity m_j. **The same eigenvalue can appear in more than one block.**

Note: The matrix A is **nonderogatory** if its JCF has only one Jordan block associated with each distinct eigenvalue.

If $T = (t_1, t_2, \ldots, t_{m_1}; t_{m_1+1}, \ldots, t_{m_2}; \ldots, t_n)$.

Then $t_1, \ldots, t_{m_1}$ must satisfy

$$At_1 = \lambda_1 t_1$$

and $At_{i+1} = \lambda_1 t_{i+1} + t_i, i = 1, 2, \ldots, m_1 - 1$.

Similarly, relations hold for the other vectors in T. The vectors t_i are called the **generalized eigenvectors** or **principal vectors** of A.

2.4.8   Positive Definite Matrix

A real symmetric matrix A is **positive definite** (positive semidefinite) if $x^T A x > 0$ (≥ 0) for every nonzero vector x.

Similarly, a complex Hermitian matrix A is positive definite (positive semidefinite) if $x^* A x > 0$ (≥ 0) for every nonzero complex vector x.

A commonly used notation for a symmetric positive definite (positive semidefinite) matrix is $A > 0 (\geq 0)$.

Unless otherwise mentioned, a real symmetric or a complex Hermitian positive definite matrix will be referred to as a **positive definite matrix.**

A symmetric positive definite matrix A admits the **Cholesky factorization** $A = HH^T$, where H is a lower triangular matrix with positive diagonal entries. **The most numerically efficient and stable way to check if a real symmetric matrix is positive definite is to compute its Cholesky factorization and see if the diagonal entries of the Cholesky factor are all positive.** See Chapter 3 (Section 3.4.2) for details.

2.4.9   Block Matrices

A matrix whose each entry is a matrix is called a **block matrix**. A **block diagonal matrix** is a diagonal matrix whose each entry is a matrix. A **block triangular matrix** is similarly defined.

The JCF is an example of a **block diagonal matrix.**

Suppose A is partitioned in the form

$$A = \begin{pmatrix} A_{11} & A_{12} \\ A_{21} & A_{22} \end{pmatrix},$$

then A is nonsingular if and only if $A_S = A_{22} - A_{21} A_{11}^{-1} A_{12}$, called the **Schur-Complement** of A, is nonsingular (assuming that A_{11} is nonsingular) and in this case, the inverse of A is given by:

$$A^{-1} = \begin{pmatrix} A_{11}^{-1} + A_{11}^{-1} A_{12} A_S^{-1} A_{21} A_{11}^{-1} & -A_{11}^{-1} A_{12} A_S^{-1} \\ -A_S^{-1} A_{21} A_{11}^{-1} & A_S^{-1} \end{pmatrix}.$$

2.5 VECTOR AND MATRIX NORMS

2.5.1 Vector Norms

Let

$$x = \begin{pmatrix} x_1 \\ x_2 \\ \vdots \\ x_n \end{pmatrix}$$

be an n-vector in $\mathbb{C}^n$. Then, a vector norm, denoted by the symbol $\|x\|$, is a real-valued **continuous** function of the components $x_1, x_2, \ldots, x_n$ of x, satisfying the following properties:

1. $\|x\| > 0$ for every nonzero x. $\|x\| = 0$ if and only if x is the zero vector.
2. $\|\alpha x\| = |\alpha| \|x\|$ for all x in $\mathbb{C}^n$ and for all scalars α.
3. $\|x + y\| \leq \|x\| + \|y\|$ for all x and y in $\mathbb{C}^n$.

The last property is known as the **Triangle Inequality**.

Note: $\|-x\| = \|x\|$ and $\big| \|x\| - \|y\| \big| \leq \|x - y\|$. It is simple to verify that the following are vector norms.

Some Commonly Used Vector Norms

1. $\|x\|_1 = |x_1| + |x_2| + \cdots + |x_n|$ (sum norm or 1-norm)
2. $\|x\|_2 = \sqrt{x_1^2 + x_2^2 + \cdots + x_n^2}$ (Euclidean norm or 2-norm)
3. $\|x\|_\infty = \max_i |x_i|$ (maximum or ∞-norm)

The above three are special case of the **p-norm** or **Hölder norm** defined by $\|x\|_p = (|x_1|^p + \cdots + |x_n|^p)^{1/p}$ for any $p \geq 1$.
Unless otherwise stated, by $\|x\|$ we will mean $\|x\|_2$.

Example 2.5.1. Let $x = (1, 1, -2)^\mathsf{T}$. Then $\|x\|_1 = 4$, $\|x\|_2 = \sqrt{1^2 + 1^2 + (-2)^2} = \sqrt{6}$, and $\|x\|_\infty = 2$.

An important property of the Hölder norm is the **Hölder inequality**

$$|x^*y| \leq \|x\|_p \|y\|_q, \quad \frac{1}{p} + \frac{1}{q} = 1.$$

A special case of the Hölder inequality is the **Cauchy–Schwartz** inequality: $|x^*y| \leq \|x\|_2\|y\|_2$.

Equivalence Property of the Vector norms

All vector norms are **equivalent** in the sense that there exist positive constants α and β such that $\alpha\|x\|_\mu \leq \|x\|_\nu \leq \beta\|x\|_\mu$, for all x, where μ and ν specify the nature of norms.

For the 2, 1, or ∞ norms, we can compute α and β easily and have the following inequalities:

Theorem 2.5.1. *Let x be in $\mathbb{C}^n$. Then*

1. $\|x\|_2 \leq \|x\|_1 \leq \sqrt{n}\|x\|_2$
2. $\|x\|_\infty \leq \|x\|_2 \leq \sqrt{n}\|x\|_\infty$
3. $\|x\|_\infty \leq \|x\|_1 \leq n\|x\|_\infty$

2.5.2   Matrix Norms

Let A be an $m \times n$ matrix. Then, analogous to the vector norm, we define the matrix norm for $\|A\|$ in $\mathbb{C}^{m \times n}$ with the following properties:

1. $\|A\| \geq 0$; $\|A\| = 0$ only if A is the zero matrix
2. $\|\alpha A\| = |\alpha|\|A\|$ for any scalar α
3. $\|A + B\| \leq \|A\| + \|B\|$, where B is also an $m \times n$ matrix.

Subordinate Matrix Norms

Given a matrix A and a vector norm $\|\cdot\|_p$ on $\mathbb{C}^n$, a nonnegative number defined by:

$$\|A\|_p = \max_{x \neq 0} \frac{\|Ax\|_p}{\|x\|_p}$$

satisfies all the properties of a matrix norm. This norm is called the matrix norm **subordinate** to (or induced by) the p-norm.

A very useful and frequently used property of a subordinate matrix norm $\|A\|_p$ (**we shall sometimes call it the p-norm of a matrix A**) is

$$\|Ax\|_p \leq \|A\|_p \|x\|_p.$$

Two important p-norms of an $m \times n$ matrix are: (i) $\|A\|_1 = \max\limits_{1 \leq j \leq n} \sum_{i=1}^{m} |a_{ij}|$ (**maximum column sum norm**) and (ii) $\|A\|_\infty = \max\limits_{1 \leq i \leq m} \sum_{j=1}^{n} |a_{ij}|$ (**maximum row sum norm**).

The Frobenius Norm

An important matrix norm is the Frobenius norm:

$$\|A\|_F = \left[\sum_{i=1}^{m} \sum_{j=1}^{n} |a_{ij}|^2 \right]^{1/2}.$$

A matrix norm $\|\cdot\|_M$ and a vector norm $\|\cdot\|_v$ are consistent if for all matrices A and vectors x, the following inequality holds:

$$\|Ax\|_v \leq \|A\|_M \|x\|_v.$$

Consistency Property of the Matrix Norm

A matrix norm is consistent if, for any two matrices A and B compatible for matrix multiplication, the following property is satisfied:

$$\|AB\| \leq \|A\| \|B\|.$$

The Frobenius norm and all subordinate norms are consistent.

Notes

1. For the identity matrix I, $\|I\|_F = \sqrt{n}$, whereas $\|I\|_1 = \|I\|_2 = \|I\|_\infty = 1$.
2. $\|A\|_F^2 = \text{trace}(A^*A)$, where trace (A) is defined as the sum of the diagonal entries of A, that is, if $A = (a_{ij})$, then trace $(A) = a_{11} + a_{22} + \cdots + a_{nn}$. The trace of A will, sometimes, be denoted by $\text{Tr}(A)$ or $\text{tr}(A)$.

Equivalence Property of Matrix Norms

As in the case of vector norms, the matrix norms are also related. There exist scalars α and β such that: $\alpha \|A\|_\mu \leq \|A\|_v \leq \beta \|A\|_\mu$. In particular, the following inequalities relating various matrix norms are true and are used very frequently in practice. We state the theorem without proof. For a proof, see Datta (1995, pp. 28–30).

Theorem 2.5.2.   *Let A be* $m \times n$. *Then,*

1. $\frac{1}{\sqrt{n}}\|A\|_\infty \leq \|A\|_2 \leq \sqrt{m}\|A\|_\infty$.
2. $\|A\|_2 \leq \|A\|_F \leq \sqrt{n}\|A\|_2$.
3. $\frac{1}{\sqrt{m}}\|A\|_1 \leq \|A\|_2 \leq \sqrt{n}\|A\|_1$.
4. $\|A\|_2 \leq \sqrt{\|A\|_1\|A\|_\infty}$.

2.6 NORM INVARIANT PROPERTIES UNDER UNITARY MATRIX MULTIPLICATION

We conclude the chapter by listing some very useful norm properties of unitary matrices that are often used in practice.

Theorem 2.6.1.   *Let U be an unitary matrix. Then,*

$$\|U\|_2 = 1.$$

Proof. $\|U\|_2 = \sqrt{\rho(U^*U)} = \sqrt{\rho(I)} = 1$. (Recall that $\rho(A)$ denotes the spectral radius of A.)   ∎

The next two theorems show that **2-norm and the Frobenius norm are invariant under multiplication by a unitary matrix**.

Theorem 2.6.2.   *Let U be an unitary matrix and AU be defined. Then,*

1. $\|AU\|_2 = \|A\|_2$
2. $\|AU\|_F = \|A\|_F$

Proof.

1. $\|AU\|_2 = \sqrt{\rho(U^*A^*AU)} = \sqrt{\rho(A^*A)} = \|A\|_2$ (Note that $U^* = U^{-1}$, and two similar matrices have the same eigenvalues).
2. $\|AU\|_F = \mathrm{trace}(U^*A^*AU) = \mathrm{trace}(A^*A) = \|A\|_F^2$ (Note that the trace of a matrix remains invariant under similarity transformation).

Thus $\|AU\|_F = \|A\|_F$.   ∎

Similarly, if UA is defined, then we have

Theorem 2.6.3.

1. $\|UA\|_2 = \|A\|_2$
2. $\|UA\|_F = \|A\|_F$

Proof.   The proof is similar to Theorem 2.6.2.   ∎

2.7   KRONECKER PRODUCT, KRONECKER SUM, AND VEC OPERATION

Let $A \in \mathbb{C}^{m \times n}$ and $B \in \mathbb{C}^{r \times s}$, then the $mr \times ns$ matrix defined by:

$$A \otimes B = \begin{pmatrix} a_{11}B & a_{12}B & \cdots & a_{1n}B \\ a_{21}B & a_{22}B & \cdots & a_{2n}B \\ \vdots & & & \vdots \\ a_{m1}B & a_{m2}B & \cdots & a_{mn}B \end{pmatrix}$$

is called the **Kronecker product** of A and B.

If A and B are invertible, then $A \otimes B$ is invertible and $(A \otimes B)^{-1} = A^{-1} \otimes B^{-1}$.

The Eigenvalues of the Kronecker Product and Kronecker Sum

Let $\lambda_1, \ldots, \lambda_n$ be the eigenvalues of $A \in \mathbb{C}^{n \times n}$, and $\mu_1, \ldots, \mu_m$ be the eigenvalues of $B \in \mathbb{C}^{m \times m}$. Then it can be shown that the eigenvalues of $A \otimes B$ are the mn numbers $\lambda_i \mu_j$, $i = 1, \ldots, n$; $j = 1, \ldots, m$, and the eigenvalues of $A \oplus B$ are the mn numbers $\lambda_i + \mu_j$, $i = 1, \ldots, n$; $j = 1, \ldots, m$.

Vec Operation

Let $X \in \mathbb{C}^{m \times n}$ and $X = (x_{ij})$.

Then the vector obtained by stacking the columns of X in one vector is denoted by $\text{vec}(X)$:

$$\text{vec}(X) = (x_{11}, \ldots, x_{m1}, x_{12}, \ldots, x_{m2}, \ldots, x_{1n}, \ldots, x_{mn})^{\mathrm{T}}.$$

If $A \in \mathbb{C}^{m \times m}$ and $B \in \mathbb{C}^{n \times n}$, then it can be shown that $\text{vec}(AX + XB) = ((I_n \otimes A) + (B^{\mathrm{T}} \otimes I_m))\text{vec } X$.

The Kronecker products and vec operations are useful in the study of the existence, uniqueness, sensitivity, and numerical solutions of the Lyapunov and Sylvester equations (**see Chapter 8**).

2.8   CHAPTER NOTES AND FURTHER READING

Most of the material in this chapter can be found in standard linear algebra text books. Some such books are cited below.

For further reading of material of Section 2.7, the readers are referred to Horn and Johnson (1985).

References

Anton H. and Rorres C. *Elementary Linear Algebra with Applications*, John Wiley, New York, 1987.

Dattta B.N. *Numerical Linear Algebra and Applications*, Brooks/Cole Publishing Company, Pacific Grove, 1995 (*Custom published by Brooks/Cole, 2003*).

Horn R.A. and Johnson C.R. *Matrix Analysis*, Cambridge University Press, Cambridge, UK, 1985.

Lay D.C. *Linear Algebra and Its Applications*, Addison-Wesley, Reading, MA, 1994.

Leon S.J. *Linear Algebra with Applications*, 4th edn, Macmillan, New York, 1994.

Noble B. and Daniel J.W. *Applied Linear Algebra*, Prentice-Hall, Englewood Cliffs, NJ, 1977.

Strang G. *Linear Algebra and Its Applications*, 3rd edn, Academic Press, New York, 1988.

CHAPTER 3

SOME FUNDAMENTAL TOOLS AND CONCEPTS FROM NUMERICAL LINEAR ALGEBRA

Topics covered

- Floating Point Numbers and Errors in Computations
- *LU* Factorization Using Gaussian Elimination
- *QR* Factorization Using Householder and Givens Matrices
- Numerical Solution of the Algebraic Linear Systems and Least-Squares Problems
- The Singular Value Decomposition (SVD)

3.1 INTRODUCTION

In this chapter, we introduce some fundamental concepts and techniques of numerical linear algebra which, we *believe, are essential for in-depth understanding of computational algorithms for control problems, discussed in this book.* The basic concepts of floating point operations, numerical stability of an algorithm, conditioning of a computational problem, and their effects on the accuracy of a solution obtained by a certain algorithm are introduced first.

Three important matrix factorizations: *LU*, *QR*, and the **singular value decomposition** (SVD), and their applications to solutions of algebraic linear systems, linear least-squares problems, and eigenvalue problems are next described in details.

The method of choice for the linear system problem is the *LU* factorization technique obtained by Gaussian elimination with partial pivoting (Section 3.4). The method of choice for the symmetric positive definite system is the **Cholesky factorization technique (Algorithm 3.4.1).**

The QR factorization of a matrix is introduced in the context of the least-squares solution of a linear system; however, it also forms the core of the **QR iteration technique, which is the method of choice for eigenvalue computation**. The QR iteration technique itself for eigenvalue computation is described in **Chapter 4**. Two numerically stable methods for the QR factorization, namely, **Householder's and Givens' methods are described in Section 3.6**. Householder's method is slightly cheaper than Givens' method for sequential computations, but the latter has computational advantages in parallel computation setting.

The SVD has nowadays become an essential tool for determining the **numerical rank**, the **distance of a matrix from a matrix of immediate lower rank, finding the orthonormal basis and projections**, etc. This important matrix factorization is described in **Section 3.9**. **The SVD is also a reliable tool for computing the least-squares solution to $Ax = b$**.

A reliable and widely used computational technique for computing the SVD of a matrix is described in **Chapter 4**.

3.2 FLOATING POINT NUMBERS AND ERRORS IN COMPUTATIONS

3.2.1 Floating Point Numbers

Most scientific and engineering computations on a computer are performed using **floating point arithmetic**. Computers may have different bases, though base 2 is most common.

A t-digit **floating point** number in base β has the form:

$$x = m \cdot \beta^e,$$

where m is a t-digit fraction called **mantissa** and e is called **exponent**.

If the first digit of the mantissa is different from zero, then the floating point number is called **normalized**. Thus, 0.3457×10^5 is a 4-digit normalized decimal floating number, whereas 0.03475×10^6 is a five-digit unnormalized decimal floating point number.

The number of digits in the mantissa is called **precision**. On many computers, it is possible to manipulate floating point numbers so that a number can be represented with about twice the usual precision. Such a precision is called **double precision**.

Most computers nowadays conform to the IEEE floating point standard (ANSI/IEEE standard 754-1985). For a single-precision, IEEE standard recommends about 24 binary digits and for a double precision, about 53 binary digits. Thus, IEEE standard for **single precision provides approximately 7 decimal digits of accuracy**, since $2^{-23} \cong 1.2 \times 10^{-7}$, and **double precision provides approximately 16 decimal digits of accuracy**, since $2^{-52} \approx 2.2 \times 10^{-16}$.

Note: Although computations with double precision increase accuracy, they require more computer time and storage.

On each computer, there is an allowable range of the exponent e: L, the minimum; U, the maximum. L and U **vary from computer to computer**.

If, during computations, the computer produces a number whose exponent is too large (too small), that is, it is outside the permissible range, then we say that an **overflow** (**underflow**) has occurred.

Overflow is a serious problem; for most systems, the result of an overflow is $\pm\infty$. Underflow is usually considered less serious. On most computers, when an underflow occurs, the computed value is set to zero, and then computations proceed. **Unless otherwise stated, we will use only decimal arithmetic.**

3.2.2   Rounding Errors

If a computed result of a given real number is not machine representable, then there are two ways it can be represented in the machine. Consider the machine representation of the number

$$\pm \cdot d_1 \cdots d_t d_{t+1} \cdots .$$

Then the first method, **chopping**, is the method in which the digits from d_{t+1} on are simply chopped off. The second method is **rounding**, in which the digits d_{t+1} through the rest are not only chopped off, but the digit d_t is also rounded up or down depending on whether $d_{t+1} \geq 5$ or $d_{t+1} < 5$.

We will denote the floating point representation of a real number x by $\mathrm{fl}(x)$.

Example 3.2.1. (*Rounding*)   Let $x = 3.141596$.

$$t = 2: \mathrm{fl}(x) = 3.1,$$
$$t = 3: \mathrm{fl}(x) = 3.14,$$
$$t = 4: \mathrm{fl}(x) = 3.142.$$

A useful measure of error in computation is the **relative error**.

Definition 3.2.1.   *Let $\hat{x}$ denote an approximation of x. Then the relative error is $|\hat{x} - x|/|x|$, $x \neq 0$.*

We now give an expression for the relative error in representing a real number x by its floating point representation $\mathrm{fl}(x)$. Proof of Theorem 3.2.1 can be found in Datta (1995, p. 47).

Theorem 3.2.1. *Let* $\mathrm{fl}(x)$ *denote the floating point representation of a real number* x *in base* β. *Then,*

$$\frac{|\mathrm{fl}(x) - x|}{|x|} \leq \mu = \begin{cases} \frac{1}{2}\beta^{1-t} \text{ for rounding,} \\ \beta^{1-t} \text{ for chopping.} \end{cases} \qquad (3.2.1)$$

Definition 3.2.2. *The number* μ *in the above theorem is called the* **machine precision, computer epsilon,** *or* **unit roundoff error**. *It is the smallest positive floating point number such that*

$$\mathrm{fl}(1 + \mu) > 1.$$

The number μ **is usually of the order** 10^{-16} **and** 10^{-7} *(on most machines) for double and single precisions computations, respectively. For example, for the IBM 360 and 370,* $\beta = 16, t = 6, \mu = 4.77 \times 10^{-7}$.

Definition 3.2.3. *The* **significant digits** *in a number are the number of digits starting with the first nonzero digit.*

For example, the number 1.5211 *has five significant digits, whereas the number* 0.0231 *has only three.*

3.2.3   Laws of Floating Point Arithmetic

The formula (3.2.1) can be written as

$$\mathrm{fl}(x) = x(1 + \delta),$$

where $|\delta| \leq \mu$.

Assuming that the IEEE standard holds, we can easily derive the following simple **laws of floating point arithmetic**.

Theorem 3.2.2. *Laws of Floating Point Arithmetic. Let* x *and* y *be two floating point numbers, and let* $\mathrm{fl}(x + y)$, $\mathrm{fl}(x - y)$, $\mathrm{fl}(xy)$, *and* $\mathrm{fl}(x/y)$ *denote, respectively, the computed sum, difference, product, and quotient. Then,*

1. $\mathrm{fl}(x \pm y) = (x \pm y)(1 + \delta)$, *where* $|\delta| \leq \mu$.
2. $\mathrm{fl}(xy) = (xy)(1 + \delta)$, *where* $|\delta| \leq \mu$.
3. *if* $y \neq 0$, *then* $\mathrm{fl}(x/y) = (x/y)(1 + \delta)$, *where* $|\delta| \leq \mu$.

On computers that do not use the IEEE standard, the following floating point law of addition might hold:

4. $\mathrm{fl}(x + y) = x(1 + \delta_1) + y(1 + \delta_2)$, *where* $|\delta_1| \leq \mu$ *and* $|\delta_2| \leq \mu$.

Example 3.2.2. Let $\beta = 10, t = 4$.

$$x = 0.1112, \; y = 0.2245 \times 10^5,$$
$$xy = 0.24964 \times 10^4,$$
$$\mathrm{fl}(xy) = 0.24960 \times 10^4.$$

Thus, $|\mathrm{fl}(xy) - xy| = 0.4000$ and $|\delta| = 1.7625 \times 10^{-4} < \frac{1}{2} \times 10^{-3}$.

3.2.4 Catastrophic Cancellation

A phenomenon, called **catastrophic cancellation**, occurs when two numbers of approximately the same size are subtracted. Very often significant digits are lost in the process.

Consider the example of computing $f(x) = e^x - 1 - x$ for $x = 0.01$. In five digit arithmetic $a = e^x - 1 = 1.0101 - 1 = 0.0101$. Then the computed value of $f(x) = a - x = 0.0001$, whereas the true answer is 0.000050167.

Note that even though the subtraction was done accurately, the final result was wrong. Indeed, subtractions in most cases can be done exactly, cancellation only signals that the error must have occurred in previous steps. **Fortunately, often cancellation can be avoided by rearranging computations**. For the example under consideration, if e^x were computed using the convergent series $e^x = 1 + x + x^2/2! + x^3/3! + \cdots$, then the result would have been 0.000050167, which is correct up to five significant digits.

For details and examples, see Datta (1995, pp. 43–61). See also Stewart (1998, pp. 136–138) for an illuminating discussion on this topic.

3.3 CONDITIONING, EFFICIENCY, STABILITY, AND ACCURACY

3.3.1 Algorithms and Pseudocodes

Definition 3.3.1. *An **algorithm** is an ordered set of operations, logical and arithmetic, which when applied to a computational problem defined by a given set of data, called the **input data**, produces a solution to the problem. A solution comprises of a set of data called the **output data**.*

In this book, for the sake of convenience and simplicity, we will very often describe algorithms by means of **pseudocodes**, which can easily be translated into computer codes. Here is an illustration.

3.3.2 Solving an Upper Triangular System

Consider the system
$$Ty = b,$$

where $T = (t_{ij})$ is a nonsingular upper triangular matrix and $y = (y_1, y_2, \ldots, y_n)^T$ and $b = (b_1, \ldots, b_n)^T$.

Algorithm 3.3.1.   *Back Substitution Method for Upper Triangular System*
   Input. T—*An $n \times n$ nonsingular upper triangular matrix, b—An n-vector.*
   Output. *The vector $y = (y_1, \ldots, y_n)^T$ such that $Ty = b$.*
Step 1. *Compute* $y_n = \dfrac{b_n}{t_{nn}}$
Step 2. *For* $i = n - 1, n - 2, \ldots, 2, 1$ *do*

$$y_i = \frac{1}{t_{ii}} \left(b_i - \sum_{j=i+1}^{n} t_{ij} y_j \right)$$

   End

Note: When $i = n$, the summation $\left(\sum \right)$ is skipped.

3.3.3   Solving a Lower Triangular System

A lower triangular system $Ly = b$ can be solved in an analogous manner. The process is known as the **forward substitution method**. Let $L = (l_{ij})$, and $b = (b_1, b_2, \ldots, b_n)^T$. Then starting with y_1, y_2 through y_n are computed recursively.

3.3.4   Efficiency of an Algorithm

Two most desirable properties of an algorithm are: **Efficiency and Stability**.

The *efficiency of an algorithm* is measured by the amount of computer time consumed in its implementation.

A theoretical and very crude measure of efficiency is the number of floating point operations (**flops**) needed to implement the algorithm. Too much emphasis should not be placed on exact flop-count when comparing the efficiency of two algorithms.

Definition 3.3.2.   *A floating point operation of* flop *is a floating point operation:* $+, -, *,$ *or* $/$.

The Big O Notation

An algorithm will be called an $O(n^p)$ algorithm if the dominant term in the operations count of the algorithm is a multiple of n^p. Thus, the solution of a triangular system is an $O(n^2)$ algorithm; because it requires n^2 flops.

Notation for Overwriting and Interchange

We will use the notation:

$$a \equiv b$$

to denote that "**b overwrites a**". Similarly, if two computed quantities a and b are interchanged, they will be written symbolically

$$a \leftrightarrow b.$$

3.3.5 The Concept of Numerical Stability

The accuracy or the inaccuracy of the computed solution of a problem usually depends upon two important factors: **the stability or the instability of the algorithm used to solve the problem** and **the conditioning of the problem** (i.e., how sensitive the problem is to small perturbations).

We first define the concept of stability of an algorithm. In the next section, we shall talk about the conditioning of a problem.

The study of stability of an algorithm is done by means of **roundoff error analysis**. There are two types: **backward error analysis** and **forward error analysis**.

In forward analysis, an attempt is made to see how the computed solution obtained by the algorithm differs from the exact solution based on the same data.

On the other hand, backward analysis relates the error to the data of the problem rather than to the problem's solution.

Definition 3.3.3. *An algorithm is called* **backward stable** *if it produces an exact solution to a* **nearby** *problem; that is, a backward algorithm exactly solves a problem whose data are close to the original data.*

Backward error analysis, popularized in the literature by J.H. Wilkinson (1965), is now widely used in matrix computations and using this analysis, the stability (or instability) of many algorithms in numerical linear algebra has been established in recent years. **In this book, by "stability" we will imply "backward stability," unless otherwise stated**.

As a simple example of backward stability, consider again the problem of computing the sum of two floating point numbers x and y. We have seen before that

$$\mathrm{fl}(x + y) = (x + y)(1 + \delta) = x(1 + \delta) + y(1 + \delta) = x' + y'.$$

Thus, the computed sum of two floating point numbers x and y is the exact sum of another two floating point numbers x' and y'. Because $|\delta| \leq \mu$, both x' and y' are close to x and y, respectively. Thus, we conclude that **the operation of adding two floating-point numbers is stable**. Similarly, *floating-point subtraction, multiplication, and division are also backward stable.*

Example 3.3.1. (*A Stable Algorithm for Linear Systems*)   Solution of an upper triangular system by Back substitution.

The back-substitution method for solving an upper triangular system $Tx = b$ **is backward stable**. It can be shown that the computed solution $\hat{x}$ satisfies

$$(T + E)\hat{x} = b,$$

where $|e_{ij}| \leq c\mu|t_{ij}|, i, j = 1, \ldots, n$ and c is a constant of order unity. Thus, the computed solution $\hat{x}$ solves exactly a nearby system. The back-substitution process is, therefore, backward stable.

Remark

- The forward substitution method for solving a lower triangular system has the same numerical stability property as above. **This algorithm is also stable**.

Example 3.3.2. (*An Unstable Algorithm for Linear Systems*) Gaussian elimination without pivoting.

It can be shown (**see Section 3.5.2**) that Gaussian elimination without pivoting applied to the linear system $Ax = b$ produces a solution $\hat{x}$ such that

$$(A + E)\hat{x} = b$$

with $\|E\|_\infty \leq cn^3\rho\|A\|_\infty\mu$. The number ρ above, called the **growth factor**, can be arbitrarily very large. When it happens, the computed solution $\hat{x}$ does not solve a nearby problem.

Example 3.3.3. (*An Unstable Algorithm for Eigenvalue Computations*)   Finding the eigenvalues of a matrix via its characteristic polynomial. **The process is numerically unstable**.

There are two reasons: First, the characteristic polynomial of a matrix may not be obtained in a numerically stable way (**see Chapter 4**); second, the zeros of a polynomial can be extremely sensitive to small perturbations of the coefficients.

A well-known example of zero-sensitivity is the Wilkinson polynomial $P_n(x) = (x - 1)(x - 2) \cdots (x - 20)$. A small perturbation of 2^{-23} to the coefficient of x^{19} changes some of the zeros significantly: some of them even become complex. See Datta (1995, pp. 81–82) for details.

Remark

- This example shows that the **eigenvalues of a matrix should not be computed by finding the roots of its characteristic polynomial**.

3.3.6   Conditioning of the Problem and Perturbation Analysis

From the preceding discussion, we should not form the opinion that if a stable algorithm is used to solve a problem, then the computed solution will be accurate.

As said before, a property of the problem called **conditioning** also contributes to the accuracy or inaccuracy of the computed result.

The conditioning of a problem is a property of the problem itself. It is concerned with how the solution of the problem will change if the input data contains some impurities. This concern arises from the fact that in practical applications, the data very often come from some experimental observations where the measurements can be subjected to disturbances (or "noises") in the data. There are other sources of error also, such as **roundoff errors, discretization errors**, and so on. Thus, when a numerical analyst has a problem in hand to solve, he or she must frequently solve the problem not with the original data, but with data that have been perturbed. The question naturally arises: **What effects do these perturbations have on the solution?**

A theoretical study done by numerical analysts to investigate these effects, which is independent of the particular algorithm used to solve the problem, is called **perturbation analysis**. This study helps us detect whether a given problem is "**bad**" or "**good**" in the sense of whether small perturbations in the data will create a large or small change in the solution. Specifically we use the following standard definition.

Definition 3.3.4. *A problem (with respect to a given set of data) is called an* **ill-conditioned** *or* **badly conditioned** *problem if a small relative error in data can cause a large relative error in the computed solution, regardless of the method of solution. Otherwise, it is called* **well-conditioned**.

Suppose a problem P is to be solved with an input c. Let $P(c)$ denote the value of the problem with the input c. Let δ_c denote the perturbation in c. Then P is said to be ill-conditioned for the input data c if the relative error $|P(c + \delta_c) - P(c)|/|P(c)|$ is much larger than the relative error in the data $|\delta_c|/|c|$

Note: The definition of conditioning is data-dependent. Thus, a problem that is ill-conditioned for *one* set of data could be well-conditioned for *another* set.

3.3.7   Conditioning of the Problem, Stability of the Algorithm, and Accuracy of the Solution

As stated in the previous section, the conditioning of a problem is a property of the problem itself, and has nothing to do with the algorithm used to solve the problem. To a user, of course, the accuracy of the computed solution is of primary importance. However, the accuracy of a computed solution by a given algorithm is directly connected with both the stability of the algorithm and the conditioning of the problem. **If the problem is ill-conditioned, no matter how stable the algorithm is, the accuracy of the computed result cannot be guaranteed.**

On the other hand, if a backward stable algorithm is applied to a well-conditioned problem, the computed result will be accurate.

Backward Stability and Accuracy

Stable Algorithm $\rightarrow$ Well-conditioned Problem $\equiv$ Accurate Result.
Stable Algorithm $\rightarrow$ Ill-conditioned Problem $\equiv$ Possibly Inaccurate Result (inaccuracy depends upon how ill-conditioned the problem is).

3.3.8   Conditioning of the Linear System and Eigenvalue Problems

The Condition Number of a Problem

Numerical analysts usually try to associate a number called the **condition number** with a problem. The condition number indicates whether the problem is ill- or well-conditioned. More specifically, the condition number gives a bound for the relative error in the solution when a small perturbation is applied to the input data.

We will now give results on the conditions of the linear system and eigenvalue problems.

Theorem 3.3.1.   *General Perturbation Theorem. Let $\triangle A$ and δb, be the perturbations, respectively, of the data A and b, and δx be the error in x. Assume that A is nonsingular and $\| \triangle A \| < 1/ \| A^{-1} \|$. Then,*

$$\frac{\| \delta x \|}{\| x \|} \leq \frac{\| A \| \| A^{-1} \|}{(1- \| \triangle A \| \| A^{-1} \|)} \left(\frac{\| \triangle A \|}{\| A \|} + \frac{\| \delta b \|}{\| b \|} \right).$$

Interpretation of the theorem: The above theorem says that if the relative perturbations in A and b are small, then the number $\| A \| \| A^{-1} \|$ is the dominating factor in determining how large the relative error in the solution can be.

Definition 3.3.5.   *The number $\| A \| \| A^{-1} \|$ is called the **condition number** of the linear system problem $Ax = b$ or just the condition number of A, and is denoted by* Cond(A).

From the theorem above, it follows that **if Cond(A) is large, then the system** $Ax = b$ **is ill-conditioned**; otherwise it is **well-conditioned**.

The condition number of a matrix certainly depends upon the norm of the matrix. However, roughly, if a matrix is ill-conditioned in one type of norm, it is ill-conditioned in other types as well. This is because the condition numbers in different norms are related. For example, for an $n \times n$ real matrix A, one can show that

$$\frac{1}{n}\text{Cond}_2(A) \le \text{Cond}_1(A) \le n\text{Cond}_2(A),$$

$$\frac{1}{n}\text{Cond}_\infty(A) \le \text{Cond}_2(A) \le n\text{Cond}_\infty(A),$$

$$\frac{1}{n^2}\text{Cond}_1(A) \le \text{Cond}_\infty(A) \le n^2\text{Cond}_1(A),$$

where $\text{Cond}_p(A)$, $p = 1, 2, \infty$ denotes the condition number in p-norm.

Next, we present the proof of the above theorem in the case $\triangle A = 0$. For the proof in the general case, see Datta (1995, pp. 249–250). We first restate the theorem in this special case.

Theorem 3.3.2. *Right Perturbation Theorem. If δb and δx, are, respectively, the perturbations of b and x in the linear system $Ax = b$, and, A is assumed to be nonsingular and $b \ne 0$, then*

$$\frac{\|\delta x\|}{\|x\|} \le \text{Cond}(A)\,\frac{\|\delta b\|}{\|b\|}.$$

Proof. We have

$$Ax = b \quad \text{and} \quad A(x + \delta x) = b + \delta b.$$

The last equation can be written as $Ax + A\delta x = b + \delta b$, or

$$A\delta x = \delta b \quad (\text{since } Ax = b) \quad \text{that is, } \delta x = A^{-1}\delta b.$$

Taking a subordinate matrix-vector norm, we get

$$\|\delta x\| \le \|A^{-1}\|\|\delta b\|. \tag{3.3.1}$$

Again, taking the same norm on both sides of $Ax = b$, we get $\|Ax\| = \|b\|$ or

$$\|b\| = \|Ax\| \le \|A\|\|x\|. \tag{3.3.2}$$

Combining (3.3.1) and (3.3.2), we have

$$\frac{\|\delta x\|}{\|x\|} \le \|A\|\|A^{-1}\|\frac{\|\delta b\|}{\|b\|}. \quad\blacksquare \tag{3.3.3}$$

Interpretation of Theorem 3.3.2: Theorem 3.3.2 says that a relative error in the solution can be as large as $\text{Cond}(A)$ multiplied by the relative perturbation in the vector b. Thus, if the condition number is not too large, then a small perturbation in the vector b will have very little effect on the solution. **On the other hand, if the condition number is large, then even a small perturbation in b might change the solution drastically.**

Example 3.3.4. (*An Ill-Conditioned Linear System Problem*)

$$A = \begin{pmatrix} 1 & 2 & 1 \\ 2 & 4.0001 & 2.002 \\ 1 & 2.002 & 2.004 \end{pmatrix}, \quad b = \begin{pmatrix} 4 \\ 8.0021 \\ 5.006 \end{pmatrix}.$$

The exact solution $x = \begin{pmatrix} 1 \\ 1 \\ 1 \end{pmatrix}$. Change b to $b' = \begin{pmatrix} 4 \\ 8.0020 \\ 5.0061 \end{pmatrix}$.

Then the relative perturbation in b:

$$\frac{\|b' - b\|}{\|b\|} = \frac{\|\delta b\|}{\|b\|} = 1.379 \times 10^{-5} \text{ (small)}.$$

If we solve the system $Ax' = b'$, we get

$$x' = x + \delta x = \begin{pmatrix} 3.0850 \\ -0.0436 \\ 1.0022 \end{pmatrix}.$$

(**x' is completely different from x**).

Note that the relative error in x: $\frac{\|\delta x\|}{\|x\|} = 1.3461$ (quite large!).

It is easily verified that the inequality in the above theorem is satisfied:

$$\text{Cond}(A) \cdot \frac{\|\delta b\|}{\|b\|} = 4.4434, \quad \text{Cond}(A) = 3.221 \times 10^5.$$

However, the predicted change is overly estimated.

Conditioning of Eigenvalues

Like the linear system problem, the eigenvalues and the eigenvectors of a matrix A can be ill-conditioned too.

The following result gives an overall sensitivity of the eigenvalues due to perturbations in the entries of A. For a proof, see Datta (1995) or Golub and Van Loan (1996).

Theorem 3.3.3. *Bauer-Fike. Let* $X^{-1}AX = D = \text{diag}(\lambda_1, \ldots, \lambda_n)$. *Then for any eigenvalue* λ *of* $A + E \in \mathbb{C}^{n \times n}$, *we have*

$$\min |\lambda_i - \lambda| \leq \text{Cond}_p(X) \| E \|,$$

where $\| \cdot \|_p$ *is a p-norm.*

*The result says that the eigenvalues of A **might be sensitive to small perturbations of the entries of A if the transforming matrix X is ill-conditioned.***

Analysis of the conditioning of the individual eigenvalues and eigenvectors are rather involved. We just state here the conditioning of simple eigenvalues of a matrix.

Let λ_i be a **simple** eigenvalue of A. Then the condition number of λ_i, denoted by Cond(λ_i), is defined to be: Cond(λ_i) $= 1/|y_i^T x_i|$, where y_i and x_i are, respectively, the unit **left** and **right** eigenvectors associated with λ_i.

A well-known example of eigenvalue sensitivity is the **Wilkinson bidiagonal matrix**:

$$A = \begin{pmatrix} 20 & & 20 & & & \\ & 19 & & 20 & & 0 \\ 0 & & \ddots & & \ddots & \\ & & & & \ddots & 20 \\ & & & & & 1 \end{pmatrix}.$$

The eigenvalues of A are $1, 2, \ldots, 20$.

A small perturbation E of the $(20, 1)$th entry of A (say $E = 10^{-10}$) changes some of the eigenvalues drastically: **they even become complex** (see Datta (1995, pp. 84–85)).

The above matrix A is named after the famous British numerical analyst **James H. Wilkinson,** who computed the condition numbers of the above eigenvalues and found that some of the condition numbers were quite large, explaining the fact why they changed so much due to a small perturbation of just one entry of A.

Note: Though the eigenvalues of a nonsymmetric matrix can be ill-conditioned- the **eigenvalues of a symmetric matrix are well-conditioned (see Datta (1995, pp. 455–456)).**

3.4 LU FACTORIZATION

In this section, we describe a well-known matrix factorization, called the **LU factorization of a matrix** and in the next section, we will show how the LU factorization is used to solve an algebraic linear system.

3.4.1 LU Factorization using Gaussian Elimination

An $n \times n$ matrix A having nonsingular principal minors can be factored into LU: $A = LU$, where L is a lower triangular matrix with 1s along the diagonal (unit lower triangular) and U is an $n \times n$ upper triangular matrix. This factorization is known as an **LU factorization of A**. A classical elimination technique, called **Gaussian elimination,** is used to achieve this factorization.

If an LU factorization exists and A is nonsingular, then the LU factorization is unique (see Golub and Van Loan (1996), pp. 97–98).

Gaussian Elimination

There are $(n - 1)$ steps in the process. Beginning with $A^{(0)} = A$, the matrices $A^{(1)}, \ldots, A^{(n-1)}$ are constructed such that $A^{(k)}$ has zeros below the diagonal in the kth column. The final matrix $A^{(n-1)}$ will then be an upper triangular matrix U. Denote $A^{(k)} = \left(a_{ij}^{(k)}\right)$. The matrix $A^{(k)}$ is obtained from the previous matrix $A^{(k-1)}$ by multiplying the entries of the row k of $A^{(k-1)}$ with $m_{ik} = -\left(a_{ik}^{(k-1)}\right)/\left(a_{kk}^{(k-1)}\right)$, $i = k+1, \ldots, n$ and adding them to those of $(k+1)$ through n. In other words,

$$a_{ij}^{(k)} = a_{ij}^{(k-1)} + m_{ik}a_{kj}^{(k-1)}, \quad i = k+1, \ldots, n; \quad j = k+1, \ldots, n. \quad (3.4.1)$$

The entries m_{ik} are called **multipliers**. The entries $a_{kk}^{(k-1)}$ are called the **pivots**.

To see how an LU factorization, when it exists, can be obtained, we note (which is easy to see using the above relations) that

$$A^{(k)} = M_k A^{(k-1)}, \quad\quad\quad (3.4.2)$$

where M_k is a unit lower triangular matrix formed out of the multipliers. The matrix M_k is known as the **elementary lower triangular matrix**. The matrix M_k can be written as:

$$M_k = I + m_k e_k^T,$$

where e_k is the kth unit vector, $e_i^T m_k = 0$ for $i \leq k$, and $m_k = (0, \ldots, 0, m_{k+1,k}, \ldots, m_{n,k})^T$.

Furthermore, $M_k^{-1} = I - m_k e_k^T$.

Using (3.4.2), we see that

$$U = A^{(n-1)} = M_{n-1} A^{(n-2)} = M_{n-1} M_{n-2} A^{(n-3)}$$
$$= \cdots = M_{n-1} M_{n-2} \cdots M_2 M_1 A$$

Thus, $A = (M_{n-1} M_{n-2} \cdots M_2 M_1)^{-1} U = LU$, where $L = (M_{n-1} M_{n-2} \cdots M_2 M_1)^{-1}$.

Since each of the matrices M_1 through M_{n-1} is a unit upper triangular matrix, so is L (*Note:* The product of two unit upper triangular matrix is an upper triangular matrix and the inverse of a unit upper triangular matrix is an upper triangular matrix).

Constructing L: The matrix L can be formed just from the multipliers, as shown below. **No explicit matrix inversion is needed**.

$$L = \begin{pmatrix} 1 & 0 & 0 & \cdots & \cdots & 0 \\ -m_{21} & 1 & 0 & \cdots & \cdots & 0 \\ -m_{31} & -m_{32} & 1 & \cdots & \cdots & 0 \\ \vdots & \vdots & & \ddots & \ddots & \vdots \\ \vdots & \vdots & & & \ddots & 0 \\ -m_{n1} & -m_{n2} & -m_{n3} & \cdots & -m_{n,n-1} & 1 \end{pmatrix}.$$

Difficulties with Gaussian Elimination without Pivoting

Gaussian elimination, as described above, fails if any of the pivots is zero, **it is worse yet if any pivot becomes close to zero**. In this case, the method can be carried to completion, but the obtained results may be totally wrong.

Consider the following simple example: Let Gaussian elimination without pivoting be applied to

$$A = \begin{pmatrix} 0.0001 & 1 \\ 1 & 1 \end{pmatrix},$$

using three decimal digit floating point arithmetic.

There is only one step. The multiplier $m_{21} = -1/10^{-4} = -10^4$. Let $\hat{L}$ and $\hat{U}$ be the computed versions of L and U. Then,

$$\hat{U} = A^{(1)} = \begin{pmatrix} 0.0001 & 1 \\ 0 & 1 - 10^4 \end{pmatrix} = \begin{pmatrix} 0.0001 & 1 \\ 0 & -10^4 \end{pmatrix}.$$

(Note that $(1 - 10^4)$ gives -10^4 in three-digit arithmetic). The matrix $\hat{L}$ formed out the multiplier m_{21} is

$$\hat{L} = \begin{pmatrix} 1 & 0 \\ 10^4 & 1 \end{pmatrix}.$$

The product of the computed $\hat{L}$ and $\hat{U}$ is:

$$\hat{L}\hat{U} = \begin{pmatrix} 0.0001 & 1 \\ 1 & 0 \end{pmatrix},$$

which is different from A.

Note that the pivot $a_{11}^{(1)} = 0.0001$ is very close to zero (in three-digit arithmetic). This small pivot gave a large multiplier. This large multiplier, when used to update the entries of A, the number 1, which is much smaller compared to 10^4, got wiped out in the subtraction of $1 - 10^4$ and the result was -10^4.

Gaussian Elimination with Partial Pivoting

The above example suggests that disaster in Gaussian elimination without pivoting in the presence of a small pivot can perhaps be avoided by identifying a **"good pivot"** (a pivot as large as possible) at each step, before the process of elimination is applied. The good pivot may be located among the entries in a column or among all the entries in a submatrix of the current matrix. In the former case, since the search is only partial, the method is called **partial pivoting**; in the latter case, the method is called **complete pivoting**. *It is important to note that the purpose of pivoting is to prevent large growth in the reduced matrices, which can wipe out the original data.* One way to do this is to keep multipliers less than 1 in magnitude, and this is exactly what is accomplished by pivoting.

We will discuss here only Gaussian elimination with partial pivoting, which also consists of $(n - 1)$ steps.

In fact, the process is just a slight modification of Gaussian elimination in the following sense: At each step, the largest entry (in magnitude) is identified among all the entries in the pivot column. This entry is then brought to the diagonal position of the current matrix by interchange of suitable rows and then, using that entry as "pivot," the elimination process is performed.

Thus, if we set $A^{(0)} = A$, at step k $(k = 1, 2, \ldots, n - 1)$, first, the largest entry (in magnitude) $a_{r_k,k}^{(k-1)}$ is identified among all the entries of the column k (below the row $(k-1)$) of the matrix $A^{(k-1)}$, this entry is then brought to the diagonal position by interchanging the rows k and r_k, and then the elimination process proceeds with $a_{r_k,k}^{(k-1)}$ as the pivot.

LU Factorization from Gaussian Elimination with Partial Pivoting

Since the interchange of two rows of a matrix is equivalent to premultiplying the matrix by a permutation matrix, the matrix $A^{(k)}$ is related to $A^{(k-1)}$ by the following relation:

$$A^{(k)} = M_k P_k A^{(k-1)}, \quad k = 1, 2, \ldots, n - 1,$$

where P_k is the permutation matrix obtained by interchanging the rows k and r_k of the identity matrix, and M_k is an elementary lower triangular matrix resulting from the elimination process. So,

$$U = A^{(n-1)} = M_{n-1} P_{n-1} A^{(n-2)} = M_{n-1} P_{n-1} M_{n-2} P_{n-2} A^{(n-3)}$$
$$= \cdots = M_{n-1} P_{n-1} M_{n-2} P_{n-2} \cdots M_2 P_2 M_1 P_1 A.$$

Setting $M = M_{n-1} P_{n-1} M_{n-2} P_{n-2} \cdots M_2 P_2 M_1 P_1$, we have the following factorization of A:

$$U = MA.$$

The above factorization can be written in the form: $PA = LU$, where $P = P_{n-1} P_{n-2} \cdots P_2 P_1$, $U = A^{(n-1)}$, and the matrix L is a unit lower triangular matrix formed out of the multipliers. For details, see Golub and Van Loan (1996, pp. 99).

For $n = 4$, the reduction of A to the upper triangular matrix U can be schematically described as follows:

$$1. \quad A \xrightarrow{P_1} P_1 A \xrightarrow{M_1} M_1 P_1 A = \begin{pmatrix} \times & \times & \times & \times \\ 0 & \times & \times & \times \\ 0 & \times & \times & \times \\ 0 & \times & \times & \times \end{pmatrix} = A^{(1)}.$$

2. $A^{(1)} \overset{P_2}{\to} P_2 A^{(1)} \overset{M_2}{\to} M_2 P_2 A^{(1)} = M_2 P_2 M_1 P_1 A = \begin{pmatrix} \times & \times & \times & \times \\ 0 & \times & \times & \times \\ 0 & 0 & \times & \times \\ 0 & 0 & \times & \times \end{pmatrix} = A^{(2)}.$

3. $A^{(2)} \overset{P_3}{\to} P_3 A^{(2)} \overset{M_3}{\to} M_3 P_3 A^{(2)} = M_3 P_3 M_2 P_2 M_1 P_1 A = \begin{pmatrix} \times & \times & \times & \times \\ 0 & \times & \times & \times \\ 0 & 0 & \times & \times \\ 0 & 0 & 0 & \times \end{pmatrix}$

$$= A^{(3)} = U.$$

The only difference between L here and the matrix L from Gaussian elimination without pivoting is that the multipliers in the kth column are now permuted according to the permutation matrix $\tilde{P}_k = P_{n-1} P_{n-2} \cdots P_{k+1}$.

Thus, to construct L, again no explicit products or matrix inversions are needed. We illustrate this below.

Consider the case $n = 4$, and suppose P_2 interchanges rows 2 and 3, and P_3 interchanges rows 3 and 4.

The matrix L is then given by:

$$L = \begin{pmatrix} 1 & 0 & 0 & 0 \\ -m_{31} & 1 & 0 & 0 \\ -m_{21} & -m_{42} & 1 & 0 \\ -m_{41} & -m_{32} & -m_{34} & 1 \end{pmatrix}.$$

Example 3.4.1.

$$A = \begin{pmatrix} 1 & 2 & 4 \\ 4 & 5 & 6 \\ 7 & 8 & 9 \end{pmatrix}.$$

$k = 1$

1. The pivot entry is 7: $r_1 = 3$.
2. Interchange rows 3 and 1.

$$P_1 = \begin{pmatrix} 0 & 0 & 1 \\ 0 & 1 & 0 \\ 1 & 0 & 0 \end{pmatrix}, \qquad P_1 A = \begin{pmatrix} 7 & 8 & 9 \\ 4 & 5 & 6 \\ 1 & 2 & 4 \end{pmatrix}.$$

3. Form the multipliers: $a_{21} \equiv m_{21} = -\frac{4}{7}$, $a_{31} \equiv m_{31} = -\frac{1}{7}$.

4. $A^{(1)} = M_1 P_1 A = \begin{pmatrix} 1 & 0 & 0 \\ -\frac{4}{7} & 1 & 0 \\ -\frac{1}{7} & 0 & 1 \end{pmatrix} \begin{pmatrix} 7 & 8 & 9 \\ 4 & 5 & 6 \\ 1 & 2 & 4 \end{pmatrix} \equiv \begin{pmatrix} 7 & 8 & 9 \\ 0 & \frac{3}{7} & \frac{6}{7} \\ 0 & \frac{6}{7} & \frac{19}{7} \end{pmatrix}.$

$k = 2$

1. The pivot entry is $\frac{6}{7}$, $r_2 = 3$.

2. Interchange rows 2 and 3.

$$P_2 = \begin{pmatrix} 1 & 0 & 0 \\ 0 & 0 & 1 \\ 0 & 1 & 0 \end{pmatrix}, \qquad P_2 A^{(1)} = \begin{pmatrix} 7 & 8 & 9 \\ 0 & \frac{6}{7} & \frac{19}{7} \\ 0 & \frac{3}{7} & \frac{6}{7} \end{pmatrix}.$$

3. Form the multiplier: $m_{32} = -\frac{1}{2}$

$$A^{(2)} = M_2 P_2 A^{(1)} = \begin{pmatrix} 1 & 0 & 0 \\ 0 & 1 & 0 \\ 0 & -\frac{1}{2} & 1 \end{pmatrix} \begin{pmatrix} 7 & 8 & 9 \\ 0 & \frac{6}{7} & \frac{19}{7} \\ 0 & \frac{3}{7} & \frac{6}{7} \end{pmatrix} = \begin{pmatrix} 7 & 8 & 9 \\ 0 & \frac{6}{7} & \frac{19}{7} \\ 0 & 0 & -\frac{1}{2} \end{pmatrix}.$$

$$\text{Form } L = \begin{pmatrix} 1 & 0 & 0 \\ -m_{31} & 1 & 0 \\ -m_{21} & -m_{32} & 1 \end{pmatrix} = \begin{pmatrix} 1 & 0 & 0 \\ \frac{1}{7} & 1 & 0 \\ \frac{4}{7} & \frac{1}{2} & 1 \end{pmatrix}.$$

$$P = P_2 P_1 = \begin{pmatrix} 0 & 0 & 1 \\ 1 & 0 & 0 \\ 0 & 1 & 0 \end{pmatrix}.$$

$$\text{Verify. } PA = \begin{pmatrix} 7 & 8 & 9 \\ 1 & 2 & 4 \\ 4 & 5 & 6 \end{pmatrix} = LU.$$

Flop-count. Gaussian elimination with partial pivoting requires only $\frac{2}{3}n^3$ flops. Furthermore, the process with partial pivoting requires at most $O(n^2)$ comparisons for identifying the pivots.

Stability of Gaussian Elimination

The stability of Gaussian elimination algorithms is better understood by measuring the growth of the elements in the reduced matrices $A^{(k)}$. **(Note that although pivoting keeps the multipliers bounded by unity, the elements in the reduced matrices still can grow arbitrarily.)**

Definition 3.4.1. *The* **growth factor** ρ *is the ratio of the largest element (in magnitude) of A, $A^{(1)}, \ldots, A^{(n-1)}$ to the largest element (in magnitude) of A: $\rho = (\max(\alpha, \alpha_1, \alpha_2, \ldots, \alpha_{n-1}))/\alpha$, where $\alpha = \max_{i,j} |a_{ij}|$, and $\alpha_k = \max_{i,j} |a_{ij}^{(k)}|$.*

The growth factor ρ **can be arbitrarily large for Gaussian elimination without pivoting.** Note that ρ for the matrix

$$A = \begin{pmatrix} 0.0001 & 1 \\ 1 & 1 \end{pmatrix}$$

without pivoting is 10^4.

Thus, Gaussian elimination without pivoting is, in general, unstable.

Note: Though Gaussian elimination without pivoting is unstable for arbitrary matrices, there are two classes of matrices, the **diagonally dominant matrices** and the **symmetric positive definite matrices**, for which the process can be shown to be stable. The growth factor of a diagonally dominant matrix is bounded by 2 and that of a symmetric positive definite matrix is 1.

The next question is: How large can the growth factor be for Gaussian elimination with partial pivoting?

The growth factor ρ for Gaussian elimination with partial pivoting can be as large as 2^{n-1}: $\rho \leq 2^{n-1}$.

Though matrices for which this bound is attained can be constructed (see Datta 1995), such matrices are rare in practice. Indeed, in many practical examples, the elements of the matrices $A^{(k)}$ very often continue to decrease in size. **Thus, Gaussian elimination with partial pivoting is not unconditionally stable in theory; in practice, however, it can be considered as a stable algorithm.**

MATLAB note: The MATLAB command $[L, U, P] = \mathrm{lu}\ (A)$ returns lower triangular matrix L, upper triangular matrix U, and permutation matrix P such that $PA = LU$.

3.4.2 The Cholesky Factorization

Every symmetric positive definite matrix A can be factored into

$$A = HH^{\mathrm{T}},$$

where H is a lower triangular matrix with positive diagonal entries.

This factorization of A is known as the **Cholesky factorization**. Since, the growth factor for Gaussian elimination of a symmetric positive definite matrix is 1, **Gaussian elimination can be safely used to compute the Cholesky factorization of a symmetric positive definite matrix.** Unfortunately, no advantage of symmetry of the matrix A can be taken in the process.

In practice, the entries of the lower triangular matrix H, called the **Cholesky factor**, are computed directly from the relation $A = HH^{\mathrm{T}}$. The matrix H is computed row by row. The algorithm is known as the Cholesky algorithm. See Datta (1995, pp. 222–223) for details.

Algorithm 3.4.1. *The Cholesky Algorithm*
  Input. *A—A symmetric positive definite matrix*
  Output. *H—The Cholesky factor*
  For $k = 1, 2, \ldots, n$ do
    For $i = 1, 2, \ldots, k - 1$ do

$$h_{ki} = \frac{1}{h_{ii}} \left(a_{ki} - \sum_{j=1}^{i-1} h_{ij}h_{kj} \right)$$

$$h_{kk} = \sqrt{ a_{kk} - \sum_{j=1}^{k-1} h_{kj}^2 }$$

End
End

Flop-count and numerical stability. Algorithm 3.4.1 requires only $n^3/3$ flops. The algorithm is **numerically stable**.

MATLAB and MATCOM notes: Algorithm 3.4.1 has been implemented in MATCOM program **choles**. MATLAB function **chol** also can be used to compute the Cholesky factor. However, note that $L = \mathbf{chol}(A)$ computes an upper triangular matrix R such that $A = R^TR$.

3.4.3 LU Factorization of an Upper Hessenberg Matrix

Recall that $H = (h_{ij})$ is an upper Hessenberg matrix if $h_{ij} = 0$ whenever $i > j+1$. Thus, Gaussian elimination scheme applied to an $n \times n$ upper Hessenberg matrix requires zeroing of only the nonzero entries on the subdiagonal. This means at each step, after a possible interchange of rows, just a multiple of the row containing the pivot has to be added to the next row.

Specifically, Gaussian elimination scheme with partial pivoting for an $n \times n$ upper Hessenberg matrix $H = (h_{ij})$ is as follows:

Algorithm 3.4.2. *LU Factorization of an Upper Hessenberg Matrix*
   Input. *H—An $n \times n$ upper Hessenberg matrix*
   Output. *U—The upper triangular matrix U of LU factorization of H, stored over the upper part of H. The subdiagonal entries of H contain the multipliers. For $k = 1, 2, \ldots, n - 1$ do*

1. *Interchange $h_{k,j}$ and $h_{k+1,j}$, if $|h_{k,k}| < |h_{k+1,k}|$, $j = k, \ldots, n$.*
2. *Compute the multiplier and store it over $h_{k+1,k}$: $h_{k+1,k} \equiv -\dfrac{h_{k+1,k}}{h_{k,k}}$.*
3. *Update $h_{k+1,j}$: $h_{k+1,j} \equiv h_{k+1,j} + h_{k+1,k} \cdot h_{k,j}$, $j = k + 1, \ldots, n$.*

End.

Flop-count and stability. The above algorithm requires n^2 flops.

It can be shown Wilkinson (1965, p. 218); Higham (1996, p. 182), that the growth factor ρ of a Hessenberg matrix for Gaussian elimination with partial pivoting is less than or equal to n. **Thus, computing LU factorization of a Hessenberg** matrix using Gaussian elimination with partial pivoting is an efficient and a numerically stable procedure.

3.5   NUMERICAL SOLUTION OF THE LINEAR SYSTEM $Ax = b$

> Given an $n \times n$ matrix A and the n-vector b, the algebraic linear system problem is the problem of finding an n-vector x such that $Ax = b$.

The principal uses of the LU factorization of a matrix A are: **solving the algebraic linear system** $Ax = b$, **finding the determinant of a matrix**, and **finding the inverse of** A.

We will discuss first how $Ax = b$ can be solved using the LU factorization of A.

The following theorem gives results on the existence and uniqueness of the solution x of $Ax = b$. Proof can be found in any linear algebra text.

Theorem 3.5.1.   *Existence and Uniqueness Theorem. The system $Ax = b$ has a solution if and only if rank $(A) = rank(A, b)$. The solution is unique if and only if A is invertible.*

3.5.1   Solving $Ax = b$ using the Inverse of A

The above theorem suggests that the unique solution x of $Ax = b$ be computed as $x = A^{-1}b$.

Unfortunately, **computationally this is not a practical idea. It generally involves more computations and gives less accurate answers**.

This can be illustrated by the following trivial example:

Consider solving $3x = 27$.

The exact answer is: $x = 27/3 = 9$. Only one flop (one division) is needed in this process. On the other hand, if the problem is solved by writing it in terms of the inverse of A, we then have $x = \frac{1}{3} \times 27 = 0.3333 \times 27 = 8.9991$ (in four digit arithmetic), a less accurate answer. Moreover, the process will need two flops: one division and one multiplication.

3.5.2   Solving $Ax = b$ using Gaussian Elimination with Partial Pivoting

Since Gaussian elimination without pivoting does not always work and, even when it works, might give an unacceptable answer in certain instances, we only discuss solving $Ax = b$ using Gaussian elimination with partial pivoting.

We have just seen that Gaussian elimination with partial pivoting, when used to triangularize A, yields a factorization $PA = LU$. In this case, the system $Ax = b$ is equivalent to the two triangular systems:

$$Ly = Pb = b' \quad \text{and} \quad Ux = y.$$

Thus, to solve $Ax = b$ using Gaussian elimination with partial pivoting, the following two steps need to be performed in the sequence.

Step 1. Find the factorization $PA = LU$ using Gaussian eliminating with partial pivoting.

Step 2. Solve the lower triangular system: $Ly = Pb = b'$ first, followed by the upper triangular system: $Ux = y$.

Forming the vector b'. The vector b' is just the permuted version of b. So, to obtain b', all that needs to be done is to permute the entries of b in the same way as the rows of the matrices $A^{(k)}$ have been interchanged. This is illustrated in the following example.

Example 3.5.1. *Solve the following system using Gaussian elimination with partial pivoting*:

$$x_1 + 2x_2 + 4x_3 = 7,$$
$$4x_1 + 5x_2 + 6x_3 = 15,$$
$$7x_1 + 8x_2 + 9x_3 = 24.$$

Here

$$A = \begin{pmatrix} 1 & 2 & 4 \\ 4 & 5 & 6 \\ 7 & 8 & 9 \end{pmatrix}, \quad b = \begin{pmatrix} 7 \\ 15 \\ 24 \end{pmatrix}.$$

Using the results of Example 3.4.1, we have

$$L = \begin{pmatrix} 1 & 0 & 0 \\ \frac{1}{7} & 1 & 0 \\ \frac{4}{7} & \frac{1}{2} & 1 \end{pmatrix}, \quad U = \begin{pmatrix} 7 & 8 & 9 \\ 0 & \frac{6}{7} & \frac{19}{7} \\ 0 & \frac{6}{7} & \frac{19}{7} \\ 0 & 0 & -\frac{1}{2} \end{pmatrix}.$$

Since $r_1 = 3$, and $r_2 = 3$,

$$b' = \begin{pmatrix} 24 \\ 7 \\ 15 \end{pmatrix}.$$

Note that to obtain b', first the 1st and 3rd components of b were permuted, according to $r_1 = 3$ (which means the interchange of rows 1 and 3), followed by the permutation of the components 2 and 3, according to $r_2 = 3$ (which means the interchange of the rows 2 and 3). $Ly = b'$ gives

$$y = \begin{pmatrix} 24 \\ 3.5714 \\ -0.5000 \end{pmatrix},$$

and $Ux = y$ gives

$$x = \begin{pmatrix} 1 \\ 1 \\ 1 \end{pmatrix}.$$

Flop-count. The factorization process requires about $\frac{2}{3}n^3$ flops. The solution of each of the triangular systems $Ly = b'$ and $Ux = y$ requires n^2 flops. Thus, the solution of the linear system $Ax = b$ using Gaussian elimination with partial pivoting requires about $\frac{2}{3}n^3 + O(n^2)$ flops. Also, the process requires $O(n^2)$ comparisons for pivot identifications.

Stability of Gaussian Elimination Scheme for $Ax = b$

We have seen that the growth factor ρ determines the stability of the triangularization procedure. Since solutions of triangular systems are numerically stable procedures, the growth factor is still the dominating factor for solving linear systems with Gaussian elimination.

The large growth factor ρ for Gaussian elimination with partial pivoting is rare in practice. **Thus, for all practical purposes, Gaussian elimination with partial pivoting for the linear system** $Ax = b$ **is a numerically stable procedure.**

3.5.3 Solving a Hessenberg Linear System

Certain control computations such as **computing the frequency response of a matrix** (see Chapter 5) require solution of a Hessenberg linear algebraic system. We have just seen that the LU factorization of a Hessenberg matrix requires only $O(n^2)$ flops and Gaussian elimination with partial pivoting is safe, because, the growth factor in this case is at most n. Thus, **a Hessenberg system can be solved using Gaussian elimination with partial pivoting using** $O(n^2)$ **flops and in a numerically stable way**.

3.5.4 Solving $AX = B$

In many practical situations, one faces the problem of solving multiple linear systems: $AX = B$. Here A is $n \times n$ and nonsingular and B is $n \times p$. Since each of the systems here has the same coefficient matrix A, to solve $AX = B$, we need to factor A just once. The following scheme, then, can be used.

Partition $B = (b_1, \ldots, b_p)$.

Step 1. Factorize A using Gaussian elimination with partial pivoting: $PA = LU$

Step 2. For $k = 1, \ldots, p$ do

        Solve $Ly = Pb_k$

        Solve $Ux_k = y$

        End

Step 3. Form $X = (x_1, \ldots, x_p)$.

3.5.5 Finding the Inverse of A

The inverse of an $n \times n$ nonsingular matrix A can be obtained as a special case of the above method. Just set $B = I_{n \times n}$. Then, $X = A^{-1}$.

3.5.6 Computing the Determinant of A

The determinant of matrix A can be immediately computed, once the LU factorization of A is available. Thus, if Gaussian elimination with partial pivoting is used giving $PA = LU$, then $\det(A) = (-1)^r \prod_{i=1}^{n} u_{ii}$, where r is the number of row interchanges in the partial pivoting process.

3.5.7 Iterative Refinement

Once the system $Ax = b$ is solved using Gaussian elimination, it is suggested that the computed solution be refined iteratively to a desired accuracy using the following procedure. The procedure is fairly inexpensive and requires only $O(n^2)$ flops for each iteration.

Let x be the computed solution of $Ax = b$ obtained by using Gaussian elimination with partial pivoting factorization: $PA = LU$.

For $k = 1, 2, \ldots$, do until desired accuracy.

    1. Compute the residual $r = b - Ax$ (in double precision).

    2. Solve $Ly = Pr$ for y.

    3. Solve $Uz = y$ for z.

    4. Update the solution $x \equiv x + z$.

3.6 THE QR FACTORIZATION

Recall that a square matrix O is said to be an **orthogonal matrix** if $OO^{\mathrm{T}} = O^{\mathrm{T}}O = I$. Given an $m \times n$ matrix A there exist an $m \times m$ orthogonal matrix Q

and an $m \times n$ upper triangular matrix R such that $A = QR$. Such a factorization of A is called the **QR factorization**. If $m \geq n$, and if the matrix Q is partitioned as $Q = [Q_1, Q_2]$, where Q_1 is the matrix of the first n columns of Q, and if R_1 is defined by

$$R = \begin{pmatrix} R_1 \\ 0 \end{pmatrix},$$

where R_1 is $n \times n$ upper triangular, then $A = Q_1 R_1$. This QR factorization is called the **"economy size"** or the **"thin"** QR factorization of A. The following theorem gives condition for uniqueness of the "thin" QR factorization. For a proof of the theorem, see Golub and Van Loan (1996, p. 230).

Theorem 3.6.1. *Let $A \in \mathbb{R}^{m \times n}, m \geq n$ have full rank. Then the thin QR factorization*

$$A = Q_1 R_1$$

is unique. Furthermore, the diagonal entries of R_1 are all positive.

There are several ways to compute the QR factorization of a matrix. Householder's and Givens' methods can be used to compute both types of QR factorizations. On the other hand, the classical Gram–Schmidt (CGS) and the modified Gram–Schmidt (MGS) compute $Q \in \mathbb{R}^{m \times n}$ and $R \in \mathbb{R}^{n \times n}$ such that $A = QR$.

The MGS has better numerical properties than the CGS. We will not discuss them here. The readers are referred to the book Datta (1995, pp. 339–343). We will discuss Householder's and Givens' methods in the sequel.

3.6.1 Householder Matrices

Definition 3.6.1. *A matrix of the form $H = I - 2uu^T/u^T u$, where u is a nonzero vector, is called a **Householder matrix**, after the celebrated American numerical analyst Alston Householder.*

A Householder matrix is also known as an **Elementary Reflector** or a **Householder transformation**.

It is easy to see that a Householder matrix H is **symmetric** and **orthogonal**.

A Householder matrix H is an important tool to create zeros in a vector:

Given $x = (x_1, x_2, \ldots, x_n)^T$, the Householder matrix $H = I - 2(uu^T/u^T u)$, where $u = x + \text{sgn}(x_1) \| x \|_2 e_1$ is such that $Hx = (\sigma, 0, \ldots, 0)^T$, where $\sigma = -\text{sgn}(x_1) \| x \|_2$.

Schematically, $x \xrightarrow{H} Hx = (\sigma, 0, \ldots, 0)^T$.

Forming Matrix–Vector and Matrix–Matrix Products With a Householder Matrix

A remarkable computational advantage involving Householder matrices is that neither a matrix–vector product with a Householder matrix H nor the matrix product HA (or AH) needs to be explicitly formed, as can be seen from the followings:

1. $Hx = \left(I - 2\dfrac{uu^T}{u^Tu}\right)x = x - \beta u(u^Tx)$, where $\beta = \dfrac{2}{u^Tu}$.
2. $HA = (I - \beta uu^T)A = A - \beta uu^TA = A - \beta uv^T$, where $v = A^Tu$.
3. $AH = A(I - \beta uu^T) = A - \beta vu^T$, where $v = Au$.

From above, we immediately see that the matrix product HA or AH requires only $O(n^2)$ flops, a substantial saving compared to $2n^3$ flops that are required to compute the product of two arbitrary matrices.

3.6.2   The Householder QR Factorization

Householder's method for the QR factorization of matrix $A \in \mathbb{R}^{m \times n}$ with $m \geq n$, consists of constructing Householder matrices $H_1, H_2, \ldots, H_n$ successively such that

$$H_n H_2 \cdots H_1 A = R$$

is an $m \times n$ upper triangular matrix. If $H_1 H_2 \cdots H_n = Q$, then Q is an orthogonal matrix (since each H_i is orthogonal) and from above, we have $Q^TA = R$ or $A = QR$. Note that

$$R = \begin{pmatrix} R_1 \\ 0 \end{pmatrix},$$

where $R_1 \in \mathbb{R}^{n \times n}$ and is upper triangular. The matrices H_i are constructed such that $A^{(i)} = H_i A^{(i-1)}$ (with $A^{(0)} = A$) has zeros below the diagonal in the ith column (see **Example 3.6.1**).

Flop-count. The Householder QR factorization method requires approximately $2n^2(m - (n/3))$ flops just to compute the triangular matrix R.

Note: The matrix Q can be computed, if required, as $Q = H_1 \cdots H_n$ by forming the product implicitly, as shown in Section 3.6.1.

It should be noted that in a majority of practical applications, it is sufficient to have Q in this factored form; in many applications, Q is not needed at all. If Q is needed explicitly, about another $4(m^2n - mn^2 + (n^3/3))$ flops will be required.

Numerical stability: The Householder QR factorization method computes the QR factorization of a slightly perturbed matrix. Specifically, it can be shown Wilkinson (1965, p. 236) that, if $\hat{R}$ denotes the computed R, then there exists an orthogonal $\hat{Q}$ such that

$$A + E = \hat{Q}\hat{R}, \quad \text{where } \|E\|_2 \approx \mu \|A\|_2 .$$

The algorithm is thus stable.

MATLAB notes: $[Q, R] = \boldsymbol{qr}(A)$ computes the QR factorization of A, using Householder's method.

Example 3.6.1.

$$A = \begin{pmatrix} 1 & 1 \\ 0.0001 & 0 \\ 0 & 0.0001 \end{pmatrix}$$

$k = 1$
  Form H_1:

$$u_1 = \begin{pmatrix} 1 \\ 0.0001 \\ 0 \end{pmatrix} + \sqrt{1 + (0.0001)^2} \begin{pmatrix} 1 \\ 0 \\ 0 \end{pmatrix} = \begin{pmatrix} 2 \\ 0.0001 \\ 0 \end{pmatrix}.$$

$$\text{Update } A \equiv A^{(1)} = H_1 A = \left(I - \frac{2u_1 u_1^T}{u_1^T u_1} \right) A = \begin{pmatrix} -1 & -1 \\ 0 & -0.0001 \\ 0 & 0.0001 \end{pmatrix}.$$

$k = 2$
  Form H_2:

$$u_2 = \begin{pmatrix} -0.0001 \\ 0.0001 \end{pmatrix} - \sqrt{(-0.0001)^2 + (0.0001)^2} \begin{pmatrix} 1 \\ 0 \end{pmatrix} = 10^{-4} \begin{pmatrix} -2.4141 \\ 0.1000 \end{pmatrix}.$$

$$\hat{H}_2 = \begin{pmatrix} 1 & 0 \\ 0 & 1 \end{pmatrix} - 2\frac{u_2 u_2^T}{u_2^T u_2} = \begin{pmatrix} -0.7071 & 0.7071 \\ 0.7071 & 0.7071 \end{pmatrix},$$

$$H_2 = \begin{pmatrix} 1 & 0 & 0 \\ 0 & -0.7071 & 0.7071 \\ 0 & 0.7071 & 0.7071 \end{pmatrix}.$$

$$\text{Update } A \equiv A^{(2)} = H_2 A^{(1)} = \begin{pmatrix} -1 & -1 \\ 0 & 0.0001 \\ 0 & 0 \end{pmatrix}.$$

Form Q and R:

$$Q = H_1 H_2 = \begin{pmatrix} -1 & 0.0001 & -0.0001 \\ -0.0001 & -0.7071 & 0.7071 \\ 0 & 0.7071 & 0.7071 \end{pmatrix}.$$

$$R = \begin{pmatrix} -1 & -1 \\ 0 & 0.0001 \\ 0 & 0 \end{pmatrix} = \begin{pmatrix} R_1 \\ 0 \end{pmatrix}, \quad \text{where } R_1 = \begin{pmatrix} -1 & -1 \\ 0 & 0.0001 \end{pmatrix}.$$

Complex *QR* Factorization

If $x \in \mathbb{C}^n$ and $x_1 = re^{i\theta}$, then it is easy to see that the Householder matrix $H = I - \beta v v^*$, where $v = x \pm e^{i\theta} \| x \|_2 e_1$ and $\beta = 2/v^*v$, is such that $Hx = \mp v e^{i\theta} \| x \|_2 e_1$.

Using the above formula, the Householder QR factorization method for a real matrix A, described in the last section, can be easily adapted to a complex matrix. The details are left to the readers.

The process of complex QR factorization of an $m \times n$ matrix, $m \geq n$, using Householder's method requires $8n^2(m - (n/3))$ real flops.

3.6.3   Givens Matrices

Definition 3.6.2. *A matrix of the form*

$$J(i, j, c, s) = \begin{pmatrix} 1 & 0 & 0 & \cdots & \cdots & \cdots & \cdots & \cdots & 0 \\ 0 & 1 & 0 & \cdots & \cdots & \cdots & \cdots & \cdots & 0 \\ \vdots & \vdots & \ddots & & & & & & \vdots \\ \vdots & \vdots & & & & & & \cdots & \vdots \\ 0 & 0 & 0 & \cdots & c & \cdots & s & \cdots & 0 \\ \vdots & \vdots & \vdots & & & & & \cdots & \vdots \\ 0 & 0 & 0 & \cdots & -s & \cdots & c & \cdots & 0 \\ \vdots & \vdots & \vdots & & & & & \ddots & \vdots \\ 0 & 0 & 0 & \cdots & \cdots & \cdots & 0 & \cdots & 1 \end{pmatrix} \quad \begin{matrix} \\ \\ \\ \\ \leftarrow i\text{th} \\ \\ \leftarrow j\text{th} \\ \\ \\ \end{matrix}$$

with *i*th and *j*th columns indicated above,

where $c^2 + s^2 = 1$, is called a **Givens matrix,** *after the name of the numerical analyst Wallace Givens.*

Since one can choose $c = \cos\theta$ and $s = \sin\theta$ for some θ, the above Givens matrix can be conveniently denoted by $J(i, j, \theta)$. Geometrically, the matrix $J(i, j, \theta)$ rotates a pair of coordinate axes (ith unit vector as its x-axis and the jth unit vector as its y-axis) through the given angle θ in the (i, j) plane. That is why, the Givens matrix $J(i, j, \theta)$ is commonly known as a **Givens Rotation** or **Plane Rotation in the** (i, j) **plane.**

Thus, when an n-vector $x = (x_1, x_2, \ldots, x_n)^{\mathrm{T}}$ is premultiplied by the Givens rotation $J(i, j, \theta)$, only the ith and jth components of x are affected; the other components remain unchanged.

Also, note that since $c^2 + s^2 = 1$; $J(i, j, \theta) \cdot J(i, j, \theta)^{\mathrm{T}} = I$. **So, the Givens matrix** $J(i, j, \theta)$ **is orthogonal.**

Zeroing the Entries of a 2×2 Vector Using a Givens Matrix

If

$$x = \begin{pmatrix} x_1 \\ x_2 \end{pmatrix}$$

is a vector, then it is a matter of simple verification that, with

$$c = \frac{x_1}{\sqrt{x_1^2 + x_2^2}} \quad \text{and} \quad s = \frac{x_2}{\sqrt{x_1^2 + x_2^2}},$$

the Givens rotation

$$J(1, 2, \theta) = \begin{pmatrix} c & s \\ -s & c \end{pmatrix}$$

is such that

$$J(1, 2, \theta)x = \begin{pmatrix} * \\ 0 \end{pmatrix}.$$

The preceding formula for computing c and s might cause some **underflow** or **overflow**. However, the following simple rearrangement of the formula might prevent that possibility.

1. If $|x_2| \geq |x_1|$, compute $t = x_1/x_2$, $s = 1/\sqrt{1 + t^2}$, and take $c = st$.
2. If $|x_2| < |x_1|$, compute $t = x_2/x_1$, $c = 1/\sqrt{1 + t^2}$, and take $s = ct$.

Implicit Construction of JA

If A is $\mathbb{R}^{m \times n}$ and $J(i, j, c, s) \in \mathbb{R}^{m \times m}$, then the update $A \equiv J(i, j, c, s)A$ can be computed implicitly as follows:

For $k = 1, \ldots, n$ do
$a \quad \equiv a_{ik}$
$b \quad \equiv a_{jk}$
$a_{ik} \equiv ac + bs$
$a_{jk} \equiv -as + bc$
End

MATCOM note: The above computation has been implemented in MATCOM program PGIVMUL.

3.6.4   The QR Factorization Using Givens Rotations

Assume that $A \in \mathbb{R}^{m \times n}$, $m \geq n$. The basic idea is just like Householder's: Compute orthogonal matrices $Q_1, Q_2, \ldots, Q_n$, using Givens rotations such that $A^{(1)} = Q_1 A$ has zeros below the $(1, 1)$ entry in the first column, $A^{(2)} = Q_2 A^{(1)}$ has zeros below the $(2, 2)$ entry in the second column, and so on. Each Q_i is generated as a product of Givens rotations. One way to form $\{Q_i\}$ is:

$$Q_1 = J(1, m, \theta)J(1, m-1, \theta) \cdots J(1, 2, \theta),$$
$$Q_2 = J(2, m, \theta)J(2, m-1, \theta) \cdots J(2, 3, \theta),$$

and so on.
Then,

$$R = A^{(n)} = Q_n A^{(n-1)} = Q_n Q_{n-1} A^{(n-2)} = \cdots$$
$$= Q_n Q_{n-1} \cdots Q_2 Q_1 A = Q^T A, \quad \text{where } Q = Q_1^T Q_2^T \cdots Q_n^T.$$

Algorithm 3.6.1.   *Givens QR Factorization*
Input. *A—An $m \times n$ matrix*
Outputs. *R—An $m \times n$ upper triangular matrix stored over A.*
Q—An $m \times m$ orthogonal matrix in factored form defined by the Givens parameters c, s, and the indices k and l.

Step 1. For $k = 1, 2, \ldots, n$ do

　　　For $l = k + 1, \ldots, m$ do

　　　1.1. Find c and s using the formulas given in **Section 3.6.3** so that

$$\begin{pmatrix} c & s \\ -s & c \end{pmatrix} \begin{pmatrix} a_{kk} \\ a_{lk} \end{pmatrix} = \begin{pmatrix} * \\ 0 \end{pmatrix}.$$

　　　1.2. Save the indices k and l and the numbers c and s

　　　1.3. Update A using the implicit construction as shown above:

　　　　　$A \equiv J(i, j, c, s) A$

　　　End

　　End

Step 2. Set $R \equiv A$.

Forming the matrix Q. If the orthogonal matrix Q is needed explicitly, then it can be computed from the product $Q = Q_1^T Q_2^T \cdots Q_n^T$, where each Q_i is the product of $m - i$ Givens rotations: $Q_i = J(i, m, \theta) J(i, m-1, \theta) \cdots J(i, i+1, \theta)$.

Flop-count. The algorithm requires $3n^2(m - n/3)$ flops; $m \geq n$. This count, of course, does not include computation of Q.

Numerical stability. **The algorithm is stable.** It can be shown Wilkinson (1965, p. 240) that for $m = n$, the computed $\hat{Q}$ and $\hat{R}$ satisfy $\hat{R} = \hat{Q}^T(A + E)$, where $\| E \|_F$ is small.

MATCOM note: The above algorithm has been implemented in MATCOM program GIVQR.

Q and R have been explicitly computed.

3.6.5　The QR Factorization of a Hessenberg Matrix Using Givens Matrices

From the structure of an upper Hessenberg matrix H, it is easy to see that the **QR factorization of H takes only** $O(n^2)$ **flops** either by Householder's or Givens' method, compared to $O(n^3)$ procedure for a full matrix. This is because only one entry from each column has to be made zero. Try this yourself using Algorithm 3.6.1.

3.7　ORTHONORMAL BASES AND ORTHOGONAL PROJECTIONS USING QR FACTORIZATION

The QR factorization of A can be used to compute the orthonormal bases and orthogonal projections associated with the subspaces $R(A)$ and $N(A^T)$. Let A be $m \times n$, where $m \geq n$ and have full rank. Suppose $Q^T A = R = \binom{R_1}{0}$. Partition $Q = (Q_1, Q_2)$, where Q_1 has n columns. **Then the columns of Q_1 form an orthonormal basis for $R(A)$.** Similarly, **the columns of Q_2 form an orthonormal**

basis for the orthogonal complement of $R(A)$. Thus, the matrix $P_A = Q_1 Q_1^T$
is the **orthogonal projection onto** $R(A)$ and the matrix $P_A^\perp = Q_2 Q_2^T$ is the
projection onto the orthogonal complement of $R(A)$. The above projections
can also be computed using the SVD (see **Section 3.9.2**).

MATLAB note: MATLAB function orth(A) computes the orthonormal basis
for $R(A)$.

QR Factorization with Column Pivoting

If A is rank-deficient, then QR factorization cannot be used to find a basis for $R(A)$.
In this case, one needs to use a modification of the QR factorization process, called
QR factorization with column pivoting.

We shall not discuss this here. The process finds a permutation matrix P, and
the matrices Q and R such that $AP = QR$. The details are given in Golub and
Van Loan (1996, pp. 248–250).

MATLAB function $[Q, R, P] = QR(A)$ can be used to compute the QR
factorization with column pivoting.

Also, $[Q, R, E] = QR(A, 0)$ produces an economy size QR factorization in
which E is a permutation vector so that $Q^* R = A(:, E)$.

3.8   THE LEAST-SQUARES PROBLEM

One of the most important applications of the QR factorization of a matrix A is
that it can be effectively used to solve the **least-squares problem** (LSP).

The **linear** LSP is defined as follows:

Given an $m \times n$ matrix A and a real vector b, find a real vector x such that
the function:
$$\| r(x) \|_2 = \| Ax - b \|_2$$
is minimized.

If $m > n$, the problem is called an **overdetermined LSP**, if $m < n$, it is called
an **underdetermined problem**.
We will discuss here only the overdetermined problem.

Theorem 3.8.1.   *Theorem on Existence and Uniqueness of the LSP. The least-
squares solution to $Ax = b$ always exists. The solution is unique if and only if
A has full rank. Otherwise, it has infinitely many solutions. The unique solution
x is obtained by solving $A^T Ax = A^T b$.*

Proof.   See Datta (1995, p. 318).   ■

3.8.1 Solving the Least-Squares Problem Using Normal Equations

The expression of the unique solution in Theorem 3.8.1 immediately suggests the following procedure, called the **Normal Equations** method, for solving the LSP:

1. Compute the **symmetric positive definite matrix** $A^T A$ (Note that if A has full rank, $A^T A$ is symmetric positive definite).
2. Solve for x: $A^T A x = A^T b$.

Computational remarks. The above procedure, though simple to understand and implement, has **serious numerical difficulties**. First, some significant figures may be lost during the explicit formation of the matrix $A^T A$. Second, **the matrix $A^T A$ will be more ill-conditioned, if A is ill-conditioned**. In fact, it can be shown that $\mathrm{Cond}_2(A^T A) = (\mathrm{Cond}_2(A))^2$. The following simple example illustrates the point. Let

$$A = \begin{pmatrix} 1 & 1 \\ 10^{-4} & 0 \\ 0 & 10^{-4} \end{pmatrix}.$$

If eight-digit arithmetic is used, then $A^T A = \begin{pmatrix} 1 & 1 \\ 1 & 1 \end{pmatrix}$, which is **singular**, though the columns of A are linearly independent.

A computationally effective method via the QR factorization of A is now presented below.

3.8.2 Solving the Least-Squares Problem Using QR Factorization

Let $Q^T A = R = \begin{pmatrix} R_1 \\ 0 \end{pmatrix}$ be the QR decomposition of the matrix A. Then, since the length of a vector is preserved by an orthogonal matrix multiplication, we have

$$\| Ax - b \|_2^2 = \| Q^T Ax - Q^T b \|_2^2$$

$$= \| R_1 x - c \|_2^2 + \| d \|_2^2, \quad \text{where } Q^T b = \begin{pmatrix} c \\ d \end{pmatrix}.$$

Thus, $\| Ax - b \|_2^2$ will be minimized if x is chosen so that $R_1 x - c = 0$. The corresponding residual norm then is given by $\| r \|_2 = \| Ax - b \|_2 = \| d \|_2$. This observation immediately suggests the following QR algorithm for solving the LSP:

Algorithm 3.8.1. *Least Squares Solution Using QR Factorization of A*
Inputs. *A—An $m \times n$ matrix $(m \geq n)$*
b—An m-vector.

Output. *The least-squares solution x to the linear system* $Ax = b$.
Step 1. *Decompose* $A = QR$, *where* $Q \in \mathbb{R}^{m \times m}$ *and* $R \in \mathbb{R}^{m \times n}$.
Step 2. *Form* $Q^T b = \binom{c}{d}$, $c \in \mathbb{R}^{n \times 1}$.
Step 3. *Obtain x by solving the upper triangular system:* $R_1 x = c$ *where*
$R = \binom{R_1}{0}$.
Step 4. *Obtain the residual norm:* $\| r \|_2 = \| d \|_2$.

Example 3.8.1.   Solve $Ax = b$ for x with

$$A = \begin{pmatrix} 1 & 2 \\ 2 & 3 \\ 3 & 4 \end{pmatrix}, \qquad b = \begin{pmatrix} 3 \\ 5 \\ 9 \end{pmatrix}.$$

Step 1. Find the QR factorization of $A : A = QR$

$$Q = \begin{pmatrix} -0.2673 & 0.8729 & 0.4082 \\ -0.5345 & 0.2182 & -0.8165 \\ -0.8018 & -0.4364 & 0.4082 \end{pmatrix},$$

$$R = \begin{pmatrix} -3.7417 & -5.3452 \\ 0 & 0.6547 \\ \hline 0 & 0 \end{pmatrix} = \begin{pmatrix} R_1 \\ 0 \end{pmatrix}.$$

Step 2. Form

$$Q^T b = \begin{pmatrix} c \\ d \end{pmatrix} = \begin{pmatrix} -10.6904 \\ -0.2182 \\ 0.8165 \end{pmatrix}.$$

Step 3. Obtain x by solving $R_1 x = c$: $x = \begin{pmatrix} 3.3532 \\ -0.3333 \end{pmatrix}$.

Step 4. $\| r \|_2 = \| d \|_2 = 0.8165$.

Use of Householder Matrices

Note that if the Householder's or Givens' method is used to compute the QR decomposition of A, then the product $Q^T b$ can be formed from the factored form of Q without explicitly computing the matrix Q.

MATCOM and MATLAB notes: MATCOM function **lsfrqrh** implements the QR factorization method for the full-rank least-squares problem using Householder's method. Alternatively, one can use the MATLAB operator: \. The command $x = A \backslash b$ gives the least-squares solution to $Ax = b$.

Flop-count and numerical stability: The least-squares method, using Householder's QR factorization, requires about $2(mn^2 - (n^3/3))$ flops. **The algorithm is numerically stable** in the sense that the computed solution satisfies a **"nearby"** LSP.

3.9   THE SINGULAR VALUE DECOMPOSITION (SVD)

We have seen two factorizations (decompositions) of A: LU and QR.

In this section we shall study another important decomposition, called the **singular value decomposition** or in short the **SVD** of A. Since $m \geq n$ is the case mostly arising in applications, we **will assume throughout this section that $m \geq n$.**

Theorem 3.9.1. *The SVD Theorem. Given $A \in \mathbb{R}^{m \times n}$, there exist orthogonal matrices $U \in \mathbb{R}^{m \times m}$ and $V \in \mathbb{R}^{n \times n}$, and a diagonal matrix $\Sigma \in \mathbb{R}^{m \times n}$ with nonnegative diagonal entries such that*

$$A = U \Sigma V^{\mathrm{T}}.$$

Proof.   See Datta (1995, pp. 552–554).   ■

The diagonal entries of Σ are called the **singular values** of A.

The columns of U are called the **left singular vectors**, and those of V are called the **right singular vectors. The singular values are unique, but U and V are not unique.**

The number of nonzero singular values is equal to the rank of the matrix A.

A convention. The n singular values $\sigma_1, \sigma_2, \ldots, \sigma_n$ of A can be arranged in nondecreasing order: $\sigma_1 \geq \sigma_2 \geq \cdots \geq \sigma_n$. The largest singular value σ_1 is denoted by $\sigma_{\max}$. Similarly, the smallest singular value σ_n is denoted by $\sigma_{\min}$.

The thin SVD. Let $U = (u_1, \ldots, u_m)$.

If $A = U \Sigma V^{\mathrm{T}}$ be the SVD of $A \in \mathbb{R}^{m \times n}$ and if $U_1 = (u_1, \ldots, u_n) \in \mathbb{R}^{m \times n}$, $\Sigma_1 = \mathrm{diag}(\sigma, \ldots, \sigma_n)$, then $A = U_1 \Sigma_1 V^{\mathrm{T}}$.

This factorization is known as the **thin SVD of A.** For obvious reasons, the thin SVD is also referred to as the **economic SVD.**

Relationship between eigenvalues and singular values. It can be shown that (see Datta (1995, pp. 555–557)).

1. The singular values $\sigma_1, \ldots, \sigma_n$ of A are the nonnegative square roots of the eigenvalues of the symmetric positive semidefinite matrix $A^{\mathrm{T}}A$.
2. The right singular vectors are the eigenvectors of the matrix $A^{\mathrm{T}}A$, and the left singular vectors are the eigenvectors of the matrix AA^{T}.

Sensitivity of the singular values. A remarkable property of the singular values is that **they are insensitive to small perturbations.** In other words, the **singular values are well-conditioned.** Specifically, the following result holds.

Theorem 3.9.2. *Insensitivity of the Singular Values. Let A be an $m \times n$ ($m \geq n$) matrix with the singular values $\sigma_1, \ldots, \sigma_n$, and $B = A+E$ be another slightly perturbed matrix with the singular values $\bar{\sigma}_1, \ldots, \bar{\sigma}_n$, then $| \bar{\sigma}_i - \sigma_i | \leq \| E \|_2$, $i = 1, \ldots, n$.*

Proof.   See Datta (1995, pp. 560–561).   ■

Example 3.9.1.   Let

$$A = \begin{pmatrix} 1 & 2 & 3 \\ 3 & 4 & 5 \\ 6 & 7 & 8 \end{pmatrix}, \qquad E = \begin{pmatrix} 0 & 0 & 0 \\ 0 & 0 & 0 \\ 0 & 0 & 2 \times 10^{-4} \end{pmatrix}.$$

The singular values of A:

$$\sigma_1 = 14.5576, \qquad \sigma_2 = 1.0372, \qquad \sigma_3 = 0.$$

The singular values of $A + E$:

$$\bar{\sigma}_1 = 14.5577, \qquad \bar{\sigma}_2 = 1.0372, \qquad \bar{\sigma}_3 = 2.6492 \times 10^{-5}.$$

It is easily verified that the inequalities in the above theorem are satisfied.

3.9.1   The Singular Value Decomposition and the Structure of a Matrix

The SVD is an effective tool in handling several computationally sensitive computations, such as the **rank** and **rank-deficiency** of matrix, the **distance of a matrix from a matrix of immediate lower rank, the orthogonormal basis and projections**, etc. It is also a reliable and numerically stable way of computing the least-squares solution to a linear system. Since these computations need to be performed routinely in control and systems theory, we now discuss them briefly in the following. The results of Theorem 3.9.3 can be easily proved.

Theorem 3.9.3.   *Let $\sigma_1 \geq \sigma_2 \geq \cdots \geq \sigma_n$ be the n singular values of an $m \times n$ matrix $A(m \geq n)$. Then,*

1. $\|A\|_2 = \sigma_1 = \sigma_{\max}$,
2. $\|A\|_F = (\sigma_1^2 + \sigma_2^2 + \cdots + \sigma_n^2)^{1/2}$,
3. $\|A^{-1}\|_2 = \dfrac{1}{\sigma_n}$, *when A is $n \times n$ and nonsingular,*
4. $\mathrm{Cond}_2(A) = \|A\|_2 \|A^{-1}\|_2 = \dfrac{\sigma_1}{\sigma_n} = \dfrac{\sigma_{\max}}{\sigma_{\min}}$, *when A is $n \times n$ and nonsingular.*

The Condition Number of a Rectangular Matrix

The condition number (with respect to 2-norm) of a rectangular matrix A of order $m \times n$ ($m \geq n$) with full rank is defined to be

$$\text{Cond}_2(A) = \frac{\sigma_{\max}(A)}{\sigma_{\min}(A)},$$

where $\sigma_{\max}(A)$ and $\sigma_{\min}(A)$ denote, respectively, the largest and smallest singular value of A.

Remark

- When A is rank-deficient, $\sigma_{\min} = 0$, and we say that $\text{Cond}(A)$ is **infinite**.

3.9.2 Orthonormal Bases and Orthogonal Projections

Let r be the rank of A, that is,

$$\sigma_1 \geq \sigma_2 \geq \cdots \geq \sigma_r > 0,$$
$$\sigma_{r+1} = \cdots = \sigma_n = 0.$$

Let u_j and v_j be the jth columns of U and V in the SVD of A. **Then the set of columns $\{v_j\}$ corresponding to the zero singular values of A form an orthonormal basis for the null-space of** A. This is because, when $\sigma_j = 0$, v_j satisfies $Av_j = 0$ and is therefore in the null-space of A. Similarly, **the set of columns $\{u_j\}$ corresponding to the nonzero singular values is an orthonormal basis for the range of** A. The orthogonal projections now can be easily computed.

Orthogonal Projections

Partition U and V as

$$U = (U_1, U_2), \qquad V = (V_1, V_2),$$

where U_1 and V_1 consist of the first r columns of U and V, then

1. Projection onto $R(A) = U_1 U_1^T$.
2. Projection onto $N(A) = V_2 V_2^T$.
3. Projection onto the orthogonal complement of $R(A) = U_2 U_2^T$.
4. Projection onto the orthogonal complement of $N(A) = V_1 V_1^T$.

Example 3.9.2.

$$A = \begin{pmatrix} 1 & 2 & 3 \\ 3 & 4 & 5 \\ 6 & 7 & 8 \end{pmatrix}.$$

$$\sigma_1 = 14.5576, \qquad \sigma_2 = 1.0372, \qquad \sigma_3 = 0.$$

$$U = \begin{pmatrix} 0.2500 & 0.8371 & 0.4867 \\ 0.4852 & 0.3267 & -0.8111 \\ 0.8378 & -0.4379 & 0.3244 \end{pmatrix}.$$

$$V = \begin{pmatrix} 0.4625 & -0.7870 & 0.4082 \\ 0.5706 & -0.0882 & -0.8165 \\ 0.6786 & -0.6106 & 0.4082 \end{pmatrix}.$$

An orthonormal basis for the null-space of A is:

$$V_2 = \left\{ \begin{array}{c} 0.4082 \\ -0.8165 \\ 0.4082 \end{array} \right\}.$$

An orthonormal basis for the range of A is:

$$U_1 = \left\{ \begin{array}{cc} 0.2500 & 0.8371 \\ 0.4852 & 0.3267 \\ 0.8370 & -0.4379 \end{array} \right\}.$$

(Now compute the four orthogonal projections yourself.)

3.9.3   The Rank and the Rank-Deficiency of a Matrix

The most obvious and the least expensive way of determining the rank of a matrix is, of course, to triangularize the matrix using Gaussian elimination and then to find the rank of the reduced upper triangular matrix. Finding the rank of a triangular matrix is trivial; one can just read it off from the diagonal. Unfortunately, however, this is not a very reliable approach in floating point arithmetic. Gaussian elimination method which uses elementary transformations, may transform a rank-deficient matrix into one having full rank, due to numerical round-off errors. Thus, in practice, it is more important, **to determine if the given matrix is near a matrix of a certain rank and in particular, to know if it is near a rank-deficient matrix. The most reliable way to determine the rank and nearness to rank-deficiency is to use the SVD.**

Suppose that A has rank r, that is, $\sigma_1 \geq \sigma_2 \geq \cdots \geq \sigma_r > 0$ and $\sigma_{r+1} = \cdots = \sigma_n = 0$. Then the question is: *How far is A from a matrix of rank $k < r$.* The following theorem can be used to answer the question. We state the theorem below, without proof. For proof, see Datta (1995, pp. 565–566).

Theorem 3.9.4.   *Distance to Rank-Deficient Matrices. Let $A = U\Sigma V^T$ be the SVD of A, and let* $\text{rank}(A) = r > 0$. *Let $k < r$. Define $A_k = U\Sigma_k V^T$,* *where*

$$\Sigma_k = \begin{pmatrix} \sigma_1 & & 0 & \\ & \ddots & & 0 \\ 0 & & \sigma_k & \\ \hline & 0 & & 0 \end{pmatrix}, \quad (\sigma_1 \geq \sigma_2 \cdots \geq \sigma_k > 0).$$

1. *Then out of all the matrices of rank $k(k < r)$, the matrix A_k is closest to A.*
2. *Furthermore, the distance of A_k from A: $\|A - A_k\|_2 = \sigma_{k+1}$.*

Corollary 3.9.1. *The relative distance of a nonsingular matrix A to the nearest singular matrix B is $1/\text{Cond}_2(A)$. That is, $\|B - A\|_2/\|A\|_2 = 1/\text{Cond}_2(A)$.*

Implication of the Above Results

Distance of a Matrix to the Nearest Matrix of Lower Rank

The above result states that the smallest nonzero singular value of A gives the 2-norm distance of A to the nearest matrix of lower rank. In particular, for a nonsingular $n \times n$ matrix A, σ_n gives the measures of the distance of A to the nearest singular matrix.

Thus, in order to know if a matrix A of rank r is close enough to a matrix of lower rank, look into the smallest nonzero singular value σ_r. If this is very small, then the matrix is very close to a matrix of rank $r - 1$, because there exists a perturbation of size as small as $|\sigma_r|$ which will produce a matrix of rank $r - 1$. In fact, one such perturbation is $u_r\sigma_r v_r^T$.

Example 3.9.3. Let

$$A = \begin{pmatrix} 1 & 0 & 0 \\ 0 & 2 & 0 \\ 0 & 0 & 4 \times 10^{-7} \end{pmatrix},$$

$$\text{Rank}(A) = 3, \qquad \sigma_3 = 0.0000004, \qquad u_3 = v_3 = (0, 0, 1)^T.$$

$$A' = A - u_3\sigma_3 v_3^T = \begin{pmatrix} 1 & 0 & 0 \\ 0 & 2 & 0 \\ 0 & 0 & 0 \end{pmatrix}, \qquad \text{rank}(A') = 2.$$

The required perturbation $u_3\sigma_3 v_3^{\mathrm{T}}$ to make A singular is:

$$10^{-7} \begin{pmatrix} 0 & 0 & 0 \\ 0 & 0 & 0 \\ 0 & 0 & 4 \end{pmatrix}.$$

3.9.4   Numerical Rank

The above discussions prompt us to define the concept of **"Numerical Rank"** of a matrix. A has **"numerical rank"** r if the computed singular values $\tilde{\sigma}_1, \tilde{\sigma}_2, \ldots, \tilde{\sigma}_n$ satisfy:

$$\tilde{\sigma}_1 \geq \tilde{\sigma}_2 \geq \cdots \geq \tilde{\sigma}_r > \delta \geq \tilde{\sigma}_{r+1} \geq \cdots \geq \tilde{\sigma}_n, \tag{3.9.1}$$

where δ is an error tolerance.

   Thus to determine the numerical rank of a matrix A, count the "large" singular values only. If this number is r, then A has numerical rank r.

Remark

- Note that finding the numerical rank of a matrix will be "tricky" if there is no suitable gap between a set of singular values.

3.9.5   Solving the Least-Squares Problem Using the Singular Value Decomposition

The SVD is also an **effective tool** to solve the LSP, both in the full rank and rank-deficient cases.

   Recall that the **linear LSP** is: Find x such that $\|r\|_2 = \|Ax - b\|_2$ is minimum.

   Let $A = U\Sigma V^{\mathrm{T}}$ be the SVD of A. Then since U is orthogonal and a vector length is preserved by orthogonal multiplication, we have

$$\|r\|_2 = \|(U\Sigma V^{\mathrm{T}}x - b)\|_2 = \|U(\Sigma V^{\mathrm{T}}x - U^{\mathrm{T}}b)\|_2 = \|\Sigma y - b'\|_2,$$

where $V^{\mathrm{T}}x = y$ and $U^{\mathrm{T}}b = b'$. **Thus, the use of the SVD of A reduces the LSP for a full matrix A to one with a diagonal matrix Σ,** which is almost trivial to solve, as shown in the following algorithm.

   Algorithm 3.9.1.   *Least Squares Solutions Using the SVD*
   Inputs. *A—An $m \times n$ matrix,*
   b—An m-vector
   Output. *x—The least-squares solution of the system $Ax = b$.*
   Step 1. *Find the SVD of A: $A = U\Sigma V^{\mathrm{T}}$.*

Step 2. *Form* $b' = U^T b = \begin{pmatrix} b'_1 \\ b'_2 \\ \vdots \\ b'_m \end{pmatrix}.$

Step 3. *Compute*

$$y = \begin{pmatrix} y_1 \\ \vdots \\ y_n \end{pmatrix}$$

choosing

$$y_i = \left\{ \begin{array}{l} \dfrac{b'_i}{\sigma_i}, \text{ when } \sigma_i \neq 0 \\ \text{arbitrary, when } \sigma_i = 0. \end{array} \right\}.$$

Step 4. *Compute the family of least squares solutions:* $x = Vy$. (**Note that in the full-rank case, the family has just one number**).

Flop-count. Using the SVD, it takes about $4mn^2 + 8n^3$ flops to solve the LSP, when A is $m \times n$ and $m \geq n$.

An Expression for the Minimum Norm Least Squares Solution

Since a rank-deficient LSP has an infinite number of solutions, it is practical to look for the one that has minimum norm. Such a solution is called the **minimum norm least square solution.**

It is clear from Step 3 above that in the **rank-deficient case**, the minimum 2-norm least squares solution is the one that is obtained by setting $y_i = 0$, whenever $\sigma_i = 0$. Thus, from above, we have the following expression for the minimum 2-norm solution:

$$x = \sum_{i=1}^{k} \frac{u_i^T b'_i}{\sigma_i} v_i, \tag{3.9.2}$$

where k is the **numerical rank** of A, and u_i and v_i, respectively, are the ith columns of U and V.

Example 3.9.4.

$$A = \begin{pmatrix} 1 & 2 & 3 \\ 2 & 3 & 4 \\ 1 & 2 & 3 \end{pmatrix}, \qquad b = \begin{pmatrix} 6 \\ 9 \\ 6 \end{pmatrix}.$$

Step 1. $\sigma_1 = 7.5358$, $\sigma_2 = 0.4597$, $\sigma_3 = 0$.

A is rank-deficient.

$$U = \begin{pmatrix} 0.4956 & 0.5044 & 0.7071 \\ 0.7133 & -0.7008 & 0.0000 \\ 0.4956 & 0.5044 & -0.7071 \end{pmatrix},$$

$$V = \begin{pmatrix} 0.3208 & -0.8546 & 0.4082 \\ 0.5470 & -0.1847 & -0.8165 \\ 0.7732 & 0.4853 & 0.4082 \end{pmatrix}.$$

Step 2. $b' = U^{T}b = (12.3667, -0.2547, 0)^{T}$.
Step 3. $y = (1.6411, -0.5541, 0)$.
The minimum 2-norm least-squares solution is $Vy = (1, 1, 1)^{T}$.

Computing the SVD of A

Since the singular values of a matrix A are the nonnegative square roots of the eigenvalues of $A^{T}A$, it is natural to think of computing the singular values and the singular vectors, by finding the eigendecomposition of $A^{T}A$. However, **this is not a numerically effective procedure**.

Some vital information may be lost during the formation of the matrix $A^{T}A$, as the following example shows.

Example 3.9.5.

$$A = \begin{pmatrix} 1.0001 & 1.000 \\ 1.000 & 1.0001 \end{pmatrix}.$$

The singular values of A are 2.0001 and 0.0001.

$$A^{T}A = \begin{pmatrix} 2.0002 & 2.0002 \\ 2.0002 & 2.0002 \end{pmatrix}.$$

The eigenvalues of $A^{T}A$ are 0 and 4.0004 (in four-digit arithmetic). Thus, the singular values computed from the eigenvalues of $A^{T}A$ are 0 and 2.0002.

A standard algorithm for computing the SVD of A is the **Golub–Kahan–Reinsch** algorithm. The algorithm will be described later in the book in **Chapter 4**.

MATLAB and MATCOM notes: MATLAB function **svd** can be used to compute the SVD. $[U, S, V] = svd(A)$ produces a diagonal matrix S, of the same dimension as A and with nonnegative diagonal entries in decreasing order, and unitary matrices U and V such that $A = USV^{*}$.

Algorithm 3.9.1 has been implemented in MATCOM program **lsqrsvd**. Also, MATCOM has a program called **minmsvd** to compute the minimum 2-norm least-squares solution using the SVD.

3.10   SUMMARY AND REVIEW

Floating Point Numbers and Errors

1. *Floating-point numbers.*   A t-digit floating point number has the form:

$$x = m\beta^e,$$

where e is called exponent, m is a t-digit fraction, and β is the base of the number system.

2. *Errors.*   The errors in a computation are measured either by absolute error or relative error. **The relative errors make more sense than absolute errors**. The relative error gives an indication of the number of significant digits in an approximate answer. The relative error in representing a real number x by its floating-point representation $fl(x)$ is bounded by a number μ, called the **machine precision (Theorem 3.2.1)**.

3. *Laws of floating-point arithmetic:*

$$fl(x * y) = (x * y)(1 + \delta).$$

Conditioning, Stability, and Accuracy

1. *Conditioning of the problem.*   The conditioning of the problem is a property of the problem. A problem is said to be **ill-conditioned** if a small change in the data can cause a large change in the solution, otherwise it is **well-conditioned**. The conditioning of a problem is data-dependent. A problem can be ill-conditioned with respect to one set of data but can be quite well-conditioned with respect to another set.

    The condition number of a nonsingular matrix, $\text{Cond}(A) = \|A\| \, \|A^{-1}\|$ is an indicator of the conditioning of the associated linear system problem: $Ax = b$. If $\text{Cond}(A)$ is large, then the linear system $Ax = b$ is ill-conditioned.

    The well-known examples of ill-conditioned problems are the **Wilkinson polynomial** for the root-finding problem, the **Wilkinson bidiagonal matrix** for the eigenvalue problem, the **Hilbert matrix** for the algebraic linear system problem, and so on.

2. *Stability of an algorithm.*   An algorithm is said to be a *backward stable algorithm* if it computes the exact solution of a nearby problem. Some examples of stable algorithms are the methods of back substitution and forward elimination for triangular systems, the QR factorization using Householder and Givens matrices transformations, and the QR iteration algorithm for eigenvalue computations.

The Gaussian elimination algorithm without row changes is unstable for arbitrary matrices. However, Gaussian elimination with partial pivoting can be considered as a stable algorithm in practice.

3. *Effects of conditioning and stability on the accuracy of the solution.* The conditioning of the problem and the stability of the algorithm both have effects on accuracy of the solution computed by the algorithm.

   If a stable algorithm is applied to a well-conditioned problem, it should compute an accurate solution. On the other hand, if a stable algorithm is applied to an ill-conditioned problem, there is no guarantee that the computed solution will be accurate. However, if a stable algorithm is applied to an ill-conditioned problem, it should not introduce more errors than that which the data warrants.

Matrix Factorizations

There are three important matrix factorizations: LU, QR, and SVD.

1. *LU factorization.* A factorization of a matrix A in the form $A = LU$, where L is unit lower triangular and U is upper triangular, is called an LU factorization of A. An LU factorization of A exists if all of its leading principal minors are nonsingular.

   A classical elimination scheme, called **Gaussian elimination**, is used to obtain an LU factorization of A (**Section 3.4.1**).

   The stability of Gaussian elimination is determined by the **growth factor**

   $$\rho = \frac{\max(\alpha, \alpha_1, \ldots, \alpha_{n-1})}{\alpha},$$

   where $\alpha = \max_{i,j} |a_{ij}|$ and $\alpha_k = \max_{i,j} |a_{ij}^{(k)}|$.

   If no pivoting is used in Gaussian elimination, ρ can be arbitrarily large. Thus, **Gaussian elimination without pivoting is, in general, an unstable process**.

   If partial pivoting is used, then Gaussian elimination yields the factorization of A in the form $PA = LU$, where P is a perturbation matrix.

   The growth factor ρ for Gaussian elimination with partial pivoting can be as large as 2^{n-1}; however, such a growth is extremely rare in practice. Thus, **Gaussian elimination with partial pivoting is considered to be a stable process in practice**.

2. *The QR factorization.* Given an $m \times n$ matrix A, there exists an orthogonal matrix Q and an upper triangular matrix R such that $A = QR$.

The QR factorization of A can be obtained using **Householder's method, Givens' method, the Gram–Schmidt processes (the CGS and MGS)**.

The Gram–Schmidt processes do not have favorable numerical properties. **Both Householder's and Givens' methods are numerically stable procedures for QR factorization.** They are discussed, respectively, in **Section 3.6.2 and Section 3.6.4 (Algorithm 3.6.1)**. Householder's method is slightly more efficient than Givens' method.

The Algebraic Linear System Problem $Ax = b$

The method of practical choice for the linear system problem $Ax = b$ is Gaussian elimination with partial pivoting (**Section 3.5.2**) followed by iterative refinement procedure (**Section 3.5.7**). A symmetric positive definite system should be solved by computing its Cholesky factor (**Algorithm 3.4.1**) R followed by solving two triangular systems: $Ry = b$ and $R^T x = y$ (**Algorithm 3.3.1** and **Section 3.3.3**).

The Least-Squares Problem

Given an $m \times n$ matrix A, the LSP is the problem of finding a vector x such that $\| Ax - b \|_2$ is minimized. The LSP can be solved using:

- The normal equations method (**Section 3.8.1**): $A^T Ax = A^T b$
- The QR factorization method (**Algorithm 3.8.1**)
- The SVD method (**Algorithm 3.9.1**).

The normal equations method might give numerical difficulties, and should not be used in practice without looking closely at the condition number. Both the QR and SVD methods for the LSP are numerically stable. Though the SVD is more expensive than the QR method, **the SVD method is most reliable and can handle both rank-deficient and full-rank cases very effectively.**

The Singular Value Decomposition

1. *Existence and uniqueness of the SVD.* The SVD of a matrix A always exists (**Theorem 3.9.1**):
   Let $A \in \mathbb{R}^{m \times n}$. Then $A = U \Sigma V^T$, where $U \in \mathbb{R}^{m \times m}$, $V \in \mathbb{R}^{n \times n}$ are orthogonal and Σ is an $m \times n$ diagonal matrix.
   The singular values (the diagonal entries of Σ) are unique, but U and V are not unique.

2. *Relationship between the singular values and the eigenvalues.* The singular values of A are the nonnegative square roots of the eigenvalues of $A^T A$ (or of $A A^T$).

3. *Sensitivity of the singular values.*   The singular values are insensitive to small perturbations (**Theorem 3.9.2**).

4. *Applications of the SVD.*   The singular values and the singular vectors of a matrix A are useful and are the most reliable tools for determining the (numerical) rank and the rank-deficiency of A; finding the orthonormal bases for range and the null space of A; finding the distance of A from another matrix of lower rank (in particular, the nearness to singularity of a nonsingular matrix); solving both full-rank and the rank-deficient LSPs.

These remarkable abilities and the fact that the singular values are insensitive to small perturbations have made the SVD an indispensable tool for a wide variety of problems in control and systems theory, as we will see throughout the book.

3.11   CHAPTER NOTES AND FURTHER READING

Material of this chapter has been taken from the recent book of the author (Datta 1995). For the advanced topics on numerical linear algebra, see Golub and Van Loan (1996). The details about stability of various algorithm and sensitivities of problems described in this chapter can be found in the book by Higham (1996). Stewart's (1998) recent book is also an excellent source of knowledge in this area. For details of various MATLAB functions, see MATLAB Users' Guide (1992). *MATCOM* is a MATLAB-based toolbox implementing all the major algorithms of the book "*Numerical Linear Algebra and Applications*" by Datta (1995). MATCOM can be obtained from the book's web page on the web site of MATH-WORKS: http://www.mathworks.com/support/books/book1329.jsp. The software (MATCOM) is linked at the bottom.

References

Datta B.N. *Numerical Linear Algebra and Applications*, Brooks/Cole Publishing Company, Pacific Grove, CA, 1995 (*Custom published by Brooks/Cole, 2003*).
Golub G.H. and Van Loan C.F. *Matrix Computations*, 3rd edn, Johns Hopkins University Press, Baltimore, MD, 1996.
Higham N.J. *Accuracy and Stability of Numerical Algorithms*, SIAM, Philadelphia, 1996.
MATCOM web site: http://www.mathworks.com/support/books/book1329.jsp.
MATLAB *User's Guide*, The Math Works, Inc., Natick, MA, 1992.
Stewart G.W. *Matrix Algorithms Volume 1: Basic Decompositions*, SIAM, Philadelphia, 1998.
Wilkinson J.H. *The Algebraic Eigenvalue Problem*, Clarendon Press, Oxford, 1965.

CHAPTER 4

CANONICAL FORMS OBTAINED VIA ORTHOGONAL TRANSFORMATIONS

Topics covered

- Numerical Instabilities in obtaining the Jordan and Companion Matrices
- Hessenberg Reduction of a Matrix
- The Double-Shift Implicit QR Iteration for the Real Schur Form (RSF) of a Matrix
- Invariant Subspace Computation from the RSF
- The QZ Algorithm for the Generalized RSF of the Matrix Pencil $A - \lambda B$
- The SVD Computation

4.1 IMPORTANCE AND SIGNIFICANCE OF USING ORTHOGONAL TRANSFORMATIONS

The Jordan and companion matrices have special structures that can be conveniently exploited to solve many control problems. **Unfortunately, however, these forms in general, cannot be obtained in a numerically stable way.**

We examine this fact here in some detail below.

Suppose that X is a nonsingular matrix and consider the computation of $X^{-1}AX$ in floating point arithmetic. It can be shown that

$$fl(X^{-1}AX) = X^{-1}AX + E,$$

where $\|E\|_2 \approx \mu \mathrm{Cond}(X)\|A\|_2$, μ is the machine precision.

Thus, when X is ill-conditioned, there will be large errors in computing $X^{-1}AX$.

For the Jordan canonical form (JCF), the transforming matrix X is highly ill-conditioned, whenever A has defective or nearly defective eigenvalue.

The reduction of a matrix A to an upper companion matrix C (**Section 2.4.5**) involves the following steps:

Step 1. A is transformed to an upper Hessenberg matrix $H_u = (h_{ij})$ by orthogonal similarity: $P^T A P = H_u$.

Step 2. Assuming that H_u is unreduced, that is, $h_{i+1,i} \neq 0, i = 1, 2, \ldots, n-1$, then H_u is further reduced to the companion matrix C by similarity. Thus, if $Y = (e_1, H_u e_1, \ldots, H_u^{n-1} e_1)$, it is easy to see that $Y^{-1} H_u Y = C$.

A numerically stable algorithm to implement Step 1 is given in the next section; however, the matrix Y in Step 2 **can be highly ill-conditioned if H_u has small subdiagonal entries.**

(Note that Y is a lower triangular matrix with $1, h_{21}h_{32}, \ldots, h_{21}h_{32} \ldots h_{n,n-1}$ as the diagonal entries).

Thus, Step 2, in general, cannot be implemented in a numerically effective manner.

The above discussions clearly show that it is important from a numerical computation viewpoint to have canonical forms which can be achieved using only well-conditioned transforming matrices, such as orthogonal matrices.

Indeed, **if a matrix A is transformed to a matrix B using an orthogonal similarity transformation, then a perturbation in A will result in a perturbation in B of the same magnitude.** That is, if

$$B = U^T A U \quad \text{and} \quad U^T(A + \Delta A)U = B + \Delta B,$$

then $\|\Delta B\|_2 \approx \|\Delta A\|_2$.

In this chapter, we show that two very important canonical forms: the **Hessenberg** form and the **Real Schur Form** (RSF) of a matrix A, can be obtained using orthogonal similarity transformations. (Another important canonical form, known as the **generalized real Schur form**, can be obtained using orthogonal equivalence.)

We will see in the rest of the book that **these canonical forms form important tools in the development of numerically effective algorithms for control problems.**

Applications of **Hessenberg** and **real Schur** forms include:

1. Computation of frequency response matrix (**Chapter 5**)
2. Solutions of Lyapunov and Sylvester equations (**Chapter 8**), Algebraic Riccati equations (**Chapter 13**), Sylvester-observer equation (**Chapter 12**).
3. Solutions of eigenvalue assignment (**Chapter 11**), feedback stabilization problems (**Chapter 10**), stability and inertia computations (**Chapter 7**).

Applications of generalized real Schur form include:

1. Solutions of certain algebraic Riccati equations **(Chapter 13)**.
2. Solution of any descriptor control problem.
3. Computations of frequencies and modes of vibrating systems.

Besides these two forms, there are two other important canonical forms, namely, the **controller-Hessenberg** and **observer-Hessenberg** forms. These forms can also be obtained in a numerically effective way and will be used throughout the book. Methods for obtaining these two forms are described in **Chapter 6.**

4.2   HESSENBERG REDUCTION OF A MATRIX

Recall that a matrix $H = (h_{ij})$ is said to be an **upper Hessenberg** matrix if $h_{ij} = 0$ for $i > j + 1$.

An $n \times n$ matrix A can always be transformed to an upper Hessenberg matrix H_u by orthogonal similarity. That is, **given an $n \times n$ matrix A, there exists an orthogonal matrix P such that $PAP^T = H_u$.**

Again, Householder and Givens matrices, being orthogonal, can be employed to obtain H_u from A.

We will discuss only Householder's method here.

Reduction to Hessenberg Form using Householder Matrices

The idea is to extend the QR factorization process using Householder matrices described in Chapter 3 to obtain P and H_u, such that $PAP^T = H_u$ is an upper Hessenberg matrix and P is orthogonal.

The matrix P is constructed as the product of $(n - 2)$ Householder matrices P_1 through P_{n-2}. The matrix P_1 is constructed to create zeros in the first column of A below the entry $(2, 1)$; P_2 is constructed to create zeros below the entry $(3, 2)$ of the second column of the matrix $P_1 A P_1^T$, and so on.

The process consists of $(n - 2)$ steps. (Note that an $n \times n$ Hessenberg matrix contains at least $(n - 2)(n - 1)/2$ zeros.)

At the end of $(n-2)$th step, the matrix $A^{(n-2)}$ is an upper Hessenberg matrix H_u. The Hessenberg matrix H_u is orthogonally similar to A. This is seen as follows:

$$H_u = A^{(n-2)} = P_{n-2}A^{(n-3)}P_{n-2}^T = P_{n-2}(P_{n-3}A^{(n-4)}P_{n-3}^T)P_{n-2}^T$$
$$= \cdots = (P_{n-2}P_{n-3}\ldots P_1)A(P_1^T P_2^T \ldots P_{n-3}^T P_{n-2}^T). \quad (4.2.1)$$

Set

$$P = P_{n-2}P_{n-3}\ldots P_1. \quad (4.2.2)$$

We then have $H_u = PAP^T$. Since each Householder matrix P_i is orthogonal, the matrix P which is the product of $(n - 2)$ Householder matrices, is also orthogonal.

For $n = 4$, schematically, we can represent the reduction as follow. Set $A^{(0)} \equiv A$. Then,

$$A \xrightarrow{P_1} P_1 A P_1^T = \begin{pmatrix} * & * & * & * \\ * & * & * & * \\ 0 & * & * & * \\ 0 & * & * & * \end{pmatrix} = A^{(1)}.$$

$$A^{(1)} \xrightarrow{P_2} P_2 A^{(1)} P_2^T = \begin{pmatrix} * & * & * & * \\ * & * & * & * \\ 0 & * & * & * \\ 0 & 0 & * & * \end{pmatrix} = A^{(2)} = H_u.$$

Notes

1. Multiplication by P_i^T to the right does not destroy the zeros already present in $P_i A^{(i-1)}$.
2. The product $P_i A^{(i-1)} P_i^T$ can be **implicitly** formed as shown in Chapter 3 (**Section 3.6.1**).

Flop-count. The process requires $\frac{10}{3} n^3$ flops to compute H_u. This count does not include the explicit computation of P, which is stored in factored form. If P is computed explicitly, another $\frac{4}{3} n^3$ flops are required. However, when n is large, the storage required to form P is prohibitive.

Roundoff property. **The process is numerically stable.** It can be shown (Wilkinson (1965, p. 351) that the computed H_u is orthogonally similar to a nearby matrix $A + E$, where

$$\|E\|_F \le cn^2 \mu \|A\|_F.$$

Here c is a constant of order unity.

MATLAB note: The MATLAB Command $[P, H] = $ **hess** (A) computes an orthogonal matrix P and an upper Hessenberg matrix H such that $PAP^T = H$.

4.2.1 Uniqueness in Hessenberg Reduction: The Implicit Q Theorem

We just described Householder's method for Hessenberg reduction. However, this form could also have been obtained using **Givens matrices** as well (see Datta (1995, pp. 163–165). The question, therefore, arises **how unique is the Hessenberg form?**

The question is answered in the following theorem, known as the **Implicit Q Theorem.** The proof can be found in Golub and Van Loan (1996, p. 347).

Theorem 4.2.1. *The Implicit Q Theorem. Let* $P = (p_1, p_2, \ldots, p_n)$ *and* $Q = (q_1, q_2, \cdots, q_n)$ *be orthogonal matrices such that* $P^T A P = H_1$ *and*

$Q^{T}AQ = H_2$ are two **unreduced** upper Hessenberg matrices. Suppose that $p_1 = q_1$. Then H_1 and H_2 are essentially the same in the sense that $H_2 = D^{-1}H_1 D$, where $D = \text{diag}(\pm 1, \ldots, \pm 1)$. Furthermore, $p_i = \pm q_i$, $i = 2, \ldots, n$.

4.3 THE REAL SCHUR FORM OF A: THE QR ITERATION METHOD

In this section, we describe how to obtain the RSF of a matrix. The RSF of a matrix A displays the eigenvalues of A. It is obtained by using the well-known QR iteration method. **This method is nowadays a standard method for computing the eigenvalues of a matrix.** First, we state a well-known classical result on this subject.

Theorem 4.3.1. *The Schur Triangularization Theorem. Let A be an n × n complex matrix, then there exists an n × n unitary matrix U such that*

$$U^{*}AU = T,$$

where T is an n × n upper triangular matrix and the diagonal entries of T are the eigenvalues of A.

Proof. See Datta (1995, pp. 433–439).

Since a real matrix can have complex eigenvalues (occurring in complex conjugate pairs), even for a real matrix A, U and T in the above theorem can be complex. However, we can choose U to be real orthogonal if T is replaced by a **quasi-triangular matrix** R, known as the RSF of A, as the following theorem shows. The proof can be found in Datta (1995, p. 434) or in Golub and Van Loan (1996, pp. 341–342). ∎

Theorem 4.3.2. *The Real Schur Triangularization Theorem. Let A be an n × n real matrix. Then there exists an n × n orthogonal matrix Q such that*

$$Q^{T}AQ = R = \begin{pmatrix} R_{11} & R_{12} & \cdots & R_{1k} \\ 0 & R_{22} & \cdots & R_{2k} \\ \vdots & & \ddots & \vdots \\ 0 & \cdots & 0 & R_{kk} \end{pmatrix}, \tag{4.3.1}$$

where each R_{ii} is either a scalar or a 2 × 2 matrix. The scalars diagonal entries correspond to real eigenvalues, and each 2 × 2 matrix on the diagonal has a pair of complex conjugate eigenvalues.

Definition 4.3.1. *The matrix R in Theorem 4.3.2 is known as the RSF of A.*

Remarks

- The 2×2 matrices on the diagonal are usually referred to as **"Schur bumps."**
- The columns of Q are called the **Schur vectors. For each $k = 1, 2, \ldots, n$, the first k columns of Q form an orthonormal basis for the invariant subspace corresponding to the first k eigenvalues.**

We present below a method, known as the QR iteration method, for computing the RSF of A. **A properly implemented QR method is widely used nowadays for computing the eigenvalues of an arbitrary matrix.** As the name suggests, the method is based on the QR factorization and is iterative in nature. Since the roots of a polynomial equation of degree higher than four cannot be found in a finite number of steps, **any numerical method to compute the eigenvalues of a matrix of order higher than four has to be iterative in nature.** The QR iteration method was proposed in algorithmic form by J.G. Francis (1961), though its roots can be traced to a work of Rutishauser (1958). The method was also independently discovered by the Russian mathematician Kublanovskaya (1961).

For references of these papers, see Datta (1995) or Golub and Van Loan (1996).

4.3.1   The Basic QR Iteration

We first present the basic QR iteration method.

Set $A_0 \equiv A$.
Compute now a sequence of matrices $\{A_k\}$ as follows:
    For $k = 1, 2, \ldots$ do
        Find the QR factorization of A_{k-1}: $A_{k-1} = Q_k R_k$
        Compute $A_k = R_k Q_k$.
    End

The matrices in the sequence $\{A_k\}$ have a very interesting property: **Each matrix in the sequence is orthogonally similar to the previous one and is, therefore, orthogonally similar to the original matrix.** It is easy to see this. For example,

$$A_1 = R_1 Q_1 = Q_1^T A_0 Q_1 \text{ (since } R_1 = Q_1^T A_0),$$
$$A_2 = R_2 Q_2 = Q_2^T A_1 Q_2 \text{ (since } R_2 = Q_2^T A_1).$$

Thus, A_1 is orthogonally similar to A and A_2 is orthogonally similar to A_1. Therefore, A_2 is orthogonally similar to A, as the following computation shows:

$$A_2 = Q_2^T A_1 Q_2 = Q_2^T (Q_1^T A_0 Q_1) Q_2 = (Q_1 Q_2)^T A_0 (Q_1 Q_2).$$

Since each matrix A_k is orthogonally similar to the original matrix A, it has the same eigenvalues as A. It can then be shown (Wilkinson (1965, pp. 518–519) that under certain conditions, the sequence $\{A_k\}$ converges to the RSF or to the Schur form of A.

4.3.2   The Hessenberg QR Iteration and Shift of Origin

The QR iteration method as presented above is not practical if the matrix A
is full and dense. This is because, as we have seen before, the QR factorization of
a matrix A requires $O(n^3)$ flops and thus n iterations will consume $O(n^4)$ flops,
making the method impractical.

Fortunately, something simple can be done:

Reduce the matrix A **to a Hessenberg matrix by orthogonal similarity**
before starting the QR iterations. An interesting practical consequence of
this is that if $A = A_0$ is initially reduced to an upper Hessenberg matrix H and
is assumed to be unreduced, then each member of the matrix sequence$\{H_k\}$
obtained by applying QR iteration to H is also upper Hessenberg. Since the
QR factorization of a Hessenberg matrix requires $O(n^2)$ flops, the whole
iteration process then becomes $O(n^3)$ method.

However, the convergence of the subdiagonal entries of H, in the presence of two
or more nearly equal (in magnitude) eigenvalues, can be painfully slow.

Fortunately, the rate of convergence can be significantly improved by using a
suitable shift.

The idea is to apply the QR iteration to the shifted matrix $\hat{H} = H - \hat{\lambda}_i I$, where
$\hat{\lambda}_i$ is an approximate eigenvalue. This is known as the **single shift** QR **iteration.**

However, since the complex eigenvalues of a real matrix occur in conjugate
pairs, in practice, the QR iteration is applied to the matrix H with double shifts.
The process then is called the **double shift** QR **iteration** method.

4.3.3   The Double Shift QR Iteration

The Hessenberg double shift QR iteration scheme can be written as follows:

For $i = 1, 2, \ldots$ do
  Choose the two shifts k_1 and k_2
  Find the QR Factorization: $H - k_1 I = Q_1 R_1$
  Form: $H_1 = R_1 Q_1 + k_1 I$
  Find the QR factorization: $H_1 - k_2 I = Q_2 R_2$
  Form: $H_2 = R_2 Q_2 + k_2 I$
End

The shifts k_1 and k_2 at each iteration are chosen as the eigenvalues of the 2×2
trailing principal submatrix at that iteration. The process is called the **explicit**
double-shift QR iteration process.

The above explicit scheme requires complex arithmetic (since k_1 and k_2 are complex) to implement, and furthermore, the matrices $H - k_1 I$ and $H_1 - k_2 I$ need to be formed explicitly. In practice, an equivalent implicit version, known as the **double shift implicit QR iteration scheme**, is used. We state one step of this process in the following.

The Double Shift Implicit QR Step

1. Compute the first column n_1 of the matrix $N = (H - k_1 I)(H - k_2 I) = H^2 - (k_1 + k_2)H + k_1 k_2 I$.
2. Find a Householder matrix P_0 such that $P_0 n_1$ is a multiple of e_1.
3. Find Householder matrices P_1 through P_{n-2} such that $H_2 = (P_{n-2}^T \cdots P_1^T P_0^T) H (P_0 P_1 \ldots P_{n-2})$ is an upper Hessenberg matrix.

It can be shown by using the **Implicit Q Theorem** (Theorem 4.2.1) that **the upper Hessenberg matrix H_2 obtained by the double shift implicit QR step is essentially the same as H_2 obtained by one step of the explicit scheme.** Furthermore, the first column n_1 of N can be computed without explicitly computing the matrix N and, the computation of H_2 from H can be done only in $O(n^2)$ flops. For details see Datta (1995, pp. 444–447).

4.3.4 Obtaining the Real Schur Form A

1. Transform the matrix A to Hessenberg form.
2. Iterate with the double shift implicit QR step.

Typically, after two to three iteration steps of the double shift implicit QR method, one or two (and sometimes more) subdiagonal entries from the bottom of the Hessenberg matrix converge to zero. This then will give us a real or a pair of complex conjugate eigenvalues.

Once a real or a pair of complex conjugate eigenvalues is computed, the last row and the last column in the first case, or the last two rows and the last two columns in the second case, are deleted and the computation of the other eigenvalues is continued with the submatrix.

This process is known as **deflation**.

Note that the eigenvalues of the deflated submatrix are also the eigenvalues of the original matrix. For, suppose, immediately before deflation, the matrix has the form:

$$H_k = \begin{pmatrix} A' & C' \\ 0 & B' \end{pmatrix},$$

where B' is the 2×2 trailing submatrix or a 1×1 matrix. Then the characteristic polynomial of H_k is: $\det(\lambda I - H_k) = \det(\lambda I - A') \det(\lambda I - B')$. Thus, the eigenvalues of H_k are the eigenvalues of A' together with those of B'. But H_k is orthogonally similar to the original matrix A and therefore has the same eigenvalues as A.

Example 4.3.1.   Find the RSF of

$$H = \begin{pmatrix} 0.2190 & -0.0756 & 0.6787 & -0.6391 \\ -0.9615 & 0.9032 & -0.4571 & 0.8804 \\ 0 & -0.3822 & 0.4526 & -0.0641 \\ 0 & 0 & -0.1069 & -0.0252 \end{pmatrix}.$$

Iteration	h_{21}	h_{32}	h_{43}
1	0.3860	-0.5084	-0.0084
2	-0.0672	-0.3773	0.0001
3	0.0089	-0.3673	0
4	-0.0011	-0.3590	0
5	0.0001	-0.3905	0
...			

The computed RSF is

$$H = \begin{pmatrix} 1.4095 & 0.7632 & -0.1996 & 0.8394 \\ 0.0001 & 0.1922 & 0.5792 & 0.0494 \\ 0 & -0.3905 & 0.0243 & -0.4089 \\ 0 & 0 & 0 & -0.0763 \end{pmatrix}.$$

The eigenvalues of $\begin{pmatrix} 0.1922 & 0.5792 \\ -0.3905 & 0.0243 \end{pmatrix}$ are $0.1082 \pm 0.4681 j$.

Balancing

It is advisable to balance the entries of the original matrix A, if they vary widely, before starting the QR process.

The balancing is equivalent to transforming the matrix A to $D^{-1}AD$, where the diagonal matrix D is chosen so that the transformed matrix has approximately equal row and column norms.

In general, preprocessing the matrix by balancing improves the accuracy of the QR iteration method. **Note that no round-off error is involved in this computation and it takes only $O(n^2)$ flops.**

MATLAB note: The MATLAB command $[T, B] = $ **balance**(A) finds a diagonal matrix T such that $B = T^{-1}AT$ has approximately the equal row and column norms. See MATLAB User's Guide (1992).

Flop-count of the QR iteration method: Since the QR iteration method is an iterative method, it is hard to give an exact flop-count for this method. However, empirical observations have established that it takes about two QR iterations per eigenvalue. Thus, it will require about $12n^3$ flops to compute all the eigenvalues. If the transforming matrix Q and the final quasitriangular matrix T are also needed, then the cost will be about $26n^3$ flops.

Numerical stability property of the QR iteration process: The QR iteration method is quite stable. An analysis of the round-off property of the algorithm shows that the computed RSF $\hat{T}$ is orthogonally similar to a nearby matrix $A + E$. Specifically,

$$Q^T(A + E)Q = \hat{T}, \quad \text{where} \| E \|_F \le \phi(n)\mu \| A \|_F,$$

where $\phi(n)$ is a slowly growing function of n and μ is the machine precision. The computed orthogonal matrix Q can also be shown to be nearly orthogonal.

MATLAB notes: The MATLAB function **schur** in the following format: $[U, T] = $ **schur**(A) produces a Schur matrix T and an unitary matrix U such that $A = UTU^*$.

By itself, **schur**(A) returns T. If A is real, the RSF is returned.

The RSF has the real eigenvalues on the diagonal and the complex eigenvalues in 2×2 blocks on the diagonal.

4.3.5   The Real Schur Form and Invariant Subspaces

The RSF of A displays information on the invariant subspaces.

Basis of an Invariant Subspace from RSF

Let

$$Q^T A Q = R = \begin{pmatrix} R_{11} & R_{12} \\ 0 & R_{22} \end{pmatrix}$$

and let's assume that R_{11} and R_{22} do not have eigenvalues in common. **Then the first p columns of Q, where p is the order of R_{11}, form a basis for the invariant subspace associated with the eigenvalues of R_{11}.**

In many applications, such as in the **solution of algebraic Riccati equations** (see Chapter 13), in constructing a reduced-order model, etc., one needs to compute an orthonormal basis of an invariant subspace associated with a selected number of eigenvalues. Unfortunately, the RSF obtained by QR iteration will not, in general, give the eigenvalues in some desired order. Thus, if the eigenvalues are not in a desired order, one wonders if some extra work can be done to bring them into that order. That this can indeed be done, is seen from the following simple discussion. Let A be 2×2.

Let

$$Q_1^T A Q_1 = \begin{pmatrix} \lambda_1 & r_{12} \\ 0 & \lambda_2 \end{pmatrix}, \quad \lambda_1 \ne \lambda_2.$$

If λ_1 and λ_2 are not in right order, all we need to do to reverse the order is to form a Givens rotation $J(1, 2, \theta)$ such that

$$J(1, 2, \theta) \begin{pmatrix} r_{12} \\ \lambda_2 - \lambda_1 \end{pmatrix} = \begin{pmatrix} * \\ 0 \end{pmatrix}.$$

Then $Q = Q_1 J(1, 2, \theta)^T$ is such that

$$Q^T A Q = \begin{pmatrix} \lambda_2 & r_{12} \\ 0 & \lambda_1 \end{pmatrix}.$$

The above simple process can be easily extended to achieve any desired ordering of the eigenvalues in the RSF. For a Fortran program, see Stewart (1976).

Example 4.3.1.

$$A = \begin{pmatrix} 1 & 2 \\ 2 & 3 \end{pmatrix},$$

$$Q_1 = \begin{pmatrix} 0.8507 & 0.5257 \\ -0.5257 & 0.8507 \end{pmatrix},$$

$$Q_1^T A Q_1 = \begin{pmatrix} -0.2361 & 0.0000 \\ 0.0000 & 4.2361 \end{pmatrix}.$$

Suppose we now want to reverse the orders of -0.2361, and 4.2361.

Form: $J(1, 2, \theta) = \begin{pmatrix} 0 & -1 \\ -1 & 0 \end{pmatrix}.$

Then, $J(1, 2, \theta) \begin{pmatrix} 0 \\ 4.4722 \end{pmatrix} = \begin{pmatrix} 4.4722 \\ 0 \end{pmatrix}.$

Form: $Q = Q_1 J(1, 2, \theta)^T = \begin{pmatrix} -0.5257 & -0.8507 \\ -0.8507 & 0.5257 \end{pmatrix}.$

Then, $Q^T A Q = \begin{pmatrix} 4.2361 & 0.00 \\ 0.00 & -0.2361 \end{pmatrix}.$

Flop-count and numerical stability. **The process is quite inexpensive.** It requires only $k(12n)$ flops, where k is the number of interchanges required to achieve the desired order. **The process is also numerically stable.**

MATCONTROL note: The routine **ordersch** in MATCONTROL can be used to order the eigenvalues in the RSF of the matrix.

Fortran Routine: The Fortran routine STRSYL in LAPACK (Anderson *et al.* (1999)) reorders the Schur decomposition of a matrix in order to find an orthonormal basis of a right invariant subspace corresponding to selected eigenvalues.

Invariant Subspace Sensitivity: Sep-Function

Let $Q^* A Q = \begin{pmatrix} T_{11} & T_{12} \\ 0 & T_{22} \end{pmatrix}$ be the Schur decomposition of A. Let $Q = (Q_1, Q_2)$.
Define

$$\text{sep}(T_{11}, T_{22}) = \min_{X \neq 0} \frac{\| T_{11} X - X T_{22} \|_F}{\| X \|_F}.$$

Then it can be shown (Golub and Van Loan (1996, pp. 325)) that **the reciprocal of sep(T_{11}, T_{22}) is a good measure of the sensitivity of the invariant subspace spanned by the columns of Q.**

4.3.6   Inverse Iteration

The inverse iteration is a commonly used procedure to compute a selected number of eigenvectors of a matrix.

Since A is initially reduced to a Hessenberg matrix H for the QR iteration process, then it is natural to take advantage of the structure of the Hessenberg matrix H in the process of inverse iteration. The **Hessenberg inverse iteration** can then be stated as follows:

Step 1. Reduce the matrix A to an upper Hessenberg matrix $H : PAP^T = H$.

Step 2. Compute an eigenvalue λ, whose eigenvector x is sought, using the implicit QR iteration method described in the previous section.

Step 3. Choose a unit-length vector $y_0 \in \mathbb{C}^n$.

> For $k = 1, 2, \ldots$ do until convergence
>> Solve for $z^{(k)} : (H - \lambda I) z^{(k)} = y^{(k-1)}$
>> Compute $y^{(k)} = z^{(k)} / \| z^{(k)} \|$
> End

Step 4. Recover the eigenvector x of the matrix $A : x = P^T y^{(k)}$, where $y^{(k)}$ is an approximation of the eigenvector y obtained at the end of Step 3.

Note: If y is an eigenvector of H, then $x = P^T y$ is the corresponding eigenvector of A.

Convergence and efficiency: The Hessenberg inverse iteration is very inexpensive. Once an eigenvalue is computed, the whole process requires only $O(n^2)$ flops. It typically requires only 1 to 2 iterations to obtain an approximate acceptable eigenvector.

4.4 COMPUTING THE SINGULAR VALUE DECOMPOSITION (SVD)

The following algorithm known as the **Golub–Kahan–Reinsch algorithm** is nowadays a standard computational algorithm for computing the SVD. The algorithm comes in two stages:

Stage I. The $m \times n$ matrix A $(m \geq n)$ is transformed to an upper $m \times n$ bidiagonal matrix by orthogonal equivalence:

$$U_0^T A \, V_0 = \begin{pmatrix} B \\ 0 \end{pmatrix}, \tag{4.4.1}$$

where B is the $n \times n$ upper bidiagonal matrix given by

$$B = \begin{pmatrix} b_1 & * & & & 0 \\ & \ddots & * & & \\ & & \ddots & \ddots & \\ & & & & * \\ 0 & & & & b_n \end{pmatrix}$$

Stage II. The transformed bidiagonal matrix B is further reduced by orthogonal equivalence to a diagonal matrix Σ using the QR iteration method; that is, orthogonal matrices U_1 and V_1 are constructed such that

$$U_1^T B V_1 = \Sigma = \mathrm{diag}(\sigma_1, \ldots, \sigma_n). \tag{4.4.2}$$

The matrix Σ is the matrix of singular values. The singular vector matrices U and V are given by $U = U_0 U_1$, $V = V_0 V_1$.

We will briefly describe Stage I here. For a description of Stage II, see Golub and Van Loan (1996, pp. 452–457).

Reduction to Bidiagonal Form

We show how Householder matrices can be employed to construct U_0 and V_0 in Stage I.

The matrices U_0 and V_0 are constructed as the product of Householder matrices as follows: $U_0 = U_1 U_2 \ldots U_n$, and $V_0 = V_1 V_2 \ldots V_{n-2}$. Let's illustrate construction of U_1, V_1 and U_2, V_2, and their role in the bidiagonalization process with $m = 5$ and $n = 4$.

First, a Householder matrix U_1 is constructed such that

$$A^{(1)} = U_1 A = \begin{pmatrix} * & * & * & * \\ 0 & * & * & * \\ 0 & * & * & * \\ 0 & * & * & * \\ 0 & * & * & * \end{pmatrix}.$$

Next, a Householder matrix V_1 is constructed such that

$$A^{(2)} = A^{(1)} V_1 = \begin{pmatrix} * & * & 0 & 0 \\ 0 & * & * & * \\ 0 & * & * & * \\ 0 & * & * & * \\ 0 & * & * & * \end{pmatrix} = \left(\begin{array}{c|ccc} * & * & 0 & 0 \\ \hline 0 & & & \\ 0 & & A' & \\ 0 & & & \\ 0 & & & \end{array} \right).$$

The process is now repeated with $A^{(2)}$; that is, Householder matrices U_2 and V_2 are constructed so that

$$U_2 A^{(2)} V_2 = \begin{pmatrix} * & * & 0 & 0 \\ 0 & * & * & 0 \\ 0 & 0 & * & * \\ 0 & 0 & * & * \\ 0 & 0 & * & * \end{pmatrix}.$$

Of course, in this step, we will work with the 4×3 matrix A' rather than the matrix $A^{(2)}$. Thus, first the orthogonal matrices U'_2 and V'_2 will be constructed such that

$$U'_2 A' V'_2 = \begin{pmatrix} * & * & 0 \\ 0 & * & * \\ 0 & * & * \\ 0 & * & * \end{pmatrix},$$

then U_2 and V_2 will be constructed from U'_2 and V'_2 in the usual way, that is, by embedding them in identity matrices of appropriate orders. The process is continued until the bidiagonal matrix B is obtained.

Example 4.4.1. Let

$$A = \begin{pmatrix} 1 & 2 & 3 \\ 3 & 4 & 5 \\ 6 & 7 & 8 \end{pmatrix}.$$

Step 1.

$$U_1 = \begin{pmatrix} -0.1474 & -0.4423 & -0.8847 \\ -0.4423 & 0.8295 & -0.3410 \\ -0.8847 & -0.3410 & 0.3180, \end{pmatrix},$$

$$A^{(1)} = U_1 A = \begin{pmatrix} -6.7823 & -8.2567 & -9.7312 \\ 0 & 0.0461 & 0.0923 \\ 0 & -0.9077 & -1.8154 \end{pmatrix}.$$

Step 2.

$$V_1 = \begin{pmatrix} 1 & 0 & 0 \\ 0 & -0.6470 & 0.7625 \\ 0 & -0.5571 & 0.6470 \end{pmatrix},$$

$$A^{(2)} = A^{(1)} V_1 = \begin{pmatrix} -6.7823 & 12.7620 & 0 \\ 0 & -1.0002 & 0.0245 \\ 0 & 1.9716 & -0.4824 \end{pmatrix}.$$

Step 3.

$$U_2 = \begin{pmatrix} 1 & 0 & 0 \\ 0 & -0.0508 & 0.9987 \\ 0 & 0.9987 & 0.0508 \end{pmatrix}$$

$$B = U_2 A^{(2)} = U_2 A^{(1)} V_1 = U_2 U_1 A V_1 = \begin{pmatrix} -6.7823 & 12.7620 & 0 \\ 0 & -1.0081 & -1.8178 \\ 0 & 0 & 0 \end{pmatrix}$$

Note that from the above expression of B, it immediately follows that zero is a singular value of A.

Flop-count: The Householder bidiagonalization algorithm requires $4mn^2 - 4n^3/3$ flops.

Stage II, that is, the process of iterative reduction of the bidiagonal matrix to a diagonal matrix containing the singular values requires $30n$ flops and $2n$ square roots. The matrices U and V can be accumulated with $6mn$ and $6n^2$ flops, respectively.

Stability: The Golub–Kahan–Reinsch algorithm is **numerically stable**. It can be shown that the process will yield orthogonal matrices U and V and a diagonal matrix Σ such that $U^T A V = \Sigma + E$, where $\|E\|_2 \approx \mu \|A\|_2$.

4.5   THE GENERALIZED REAL SCHUR FORM:
THE QZ ALGORITHM

In this section, we describe two canonical forms for a pair of matrices (A, B): The **Hessenberg-triangular** and the **Generalized RSF**. The Generalized RSF displays the eigenvalues of the matrix pencil $A - \lambda B$, as the RSF does for the matrix A.

Given $n \times n$ matrices A and B, a scalar λ and a nonzero vector x satisfying

$$Ax = \lambda Bx$$

are respectively called an **eigenvalue** and **eigenvector** for the pencil $A - \lambda B$. The eigenvalue problem itself is called **generalized eigenvalue problem**. The eigenvalues and eigenvectors of the generalized eigenvalue problem are often called **generalized eigenvalues** and **generalized eigenvectors**. The matrix pencil $A - \lambda B$ is often conveniently denoted by the pair (A, B).

The pair (A, B) is called **regular** if $\det(A - \lambda B)$ is not identically zero. Otherwise, it is **singular**. **We will consider only regular pencil here**. If B is nonsingular, then the eigenvalues **of the regular** pair (A, B) are finite and are the same as those of AB^{-1} or $B^{-1}A$.

If B is singular, and if the degree of $\det(A - \lambda B)$ is $r (< n)$, then $n - r$ eigenvalues of (A, B) are ∞, and the remaining ones are the zeros of $\det(A - \lambda B)$.

As we will see later, the generalized RSF is an important tool in the numerical solutions of the **discrete algebraic Riccati equation** and **the Riccati equations with singular and ill-conditioned control weighting matrices** (Chapter 13).

The QZ algorithm

Assume that B is nonsingular. Then the basic idea is to apply the QR iteration algorithm to the matrix $C = B^{-1}A$ (or to AB^{-1}), without explicitly forming the matrix C. For if B is nearly singular, then it is not desirable to form B^{-1}. In this case the entries of C will be much larger than those of A and B, and the eigenvalues of C will be computed inaccurately. (Note that the eigenvalues of $B^{-1}A$ are the same as those of AB^{-1}, because $AB^{-1} = B(B^{-1}A)B^{-1}$). If AB^{-1} or $B^{-1}A$ is not to be computed explicitly, then the next best alternative, of course, is to transform A and B simultaneously to some reduced forms such as the triangular forms and then extract the generalized eigenvalues from these reduced forms. The simultaneous reduction of A and B to triangular forms by equivalence is guaranteed by the following theorem:

Theorem 4.5.1.   *The Generalized Real Schur Decomposition. Given two $n \times n$ real matrices A and B, there exist orthogonal matrices Q and Z such that $Q^T AZ$ is an upper real Schur matrix and $Q^T BZ$ is upper triangular:*

$$Q^T AZ \equiv A', \text{ an upper real Schur matrix,}$$

$$Q^T BZ \equiv B', \text{ an upper triangular matrix.}$$

The pair (A', B') is said to be in **generalized RSF**.

The reduction to the generalized RSF is achieved in two stages.

Stage I. The matrices A and B are reduced to an upper Hessenberg and an upper triangular matrix, respectively, by simultaneous orthogonal equivalence:

$$A \equiv Q^T A Z, \text{ an upper Hessenberg matrix,}$$

$$B \equiv Q^T B Z, \text{ an upper triangular matrix.}$$

Stage II. The Hessenberg-triangular pair (A, B) is further reduced to the **generalized RSF** by applying **implicit QR iteration** to AB^{-1}.

This process is known as the **QZ Algorithm**.

We will now briefly sketch these two stages in the sequel.

4.5.1   Reduction to Hessenberg-Triangular Form

Let A and B be two $n \times n$ matrices. Then,

Step 1. Find an orthogonal matrix U such that

$$B \equiv U^T B$$

is an upper triangular matrix by finding the QR factorization of B.

Form

$$A \equiv U^T A$$

(**in general, A will be full**).

Step 2. Reduce A to Hessenberg form while preserving the triangular structure of B.

Step 2 is achieved as follows:

To start with, we have

$$A \equiv U^T A = \begin{pmatrix} * & * & \cdots & * \\ * & * & \cdots & * \\ \vdots & & & \\ * & * & \cdots & * \\ * & * & \cdots & * \end{pmatrix},$$

$$B \equiv U^T B = \begin{pmatrix} * & * & \cdots & \cdots & * \\ 0 & * & \cdots & \cdots & * \\ 0 & 0 & \ddots & \cdots & * \\ \vdots & & & \ddots & \vdots \\ 0 & 0 & 0 & \cdots 0 & * \end{pmatrix}.$$

First, the $(n, 1)$th entry of A is made zero by applying a Givens rotation $Q_{n-1,n}$ in the $(n - 1, n)$ plane:

$$A \equiv Q_{n-1,n} A = \begin{pmatrix} * & * & \cdots & * \\ * & * & \cdots & * \\ \vdots & & & \\ * & * & \cdots & * \\ 0 & * & \cdots & * \end{pmatrix}.$$

This transformation, when applied to B from the left, will give a fill-in in the $(n, n - 1)$ position:

$$B \equiv Q_{n-1,n} B = \begin{pmatrix} * & * & \cdots & \cdots & * \\ 0 & * & \cdots & \cdots & * \\ 0 & 0 & * & \cdots & * \\ \vdots & & & \ddots & \vdots \\ 0 & 0\cdots & 0 & * & * \end{pmatrix}.$$

The Givens rotation $Z_{n-1,n} = J(n - 1, n, \theta)$ is now applied to the right of B to make the $(n, n - 1)$ entry of B zero. Fortunately, this rotation, when applied to the right of A, does not destroy the zero produced earlier. Schematically, we have

$$B \equiv B Z_{n-1,n} = \begin{pmatrix} * & * & * & \cdots & * \\ 0 & * & * & \cdots & * \\ 0 & 0 & * & \cdots & * \\ \vdots & & \ddots & \ddots & \\ 0 & 0 & \cdots & 0 & * \end{pmatrix},$$

$$A \equiv A Z_{n-1,n} = \begin{pmatrix} * & * & * & \cdots & * \\ * & * & * & \cdots & * \\ * & * & * & \cdots & * \\ \vdots & \vdots & \vdots & & \vdots \\ * & * & * & \cdots & * \\ 0 & * & * & \cdots & * \end{pmatrix}.$$

The entries $(n - 1, 1)$, $(n - 2, 1)$, ..., $(3, 1)$ of A are now successively made zero, each time applying an appropriate rotation to the left of A, followed by another appropriate Givens rotation to the right of B to zero out the undesirable fill-in in B. At the end of the first step, the matrix A is Hessenberg in its first column, while

B remains upper triangular:

$$
A = \begin{pmatrix} * & * & \cdots & * \\ * & * & \cdots & * \\ 0 & * & \cdots & * \\ \vdots & \vdots & \ddots & \vdots \\ 0 & * & \cdots & * \end{pmatrix}, \qquad
B = \begin{pmatrix} * & * & \cdots & \cdots & * \\ * & * & \cdots & \cdots & * \\ 0 & \ddots & \ddots & & \vdots \\ \vdots & \ddots & \ddots & \ddots & * \\ 0 & \cdots & 0 & * & * \end{pmatrix}.
$$

The zeros are now produced on the second column of A in the appropriate places while retaining the triangular structure of B in an analogous manner.

The process is continued until the matrix A is an upper Hessenberg matrix while keeping B in upper triangular form.

Example 4.5.1.

$$
A = \begin{pmatrix} 1 & 2 & 3 \\ 1 & 3 & 4 \\ 1 & 3 & 3 \end{pmatrix}, \qquad B = \begin{pmatrix} 1 & 1 & 1 \\ 0 & 1 & 2 \\ 0 & 0 & 2 \end{pmatrix}.
$$

1. Form the Givens rotation Q_{23} to make a_{31} zero:

$$
Q_{23} = \begin{pmatrix} 1 & 0 & 0 \\ 0 & 0.7071 & 0.7071 \\ 0 & -0.7071 & 0.7071 \end{pmatrix},
$$

$$
A \equiv A^{(1)} = Q_{23}A = \begin{pmatrix} 1 & 2 & 3 \\ 1.4142 & 4.2426 & 4.9497 \\ 0 & 0 & -0.7071 \end{pmatrix}.
$$

2. Update B:

$$
B \equiv B^{(1)} = Q_{23}B = \begin{pmatrix} 1 & 1 & 1 \\ 0 & 0.7071 & 2.8284 \\ 0 & -0.7071 & 0 \end{pmatrix}.
$$

3. Form the Givens rotation Z_{23} to make b_{32} zero:

$$
Z_{23} = \begin{pmatrix} 1 & 0 & 0 \\ 0 & 0 & -1 \\ 0 & 1 & 0 \end{pmatrix},
$$

$$
B \equiv B^{(1)}Z_{23} = Q_{23}BZ_{23} = \begin{pmatrix} 1 & 1 & -1 \\ 0 & 2.8284 & -0.7071 \\ 0 & 0 & 0.7071 \end{pmatrix}.
$$

4. Update A:

$$A \equiv A^{(1)}Z_{23} = Q_{23}AZ_{23} = \begin{pmatrix} 1 & 3 & -2 \\ 1.4142 & 4.9497 & -4.2426 \\ 0 & -0.7071 & 0 \end{pmatrix}.$$

Now A is an upper Hessenberg and B is in upper triangular form.

4.5.2   Reduction to the Generalized Real Schur Form

At the beginning of this process, we have A and B as an upper Hessenberg and an upper triangular matrix, respectively, obtained from Stage 1. We can assume without loss of generality that the matrix A is an unreduced upper Hessenberg matrix. **The basic idea now is to apply an implicit QR step to AB^{-1} without ever forming this matrix explicitly.** We sketch just the basic ideas here. For details, see Datta (1995, pp. 500–504).

Thus a QZ step, analogous to an implicit QR step, will be as follows:

1. Compute the first column n_1 of $N = (C - \alpha_1 I)(C - \alpha_2 I)$, where $C = AB^{-1}$ and α_1 and α_2 are suitably chosen shifts, without explicitly forming the matrix AB^{-1}.
    (Note that n_1 has only three nonzero entries and the rest are zero).
2. Find a Householder matrix Q_1, such that $Q_1 n_1$ is a multiple of e_1.
3. Form $Q_1 A$ and $Q_1 B$.
4. Simultaneously transform $Q_1 A$ to an upper Hessenberg matrix A_1, and $Q_1 B$ to an upper triangular matrix B_1:

$$A_1 \equiv Q^T(Q_1 A)Z : \text{ an upper Hessenberg;}$$
$$B_1 \equiv Q^T(Q_1 B)Z : \text{ an upper triangular.}$$

Using the implicit Q theorem (Theorem 4.2.1) we can show that the matrix $A_1 B_1^{-1}$ is essentially the same as that would have been obtained by applying an implicit QR step directly to AB^{-1}.

Applications of a few QZ steps in sequence will then yield a quasi-triangular matrix $R = Q^T A Z$ and an upper triangular $T = Q^T B Z$, from which the generalized eigenvalues can be easily extracted.

Choosing the Shifts

The double shifts α_1 and α_2 at a QZ step can be taken as the eigenvalues of the lower 2×2 submatrix of $C = AB^{-1}$. **The 2×2 lower submatrix of C again can be computed without explicitly forming B^{-1}** (see Datta (1995, p. 501)).

Algorithm 4.5.1. *The Complete QZ Algorithm for Reduction to Generalized Schur Form*

Inputs: Real $n \times n$ matrices A and B.
Outputs: The pair (R, T) of the generalized RSF of the pencil $A - \lambda B$. The matrix R is Quasi-triangular and T is upper triangular.

1. Transform (A, B) to a Hessenberg-triangular pair by orthogonal equivalence:

$$A \equiv Q^{\mathrm{T}} A Z, \text{ an upper Hessenberg,}$$

$$B \equiv Q^{\mathrm{T}} B Z, \text{ an upper triangular.}$$

2. Apply a sequence of the QZ steps to the Hessenberg-triangular pair (A, B) to produce $\{A_k\}$ and $\{B_k\}$, with properly chosen shifts.
3. Monitor the convergence of the sequences $\{A_k\}$ and $\{B_k\}$:

$$\{A_k\} \longrightarrow R, \text{ quasi-triangular (in RSF),}$$

$$\{B_k\} \longrightarrow T, \text{ upper triangular.}$$

Flop-count: The implementation of (1)–(3) requires about $30n^3$ flops. The formation of Q and Z, if required, needs, respectively, another $16n^3$ and $20n^3$ flops (from experience it is known that about two QZ steps per eigenvalue are adequate).

Numerical Stability Properties: The QZ iteration algorithm is as **stable** as the QR iteration algorithm. It can be shown that the computed $\hat{R}$ and $\hat{S}$ satisfy

$$Q_0^{\mathrm{T}}(A + E)Z_0 = \hat{R}, \qquad Q_0^{\mathrm{T}}(B + F)Z_0 = \hat{S}.$$

Here Q_0 and Z_0 are orthogonal, $\|E\| \cong \mu \|A\|$ and $\|F\| \cong \mu \|B\|$; μ is the machine precision.

4.6   COMPUTING OF THE EIGENVECTORS OF THE PENCIL $A - \lambda B$

Once an approximate generalized eigenvalue λ is computed, the corresponding eigenvector v of the pencil $A - \lambda B$ can be computed using the **generalized inverse iteration** as before.
Step 1. Choose an initial eigenvector v_0.
Step 2. For $k = 1, 2, \ldots$ do until convergence

$$\text{Solve } (A - \lambda B)\hat{v}_k = B v_{k-1};$$

$$v_k = \hat{v}_k / \|\hat{v}_k\|_2. \tag{4.6.1}$$

A Remark on Solving $(A - \lambda B)\hat{v}_k = B v_{k-1}$

In solving $(A - \lambda B)\hat{v}_k = B v_{k-1}$, substantial savings can be made by exploiting the Hessenberg-triangular structure to which the pair (A, B) is reduced as a part of the QZ algorithm. Note that in this case for a given λ, the matrix $A - \lambda B$ is also a Hessenberg matrix. Thus, at each iteration, only a Hessenberg system needs to be solved, which requires only $O(n^2)$ flops, compared to $O(n^3)$ flops required for a system with a full matrix.

Example 4.6.1.

$$A = 10^9 \begin{pmatrix} 3 - 1.50 \\ -1.53 - 1.5 \\ 0 - 1.51.5 \end{pmatrix}, \qquad B = 10^3 \begin{pmatrix} 200 \\ 030 \\ 004 \end{pmatrix}.$$

$\lambda_1 =$ a generalized eigenvalue of $(A - \lambda B) = 1950800$.

$$v_0 = \begin{pmatrix} 1 \\ 1 \\ 1 \end{pmatrix}.$$

$k = 1$: Solve for v_1:

$$\text{Solve:} \ (A - \lambda_1 B)\hat{v}_1 = Bv_0$$

$$\hat{v}_1 = \begin{pmatrix} 0.0170 \\ -0.0102 \\ 0.0024 \end{pmatrix}, \qquad v_1 = \hat{v}_1/\|\hat{v}_1\| = \begin{pmatrix} 0.8507 \\ -0.5114 \\ 0.1217 \end{pmatrix}.$$

MATLAB and MATCOM notes: The MATLAB function **qz** in the form: $[AA, BB, Q, Z, V] = \mathbf{qz}(A, B)$ produces upper triangular matrices AA and BB, and the orthogonal matrices Q and Z such that $QAZ = AA$, $QBZ = BB$.

The matrix V contains the eigenvectors. The generalized eigenvalues are obtained by taking the ratios of the corresponding diagonal entries of AA and BB. The MATLAB function **eig** (A, B) gives only the generalized eigenvalues of the pencil $A - \lambda B$ from the generalized Schur decomposition. MATCOM functions HESSTRI and INVITRGN compute, respectively, the Hessenberg-triangular reduction of the pair (A, B) and the eigenvectors of the pencil $A - \lambda B$ using inverse iteration.

Deflating Subspace for the Pencil $A - \lambda B$

A k-dimensional subspace $S \in \mathbb{R}^n$ is a **deflating subspace** of the pencil $A - \lambda B$ if the subspace $\{Ax + By \mid x, y \in S\}$ has dimension k or less. It can be easily seen that the **columns of Z in the generalized Schur decomposition form a family of deflating subspaces.** Also, span$\{Az_1, \ldots, Az_k\}$ and span$\{Bz_1, \ldots, Bz_k\}$ belong to span$\{q_1, \ldots, q_k\}$, where z_i and q_i are, respectively, the columns of Z and Q.

Remark

- In solving algebraic Riccati equations, deflating subspaces with specified spectrum need to be computed. There exist Fortran routines for computing such deflating subspaces developed by Van Dooren (1982).

4.7 SUMMARY AND REVIEW

Numerical Instability in Obtaining Jordan and Companion Matrices

The JCF and a companion form of a matrix, because of their rich structures, are important theoretical tools. Using these two decompositions, many important results in control theory have been established (see Kailath 1980).

Unfortunately, however, these two forms cannot be obtained in a numerically stable way in general. Since it is necessary to use non-orthogonal transformations to achieve these forms, the transforming matrices can be highly ill-conditioned. Some discussions to this effect have been given in **Section 4.1**. Because of possible numerical instabilities in reduction of A to a companion matrix, and the fact that the zeros of a polynomial can be extremely sensitive to small perturbations, **it is not advisable to compute the eigenvalues of a matrix by finding the zeros of its characteristic polynomial.**

Hessenberg and Real Schur Forms

Both Hessenberg and RCFs can be obtained via orthogonal similarity transformations. These two forms, thus, are extremely valuable tools in numerical computations. In fact, many of the numerically effective algorithms for control problems described in this book, are based on these two forms.

Reduction to Hessenberg form. A Hessenberg form, via orthogonal similarity transformation, is obtained using either Householder or Givens transformations. The Householder method for Hessenberg reduction is described in Section 4.2. For a description of Givens Hessenberg reduction, see Datta (1995) or Golub and Van Loan (1996). The **implicit Q theorem (Theorem 4.2.1)** guarantees that the Hessenberg forms obtained by two different methods are essentially the same, provided that the transforming matrices have the same first column.

Real Schur form: Computing the eigenvalues, eigenvectors, and orthonormal bases for invariant subspaces. The RSF of a matrix is a **quasi-triangular matrix** whose diagonal entries are either scalars or 2×2 matrices. Every real matrix A can be transformed to RSF by an orthogonal similarity. Since the RSF of a matrix A displays the eigenvalues of A, any numerical method for obtaining the RSF of order higher than four X has to be iterative in nature. The standard method for obtaining the RSF is the QR iteration method with implicit double shift. This method is described in some detail in Sections 4.3.1–4.3.4. **The double shift implicit QR iteration method is nowadays the standard method for finding the eigenvalues of a matrix.**

An orthonormal basis for the invariant subspace associated with a given set of eigenvalues can also be found by reordering the eigenvalues in RSF in a suitable way. This is discussed in **Section 4.3.5.**

Once the RSF is found, it can be employed to compute the eigenvectors of A. This is not discussed here. Interested readers are referred to Datta (1995, pp. 452–455). Instead, a commonly used procedure for computing selected eigenvectors, called the **inverse iteration method, is described in Section 4.3.**

Computing the SVD of a Matrix

The standard method for computing the SVD, called the **Golub–Kahan–Reinsch algorithm**, is described in Section 4.4. The method comes in two stages:

Stage I. Reduction of the matrix A to a bidiagonal form.

Stage II. Further reduction of the bidiagonal matrix obtained in Stage I to a diagonal matrix using implicit QR iteration.

The detailed discussion of Stage II is omitted here. The readers are referred to Golub and Van Loan (1996, pp. 452–456).

The Generalized Real Schur Form

The generalized RSF of a pair of matrices (A, B) is a matrix-pair (A', B'), where A' is an upper real Schur matrix and B' is an upper triangular matrix **(Theorem 4.5.1).**

The standard method for computing the general RSF is the **QZ iteration algorithm.** The QZ algorithm also comes in two stages:

Stage I. Reduction of (A, B) to Hessenberg-triangular form.

Stage II. Further reduction of the Hessenberg-triangular form obtained in Stage I to the generalized RSF.

Stage I is a finite procedure. Again, the Householder or Givens transformations can be used. The Householder procedure is described in **Section 4.5.1.** Stage II is an iterative procedure. Only a brief sketch of the procedure is presented here in **Section 4.5.2.** For details, readers are referred to Datta (1995, pp. 500–504).

The generalized RSF displays the eigenvalues (called generalized eigenvalues) of the linear pencil $A - \lambda B$. Once the eigenvalues are obtained, the selected eigenvectors can be computed using **generalized inverse iteration (Section 4.6).**

4.8   CHAPTER NOTES AND FURTHER READING

The material of this chapter has been taken from the recent book of the author (Datta 1995). For advanced readings of the topics dealt with in this chapter, consult the book by Golub and Van Loan (1996) and Stewart (2001). For a description of the toolbox MATCOM and how to obtain it, see the section on Chapter Notes and Further Reading of Chapter 3 (**Section 3.11**). For MATLAB functions and LAPACK routines, see the respective user's guides; Anderson *et al.* (1995) and MATLAB User's Guide (1992)

References

Anderson E., Bai Z., Bischof C., Blackford S., Demmel J., Dongarra J., Du Croz J., Greenbaum A., Hammarling S., McKenney A., and Sorensen D. *LAPACK Users' Guide*, 2nd edn, SIAM, Philadelphia, 1999.

Datta B.N. *Numerical Linear Algebra and Applications*, Brooks/Cole Publishing Company, Pacific Grove, CA. 1995.

Golub G.H. and Van Loan C.F. *Matrix Computations*, 3rd edn, The Johns Hopkins University Press, Baltimore, MD, 1996.

Kailath T. *Linear Systems*, Prentice Hall, Englewood Cliffs, N.J, 1980.

MATLAB *User's Guide*, The Math Works, Inc., Natick, MA, 1992.

Stewart G.W. "Algorithm 406. HWR3 and EXCHNG: FORTRAN programs for calculating the eigensystems of a real upper Hessenberg matrix in a prescribed order," *ACM Trans. Math. Soft.* Vol. 2, pp. 275–280, 1976.

Stewart G.W. *Matrix Algorithms, Vol. II: Eigen Systems*, SIAM, Philadelphia, 2001.

Van Dooren P. "Algorithm 590–DSUBSP and EXCHQZ: Fortran subroutines for computing deflating subspaces with specified spectrum," *ACM Trans. Math. Soft.*, Vol. 8, pp. 376–382, 1982.

Wilkinson J.H. *The Algebraic Eigenvalue Problem*, Clarendon Press, Oxford, England, 1965.

CONTROL SYSTEMS ANALYSIS

CHAPTER 5

LINEAR STATE-SPACE MODELS AND SOLUTIONS OF THE STATE EQUATIONS

Topics covered

- State-Space Models
- Solutions of the State Equations
- System Responses
- Sensitivity Analysis of the Matrix Exponential Problem
- Numerical Methods for Computing the Matrix Exponential and the Integral involving an Matrix Exponential
- Computation of the Frequency Response Matrix

5.1 INTRODUCTION

A finite-dimensional time-invariant linear continuous-time dynamical system may be described using the following system of first-order ordinary differential equations:

$$\dot{x}(t) = Ax(t) + Bu(t),$$
$$y(t) = Cx(t) + Du(t).$$

The input and the output of the system are defined in continuous-time over the interval $[0, \infty)$. The system is, therefore, known as a **continuous-time system.** The discrete-time analog of this system is the system of difference equations:

$$x(k+1) = Ax(k) + Bu(k),$$
$$y(k) = Cx(k) + Du(k).$$

We will consider in this book only time-invariant systems, that is, the matrices A, B, C, and D will be assumed constant matrices throughout the book.

It is first shown in Section 5.2 how some simple familiar physical systems can be described in state-space forms. Very often the mathematical model of a system is not obtained in first-order form; it may be a system of nonlinear equations, a system of **second-order differential equations** or **partial differential equations.** It is shown how such systems can be reduced to the standard first-order state-space forms. The computational methods for the state equations are then considered both in time and frequency domain.

The major computational component of the time-domain solution of a continuous-time system is the matrix exponential e^{At}. Some results on the sensitivity of this matrix and various well-known methods for its computation: the Taylor series method, the Padé approximation method, the methods based on decompositions of A, the ordinary-differential equation methods, etc., are described in Section 5.3. A comparative study of these methods is also included. **The Padé method (Algorithm 5.3.1) (with scaling and squaring)** and the **method, based on the Real Schur decomposition of A (Algorithm 5.3.2), are recommended for practical use.** This section concludes with an algorithm for numerically computing an integral with an matrix exponential (**Algorithm 5.3.3**).

Section 5.4 describes the state-space solution of a **discrete-time system**. The major computational task here is computation of various powers of A.

In **Section 5.5**, the problem of computing the **frequency response matrix** for many different values of the frequencies is considered. The computation of the frequency response matrix is necessary to study various system responses in frequency domain. A widely used method **(Algorithm 5.5.1)**, based on the one-time reduction of the state matrix A to a Hessenberg matrix, is described in detail and the references to the other recent methods are given.

Reader's Guide for Chapter 5

The readers familiar with basic concepts and results of modeling and state-space systems can skip Sections 5.2, 5.4, and 5.5.1.

5.2   STATE-SPACE REPRESENTATIONS OF CONTROL SYSTEMS

5.2.1   Continuous-Time Systems

Consider the dynamical system represented by means of the following system of ordinary first-order differential equations:

$$\dot{x}(t) = Ax(t) + Bu(t), \quad x(t_0) = x_0, \tag{5.2.1}$$

$$y(t) = Cx(t) + Du(t). \tag{5.2.2}$$

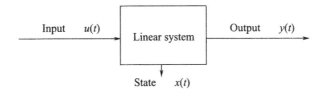

FIGURE 5.1:   Representation of a continuous-time state-space model.

In this description,

> $x(t)$ is an n-dimensional vector, called the system **state**,
> $u(t)$ is an m-dimensional vector ($m \leq n$), called the system **input**,
> $y(t)$ is an r-dimensional vector, called the system **output**.

The vector $x(t_0)$ is the **initial condition** of the system. The components of $x(t)$ are called **state variables.**

The matrices A, B, C, and D are **time-invariant matrices**, respectively, of dimensions $n \times n, n \times m, r \times n$, and $r \times m$. The above representation is known as a time-invariant **continuous-time state-space model** of a dynamical system.

Schematically, the model is represented in Figure 5.1.

Clearly, at a given time t, the variables arriving at the system would form the input, those internal to the system form the state, while the others that can be measured directly comprise the output.

The space $X \subseteq \mathbb{R}^n$, where all the states lie for all $t \geq 0$ is called the **state-space**, the Eq. (5.2.1) is called the **state equation** and the Eq. (5.2.2) is called the **output equation**. If $m = r = 1$, the system is said to be a **single-input single-output (SISO) system. A multi-input multi-output (MIMO) system** is similarly defined. If a system has more than one input or more than one output it is referred to be a **multivariable** system. *The system represented by the Eqs. (5.2.1) and (5.2.2) is sometimes written compactly as (A, B, C, D) or as (A, B, C), in case D is not used in modeling. Sometimes $\dot{x}(t)$ and $x(t)$ will be written just as $\dot{x}$ and x for the sake of convenience.* Similarly, $u(t)$ and $y(t)$ will be written as u and y, respectively.

We provide below a few examples to illustrate the state-space representations of some simple systems.

Example 5.2.1 (A Parallel RLC Circuit).   Consider a parallel RLC circuit excited by the current source $u(t)$ and with output $y(t)$ (Figure 5.2).

The current and voltage equations governing the circuit are:

$$u = i_R + i_L + i_C; \qquad i_C = C\frac{de_C}{dt}; \qquad e_C = L\frac{di_L}{dt} = Ri_R.$$

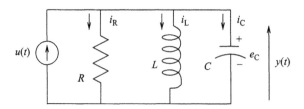

FIGURE 5.2:   A parallel RLC circuit.

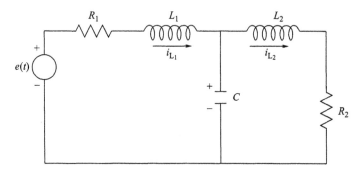

FIGURE 5.3:   An expanded RLC circuit.

Defining the states by $x_1 := i_L$ and $x_2 := e_C$, the state and output equations are, respectively:

$$\dot{x} = Ax + bu \quad \text{and} \quad y = cx,$$

where $x = [x_1, x_2]^T$,

$$A = \begin{bmatrix} 0 & 1/L \\ -1/C & -1/RC \end{bmatrix}, \qquad b = \begin{bmatrix} 0 \\ 1/C \end{bmatrix}, \qquad c = [0\ 1].$$

Example 5.2.2.   Consider again another electric circuit, as shown in Figure 5.3: The state variables here are taken as voltage across the capacitor and the current through the inductor. The state equations are

$$L_1 \frac{di_{L_1}(t)}{dt} = -R_1 i_{L_1}(t) - e_C(t) + e(t),$$

$$L_2 \frac{di_{L_2}(t)}{dt} = -R_2 i_{L_2}(t) + e_C(t),$$

$$C \frac{de_C(t)}{dt} = i_{L_1}(t) - i_{L_2}(t).$$

Setting $x_1 = i_{L_1}, x_2 = i_{L_2}, x_3 = e_C, b = \begin{pmatrix} 1/L_1 \\ 0 \\ 0 \end{pmatrix}, u = e(t)$, the matrix form of the state-space representation of the above system is given by:

$$\dot{x}(t) = \begin{pmatrix} \dot{x}_1(t) \\ \dot{x}_2(t) \\ \dot{x}_3(t) \end{pmatrix} = \begin{pmatrix} -\dfrac{R_1}{L_1} & 0 & \dfrac{1}{L_1} \\ 0 & \dfrac{R_2}{L_2} & \dfrac{1}{L_2} \\ \dfrac{1}{C} & \dfrac{1}{C} & 0 \end{pmatrix} \begin{pmatrix} x_1(t) \\ x_2(t) \\ x_3(t) \end{pmatrix} + bu(t).$$

First-Order State-Space Representation of Second-Order Systems

Mathematical models of several practical problems, especially those arising in vibrations of structures, are second-order differential equations.

We show by means of a simple example of a spring-mass system how the equations of motion represented by a second-order differential equation can be converted to a first-order state-space representation.

Example 5.2.3. (A Spring-Mass System). Consider the spring-mass system shown in Figure 5.4 with equal spring constants k and masses m_1 and m_2. Let force u_1 be applied to mass m_1 and u_2 be applied to mass m_2.

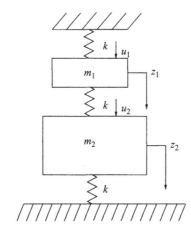

FIGURE 5.4:   A spring-mass system.

The equations of motion for the system are:

$$m_1\ddot{z}_1 + k(z_1 - z_2) + kz_1 = u_1,$$
$$m_2\ddot{z}_2 - k(z_1 - z_2) + kz_2 = u_2,$$

(5.2.3)

or in matrix form:

$$\begin{pmatrix} m_1 & 0 \\ 0 & m_2 \end{pmatrix} \begin{pmatrix} \ddot{z}_1 \\ \ddot{z}_2 \end{pmatrix} + \begin{pmatrix} 2k & -k \\ -k & 2k \end{pmatrix} \begin{pmatrix} z_1 \\ z_2 \end{pmatrix} = \begin{pmatrix} u_1 \\ u_2 \end{pmatrix}.$$

(5.2.4)

Set

$$z = \begin{pmatrix} z_1 \\ z_2 \end{pmatrix}.$$

Then, we have

$$M\ddot{z} + Kz = u,$$

(5.2.5)

where

$$M = \mathrm{diag}(m_1, m_2), \qquad K = \begin{pmatrix} 2k & -k \\ -k & k \end{pmatrix}, \quad \text{and} \quad u = \begin{pmatrix} u_1 \\ u_2 \end{pmatrix}.$$

Let us make a change of variables from z to x as follows:
Set

$$x_1 = z \quad \text{and} \quad x_2 = \dot{z}.$$

Then, in terms of the new variables, the equations of motion become

$$\dot{x}_1 = x_2,$$
$$M\dot{x}_2 = -Kx_1 + u,$$

(5.2.6)

or

$$\dot{x} = \begin{pmatrix} 0 & I \\ -M^{-1}K & 0 \end{pmatrix} x + \begin{pmatrix} 0 \\ M^{-1} \end{pmatrix} u,$$

(5.2.7)

where

$$x = (x_1, x_2)^{\mathrm{T}} = (z, \dot{z})^{\mathrm{T}}.$$

Equation (5.2.7) is a first-order representation of the second-order system (5.2.4).

State-Space Representations of Nonlinear Systems

Mathematical models of many real-life applications are nonlinear systems of differential equations. Very often it is possible to linearize a nonlinear system, and then after linearization, the first-order state-space representation of transformed linear system can be obtained. We will illustrate this by means of the following well-known examples (see Luenberger 1979; Chen 1984; Szidarovszky and Bahill 1991; etc.).

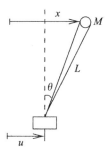

FIGURE 5.5:   Balancing of a stick.

Example 5.2.4. (Balancing a Stick).   Consider the simple problem of balancing a stick on your hand as shown in the Figure 5.5:

Here L is the length of the stick, and M is the mass of the stick concentrated on the top. The input $u(t)$ is the position of the hand. Then, the position of the top of the stick is

$$x(t) = L \sin \theta(t) + u(t).$$

The torque due to gravity acting on the mass is $MgL \sin \theta(t)$. The rotational inertia of the mass on the stick is $ML^2 \ddot{\theta}(t)$. The shift of the inertial term down to the pivot point is $\ddot{u}(t)ML \cos \theta(t)$. Thus, we have:

$$MgL \sin \theta(t) = ML^2 \ddot{\theta}(t) + \ddot{u}(t)ML \cos \theta(t).$$

The above equations are clearly nonlinear. We now linearize these equations by assuming that θ is small. We then can take $\cos \theta = 1$, $\sin \theta = \theta$.
This gives us

$$x(t) = L\theta(t) + u(t)$$

and

$$MgL\theta(t) = ML^2 \ddot{\theta}(t) + \ddot{u}(t)ML.$$

Eliminating $\theta(t)$ from these two equations, we obtain

$$\ddot{x}(t) = (g/L)(x(t) - u(t)).$$

We can now write down the first-order state-space representation by setting $v(t) = \dot{x}(t)$. The first-order system is then:

$$\begin{pmatrix} \dot{x}(t) \\ \dot{v}(t) \end{pmatrix} = \begin{pmatrix} 0 & 1 \\ \dfrac{g}{L} & 0 \end{pmatrix} \begin{pmatrix} x(t) \\ v(t) \end{pmatrix} + \begin{pmatrix} 0 \\ -\dfrac{g}{L} \end{pmatrix} u(t).$$

Example 5.2.5. (A Cart with an Inverted Pendulum).   Next we consider a similar problem (Figure 5.6), but this time with some more forces exerted (taken from Chen (1984, pp. 96–98)).

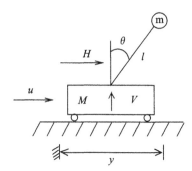

FIGURE 5.6:   A cart with an inverted pendulum.

In Figure 5.6, a cart is carrying an inverted pendulum with mass m and length l. Let M be the mass of the cart. Let H and V be, respectively, the horizontal and vertical forces exerted by the cart on the pendulum. Newton's law applied to the linear movements gives:

$$M \ddot{y}(t) = u - H,$$

$$H = m\ddot{y} + ml \cos \theta \ddot{\theta} - ml \sin \theta (\dot{\theta})^2,$$

and
$$mg - V = ml \left(-\sin \theta \ddot{\theta} - \cos \theta \, (\dot{\theta})^2 \right).$$

Newton's law applied to the rotational movement of the pendulum gives:

$$ml^2 \ddot{\theta} = mgl \sin \theta + Vl \sin \theta - Hl \cos \theta.$$

These are nonlinear equations. We now linearize them by making the same assumptions as before; that is, we assume that θ is small so that we can take $\sin \theta = \theta$, $\cos \theta = 1$. Dropping the terms involving θ^2, $\dot{\theta}^2$, $\theta \dot{\theta}$, and $\theta \ddot{\theta}$, and setting $\sin \theta = \theta$, and $\cos \theta = 1$, we obtain, from above, by eliminating V and H

$$(M + m)\ddot{y} + ml\ddot{\theta} = u$$

and
$$2l\ddot{\theta} - 2g\theta + \ddot{y} = 0.$$

Solving for $\ddot{y}$ and $\ddot{\theta}$, we obtain

$$\ddot{y} = -\frac{2gm}{2M + m}\theta + \frac{2}{2M + m}u,$$

$$\ddot{\theta} = \frac{2g(M + m)\theta}{(2M + m)l} - \frac{1}{(2M + m)l}u.$$

The state-space representations of these linear equations can now be written down by setting $x_1 = y$, $x_2 = \dot{y}$, $x_3 = \theta$, and $x_4 = \dot{\theta}$, as follows:

$$
\begin{pmatrix} \dot{x}_1 \\ \dot{x}_2 \\ \dot{x}_3 \\ \dot{x}_4 \end{pmatrix} = \begin{pmatrix} 0 & 1 & 0 & 0 \\ 0 & 0 & \dfrac{-2gm}{2M+m} & 0 \\ 0 & 0 & 0 & 1 \\ 0 & 0 & \dfrac{2g(M+m)}{(2M+m)l} & 0 \end{pmatrix} \begin{pmatrix} x_1 \\ x_2 \\ x_3 \\ x_4 \end{pmatrix} + \begin{pmatrix} 0 \\ \dfrac{2}{2M+m} \\ 0 \\ -\dfrac{1}{(2M+m)l} \end{pmatrix} u,
$$

$$
y = (1, 0, 0, 0)x.
$$

The nonlinear equations in Examples 5.2.4 and 5.2.5 are special cases of the general nonlinear equations of the form:

$$
\dot{\tilde{x}}(t) = f(\tilde{x}(t), \tilde{u}(t), t), \qquad \tilde{x}(t_0) = \tilde{x}_0,
$$
$$
\tilde{y}(t) = h(\tilde{x}(t), \tilde{u}(t), t),
$$

where f and h are vector functions. We will now show how these equations can be written in the standard first-order state-space form (Sayed 1994).
Assume that the nonlinear differential equation:

$$
\dot{\tilde{x}}(t) = f(\tilde{x}(t), \tilde{u}(t), t), \qquad \tilde{x}(t_0) = x_0
$$

has a unique solution and this unique solution is also continuous with respect to the initial condition.

Let $\tilde{x}_{\text{nom}}(t)$ denote the unique solution corresponding to the given input $\tilde{u}_{\text{nom}}(t)$ and the given initial condition $\tilde{x}_{\text{nom}}(t_0)$.

Let the nominal data $\{\tilde{u}_{\text{nom}}(t), \tilde{x}_{\text{nom}}(t)\}$ be perturbed so that

$$
\tilde{u}(t) = \tilde{u}_{\text{nom}}(t) + u(t)
$$

and $\quad \tilde{x}(t_0) = \tilde{x}_{\text{nom}}(t_0) + x(t_0)$, where $\quad \|u(t)\| \quad$ and $\quad \|x(t_0)\| \quad$ are small; $\|u(t)\| = \sup\limits_{t} \|u(t)\|_2$.

Assume further that

$$
\tilde{x}(t) = \tilde{x}_{\text{nom}}(t) + x(t), \qquad \tilde{y}(t) = \tilde{y}_{\text{nom}}(t) + y(t),
$$

where $\|x\|$ and $\|y\|$ are small.

These nonlinear equations can then be linearized (assuming that f and h are smooth enough) by expanding f and h around $(\tilde{u}_{\text{nom}}(t),\ \tilde{x}_{\text{nom}}(t))$, giving rise to a time-invariant linear state-space model of the form:

$$\dot{x}(t) = Ax(t) + Bu(t), \quad x(t_0) = x_0,$$
$$y(t) = Cx(t) + Du(t),$$

where

$$A = \left.\frac{\partial f}{\partial \tilde{x}}\right|_{(\tilde{x}_{\text{nom}}(t),\tilde{u}_{\text{nom}}(t))}, \qquad B = \left.\frac{\partial f}{\partial \tilde{u}}\right|_{(\tilde{x}_{\text{nom}}(t),\tilde{u}_{\text{nom}}(t))},$$

$$C = \left.\frac{\partial h}{\partial \tilde{x}}\right|_{(\tilde{x}_{\text{nom}}(t),\tilde{u}_{\text{nom}}(t))}, \qquad D = \left.\frac{\partial h}{\partial \tilde{u}}\right|_{(\tilde{x}_{\text{nom}}(t),\tilde{u}_{\text{nom}}(t))}.$$

Example 5.2.6. (The Motion of a Satellite (Sayed 1994)). Suppose that a satellite of unit mass orbits the earth at a distance $d(t)$ from its center (figure 5.7). Let $\theta(t)$ be the angular position of the satellite at time t, and the three forces acting on the satellite are: a radial force $u_1(t)$, a tangential force $u_2(t)$, and an attraction force $\alpha/d^2(t)$, where α is a constant.

The equations of motion are given by

$$\ddot{d}(t) = d(t)\dot{\theta}^2(t) - \frac{\alpha}{d^2(t)} + u_1(t),$$
$$\ddot{\theta}(t) = \frac{-2\dot{d}(t)\dot{\theta}(t)}{d(t)} + \frac{u_2(t)}{d(t)}.$$

Let's define the state variable as

$$\tilde{x}_1(t) = d(t), \qquad \tilde{x}_2(t) = \dot{d}(t), \qquad \tilde{x}_3(t) = \theta(t), \qquad \tilde{x}_4(t) = \dot{\theta}(t)$$

and the output variables as

$$\tilde{y}_1(t) = d(t), \qquad \tilde{y}_2(t) = \theta(t).$$

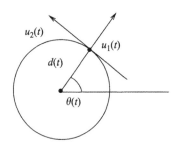

FIGURE 5.7:   The motion of a satellite.

The state-space model is then given by:

$$
\begin{pmatrix} \dot{\tilde{x}}_1(t) \\ \dot{\tilde{x}}_2(t) \\ \dot{\tilde{x}}_3(t) \\ \dot{\tilde{x}}_4(t) \end{pmatrix} = \begin{pmatrix} \tilde{x}_2(t) \\ \tilde{x}_1(t)\tilde{x}_4^2(t) - \dfrac{\alpha}{\tilde{x}_1^2(t)} + u_1(t) \\ \tilde{x}_4(t) \\ \dfrac{-2\tilde{x}_2(t)\tilde{x}_4(t)}{\tilde{x}_1(t)} + \dfrac{u_2(t)}{\tilde{x}_1(t)} \end{pmatrix},
$$

$$
\begin{pmatrix} \tilde{y}_1(t) \\ \tilde{y}_2(t) \end{pmatrix} = \begin{pmatrix} 1 & 0 & 0 & 0 \\ 0 & 0 & 1 & 0 \end{pmatrix} \begin{pmatrix} \tilde{x}_1(t) \\ \tilde{x}_2(t) \\ \tilde{x}_3(t) \\ \tilde{x}_4(t) \end{pmatrix}
$$

and the initial conditions are:

$$
\tilde{x}_0 = \tilde{x}(0) = \begin{pmatrix} \tilde{x}_1(0) \\ \tilde{x}_2(0) \\ \tilde{x}_3(0) \\ \tilde{x}_4(0) \end{pmatrix} = \begin{pmatrix} d(0) \\ 0 \\ \theta(0) \\ \omega_0 \end{pmatrix} = \begin{pmatrix} d_0 \\ 0 \\ \theta_0 \\ \omega_0 \end{pmatrix}.
$$

The above is still a nonlinear model of the form:

$$
\dot{\tilde{x}}(t) = f(\tilde{x}(t), \tilde{u}(t), t), \qquad \tilde{x}(t_0) = \tilde{x}_0,
$$
$$
\tilde{y}(t) = h(\tilde{x}(t), \tilde{u}(t), t).
$$

Linearizing this nonlinear model around the initial point $(\tilde{x}(0), \tilde{u}(0))$, where $\tilde{u}(0) = \begin{pmatrix} 0 \\ 0 \end{pmatrix}$, we obtain the linear model:

$$
\dot{x}(t) = Ax(t) + Bu(t),
$$
$$
y(t) = Cx(t),
$$

where

$$
A = \left. \frac{\partial f}{\partial \tilde{x}} \right|_{(\tilde{x}(0),\, \tilde{u}(0))}
$$

$$
= \begin{pmatrix} 0 & 1 & 0 & 0 \\ 3\omega_0^2 & 0 & 0 & 2d_0\omega_0 \\ 0 & 0 & 0 & 1 \\ 0 & -2\dfrac{\omega_0}{d_0} & 0 & 0 \end{pmatrix},
$$

$$B = \frac{\partial f}{\partial \tilde{u}}\bigg|_{(\tilde{x}(0),\tilde{u}(0))} = \begin{pmatrix} 0 & 0 \\ 1 & 0 \\ 0 & 0 \\ 0 & \frac{1}{d_0} \end{pmatrix},$$

and

$$C = \frac{\partial h}{\partial \tilde{x}}\bigg|_{(\tilde{x}(0),\tilde{u}(0))} = \begin{pmatrix} 1 & 0 & 0 & 0 \\ 0 & 0 & 1 & 0 \end{pmatrix}.$$

State-Space Representation of Systems Modeled by Partial Differential Equations

Mathematical models of many engineering problems such as those arising in fluid dynamics, mechanical systems, heat transfer, etc., are partial differential equations. The discretizations of these equations naturally lead to state-space models. We illustrate the idea by means of the following example.

Example 5.2.7.   Consider the partial differential equation

$$\frac{\partial^4 y}{\partial x^4} + \frac{P}{EI} \frac{\partial^2 y}{\partial t^2} = \frac{1}{EI} F(x, t),$$

which models the **deflection of a prismatic beam** (Soong (1990, pp. 180–181)). Let $y(x, t)$ be the transverse displacement of a typical segment of the beam that is located at a distance x from the end, and $F(x, t)$ be the applied force. EI is the flexural rigidity, and P is the density of the material of the beam per unit length. Let L be the length of the beam.

Assume that the solution $y(x, t)$ can be written as

$$y(x, t) = \sum_{j=1}^{n} v_j(x) p_j(t)$$

($n = \infty$ in theory, but in practice it is large but finite). Also assume that

$$F(x, t) = \sum_{j=1}^{r} \delta(x - a_j) u_j(t),$$

where $\delta(\cdot)$ is the Dirac delta function.

That is, we assume that the force is point-wise and is exerted at r points of the beam.

Substituting these expressions of $y(x, t)$ and $F(x, t)$ in the partial differential equation, it can be shown that the state equation for the beam in the standard form is:

$$\dot{z}(t) = \mathbf{A}z(t) + Bu(t),$$

where $z(t) = (p_1, \dot{p}_1, p_2, \dot{p}_2, \ldots, p_n, \dot{p}_n)^{\mathrm{T}}$ is the $2n$-dimensional state vector,

$$\mathbf{A} = \mathrm{diag}\,(\Lambda_1, \Lambda_2, \ldots, \Lambda_n), \qquad B = \begin{pmatrix} B_1 \\ B_2 \\ \vdots \\ B_n \end{pmatrix},$$

The matrices B_j and the vector u are defined by:

$$B_j = \frac{1}{\mathrm{EI}} \begin{pmatrix} 0 & 0 & \cdots & 0 \\ v_j(a_1) & v_j(a_2) & \cdots & v_j(a_r) \end{pmatrix}_{n \times r} \quad \text{and} \quad u = \begin{pmatrix} u_1 \\ u_2 \\ \vdots \\ u_r \end{pmatrix}.$$

5.2.2 Discrete-Time Systems

A linear time-invariant discrete-time system may be represented by means of a system of difference equations:

$$x(k+1) = Ax(k) + Bu(k), \tag{5.2.8}$$
$$y(k) = Cx(k) + Du(k). \tag{5.2.9}$$

As before, $x(k)$ is the n-dimensional **state vector**, $u(k)$ is the m-dimensional **input vector**; and A, B, C and D are time-invariant matrices of dimensions $n \times n$, $n \times m$, $r \times n$, and $r \times m$, respectively. The inputs and outputs of a discrete-time system are defined only at discrete time instants.

Sometimes we will write the above equations in the form:

$$x_{k+1} = Ax_k + Bu_k, \tag{5.2.10}$$
$$y_k = Cx_k + Du_k. \tag{5.2.11}$$

5.2.3 Descriptor Systems

The system represented by Eqs. (5.2.1) and (5.2.2) is a special case of a more general system, known as the **descriptor system**.

A continuous-time linear descriptor system has the form:

$$E\dot{x}(t) = Ax(t) + Bu(t), \tag{5.2.12}$$
$$y(t) = Cx(t) + Du(t). \tag{5.2.13}$$

Similarly, a discrete-time linear descriptor system has the form:

$$Ex(k + 1) = Ax(k) + Bu(k), \tag{5.2.14}$$
$$y(k) = Cx(k) + Du(k). \tag{5.2.15}$$

If the matrix E is nonsingular, then, of course, the descriptor system represented by (5.2.12) and (5.2.13) is reduced to the standard form (5.2.1)–(5.2.2). Similarly, for the system (5.2.14)–(5.2.15). However, the case when E is singular or nearly singular is more interesting. A book devoted to singular systems of differential equations is by Campbell (1980). We will now give an example to show how a singular descriptor systems arises in practice.

Example 5.2.8. (Simplified Samuelson's Model of Economics).   Let NI(k), CS(k), IV(k), and GE(k), denote, respectively, the national income, consumption, investment, and government expenditure of a country at a given year k.

The economist P.A. Samuelson proposed a model of the national economy of a country, which is based on the following assumptions:

1. National income NI(k) is equal to the sum of the consumption CS(k), investment IV(k), and the government expenditure GE(k) at a given year k.
2. Consumption CS($k + 1$) at the year $k + 1$ is proportional to the national income NI(k) at the previous year k.
3. Investment IV($k + 1$) at the year $k + 1$ is proportional to the difference of the consumer spending CS($k + 1$) at that year and that of the previous year CS(k).

The state-space representation of Samuelson's model, then, can be written in the form:

$$\text{NI}(k) = \text{CS}(k) + \text{IV}(k) + \text{GE}(k),$$
$$\text{CS}(k + 1) = \alpha \text{NI}(k), \tag{5.2.16}$$
$$\text{IV}(k + 1) = \beta \left[\text{CS}(k + 1) - \text{CS}(k) \right].$$

These equations in matrix form are:

$$\begin{pmatrix} 0 & 0 & 0 \\ 0 & 1 & 0 \\ 1 & -\beta & 0 \end{pmatrix} \begin{pmatrix} \text{IV}(k+1) \\ \text{CS}(k+1) \\ \text{NI}(k+1) \end{pmatrix} = \begin{pmatrix} 1 & 1 & -1 \\ 0 & 0 & \alpha \\ 0 & -\beta & 0 \end{pmatrix} \begin{pmatrix} \text{IV}(k) \\ \text{CS}(k) \\ \text{NI}(k) \end{pmatrix} + \begin{pmatrix} 1 \\ 0 \\ 0 \end{pmatrix} \text{GE}(k).$$

or

$$Ex(k + 1) = Ax(k) + Bu(k),$$

where $x(k) = \begin{pmatrix} IV(k) \\ CS(k) \\ NI(k) \end{pmatrix}$, $B = \begin{pmatrix} 1 \\ 0 \\ 0 \end{pmatrix}$, $u(k) = \text{GE}(k)$.

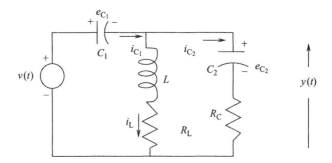

FIGURE 5.8: An electric circuit for a descriptor system.

(Note that E is singular).
For details, see Luenberger (1979, pp. 122–123).

Example 5.2.9.   Consider another electric circuit given in Figure 5.8 driven by a
voltage source $v(t)$.

The state variables are taken as $x_1 := e_{C_1}$, $x_2 := i_L$, and $x_3 := e_{C_2}$. By applying
Kirchhoff's current and voltage laws we have:

$$i_{C_1} = i_L + i_{C_2}, \qquad i_{C_1} = C_1 \frac{de_{C_1}}{dt}, \qquad i_{C_2} = C_2 \frac{de_{C_2}}{dt},$$

$$v(t) = e_{C_1} + L\frac{di_L}{dt} + R_L i_L = e_{C_1} + e_{C_2} + R_C i_{C_2}.$$

Manipulating these equations we have the state equation:

$$E\dot{x} = Ax + bu,$$

where $u := v$, $E = \operatorname{diag}(R_C C_1, L, R_C C_2)$

$$x = [x_1, x_2, x_3]^{\mathrm{T}}$$

$$A = \begin{bmatrix} -1 & R_C & -1 \\ -1 & -R_L & 0 \\ -1 & 0 & -1 \end{bmatrix}, \qquad b = \begin{bmatrix} 1 \\ 1 \\ 1 \end{bmatrix}.$$

The output is defined by $y(t) = e_{C_2} + R_C i_{C_2}$, so the output equation becomes

$$y = cx + du,$$

where $c = (-1\ 0\ 0)$, $d = 1$.

5.3 SOLUTIONS OF A CONTINUOUS-TIME SYSTEM: SYSTEM RESPONSES

Theorem 5.3.1. *Continuous-Time State-Space Solution. The solutions of the continuous-time dynamical equations*

$$\dot{x}(t) = Ax(t) + Bu(t), \quad x(t_0) = x_0. \tag{5.3.1}$$

$$y(t) = Cx(t) + Du(t) \tag{5.3.2}$$

are given by

$$x(t) = e^{A(t-t_0)}x_0 + \int_{t_0}^{t} e^{A(t-s)}Bu(s)\,ds, \tag{5.3.3}$$

$$y(t) = Ce^{A(t-t_0)}x_0 + \int_{t_0}^{t} Ce^{A(t-s)}Bu(s)\,ds + Du(t). \tag{5.3.4}$$

Remark

- If $u(t) = 0$, then $x(t) = e^{A(t-t_1)}x(t_1)$ for every $t \geq t_0$ and any $t_1 \geq t_0$.

Definition 5.3.1. *The matrix $e^{A(t-t_1)}$ is called the **state transition matrix**.*

Since the state at any time can be obtained from that of any other time through the state transition matrix, **it will be assumed, without any loss of generality, that $t_0 = 0$, unless otherwise mentioned.**
Assuming $t_0 = 0$, the Eqs. (5.3.3) and (5.3.4) will, respectively, be reduced to

$$x(t) = e^{At}x_0 + \int_0^t e^{A(t-s)}Bu(s)\,ds \tag{5.3.5}$$

and

$$y(t) = Ce^{At}x_0 + \int_0^t Ce^{A(t-s)}Bu(s)\,ds + Du(t). \tag{5.3.6}$$

Definition 5.3.2. *The matrix e^{At} defined above has the form:*

$$e^{At} = \sum_{k=0}^{\infty} \frac{(At)^k}{k!}$$

*and is called the **matrix exponential**.*

Proof. Proof of (5.3.5) and (5.3.6): Noting that $(d/dt)(e^{At}) = Ae^{At}$ (see **Section 5.3.1**), we first verify that the expression (5.3.5) satisfies (5.3.1) with

$t = 0$. Differentiating (5.3.5), we have

$$\dot{x}(t) = Ae^{At}x_0 + Bu(t) + \int_0^t \frac{d}{dt} e^{A(t-s)} Bu(s) \, ds,$$

$$= Ae^{At}x_0 + Bu(t) + A \int_0^t e^{A(t-s)} Bu(s) \, ds,$$

$$= A \left[e^{At}x_0 + \int_0^t e^{A(t-s)} Bu(s) \, ds \right] + Bu(t),$$

$$= Ax(t) + Bu(t).$$

Also, note that at $t = 0$,

$$x(0) = x_0.$$

Thus, the solution $x(t)$ also satisfies the initial condition.

The expression for $y(t)$ in (5.3.6) follows immediately by substituting the expression for $x(t)$ from (5.3.5) into $y(t) = Cx(t) + Dx(t)$. ■

Free, Forced, and Steady-State Responses

Given the initial condition x_0 and the control input $u(t)$, the vectors $x(t)$ and $y(t)$ determine the **system time responses** for $t \geq 0$. The system time responses determine the behavior of the dynamical system for particular classes of inputs. System characteristics such as **overshoot, rise-time, settling time,** etc., can be determined from the time responses.

In the expression (5.3.6), the first term $Ce^{At}x_0$ represents the response of the system due to the initial condition x_0 with zero input **(zero-input response)**. This is called the **free response** of the system.

On the other hand, the second term in (5.3.6) determines, what is known, as the **forced response** of the system. It is due to the forcing function $u(t)$ applied to the system with zero initial conditions. A special case of the forced response is known as the **impulse response** which is obtained from (5.3.6) by setting $x_0 = 0$ and $u(t) = \delta(t)$, where $\delta(t)$ is the unit impulse or **Dirac delta function**. Then,

$$y(t) = \int_0^t (Ce^{A(t-s)} B + D\delta(t - s))u(s) \, ds,$$

$$= \int_0^t H(t - s)u(s) \, ds, \tag{5.3.7}$$

where the matrix $H(t)$, **the impulse response matrix**, is defined by

$$H(t) := Ce^{At} B + D\delta(t). \tag{5.3.8}$$

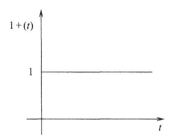

FIGURE 5.9:   A unit step function.

In fact, if $x_0 = 0$, then the (i, j)th element of the impulse response matrix $H(t)$ of the system (5.3.1)–(5.3.2) is the response at time t at the output i due to a unit impulse applied at time $t = 0$ at the input j of the system, while all other inputs are kept at zero. Similarly, the **unit step response** is defined to be the output using the input as the unit step function in the manner done for an impulse response, assuming again that the initial state is zero; that is, $x_0 = 0$.

A unit step function $1_+(t)$ (Figure 5.9) is given by

$$1_+(t) = \begin{cases} 1, & t \geq 0, \\ 0, & t < 0. \end{cases}$$

For any finite value of time t, the response $y(t)$, that is, the right-hand side of (5.3.6), will contain terms consisting of $e^{\alpha_i t} e^{j\omega_i t}$ if $\lambda_i = \alpha_i + j\omega_i$, $j = \sqrt{-1}$, is a **simple eigenvalue** of A, and the other terms are governed by the nature of the input function $u(t)$.

When t is finite, the part of the response in $y(t)$ which is governed by $e^{\alpha_i t} e^{j\omega_i t}$ is called the **transient response** of the system. As t tends to infinity, this transient part of the response tends to zero if all α_is are negative or it grows to become unbounded if at least one of α_is is positive. Thus, $y_{ss}(t) \equiv \lim_{t \to \infty} y(t)$ will be called the **steady-state response** of the system. The speed with which the response $y(t)$ will reach the steady-state value $y_{ss}(t)$ will be determined by the largest value of αs.

MATLAB note: MATLAB functions **step, impulse**, and **initial** in MATLAB CONTROL TOOLBOX can be used to generate plots for step, impulse, and initial condition responses, respectively. Their syntax are:

$$\text{step (sys)},$$
$$\text{impulse (sys)},$$
$$\text{initial (sys, } x_0).$$

Example 5.3.1. (Baldwin (1961, pp. 29–44)).   The dynamic behavior of a moving coil galvanometer, see Figure 5.10,

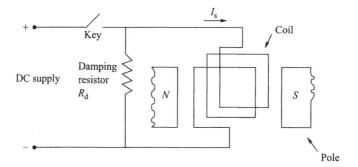

FIGURE 5.10:   Basic Circuit of a Moving Coil Galvanometer.

is governed by

$$J\frac{d^2\theta}{dt^2} + D\frac{d\theta}{dt} + C\theta = GI_s, \tag{5.3.9}$$

where

J = the moment of inertia about the axis of rotation of moving parts of the instrument,

D = the damping constant,

C = stiffness of suspension,

G = galvanometer constant which represents the electromagnetic torque developed on the coil by 1 A of current flowing through it,

θ = the deflection of the coil about its axis of rotation,

I_s = the steady-state current flowing through the galvanometer coil, and

R_g = galvanometer resistance.

It can be shown that D is given by

$$D = \frac{G^2}{R_g + R_d} + D_{air},$$

where

R_g = resistance of galvanometer coil,

R_d = damping resistor,

D_{air} = damping to the coil due to air.

If the key is opened interrupting supply current I_s to the galvanometer, the response of the coil is determined by (5.3.9) with $I_s = 0$ and is shown in Figure 5.11 where θ_0 is the steady-state deflection with I_s flowing through the coil.

A galvanometer with a very low damping constant is known as a ballistic galvanometer. If a charged capacitor is discharged through a ballistic galvanometer such that

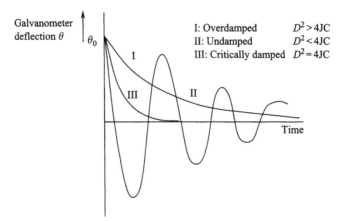

FIGURE 5.11:   Step-response of the galvanometer.

the whole charge should have passed before the coil has time to move appreciably, we have the **torque impulse** due to the whole charge equal to $\int Gi\, dt = GQ$, the integral being taken over the time of passage of the charge Q and i is the instantaneous current flowing through galvanometer coil. The subsequent time response of the galvanometer will be similar to that shown in Figure 5.11 but will differ in the fact that the response now starts from the origin. The responses in three cases: damped, undamped, and critically damped, are shown in Figure 5.11.

Causality: If the output of the system at time t does not depend upon the input applied to the system after time t, but depends only upon the present and the past inputs, the system is said to be **causal.**

In this book, all systems will be assumed to be causal.

5.3.1   Some Important Properties of the Matrix e^{At}

Since the matrix exponential e^{At} plays a fundamental role in the solution of the state equations, we will now discuss the various methods for computing this matrix. Before doing that, we list some important properties of this matrix. These properties are easily verifiable and left as Exercises (5.8–5.10) for the readers.

1. $e^{A(t+s)} = e^{At} \cdot e^{As}$
2. $d/dt\,(e^{At}) = Ae^{At} = e^{At}A$
3. $e^{(A+B)t} = e^{At} \cdot e^{Bt}$, if and only if A and B commute; that is, if and only if $AB = BA$
4. e^{At} is nonsingular and $(e^{At})^{-1} = e^{-At}$
5. $(e^{A/m})^{m} = e^{A}$, where m is an integer
6. $e^{P^{-1}APt} = P^{-1}e^{At}P$.

5.3.2 Sensitivity of e^{At}

We know that the accuracy of a solution obtained by an algorithm is greatly influenced by the sensitivity of the problem. We will, therefore, consider the sensitivity aspect of the problem of computing e^{At} first. We just state a result due to Van Loan (1977) without proof, which will help identify the condition number of the problem.

Let E be a perturbation matrix. We are interested in knowing how large the **relative error**

$$\rho = \frac{\|e^{(A+E)t} - e^{At}\|_2}{\|e^{At}\|_2}$$

can be.

Differentiating $e^{(A+E)s}e^{A(t-s)}$ with respect to s, we obtain

$$e^{(A+E)t} - e^{At} = \int_0^t e^{A(t-s)} E e^{(A+E)s} ds.$$

It then follows that

$$\rho \leq \frac{\|E\|_2}{\|e^{At}\|_2} \int_0^t \|e^{A(t-s)}\|_2 \|e^{(A+E)s}\|_2 ds$$

Further simplification of this result is possible.

Van Loan (1977) has shown that, for a given t, there exists a perturbation matrix E such that

$$\rho = \frac{\|e^{(A+E)t} - e^{At}\|_2}{\|e^{At}\|_2} \approx \kappa(A, t) \frac{\|E\|_2}{\|A\|_2},$$

where

$$\kappa(A, t) = \max_{\|E\|_2 \leq 1} \left\| \int_0^t e^{A(t-s)} E e^{As} ds \right\|_2 \frac{\|A\|_2}{\|e^{At}\|_2}.$$

This result shows that $\kappa(A, t)$ **is the condition number for the problem** e^{At}. **If this number is large, then a small change in A can cause a large change in** e^{At}, **for a given** t.

Though determining $\kappa(A, t)$ involves computation of a matrix integral, it can be easily verified that

$$\kappa(A, t) \geq t \|A\|_2,$$

with equality holding for all nonnegative t if and only if A is a normal matrix, that is, if $A^T A = A A^T$. **"When A is not normal, $\kappa(A, t)$ can grow like a high degree polynomial in t"** (Moler and Van Loan 1978).

Example 5.3.2. Consider computing e^A, where

$$A = \begin{pmatrix} -1 & 1000 \\ 0 & -1 \end{pmatrix}.$$

Since $\|A\|_2 = 10^3$, the problem of computing the matrix exponential e^A is expected to be ill-conditioned.

Let's verify this as follows. The matrix e^A computed by using the MATLAB function **expm** is

$$e^A = \begin{pmatrix} 0.3679 & 367.8794 \\ 0 & 0.3679 \end{pmatrix}.$$

Now change the $(2, 1)$ entry of A to 10^{-8} and call this perturbed matrix A_{new}. We now obtain

$$e^{A_{\text{new}}} = \begin{pmatrix} 0.3679 & 367.8801 \\ 0 & 0.3679 \end{pmatrix}.$$

The relative error in the solution is

$$\frac{\|e^A - e^{A_{\text{new}}}\|_2}{\|e^A\|_2} = O(10^{-6}).$$

On the other hand, the relative error in the data is

$$\frac{\|A - A_{\text{new}}\|_2}{\|A\|_2} = O(10^{-11}).$$

(The matrix $e^{A_{\text{new}}}$ was also computed by the MATLAB function **expm**).

5.3.3 Computational Methods for e^{At}

There is a wide variety of methods to compute the matrix exponential. Several of these methods have been carefully analyzed, with respect to efficiency, and numerical stability, in an authoritative paper on the subject by Moler and Van Loan (1978). We discuss the following ones briefly.

- **The eigenvalue–eigenvector method**
- **Series methods**
- **Ordinary differential equations (ODE) Methods**
- **Matrix decomposition methods.**

The Eigenvalue–Eigenvector Method

We have seen that the solution of the unforced system:

$$\dot{x}(t) = Ax(t), \qquad x(0) = x_0 \tag{5.3.10}$$

is given by

$$x(t) = e^{At}x_0. \tag{5.3.11}$$

Equation (5.3.11) shows that the ith column of the matrix e^{At} is just the vector $x(t)$ with $x(0) = e_i$, the ith unit vector.

Again, $x(t)$ can be expressed in terms of the eigenvalues and eigenvectors of A:

$$x(t) = c_1 e^{\lambda_1 t} v_1 + c_2 e^{\lambda_2 t} v_2 + \cdots + c_n e^{\lambda_n t} v_n, \tag{5.3.12}$$

where $\lambda_1, \lambda_2, \ldots, \lambda_n$ are the **simple eigenvalues of** A and v_1 through v_n are a set of linearly independent eigenvectors associated with λ_1 through λ_n. The scalar cs are computed from the given initial condition.

Thus, when the eigenvalues of A are simple, the matrix e^{At} is completely determined by the eigenvalues and eigenvectors of the matrix A. The same can be shown to be true when A has multiple eigenvalues.

A difficulty with this approach arises when A has some nearly equal eigenvalues. This can be seen from the following theorem (Moler and Van Loan 1978).

Theorem 5.3.2. *Let A be an $n \times n$ nondefective matrix; that is, it has a set of n linearly independent eigenvectors. Let $X^{-1}AX = \mathrm{diag}(\lambda_1, \lambda_2, \ldots, \lambda_n)$, where $\lambda_1, \lambda_2, \ldots, \lambda_n$ are the eigenvalues of A. Then,*

$$\| fl(e^{At}) - e^{At} \|_2 \le n\mu e^{\rho(A)t} \mathrm{Cond}_2(X),$$

where $\rho(A) = \max |\lambda_i|$ is the spectral radius of A.

Interpretation of Theorem 5.3.2. Theorem 5.3.2 shows that **there might be a large error in computing e^{At} whenever there is a coalescence of eigenvalues of A,** because, in this case, $\mathrm{Cond}_2(X)$ will be large.

The following simple 2×2 example taken from Moler and Van Loan (1978) illustrates the difficulty.

Let

$$A = \begin{pmatrix} \lambda & \alpha \\ 0 & \mu \end{pmatrix}.$$

Then,

$$e^{At} = \begin{pmatrix} e^{\lambda t} & \alpha \dfrac{e^{\lambda t} - e^{\mu t}}{\lambda - \mu} \\ 0 & e^{\mu t} \end{pmatrix}.$$

Clearly, the result will be inaccurate if λ is close to μ, but not exactly equal to μ. A large round-off error can be expected in this case.

Series Method for Computing the Matrix Exponential

In this section, we briefly state two series methods: **The Taylor Series Method** and the **Padé Approximation Method.** When properly implemented, these methods become numerically effective for computing the matrix exponential.

The Taylor Series Method
An obvious way to approximate e^A is to evaluate a finite-series sum by truncating the Taylor series:

$$e^A = I + A + \frac{A^2}{2} + \frac{A^3}{6} + \cdots$$

after k terms. Thus, if

$$T_k(A) = \sum_{j=0}^{k} \frac{A^j}{j!}$$

and if $\text{fl}(T_k(A))$ denotes the floating point evaluation of $T_k(A)$, then it is natural to choose k so that $\text{fl}(T_k(A)) = \text{fl}(T_{k+1}(A))$. **The drawbacks to this method are that a large number of terms is needed for convergence, and even when convergence occurs, the answer can be totally wrong.**
   Consider the following example from Moler and Van Loan (1978).

Example 5.3.3.   Consider computing e^A using the Taylor series methods with the following 2×2 matrix A and a relative accuracy of about 10^{-5}.

$$A = \begin{pmatrix} -49 & 24 \\ -64 & 31 \end{pmatrix}.$$

A total of $k = 59$ terms were required for convergence and the computed output was

$$e^A \approx \begin{pmatrix} -22.25880 & -1.432766 \\ -61.49931 & -3.474280 \end{pmatrix},$$

which is nowhere close to the true answer (to 6 decimal places)

$$e^A \approx \begin{pmatrix} -0.735759 & 0.551819 \\ -1.471518 & 1.103638 \end{pmatrix}.$$

The source of error here is the catastrophic cancellation that took place in the evaluation of $(A^{16}/16!) + (A^{17}/17!)$, using finite-precision arithmetic (see **Chapter 3**). These two terms have almost the same magnitude but are of opposite signs, as seen from the following expressions of $(A^{16}/16!)$ and $(A^{17}/17!)$:

$$\frac{A^{16}}{16!} = 10^6 \begin{pmatrix} 6.9773 & -3.4886 \\ 9.3030 & -4.6515 \end{pmatrix},$$

$$\frac{A^{17}}{17!} = 10^6 \begin{pmatrix} -6.9772 & 3.4885 \\ -9.3030 & 4.6515 \end{pmatrix}.$$

For relative accuracy of 10^{-5}, "the elements of these intermediate results have absolute errors larger than the final result".

The Padé Approximation Method
Suppose that $f(x)$ is a power series represented by

$$f(x) = f_0 + f_1 x + f_2 x^2 + \cdots .$$

Then the (p, q) Padé approximation to $f(x)$ is defined as

$$f(x) \approx \frac{c(x)}{d(x)} = \frac{\sum_{k=0}^{p} c_k x^k}{\sum_{k=0}^{q} d_k x^k}.$$

The coefficients $c_k, k = 0, \ldots, p$, and $d_k, k = 0, \ldots, q$ are chosen such that the terms containing $x^0, x^1, x^2, \ldots, x^{p+q}$ are cancelled out in $f(x)d(x) - c(x)$. The order of the Padé approximation is $p + q$. The ratio $(c(x)/d(x))$ is unique and the coefficients $c_0, c_1, \ldots, c_p$ and $d_0, d_1, \ldots, d_q$ always exist. The (p, q) Padé approximation to e^A is given by

$$R_{pq}(A) = \left[D_{pq}(A) \right]^{-1} N_{pq}(A),$$

where

$$D_{pq}(A) = \sum_{j=0}^{q} \frac{(p+q-j)!q!}{(p+q)!j!(q-j)!}(-A)^j$$

and

$$N_{pq}(A) = \sum_{j=0}^{p} \frac{(p+q-j)!p!}{(p+q)!j!(p-j)!} A^j.$$

It can be shown (**Exercise 5.16**) that if p and q are large, or if A is a matrix having all its eigenvalues **with negative real parts**, then $D_{pq}(A)$ is nonsingular.

    Round-off errors due to catastrophic cancellation is again a major concern for this method. It is less when $\|A\|$ is not too large and the diagonal approximants $(p = q)$ are used.

Scaling and Squaring in Taylor Series and Padé Approximations
The difficulties with the round-off errors in Taylor series and Padé approximation methods can somehow be controlled by a technique known as **Scaling and Squaring**. Since $e^A = (e^{A/m})^m$, the idea here is to choose m to be a power of 2 so that $e^{A/m}$ can be computed rather efficiently and accurately using the method under consideration, and then to form the matrix $(e^{A/m})^m$ by repeated squaring.

Moler and Van Loan (1978) have remarked **"When properly implemented, the resulting algorithm is one of the most effective we know."** The method may fail but it is **reliable** in the sense that it tells the user when it does.

The method has favorable numerical properties when $p = q$. We will, therefore, describe the algorithm for this case only.

Let $m = 2^j$ be chosen such that $\|A\|_\infty \le 2^{j-1}$, then Moler and Van Loan (1978) have shown that there exists an E such that

$$\left[R_{pp}(A/2^j) \right]^{2^j} = e^{A+E},$$

where $\|E\|_\infty \le \epsilon \|A\|_\infty$, with

$$\epsilon = 2^{3-2p} \left(\frac{(p!)^2}{(2p)!(2p+1)!} \right).$$

Given an error-tolerance δ, the above expression, therefore, gives a criterion to choose p such that $\|E\|_\infty \le \delta \|A\|_\infty$.

The above discussion leads to the following algorithm.

Algorithm 5.3.1.   *Padé Approximation to e^A using Scaling and Squaring.*
   Input. *$A \in \mathbb{R}^{n \times n}$, $\delta > 0$, an error-tolerance.*
   Output. *$F = e^{A+E}$ with $\|E\|_\infty \le \delta \|A\|_\infty$.*
   Step 1. *Choose j such that $\|A\|_\infty \le 2^{j-1}$. Set $A \equiv A/2^j$.*
   Step 2. *Find p such that p is the smallest nonnegative integer satisfying*

$$\left(\frac{8}{2^{2p}} \right) \frac{(p!)^2}{(2p)!(2p+1)!} \le \delta.$$

   Step 3. *Set $D \equiv I$, $N \equiv I$, $Y \equiv I$, $c = 1$.*
   Step 4. *For $k = 1, 2, \ldots, p$ do*

$$c \equiv c(p - k + 1)/[(2p - k + 1)k]$$

$$Y \equiv AY, N \equiv N + cY, D = D + (-1)^k cY.$$

   End
   Step 5. *Solve for F: $DF = N$.*

Step 6. *For $k = 1, 2, \ldots, j$ do*

$$F \equiv F^2.$$

End

Flop-count: · The algorithm requires about $2(p + j + (1/3)n^3$ flops.

Numerical Stability Property: **The algorithm is numerically stable when A is normal. When A is non-normal,** an analysis of the stability property becomes difficult, because, in this case e^A **may grow before it decays during the squaring process;** which is known as **"hump"** phenomenon. For details, see Golub and Van Loan (1996, p. 576).

MATLAB note: The MATLAB function **expm** computes the exponential of a matrix A, using Algorithm 5.3.1.

MATCONTROL note: Algorithm 5.3.1 has been implemented in MATCON-TROL function **expmpade**.

Example 5.3.4. Consider computing e^A using Algorithm *5.3.1* with

$$A = \begin{pmatrix} 5 & 1 & 0 \\ 0 & 2 & 0 \\ 2 & 3 & 1 \end{pmatrix}.$$

Set $\delta = 0.50$.
Step 1. Since $\|A\|_\infty = 7$, choose $j = 4$.
Then,

$$A \equiv \frac{A}{2^4} = \begin{pmatrix} 0.3125 & 0.8625 & 0 \\ 0 & 0.1750 & 0 \\ 0.1250 & 0.1875 & 0.6625 \end{pmatrix}.$$

Step 2. $p = 1$.
Step 3. $D = N = Y = I$.
Step 4. $k = 1, c = 0.5$.

$$Y = \begin{pmatrix} 0.3125 & 0.0625 & 0 \\ 0 & 0.1250 & 0 \\ 0.1250 & 0.7875 & 0.0625 \end{pmatrix},$$

$$N = \begin{pmatrix} 1.1563 & 0.0313 & 0 \\ 0 & 1.0625 & 0 \\ 0.0625 & 0.09838 & 1.0313 \end{pmatrix}, \quad D = \begin{pmatrix} 0.8438 & -0.0313 & 0 \\ 0 & 0.9375 & 0 \\ -0.0625 & -0.0938 & 0.9638 \end{pmatrix},$$

$$F = \begin{pmatrix} 1.3704 & 0.0790 & 0 \\ 0 & 1.1333 & 0 \\ 0 & 0.2115 & 1.0645 \end{pmatrix}.$$

Step 5. $F \equiv F^2 = \begin{pmatrix} 154.6705 & 49.0874 & 0 \\ 0 & 7.4084 & 0 \\ 75.9756 & 36.2667 & 2.7192 \end{pmatrix}.$

The matrix e^A given by F in Step 5 is different from what would have been obtained by using MATLAB function $expm(A)$, which is:

$$e^A = \begin{pmatrix} 148.4132 & 47.0080 & 0 \\ 0 & 7.3891 & 0 \\ 72.8474 & 35.1810 & 2.7183 \end{pmatrix}.$$

This is because MATLAB uses much smaller tolerance than what was prescribed for this example.

ODE Methods

Consider solving

$$\dot{x}(t) = f(x, t), \qquad x(0) = x_0$$

with

$$f(x, t) = Ax(t).$$

Then the kth column of e^{At} becomes equal to the solution $x(t)$ by setting $x(0)$ as the kth column of the identity matrix. Thus, any ODE solver can be used to compute e^{At}.

However, computing e^{At} using a general-purpose ODE solver will be rather expensive, because such a method does not take advantage of the special form of $f(x, t) = Ax(t)$. **An ODE routine will treat $Ax(t)$ as any function $f(x, t)$ and the computation will be carried on.**

However, a single-step ODE method such as the fourth order **Taylor** or **Runge–Kutta** method and a multistep solver such as the **Adams formulas** with variable step size, could be rather computationally practically feasible (and also reliable and stable) for the matrix vector problem of computing $e^{At}x(0)$, when such a problem is to be solved for many different values of t and also when A is large and sparse.

Matrix Decomposition Methods

The basic idea behind such a method is to first transform the matrix A to a suitable canonical form R so that e^R can be easily computed, and then compute e^A from e^R. Thus, if P is the transforming matrix such that

$$P^{-1}AP = R,$$

then $e^A = Pe^R P^{-1}$.

The convenient choices for R include the **Jordan canonical** form (JCF), **the companion form, the Hessenberg form, and the real Schur form (RSF) of a matrix.** Because of the difficulties in using the JCF and the companion form, stated in Section 4.1, we will not discuss computing e^A via these forms here.

Though the Hessenberg form can be obtained in a numerically stable way, computation of e^H via an upper Hessenberg matrix H will involve divisions by the superdiagonal entries, and if the product of these entries is small, large round-off errors can contaminate the computation (**see Exercise 5.12**). Thus, **our choice for R here is the RSF.**

Computing e^A via the Real Schur Form
Let A be transformed to a real Schur matrix R using an orthogonal similarity transformation:

$$P^T A P = R,$$

then $e^A = P e^R P^T$.

The problem is now how to compute e^R. Parlett (1976) has given an elegant formula to compute $f(R)$ for an analytic function $f(x)$, where R is **upper triangular**. The formula is derived from the commutativity property: $Rf(R) = f(R)R$.

Adapting this formula for computing $F = e^R$, we have the following algorithm when R is an upper triangular matrix. The algorithm needs to be modified when R is a **quasi-triangular matrix (Exercise 5.21)**.

Algorithm 5.3.2. *The Schur Algorithm for e^A.*
Input. $A \in \mathbb{R}^{n \times n}$
Output. e^A.
Step 1. *Transform A to R, an **upper triangular** matrix, using the QR iteration algorithm (Chapter 4):*

$$P^T A P = R.$$

(Note that when the eigenvalues of A are all real, the RSF is upper triangular.)
Step 2. *Compute $e^R = G = (g_{ij})$:*
For $i = 1, \ldots, n$ do
$g_{ii} = e^{r_{ii}}$
End
For $k = 1, 2, \ldots, n - 1$ do
  For $i = 1, 2, \ldots, n - k$ do

  Set $j = i + k$

  $$g_{ij} = \frac{1}{(r_{ii} - r_{jj})} \left[r_{ij}(g_{ii} - g_{jj}) + \sum_{p=i+1}^{j-1} (g_{ip} r_{pj} - r_{ip} g_{pj}) \right].$$

  End

End
Step 3. *Compute $e^A = P e^R P^T$.*

Flop-count: Computation of e^R in Step 2 requires about $(2n^3/3)$ flops.

MATCONTROL note: Algorithm 5.3.2 has been implemented in MATCONTROL function **expmschr**.

Example 5.3.5.   Consider computing e^A using Algorithm *5.3.2* with the matrix A of Example 5.3.3.

$$A = \begin{pmatrix} 5 & 1 & 0 \\ 0 & 2 & 0 \\ 2 & 3 & 1 \end{pmatrix}.$$

Using MATLAB function $[P, R] = \mathbf{schur}(A)$, we obtain

$$P = \begin{pmatrix} 0 & 1 & 0 \\ 0 & 0 & 1 \\ 1 & 0 & 0 \end{pmatrix} \quad \text{and} \quad R = \begin{pmatrix} 1 & 2 & 3 \\ 0 & 5 & 1 \\ 0 & 0 & 2 \end{pmatrix}.$$

$g_{11} = e^{r_{11}} = 2.7183,$ $\qquad g_{22} = e^{r_{22}} = 148.4132,$ $\qquad g_{33} = e^{r_{33}} = 7.3891,$

$k = 1, i = 1, j = 2:$ $\qquad g_{12} = \dfrac{1}{(r_{11} - r_{22})}[r_{12}(g_{11} - g_{22})] = 72.8474,$

$k = 1, i = 2, j = 3:$ $\qquad g_{23} = \dfrac{1}{(r_{22} - r_{33})}[r_{23}(g_{22} - g_{33})] = 47.0090,$

$k = 2, i = 1, j = 3:$ $\qquad g_{13} = \dfrac{1}{(r_{11} - r_{33})}[r_{13}(g_{11} - g_{33}) + (g_{12}r_{23} - r_{12}g_{23})]$

$$= 35.1810.$$

So,

$$e^R = \begin{pmatrix} 2.7183 & 72.8474 & 35.1810 \\ 0 & 148.4132 & 47.0080 \\ 0 & 0 & 7.3891 \end{pmatrix}.$$

Thus,

$$e^A = Pe^R P^T = \begin{pmatrix} 148.4132 & 47.0080 & 0 \\ 0 & 7.3891 & 0 \\ 72.8474 & 35.1810 & 2.7183 \end{pmatrix}.$$

Note that e^A obtained here is exactly the same (in four-digit arithmetic) as given by MATLAB function **expm** (A), which is based on Algorithm 5.3.1.

Numerical stability of the schur algorithm: Numerical difficulties clearly arise when A has equal or nearly equal confluent eigenvalues, even though the transformation matrix P is orthogonal.

5.3.4 Comparison of Different Methods for Computing the Exponential Matrix

From our discussions in previous sections, it is clear that the **Padé approximation method** (with scaling and squaring) and the **Schur method** should, in general, be attractive from computational viewpoints.

The Taylor series method and the methods based on reduction of A to a companion matrix or to the JCF should be avoided for the reason of numerical instability. The ODE techniques should be preferred when the matrix A is large and sparse.

5.3.5 Evaluating an Integral with the Matrix Exponential

We have discussed methods for computing the matrix exponential. We now present a method due to Van Loan (1978) to compute integrals involving exponential matrices.

The method can be used, in particular, to compute the state-space solution (5.3.3) of the Eq. (5.3.1), and the **controllability** and **observability Grammians**, which will be discussed in the next chapter.

The method uses diagonal Padé approximation discussed in the last section.

Let

$$H(\Delta) = \int_0^\Delta e^{As} B \, ds, \qquad M(\Delta) = \int_0^\Delta e^{A^T s} Q H(s) \, ds,$$

$$N(\Delta) = \int_0^\Delta e^{A^T s} Q e^{As} ds, \qquad W(\Delta) = \int_0^\Delta H(s)^T Q H(s) \, ds$$

where A and B are matrices of order n and $n \times m$, respectively, and Q is a symmetric positive semidefinite matrix.

Algorithm 5.3.3. *An Algorithm for Computing Integrals involving Matrix Exponential.*

Inputs.

> A—The $n \times n$ state matrix
>
> B—An $n \times m$ matrix
>
> Q—A symmetric positive semidefinite matrix.

Outputs. F, H, Q, M, and W which are, respectively, the approximations to $e^{A\Delta}$, $H(\Delta)$, $N(\Delta)$, $M(\Delta)$, and $W(\Delta)$.

Step 1. *Form the* $(3n + m) \times (3n + m)$ *matrix*

$$\hat{C} = \begin{pmatrix} -A^\mathrm{T} & I & 0 & 0 \\ 0 & -A^\mathrm{T} & Q & 0 \\ 0 & 0 & A & B \\ 0 & 0 & 0 & 0 \end{pmatrix}.$$

Find the smallest positive integer j *such that* $(\|\hat{C}\Delta\|_\mathrm{F}/2^j) \leq \frac{1}{2}$. *Set* $t_0 = (\Delta/2^j)$.

Step 2. *For some* $q \geq 1$, *compute*

$$Y_0 = R_{qq}\left(\frac{\hat{C}\Delta}{2^j}\right),$$

where R_{qq} *is the* (q, q) *Padé approximant to* e^z:

$$R_{qq}(z) = \frac{\sum_{k=0}^z c_k z^k}{\sum_{k=0}^q c_k (-z)^k}, \quad \text{where } c_k = \frac{(2q - k)!q!}{(2q)!k!(q - k)!}.$$

Write

$$Y_0 = \begin{pmatrix} F_1(t_0) & G_1(t_0) & H_1(t_0) & K_1(t_o) \\ 0 & F_2(t_0) & G_2(t_0) & H_2(t_0) \\ 0 & 0 & F_3(t_0) & G_3(t_0) \\ 0 & 0 & 0 & F_4(t_0) \end{pmatrix}$$

and set

$$F_0 = F_3(t_0) \qquad M_0 = F_3(t_0)^\mathrm{T} H_2(t_0)$$
$$H_0 = G_3(t_0) \qquad W_0 = [B^\mathrm{T} F_3(t_0)^\mathrm{T} K_1(t_0)] + (B^\mathrm{T} F_3(t_0)^\mathrm{T} K_1(t_0)]^\mathrm{T}.$$
$$Q_0 = F_3(t_0)^\mathrm{T} G_2(t_0).$$

Step 3. *For* $k = 0, 1, \ldots, j - 1$ *do*

$$W_{k+1} = 2W_k + H_k^\mathrm{T} M_k + M_k^\mathrm{T} H_k + H_k^\mathrm{T} Q_k H_k$$
$$M_{k+1} = M_k + F_k^\mathrm{T}[Q_k H_k + M_k]$$
$$Q_{k+1} = Q_k + F_k^\mathrm{T} Q_k F_k$$
$$H_{k+1} = H_k + F_k H_k$$
$$F_{k+1} = F_k^2$$

End

Step 4. *Set* $F \equiv F_j$, $H \equiv H_j$, $Q \equiv Q_j$, $M \equiv M_j$, *and* $W \equiv W_j$.

Remark

- It has been proved by Van Loan (1978) that the accuracy of the algorithm can be controlled by selecting q properly. For **"how to choose q properly"** and other computational details, see the paper of Van Loan (1978).

MATCONTROL note: Algorithm 5.3.3 has been implemented in MATCON-TROL function **intmexp**.

5.4 STATE-SPACE SOLUTION OF THE DISCRETE-TIME SYSTEM

In this section, we state a discrete-analog of Theorem 5.3.1, and then discuss how to compute the discrete-time solution.

Theorem 5.4.1. *Discrete-Time State-Space Solution. The solutions to the linear time-invariant discrete-time system of equations*

$$x(k + 1) = Ax(k) + Bu(k), \quad x(0) = x_0 \tag{5.4.1}$$

and

$$y(k) = Cx(k) + Du(k) \tag{5.4.2}$$

are

$$x(k) = A^k x_0 + \sum_{i=0}^{k-1} A^{k-1-i} Bu(i) \tag{5.4.3}$$

and

$$y(k) = CA^k x_0 + \left\{ \sum_{i=0}^{k-1} CA^{k-i-1} Bu(i) \right\} + Du(k). \tag{5.4.4}$$

Proof. From

$$x(k + 1) = Ax(k) + Bu(k), \tag{5.4.5}$$

we have

$$
\begin{aligned}
x(k) &= A[Ax(k - 2) + Bu(k - 2)] + Bu(k - 1), \\
&= A^2 x(k - 2) + ABu(k - 2) + Bu(k - 1), \\
&= A^2[Ax(k - 3) + Bu(k - 3)] + ABu(k - 2) + Bu(k - 1), \\
&\vdots \\
&= A^k x_0 + \sum_{i=0}^{k-1} A^{k-1-i} Bu(i).
\end{aligned}
$$

This proves (5.4.3).

Equation (5.4.4) is now obtained by substituting the expression of $x(t)$ from (5.4.3) into (5.4.2). ■

Computing the Powers of a Matrix

Theorem 5.4.1 shows that, to find the state-space solution of a discrete-time system, one needs to compute the various powers of A. The powers of a matrix A are more easily computed if A is first decomposed into a condensed form by similarity. Thus, if $P^T A P = R$, where P is orthogonal, then $A^s = P R^s P^T$. **For the sake of numerical stability, only those condensed forms such as the Hessenberg form or the RSF of a matrix should be considered.** The matrix R^s can be easily calculated by exploiting the Hessenberg or Schur structure of R. Furthermore, the reduction to R can be achieved using a numerically stable procedure such as Householder's or Givens' method (**Chapter 4**).

5.5   TRANSFER FUNCTION AND FREQUENCY RESPONSE

In this section, we introduce the important concepts of **transfer function** and **frequency response** matrices and describe a numerical algorithm for computing the frequency response matrix.

5.5.1   Transfer Function

Consider

$$\dot{x}(t) = Ax(t) + Bu(t), \quad x(0) = x_0, \tag{5.5.1}$$

$$y(t) = Cx(t) + Du(t). \tag{5.5.2}$$

Let $\hat{x}(s)$, $\hat{y}(s)$, and $\hat{u}(s)$, respectively, denote the Laplace transforms of $x(t)$, $y(t)$, and $u(t)$. Then taking the Laplace transform of (5.5.1) and (5.5.2), we obtain

$$s\hat{x}(s) - x_0 = A\hat{x}(s) + B\hat{u}(s), \tag{5.5.3}$$

$$\hat{y}(s) = C\hat{x}(s) + D\hat{u}(s). \tag{5.5.4}$$

From (5.5.3) and (5.5.4), we have

$$\hat{x}(s) = R(s)x(0) + R(s)B\hat{u}(s) \tag{5.5.5}$$

$$\hat{y}(s) = CR(s)x(0) + G(s)\hat{u}(s), \tag{5.5.6}$$

where

$$R(s) = (sI - A)^{-1} \tag{5.5.7}$$

and

$$G(s) = C(sI - A)^{-1}B + D. \tag{5.5.8}$$

If $x(0) = 0$, then (5.5.6) gives

$$\hat{y}(s) = G(s)\hat{u}(s).$$

Definition 5.5.1. *The matrix $R(s)$ is called the* **resolvent** *and $G(s)$ is called the* **transfer function**.

The transfer function $G(s)$ is a matrix transfer function of dimension $r \times m$. Its (i, j)th entry denotes the transfer function from the jth input to the ith output. That is why, it is also referred to as the **transfer function matrix** or simply the **transfer matrix**.

Definition 5.5.2. *The points p at which $G(p) = \infty$ are called the* **poles** *of the system.*

If $G(\infty) = 0$, then the transfer function is called **strictly proper** *and is* **proper** *if $G(\infty)$ is a constant matrix.*

State-Space Realization

For computational convenience, the transfer function matrix $G(s)$ will be written sometimes as

$$G(s) = \left[\begin{array}{c|c} A & B \\ \hline C & D \end{array} \right].$$

The state-space model (A, B, C, D) having $G(s)$ as its transfer function matrix is called a realization of $G(s)$. For more on this topic, see **Chapter 9**. **In general,** it will be assumed that $G(s)$ **is a real-rational transfer matrix that is proper.**

Operations with Transfer Function Matrices
Let $G_1(s)$ and $G_2(s)$ be the transfer functions of the two systems S_1 and S_2. Then the transfer function matrix of the **parallel connection** of S_1 and S_2 is $G_1(s) + G_2(s)$. Using our notation above, we obtain:

$$G_1(s)+G_2(s) = \left[\begin{array}{c|c} A_1 & B_1 \\ \hline C_1 & D_1 \end{array} \right] + \left[\begin{array}{c|c} A_2 & B_2 \\ \hline C_2 & D_2 \end{array} \right] = \left[\begin{array}{cc|c} A_1 & 0 & B_1 \\ 0 & A_2 & B_2 \\ \hline C_1 & C_2 & D_1 + D_2 \end{array} \right].$$

Similarly, the transfer function matrix of the **series** or **cascade connection** of S_1 and S_2 (i.e., a system with the output of the second system as the input of the first system) is $G_1(s)G_2(s)$, given by

$$G_1(s)G_2(s) = \left[\begin{array}{c|c} A_1 & B_1 \\ \hline C_1 & D_1 \end{array}\right]\left[\begin{array}{c|c} A_2 & B_2 \\ \hline C_2 & D_2 \end{array}\right] = \left[\begin{array}{cc|c} A_2 & 0 & B_1 \\ B_1 C_2 & A_1 & B_1 D_2 \\ \hline D_1 C_2 & C_1 & D_1 D_2 \end{array}\right].$$

The **transpose** of a transfer function matrix $G(s)$ is defined as:

$$G^T(s) = B^T(sI - A^T)^{-1}C^T + D^T,$$

or equivalently,

$$G^T(s) = \left[\begin{array}{c|c} A^T & B^T \\ \hline C^T & D^T \end{array}\right].$$

The **conjugate** of $G(s)$ is defined as:

$$G^\sim(s) \equiv G^T(-s) = B^T(-sI - A^T)^{-1}C^T + D^T,$$

or equivalently,

$$G^\sim(s) = \left[\begin{array}{c|c} -A^T & -C^T \\ \hline -B^T & D^T \end{array}\right].$$

The **inverse** of a transfer function matrix $G(s)$, denoted by $\hat{G}(s)$ is such that $G(s)\hat{G}(s) = \hat{G}(s)G(s) = I$. If $G(s)$ is square and D is invertible, then,

$$\hat{G}(s) \equiv G^{-1}(s) = \left[\begin{array}{c|c} A - BD^{-1}C & -BD^{-1} \\ \hline D^{-1}C & D^{-1} \end{array}\right].$$

MATLAB notes: MATLAB functions **parallel, series, transpose, inv (ss/inv.m)** can be used to compute parallel, series, transpose, and inverse, respectively.

Transfer Function of Discrete-Time System

The transfer function matrix of the discrete-time system (5.4.1)–(5.4.2) is

$$G(z) = C(zI - A)^{-1}B + D.$$

It is obtained by taking the z-transform of the system.

5.5.2   The Frequency Response Matrix and its Computation

In this section, we describe a computationally viable approach for computing the **frequency response matrix**.

Definition 5.5.3.   *For the continuous-time state-space model (5.5.1–5.5.2), the matrix*

$$G(j\omega) = C(j\omega I - A)^{-1}B + D \qquad (5.5.9)$$

is called the **frequency response matrix***;* $\omega \in \mathbb{R}$ *is called frequency.*

The frequency response matrix of a discrete-time system is similarly defined by using the transformation:

$$z = e^{j\omega T},$$

where T is the sample time. This transformation maps the frequencies ω to points on the unit circle. The frequency response matrix is then evaluated at the resulting z values.

Computing the Frequency Response Matrix

In order to study the different responses of the system, it is necessary to compute $G(j\omega)$ for many different values of ω. Furthermore, the singular values of the **return difference matrix** $I + L(j\omega)$ and of the inverse return difference matrix $I + L^{-1}(j\omega)$, where $L(j\omega)$ is square and of the form $L = KGM$ for appropriate K and M, **provide robustness measure of a linear system with respect to stability, noise, disturbance attenuation, sensitivity, etc.** (Safonov *et al.* 1981).

We therefore describe a numerical approach to compute $G(j\omega)$. **For simplicity, since D does not have any computational role in computing $G(j\omega)$, we will assume that $D = 0$.**

The computation of $(j\omega I - A)^{-1}B$ is equivalent to solving m systems:

$$(j\omega I - A)X = B,$$

A usual scheme for computing the frequency response matrix is:

Step 1. Solve the m systems for m columns $x_1, x_2, \ldots, x_m$ of X:

$$(j\omega I - A)x_i = b_i, \qquad i = 1, 2, \ldots, m,$$

where b_i is the ith column of B.

Step 2. Compute CX.

If A is $n \times n$, B is $n \times m$ ($m \le n$), and C is $r \times n$ ($r \le n$), and if LU factorization **(Chapter 3)** is used to solve the systems $(j\omega I - A)x_i = b_i, i = 1, 2, \ldots, m$, then the total flop-count **(complex) for each** ω is about $\frac{2}{3}n^3 + 2mn^2 + 2mnr$. Note that, since the matrix $j\omega I - A$ is the same for each linear system for a given ω, the matrix $j\omega I - A$ has to be factored into LU only once, and the same factorization can be used to solve all the m systems. Since the matrix $G(j\omega)$ needs to be computed for many different values of ω, the computation in this way will be fairly expensive.

Laub (1981) noted that the computational cost can be reduced significantly when $n > m$ (which is normally the case in practice), if the matrix A is initially transformed to an upper Hessenberg matrix H by orthogonal similarity, before the frequency response computation begins. The basic observation here is that if $P^T A P = H$, an upper Hessenberg matrix, then

$$\begin{aligned}
G(j\omega) &= C(j\omega I - A)^{-1} B, \\
&= C(j\omega I - PHP^T)^{-1} B, \\
&= C(P(j\omega I - H)P^T)^{-1} B, \\
&= CP(j\omega I - H)^{-1} P^T B.
\end{aligned}$$

Thus, if A is transformed initially to an upper Hessenberg matrix H, and $CP = C'$ and $P^T B = B'$ are computed once for all, then for each ω, we have the following algorithm.

Algorithm 5.5.1. *A Hessenberg Algorithm for the Frequency Response Matrix.*

Input. *A—The $n \times n$ state matrix*
ω—Frequency, a real number
B—The $n \times m$ input matrix
C—The $r \times n$ output matrix.
Output. *The Frequency Response Matrix $G(j\omega) = C(j\omega I - A)^{-1} B$.*
Step 1. *Transform A to an upper Hessenberg matrix H (Section 3.5.3):*
$P^T A P = H$.
Step 2. *Compute $B' = P^T B$ and $C' = CP$*
Step 3. *Solve the m Hessenberg systems:*

$$(j\omega I - H) x_i = b'_i, \quad i = 1, \ldots, m,$$

where b'_i is the ith column of B'.

Step 4. *Compute $C'X$.*

Flop-Count: Since the system matrices A, B, and C are real, Steps 1 and 2 can be done using real arithmetic and require approximately $\frac{10}{3}n^3 + 2mn^2 + 2rn^2$ (real) flops. Steps 3 and 4 require complex arithmetic and require approximately $2mn^2 + 2rnm$ **complex** flops.

Comparison: For N values of ω, the Hessenberg method require $\frac{10}{3}n^3 + 2(m + r)n^2$ real $+[2mn^2 + 2rnm]N$ complex flops compared to $[\frac{2}{3}n^3 + 2mn^2 + +2mnr]N$ complex flops, required by the non-Hessenberg scheme.

Numerical stability: It is possible to show (Laub and Linnemann 1986) that if the data is well-conditioned, then the frequency response of the computed Hessenberg form is $(C + \triangle C)(j\omega I - A - \triangle A)^{-1}(B + \triangle B)$, where $\triangle A$, $\triangle B$, and $\triangle C$ are small. Thus, the **Hessenberg method is stable.**

MATCONTROL note: Algorithm 5.5.1 has been implemented in MATCONTROL function **freqresh**.

Example 5.5.1. Compute the frequency response matrix with

$$A = \begin{pmatrix} 1 & 2 & 3 \\ 2 & 3 & 4 \\ 0 & 1 & 1 \end{pmatrix}, \qquad B = \begin{pmatrix} 1 & 1 \\ 1 & 1 \\ 1 & 1 \end{pmatrix}, \qquad C = \begin{pmatrix} 1 & 1 & 1 \\ 1 & 1 & 1 \end{pmatrix}, \qquad \omega = 1.$$

Since A is already in upper Hessenberg form, Steps 1 and 2 are skipped.
   Step 3. Solve for x_1: $(j\omega I - H)x_1 = b'_1 = b_1$,

$$x_1 = \begin{pmatrix} -5.000 - 0.0000i \\ 0.4000 - 0.8000i \\ +0.1000 - 0.7000i \end{pmatrix}.$$

Solve for x_2: $(j\omega I - H)x_2 = b'_2 = b_2$,

$$x_2 = \begin{pmatrix} -0.5000 - 0.0000i \\ 0.4000 + 0.8000i \\ 0.1000 - 0.7000i \end{pmatrix}.$$

Step 4. Compute the frequency response matrix:

$$G(j\omega) = C'X = CX,$$

$$= \begin{pmatrix} -0.8000 + 0.1000i & -0.8000 + 0.1000i \\ -0.8000 + 0.1000i & -0.8000 + 0.1000i \end{pmatrix}.$$

Other Methods for Frequency Response Computation

1. A method based on a determinant identity for the computation of the frequency response matrix has been proposed by Misra and Patel (1988). The method seems to be considerably faster and at least as accurate as the Hessenberg method just described. The method uses the controller-Hessenberg and observer-Hessenberg forms which will be described in the next chapter. The Misra–Patel method for computing the frequency response matrix is based on the following interesting observation:

$$g_{lk}(j\omega) = \frac{\det(j\omega I - A + b_k c_l^T)}{\det(j\omega I - A)} - 1, \qquad (5.5.10)$$

   where b_k and c_l denote, respectively, the kth and lth columns of the matrices B and C.

2. An alternative method also based on the reduction to controller-Hessenberg form, has been given by Laub and Linnemann (1986).

3. Another method for the problem has been proposed by Kenney *et al.* (1993). The method uses matrix interpolation techniques and seems to have better efficiency than the Hessenberg method.

4. The frequency response of a discrete time system is obtained by evaluating the transfer function $H(z)$ in (5.5.8) on the unit circle.

Bode Diagram and the Singular Value Plot

Traditionally, Bode diagram is used to measure the magnitude and angle of frequency response of an SISO system. Thus expressing the frequency response function in polar coordinates, we have

$$G(j\omega) = M(\omega)e^{j\sigma(\omega)}.$$

$M(\omega)$ is the **magnitude** and $\sigma(\omega)$ is the **angle**. It is customary to express $M(\omega)$ in decibels (abbreviated by dB). Thus,

$$M(\omega)|_{dB} = 20\log_{10} M(\omega).$$

The angle is measured in degrees.

The **Singular Value Plot** is the plot of singular values of $H(j\omega)$ as a function of the frequency ω. If the system is an SISO system, the singular value plot is the same as the Bode diagram. The singular value plot is a useful tool in robustness analysis.

MATLAB note: MATLAB functions **bode** and **sigma** can be used to draw the Bode plot and the singular value plot, respectively. For the Bode plot of multi-input, multi-output (MIMO) system, the system is treated as arrays of SISO systems and the magnitudes and phases are computed for each SISO entry h_{ij} independently.

MATLAB function **freqresp** can be used to compute frequency response at some specified individual frequencies or over a grid of frequencies. When numerically safe, the frequency response is computed by diagonalizing the matrix A; otherwise, Algorithm 5.5.1 is used.

5.6   SOME SELECTED SOFTWARE

5.6.1   Matlab Control System Toolbox

Time response

step	Step response
impulse	Impulse response
initial	Response of state-space system with given initial state

lsim	Response to arbitrary inputs
ltiview	Response analysis GUI
gensig	Generate input signal for LSIM
stepfun	Generate unit-step input.

Frequency response

bode	Bode plot for the frequency response
sigma	Singular value frequency plot
nyquist	Nyquist plot
nichols	Nichols chart
ltiview	Response analysis GUI
evalfr	Evaluate frequency response at given frequency
freqresp	Frequency response over a frequency grid
margin	Gain and phase margins

5.6.2 MATCONTROL

EXPMPADE	The Padé approximation to the exponential of a matrix
EXPMSCHR	Computing the exponential of a matrix using Schur decomposition
FREQRESH	Computing the frequency response matrix using Hessenberg decomposition
INTMEXP	Computing an integral involving a matrix exponentials.

5.6.3 SLICOT

MB05MD	Matrix exponential for a real non-defective matrix
MB05ND	Matrix exponential and integral for a real matrix
MB05OD	Matrix exponential for a real matrix with accuracy estimate.

State-space to rational matrix conversion
| TB04AD | Transfer matrix of a state-space representation. |

State-space to frequency response
TB05AD	Frequency response matrix of a state-space representation
TF	Time response
TF01MD	Output response of a linear discrete-time system
TF01ND	Output response of a linear discrete-time system (Hessenberg matrix).

5.6.4   MATRIX$_X$

Purpose: Gain and phase plots of discrete-time systems.

Syntax:
[GAIN, DB, PHASE]=DBODE (SD, NS, OMEGANMIN, OMEGANMAX, NPTS, 'OPT') OR
[GAIN, DB, PHASE]=DBODE (DNUM, DDEN, OMEGANMIN, OMEGAN-MAX, NPTS) OR
[GIAN, DB, PHASE]=DBODE (SD, NS, OMEGAN)

Purpose: Initial value response of discrete-time dynamic system.

Syntax: [N, Y]=DINITIAL (SD, NS, XO, NPTS).

Purpose: Step-response of discrete-time system.

Syntax: [N, Y]=DSTEP (SD, NS, NPTS) OR
[N, Y]=DSTEP (DNUM, DDEN, NPTS)

Purpose: Frequency response of dynamic system. FREQ transforms the *A* matrix to Hessenberg form prior to finding the frequency response.

Syntax: [OMEGA, H]=FREQ (S, NS, RANGE, option) OR
H=FREQ (S, NS, OMEGA, 'options')

Purpose: Compute the impulse response of a linear continuous-time system.
Syntax: [T, Y]=IMPULSE (S, NS, TMAX, NPTS) OR
[T, Y]=IMPULSE (NUM, DEN, TMAX, NPTS)

Purpose: Initial value response of continuous-time dynamic system.

Syntax: [T, Y]=INITIAL (S, NS, XO, TMAX, NPTS)

Purpose: Response of continuous-time system to general inputs.

Syntax: [T, Y]=LSIM (S, NS, U, DELTAT, X0)

Purpose: Step response of continuous-time system.

Syntax: [T, Y]=STEP (S, NS, TMAX, NPTS) OR
[T, Y]=STEP (NUM, DEN, TMAX, NPTS)

Purpose: Gives transfer function form of a state-space system.

Syntax: [NUM, DEN]=TFORM (S, NS)

Purpose: Impulse response of continuous-time system.

Syntax: [T, Y]=TIMR (S, NS, RANGE, 'MODE')

5.7 SUMMARY AND REVIEW

State-Space Representations

A **continuous-time** linear time-invariant control system may be represented by systems of differential equations of the form (5.2.1)–(5.2.2)

$$\dot{x}(t) = Ax(t) + Bu(t),$$
$$y(t) = Cx(t) + Du(t),$$

where $x(t)$ is the **state vector,** $u(t)$ is the **input vector,** $y(t)$ is the **output vector.**
  The matrices A, B, C, and D are **time-invariant matrices** known, respectively, as the **state matrix**, the **input matrix**, the **output matrix**, and the **direct transmission matrix**.
  A **discrete-time** linear time-invariant control system may analogously be represented by systems of difference equations (5.4.1)–(5.4.2).

$$x(t+1) = Ax(t) + Bu(t),$$
$$y(t) = Cx(t) + Du(t).$$

where $x(t), u(t), y(t)$, and A, B, C, and D have the same meaning as above.

Solutions of the Dynamical Equations

The solutions of the equations representing the continuous-time system in state-space form are given by (assuming $t_0 = 0$):

$$x(t) = e^{At}x_0 + \int_0^t e^{A(t-s)} Bu(s)\, ds$$

and

$$y(t) = Ce^{At}x_0 + \int_0^t Ce^{A(t-s)} Bu(s)\, ds + Du(t).$$

where x_0 is the value of $x(t)$ at $t = 0$. The matrix e^{At} is the **state-transition matrix in this case.**

The solutions of the equations representing the discrete-time system are given by:

$$x(k) = A^k x(0) + \sum_{i=0}^{k-1} A^{k-1-i} Bu(i)$$

and

$$y(k) = C A^k x(0) + \left\{ \sum_{i=0}^{k-1} C A^{k-i-1} Bu(i) \right\} + Du(k).$$

Computing e^{At}

There exist several methods for computing the state-transition matrix e^{At}. These include: The Taylor series method, the Padé approximation method, ODE methods, the eigenvalue–eigenvector method, and the matrix decomposition methods.

Among these, **the Padé approximation method with scaling and squaring (Algorithm 5.3.1)** and the **method based on the Schur decomposition of A (Algorithm 5.3.2) are the ones that are recommended for use in practice.** If the problem is ill-conditioned, these methods, however, might not give accurate results. The **ODE methods (Section 5.3.3) should be preferred if A is large and sparse.**

Computing Integrals Involving Matrix Exponentials

An algorithm **(Algorithm 5.3.3)** is presented for computing integrals involving matrix exponentials.

Transfer Function Matrix

If $\hat{u}(s)$ and $\hat{y}(s)$ are the Laplace transforms of $u(t)$ and $y(t)$, then assuming zero initial condition, we obtain:

$$\hat{y}(s) = G(s)\hat{u}(s),$$

where

$$G(s) = C(sI - A)^{-1} B + D.$$

The matrix $G(s)$ is called the transfer function matrix and is conveniently written as:

$$G(s) \equiv \left[\begin{array}{c|c} A & B \\ \hline C & D \end{array} \right].$$

The Frequency Response Matrix

The matrix $G(j\omega) = C(j\omega I - A)^{-1}B + D$ is called the **frequency response matrix.**

The frequency response plot for different values of ω is important in the study of different responses of a control system. For this the frequency response matrix needs to be computed.

An efficient method (**Algorithm 5.5.1**), based on transformation of A to Hessenberg form, is described. **The Hessenberg method is nowadays widely used in practice.**

5.8 CHAPTER NOTES AND FURTHER READING

The examples on state-space model in Section 5.2.1 have been taken from various sources. These include the books by Brogan (1991), Chen (1984), Kailath (1980), Luenberger (1979), Szidarovszky and Bahill (1991), Zhou with Doyle (1998). Discussions on system responses can be found in any standard text books. The books mentioned above and also the books by DeCarlo (1989), Dorf and Bishop (2001), Friedland (1986), Patel and Munro (1982), etc., can be consulted. For various ways of computing the matrix exponential e^{At}, the paper by Moler and Van Loan (1978) is an excellent source. Some computational aspects of the matrix exponential have also been discussed in DeCarlo (1989).

The frequency response algorithm is due to Laub (1981). For discussions on applications of frequency response matrix, see Safonov *et al.* (1981). For alternative algorithms for frequency response computation, see Misra and Patel (1988), Kenney *et al.* (1993). The algorithm for computing integrals (**Algorithm 5.3.3**) involving matrix exponential has been taken from the paper of Van Loan (1978). The sensitivity analysis of the matrix e^{At} is due to Van Loan (1977). See also Golub and Van Loan (1996).

Exercises

5.1   (a)   Consider the electric circuit

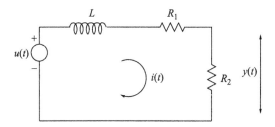

(i) Show that the state-space representation of this circuit is given by

$$L\frac{di(t)}{dt} + (R_1 + R_2)i(t) = u(t), \quad y(t) = R_2 i(t).$$

(ii) Give an explicit expression for the solution of the state equation.

(b) Write the state equations for the following electric network:

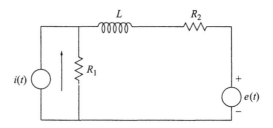

5.2 (a) Write down the equations of motion of the following spring-mass system in state-space form:

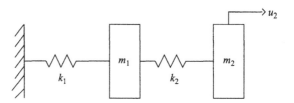

(b) Write down the state-space representation of the following spring-mass system:

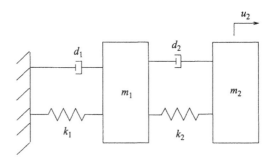

5.3   Consider the following diagram of a cart of mass M carrying two sticks of masses M_1 and M_2, and lengths l_1 and l_2.

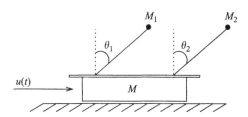

(a)   Write down the equations of motion, in terms of the velocity v of the cart, u and $\theta_1, \theta_2, \dot{\theta}_1, \ddot{\theta}_2$.

(b)   Assuming θ_1 and θ_2 are small, linearize the equations of motion and then write down a first-order state-space representation.

5.4   (Saad 1988) Consider the differential equation

$$\frac{\partial u}{\partial t} = \frac{\partial^2 u}{\partial x^2} + \frac{\partial^2 u}{\partial y^2} + \beta\frac{\partial u}{\partial x} + vu + F(x, y, t)$$

on the unit square $\Omega = (0, 1) \times (0, 1)$, that models a chemical reaction process, where u represents the concentration of a chemical component that diffuses and convects. Let the boundary conditions $u(x, y, t) = 0$ for every t. Assume that $F(x, y, t)$ has the form:

$$F(x, y, t) = F(x, y)g(t).$$

The term vu simulates a chemical reaction that results in an increase of the concentration that is proportional to u.

(a)   Discretize the region with n interior points in the x-direction and m interior points in the y-direction and show that this leads to the state-space representation of the form:

$$\dot{u} = Au + bg,$$

where the dimension of A is nm.

(b)   Solve the above problem on a computer with $\beta = 20, v = 180, n = 20, m = 10$.

5.5   (**Lanchester War Model**) The following simple model of Warfare was developed by F. Lanchester in 1916.

Let x_1 and x_2 be the number of units in the two forces which are engaged in a war. Suppose that each unit of the first force has the "hitting" power h_1 and that of the second force has the "hitting" power h_2.

The "hitting" power is defined to be the number of casualties per unit time that one member can inflict on the other.

Suppose further that the hitting power of each force is directed uniformly against all units of the other force.
(a) Develop a state-space representation of the model.
(b) Show that

$$x_1(t) = c_1 e^{t\sqrt{h_1 h_2}} + c_2 e^{-t\sqrt{h_1 h_2}},$$

$$x_2(t) = -c_1 \sqrt{\frac{h_1}{h_2}} e^{t\sqrt{h_1 h_2}} + c_2 \sqrt{\frac{h_1}{h_2}} e^{-t\sqrt{h_1 h_2}},$$

where c_1 and c_2 are constants to be determined from initial conditions.

5.6 (a) Find an explicit expression for the solution of the initial value problem

$$\dot{x}(t) = \begin{pmatrix} 0 & \lambda \\ -\lambda & 0 \end{pmatrix} x(t) + \begin{pmatrix} 0 \\ 1 \end{pmatrix}, \qquad x(0) = \begin{pmatrix} 1 \\ 0 \end{pmatrix}.$$

(b) Find the free response of the system.

5.7 Find an explicit expression for the solution of the homogeneous discrete-time system

$$x(t+1) = \begin{pmatrix} 0 & 1 \\ 1 & 1 \end{pmatrix} x(t), \qquad x(0) = \begin{pmatrix} 0 \\ 1 \end{pmatrix}.$$

5.8 Prove the following properties of e^{At}:
(a) $e^{A(t+s)} = e^{At} \cdot e^{As}$,
(b) $e^{(A+B)t} = e^{At} \cdot e^{Bt}$ if and only if A and B commute,
(c) e^{At} is nonsingular and $(e^{At})^{-1} = e^{-At}$,
(d) $\left(e^{A/m}\right)^m = e^A$, where m is an integer,
(e) $e^{P^{-1}APt} = P^{-1}e^{At}P$.

5.9 Prove that the infinite series

$$e^{At} = \sum_{k=0}^{\infty} \frac{1}{k!} A^k t^k$$

converges uniformly and absolutely for t in any bounded interval.

5.10 Prove that $(d/dt)(e^{At}) = Ae^{At}$
$\left(Hint: \text{ Differentiate } e^{At} = \sum_{k=0}^{\infty} 1/k! A^k t^k \text{ term by term}\right).$

5.11 Illustrate the difficulty with the eigenvalue–eigenvector method for computing e^{At} with the matrix:

$$A = \begin{pmatrix} \lambda & \alpha \\ 0 & \mu \end{pmatrix}$$

by choosing λ, μ, and α appropriately.

5.12 Let $R = (r_{ij})$ be an unreduced lower Hessenberg matrix and let

$$e^R = \begin{pmatrix} f_1 \\ f_2 \\ \vdots \\ f_r \end{pmatrix} \quad \text{and} \quad B_i = R - r_{ii}I.$$

Then prove that

$$f_2 = \frac{1}{r_{12}} f_1 B_1$$

and

$$f_{i+1} = \frac{1}{r_{i-1,i}} \left(f_i B_i - \sum_{j=1}^{i-1} r_{ij} f_j \right), \quad i = 2, 3, \ldots, n-1.$$

(Consult Datta and Datta (1976), if necessary.) What difficulties do you foresee in computing e^R using these formulas? Give an illustrative example.

5.13 Compute e^A for

$$A = \begin{pmatrix} 1 & 1 & 0 \\ 1 & 1 & 1 \\ 1 & 1 & 2 \end{pmatrix}$$

using
  (a) a fifth-order variable-step Runge–Kutta method;
  (b) the Adams–Moulton predictor correct formulas of variable order and variable step;
  (c) a general purpose ODE solver. (**Use error tolerance** 10^{-6}.)
  Compare the result in each case with that obtained by MATLAB function **expm**.

5.14 (a) Write an algorithm based on the block diagonal decomposition of A to compute e^{At}:

$$A = X \operatorname{diag}(T_1, T_2, \ldots, T_p)X^{-1}.$$

  (b) Determine the flop-count of this algorithm.
  (c) What numerical difficulty do you expect out of this algorithm?

5.15 Prove that the matrix $D_{pq}(A)$ in the Padé approximation method is nonsingular if all the eigenvalues of A have negative real parts or if p and q are sufficiently large.

5.16 Develop an algorithm to compute A^s, where s is a positive integer, and A is an unreduced lower Hessenberg matrix. Apply your algorithm to compute A^{10}, where $A =$

$$(i) \begin{pmatrix} 1 & 2 & 0 \\ 1 & 1 & 0.0001 \\ 1 & 1 & 1 \end{pmatrix}, \quad (ii) \begin{pmatrix} 0 & 1 & 0 & 0 \\ 0 & 0 & 1 & 0 \\ 0 & 0 & 0 & 1 \\ 1 & 2 & 3 & 4 \end{pmatrix}, \quad (iii) \begin{pmatrix} 1 & 10^{-3} & 0 \\ 1 & 1 & 10^{-4} \\ 1 & 1 & 1 \end{pmatrix}.$$

(Consult Datta and Datta 1976.)

5.17 Let $A = X \Lambda X^{-1}$, where $\Lambda = \text{diag}\,(\lambda_1, \ldots, \lambda_n)$. Prove that $A^k = X \Lambda^k X^{-1}$, for each k. Under what conditions on $\lambda_1, \ldots, \lambda_n$ does the infinite series of matrices $\sum c_k A^k$ converge?

5.18 Develop an algorithm to compute A^s, where A is an upper real Schur matrix, and s is a positive integer. Apply your algorithm to compute A^5, where

$$A = \begin{pmatrix} 1 & 2 & 3 & 4 \\ 0 & 0.99 & 1 & 1 \\ 0 & 0 & 2 & 1 \\ 0 & 0 & 0 & 1.99 \end{pmatrix}.$$

5.19 Prove that the Laplace transform of $y(t) = e^{At}$ is $y(s) = (sI - A)^{-1}$.

5.20 Modify Algorithm 5.3.2 to compute e^A, where A is in RSF.

5.21 Show that the transformation $\tilde{x} = Tx$, where T is nonsingular, preserves the transfer function.

5.22 Show that the transfer function of the system:

$$\dot{x}(t) = \begin{pmatrix} 0 & \omega \\ -\omega & 0 \end{pmatrix} x + \begin{pmatrix} 0 \\ 1 \end{pmatrix} u, \qquad x(0) = \begin{pmatrix} 1 \\ 0 \end{pmatrix},$$

$$y = (1, 1)x.$$

is $H(s) = \dfrac{s + \omega}{s^2 + \omega^2}$.

5.23 Modify the Hessenberg algorithm (Algorithm 5.5.1) for computation of the frequency response matrix that uses only real arithmetic. Give a flop-count of this modified algorithm.

5.24 Develop an algorithm for computing the frequency response matrix using formula (5.5.10) and the fact that the determinant of a matrix A is just the product of the diagonal entries of the matrix U in its LU factorization. (Consult Misra and Patel 1988.)

5.25 Develop an algorithm for computing the frequency response matrix based on the reduction of A to RSF. Compare the flop-count of this algorithm with that of the Hessenberg algorithm (Algorithm 5.5.1).

5.26 Develop an algorithm for computing the frequency response matrix of the descriptor model:

$$E\dot{x} = Ax + Bu$$

based on the initial reduction of the pair (A, B) to Hessenberg-triangular form described in Chapter 4. (Consult Misra 1989.)

References

Ali Sayed, *Lecture Notes on Dynamical Systems*, University of California, Los Angeles, 1994.

Baldwin C.T. *Fundamentals of Electrical Measurements*, George G. Harp and Co. Ltd., London, 1961.

Brogan W.L. *Modern Control Theory*, 3rd edn, Prentice Hall, Englewood Cliffs, NJ, 1991.

Campbell S.L. *Singular Systems of Differential Equations,* Pitman, Marshfield, MA, 1980.

Chen C.-T. *Linear System Theory and Design*, CBS College Publishing, New York, 1984.

Datta B.N. and Datta, K. "An algorithm to compute the powers of a Hessenberg matrix and applications," *Lin. Alg. Appl.* Vol. 14, pp. 273–284, 1976.

DeCarlo R.A. *Linear Systems: A State Variable Approach with Numerical Implementation,* Prentice Hall, Englewood Cliffs, NJ, 1989.

Dorf R.C. and Bishop R.H. *Modern Control Systems*, 9th ed, Prentice Hall, Upper Saddle River, NJ, 2001

Friedland B. *Control Systems Design: An Introduction to State-Space Methods*, McGraw-Hill, New York, 1986.

Golub G.H. and Van Loan C.F. *Matrix Computations*, 3rd edn, Johns Hopkins University Press, Baltimore, MD, 1996.

Kailath T. *Linear Systems*, Prentice Hall, Englewood Cliffs, NJ, 1980.

Kenney C.S., Laub A.J., and Stubberud, S.C. "Frequency response computation via rational interpolation," *IEEE Trans. Autom. Control*, Vol. 38, pp. 1203–1213, 1993.

Laub A.J. "Efficient multivariable frequency response computations," *IEEE Trans. Autom. Control*, Vol. AC-26, pp. 407–408, 1981.

Laub A.J. and Linnemann A. "Hessenberg and Hessenberg/triangular forms in linear systems theory," *Int. J. Control*, Vol. 44, pp. 1523–1547, 1986.

Luenberger D.G. *Introduction to Dynamic Systems: Theory, Methods, & Applications*, John Wiley & Sons, New York, 1979.

Misra P. "Hessenberg-triangular reduction and transfer function matrices of singular systems," *IEEE Trans. Circuits Syst.*, Vol. CAS-36, pp. 907–912, 1989.

Misra P. and Patel R.V. "A determinant identity and its application in evaluating frequency response matrices," *SIAM J. Matrix Anal. Appl.*, Vol. 9, pp. 248–255, 1988.

Moler C.B. and Van Loan C.F. "Nineteen dubious ways to compute the exponential of a matrix," *SIAM Rev.*, Vol. 20, pp. 801–836, 1978.

Parlett B.N. "A recurrence among the elements of functions of triangular matrices," *Lin. Alg. Appl.*, Vol. 29, pp. 323–346, 1976.

Patel R.V. and Munro N. *Multivariable Systems Theory and Design,* Pergamon Press, Oxford, UK, 1982.

Safonov M.G., Laub A.J. and Hartman, G.L. "Feedback properties of multivariable systems: the role and use of the return difference matrix," *IEEE Trans. Autom. Control*, Vol. AC-26, pp. 47–65, 1981.

Saad Y. "Projection and deflation methods for partial pole assignment in state feedback," IEEE Trans. Automat. Control, Vol. 33, No. 3, pp. 290–297, 1988.

Soong T.T. *Active Structural Control: Theory and Practice*, Longman Scientific and Technical, Essex, UK, 1990.

Szidarovszky F. and Terry Bahill, A. *Linear Systems Theory*, CRC Press, Boca Raton, 1991.

Van Loan C.F. "The sensitivity of the matrix exponential," *SIAM J. Numer. Anal.*, Vol. 14, pp. 971–981, 1977.

Van Loan C.F. "Computing integrals involving the matrix exponential," *IEEE Trans. Autom. Control*, Vol. AC-23, pp. 395–404, 1978.

Zhou K. (with Doyle J.), *Essentials of Robust Control*, Prentice Hall, Upper Saddle River, NJ, 1998.

CHAPTER **6**_____

CONTROLLABILITY, OBSERVABILITY, AND DISTANCE TO UNCONTROLLABILITY

Topics covered

- Controllability and Observability Criteria
- Controller- and Observer-Companion Forms
- Kalman Decomposition
- Controller and Observer Hessenberg Forms
- Distance to Uncontrollability

6.1 INTRODUCTION

This chapter deals with discussions on the two most fundamental notions, **controllability** and **observability**, and related concepts. The well-known criteria of controllability and observability are stated and proved in Theorem 6.2.1.

These theoretically important criteria, unfortunately, do not yield numerically effective tests of controllability. This is demonstrated by means of some examples and discussions in **Section 6.6**. **Numerically effective tests**, based on reduction of the pairs (A, B) and (A, C), respectively, to the **controller-Hessenberg** and **observer-Hessenberg** pairs, achieved by means of orthogonal similarly, are described in **Sections 6.7 and 6.8**.

Controllability and observability are generic concepts. What is more important in practice is to know when a controllable system is close to an uncontrollable one. To this end, a measure of the distance to uncontrollability is introduced in **Section 6.9** and a characterization (**Theorem 6.9.1**) in terms of the minimum singular value of a certain matrix is stated and proved. Finally, two algorithms (**Algorithms 6.9.1 and 6.9.2**) are described to measure the distance to uncontrollability.

The chapter concludes with a brief discussion (**Section 6.10**) on the relationship between the distance to uncontrollability and the singular values of the controllability matrix. *The important message here is that the singular values of the controllability matrix as such cannot be used to make a prediction of how close the system is to an uncontrollable system.* It is the largest gap between two singular values that should be considered.

Reader's Guide for Chapter 6

The readers having knowledge of basic concepts and results on controllability and observability, can skip Sections 6.2–6.5.

6.2 CONTROLLABILITY: DEFINITIONS AND BASIC RESULTS

In this section, we introduce the basic concepts and some algebraic criteria of controllability.

6.2.1 Controllability of a Continuous-Time System

Definition 6.2.1. *The system:*

$$\dot{x}(t) = Ax(t) + Bu(t),$$
$$y(t) = Cx(t) + Du(t) \tag{6.2.1}$$

is said to be **controllable**, *if starting from any initial state $x(0)$, the system can be driven to any final state $x_1 = x(t_1)$ in some finite time t_1, choosing the input variable $u(t), 0 \le t \le t_1$ appropriately.*

Remark

- The controllability of the system (6.2.1) is often referred to as the controllability of the pair (A, B), the reason for which will be clear in the following theorem.

Theorem 6.2.1. *Criteria for Continuous-Time Controllability. Let $A \in \mathbb{R}^{n \times n}$ and $B \in \mathbb{R}^{n \times m} (m \le n)$.*

The following are equivalent:

(i) *The system (6.2.1) is controllable.*

(ii)   *The $n \times nm$ matrix*

$$C_M = (B, AB, A^2B, \ldots, A^{n-1}B)$$

has full rank n.

(iii)   *The matrix*

$$W_C = \int_0^{t_1} e^{At} BB^T e^{A^T t}\, dt$$

is nonsingular for any $t_1 > 0$.

(iv)   *If (λ, x) is an eigenpair of A^T, that is, $x^T A = \lambda x^T$, then $x^T B \neq 0$*

(v)   *Rank $(A - \lambda I, B) = n$ for every eigenvalue λ of A.*

(vi)   *The eigenvalues of $A - BK$ can be arbitrarily assigned (assuming that the complex eigenvalues occur in conjugate pairs) by a suitable choice of K.*

Proof.   Without loss of generality, we can assume that $t_0 = 0$. Let $x(0) = x_0$.
(i) $\Rightarrow$ (ii). Suppose that the rank of C_M is not n. From Chapter 5, we know that

$$x(t_1) = e^{At_1} x_0 + \int_0^{t_1} e^{A(t_1-t)} Bu(t)\, dt. \tag{6.2.2}$$

That is,

$$x(t_1) - e^{At_1} x_0 = \int_0^{t_1} \left\{ I + A(t_1 - t) + \frac{A^2}{2!}(t_1 - t)^2 + \cdots \right\} Bu(t)\, dt$$

$$= B \int_0^{t_1} u(t)\, dt + AB \int_0^{t_1} (t_1 - t)u(t)\, dt + A^2 B \int_0^{t_1} \frac{(t_1 - t)^2}{2!} u(t)\, dt + \cdots$$

From the Cayley–Hamilton Theorem (see **Chapter 1**), it then follows that the vector $x(t_1)$ is a linear combination of the columns of $B, AB, A^2B, \ldots, A^{n-1}B$.

Since C_M does not have rank n, these columns vectors cannot form a basis of the state-space and therefore for some t_1, $x(t_1) = x_1$ cannot be attained, implying that (6.2.1) is not controllable.

(ii) $\Rightarrow$ (iii). Suppose that the matrix C_M has rank n, but the matrix:

$$W_C = \int_0^{t_1} e^{At} BB^T e^{A^T t}\, dt \tag{6.2.3}$$

is singular.

Let v be a nonzero vector such that $W_C v = 0$. Then, $v^T W_C v = 0$. That is, $\int_0^{t_1} v^T e^{At} BB^T e^{A^T t} v\, dt = 0$. The integrand is always nonnegative, since it is of the form $c^T(t)c(t)$, where $c(t) = B^T e^{A^T t} v$. Thus, for the above integral to be equal to zero, we must have:

$$v^T e^{At} B = 0, \quad \text{for } 0 \le t \le t_1.$$

From this we obtain (evaluating the successive derivatives with respect to t, at $t = 0$):

$$v^T A^i B = 0, \quad i = 1, 2, \ldots, n - 1.$$

That is, v is orthogonal to all columns of the matrix C_M. Since it has been assumed that the matrix C_M has rank n, this means that $v = 0$, which is a contradiction.

(iii) $\Rightarrow$ **(i)**. We show that $x(t_1) = x_1$. Let us now choose the vector $u(t)$ as

$$u(t) = B^T e^{A^T(t_1-t)} W_C^{-1}(-e^{At_1}x_0 + x_1).$$

Then from (6.2.2), it is easy to see that $x(t_1) = x_1$. This implies that the system (6.2.1) is controllable.

(ii) $\Rightarrow$ **(iv)**. Let x be an eigenvector of A^T corresponding to an eigenvalue λ; that is, $x^T A = \lambda x^T$. Suppose that $x^T B = 0$. Then,

$$x^T C_M = (x^T B, \lambda x^T B, \lambda^2 x^T B, \ldots, \lambda^{n-1} x^T B) = 0.$$

Since the matrix C_M has full rank, $x = 0$, which is a contradiction.

(iv) $\Rightarrow$ **(ii)**. Assume that none of the eigenvectors of A^T is orthogonal to the columns of B, but rank $(C_M) = k < n$.

We will see later in this chapter (Theorem 6.4.1) that, in this case, there exists a nonsingular matrix T such that

$$\bar{A} = T A T^{-1} = \begin{pmatrix} \bar{A}_{11} & \bar{A}_{12} \\ 0 & \bar{A}_{22} \end{pmatrix}, \qquad \bar{B} = T B = \begin{pmatrix} \bar{B}_1 \\ 0 \end{pmatrix}, \tag{6.2.4}$$

where $\bar{A}_{22}$ is of order $(n - k)$, and $k = \text{rank}(C_M)$.

Let v_2 be an eigenvector of $(\bar{A}_{22})^T$ corresponding to an eigenvalue λ. Then,

$$(\bar{A})^T \begin{pmatrix} 0 \\ v_2 \end{pmatrix} = \begin{pmatrix} \bar{A}_{11}^T & 0 \\ \bar{A}_{12}^T & \bar{A}_{22}^T \end{pmatrix} \begin{pmatrix} 0 \\ v_2 \end{pmatrix} = \begin{pmatrix} 0 \\ \bar{A}_{22}^T v_2 \end{pmatrix} = \lambda \begin{pmatrix} 0 \\ v_2 \end{pmatrix}.$$

Furthermore,

$$(0, v_2^T)\bar{B} = (0, v_2^T) \begin{pmatrix} \bar{B}_1 \\ 0 \end{pmatrix} = 0.$$

That is, there is an eigenvector, namely $\begin{pmatrix} 0 \\ v_2 \end{pmatrix}$ of $(\bar{A})^T$ such that it is orthogonal to the columns of $\bar{B}$. This means that the pair $(\bar{A}, \bar{B})$ is not controllable.

This is a contradiction because a similarity transformation does not change controllability.

(ii) $\Rightarrow$ (v). Rank$(\lambda I - A, B) < n$ if and only if there exists a nonzero vector v such that $v^{\mathrm{T}}(\lambda I - A, B) = 0$.

This equation is equivalent to:

$$A^{\mathrm{T}}v = \lambda v \quad \text{and} \quad v^{\mathrm{T}}B = 0.$$

This means that v is an eigenvector of A^{T} corresponding to the eigenvalue λ and it is orthogonal to the columns of B. The system (A, B) is, therefore, not controllable by (iv).

(v) $\Rightarrow$ (ii). If (v) were not true, then from (iv), we would have had

$$x^{\mathrm{T}}(B, AB, \ldots, A^{n-1}B) = 0,$$

meaning that rank(C_{M}) is less than n.

(vi) $\Rightarrow$ (i). Suppose that (vi) holds, but not (i). Then the system can be decomposed into (6.2.4) such that a subsystem corresponding to $\bar{A}_{22}$ is uncontrollable, whose eigenvalues, therefore, cannot be changed by the control. This contradicts (vi).

(i) $\rightarrow$ (vi). The proof of this part will be given in Chapter 10 (**Theorem 10.4.1**). It will be shown there that if (A, B) is controllable, then a matrix K can be constructed such that the eigenvalues of the matrix $(A - BK)$ are in desired locations. ■

Definition 6.2.2. *The matrix*

$$C_{\mathrm{M}} = (B, AB, A^2B, \ldots, A^{n-1}B) \tag{6.2.5}$$

is called the **controllability matrix**.

Remark

- The eigenvector criterion (iv) and the eigenvalue criterion (v) are popularly known as the **Popov–Belevitch–Hautus (PBH)** criteria of controllability (see Hautus 1969). For a historical perspective of this title, see Kailath (1980, p. 135).

Component controllability. The controllability, as defined in the Definition 6.2.1, is often referred to as the **complete controllability** implying that all the states are controllable.

However, if only one input, say $u_j(t)$, from $u(t) = (u_1(t), \ldots, u_m(t))^{\mathrm{T}}$ is used, then the rank of the corresponding $n \times n$ controllability matrix $C_{\mathrm{M}}^j = (b_j, Ab_j, \ldots, A^{n-1}b_j)$, where b_j is the jth column of B, determines the number of states that are controllable using the input $u_j(t)$. This is illustrated in the following example.

Consider Example 5.2.6 on the motions of an orbiting satellite with $d_0 = 1$.

It is easy to see that C_M has rank 4, so that all states are controllable using both inputs. However, if only the first input $u_1(t)$ is used, then

$$C_M^1 = (b_1, Ab_1, A^2b_1, A^3b_1) = \begin{pmatrix} 0 & 1 & 0 & -\omega^2 \\ 1 & 0 & -\omega^2 & 0 \\ 0 & 0 & -2\omega & 0 \\ 0 & -2\omega & 0 & 2\omega^3 \end{pmatrix},$$

which is singular.

Thus, one of the states is not controllable by using only the radial force $u_1(t)$.

However, it can be easily verified that all the states would be controllable if the tangential force $u_2(t)$ were used instead of $u_1(t)$.

Controllable and Uncontrollable Modes

From the eigenvalue and eigenvector criteria above, it is clear that the controllability and uncontrollability of the pair (A, B) are tied with the eigenvalues and eigenvectors of the system matrix A.

Definition 6.2.3. *A mode of the system (6.2.1) or, equivalently, an eigenvalue λ of A is controllable if the associated left eigenvector (i.e., the eigenvector of A^T associated with λ) is not orthogonal to the columns of B. Otherwise, the mode is uncontrollable.*

6.2.2   Controllability of a Discrete-Time System

Definition 6.2.4. *The discrete-time system*

$$\begin{aligned} x_{k+1} &= Ax_k + Bu_k, \\ y_k &= Cx_k + Du_k \end{aligned} \tag{6.2.6}$$

is said to be **controllable** *if for any initial state x_0 and any final state x_1, there exists a finite sequence of inputs $\{u_0, u_1, \cdots, u_{N-1}\}$ that transfers x_0 to x_1; that is, $x(N) = x_1$.*

In particular, if $x_0 = 0$ and the system (6.2.6) is controllable, then it is called **reachable** *(see Chen 1984). It is also known as controllability from the origin.*

Note: To avoid any confusion, we will assume (without any loss of generality) that $x_0 = 0$. So, in our case, the notion of controllability and reachability are equivalent.

Most of the criteria on controllability in the continuous-time case also hold in the discrete-time case. Here we will state and prove only one criterion analogous to (ii) of Theorem 6.2.1.

Theorem 6.2.2. *The discrete-time system* (6.2.6) *or the pair* (A, B) *is controllable if and only if the rank of the controllability matrix*

$$C_M = (B, AB, \ldots, A^{n-1}B)$$

is n.

Proof. From Theorem 5.4.1, we know that the general solution of the discrete-time systems is

$$x_N = A^{N-1}Bu_0 + A^{N-2}Bu_1 + \cdots + Bu_{N-1}.$$

Thus, x_N can be expressed as a linear combination of $A^{k-1}B$, $k = N, \ldots, 1$.

So, it is possible to choose u_0 through u_{N-1} for arbitrary x_N if and only if the sequence $\{A^i B\}$ has a finite number of columns that span $\mathbb{R}^n$; and this is possible, if and only if the controllability matrix C_M has rank n. ∎

6.3 OBSERVABILITY: DEFINITIONS AND BASIC RESULTS

In this section we state definitions and basic results of observability. The results will not be proved here because they can be easily proved by duality of the results on controllability proved in the previous section.

6.3.1 Observability of a Continuous-Time System

The concept of observability is dual to the concept of controllability.

Definition 6.3.1. *The continuous-time system* (6.2.1) *is said to be* **observable** *if there exists* $t_1 > 0$ *such that the initial state* $x(0)$ *can be uniquely determined from the knowledge of* $u(t)$ *and* $y(t)$ *for all* $t, 0 \leq t \leq t_1$.

Remark

- The observability of the system (6.2.1) is often referred to as the observability of the pair (A, C).

Analogous to the case of controllability, we state the following criteria of observability.

Theorem 6.3.1. Criteria for Continuous-Time Observability. *The following are equivalent:*

(i) *The system* (6.2.1) *is observable.*

(ii) *The observability matrix*

$$O_M = \begin{pmatrix} C \\ CA \\ CA^2 \\ \vdots \\ CA^{n-1} \end{pmatrix}$$

has full rank n.
(iii) *The matrix*

$$W_O = \int_0^{t_1} e^{A^T\tau} C^T C e^{A\tau} d\tau$$

is nonsingular for any $t_1 > 0$.
(iv) *The matrix*

$$\begin{bmatrix} \lambda I - A \\ C \end{bmatrix}$$

has rank n for every eigenvalue λ of A.
(v) *None of the eigenvectors of A is orthogonal to the rows of C, that is, if (λ, y) is an eigenpair of A, then $Cy \neq 0$.*
(vi) *There exists a matrix L such that the eigenvalues of $A + LC$ can be assigned arbitrarily, provided that the complex eigenvalues occur in conjugate pairs.*

We only prove (iii) $\Longleftrightarrow$ (i) *and leave the others as an exercise (**Exercise 6.6**).*

Theorem 6.3.2. *The pair (A, C) is observable if and only if the matrix W_O is nonsingular for any $t_1 > 0$.*

Proof. First suppose that the matrix W_O is nonsingular. Since $y(t)$ and $u(t)$ are known, we can assume, without any loss of generality, that $u(t) = 0$ for every t. Thus,

$$y(t) = Ce^{At}x(0).$$

This gives

$$W_O x(0) = \int_0^{t_1} e^{A^T\tau} C^T y(\tau) \, d\tau.$$

Thus, $x(0)$ is uniquely determined and is given by $x(0) = W_O^{-1} \int_0^{t_1} e^{A^T\tau} C^T y(\tau) d\tau$.

Conversely, if W_O is singular, then there exists a nonzero vector z such that $W_O z = 0$, which in turn implies that $Ce^{A\tau} z = 0$. So, $y(\tau) = Ce^{A\tau}(x(0) + z) = Ce^{A\tau} x(0)$.

Thus, $x(0)$ cannot be determined uniquely, implying that (A, C) is not observable. ■

Component observability. As in the case of controllability, we can also speak of component observability when all the states are not observable by certain output. The rank of the observability matrix

$$C_M^j = \begin{pmatrix} c_j \\ c_j A \\ \vdots \\ c_j A^{n-1} \end{pmatrix},$$

where c_j, the jth row of the output matrix C, determines the number of states that are observable by the output $y_j(t)$.

6.3.2 Observability of a Discrete-Time System

Definition 6.3.2. *The discrete-time system (6.2.6) is said to be observable if there exists an index N such that the initial state x_0 can be completely determined from the knowledge of inputs $u_0, u_1, \ldots, u_{N-1}$, and the outputs $y_0, y_1, \ldots, y_N$.*

The criteria of observability in the discrete-time case are the same as in the continuous-time case, and therefore, will not be repeated here.

6.4 DECOMPOSITIONS OF UNCONTROLLABLE AND UNOBSERVABLE SYSTEMS

Suppose that the pair (A, B) is not controllable. Let the rank of the controllability matrix be $k < n$. Then the following theorem shows that the system can be decomposed into controllable and uncontrollable parts.

Theorem 6.4.1. *Decomposition of Uncontrollable System. If the controllability matrix C_M has rank k, then there exists a nonsingular matrix T such that*

$$\bar{A} = TAT^{-1} = \begin{pmatrix} \bar{A}_{11} & \bar{A}_{12} \\ 0 & \bar{A}_{22} \end{pmatrix}, \qquad \bar{B} = TB = \begin{pmatrix} \bar{B}_1 \\ 0 \end{pmatrix}, \tag{6.4.1}$$

where $\bar{A}_{11}, \bar{A}_{12}$, and $\bar{A}_{22}$ are, respectively, of order $k \times k, k \times (n-k)$, and $(n-k) \times (n-k)$, and $\bar{B}_1$ has k rows. Furthermore, $(\bar{A}_{11}, \bar{B}_1)$ is controllable.

Proof. Let $v_1, \ldots, v_k$ be the independent columns of the controllability matrix C_M. We can always choose a set of $n - k$ vectors $v_{k+1}, \ldots, v_n$ so that the vectors $(v_1, v_2, \ldots, v_k, v_{k+1}, \ldots, v_n)$ form a basis of $\mathbb{R}^n$.

Then it is easy to see that the matrix $T^{-1} = (v_1, \ldots, v_n)$ is such that TAT^{-1} and TB will have the above forms.

To prove that $(\bar{A}_{11}, \bar{B}_1)$ is controllable, we note that the controllability matrix of the pair $(\bar{A}, \bar{B})$ is

$$
\begin{pmatrix}
\bar{B}_1 & \bar{A}_{11}\bar{B}_1 & \cdots & (\bar{A}_{11})^{k-1}\bar{B}_1 & \cdots & (\bar{A}_{11})^{n-1}\bar{B}_1 \\
0 & 0 & & 0 & \cdots & 0
\end{pmatrix}.
$$

Since, for each $j \geq k$, $(\bar{A}_{11})^j$ is a linear combination of $(\bar{A}_{11})^i$, $i = 0, 1, \ldots$, $(k - 1)$, by the Cayley–Hamilton Theorem (see Chapter 1), we then have rank$(\bar{B}_1, \bar{A}_{11}\bar{B}_1, \ldots, \bar{A}_{11}^{k-1}\bar{B}_1) = k$, proving that $(\bar{A}_{11}, \bar{B}_1)$ is controllable. ∎

Note: If we define $\bar{x} = Tx$, then the state vector $\bar{x}$ corresponding to the system defined by $\bar{A}$ and $\bar{B}$ is given by $\bar{x} = \begin{pmatrix} \bar{x}_1 \\ \bar{x}_2 \end{pmatrix}$.

Remark (Choosing T Orthogonal)

• Note that T in Theorem 6.4.1 can be chosen to be orthogonal by finding the QR factorization of the controllability matrix C_M. Thus, if $C_M = QR$, then $T = Q^T$.

Using duality, we have the following decomposition for an unobservable pair. The proof is left as an exercise (**Exercise 6.7**).

Theorem 6.4.2. *Decomposition of Unobservable System. If the observability matrix O_M has rank $k' < n$, then there exists a nonsingular matrix T such that*

$$
\bar{A} = TAT^{-1} = \begin{pmatrix} \bar{A}_{11} & \bar{A}_{12} \\ 0 & \bar{A}_{22} \end{pmatrix}, \qquad \bar{C} = CT^{-1} = (0, \bar{C}_1) \qquad (6.4.2)
$$

with $(\bar{A}_{11}, \bar{C}_1)$ observable and $\bar{A}_{11}$ is of order k'.

The Kalman Decomposition

Combining Theorems 6.4.1 and 6.4.2, we obtain (after some reshuffling of the coordinates) the following decomposition, known as the **Kalman Canonical Decomposition**. The proof of the theorem is left as an exercise (**Exercise 6.8**), and can also be found in any standard text on linear systems theory.

Theorem 6.4.3. *The Kalman Canonical Decomposition Theorem. Given the system (6.2.1) there exists a nonsingular coordinate transformation $\bar{x} = Tx$*

such that

$$\begin{pmatrix} \dot{\bar{x}}_{c\bar{o}} \\ \dot{\bar{x}}_{co} \\ \dot{\bar{x}}_{\bar{c}\bar{o}} \\ \dot{\bar{x}}_{\bar{c}o} \end{pmatrix} = \begin{pmatrix} \bar{A}_{c\bar{o}} & \bar{A}_{12} & \bar{A}_{13} & \bar{A}_{14} \\ 0 & \bar{A}_{co} & 0 & \bar{A}_{24} \\ 0 & 0 & \bar{A}_{\bar{c}\bar{o}} & A_{34} \\ 0 & 0 & 0 & A_{\bar{c}0} \end{pmatrix} \begin{pmatrix} \bar{x}_{c\bar{o}} \\ \bar{x}_{co} \\ \bar{x}_{\bar{c}\bar{o}} \\ \bar{x}_{\bar{c}o} \end{pmatrix} + \begin{pmatrix} \bar{B}_{c\bar{o}} \\ \bar{B}_{co} \\ 0 \\ 0 \end{pmatrix}, \qquad (6.4.3)$$

$$y = (0, \bar{C}_{co}, 0, \bar{C}_{\bar{c}o}) \begin{pmatrix} \bar{x}_{c\bar{o}} \\ \bar{x}_{co} \\ \bar{x}_{\bar{c}\bar{o}} \\ \bar{x}_{\bar{c}o} \end{pmatrix} + Du,$$

$\bar{x}_{c\bar{o}} \equiv$ *states which are controllable but not observable.*
$\bar{x}_{co} \equiv$ *states which are both controllable and observable*
$\bar{x}_{\bar{c}\bar{o}} \equiv$ *states which are both uncontrollable and unobservable*
$\bar{x}_{\bar{c}o} \equiv$ *states which are uncontrollable but observable.*
Moreover, the transfer function matrix from u to y is given by

$$G(s) = \bar{C}_{co}(sI - \bar{A}_{co})^{-1}\bar{B}_{co} + D.$$

Remark

- It is interesting to observe that the uncontrollable and/or unobservable parts of the system do not appear in the description of the transfer function matrix.

6.5 CONTROLLER- AND OBSERVER-CANONICAL FORMS

An important property of a linear system is that **controllability and observability remain invariant under certain transformations**. We will state the result for controllability without proof. A similar result, of course, holds for observability. Proof is left as an **Exercise (6.9)**.

Theorem 6.5.1. *Let T be a nonsingular matrix such that $TAT^{-1} = \tilde{A}$, and $TB = \tilde{B}$, then (A, B) is controllable if and only if $(\tilde{A}, \tilde{B})$ is controllable.*

The question naturally arises if the matrix T can be chosen so that $\tilde{A}$ and $\tilde{B}$ will have simple forms, from where conclusions about controllability or observability can be easily drawn. Two such forms, **controller-canonical form** (or the **controller-companion form**) and the **Jordan Canonical Form** (JCF) are well known in control text books (Luenberger 1979; Kailath 1980; Chen 1984; Szidarovszky and Bahill 1991 etc.). **Unfortunately, neither of these two forms, in general, can be obtained in a numerically stable way, because, T, not being an orthogonal matrix in general, can be highly ill-conditioned.**

The controller- and observer-canonical forms are, however, very valuable theoretical tools in establishing many theoretical results in control and systems theory (e.g., the proof of the eigenvalue assignment (EVA) theorem by state feedback (**Theorem 10.4.1 in Chapter 10**)). We just state these two forms here for later uses as theoretical tools. First consider the single-input case.

Let (A, b) be controllable and let C_M be the controllability matrix. Let s_n be the last row of C_M^{-1}. Then the matrix T defined by

$$T = \begin{pmatrix} s_n \\ s_n A \\ \vdots \\ s_n A^{n-1} \end{pmatrix} \tag{6.5.1}$$

is such that

$$\tilde{A} = TAT^{-1} = \begin{pmatrix} 0 & 1 & 0 & \cdots & 0 \\ 0 & 0 & 1 & \cdots & 0 \\ \vdots & & & \ddots & 1 \\ -a_1 & -a_2 & -a_3 & \cdots & -a_n \end{pmatrix}, \qquad \tilde{b} = Tb = \begin{pmatrix} 0 \\ 0 \\ \vdots \\ 0 \\ 1 \end{pmatrix}. \tag{6.5.2}$$

Similarly, it is possible to show that if (A, b) is controllable, then there exists a nonsingular matrix P such that

$$PAP^{-1} = \begin{pmatrix} -a_n & -a_{n-1} & \cdots & -a_2 & -a_1 \\ 1 & 0 & \cdots & 0 & 0 \\ 0 & 1 & \cdots & 0 & 0 \\ \cdots & \cdots & \cdots & \cdots & \cdots \\ 0 & 0 & \cdots & 1 & 0 \end{pmatrix}, \qquad Pb = \begin{pmatrix} 1 \\ 0 \\ 0 \\ \vdots \\ 0 \end{pmatrix}. \tag{6.5.3}$$

The forms (6.5.1) and (6.5.3) are, respectively, known as **lower** and **upper companion (or controller) canonical forms**. By duality, the **observer-canonical forms** (in lower and upper companion forms) can be defined. Thus, the pair $(\tilde{A}, \tilde{c})$ given by

$$\tilde{A} = \begin{pmatrix} 0 & 0 & \cdots & 0 & -a_1 \\ 1 & 0 & \cdots & 0 & -a_2 \\ 0 & 1 & \cdots & 0 & -a_3 \\ \vdots & & & & \vdots \\ 0 & 0 & \cdots & 1 & -a_n \end{pmatrix}, \qquad \tilde{c} = (0, 0, \ldots, 0, 1).$$

is an **upper observer-canonical form**.

MATCONTROL note: The MATCONTROL function **cntrlc** can be used to find a controller-canonical form.

The Luenberger Canonical Form

In the multi-input case, the controllable pair (A, B) can be reduced to the pair $(\tilde{A}, \tilde{B})$, where $\tilde{A} = TAT^{-1}$ is a block matrix, *in which each diagonal block matrix is a companion matrix* of the form given in (6.5.2), and $\tilde{B}$ is also a block matrix with each block as a companion matrix of the form (6.5.2) having nonzero entries only on the last row.

The number of diagonal blocks of $\tilde{A}$ is equal to the rank of B. Such a form is known as the **Luenberger controller-canonical form**. Similarly, by duality, the **Luenberger observer-canonical form** can be written down.

Numerical instability: In general, like a controller-companion form, the Luenberger canonical form also cannot be obtained in a numerically stable way.

6.6 NUMERICAL DIFFICULTIES WITH THEORETICAL CRITERIA OF CONTROLLABILITY AND OBSERVABILITY

Each of the algebraic criterion of controllability (observability) described in Section 6.2 (Section 6.3) suggests a test of controllability (observability). Unfortunately, most of them do not lead to numerically viable tests as the following discussions show. First, let's look into the controllability criterion (ii) of Theorem 6.2.1.

This criterion requires successive matrix multiplications and determination of the rank of an $n \times nm$ matrix.

It is well known that matrix multiplications involving nonorthogonal matrices can lead to a severe loss of accuracy (see Chapter 3). The process may transform the problem to a more sensitive one. To illustrate this, consider the following illuminating example from Paige (1981).

Example 6.6.1

$$A = \begin{pmatrix} 1 & & & \\ & 2^{-1} & & \\ & & \ddots & \\ & & & 2^{-9} \end{pmatrix}_{10 \times 10} \quad , \quad B = \begin{pmatrix} 1 \\ 1 \\ 1 \\ \cdots \\ 1 \end{pmatrix}.$$

The pair (A, B) is clearly controllable. The controllability matrix $(B, AB, \ldots, A^9 B)$ can be computed easily and stored accurately. Note that the (i, j)-th entry of this matrix is $2^{(-i+1)(j-1)}$. This matrix has three smallest singular values $0.613 \times 10^{-12}, 0.364 \times 10^{-9}$, and 0.712×10^{-7}. Thus, on a computer with machine precision no smaller than 10^{-12}, one will conclude that the *numerical rank of this matrix* is less than 10, indicating that the system is uncontrollable. (Recall that matrix A is said to have a *numerical rank r* if the computed singular values $\tilde{\sigma}_i, i = 1, \ldots, n$ satisfy $\tilde{\sigma}_1 \geq \tilde{\sigma}_2 \geq \cdots \geq \tilde{\sigma}_r \geq \delta \geq \tilde{\sigma}_{r+1} \geq \cdots \geq \tilde{\sigma}_n$, where δ is a tolerance) (**Section 3.9.4**).

Note that determining the rank of a matrix using the singular values is the most effective way from a numerical viewpoint.

The criteria (iv)–(vi) in Theorem 6.2.1 are based on eigenvalues and eigenvectors computations. We know that the eigenvalues and eigenvectors of certain matrices can be very ill-conditioned and that the ill-conditioning of the computed eigenvalues will again lead to inaccuracies in computations. For example, by criteria (vi) of controllability in Theorem 6.2.1, it is possible, when (A, B) is controllable, to find a matrix K such that $A + BK$ and A have disjoint spectra. Computationally, it is indeed a difficult task to decide if two matrices have a common eigenvalue if that eigenvalue is ill-conditioned. This can be seen from the following discussion.

Let λ and δ be the eigenvalues of A and $A + BK$, respectively. We know that a computed eigenvalue $\tilde{\lambda}$ of A is an eigenvalue of the perturbed matrix $\tilde{A} = A + \Delta A$, where $\|\Delta A\|_2 \leq \mu\|A\|_2$. Similarly, a computed eigenvalue $\tilde{\delta}$ of $A + BK$ is an eigenvalue of $A + BK + \Delta A'$, where $\|\Delta A'\|_2 \leq \mu\|A + BK\|_2$. Thus, even if $\lambda = \delta$, $\tilde{\lambda}$ can be very different from λ and $\tilde{\delta}$ very different from δ, implying that $\tilde{\lambda}$ and $\tilde{\delta}$ are different.

Example 6.6.2. Consider the following example due to Paige (1981) where the matrices A and B are taken as

$$A = Q^{\mathrm{T}}\hat{A}Q, \qquad B = Q^{\mathrm{T}}\hat{B}$$

Here $\hat{A}$ is the well-known 20×20 Wilkinson bidiagonal matrix

$$\hat{A} = \begin{pmatrix} 20 & 20 & & & 0 \\ & 19 & 20 & & \\ & & \ddots & \ddots & \\ & & & \ddots & 20 \\ 0 & & & & 1 \end{pmatrix},$$

$$\hat{B} = (1, 1, \ldots, 1, 0)^{\mathrm{T}},$$

and Q is the Q-matrix of the QR factorization of a randomly generated 20×20 arbitrary matrix whose entries are uniform random numbers on $(-1, 1)$.

Clearly, the pair $(\hat{A}, \hat{B})$, and therefore, the pair (A, B), are uncontrollable. (Note that controllability or uncontrollability is preserved by nonsingular transformations (Theorem 6.5.1)).

Now taking K as a 1×20 matrix with entries as random numbers uniformly distributed on $(-1, 1)$, the eigenvalues $\tilde{\lambda}_i$ of A and μ_i of $A + BK$ were computed and tabulated. They are displayed in the following table.

In this table, $\rho(B, A - \tilde{\lambda}_i I)$ denotes the ratio of the smallest to the largest singular value of the matrix $(B, A - \tilde{\lambda}_i I)$.

Eigenvalues $\tilde{\lambda}_i(A)$	Eigenvalues $\mu_i(A + BK)$	$\rho(B, A - \tilde{\lambda}_i I)$
$-0.32985 \pm j1.06242$	$0.99999 \pm j0$	0.002
$0.9219 \pm j3.13716$	$-8.95872 \pm j3.73260$	0.004
$3.00339 \pm j3.13716$	$-5.11682 \pm j9.54329$	0.007
$5.40114 \pm j6.17864$	$-0.75203 \pm j14.148167$	0.012
$8.43769 \pm j7.24713$	$5.77659 \pm j15.58436$	0.018
$11.82747 \pm j7.47463$	$11.42828 \pm j14.28694$	0.026
$15.10917 \pm j6.90721$	$13.30227 \pm j12.90197$	0.032
$18.06886 \pm j5.66313$	$18.59961 \pm j14.34739$	0.040
$20.49720 \pm j3.81950$	$23.94877 \pm j11.80677$	0.052
$22.06287 \pm j1.38948$	$28.45618 \pm j8.45907$	0.064

The table shows that $\tilde{\lambda}_i$ are almost unrelated to μ_i. One will, then, erroneously conclude that the pair (A, B) is controllable.

The underlying problem, of course, is the ill-conditioning of the eigenvalues of $\hat{A}$. Note that, because of ill-conditioning, the computed eigenvalues of A are different from those of $\hat{A}$, which, in theory, should have been the same because A and $\hat{A}$ are similar.

The entries of the third column of the table can be used to illustrate the difficulty with the eigenvalue criterion (Criterion (v) of Theorem 6.2.1).

Since the pair (A, B) is uncontrollable, by the eigenvalue criterion of controllability, rank$(B, A - \tilde{\lambda}_i I)$, for some $\tilde{\lambda}_i$, should be less than n; consequently, one of the entries of the third column should be identically zero. But this is not the case; **only there is an indication that some "modes" are less controllable than the others**.

To confirm the fact that ill-conditioning of the eigenvalues of $\hat{A}$ is indeed the cause of such failure, rank$(B, A - I)$, which corresponds to the exact eigenvalue 1 of $\hat{A}$ was computed and seen to be

$$\text{rank}(B, A - I) = 5 \times 10^{-8}.$$

Thus, this test would have done well if the exact eigenvalues of $\hat{A}$ were used in place of the computed eigenvalues of A, which are complex.

6.7 A NUMERICALLY EFFECTIVE TEST OF CONTROLLABILITY

A numerically effective test of controllability can be obtained through the reduction of the pair (A, B) to a *block Hessenberg form* using **orthogonal similarity transformation**. The process constructs an orthogonal matrix P such that

$$PAP^T = H, \quad \text{a block upper Hessenberg matrix}$$

$$PB = \tilde{B} = \begin{pmatrix} B_1 \\ 0 \end{pmatrix}. \tag{6.7.1}$$

The form (6.7.1) is called the **controller-Hessenberg** form of (A, B), and the pair $(H, \bar{B})$ is called the controller-Hessenberg pair of (A, B) (see Rosenbrock 1970). The reduction to this form can be done using Householder's or Givens' method. We describe the reduction here using Householder's transformations. The algorithmic procedure appears in Boley (1981), Van Dooren and Verhaegen (1985), and Paige (1981), Patel (1981), Miminis (1981) etc. The algorithm is usually known as the **staircase algorithm**.

Algorithm 6.7.1. *Staircase Algorithm. Let A be $n \times n$ and B be $n \times m$ ($m \leq n$).*
   Step 0. *Triangularize the matrix B using the QR factorization with column pivoting (Golub and Van Loan, 1996, pp. 248–250), that is, find an orthogonal matrix P_1 and a permutation matrix E_1 such that*

$$P_1 B E_1 = \begin{pmatrix} B_1 \\ 0 \end{pmatrix},$$

where B_1 is an $n_1 \times m$ upper triangular matrix and $n_1 = \text{rank}(B) = \text{rank}(B_1)$.
   Step 1. *Update A and B, that is, compute*

$$P_1 A P_1^T = H = \begin{pmatrix} H_{11}^{(1)} & H_{12}^{(1)} \\ H_{21}^{(1)} & H_{22}^{(1)} \end{pmatrix}, \qquad \tilde{B} = P_1 B = \begin{pmatrix} B_1 \\ 0 \end{pmatrix} E^T \equiv \begin{pmatrix} B_1 \\ 0 \end{pmatrix}.$$

where $H_{11}^{(1)}$ is $n_1 \times n_1$ and $H_{21}^{(1)}$ is $(n - n_1) \times n_1$, $n_1 \leq n$. If $H_{21}^{(1)} = 0$, stop.
   Step 2. *Triangularize $H_{21}^{(1)}$ using the QR factorization with column pivoting, that is, find an orthogonal matrix $\hat{P}_2$ and a permutation matrix E_2 such that*

$$\hat{P}_2 H_{21}^{(1)} E_2 = \begin{pmatrix} H_{21}^{(*)} \\ 0 \end{pmatrix},$$

where $H_{21}^{(*)}$ is $n_2 \times n_1$, $n_2 = \text{rank}(H_{21}^{(1)}) = \text{rank}(H_{21}^{(*)})$, and $n_2 \leq n_1$. If $n_1 + n_2 = n$, stop.
   Form

$$P_2 = \text{diag}\,(I_{n_1}, \hat{P}_2) = \left(\begin{array}{c|c} I_{n_1} & 0 \\ \hline 0 & \hat{P}_2 \end{array} \right),$$

where I_{n_1} is a matrix consisting of the first n_1 rows and columns of the identity matrix.
   Compute

$$H_2 = P_2 H_1 P_2^T = \left(\begin{array}{c|cc} H_{11}^{(1)} & H_{12}^{(2)} & H_{13}^{(2)} \\ \hline H_{21}^{(2)} & H_{22}^{(2)} & H_{23}^{(2)} \\ \hline 0 & H_{32}^{(2)} & H_{33}^{(2)} \end{array} \right),$$

where $H_{22}^{(2)}$ is $n_2 \times n_2$ and $H_{32}^{(2)}$ is $(n - n_1 - n_2) \times n_2$. Note that $H_{21}^{(2)} = H_{21}^{(*)} E_2^T$.

Update $P \equiv P_2 P_1$.

If $H_{32}^{(2)} = 0$, *stop.* (*Note that* $H_{11}^{(1)}$ *does not change.*)

The matrix $\tilde{B}$ **remains unchanged.**

Step 3. *Triangularize* $H_{32}^{(2)}$ *to obtain its rank. That is, find* $\hat{P}_3$ *and* E_3 *such that*

$$\hat{P}_3 H_{32}^{(2)} E_3 = \begin{pmatrix} H_{32}^{(3)} \\ 0 \end{pmatrix}.$$

Let $n_3 = \text{rank}(H_{32}^{(2)}) = \text{rank}(H_{32}^{(3)}); n_3 \leq n_2$.

If $n_1 + n_2 + n_3 = n$, *stop. Otherwise, compute* P_3, H_3, *and update* P *as above.* (**Note that** $\tilde{B}$ **remains unchanged.**)

Step 4. *Continue the process until for some integer* $k \leq n$, *the algorithm produces*

$$H \equiv \begin{pmatrix} H_{11} & H_{12} & H_{13} & \cdots & H_{1k} \\ H_{21} & H_{22} & H_{23} & \cdots & H_{2k} \\ 0 & \ddots & \ddots & & \vdots \\ \vdots & \ddots & \ddots & \ddots & \vdots \\ 0 & \cdots & 0 & H_{k,k-1} & H_{kk} \end{pmatrix}, \qquad \tilde{B} \equiv \begin{pmatrix} B_1 \\ 0 \end{pmatrix}, \qquad (6.7.2)$$

where, either $H_{k,k-1}$ *has full rank* n_k, *signifying that the pair* (A, B) *is controllable, or* $H_{k,k-1}$ *is a zero matrix signifying that the pair* (A, B) *is uncontrollable.*

(**Note that in the above expressions for** H **and** $\tilde{B}$**, the superscripts have been dropped, for convenience.** *However,* H_{21} *stands for* $H_{21}^{(2)}$, H_{32} *stands for* $H_{32}^{(3)}$, *etc. that is,* $H_{k,k-1}$ *is established at step* k).

It is easy to see that

$$(\tilde{B}, H\tilde{B}, H^2\tilde{B}, \ldots, H^{k-1}\tilde{B}) = P(B, AB, \ldots, A^{k-1}B)$$

$$= \begin{pmatrix} B_1 & \cdots & \cdots \\ \overline{H_{21}B_1} & \cdots & \cdots \\ & \overline{H_{32}H_{21}B_1} & \cdots \\ & & \ddots \\ & & \overline{H_{k,k-1}\ldots H_{21}B_1} \end{pmatrix}$$

That is, it is block triangular matrix with B_1, $H_{21}B_1$, $\ldots$, $H_{k,k-1}$, $\ldots$, $H_{21}B_1$ *on the diagonal.*

This implies that the matrix $H_{k,k-1}$ *is of full rank if the system is controllable or is a zero matrix if the system is uncontrollable.*

Theorem 6.7.1. *Controller-Hessenberg Theorem.* (*i*) *Given the pair* (A, B), *the orthogonal matrix* P *constructed by the above procedure is such that*

$PAP^T = H$ and $PB = \bar{B}$, where H and $\bar{B}$ are given by (6.7.2). (ii) If pair (A, B) is controllable, then $H_{k,k-1}$ has full rank. If it is uncontrollable, then $H_{k,k-1} = 0$.

Proof.   The proof of Theorem 6.7.1 follows from the above construction. However, we will prove here part (ii) using (v) of Theorem 6.2.1.

Obviously, rank $(B, A - \lambda I) = n$ for all λ if and only if rank $(\bar{B}, H - \lambda I) = n$ for all λ.

Now,

$$(\tilde{B}, H - \lambda I) = \begin{pmatrix} B_1 & H_{11} - \lambda I_1 & \cdots & & \cdots & H_{1k} \\ 0 & H_{21} & H_{22} - \lambda I_2 & \cdots & & H_{2k} \\ \vdots & 0 & \ddots & & \ddots & \vdots \\ \vdots & \vdots & \ddots & \ddots & & \ddots \\ 0 & 0 & \cdots & 0 & H_{k,k-1} & H_{kk} - \lambda I_k \end{pmatrix}.$$

If the system is controllable, then the matrix $(\tilde{B}, H - \lambda I)$ must have full rank and thus, the matrix $H_{k,k-1}$ has full rank. On the other hand, if the system is not controllable, then the matrix $(\tilde{B}, H - \lambda I)$ cannot have full rank implying that $H_{k,k-1}$ must be a zero matrix.   ■

Notes

1. The matrix $\bar{B}$ is not affected throughout the whole process.
2. At each step of computation, the rank of a matrix has to be determined. We have used QR factorization with column pivoting for this purpose. However, **the best way to do this is to use singular value decomposition (SVD) of that matrix**. (See Golub and Van Loan (1996) or Datta (1995)).
3. From the construction of the block Hessenberg pair $(H, \tilde{B})$, it follows that as soon as we encounter a zero block on the subdiagonal of H or if the matrix B_1 does not have full rank, we stop, concluding that (A, B) is not controllable.

Example 6.7.1.   An Uncontrollable Pair

$$A = \begin{pmatrix} 1 & 1 & 1 \\ 1 & 1 & 1 \\ 0 & 0 & 1 \end{pmatrix}, \qquad B = \begin{pmatrix} 1 & 1 \\ 1 & 1 \\ 1 & 1 \end{pmatrix}.$$

Step 0. $P_1 = \begin{pmatrix} -0.5774 & -0.5774 & -0.5774 \\ 0.8165 & -0.4082 & -0.4082 \\ 0 & -0.7071 & 0.7071 \end{pmatrix}$, $E_1 = \begin{pmatrix} 1 & 0 \\ 0 & 1 \end{pmatrix}$,

$$P_1 B E_1 = \begin{pmatrix} -1.7321 & -1.7321 \\ 0 & 0 \\ 0 & 0 \end{pmatrix}.$$

Step 1. $H_1 = P_1 A P_1^T = \begin{pmatrix} 2.3333 & 0.2357 & -0.4082 \\ -0.4714 & 0.1667 & -0.2887 \\ 0.8165 & -0.2887 & 0.5000 \end{pmatrix}$,

$\tilde{B} = \begin{pmatrix} -1.7321 & -1.7321 \\ 0 & 0 \\ 0 & 0 \end{pmatrix}.$

Step 2. $H_{21}^{(1)} = \begin{pmatrix} -0.4714 \\ 0.8165 \end{pmatrix}$

$$\hat{P}_2 = \begin{pmatrix} -0.5000 & 0.8660 \\ 0.8660 & 0.5000 \end{pmatrix}, \qquad E_2 = 1, \qquad P_2 = \text{diag}(I_1, \hat{P}_2)$$

$$H_2 = P_2 H_1 P_2^T = \begin{pmatrix} 2.3333 & -0.4714 & 0 \\ 0.9428 & 0.6667 & 0 \\ 0 & 0 & 0 \end{pmatrix}.$$

Since $H_{32}^{(2)} = 0$, we stop.

The controller-Hessenberg form is given by $(H \equiv H_2, \tilde{B})$

$$H = \begin{pmatrix} 2.3333 & -0.4714 & 0 \\ 0.9428 & 0.6667 & 0 \\ 0 & 0 & 0 \end{pmatrix}, \qquad \tilde{B} = \begin{pmatrix} -1.7321 & -1.7321 \\ 0 & 0 \\ 0 & 0 \end{pmatrix}.$$

Clearly the pair (A, B) is not controllable.

Example 6.7.2. *A Controllable Pair*

$$A = \begin{pmatrix} 0.7665 & 0.1665 & 0.9047 & 0.4540 & 0.5007 \\ 0.4777 & 0.4865 & 0.5045 & 0.2661 & 0.3841 \\ 0.2378 & 0.8977 & 0.5163 & 0.0907 & 0.2771 \\ 0.2749 & 0.9092 & 0.3190 & 0.9478 & 0.9138 \\ 0.3593 & 0.0606 & 0.9866 & 0.0737 & 0.5297 \end{pmatrix},$$

$$B = \begin{pmatrix} 0.4644 & 0.8278 \\ 0.9410 & 0.1254 \\ 0.0501 & 0.0159 \\ 0.7615 & 0.6885 \\ 0.7702 & 0.8682 \end{pmatrix}.$$

Step 0.

$$P_1 = \begin{pmatrix} -0.3078 & -0.6236 & -0.0332 & -0.5047 & -0.5104 \\ 0.5907 & -0.7058 & -0.0263 & 0.1485 & 0.3610 \\ -0.0080 & -0.0451 & 0.9989 & -0.0047 & -0.0004 \\ -0.4561 & -0.2970 & -0.0132 & 0.8208 & -0.1728 \\ -0.5901 & -0.1510 & -0.0123 & -0.2225 & 0.7611 \end{pmatrix}, \quad E_1 = \begin{pmatrix} 1 & 0 \\ 0 & 1 \end{pmatrix}.$$

$$P_1 B E_1 = \begin{pmatrix} -1.5089 & -1.1241 \\ 0 & 0.8157 \\ 0 & 0 \\ 0 & 0 \\ 0 & 0 \end{pmatrix}.$$

$n_1 = \text{rank}(B) = 2$

Step 1.

$$H_1 = P_1 A P_1^T = \begin{pmatrix} 1.8549 & -0.3935 & -1.2228 & 0.1796 & -0.0198 \\ -0.3467 & 0.3934 & 0.5690 & 0.0593 & 0.0470 \\ -0.7857 & -0.4020 & 0.4421 & -0.3453 & -0.0848 \\ -0.5876 & -0.3573 & -0.4998 & 0.3311 & 0.2607 \\ 0.6325 & -0.0777 & 0.0837 & -0.1295 & 0.2253 \end{pmatrix}$$

$$= \begin{pmatrix} H_{11}^{(1)} & H_{12}^{(1)} \\ H_{21}^{(1)} & H_{22}^{(1)} \end{pmatrix}.$$

$$\tilde{B} = \begin{pmatrix} -1.5089 & -1.1241 \\ 0 & 0.8157 \\ 0 & 0 \\ 0 & 0 \\ 0 & 0 \end{pmatrix}.$$

Step 2.

$$\hat{P}_2 = \begin{pmatrix} -0.6731 & -0.5034 & 0.5418 \\ -0.3545 & -0.4233 & -0.8337 \\ 0.6490 & -0.7533 & 0.1064 \end{pmatrix}, \quad E_2 = \begin{pmatrix} 1 & 0 \\ 0 & 1 \end{pmatrix}.$$

$$\hat{P}_2 H_{21}^{(1)} = \begin{pmatrix} H_{21}^{(2)} \\ 0 \end{pmatrix} = \begin{pmatrix} 1.1674 & 0.4084 \\ 0 & 0.3585 \\ 0 & 0 \end{pmatrix}.$$

$n_2 = \text{rank}(H_{21}^{(1)}) = \text{rank}(H_{21}^{(2)}) = 2.$

$$P_2 = \begin{pmatrix} 1 & 0 & 0 & 0 & 0 \\ 0 & 1 & 0 & 0 & 0 \\ 0 & 0 & -0.6731 & -0.5034 & 0.5418 \\ 0 & 0 & -0.3545 & -0.4233 & -0.8337 \\ 0 & 0 & 0.6490 & -0.7533 & 0.1064 \end{pmatrix},$$

$$P = \begin{pmatrix} -0.3078 & -0.6236 & -0.0332 & -0.5047 & -0.5104 \\ 0.5907 & -0.7058 & -0.0263 & 0.1485 & 0.3610 \\ -0.0847 & 0.0980 & -0.6724 & -0.5306 & 0.4996 \\ 0.6879 & 0.2676 & -0.3383 & -0.1602 & -0.5613 \\ 0.2756 & 0.1784 & 0.6570 & -0.6450 & 0.2109 \end{pmatrix}.$$

$$H_2 = P_2 H_1 P_2^T = \begin{pmatrix} 1.8549 & -0.3935 & 0.7219 & 0.3740 & -0.9310 \\ -0.3467 & 0.3934 & -0.3874 & -0.2660 & 0.3297 \\ 1.1674 & 0.4084 & 0.0286 & -0.0378 & 0.0080 \\ 0 & 0.3585 & -0.1807 & 0.1907 & -0.1062 \\ 0 & 0 & -0.3304 & 0.1575 & 0.7792 \end{pmatrix}$$

$$= \begin{pmatrix} H_{11}^{(1)} & H_{12}^{(2)} & H_{13}^{(2)} \\ H_{21}^{(2)} & H_{22}^{(2)} & H_{23}^{(2)} \\ 0 & H_{32}^{(2)} & H_{33}^{(2)} \end{pmatrix}.$$

$$\bar{B} = \begin{pmatrix} -1.5089 & -1.1242 \\ 0 & 0.8157 \\ 0 & 0 \\ 0 & 0 \\ 0 & 0 \end{pmatrix}.$$

$n_3 = \text{rank}(H_{32}^{(2)}) = 1.$

Since $n_1 + n_2 + n_3 = 2 + 2 + 1 = 5$, we stop.

The controller-Hessenberg form (A, B) is given by $(H \equiv H_2, \bar{B})$.

The pair (A, B) is controllable, because $H_{21}^{(2)}$ and $H_{32}^{(2)}$ have full rank.

The next example (Example 6.7.3) shows the uses of non-identity premutation matrices in QR factorization with column pivoting.

Example 6.7.3.

$$A = \begin{pmatrix} 0.7665 & 0.1665 & 0.9047 & 0.4540 & 0.5007 \\ 0.4777 & 0.4865 & 0.5045 & 0.2661 & 0.3841 \\ 0.2378 & 0.8977 & 0.5163 & 0.0907 & 0.2771 \\ 0.2749 & 0.9092 & 0.3190 & 0.9478 & 0.9138 \\ 0.3593 & 0.0606 & 0.9866 & 0.0737 & 0.5297 \end{pmatrix},$$

$$B = \begin{pmatrix} 0.4644 & 0.8278 \\ 0.9410 & 0.1254 \\ 0.0501 & 1.0159 \\ 0.7615 & 0.6885 \\ 0.7702 & 0.8682 \end{pmatrix}.$$

Step 0.

$$P_1 = \begin{pmatrix} -0.4811 & -0.0729 & -0.5904 & -0.4001 & -0.5046 \\ 0.0213 & -0.7765 & 0.4917 & -0.3183 & -0.2312 \\ -0.6160 & 0.4529 & 0.6231 & -0.0783 & -0.1450 \\ -0.3829 & -0.3429 & -0.0646 & 0.8375 & -0.1739 \\ -0.4919 & -0.2627 & -0.1313 & -0.1764 & 0.8004 \end{pmatrix},$$

$$E_1 = \begin{pmatrix} 0 & 1 \\ 1 & 0 \end{pmatrix}.$$

Step 1.

$$\bar{B} = P_1 B = \begin{pmatrix} -1.0149 & -1.7207 \\ -1.1166 & -0.0000 \\ -0.0000 & 0.0000 \\ 0.0000 & 0.0000 \\ 0.0000 & 0.0000 \end{pmatrix},$$

$$n_1 = 2$$

$$H_1 = P_1 A P_1^{\mathsf{T}} = \begin{pmatrix} 2.1262 & 0.5717 & -0.6203 & 0.3595 & 0.1402 \\ 0.9516 & 0.2348 & -0.0160 & -0.0300 & -0.0472 \\ 0.2778 & -0.4685 & 0.3036 & -0.2619 & -0.0998 \\ 0.0115 & -0.8294 & 0.0243 & 0.3571 & 0.2642 \\ 0.3480 & 0.5377 & 0.0873 & -0.0673 & 0.2250 \end{pmatrix}.$$

Step 2.

$$\hat{P}_2 = \begin{pmatrix} -0.4283 & -0.7582 & 0.4916 \\ 0.6685 & 0.1002 & 0.7370 \\ -0.6080 & 0.6442 & 0.4640 \end{pmatrix},$$

$$E_2 = \begin{pmatrix} 0 & 1 \\ 1 & 0 \end{pmatrix},$$

$$n_2 = 2,$$

$$P_2 = \begin{pmatrix} 1.0000 & 0 & 0 & 0 & 0 \\ 0 & 1.0000 & 0 & 0 & 0 \\ 0 & 0 & -0.4283 & -0.7582 & 0.4916 \\ 0 & 0 & 0.6685 & 0.1002 & 0.7370 \\ 0 & 0 & -0.6080 & 0.6442 & 0.4640 \end{pmatrix},$$

$$P = P_2 P_1,$$

$$P = \begin{pmatrix} -0.4811 & -0.0729 & -0.5904 & -0.4001 & -0.5046 \\ 0.0213 & -0.7765 & 0.4917 & -0.3183 & -0.2312 \\ 0.3124 & -0.0631 & -0.2824 & -0.6881 & 0.5875 \\ -0.8127 & 0.0749 & 0.3133 & -0.0985 & 0.4755 \\ -0.1004 & -0.6182 & -0.4813 & 0.5053 & 0.3475 \end{pmatrix},$$

$$H_2 = P_2 H_1 P_2^T,$$

$$H_2 = \begin{pmatrix} 2.1262 & 0.5717 & 0.0619 & -0.2753 & 0.6738 \\ 0.9516 & 0.2348 & 0.0064 & -0.0485 & -0.0315 \\ 0.0434 & 1.0938 & 0.1675 & -0.1244 & -0.0811 \\ 0.4433 & 0.0000 & 0.0895 & 0.2539 & -0.2275 \\ -0.0000 & -0.0000 & -0.0517 & 0.1971 & 0.4644 \end{pmatrix},$$

$$\bar{B} = \begin{pmatrix} -1.0149 & -1.7207 \\ -1.1166 & -0.0000 \\ -0.0000 & 0.0000 \\ 0.0000 & 0.0000 \\ 0.0000 & 0.0000 \end{pmatrix}.$$

Flop-count: Testing controllability using the constructive proof of Theorem 6.7.1 requires roughly $6n^3 + 2n^2m$ flops. The count includes the construction of the transforming matrix P (see Van Dooren and Verhaegen 1985).

Round-off error analysis and stability: The procedure is **numerically stable**. It can be shown that the computed matrices $\hat{H}$ and $\hat{B}$ are such that $\hat{H} = H + \Delta H$ and $\hat{B} = \bar{B} + \Delta B$, where $\|\Delta H\|_F \le c\mu\|H\|_F$ and $\|\Delta B\|_F \le c\mu\|\|\bar{B}\|_F$ for some small constant c. Thus, with the computed pair $(\hat{H}, \hat{B})$, we will compute the controllability of a system determined by the pair of matrices which are close to H and $\bar{B}$. Since the controllability of the pair $(H, \bar{B})$ is the same as that of the pair (A, B), **this can be considered as a backward stable method for finding the controllability of the pair** (A, B).

MATCONTROL note: Algorithm 6.7.1 has been implemented in MATCONTROL function **cntrlhs**.

The function **cntrlhst** gives block Hessenberg form with triangular subdiagonal blocks.

Controllability Index and Controller-Hessenberg Form

Let $B = (b_1, b_2, \ldots, b_m)$. Then the controllability matrix C_M can be written as

$$C_M = (b_1, b_2, \ldots, b_m; Ab_1, Ab_2, \ldots, Ab_m; \ldots; A^{n-1}b_1, A^{n-1}b_2, \ldots, A^{n-1}b_m).$$

Suppose that the linearly independent columns of the matrix C_M have been obtained in order from left to right. Reorder these independent columns to obtain:

$$C_M' = (b_1, Ab_1, \ldots, A^{\mu_1-1}b_1; b_2, Ab_2, \ldots, A^{\mu_2-1}b_2; \ldots; b_m,$$
$$Ab_m, \ldots, A^{\mu_m-1}b_m).$$

The integers $\mu_1, \ldots, \mu_m$ are called the **controllability indices** associated with $b_1, b_2, \ldots, b_m$, respectively if $\mu_1 \ge \cdots \ge \mu_m$. Note that μ_i is the number of independent columns associated with b_i.

Furthermore, $\mu = \max(\mu_1, \ldots, \mu_m)$ is called the **controllability index**. If $\mu_1 + \mu_2 + \cdots + \mu_m = n$, then the system is controllable.

It is clear that determining the controllability index is a delicate problem from the numerical view point because it is basically a rank-determination problem.

Fortunately, the block-Hessenberg pair $(H, \bar{B})$ of (A, B) not only determines if the pair (A, B) is controllable, but it also gives us the controllability index. In the block-Hessenberg pair $(H, \bar{B})$ in (6.7.2), k is the controllability index. Thus, for Example 6.7.2, the controllability index is 3.

Controllability Test in the Single-Input Case

In the single-input case, the controller-Hessenberg form of (A, b) becomes:

$$PAP^\mathsf{T} = H = \begin{pmatrix} h_{11} & h_{12} & \cdots & & h_{1n} \\ h_{21} & h_{22} & \cdots & & h_{2n} \\ 0 & h_{32} & \ddots & \cdots & h_{3n} \\ \vdots & \vdots & \ddots & \ddots & \vdots \\ 0 & \cdots & 0 & h_{n,n-1} & h_{nn} \end{pmatrix}, \qquad Pb = \bar{b} = \begin{pmatrix} b_1 \\ 0 \\ \vdots \\ 0 \end{pmatrix}.$$

$$(6.7.3)$$

Theorem 6.7.2. *(A, b) is controllable if the controller-Hessenberg pair $(H, \bar{b})$ is such that H is an unreduced upper Hessenberg, that is, $h_{i,i-1} \neq 0, i = 2, \ldots, n$, and $b_1 \neq 0$; otherwise, it is uncontrollable.*

We will give an independent proof of this test using the controllability criterion (ii) of Theorem 6.2.1.

Proof. Observe that $\text{Rank}(b, Ab, \ldots, A^{n-1}b) = \text{rank}(Pb, PAP^\mathsf{T}Pb, \ldots, PA^{n-1}P^\mathsf{T}Pb) = \text{rank}(\bar{b}, H\bar{b}, \ldots, H^{n-1}\bar{b})$.

The last matrix is a lower triangular matrix with $b_1, h_{21}b_1, h_{21}h_{32}b_1, \ldots, h_{21}, \ldots, h_{n,n-1}b_1$ as the diagonal entries.

Since $h_{i,i-1} \neq 0, i = 2, \ldots, n$ and $b_1 \neq 0$, it follows that $\text{rank}(b, Ab, \ldots, A^{n-1}b) = n$.

On the other hand, if any of $h_{i,i-1}$ or b_1 is zero, the matrix $(b, Ab, \ldots, A^{n-1}b)$ is rank deficient, and therefore, the system is uncontrollable. ∎

Example 6.7.4 (*Example 6.6.2 Revisited*). Superiority of the algorithm over the other theoretical criteria.

To demonstrate the superiority of the test of controllability given by Theorem 6.7.2 over some of the theoretical criteria that we considered in the last section, Paige applied the controller-Hessenberg test to the same ill-conditioned problem as in Example 6.6.2. The computations gave $b_1 = 4.35887$, $h_{21} = 8.299699$,

$17 < \|h_{i,i-1}\| < 22, i = 3, 4, \ldots, 9$, and $h_{20,19} = 0.0000027$. Since $h_{20,19}$ is computationally zero in a single-precision computation, the system is uncontrollable, according to the test based on Theorem 6.7.2.

6.8   A NUMERICALLY EFFECTIVE TEST OF OBSERVABILITY

Analogous to the procedure of obtaining the form $(H, \bar{B})$ from (A, B), the pair (A, C) can be transformed to $(H, \bar{C})$, where

$$H = OAO^T = \begin{pmatrix} H_{11} & H_{12} & \cdots & H_{1k} \\ H_{21} & \ddots & & \vdots \\ & \ddots & \ddots & \vdots \\ 0 & & H_{k,k-1} & H_{kk} \end{pmatrix}, \tag{6.8.1}$$

$$\bar{C} = CO^T = (0, C_1). \tag{6.8.2}$$

The pair (A, C) is observable if H is block unreduced (i.e., all the subdiagonal blocks have full rank) and the matrix C_1 has full rank.

The pair $(H, \bar{C})$ is said to be an **observer-Hessenberg pair**.

Flop-count: The construction of the observer-Hessenberg form this way requires roughly $6n^3 + 2n^2r$ flops.

Single-output case: In the single-output case, that is, when C is a row vector, the pair (A, C) is observable if H is an upper Hessenberg matrix and $\bar{C} = (0, \ldots, 0, c_1); c_1 \neq 0$.

MATCONTROL note: MATCONTROL function **obserhs** can be used to obtain the reduction (6.8.1).

6.9   DISTANCE TO AN UNCONTROLLABLE SYSTEM

The concepts of controllability and observability are generic ones. Since determining if a system is controllable depends upon whether or not a certain matrix (or matrices) has full rank, it is immediately obvious from our discussion on numerical rank of a matrix in Chapter 4 that any uncontrollable system is arbitrary close to a controllable system. To illustrate this, let us consider the following well-known example (Eising 1984):

$$A = \begin{pmatrix} -1 & -1 & \cdot & \cdot & \cdot & \cdot & -1 & -1 \\ 1 & \cdot & & & & & & -1 \\ & 1 & \cdot & & & & & -1 \\ & & & \cdot & \vdots & & & \vdots \\ & & & & \cdot & \cdot & & \vdots \\ & & & & & \cdot & \cdot & \vdots \\ 0 & & & & & \cdot & \cdot & -1 \\ & & & & & & 1 & 1 \end{pmatrix}, \quad B = \begin{pmatrix} 1 \\ 0 \\ \vdots \\ \vdots \\ 0 \end{pmatrix}. \tag{6.9.1}$$

The pair (A, b) is obviously controllable. However, it is easily verified that if we add $(-2^{1-n}, -2^{1-n}, \ldots, -2^{1-n})$ to the last row of (B, A), we obtain an uncontrollable system. Clearly, when n is large, the perturbation 2^{1-n} is small, implying that the original controllable system (A, B) is close to an uncontrollable system. **Thus, what is important in practice is knowledge of how close a controllable system is to an uncontrollable one rather than determining if a system is controllable or not.** To this end, we introduce, following Paige (1981), **a measure of the distance to uncontrollability,** denoted by $\mu(A, B)$:

$$\mu(A, B) \equiv \min\{\|\Delta A, \Delta B\|_2 \text{ such that the system}$$
$$\text{defined by } (A + \Delta A, B + \Delta B) \text{ is uncontrollable}\}.$$

Here ΔA and ΔB are allowable perturbations over a field F. If the field F is $\mathbb{R}$, then we will use the symbol $\mu_R(A, B)$ to distinguish it from $\mu(A, B)$.

The quantity $\mu(A, B)$ gives us a measure of the distance of a controllable pair (A, B) to the nearest uncontrollable pair. **If this distance is small, then the original controllable system is close to an uncontrollable system. If this distance is large, then the system is far from an uncontrollable system.**

Here is a well-known result on $\mu(A, B)$. See Miminis (1981), Eising (1984) and Kenney and Laub (1988). **Unless otherwise stated, the perturbations are assumed to be over the field of complex numbers, that is, $F = \mathbb{C}$.**

Theorem 6.9.1. *Singular Value Characterization to Distance to Uncontrollability.* $\mu(A, B) = \min \sigma_n(sI - A, B)$, *where* $\sigma_n(sI - A, B)$ *is the smallest singular value of* $(sI - A, B)$ *and s runs over all complex numbers.*

Proof. Suppose that $(A + \Delta A, B + \Delta B)$ is an uncontrollable pair. Then according to (v) of Theorem 6.2.1, we have

$$\text{rank}(A + \Delta A - \lambda I, B + \Delta B) < n, \quad \text{for some } \lambda \in \mathbb{C}.$$

Since the smallest perturbation that can make $\text{rank}(A - \lambda I, B)$ less than n is $\sigma_n(A - \lambda I, B)$ (see Section 3.9.3 of Chapter 3), we have

$$\sigma_n(A - \lambda I, B) \leq \|\Delta A, \Delta B\|_2$$

and the equality holds if

$$(\Delta A, \Delta B) = -\sigma_n u_n v_n^*,$$

where σ_n is the smallest singular value of $(A - \lambda I, B)$, and u_n and v_n are the corresponding left and right singular vectors. Taking the minimum over all $\lambda \in \mathbb{C}$, and using criterion (v) of Theorem 6.2.1, we obtain the result. ■

Algorithms for Computing $\mu(A, B)$

Based on Theorem 6.9.1, several algorithms (Miminis 1981; Eising 1984; Wicks and De Carlo 1991) have been developed in the last few years to compute $\mu(A, B)$ and $\mu_R(A, B)$.

We will briefly describe here a Newton algorithm due to Elsner and He (1991), and an algorithm due to Wicks and DeCarlo (1991).

6.9.1   Newton's and the Bisection Methods for Computing the Distance to Uncontrollability

Let's denote $\sigma_n[sI - A, B]$ by $\sigma(s)$. The problem of finding $\mu(A, B)$ is then clearly the problem of minimizing $\sigma(s)$ over the complex plane.

To this end, define

$$f(s) = v_n^*(s) \begin{pmatrix} u_n(s) \\ 0 \end{pmatrix},$$

where $u_n(s)$ and $v_n(s)$ are the normalized nth columns of U and V in the SVD of $(A - sI, B)$, that is, $(A - sI, B) = U\Sigma V^T$. The function $f(s)$ plays an important role. The first and second derivatives of $\sigma(s) = \sigma(x + jy) = \sigma(x, y)$ can be calculated using this SVD. It can be shown that if $s = x + jy$, then

$$\frac{\partial \sigma}{\partial x} = \frac{\partial \sigma(x + jy)}{\partial x} = -\text{Re}\, f(x+jy), \quad \text{and} \quad \frac{\partial \sigma}{\partial y} = \frac{\partial \sigma(x + jy)}{\partial y} = -\text{Im}\, f(x+jy).$$

Knowing the first derivatives, the second derivatives can be easily calculated. **Hence the zeros of $f(s)$ are the critical points of the function $\sigma(s)$.**

Thus, some well-established root-finding methods, such as **Newton's method**, or the **Bisection method** can be used to compute these critical points.

An interesting observation about the critical points is: **The critical points satisfy $s = u_n^*(s) A u_n(s)$, and hence they lie in the field of values of A.**

The result follows from the fact that $\sigma(s) f(s) = u_n^*(s)(A - sI)u_n(s)$, since $(A - sI, B)^* u_n(s) = \sigma(s) v_n(s)$. (For the definition of field of values, see Horn and Johnson (1985).)

To decide which critical points are local minima, one can use the following well-known criterion.

A critical point $s = x_c + jy_c$ of $\sigma(s)$ is a local minimum of $\sigma(x, y)$ if

$$\left(\frac{\partial^2 \sigma}{\partial x^2}\right)\left(\frac{\partial^2 \sigma}{\partial y^2}\right) - \left(\frac{\partial^2 \sigma}{\partial x \partial y}\right)^2 > 0 \quad \text{and} \quad \frac{\partial^2 \sigma}{\partial x^2} > 0.$$

Another sufficient condition is: If $\sigma_{n-1}(A - sI, B) > \sqrt{5}\sigma_n(A - sI, B)$, where $s = x_c + jy_c$ is a critical point, then (x_c, y_c) is a local minimum point of $\sigma(x, y)$.

Newton's method needs a starting approximation. The local minima of $\sigma(x, y)$, generally, are simple. Since all critical points s satisfy $u_n^* A u_n = s$, all minimum points $s = x + jy$ will lie in the field of values of A, and hence

$$\lambda_{\min}\left(\frac{A + A^T}{2}\right) \le x \le \lambda_{\max}\left(\frac{A + A^T}{2}\right) \quad \text{and} \quad \lambda_{\min}\left(\frac{A - A^T}{2j}\right) \le y$$

$$\le \lambda_{\max}\left(\frac{A - A^T}{2j}\right),$$

where $\lambda_{\max}(C)$ and $\lambda_{\min}(C)$ denote the largest and smallest eigenvalues of the matrix C. Furthermore, since $\sigma_n(A - sI, B) = \sigma_n(A - \bar{s}I, B)$, the search for all local minimum points can be restricted to $0 \le y \le \lambda_{\max}((A - A^T)/2j)$.

Based on the above discussion, we now state Newton's algorithms for finding $\mu(A, B)$. Denote $x_k + jy_k$ by $\begin{pmatrix} x_k \\ y_k \end{pmatrix}$.

Algorithm 6.9.1. *Newton's Algorithm For Distance to Uncontrollability*
Inputs. *The matrices A and B.*
Output. *A local minimum of $\sigma(s)$.*
Step 0. *Choose $\begin{pmatrix} x_0 \\ y_0 \end{pmatrix}$ using the above criterion.*
Step 1. *For $k = 0, 1, 2, \ldots$ do until convergence*

$$\begin{pmatrix} x_{k+1} \\ y_{k+1} \end{pmatrix} = \begin{pmatrix} x_k \\ y_k \end{pmatrix} - \theta_k \begin{pmatrix} p_{k_1} \\ p_{k_2} \end{pmatrix},$$

$$\text{where } \begin{pmatrix} p_{k_1} \\ p_{k_2} \end{pmatrix} = \begin{pmatrix} \text{Re}\,\dfrac{\partial f}{\partial x} & \text{Re}\,\dfrac{\partial f}{\partial y} \\[2mm] \text{Im}\,\dfrac{\partial f}{\partial x} & \text{Im}\,\dfrac{\partial f}{\partial y} \end{pmatrix}^{-1} \begin{pmatrix} \text{Re}\,f(x, y) \\ \text{Im}\,f(x, y) \end{pmatrix},$$

choosing θ_k such that $\sigma(x_k - \theta_k p_{k1}, y_k - \theta_k p_{k2}) = \min\limits_{-1 \le \theta \le 1} \sigma(x_k - \theta p_{k1}, y_k - \theta p_{k2})$. (see Elsner and He (1991) for formulas for computing $\partial f/\partial x$ and $\partial f/\partial y$).

End.

Step 2. *If $s_C = \begin{pmatrix} x_f \\ y_f \end{pmatrix}$ is the final point upon conclusion of Step 2, then compute the smallest singular value σ_n of the matrix $(A - s_C I, B)$, and take σ_n as the local minimum of $\sigma(s)$.*

Choosing θ_k: θ_k can be chosen using Newton's algorithm, again as follows: Define $g(\theta) = p_{k1}\text{Re}\,f(\theta) + p_{k2}\text{Im}\,f(\theta)$. Then $g'(\theta) = p_{k1}\text{Re}\,f'(\theta) + p_{k2}\text{Im}\,f'(\theta)$.

Newton's Algorithm for Computing θ_k
Step 1. *Choose $\theta_0 = 1$.*

Step 2. *For $j = 1, 2, \ldots$ do until convergence*

$$\theta_{j+1} = \theta_j - \eta_j \frac{g(\theta_j)}{g'(\theta_j)},$$

where η_j is chosen such that

$$\sigma(x_k - \theta_{j+1}p_{k1}, \; y_k - \theta_{j+1}p_{k2}) < \sigma(x_k - \theta_j p_{k1}, \; y_k - \theta_j p_{k2})$$

End.

Remark

- Numerical experiments suggest that it is necessary to compute θ_k's only a few times to get a good initial point and then as soon as it becomes close 1, it can be set to 1. **Newton's method with $\theta_k = 1$ converges quadratically.**

Example 6.9.1 Elsner and He 1991.   Let

$$A = \begin{pmatrix} 1 & 1 & 1 \\ 0.1 & 3 & 5 \\ 0 & -1 & -1 \end{pmatrix}, \qquad B = \begin{pmatrix} 1 \\ 0.1 \\ 0 \end{pmatrix}.$$

$$\lambda_{\max}\left(\frac{A + A^{\mathrm{T}}}{2}\right) = 3.9925, \qquad \lambda_{\min}\left(\frac{A + A^{\mathrm{T}}}{2}\right) = -1.85133,$$

$$\lambda_{\max}\left(\frac{A - A^{\mathrm{T}}}{2j}\right) = 3.0745, \qquad \lambda_{\min}\left(\frac{A - A^{\mathrm{T}}}{2j}\right) = -3.0745.$$

Thus all zero points lie in the rectangular region given by $-1.8513 \leq x \leq 3.9925$, $-3.0745 \leq y \leq 3.0745$. Choose $x_0 = 1.5$ and $y_0 = 1$.

Then, $s_0 = 1.5 + j$.

$\theta_0 = 0.09935$, $\theta_1 = 0.5641$, $\theta_2 = 1.0012$. Starting from here, θ_k was set to 1. The method converged in five steps. $s_C = \begin{pmatrix} x_5 \\ y_5 \end{pmatrix} = \begin{pmatrix} 0.93708 \\ 0.998571 \end{pmatrix} = 0.93708 + 0.998571j$.

The minimum singular value of $(A - s_C I, B) = 0.0392$.

Thus, $\mu(A, B) = 0.039238$.

MATLAB note: MATLAB codes for Algorithm 6.9.1 are available from the authors of the paper.

The Bisection Method (Real Case)

In the real case, the following bisection method can also be used to compute the zeros of $f(s)$.

Step 1. Find an interval $[a, b]$ such that $f(a)f(b) < 0$.

Step 2.

    2.1. Compute $c = (a + b)/2$.

2.2. If $f(c)f(b) < 0$, then set $a = c$ and return to Step 2.1.
If $f(a)f(c) < 0$, then set $b = c$ and return to Step 2.1.
Step 3. Repeat Step 2 until c is an acceptable zero point of $f(s)$.
  Note: For Example 6.9.1, $f(s)$ has only one real zero $s = 1.027337$, and $\mu_R(A, B) = 0.1725$.

6.9.2 The Wicks–DeCarlo Method for Distance to Uncontrollability

Newton's algorithm, described in Section 6.9.1 is based on minimization of $\sigma_n(sI - A, B)$ over all complex numbers s. It requires an SVD computation at each iteration.

In this section, we state an algorithm due to Wicks and DeCarlo (1991). The algorithm is also iterative in nature but "requires only two QR factorizations at each iteration without the need for searching or using a general minimization algorithm."

The algorithm is based on the following observation:

$$\mu(A, B) = \min_{u \in \mathbb{C}^n} ||(u^*A(I - uu^*)u^*B||, \tag{6.9.2}$$

subject to $u^*u = 1$. Based on this observation, they developed three algorithms for computing $\mu_R(A, B)$ and $\mu(A, B)$.

We state here why one of the algorithms (Algorithm II in Wicks and DeCarlo (1991)) is used for computing $\mu(A, B)$.

Definition 6.9.1. *Let the distance measure $d_1(A, B)$ be defined by*

$$[d_1(A, B)]^2 = ||[e_n^*(A(I - e_n e_n^*) B)]||_2^2$$

$$= \sum_{j=1}^{n-1} |a_{nj}|^2 + \sum_{j=1}^{m} |b_{nj}|^2.$$

Using the above notation, it has been shown in Wicks and DeCarlo (1991) that

$$\mu(A, B) = \min_{\substack{U \in \mathbb{C}^{n \times n} \\ U^*U = I}} d_1(U^*AU, U^*B).$$

The algorithm proposed by them constructs a sequence of unitary matrices $U_1, U_2, \ldots$, such that

1. $A_{k+1} = U_k^* A_k U_k$, $B_{k+1} = U_k^* B$
2. $d_1(A_{k+1}, B_{k+1}) < d_1(A_k, B_k)$
3. $\lim_{k \to \infty} d_1(A_k, B_k)$ is a local minimum of (6.9.2).

Algorithm 6.9.2. *An Algorithm for Computing $\mu(A, B)$*
Inputs. *The matrices A and B.*
Output. $\mu(A, B)$.
Step 0. *Set $A_1 \equiv A$, $B_1 \equiv B$.*
Step 1. *For $k = 1, 2, \ldots$ until convergence.*
   1.1. *Form $M_k = (A_k - (a_{nn})_k I \quad B_k)$.*
   1.2. *Factor $M_k = L_k V_k$, where L_k is lower triangular and V_k is unitary.*
   1.3. *Find the QR factorization of $L_k = U_k^* R_k$.*
   1.4. *Set $A_{k+1} = U_k^* A_k U_k$, $B_{k+1} = U_k^* B_k$.*
   1.5. *If $d_1(A_{k+1}, B_{k+1}) = d_1(A_k, B_k)$, stop.*
End.

Proof. The proof amounts to showing that $d_1(A_k, B_k) \geq |r_{nn}|$, where

$$
R_k = \begin{pmatrix}
r_{11} & \cdots & \cdots & r_{1n} \\
0 & r_{22} & & r_{2n} & 0 \\
\vdots & & \ddots & \vdots \\
0 & & & r_{nn}
\end{pmatrix}
$$

and as such $|r_{nn}| = d_1(U_k^* A_k U_k, U_k^* B_k)$. ∎

For details, the readers are referred to Wicks and DeCarlo (1991).

Example 6.9.2.

$$
A = \begin{pmatrix}
0.950 & 0.891 & 0.821 & 0.922 \\
0.231 & 0.762 & 0.445 & 0.738 \\
0.607 & 0.456 & 0.615 & 0.176 \\
0.486 & 0.019 & 0.792 & 0.406
\end{pmatrix},
$$

$$
B = \begin{pmatrix}
0.9350 & 0.0580 & 0.1390 \\
0.9170 & 0.3530 & 0.2030 \\
0.4100 & 0.8130 & 0.1990 \\
0.8940 & 0.0100 & 0.6040
\end{pmatrix}.
$$

Let the tolerance for stopping the iteration be: Tol $= 0.00001$.
   Define $\mu_k = d_1(A_k, B_k)$.

The algorithm produces the following converging sequence of μ_k:

k	μ_k	k	μ_k
0	1.42406916966838	10	0.41450782001833
1	0.80536738314449	11	0.41450781529413
2	0.74734006994998	12	0.41450781480559
3	0.52693889988172	13	0.41450781475487
4	0.42241562062172	14	0.41450781474959
5	0.41511102322896	15	0.41450781474904
6	0.41456112538077	16	0.41450781474899
7	0.41451290008455	17	0.41450781474898
8	0.41450831981698	18	0.41450781474898
9	0.41450786602577	19	0.41450781474898

After 19 iterations the algorithm returns $\mu = 0.41450781474898$.

MATCONTROL note: Algorithm 6.9.2 has been implemented in MATCONTROL function **discntrl**.

6.9.3   A Global Minimum Search Algorithm

The algorithms by Elsner and He (1991) and Wicks and DeCarlo (1991) are guaranteed only to converge to a local minimum rather than a global minimum. A global minimum search algorithm was given by Gao and Neumann (1993). Their algorithm is based on the observation that if $\text{rank}(B) < n$, then the minimization problem can be transformed to a minimization problem in the bounded region $\{(x, z) | x \leq \|A\|_2, |z| \leq \|A\|_2\}$ in the two-dimensional real plane.

The algorithm then progressively partitions this region into simplexes and finds lower and upper bounds for $\mu(A, B)$ by determining if the vertices (x_k, z_k) satisfy

$$z_k > \min_{y \in \mathbb{R}} \sigma_{\min}(A - (x_k + jy)I, B).$$

These bounds are close to each other if $\mu(A, B)$ is small. "If $\mu(A, B)$ is not small, then the algorithm produces a lower bound which is not small, thus leading us to a safe conclusion that (A, B) is not controllable."

For details of the algorithm, see Gao and Neumann (1993). See also **Exercise 6.26**.

6.10   DISTANCE TO UNCONTROLLABILITY AND THE SINGULAR VALUES OF THE CONTROLLABILITY MATRIX

Since the rank of the controllability matrix C_M determines whether a system is controllable or not, and the most numerically effective way to determine the rank of a matrix is via the singular values of the matrix, it is natural to wonder, what roles do the singular values of the controllability matrix play in deciding if a given controllable system is near an uncontrollable system. (Note that Theorem 6.9.1 and the associated algorithm for computing $\mu(A, B)$ use the singular values of $(A - sI, B)$).

The following result due to Boley and Lu (1986) sheds some light in that direction. We state the result without proof. Proof can be found in Boley and Lu (1986).

Theorem 6.10.1. *Let (A, B) be a controllable pair. Then,*

$$\mu(A, B) \leq \mu_R(A, B) \leq \left(1 + \frac{\|C_p\|}{\sigma_{n-1}}\right) \sigma_n,$$

where $\sigma_1 \geq \sigma_2 \geq \cdots \geq \sigma_{n-1} \geq \sigma_n$ are the singular values of the controllability matrix $C_M = (B, AB, \ldots, A^{n-1}B)$ and C_p is a companion matrix for A.

Example 6.10.1. We consider Example 6.9.1 again.
The singular values of the controllability matrix are 2.2221, 0.3971, 0.0227.
The companion matrix C_p is calculated as follows:

$$x_1 = (1, 0, 0)^T, \qquad x_2 = Ax_1 = (1, 0.1, 0)^T, \qquad x_3 = Ax_2 = (1.1, 0.4, -0.1)^T.$$

Then the matrix $X = (x_1, x_2, x_3)$ is such that

$$X^{-1}AX = C_p = \begin{pmatrix} 0 & 0 & 2 \\ 1 & 0 & -3.9 \\ 0 & 1 & 3 \end{pmatrix}.$$

According to Theorem 6.10.1, we then have

$$\mu(A, B) \leq \mu_R(A, B) \leq \left(1 + \frac{5.3919}{0.3971}\right) \times 0.02227 = 0.3309.$$

Remark

- The above theorem can be used to predict the order of perturbations needed to transform a controllable system to an uncontrollable system. *It is the largest gap between the consecutive singular values.* (**However, note that, in general, the singular values of the controllability matrix cannot be used directly to make a prediction of how close the system is to an uncontrollable system.**)

In other words, it is **not** true that one can obtain a nearly uncontrollable system by applying perturbations ΔA, ΔB, with norm bounded by the smallest nonzero singular value of the controllability matrix.

6.11   SOME SELECTED SOFTWARE

6.11.1   MATLAB Control System Toolbox

canon—State-space canonical forms.
ctrb, obsv—Controllability and observability matrices.
gram—Controllability and observability gramians.
ctrbf—Controllability staircase form.
obsvf—observability staircase form.

6.11.2   MATCONTROL

CNTRLHS—Finding the controller-Hessenberg form.
CNTRLHST—Finding the Controller-Hessenberg form with triangular
   subdiagonal blocks.
OBSERHS—Finding the observer-Hessenberg form.
CNTRLC—Find the controller-canonical form (Lower Companion).
DISCNTRL—Distance to controllability using the Wicks–DeCarlo algorithm.

6.11.3   CSP-ANM

- **Reduction to controller-Hessenberg and observer-Hessenberg forms**

   - Block controller-Hessenberg forms are computed by `controller-HessenbergForm` [*system*] and `LowercontrollerHessenbergForm` [*system*].
   - Block observer-Hessenberg forms are computed by `Observer-HessenbergForm` [*system*] and `UpperObserverHessenbergForm` [*system*].

- **Controllability and observability tests**

   - Tests of controllability and observability using block controller-Hessenberg and block observer-Hessenberg forms are performed via `Controllable` [*system*, `ControllabilityTest` → `FullRankcontrollerHessenbergBlocks`] and `Observable` [*system*, `ObservabilityTest` → `FullRankObserverHessenbergBlocks`].
   - Tests of controllability and observability of a stable system via positive definiteness of Gramians are performed via `Controllable` [*system*, `ControllabilityTest` → `PositiveDiagonalCholeskyFactorControllabilityGramian`] and `Observable` [*system*, `ObservabilityTest` → `PositiveDiagonalCholeskyFactorObservabilityGramian`].

6.11.4   SLICOT

Canonical and quasi canonical forms:

AB01MD—Orthogonal controllability form for single-input system

AB01ND—Orthogonal controllability staircase form for multi-input system

AB01OD—Staircase form for multi-input system using orthogonal transformations

TB01MD—Upper/lower controller-Hessenberg form

TB01ND—Upper/lower observer-Hessenberg form

TB01PD—Minimal, controllable or observable block Hessenberg realization

TB01UD—Controllable block Hessenberg realization for a state-space representation

TB01ZD—Controllable realization for single-input systems

6.11.5   MATRIX$_X$

Purpose: Obtain controllable part of a dynamic system.
Syntax: [SC, NSC, T]= CNTRLABLE (S, NS, TOL)

Purpose: Compute observable part of a system.
Syntax: [SOBS, NSOBS, T]= OBSERVABLE (S, NS, TOL)

Purpose: Staircase form of a system matrix.
Syntax: [SST, T, NCO]= STAIR (S, NS, TOL)

6.12   SUMMARY AND REVIEW

Algebraic Criteria of Controllability and Observability

Controllability and observability are two most fundamental concepts in control theory. The algebraic criteria of controllability and observability are summarized in **Theorems 6.2.1** and **6.3.1**, respectively.

Unfortunately, these algebraic criteria very often do not yield numerically viable tests for controllability and observability. The numerical difficulties with these criteria as practical tests of controllability are discussed and illustrated in Section 6.6. The pair (A, B) in Example 6.6.1 is a controllable pair; however, it is shown that the Example 6.6.1 is a controllable pair; however, it is shown that criterion (ii) of Theorem 6.2.1 leads to an erroneous conclusion due to a computationally small singular value of the controllability matrix. Similarly, in Example 6.6.2, it is shown how an obviously uncontrollable pair can be taken as a controllable pair by using the eigenvalue criterion of controllability (Criterion (v) of Theorem 6.2.1) as a numerical test, due to the ill-conditioning of the eigenvalues of the matrix A.

Numerically Effective Tests of Controllability and Observability

Computationally viable tests of controllability and observability are given in **Sections 6.7** and **6.8**. These tests are based on the reductions of the pairs (A, B) and (A, C), respectively, to **controller-Hessenberg** and **observer-Hessenberg** forms. These forms can be obtained by using orthogonal transformations and the tests can be shown to be numerically stable.

Indeed, when controller-Hessenberg test is applied to Example 6.6.2, it was concluded correctly that, in spite of the ill-conditioning of the eigenvalues of A, the pair (A, B) is uncontrollable.

Distance to Uncontrollability

Since determining the rank of a matrix is numerically a delicate problem and the problem is sensitive to small perturbations, **in practice, it is more important to find when a controllable system is close to an uncontrollable system.** To this end, a practical measure of the distance to uncontrollability, denoted by $\mu(A, B)$, is introduced in Section 6.9: $\mu(A, B) = \min\{\|\Delta A, \Delta B\|_2$ such that the pair $(A + \Delta A, B + \Delta B)$ is controllable.

A well-known characterization of $\mu(A, B)$ is given in **Theorem 6.9.1.** This theorem states: $\mu(A, B) = \min \sigma_n(sI - A, B)$, where $\sigma_n(sI - A, B)$ is the smallest singular value of the matrix $(sI - A, B)$ and s runs over all complex numbers.

Two algorithms (**Algorithms 6.9.1** and **6.9.2**), have been described to compute $\mu(A, B)$.

6.13   CHAPTER NOTES AND FURTHER READING

Controllability and observability are two most basic concepts in control theory. The results related to controllability and observability can be found in any standard book on linear systems (e.g., Kalman *et al.* 1969; Brockett 1970; Rosenbrock 1970; Luenberger 1979; Kailath 1980; Chen 1984; DeCarlo 1989; Brogan 1991; etc.).

For details on the staircase algorithms for finding the controller-Hessenberg and observer-Hessenberg forms, see Boley (1981), Paige (1981), Van Dooren and Verhaegen (1985), etc. For computation of the Kalman decomposition, see Boley (1980, 1991), etc. For more on the concept of the distance to uncontrollability and related algorithms, see Boley (1987), Boley and Lu (1986), Eising (1984), Wicks and DeCarlo (1991), Elsner and He (1991), Paige (1981), Miminis (1981), Kenney and Laub (1988), and Gao and Neumann (1993). For a test of controllability via real Schur form, see Varga (1979).

Exercises

6.1   Prove that (A, B) is controllable if and only if for a constant matrix F, the matrix $(A + BF, B)$ is controllable, that is, the controllability of a system does not change under state feedback. **(The concept of state feedback is defined in Chapter 10.)**

6.2 Construct an example to show that the observability of a system may change under state feedback.

6.3 A matrix A is called a **cyclic matrix** if in the JCF of A, there is only one Jordan box associated with each distinct eigenvalue.

Let A be a cyclic matrix and let the pair (A, B) be controllable. Then prove that for almost all vectors v, the pair (A, Bv) is controllable.

6.4 Give a 2×2 example to show that the cyclicity assumption is essential for the result of Problem 6.3 to hold.

6.5 Show that (A, c) is observable if and only if there exists a vector k such that

$$\left(\begin{pmatrix} c \\ k \end{pmatrix}, \ A - bk \right)$$

is observable.

6.6 Prove the parts (i), (ii), (iv)–(vi) of Theorem 6.3.1.

6.7 Prove Theorem 6.4.2.

6.8 Using Theorems 6.4.1 and 6.4.2, give a proof of Theorem 6.4.3 (**The Kalman Canonical Decomposition Theorem**).

6.9 Prove that the change of variable $\bar{x} = Tx$, where T is nonsingular, preserves the controllability and observability of the system (A, B, C).

6.10 Work out an algorithm to compute the nonsingular transforming matrix that transforms the pair (A, b) to the upper companion form. When can the matrix transforming T be highly ill-conditioned? Construct a numerical example to support your statement.

6.11 Apply the test based on the eigenvalue criterion of controllability to Example 6.6.2 and show that this test will do better than the one based on the criterion (ii) of Theorem 6.2.1.

6.12 Applying the staircase algorithm to the pair (A, b) in Example 6.6.2, show that the pair (A, b) is uncontrollable.

6.13 If the controller-Hessenberg pair $(H, \bar{B})$ of the controllable system (A, B) is such that the subdiagonal blocks of H are nearly rank-deficient, then the system may be very near to an uncontrollable system.

(a) Construct examples both in the single- and multi-input cases in support of the above statement.

(b) Construct another example to show that the converse is not necessarily true, that is, even if the subdiagonal blocks of H have robust ranks, the system may be close to an uncontrollable system.

6.14 Show that to check the controllability for the pair (A, B), where

$$A = \text{diag}(1, 2^{-1}, \ldots, 2^{1-n}) \quad \text{and} \quad B = \begin{pmatrix} 1 \\ 1 \\ \vdots \\ 1 \end{pmatrix},$$

the eigenvalue-criterion for controllability (the PBH criterion) will do better than the criterion (ii) of Theorem 6.2.1.

6.15 Let $(H = (H_{ij}), \bar{B})$ be the controller-Hessenberg form of the pair (A, B). Then prove the following:
  (a) If $H_{i,i-1} = 0$ for any i, then (A, B) is not controllable.
  (b) $\mu(A, B) \leq \min\limits_{1 \leq i \leq k} \|H_{i,i-1}\|_2$.

6.16 Prove that $\mu(A, B)$ remains invariant under an orthogonal transformation.

6.17 Let $(\tilde{A}, \tilde{b})$ be as in (6.5.2). Prove that $\mu(\tilde{A}, \tilde{b}) \leq \sin(\pi/n)$ (Kenney and Laub 1988).

6.18 Develop an algorithm for the reduction of the pair (A, C) to the observer-Hessenberg form (6.8.1), without invoking the algorithm for the controller-Hessenberg reduction to the pair (A^T, C^T). How can one obtain the observability indices of the pair (A, C) from this form?

6.19 Construct a simple example to show that the minimum which yields $\mu(A, B)$ is not achieved when s is an eigenvalue of A.

6.20 Rework Example 6.9.1 with the initial point as one of the eigenvalues of $F = \left(\begin{smallmatrix} A & B \\ C & D \end{smallmatrix} \right)$, where (C, D) is a random matrix such that F is square.

6.21 Apply the bisection method to Example 6.9.1 to find an estimate of $\mu_R(A, B)$.

6.22 Find $\mu(A, B)$ and $\mu_R(A, B)$, where A and B are given by:

$$ A = \begin{pmatrix} 0 & 1 & & & \\ & 0 & \ddots & & \\ & & \ddots & \ddots & \\ & & & 0 & 1 \\ & & & & 0 \end{pmatrix}_{10 \times 10} \quad \text{and} \quad B = \begin{pmatrix} 0 \\ 0 \\ 0 \\ \vdots \\ 0 \\ 1 \end{pmatrix}_{10 \times 1}. $$

6.23 Derive Newton's algorithm for computing $\mu_R(A, B)$.

6.24 (Laub and Linnemann 1986). Consider the controllable pair

$$ (H, \bar{b}) = \left(\begin{pmatrix} -4 & 0 & 0 & 0 \\ \alpha & -3 & 0 & 0 \\ 0 & \alpha & -2 & 0 \\ 0 & 0 & \alpha & -1 \end{pmatrix}, \begin{pmatrix} 1 \\ 0 \\ 0 \\ 0 \end{pmatrix} \right), $$

with $0 < \alpha \leq 1$. Show (experimentally or mathematically) that the pair $(H, \bar{b})$ is close to an uncontrollable pair.

6.25 Let (A, B) be controllable. Let $B = (B_1, B_2)$, with B_1 consisting of minimum number of inputs such that (A, B) is controllable. Then prove that (A, B_1) is closer to an uncontrollable system than (A, B) is; that is, prove that

$$ \min \sigma_n(B_1, A - sI) \leq \min \sigma_n(B, A - sI), \quad s \in \mathbb{C} $$

6.26 (Gao and Neumann 1993). Let $\lambda_0 \in \mathbb{C}$ and let $p \in \mathbb{C}$ be on the unit circle. Consider the straight line $\lambda = \lambda_0 + tp$, $t \in \mathbb{R}$. Then prove that

$$ \min\limits_{t \in \mathbb{R}} \sigma_{\min}(A - (\lambda_0 + tp)I, B) \leq \alpha $$

if and only if the matrix

$$ G(\alpha) = \begin{pmatrix} \bar{p}(A - \lambda_0 I) & \bar{p}(BB^* - \alpha^2 I) \\ -pI & p(A^* - \bar{\lambda}_0 I) \end{pmatrix} $$

has a real eigenvalue.

Based on the above result, derive an algorithm for computing $\mu_R(A, B)$.
(Hint: take $\lambda_0 = 0$, $p = 1$.)
Test your algorithm with Example 6.9.1.

6.27 (a) Construct a state-space representation of the following second-order model.

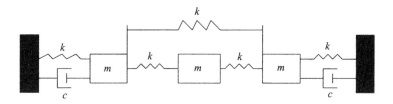

$$
\begin{bmatrix} m & & \\ & m & \\ & & m \end{bmatrix} \begin{pmatrix} \ddot{u}_1 \\ \ddot{u}_2 \\ \ddot{u}_3 \end{pmatrix} + \begin{bmatrix} c & & \\ & 0 & \\ & & c \end{bmatrix} \begin{pmatrix} \dot{u}_1 \\ \dot{u}_2 \\ \dot{u}_3 \end{pmatrix}
$$
$$
+ \begin{bmatrix} 3k & -k & -k \\ -k & 2k & -k \\ -k & -k & 3k \end{bmatrix} \begin{pmatrix} u_1 \\ u_2 \\ u_3 \end{pmatrix} = \begin{pmatrix} 0 \\ 0 \\ 0 \end{pmatrix}
$$

(b) Show that the system is not controllable for

$$
b = \begin{pmatrix} 0 \\ 1 \\ 0 \end{pmatrix}, \quad b = \begin{pmatrix} \alpha \\ \beta \\ \alpha \end{pmatrix}, \quad \text{or} \quad b = \begin{pmatrix} \beta \\ 0 \\ -\beta \end{pmatrix}.
$$

6.28 Consider Example 5.2.6 on the motion of an orbiting satellite with $d_0 = 1$. Let $x(t) = (x_1(t), x_2(t), x_3(t), x_4(t))^T$, $u(t) = (u_1(t), u_2(t))^T$, and $y(t) = (y_1(t), y_2(t))^T$.
   (a) Show that one of the states cannot be controlled by the radial force $u_1(t)$ alone, but all the states can be controlled using the tangential force $u_2(t)$.
   (b) Show that all the states are observable using both the outputs; however, one of the states cannot be observed by $y_1(t)$ alone.

6.29 (Boley 1985). Let $(H, \bar{b})$ be the controller-Hessenberg pair of the controllable pair (A, b) such that $\|H\|_2 + \|\bar{b}\|_2 \le \frac{1}{4}$. Then prove that the quantity $|\bar{b}_1 h_{21} h_{32} \cdots h_{n,n-1}|$ gives a lower bound on the perturbations needed to obtain an uncontrollable pair. Construct an example to support this.

6.30 Does the result of the preceding exercise hold in the multi-input case? Prove or disprove.

6.31 Consider the example of balancing a stick on your hand (Example 5.2.4). We know from our experience that a stick can be balanced. Verify this using a criterion of controllability. Take $L = 1$.

$$
A = \begin{pmatrix} 0 & 1 \\ \dfrac{g}{L} & 0 \end{pmatrix}, \quad b = \begin{pmatrix} 0 \\ -\dfrac{g}{L} \end{pmatrix}.
$$

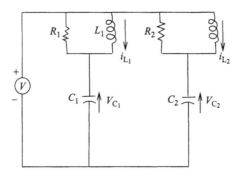

FIGURE 6.1:   Uncontrollability of an electrical network.

6.32   (An Uncontrollable System) (Szidarovszky and Bahill (1991, pp. 223–224)). Consider the electric network in Figure 6.1 with two identical circuits in parallel. Intuitively, it is clear that there cannot exist a single input that will bring one circuit to one state and the other to a different state. Verify this using a criterion of controllability. Take $L_1 = L_2 = 1$, $C_1 = C_2 = 1$, and $R_1 = R_2 = 1$.

References

Boley D. "On Kalman's procedure for the computation of the controllable/observable canonical form," *SIAM J. Control Optimiz.*, Vol. 18, pp. 624–626, 1980.

Boley D. *Computing the Controllability/Observability of a Linear Time-Invariant Dynamic System: A Numerical Approach*, Ph.D. thesis, Computer Science Department, Stanford University, Stanford, CA, June 1981.

Boley D. and Lu W.-S. "Measuring how far a controllable system is from an uncontrollable system," *IEEE Trans. Autom. Control*, Vol. AC-31, pp. 249–251, 1986.

Boley D. A. "Perturbation result for linear control problems," *SIAM J. Alg. Discr. Methods*, Vol. 6, pp. 66–72, 1985.

Boley D. "Computing the Kalman decomposition, an optimal method," *IEEE Trans. Autom. Control*, Vol. AC-29, pp. 51–53, 1984; correction in Vol. 36, p. 1341, 1991.

Brockett R. *Finite Dimensional Linear Systems*, Wiley, New York, 1970.

Brogan W.L. *Modern Control Theory*, 3rd edn, Prentice Hall, Englewood Cliffs, NJ, 1991.

Chen C.-T. *Linear Systems Theory and Design*, CBS College Publishing, New York, 1984.

Datta B.N. *Numerical Linear Algebra and Applications*, Brooks/Cole Publishing Company, Pacific Grove, CA, 1995.

DeCarlo R.A. *Linear Systems: A State Variable Approach with Numerical Implementation*, Prentice Hall, Englewood Cliffs, NJ, 1989.

Eising R. "Between controllable and uncontrollable," *Syst. Control Lett.*, Vol. 4, pp. 263–264, 1984.

Elsner L. and He C. "An algorithm for computing the distance to uncontrollability," *Syst. Control Lett.*, Vol. 17, pp. 453–464, 1991.

Gao M. and Neumann M. "A global minimum search algorithm for estimating the distance to uncontrollability," *Lin. Alg. Appl.*, Vol. 188/189, pp. 305–350, 1993.

Golub G.H. and Van Loan C.F. *Matrix Computations*, 3rd edn, Johns Hopkins University, Baltimore, MD, 1996.

Hautus M.L.J. "Controllability and observability conditions of linear autonomous systems," *Proc. Kon. Ned. Akad. Wetensh*, Ser. A. Vol. 72, pp. 443–448, 1969.

Horn R.A. and Johnson C. *Matrix Analysis*, Cambridge University Press, Cambridge, UK, 1985.

Kailath T. *Linear Systems*, Prentice Hall, Englewood Cliffs, NJ, 1980.

Kalman R.E., Flab P.L., and Arbib M.A. *Topics in Mathematical Systems Theory*, McGraw Hill, New York, 1969.

Kenney C.S. and Laub A.J. "Controllability and stability radii for companion form systems," *Math. Control, Signals Syst.*, Vol. 1, pp. 239–256, 1988.

Laub A.J. and Linnemann A. "Hessenberg and Hessenberg/triangular forms in linear systems theory," *Int. J. Control*, Vol. 44, pp. 1523–1547, 1986b.

Luenberger D.G. *Introduction to Dynamic Systems: Theory, Models & Applications*, John Wiley & Sons, New York, 1979.

Miminis G.S. *Numerical Algorithms for Controllability and Eigenvalue Allocations*, Master Thesis, School of Computer Science, McGill University, Montreal, Canada, May, 1981.

Paige C.C. "Properties of numerical algorithms related to computing controllability," *IEEE Trans. Autom. Control*, Vol. AC-26, pp. 130–138, 1981.

Patel R. "Computation of Minimal order state-space realization and observability indices using orthogonal transformation," *Int. J. Control.*, Vol. 33, pp. 227–246, 1981.

Rosenbrock M.M. *State-Space and Multivariable Theory*, John Wiley, New York, 1970.

Szidarovszky F. and Terry Bahill A. *Linear Systems Theory*, CRC Press, Boca Raton, 1991.

Van Dooren P.M. and Verhaegen M. "On the use of unitary state-space transformations, Contemporary Mathematics," in *Contemporary Mathematics Series* (Brualdi R. *et al.*, eds.), Vol. 47, pp. 447–463, Amer. Math. Soc., Providence, RI, 1985.

Varga A. "Numerically reliable algorithm to test controllability," *Electronic Lett.*, Vol. 15, pp. 452–453, 1979.

Wicks M. and DeCarlo A. "Computing the distance to an uncontrollable system," *IEEE Trans. Autom. Control*, Vol. 36, pp. 39–49, 1991.

CHAPTER 7

STABILITY, INERTIA, AND ROBUST STABILITY

> **Topics covered**
>
> - Lyapunov Stability Theory for Continuous-Time and Discrete-Time Systems
> - Controllability and Observability Grammians via Lyapunov Equations
> - Theory and Computation of the Inertia
> - Distance to Instability
> - Robust Stability
> - Stability Radius

7.1 INTRODUCTION

This chapter deals with **stability of a linear time-invariant system** and the associated aspects such as the **inertia of a matrix, distance from an unstable system, robust stability, and stability radius** and computing the H_2**-norm of a stable transfer function**. A classical approach to determine the stability and inertia is to solve a **Lyapunov equation** or to find the characteristic polynomial of the state matrix A followed by application of the **Routh–Hurwitz criterion** in the continuous-time case and the **Schur–Cohn** criteria in the discrete-time case. These approaches are historically important and were developed at a time when numerically finding the eigenvalues of a matrix, even of a modest order, was a difficult problem. However, nowadays, with the availability of the QR iteration method for eigenvalue computation which is reliable, accurate, and fast, these approaches for stability and inertia, seem to have very little practical value. Furthermore, the Lyapunov equation approach is counterproductive in a practical computational setting in the sense that the most numerically viable method, currently available for solution of the Lyapunov equation, namely, the **Schur method (described in Chapter 8)**, is based on transformation of the matrix A to a real Schur form (RSF) and the latter either explicitly displays the eigenvalues of A or the eigenvalues can be trivially found once A is transformed into this form. Also, as mentioned before, **finding the characteristic polynomial of a matrix, in general, is a numerically unstable process.** In view of the above statements, it is clear that the best way to numerically check the stability and inertia is to explicitly compute all the eigenvalues. However, by computing the eigenvalues, one gets more than stability and inertia. Furthermore, if the eigenvalues of A are very ill-conditioned, determining the stability and inertia using eigenvalues may be misleading (**see Section 7.6**). The question, therefore, arises if an approach can be developed that does not require explicit computation of the eigenvalues of the state matrix A nor solution of a Lyapunov equation. Such an implicit method (**Algorithm 7.5.1**) is developed in **Section 7.5**. **This method is about three times faster than the eigenvalue method** and, **of course, many times faster than solving Lyapunov equation in a numerically effective way using the Schur method.**

An important practical problem "**How nearly unstable is a stable system** (or equivalently a stable matrix)?" is discussed in **Section 7.6**. A simple **bisection algorithm** (**Algorithm 7.6.1**) due to Byers (1988) to measure the distance of a stable matrix A from a set of unstable matrices is provided. A brief discussion of **robust stability** is the topic of **Section 7.7**.

The concept of **stability radius** in the context of robust stability is introduced in **Section 7.8** and a recent important formula for real stability radius due to Qiu *et al.* (1995) is stated. This concept will again be revisited in **Chapter 10**, where a connection of the complex stability radius with an algebraic Riccati equation (ARE) will be made.

The relationships between the controllability and observability Grammians and the H_2-norm of an asymptotically stable system with Lyapunov equations are discussed in **Sections 7.2.3, 7.2.4, and 7.3**, and a computational algorithm (**Algorithm 7.2.1**) for computing the H_2-norm of a stable continuous-time system is described in **Section 7.2.4**.

Reader's Guide to Chapter 7

Readers familiar with the basic concepts of stability and Lyapunov stability theory can skip Sections 7.2 and 7.3.

7.2   STABILITY OF A CONTINUOUS-TIME SYSTEM

The stability of a system is defined with respect to an equilibrium state.

Definition 7.2.1.   *An **equilibrium state** of the unforced system*

$$\dot{x}(t) = Ax(t), \qquad x(0) = x_0, \qquad (7.2.1)$$

is the vector x_e satisfying

$$Ax_e = 0.$$

Clearly, $x_e = 0$ is an equilibrium state and it is the unique equilibrium state if and only if A is nonsingular.

Definition 7.2.2. *An equilibrium state x_e is **asymptotically stable** if for any initial state, the state vector $x(t)$ approaches x_e as time increases.*

The system (7.2.1) is asymptotically stable if and only if the equilibrium state $x_e = 0$ is asymptotically stable. Thus, the system (7.2.1) is asymptotically stable if and only if $x(t) \to 0$ as $t \to \infty$.

7.2.1 Eigenvalue Criterion of Continuous-Time Stability

Below we state a well-known criterion of asymptotic stability of a continuous-time system.

Theorem 7.2.1. *The system (7.2.1) is asymptotically stable if and only if all the eigenvalues of the matrix A have negative real parts.*

Proof. From Chapter 5, we know that the general solution of (7.2.1) is

$$x(t) = e^{At}x_0.$$

Thus, $x(t) \to 0$ if and only if $e^{At} \to 0$ as $t \to \infty$. We will now show that this happens if and only if all the eigenvalues of A have negative real parts.

Let $X^{-1}AX = \mathrm{diag}(J_1, J_2, \ldots, J_k)$ be the Jordan canonical form (JCF) of A. Then,

$$e^{At} = X \, \mathrm{diag}(e^{J_1 t}, e^{J_2 t}, \ldots, e^{J_k t})X^{-1}.$$

Let λ_i be the eigenvalue of A associated with J_i. Then $e^{J_i t} \to 0$ if and only if λ_i has a negative real part. Therefore, $e^{At} \to 0$ if and only if all the eigenvalues of A have negative real parts. ∎

Definition 7.2.3. *A matrix A is called a **stable matrix** if all of the eigenvalues of A have negative real parts.*

A stable matrix is also known as a **Hurwitz** matrix in control literature. In analogy, an eigenvalue with negative real part is called a **stable eigenvalue**.

Since the asymptotic stability of (7.2.1) implies that its zero-input response approaches zero exponentially, the asymptotic stability is also referred to as **exponential stability**.

Definition 7.2.4. *Let $\lambda_1, \ldots, \lambda_n$ be the eigenvalues of A. Then the distance $\min\{-Re(\lambda_i): i = 1, \ldots, n\}$ to the imaginary axis is called the **stability margin** of A.*

> In this book, the "stability" of a system means "**asymptotic stability**," and the associated matrix A will be referred to as "**stable matrix**," **not** asymptotically stable matrix.

Bounded-Input Bounded-Output Stability

The continuous-time linear system:

$$\dot{x}(t) = Ax(t) + Bu(t),$$
$$y(t) = Cx(t) \tag{7.2.2}$$

is said to be **bounded-input bounded-output** (BIBO) **stable** if for any bounded input, the output is also bounded.

The BIBO stability is governed by the poles of the transfer function $G(s) = C(sI - A)^{-1}B$. Specifically, the following result can be proved: (**Exercise 7.5**).

Theorem 7.2.2.   *The system (7.2.2) is BIBO stable if and only if every pole of $G(s)$ has a negative real part.*
Since every pole of $G(s)$ is also an eigenvalue of A, **an asymptotically stable system is also BIBO stable.** *However,* **the converse is not true.** *The following simple example illustrates this.*

Example 7.2.1.

$$\dot{x} = \begin{pmatrix} 1 & 0 \\ 0 & -1 \end{pmatrix} x + \begin{pmatrix} 0 \\ 1 \end{pmatrix} u, \qquad y = (1, 1)x.$$

$$G(s) = C(sI - A)^{-1}B = (1, 1) \begin{pmatrix} s-1 & 0 \\ 0 & s+1 \end{pmatrix}^{-1} \begin{pmatrix} 0 \\ 1 \end{pmatrix} = \frac{1}{s+1}.$$

Thus, the system is BIBO (note that the pole of $G(s)$ is -1), but not asymptotically stable.

Bounded-Input Bounded-State (BIBS) Stability

Definition 7.2.5.   *The system (7.2.2) is **BIBS** stable if, for any bounded input, the state response is also bounded.*

The following characterization of BIBS can be given in terms of eigenvalues of A and the controllability of the modes. For a proof of Theorem 7.2.2$'$, see DeCarlo (1989, pp. 416–417).

Theorem 7.2.2$'$.   *BIBS. The system (7.2.2) is BIBS stable if and only if*

(i)   *All the eigenvalues of A have nonnegative real parts.*

(ii)   *If an eigenvalue λ_i has a zero real part, then the order of the associated factor in the minimal polynomial of A must be 1.*

(iii) *The mode associated with an eigenvalue with zero real part must be uncontrollable.*

7.2.2   Continuous-Time Lyapunov Stability Theory

In this section, we present the historical Lyapunov criterion of stability. Before the advent of computers, finding the eigenvalues of a matrix A was an extremely difficult task. The early research on stability, therefore, was directed toward finding the criteria that do not require explicit computation of the eigenvalues of a matrix. In 1892, the Russian mathematician A. Lyapunov (1857–1918) developed a historical stability criterion for nonlinear systems of equations. In the linear case, this criterion may be formulated in terms of the solution of a matrix equation.

Theorem 7.2.3.   *Lyapunov Stability Theorem. The system:*

$$\dot{x}(t) = Ax(t),$$

is asymptotically stable if and only if, for any symmetric positive definite matrix M, there exists a unique symmetric positive definite matrix X satisfying the equation:

$$XA + A^{T}X = -M. \tag{7.2.3}$$

Proof.   Let's define a matrix X by

$$X = \int_{0}^{\infty} e^{A^{T}t} M e^{At} dt. \tag{7.2.4}$$

Then, we show that when the system is asymptotic stable, X is a unique solution of the equation (7.2.3) and is symmetric positive definite.

Using the expression of X in (7.2.3), we obtain

$$\begin{aligned}
XA + A^{T}X &= \int_{0}^{\infty} e^{A^{T}t} M e^{At} A \, dt + \int_{0}^{\infty} A^{T} e^{A^{T}t} M e^{At} dt \\
&= \int_{0}^{\infty} \frac{d}{dt} (e^{A^{T}t} M e^{At}) \, dt = \left[e^{A^{T}t} M e^{At} \right]_{0}^{\infty}
\end{aligned}$$

Since A is stable, $e^{A^{T}t} \rightarrow 0$ as $t \rightarrow \infty$. Thus, $XA + A^{T}X = -M$, showing that X defined by (7.2.4) satisfies the Eq. (7.2.3).

To show that X is positive definite, we have to show that $u^T X u > 0$ for any nonzero vector u. Using (7.2.4) we can write

$$u^T X u = \int_0^\infty u^T e^{A^T t} M e^{At} \, u \, dt.$$

Since the exponential matrices $e^{A^T t}$ and e^{At} are both nonsingular and M is positive definite, we conclude that $u^T X u > 0$.

To prove that X is unique, assume that there are two solutions X_1 and X_2 of (7.2.3). Then,

$$A^T (X_1 - X_2) + (X_1 - X_2)A = 0,$$

which implies that

$$e^{A^T t}(A^T (X_1 - X_2) + (X_1 - X_2)A)e^{At} = 0$$

or

$$\frac{d}{dt}\left[e^{A^T t}(X_1 - X_2)e^{At} \right] = 0,$$

and hence $e^{A^T t}(X_1 - X_2)e^{At}$ is a constant matrix for all t.

Evaluating at $t = 0$ and $t = \infty$ we conclude that $X_1 - X_2 = 0$.

We now prove the converse, that is, we prove that if X is a symmetric positive definite solution of the equation (7.2.3), then A is stable.

Let (λ, x) be an eigenpair of A. Then premultiplying the equation (7.2.3) by x^* and postmultiplying it by x, we obtain:

$$x^* X A x + x^* A^T X x = \lambda x^* X x + \bar{\lambda} x^* X x = (\lambda + \bar{\lambda})x^* X x = -x^* M x.$$

Since M and X are both symmetric positive definite, we have $\lambda + \bar{\lambda} < 0$ or $\mathrm{Re}(\lambda) < 0$. ■

Note: The matrix X defined by (7.2.4) satisfies the Eq. (7.2.3) even when M is not positive definite.

Definition 7.2.6.   *The matrix equation:*

$$X A + A^T X = -M$$

and its dual

$$A X + X A^T = -M$$

are called the **Lyapunov equations.**

Remark (Lyapunov Function)

- The Lyapunov stability theory was originally developed by Lyapunov (Liapunov (1892)) in the context of stability of a nonlinear system. The stability of a nonlinear system is determined by **Lyapunov functions.** See Luenberger (1979) for details. For the linear system:

$$\dot{x}(t) = Ax(t),$$

the function $V(x) = x^T X x$, where X is symmetric is a **Lyapunov function** if the $\dot{V}(x)$, the derivative of $V(x)$, is negative definite. This fact yields an alternative proof of Theorem 7.2.3. This can be seen as follows:

$$\dot{V}(x) = \dot{x}^T X x + x^T X \dot{x},$$
$$= x^T (A^T X + X A) x,$$
$$= x^* (-M) x.$$

Thus, $\dot{V}(x)$ is negative definite if and only if M is positive definite.

We note the following from the proof of Theorem 7.2.3.

Integral Representations of the Unique Solutions of Lyapunov Equations

Let A be a stable matrix and let M be symmetric, positive definite, or semidefinite. Then,

1. The unique solution X of the Lyapunov equation:

$$X A + A^T X = -M$$

   is given by

$$X = \int_0^\infty e^{A^T t} M e^{At} dt. \qquad (7.2.5)$$

2. The unique solution X of the Lyapunov equation

$$A X + X A^T = -M$$

   is given by

$$X = \int_0^\infty e^{At} M e^{A^T t} dt. \qquad (7.2.6)$$

As we will see later, the Lyapunov equations also arise in many other important control theoretic applications. In many of these applications, the right-hand side matrix M is positive semi-definite, rather than positive definite. The typical examples are $M = BB^T$ or $M = C^T C$, where B and C

are, respectively, the input and output matrices. The Lyapunov equations of the above types arise in finding **Grammians** of a stable system (see **Section 7.2.3**).

Theorem 7.2.4. *Let A be a stable matrix. Then the Lyapunov equation:*

$$XA + A^{\mathrm{T}}X = -C^{\mathrm{T}}C \tag{7.2.7}$$

has a unique symmetric positive definite solution X if and only if (A, C) is observable.

Proof. We first show that the observability of (A, C) and stability of A imply that X is positive definite.

Since A is stable, by (7.2.5) the unique solution X of the equation (7.2.7) is given by

$$X = \int_0^\infty e^{A^{\mathrm{T}}t} C^{\mathrm{T}} C e^{At} dt.$$

If X is not positive definite, then there exists a nonzero vector x such that $Xx = 0$. In that case

$$\int_0^\infty \|Ce^{At}x\|^2 \, dt = 0;$$

this means that $Ce^{At}x = 0$. Evaluating $Ce^{At}x = 0$ and its successive derivatives at $t = 0$, we obtain $CA^i x = 0, i = 0, 1, \ldots, n - 1$. This gives $O_M x = 0$, where O_M is the observability matrix. Since (C, A) is observable, O_M has full rank, and this implies that $x = 0$, which is a contradiction.

Hence $Ce^{At}x \neq 0$, for every t. So, X is positive definite.

Next, we prove the converse. That is, we prove that the stability of A and definiteness of X imply that (A, C) is observable. The proof is again by contradiction.

Suppose (A, C) is not observable. Then, according to criterion (v) of Theorem 6.3.1, there is an eigenvector x of A such that $Cx = 0$. Let λ be the eigenvalue corresponding to the eigenvector x. Then from the equation:

$$XA + A^{\mathrm{T}}X = -C^{\mathrm{T}}C,$$

we have $x^* X A x + x^* A^{\mathrm{T}} X x = -x^* C^{\mathrm{T}} C x$ or $(\lambda + \bar{\lambda})x^* X x = - \|Cx\|^2$.

So, $(\lambda + \bar{\lambda})x^* X x = 0$.

Since A is stable, $\lambda + \bar{\lambda} < 0$. Thus,

$$x^* X x = 0.$$

But X is positive definite, so x must be a zero vector, which is a contradiction. ∎

We next prove a necessary and sufficient condition of stability assuming that (A, C) is observable.

Theorem 7.2.5. *Let* (A, C) *be observable. Then* A *is stable if and only if there exists a unique symmetric positive definite solution matrix* X *satisfying the Lyapunov equation* (7.2.7).

Proof. We have already proved the theorem in one direction, that is, we have proved that if A is stable and (A, C) is observable, then the Lyapunov equation (7.2.7) has a unique symmetric positive definite solution X given by:

$$X = \int_0^\infty e^{A^\mathrm{T} t} C^\mathrm{T} C e^{At} dt.$$

We now prove the other direction. Let (λ, x) be an eigenpair of A. Then as before we have

$$(\lambda + \bar{\lambda}) x^* X x = - \|Cx\|^2 .$$

Since (A, C) is observable, $Cx \neq 0$, and since X is positive definite, $x^* X x > 0$. Hence $\lambda + \bar{\lambda} < 0$, which means that A is stable. ∎

For the sake of convenience, we combine the results of Theorems 7.2.4 and 7.2.5 in Theorem 7.2.6.

In the rest of this chapter, for notational convenience, a symmetric positive definite (positive semidefinite) matrix X will be denoted by the symbol $X > 0$ (≥ 0).

Theorem 7.2.6. *Let* X *be a solution of the Lyapunov equation* (7.2.7). *Then the followings hold:*

   (i)   *If* $X > 0$ *and* (A, C) *is observable, then* A *is a stable matrix.*
  (ii) *If* A *is a stable matrix and* (A, C) *is observable, then* $X > 0$.
 (iii) *If* A *is a stable matrix and* $X > 0$, *then* (A, C) *is observable.*

Since observability is a dual concept of controllability, the following results can be immediately proved by duality of Theorems 7.2.4 and 7.2.5.

Theorem 7.2.7. *Let* A *be a stable matrix. Then the Lyapunov equation:*

$$AX + XA^\mathrm{T} = -BB^\mathrm{T} \tag{7.2.8}$$

has a unique symmetric positive definite solution X *if and only if* (A, B) *is controllable.*

Theorem 7.2.8. *Let* (A, B) *be controllable. Then* A *is stable if and only if there exists a unique symmetric positive definite* X *satisfying the Lyapunov equation* (7.2.8).

Theorems 7.2.7 and 7.2.8 can also be combined, in a similar manner, as in Theorem 7.2.6, to obtain the following:

Theorem 7.2.9. *Let X be a solution of the Lyapunov equation (7.2.8). Then the followings hold:*

(i) *If $X > 0$ and (A, B) is controllable, then A is a stable matrix.*
(ii) *If A is a stable matrix and (A, B) is controllable, then $X > 0$.*
(iii) *If A is a stable matrix and $X > 0$, then (A, B) is controllable.*

7.2.3 Lyapunov Equations and Controllability and Observability Grammians

Definition 7.2.7. *Let A be a stable matrix. Then the matrix:*

$$C_G = \int_0^\infty e^{At} B B^T e^{A^T t} dt \qquad (7.2.9)$$

*is called the **controllability Grammian**, and the matrix:*

$$O_G = \int_0^\infty e^{A^T t} C^T C e^{At} dt \qquad (7.2.10)$$

*is called the **observability Grammian**.*

In view of these definitions, Theorems 7.2.7 and 7.2.4 can be, respectively, restated as follows.

Theorem 7.2.10. *Controllability Grammian and the Lyapunov Equation. Let A be a stable matrix. Then the controllability Grammian C_G satisfies the Lyapunov equation*

$$AC_G + C_G A^T = -B B^T \qquad (7.2.11)$$

and is symmetric positive definite if and only if (A, B) is controllable.

Theorem 7.2.11. *Observability Grammian and the Lyapunov Equation. Let A be a stable matrix. Then the observability Grammian O_G satisfies the Lyapunov equation*

$$O_G A + A^T O_G = -C^T C \qquad (7.2.12)$$

and is symmetric positive definite if and only if (A, C) is observable.

Example 7.2.2. Let

$$A = \begin{pmatrix} -1 & -2 & -3 \\ 0 & -2 & -1 \\ 0 & 0 & -3 \end{pmatrix}, \qquad B = \begin{pmatrix} 1 \\ 1 \\ 1 \end{pmatrix}.$$

The controllability Grammian C_G obtained by solving the Lyapunov equation (using MATLAB command **lyap**) $AX + XA^T = -BB^T$ is

$$C_G = \begin{pmatrix} 0.2917 & 0.0417 & 0.0417 \\ 0.0417 & 0.1667 & 0.1667 \\ 0.0417 & 0.1667 & 0.1667 \end{pmatrix},$$

which is clearly singular. So, (A, B) is not controllable.

Verify: The singular values of the controllability matrix C_M are 25.6766, 0.8425, and 0.

7.2.4 Lyapunov Equations and the H_2-Norm

In this section, we show how the H_2-norm of the transfer matrix of an asymptotically stable continuous-time system can be computed using Lyapunov equations.

Definition 7.2.8. *The H_2-norm of the transfer matrix $G(s)$ of an asymptotically stable continuous-time system:*

$$\dot{x} = Ax + Bu, \qquad \qquad (7.2.13)$$
$$y = Cx,$$

denoted by $\|G\|_2$, is defined by

$$\|G\|_2 = \left(\frac{1}{2\pi} \int_{-\infty}^{\infty} \text{Trace}(G(j\omega)^* G(j\omega)) \, d\omega \right)^{1/2}. \qquad (7.2.14)$$

Thus, **the H_2-norm measures the steady-state covariance of the output response $y = Gv$ to the white noise inputs v.**

Computing the H_2-Norm

By Parseval's theorem in complex analysis (Rudin 1966, p. 191), (7.2.14) can be written as

$$\|G(s)\|_2 = \left(\int_0^{\infty} \text{Trace } (h^T(t)h(t)) \, dt \right)^{1/2},$$

where $h(t)$ is the impulse response matrix:

$$h(t) = Ce^{At}B.$$

Thus,

$$\|G\|_2^2 = \text{Trace}\left(B^{\text{T}} \left(\int_0^\infty e^{A^{\text{T}}t} C^{\text{T}} C e^{At} dt \right) B \right),$$

$$= \text{Trace}(B^{\text{T}} O_G B),$$

where O_G is the observability Grammian given by (7.2.10).
   Similarly, we can show that

$$\|G\|_2^2 = \text{Trace}(C C_G C^{\text{T}}), \tag{7.2.15}$$

where C_G is the controllability Grammian given by (7.2.9).
   Since A is stable, the controllability and observability Grammians satisfy, respectively, the Lyapunov equations (7.2.11) and (7.2.12).
   Thus, a straightforward method for computing the H_2-norm is as follows:

Algorithm 7.2.1.   *Computing the H_2-Norm*
   Input. *The system matrices A, B, and C.*
   Output. *The H_2-norm of the system (A, B, C).*
   Assumption. *A is stable.*
   Step 1. *Solve the Lyapunov equation (7.2.11) or (7.2.12)*
   Step 2. *Compute either* $\text{Trace}(C C_G C^{\text{T}})$ *or* $\text{Trace}(B^{\text{T}} O_G B)$, *depending upon which of the two Lyapunov equations is solved, and take the square-root of either of these two values as the value of the H_2-norm.*

Example 7.2.3.

$$A = \begin{pmatrix} -1 & 2 & 3 \\ 0 & -2 & 2 \\ 0 & 0 & -3 \end{pmatrix}, \qquad B = \begin{pmatrix} 1 & 1 \\ 1 & 1 \\ 1 & 1 \end{pmatrix}, \qquad C = \begin{pmatrix} 1 & 1 & 1 \\ 1 & 1 & 1 \end{pmatrix}.$$

Step 1. The solution of the Lyapunov equation (7.2.11), C_G, is

$$C_G = \begin{pmatrix} 9.1833 & 2.5667 & 1.0167 \\ 2.5667 & 1.0333 & 0.5333 \\ 1.0167 & 0.5333 & 0.3333 \end{pmatrix},$$

$$C' = C C_G C^{\text{T}} = \begin{pmatrix} 18.7833 & 18.7833 \\ 18.7833 & 18.7833 \end{pmatrix}.$$

The solution of the Lyapunov equations (7.2.12), O_G, is

$$O_G = \begin{pmatrix} 1 & 1.3333 & 1.9167 \\ 1.3333 & 1.8333 & 2.7000 \\ 1.9167 & 2.7000 & 4.0500 \end{pmatrix},$$

$$B' = B^T O_G B = \begin{pmatrix} 18.7833 & 18.7833 \\ 18.7833 & 18.7833 \end{pmatrix}.$$

Step 2. H_2-norm $= \sqrt{\text{Trace}(B')} = \sqrt{\text{Trace}(C')} = \sqrt{37.5667} = 6.1292.$

MATCONTROL note: Algorithm 7.2.1 has been implemented in MATCON-TROL function **h2nrmcg** and **h2nrmog**.

MATLAB Note. MATLAB function **norm(sys)** computes the H_2-norm of a system.

7.3 STABILITY OF A DISCRETE-TIME SYSTEM

7.3.1 Stability of a Homogeneous Discrete-Time System

Consider the discrete-time system:

$$x_{k+1} = Ax_k \tag{7.3.1}$$

with initial value x_0.

A well-known mathematical criterion for asymptotic stability of the homogeneous discrete-time system now follows. The proof is analogous to that of Theorem 7.2.1 and can be found in Datta (1995).

Theorem 7.3.1. *The system* (7.3.1) *is asymptotically stable if and only if all the eigenvalues of A are inside the unit circle.*

Definition 7.3.1. *A matrix A having all its eigenvalues inside the unit circle is called a* **discrete-stable matrix,** *or a* **convergent matrix** *or a* **Schur matrix.** *We shall use the terminology* **discrete-stable** *throughout the book.*

Discrete-Time Lyapunov Stability Theory

Each of the theorems in Section 7.2 has a discrete counterpart. In the discrete case, the continuous-time Lyapunov equations $XA + A^T X = -M$ and $AX + XA^T = -M$ are, respectively, replaced by their discrete-analogs $X - A^T X A = M$ and $X - AXA^T = M$.

These discrete counterparts of the continuous-time Lyapunov equations are called the **Stein equations.** The Stein equations are also known as **discrete-Lyapunov equations** in control literature.

In the following, we state and prove a discrete analog of Theorem 7.2.3. The statements and proofs of the discrete counterparts of Theorems 7.2.4 through 7.2.9 are analogous. In fact, the Lyapunov and Stein equations are related via the matrix analogs of the well-known bilinear transformation (known as the **Cayley transformation**):

$$z = \frac{1+s}{1-s}, \qquad s = \frac{z-1}{z+1} \tag{7.3.2}$$

Note that $|z| < 1 \Leftrightarrow \text{Re}(s) < 0$ and $|z| = 1 \Leftrightarrow \text{Re}(s) = 0$.

Theorem 7.3.2. *Discrete-Time Lyapunov Stability Theorem. The discrete-time system (7.3.1) is asymptotically stable if and only if, for any positive definite matrix M, there exists a unique positive definite matrix X satisfying the discrete Lyapunov equation:*

$$X - A^{\mathrm{T}} X A = M. \tag{7.3.3}$$

Proof. We prove the theorem in one direction, that is, we prove that if A is discrete-stable, then Eq. (7.3.3) has a unique symmetric positive definite solution X. The proof of the other direction is left as an **Exercise (7.10)**.

Define the matrix

$$X = \sum_{k=0}^{\infty} (A^{\mathrm{T}})^k M A^k. \tag{7.3.4}$$

Since A is discrete-stable, the infinite series on the right-hand side converges. Furthermore, the matrix X is symmetric and positive definite.

We now show that X is the unique solution of the Eq. (7.3.3). Indeed,

$$X - A^{\mathrm{T}} X A = \sum_{k=0}^{\infty} (A^{\mathrm{T}})^k M A^k - \sum_{k=1}^{\infty} (A^{\mathrm{T}})^k M A^k = M. \tag{7.3.5}$$

Thus, X defined by (7.3.4) satisfies the Eq. (7.3.3).

To prove that X is unique, let's assume that there is another symmetric positive definite solution X_1 of (7.3.3).

Then,

$$X_1 - A^{\mathrm{T}} X_1 A = M,$$

and

$$X = \sum_{k=0}^{\infty} (A^{\mathrm{T}})^k M A^k = \sum_{k=0}^{\infty} (A^{\mathrm{T}})^k (X_1 - A^{\mathrm{T}} X_1 A) A^k,$$

$$= \sum_{k=0}^{\infty} (A^{\mathrm{T}})^k X_1 A^k - \sum_{k=1}^{\infty} (A^{\mathrm{T}})^k X_1 A^k = X_1. \quad \blacksquare$$

Remark (BIBO and BIBS Stability of a Discrete-Time System)

- Results on BIBO stability and BIBS stability, the discrete counter parts of Theorem 7.2.2 and Theorem 7.2.2′ can be obtained, for the discrete-time system:

$$x_{k+1} = Ax_k + Bu_k.$$

See **Exercises 7.7** and **7.8** and the book by DeCarlo (1989).

Definition 7.3.2. *Let A be discrete-stable. Then the matrices:*

$$C_G^D = \sum_{k=0}^{\infty} A^k B B^T (A^T)^k \tag{7.3.6}$$

and

$$O_G^D = \sum_{k=0}^{\infty} (A^T)^k C^T C A^k \tag{7.3.7}$$

*are, respectively, called the **discrete-time controllability Grammian** and* **discrete-time observability Grammians**.

The discrete counterparts of Theorems 7.2.10 and 7.2.11 are:

Theorem 7.3.3. *Discrete-Time Controllability Grammian and Lyapunov Equation. Let A be discrete-stable. Then the discrete-time controllability Grammian C_G^D satisfies the discrete Lyapunov equation*

$$C_G^D - A C_G^D A^T = B B^T \tag{7.3.8}$$

and is symmetric positive definite if and only if (A, B) is controllable.

Theorem 7.3.4. *Discrete-Time Observability Grammian and Lyapunov Equation. Let A be discrete-stable. Then the discrete-time observability Grammian O_G^D satisfies the discrete Lyapunov equation:*

$$O_G^D - A^T O_G^D A = C^T C \tag{7.3.9}$$

and is symmetric positive definite if and only if (A, C) is observable.

7.4   SOME INERTIA THEOREMS

Certain design specifications require that the eigenvalues lie in a certain region of the complex plane. Thus, finding if a matrix is stable is not enough in many practical instances. We consider the following problem, known as the inertia problem, which is concerned with counting the number of eigenvalues in a given region."

Definition 7.4.1.   *The inertia of a matrix A of order n, denoted by* In*(A), is the triplet* $(\pi(A), \nu(A), \delta(A))$, *where* $\pi(A)$, $\nu(A)$, *and* $\delta(A)$ *are, respectively, the number of eigenvalues of A with positive, negative, and zero real parts, counting multiplicities.*

Note that $\pi(A) + \nu(A) + \delta(A) = n$ and A **is a stable matrix if and only if** In$(A) = (0, n, 0)$.

The inertia, as defined above, is the **half-plane** or the **continuous-time inertia.**

The inertia with respect to the other regions of the complex plane can be defined similarly.

The **discrete-time inertia** or the **unit-circle inertia** is defined by the triplet $(\pi_O(A), \nu_O(A), \delta_O(A))$, where $\pi_O(A)$, $\nu_O(A)$, $\delta_O(A)$, are, respectively the number of eigenvalues of A outside, inside, and on the unit circle. It will be denoted by In$_O(A)$.

Unless otherwise stated, by "inertia" we will mean the "half-plane inertia."

Much work has been done on the inertia theory of matrices. We will just give here a glimpse of the existing inertia theory and then present a computational algorithm for computing the inertia. For details, we refer the curious readers to the recent survey paper of the author (Datta 1999). This paper gives an overview of the state-of-the-art theory and applications of matrix inertia and stability. The applications include new matrix theoretic proofs of several classical stability tests, applications to D-stability and to continued functions, etc. (Datta 1978a, 1978b, 1979, 1980). For other control theoretic applications of the inertia of a matrix, see Glover (1984), and the book by Zhou *et al.* (1996).

7.4.1   The Sylvester Law of Inertia

A classical law on the inertia of a symmetric matrix A is the **Sylvester Law of Inertia,** stated as follows:

Let A be a symmetric matrix and P be a nonsingular matrix. Then,

$$\text{In}(A) = \text{In}(PAP^{\text{T}}).$$

Proof.   See Horn and Johnson (1985, pp. 223–229).

Computing the Inertia of a Symmetric Matrix

If A is symmetric, then Sylvester's law of inertia provides an inexpensive and numerically effective method for computing its inertia.

A symmetric matrix A admits a triangular factorization:

$$A = UDU^T,$$

where U is a product of elementary unit upper triangular and permutation matrices, and D is a symmetric block diagonal with blocks of order 1 or 2. This is known as **diagonal pivoting factorization**. Thus, by Sylvester's law of inertia $\text{In}(A) = \text{In}(D)$. Once this diagonal pivoting factorization is obtained, the inertia of the symmetric matrix A can be obtained from the entries of D as follows:

Let D have p blocks of order 1 and q blocks of order 2, with $p + 2q = n$. Assume that none of the 2×2 blocks of D is singular. Suppose that out of p blocks of order 1, p' of them are positive, p'' of them are negative, and p''' of them are zero (i.e., $p' + p'' + p''' = p$). Then,

$$\pi(A) = p' + q,$$
$$\upsilon(A) = p'' + q,$$
$$\delta(A) = p'''.$$

The **diagonal pivoting factorization can be achieved in a numerically stable way**. It requires only $n^3/3$ flops. For details of the diagonal pivoting factorization, see Bunch (1971), Bunch and Parlett (1971), and Bunch and Kaufman (1977).

LAPACK implementation: The diagonal pivoting method has been implemented in the LAPACK routine **SSYTRF**.

7.4.2   The Lyapunov Inertia Theorems

The Sylvester Law of Inertia and the matrix formulation of the Lyapunov criterion of stability seem to have made a significant impact on the development of nonsymmetric inertia theorems. Many inertia theorems for nonsymmetric matrices have been developed over the years. These theorems attempt to find a symmetric matrix X, given a nonsymmetric matrix A, as a solution of a certain matrix equation, in such a way that, under certain conditions, the inertia of the nonsymmetric matrix A becomes equal to the inertia of the symmetric matrix X. Once the symmetric matrix X is obtained, its inertia can be computed rather cheaply by application of the Sylvester Law of Inertia to the LDL^T decomposition of X.

Theorem 7.4.1 is the **Fundamental Theorem** on the inertia of a nonsymmetric matrix and is known as the **Main Inertia Theorem (MIT)** (Taussky (1961) , and Ostrowski and Schneider (1962)). This theroem is also known as **Ostrowski-Schneider-Taussky** (OST) **Theorem**.

Theorem 7.4.1. *The Main Inertia Theorem. (i) There exists a unique symmetric matrix X such that*

$$XA + A^TX = M > 0 \qquad (7.4.1)$$

if and only if $\delta(A) = 0$.
(ii) Whenever Eq. (7.4.1) has a symmetric solution X, $\text{In}(A) = \text{In}(X)$.

Recovery of the Lyapunov Stability Theorem

As an immediate corollary of Theorem 7.4.1, we obtain the following.

Corollary 7.4.1. *A necessary and sufficient condition for A to be stable is that there exists a symmetric positive definite matrix X such that*

$$XA + A^TX = -M, \quad M > 0.$$

The Lyapunov Stability Theorem (**Theorem 7.2.3**) now follows from Corollary 7.4.1 by noting the fact that the Lyapunov equation for any given positive definite matrix M, has a unique solution if and only if $\triangle(A) = \prod_{i,j=1}^{n}(\lambda_i + \lambda_j) \neq 0$, where $\lambda_1, \lambda_2, \ldots, \lambda_n$ are the eigenvalues of A, and $\triangle(A) \neq 0$ implies that $\delta(A) = 0$ (see Chapter 8).

Theorem 7.4.2. *Continuous-Time Semidefinite Inertia Theorem. Assume that $\delta(A) = 0$ and let X be a nonsingular symmetric matrix such that*

$$XA + A^TX = M \geq 0.$$

Then $\text{In}(A) = \text{In}(X)$.

Remarks

- Theorem 7.4.2 is due to Carlson and Schneider (1963).
- For a discrete version of Theorem 7.4.1, see Wimmer (1973), and Taussky (1964).
- For a discrete version of Theorem 7.4.2, see Datta (1980).
- The condition $\delta(A) = 0$ in Theorem 7.4.2 can be shown to be equivalent to the controllability of the pair (A^T, M); see Chen (1973) and Wimmer (1974). For discrete analogue, see Wimmer and Ziebur (1975).

7.5   DETERMINING THE STABILITY AND INERTIA OF A NONSYMMETRIC MATRIX

From our discussions in the two previous sections, it is clear that the stability and inertia of a nonsymmetric matrix can be determined by solving an appropriate Lyapunov equation.

Unfortunately **this is computationally a counterproductive approach.** The reason is that the most numerically effective (and widely used) method for solving a Lyapunov equation, the Schur method **(see Chapter 8)**, is based on reduction of the matrix A to the RSF. The RSF either displays the eigenvalues of A or can be trivially obtained from there. Of course, once the eigenvalues are computed, the stability and inertia are immediately known.

An alternative classical approach (see Marden 1966) is to compute the characteristic polynomial of A, followed by application of the **Routh–Hurwitz criterion** in the continuous-time case and the **Schur–Cohn Criterion** in the discrete-time case. **This is, unfortunately, also not a numerically viable approach.** The reasons are that: (i) computing the characteristic polynomial may be a highly numerically unstable process and (ii) the coefficients of a polynomial may be extremely sensitive to small perturbations. See our discussions in Chapter 4 **(Section 4.1)**.

In view of the above considerations, the numerical analysts believe that the most numerically effective way to compute the inertia and stability of a matrix A is to explicitly compute the eigenvalues of A. However, by explicitly computing the eigenvalues of A, one gets more than what is needed, and furthermore, since the eigenvalues of a matrix can be sensitive to small perturbations, computing the inertia and stability this way may be quite misleading sometimes (see the example in *Section 7.6 which shows that a perfectly stable matrix may become unstable by a very small perturbation of a single entry of the matrix*).

It is, therefore, of interest to develop a method for inertia and stability that does not require solution of a Lyapunov equation, or explicit computation of the characteristic polynomial or the eigenvalues of A. We will now describe such a method.

Algorithm 7.5.1 is based on the **implicit solution** of a matrix equation. The algorithm constructs a symmetric matrix F which satisfies a Lyapunov matrix equation with a positive semidefinite matrix on the right-hand side, but **the Lyapunov matrix equation is not explicitly solved.** The algorithm was developed by Carlson and Datta (1979b).

Algorithm 7.5.1. *An Implicit Matrix Equation Method for Inertia and Stability*

Input. *An $n \times n$ real matrix A*

Output. *The inertia of A.*

Step 1. *Transform A to a lower Hessenberg matrix H using an orthogonal similarity. Assume that H is unreduced* **(Chapter 4)**.

Step 2. *Construct a nonsingular lower triangular matrix L such that*

$$LH + HL = R = \begin{pmatrix} 0 \\ r \end{pmatrix}$$

is a matrix whose first $(n-1)$ *rows are zero, starting with the first row* l_1 *of L as* $l_1 = (1, 0, \ldots, 0)$.

Step 3. *Having constructed L, compute the last row r of R.*

Step 4. *Construct now a matrix S such that*

$$SH = H^T S,$$

with the last row s_n *of S as the last row r of R.*

Step 5. *Compute* $F = L^T S$.

Step 6. *If F is nonsingular, compute the inertia of the symmetric matrix F, using the Sylvester law of inertia, as described in Section 7.4.1.*

Step 7. *Obtain the inertia of A:* $\mathrm{In}(A) = \mathrm{In}(F)$.

Theorem 7.5.1. (i) *If F is nonsingular, then it is symmetric and* $\mathrm{In}(A) = \mathrm{In}(F)$. (ii) *A is stable if and only if F is negative definite.*

Proof.   Proof of Part (i).

$$FH + H^T F = L^T SH + H^T L^T S = L^T H^T S + H^T L^T S,$$
$$= (L^T H^T + H^T L^T)S = R^T S = r^T r \geq 0.$$

The nonsingularity of F implies the nonsingularity of S, and it can be shown (see Datta and Datta (1987)) that S is nonsingular if and only if H and $-H$ do not have a common eigenvalue. Thus, F is a unique solution of the matrix equation (see Theorem 8.2.1):

$$FH + H^T F = r^T r \geq 0,$$

and is, therefore, necessarily symmetric. Furthermore, since H and $-H$ do not have a common eigenvalue, we have $\delta(H) = 0$. Theorem 7.4.2 now can be applied to the above matrix equation to obtain Part (i) of Theorem 7.5.1.

Proof of Part (ii). First suppose that A is stable, then we prove that F is negative definite. Since A is stable, so is H, and therefore, $\delta(H) = 0$. Again $\delta(H) = 0$ implies that H and $-H$ do not have an eigenvalue in common. Therefore, by Theorem 8.2.1 (see Chapter 8), the Lyapunov equation:

$$FH + H^T F = r^T r \geq 0$$

has a unique solution F and therefore, must be symmetric F. By Theorem 7.4.2, we then have

$$\mathrm{In}(F) = \mathrm{In}(A) = (0, n, 0).$$

Thus, F is negative definite. Conversely, let F be negative definite. Then F is nonsingular. By part (i), we then have that $\mathrm{In}(A) = \mathrm{In}(F) = (0, n, 0)$. So, A is stable. ■

Computational remarks

- *Computation of L.* Once the first row of $L = (l_{ij})$ in step 2 is pre-scribed, the diagonal entries of L are immediately known. These are: $1, -1, 1, \ldots, (-1)^{n-1}$. Having known these diagonal entries, the $n(n-1)/2$ off-diagonal entries $l_{ij} (i > j)$ of L lying below the main diagonal can now be uniquely determined by solving a lower triangular system if these entries are computed in the following order: $l_{21}; l_{31}, l_{32}; \ldots, l_{n1}, l_{n2}, \ldots, l_{n,n-1}$.

- *Computation of S.* Similar remarks hold for computing S in Step 4. Knowing the last row of the matrix S, the rows s_{n-1} through s_1 of S can be computed directly from the relation $SH = H^\mathrm{T} S$.

Notes

1. The above algorithm has been modified and made more efficient by *Datta and Datta* (1987). The modified algorithm uses the matrix-adaptation of the well-known Hyman method for computing the characteristic polynomial of a Hessenberg matrix (see *Wilkinson* 1965), which is numerically effective with proper scaling.

2. The algorithm has been extended by *Datta and Datta (1986)* to obtain information on the number of eigenvalues of a matrix in several other regions of the complex plane including strips, ellipses, and parabolas.

3. A method of this type for finding distribution of eigenvalues of a matrix with respect to the unit circle has been reported by *L.Z. Lu* (an unpublished manuscript (1987)).

4. A comparison of various methods for inertia computation, and a compu-tationally more effective version of the algorithm reported in this section appeared in the M.Sc. Thesis of *Daniel Pierce* (1983).

Flop-count of Algorithm 7.5.1 and comparisons with other methods: Algorithm 7.5.1 requires about n^3 flops once the matrix A has been transformed to the lower Hessenberg matrix H. Since it requires $\frac{10}{3}n^3$ flops to transform A to H, a total of about $\frac{13}{3}n^3$ flops is needed to determine the inertia and stability of A using Algorithm 7.5.1. This count compares very favorable with about $12n^3$ flops needed to compute the eigenvalues of A using the QR iteration algorithm described in Chapter 4. **Thus, Algorithm 7.5.1 seems to be about three times faster than the eigenvalue method.**

We have not included the Lyapunov equation approach and the characteristic polynomial approach in our comparisons here because of the numerical difficulties with the characteristic polynomial approach and the counterproductivity of the Lyapunov equation approach, as mentioned in the beginning of this section.

Example 7.5.1. We compute In(A) using Algorithm 7.5.1 with

$$A = \begin{pmatrix} 1.997 & -0.724 & 0.804 & -1.244 & -1.365 & -2.014 \\ 0.748 & 2.217 & -0.305 & 1.002 & -2.491 & -0.660 \\ -1.133 & -1.225 & -0.395 & -0.620 & 1.504 & 1.498 \\ -0.350 & 0.515 & -0.063 & 2.564 & 0.627 & 0.422 \\ -0.057 & -0.631 & 1.544 & 0.001 & 1.074 & -1.750 \\ -1.425 & -0.788 & 1.470 & -1.515 & 0.552 & -0.036 \end{pmatrix}.$$

Step 1. Reduction to Lower Hessenberg form:

$$H = \begin{pmatrix} 1.9970 & 2.9390 & & & & \\ 0.6570 & -1.0007 & 1.9519 & & & \\ 0.4272 & 1.5242 & 0.4502 & 0.8785 & & \\ -0.1321 & 1.2962 & 0.9555 & 1.4541 & 0.4940 & \\ -0.0391 & -1.5738 & 0.6601 & 0.2377 & 2.3530 & -0.4801 \\ -1.8348 & -0.5976 & 0.7595 & 0.1120 & -3.3993 & 2.1673 \end{pmatrix}.$$

Step 2. Construction of the lower triangular L such that $LH + HL = R$:

$$L = \begin{pmatrix} 1 & & & & & \\ -1.3590 & 1 & & & & \\ 0.6937 & 1.0209 & 1 & & & \\ -1.3105 & -1.6810 & -3.2933 & 1 & & \\ 16.4617 & 22.7729 & 19.3373 & 11.7433 & 1 & \\ 198.8687 & 258.5635 & 229.9657 & 128.4966 & 21.8842 & 1 \end{pmatrix}.$$

Step 3. Last row of the matrix R is

$$r = (1023.6330, \; 1293.0942, \; 1177.7393, \; 632.4162, \; 162.4031, \; -14.8420).$$

Step 4. Construction of S such that $SH = H^T S$:

$$S = \begin{pmatrix} 2.1404 & 3.3084 & 2.7775 & 1.3224 & -1.0912 & 0.4808 \\ 3.3084 & 5.0426 & 4.1691 & 2.0997 & -1.2521 & 0.6073 \\ 2.7775 & 4.1691 & 3.6050 & 1.8169 & -1.1757 & 0.5531 \\ 1.3224 & 2.0996 & 1.8169 & 0.8899 & -0.6070 & 0.2970 \\ -1.0912 & -1.2521 & -1.1757 & -0.6070 & -0.3845 & 0.0763 \\ 0.4808 & 0.6073 & 0.5531 & 0.2970 & 0.0763 & -0.0070 \end{pmatrix}.$$

Step 5. Computation of $F = L^T S$:

$$F = \begin{pmatrix} 75.4820 & 96.7596 & 87.8785 & 47.6385 & 9.4298 & -0.4808 \\ 96.7596 & 124.1984 & 112.7028 & 61.2339 & 12.0384 & -0.6073 \\ 87.8785 & 112.7028 & 102.0882 & 55.4523 & 10.9292 & -0.5531 \\ 47.6385 & 61.2339 & 55.4523 & 30.1476 & 5.8930 & -0.2970 \\ 9.4298 & 12.0384 & 10.9292 & 5.8930 & 1.2847 & -0.0763 \\ -0.4808 & -0.6073 & -0.5531 & -0.2970 & -0.0763 & 0.0070 \end{pmatrix}.$$

Step 6. Gaussian elimination with diagonal pivoting: $PFP^{\mathrm{T}} = WDW^{\mathrm{T}}$, gives

$$W = \begin{pmatrix} 1 & & & & & \\ 0.9074 & 1 & & & & \\ 0.7791 & -0.4084 & 1 & & & \\ 0.0969 & -0.0274 & 0.4082 & 1 & & \\ 0.4930 & 0.6227 & -0.8765 & 0.0102 & 1 & \\ -0.0049 & 0.0111 & -0.0651 & -0.1454 & 0.0502 & 1 \end{pmatrix},$$

$$P = \begin{pmatrix} 0 & 1 & 0 & 0 & 0 & 0 \\ 0 & 0 & 1 & 0 & 0 & 0 \\ 1 & 0 & 0 & 0 & 0 & 0 \\ 0 & 0 & 0 & 0 & 1 & 0 \\ 0 & 0 & 0 & 1 & 0 & 0 \\ 0 & 0 & 0 & 0 & 0 & 1 \end{pmatrix},$$

$$D = \begin{pmatrix} 124.1984 & 0 & 0 & 0 & 0 & 0 \\ 0 & -0.1831 & 0 & 0 & 0 & 0 \\ 0 & 0 & 0.1298 & 0 & 0 & 0 \\ 0 & 0 & 0 & 0.0964 & 0 & 0 \\ 0 & 0 & 0 & 0 & -0.0715 & 0 \\ 0 & 0 & 0 & 0 & 0 & 0.0016 \end{pmatrix}.$$

Step 7. $\mathrm{In}(A) = \mathrm{In}(F) = \mathrm{In}(D) = (4, 2, 0)$.
Verification: The eigenvalues of A are:

$$\{-2.1502, \ 0.8553, \ 3.6006, \ 2.0971, \ 3.1305, \ -0.1123\},$$

confirming that $\mathrm{In}(A) = (4, 2, 0)$.

MATCONTROL note: Algorithm 7.5.1 has been implemented in MATCON-TROL function **inertia**.

7.6   DISTANCE TO AN UNSTABLE SYSTEM

Let A be an $n \times n$ **complex stable matrix.** A natural question arises:

How "nearly unstable" is the stable matrix A?

We consider the above question in this section.

Definition 7.6.1. *Let $A \in \mathbb{C}^{n \times n}$ have no eigenvalue on the imaginary axis. Let $U \in \mathbb{C}^{n \times n}$ be the set of matrices having at least one eigenvalue on the imaginary axis. Then, with $\|\cdot\|$ as the 2-norm or the Frobenius norm, the distance from A to U is defined by*

$$\beta(A) = \min\{\|E\| \mid A + E \in U\}.$$

If A is stable, then $\beta(A)$ is the distance to the set of unstable matrices.

*The concept of "**distance to instability**" is an important practical concept. Note that a theoretically perfect stable matrix may be very close to an unstable matrix. For example, consider the following matrix (Petkov et al. 1991):*

$$
A = \begin{pmatrix}
-0.5 & 1 & 1 & 1 & 1 & 1 \\
0 & -0.5 & 1 & 1 & 1 & 1 \\
0 & 0 & -0.5 & 1 & 1 & 1 \\
0 & 0 & 0 & -0.5 & 1 & 1 \\
0 & 0 & 0 & 0 & -0.5 & 1 \\
0 & 0 & 0 & 0 & 0 & -0.5
\end{pmatrix}.
$$

Since its eigenvalues are all -0.5, it is perfectly stable. However, if the $(6, 1)$th entry is perturbed to $\epsilon = 1/324$ from zero, then the eigenvalues of this slightly perturbed matrix become:

$-0.8006, -0.7222 \pm 0.2485 j, -0.3775 \pm 0.4120 j, 0.000.$

Thus, the perturbed matrix is unstable, showing that the stable matrix A is very close to an unstable matrix.

We now introduce a measure of $\beta(A)$ in terms of singular values and describe a simple bisection algorithm to approximately measure it.

Let $\sigma_{\min}(A - j\omega I)$ be the smallest singular value of $A - j\omega I$. Then it can be shown (**Exercise 7.14**) that

$$
\beta(A) = \min_{\omega \in \mathcal{R}} \sigma_{\min}(A - j\omega I). \tag{7.6.1}
$$

So, for any real ω, $\sigma_{\min}(A - j\omega I)$ is an upper bound on $\beta(A)$, that is, $\beta(A) \leq \sigma_{\min}(A - j\omega I)$.

Based on this idea, Van Loan (1985) gave two estimates for $\beta(A)$. One of them is a **heuristic estimate**:

$$
\beta(A) \approx \min \left\{ \sigma_{\min}(A - j\operatorname{Re}(\lambda)I) \big| \lambda \in \Lambda(A) \right\}, \tag{7.6.2}
$$

where $\Lambda(A)$ denotes the spectrum of A.

Thus, using this heuristic estimate, $\beta(A)$ may be estimated by finding the singular values of the matrix $(A - j\operatorname{Re}(\lambda)I)$, for every eigenvalue λ of A. This approach was thought to give an upper bound within an order of magnitude of $\beta(A)$. However, Demmel (1987) has provided examples to show this bound can be larger than $\beta(A)$ by an arbitrary amount.

The other approach of Van Loan requires application of a general nonlinear minimization algorithm to $f(\omega) = \sigma_{\min}(A - j\omega I)$. We will not pursue these approaches here. Rather, we will describe a simple bisection method to estimate

$\beta(A)$ due to Byers (1988). The bisection algorithm estimates $\beta(A)$ within a factor of 10 or indicates that $\beta(A)$ is less than a small tolerance. This is sufficient in practice. The algorithm makes use of the crude estimate of the upper bound $\beta(A) \leq \frac{1}{2}\|A + A^*\|_2$.

To describe the algorithm, let's define a $2n \times 2n$ **Hamiltonian matrix** $H(\sigma)$, given $\sigma \geq 0$, by

$$H(\sigma) = \begin{pmatrix} A & -\sigma I \\ \sigma I & -A^* \end{pmatrix}. \qquad (7.6.3)$$

The bisection method is based on the following interesting spectral property of the matrix $H(\sigma)$. **For more on Hamiltonian matrices, see Chapters 10 and 13.**

Theorem 7.6.1. $\sigma \geq \beta(A)$ if and only if $H(\sigma)$ defined by (7.6.3) has a purely imaginary eigenvalue.

Proof. Let ω_i be a purely imaginary eigenvalue of $H(\sigma)$. Then there exist nonzero complex vectors u, v such that

$$\begin{pmatrix} A & -\sigma I \\ \sigma I & -A^* \end{pmatrix} \begin{pmatrix} u \\ v \end{pmatrix} = \omega_i \begin{pmatrix} u \\ v \end{pmatrix}. \qquad (7.6.4)$$

This gives us

$$(A - \omega_i I)u = \sigma v \qquad (7.6.5)$$

and

$$(A - \omega_i I)^* v = \sigma u \qquad (7.6.6)$$

This means that σ is a singular value of the matrix $A - \omega_i I$. Also, since $\beta(A) \leq \sigma_{\min}(A - j\omega I)$ for any real ω, we obtain $\sigma \geq \beta(A)$.

Conversely, suppose that $\sigma \geq \beta(A)$. Define

$$f(\alpha) = \sigma_{\min}(A - j\alpha I).$$

The function f is continuous and $\lim_{\alpha \to \infty} f(\alpha) = \infty$. Therefore, f has a minimum value $f(\alpha) = \beta(A) \leq \sigma$, for some real α.

By the Intermediate Value Theorem of Calculus, we have $f(\omega) = \sigma$ for some real ω.

So, σ is a singular value of $A - j\omega I = A - \omega_i I$ and there exist unit complex vectors u and v satisfying (7.6.5) and (7.6.6). This means that ω_i is a purely imaginary eigenvalue of $H(\sigma)$. ∎

Algorithm 7.6.1. *The Bisection Algorithm for Estimating the Distance to an Unstable System*

Inputs. *A—An $n \times n$ stable complex matrix*
τ—Tolerance (> 0).

Outputs. *Real numbers α and v such that either $v/10 \leq \alpha \leq \beta(A) \leq v$ or $0 = \alpha \leq \beta(A) \leq v \leq 10\tau$.*
    Step 1. *Set $\alpha \equiv 0$, $v = \frac{1}{2}\|(A + A^*)\|_2$*
    Step 2. *Do while $v > 10\max(\tau, \alpha)$*
$\sigma \equiv \sqrt{v\max(\tau, \alpha)}$
If $= H(\sigma)$ has a purely imaginary eigenvalue, then set $v \equiv \sigma$; else $\alpha \equiv \sigma$

Example 7.6.1. Consider finding $\beta(A)$ for the matrix:

$$A = \begin{pmatrix} -1 & 1 \\ 0 & -0.0001 \end{pmatrix}.$$

$$\tau = 0.00100$$

Iteration 1.
    Step 1. Initialization: $\alpha = 0$, $v = 1.2071$.
    Step 2. $10 \times \max(\tau, \alpha) = 0.0100$.

$$\sigma = 0.0347$$

$$H(\sigma) = \begin{pmatrix} -1 & 1 & -0.0347 & 0 \\ 0 & -0.0001 & 0 & -0.0347 \\ 0.0347 & 0 & 1 & 0 \\ 0 & 0.0347 & -1 & 0.0001 \end{pmatrix}$$

The eigenvalues of $H(\sigma)$ are $\pm 1, \pm 0.0491j$. Since $H(\sigma)$ has an purely imaginary eigenvalue, we set

$$v = \sigma = 0.0347.$$
$$v = 0.0347 > 10 \max(\tau, \alpha) = 0.0100,$$

the iteration continues, until $v = 0.0059$ is reached, at which point, the iteration terminates with $\beta(A) \leq 0.0059 < 10\tau$.

Conclusion: $\beta(A) \leq 0.0059 < 10\tau$.

Computational remarks:

- The bulk of the work of the Algorithm 7.6.1 is in deciding whether $H(\sigma)$ has an imaginary eigenvalue.
- Also, the decision of whether $H(\sigma)$ has an imaginary eigenvalue in a computational setting (in the presence of round-off errors) is a tricky one. Some sort of threshold has to be used. However, if that decision is made in a numerically effective way, then in the worst case, "$\beta(A)$ **might lie outside the bound given by the algorithm by an amount proportional to the precision of the arithmetic**" (Byers 1988).

- Because of the significant computational cost involved in deciding if the matrix $H(\sigma)$ at each step has an imaginary eigenvalue, **the algorithm may not be computationally feasible for large problems.**

Convergence. If $\tau = \frac{1}{2}10^{-p}\|A + A^*\|$, then at most $\log_2 p$ bisection steps are required; for example, if $\tau = \frac{1}{2} \times 10^{-8}\|A + A^*\|$, then at most three bisection steps are required.

MATCONTROL note: Algorithm 7.6.1 has been implemented in MATCONTROL function **disstabc**.

Relation to Lyapunov Equation

Since Lyapunov equations play a vital role in stability analysis of a linear system, it is natural to think that the distance to a set of unstable matrices $\beta(A)$ is also related to a solution of a Lyapunov equation. Indeed, the following result can be proved (See Malyshev and Sadkane (1999) and also Hewer and Kenney (1988)).

Theorem 7.6.2. *Distance to an Unstable System and Lyapunov Equation. Let* **A be complex stable** *and let X be the unique positive Hermitian definite solution of the Lyapunov equation:*

$$XA + A^*X = -M, \qquad (7.6.7)$$

where M is Hermitian positive definite. Then

$$\beta(A) \geq \frac{\lambda_{\min}(M)}{2\,\|X\|_2},$$

where $\lambda_{\min}(M)$ denotes the smallest eigenvalue of M.

Proof. Let $\omega \in \mathbb{R}$ and u be a unit vector such that

$$\frac{1}{\beta(A)} = \max_{Re(z)=0} \|(A - zI)^{-1}\| = \|(A - j\omega I)^{-1}u\|. \qquad (7.6.8)$$

Let $x = (A - j\omega I)^{-1}u$. Then, $\|x\|_2 = 1/\beta(A)$.

Multiplying the Lyapunov equation (7.6.7) by x^* to the left and x to the right, we have

$$x^*(XA + A^*X)x = -x^*Mx,$$
$$x^*XAx + x^*A^*Xx = -x^*Mx.$$

Then,

$$x = (A - j\omega I)^{-1}u \Rightarrow (A - j\omega I)x = u \Rightarrow Ax = u + j\omega x$$

and

$$x = (A - j\omega I)^{-1}u \Rightarrow x^*(A - j\omega I)^* = u^* \Rightarrow x^*A^* + j\omega x^*$$
$$= u^* \Rightarrow x^*A^* = u^* - j\omega x^*.$$

Therefore,

$$|x^*XAx + x^*A^*Xx| = |x^*X(u + j\omega x) + (u^* - j\omega x^*)Xx|$$
$$= 2|u^*Xx| \leq 2||X||_2||x||_2. \qquad (7.6.9a)$$

Also, by the Rayleigh quotient (see Datta (1995) or Golub and Van Loan (1996)), we have, $\lambda_{\min}(M) \leq x^*Mx/x^*x$, that is,

$$\lambda_{\min}(M)||x||_2^2 \leq x^*Mx = |-x^*Mx|. \qquad (7.6.9b)$$

Thus, combining (7.6.9a) and (7.6.9b) yields

$$\lambda_{\min}(M)||x||_2^2 \leq |x^*Mx| = |x^*(XA + A^*X)x| \leq 2||X||_2||x||_2$$

or

$$\lambda_{\min}(M)||x||_2 \leq 2||X||_2.$$

Since $||x||_2 = 1/\beta(A)$, this means that $\lambda_{\min}(M)(1/\beta(A)) \leq 2||X||_2$ or $\beta(A) \geq \lambda_{\min}/2||X||_2$. ∎

Example 7.6.2.   Consider Example 7.6.1 again.

$$A = \begin{pmatrix} -1 & 1 \\ 0 & -0.0001 \end{pmatrix}.$$

Take $M = I_2$. Then

$$X = \begin{pmatrix} 0.5 & 0.5 \\ 0.5 & 9999.5 \end{pmatrix}.$$

$$\beta(A) \approx 0.0059 \geq \frac{1}{2||x||_2} = 5.0002 \times 10^{-5}.$$

Verify: The eigenvalues of $A + 5.0002 \times 10^{-5}I$ are -0.9999 and 0.

Distance to a Discrete Unstable System

The discrete analog of $\beta(A)$ is defined to be

$$\gamma(A) = \min\{||E|| \mid \text{for some } \theta \in \mathbb{R}; \ e^{i\theta} \in \Omega(A + E)\}. \qquad (7.6.10)$$

That is, $\gamma(A)$ **measures the distance from A to the nearest matrix with an eigenvalue on the unit circle.** If A is discrete-stable, then $\gamma(A)$ is a measure of how "**nearly discrete-unstable**" A is. In above, $\Omega(M)$ denotes the spectrum of M. A discrete-analog of Theorem 7.6.1 is:

Theorem 7.6.3. *Given an $n \times n$ complex matrix A, there exists a number $\Gamma(A) \in \mathbb{R}$ such that $\Gamma(A) \geq \gamma(A)$ and for $\Gamma(A) \geq \sigma \geq \gamma(A)$, the $2n \times 2n$ Hamiltonian matrix pencil*

$$H_D(\sigma) = F(\sigma) - \lambda G(\sigma) = \begin{pmatrix} -\sigma I_n & A \\ I_n & 0 \end{pmatrix} - \lambda \begin{pmatrix} 0 & I_n \\ A^T & -\sigma I_n \end{pmatrix}$$

has a generalized eigenvalue of magnitude 1. Furthermore, if $\sigma < \gamma(A)$, then the above pencil has no generalized eigenvalue of magnitude 1.

Proof. See Byers (1988). ■

Based on the above result, Byers (1988) described the following bisection algorithm to compute $\gamma(A)$, analogous to Algorithm 7.6.1.

The algorithm estimates $\gamma(A)$ within a factor of 10 or indicates that $\gamma(A)$ is less than a tolerance. The algorithm uses a crude bound $\Gamma(A) \geq \sigma_{\min}(A - I)$.

Algorithm 7.6.2. *The Bisection Algorithm for Estimating the Distance to a Discrete-Unstable System*
  Inputs. *An $n \times n$ complex matrix A and a tolerance $\tau > 0$.*
  Outputs. *Real numbers α and δ such that $\delta/10 \leq \alpha \leq \gamma(A) \leq \delta$ or $0 = \alpha \leq \gamma(A) \leq \delta \leq 10\tau$.*
  Step 1. *Set $\alpha \equiv 0$; $\delta \equiv \sigma_{\min}(A - I)$.*
  Step 2. *Do while $\delta > 10 \max(\tau, \alpha)$*
      $\sigma \equiv \sqrt{\delta \max(\tau, \alpha)}$.
      If the pencil $F(\sigma) - \lambda G(\sigma)$, defined above, has a generalized eigenvalue of magnitude 1, then set $\delta \equiv \sigma$, else $\alpha \equiv \sigma$.
  End.

Example 7.6.3. Let $A = \begin{pmatrix} 0.9999 & 1 \\ 0 & 0.5 \end{pmatrix}$, $\tau = 10^{-8}$. The matrix A is discrete-stable.
Iteration 1:
  Step 1: $\alpha = 0$, $\delta = 4.4721 \times 10^{-5}$.
  Step 2: $\delta > 10 \max(\tau, \alpha)$ is verified, we compute $\sigma = 6.6874 \times 10^{-6}$. The eigenvalues of $H_D(\sigma)$ are 2, 1.0001, 0.9999, and 0.5000. Thus, $\alpha \equiv 6.6874 \times 10^{-6}$.

Iteration 2: Since $\delta > 10 \max(\tau, \alpha)$ is verified, we compute $\sigma = 5.4687 \times 10^{-6}$. The eigenvalues of $H_D(\sigma)$ are 2, 1.001, 0.999, 0.5000; we set $\alpha = \sigma = 5.4687 \times 10^{-6}$.
Iteration 3: $\delta < 10 \max(\tau, \alpha)$, the iteration stops, and on exit we obtain

$$\alpha = 5.4687 \times 10^{-6} \qquad \delta = 4.4721 \times 10^{-5}.$$

MATCONTROL note: Algorithm 7.6.2 has been implemented in MATCONTROL function **disstabd**.

7.7   ROBUST STABILITY

Even though a system is known to be stable, it is important to investigate if the system remains stable under certain perturbations. Note that in most physical systems, the system matrix A is not known exactly; what is known is $A + E$, where E is an $n \times n$ perturbation matrix. Thus, in this case the stability problem becomes the problem of finding if the system:

$$\dot{x}(t) = (A + E)x(t) \qquad (7.7.1)$$

remains stable, given that A is stable.

The solution of the Lyapunov equations can be used again to obtain bounds on the perturbations that guarantee that the perturbed system (7.7.1) remains stable.

In Theorem 7.7.1, $\sigma_{\max}(M)$, as usual, stands for the largest singular value of M.

We next state a general result on robust stability due to Keel *et al.* (1988). The proof can be found in Bhattacharyya *et al.* (1995, pp. 519–520). The result there is proved in the context of feedback stabilization, and we will revisit the result later in that context. The other earlier results include those of Patel and Toda (1980) and Yedavalli (1985).

Theorem 7.7.1.   *Let A be a stable matrix and let the perturbation matrix E be given by*

$$E = \sum_{i=1}^{r} p_i E_i, \qquad (7.7.2)$$

where $E_i, i = 1, \ldots, r$ are matrices determined by structure of the perturbations.

Let Q be a symmetric positive definite matrix and X be a unique symmetric positive definite solution of the Lyapunov equation:

$$XA + A^TX + Q = 0. \tag{7.7.3}$$

Then the system (7.7.1) remains stable for all p_i satisfying

$$\sum_{i=1}^{r} |p_i|^2 < \frac{\sigma_{\min}^2(Q)}{\sum_{i=1}^{r} \mu_i^2},$$

where $\sigma_{\min}(Q)$ denotes the minimum singular value of Q and μ_i is given by

$$\mu_i = \|E_i^T X + X E_i\|_2 .$$

Example 7.7.1. Let $r = 1$, $p_1 = 1$. Take

$$E = E_1 = \begin{pmatrix} 0.0668 & 0.0120 & 0.0262 \\ 0.0935 & 0.0202 & 0.0298 \\ 0.0412 & 0.0103 & 0.0313 \end{pmatrix}.$$

Let $A = \begin{pmatrix} -4.1793 & 9.712 & 1.3649 \\ 0 & -1.0827 & 0.3796 \\ 0 & 0 & -9.4673 \end{pmatrix}$

Choose $Q = 2I$.
Then, $\mu_1 = \|E_1^T X + X E_1\|_2 = 0.7199$, and the right-hand side of (7.7.3) is 7.7185. Since $|p_1^2| = 1 < 7.7185$, the matrix $A + E$ is stable.

A result similar to that stated in Theorem 7.7.1 was also proved by Zhou and Khargonekar (1987). We state the result below.

Theorem 7.7.2. *Let A be a stable matrix and let E be given by (7.7.2). Let X be the unique symmetric positive definite solution of (7.7.3). Define*

$$X_i = (E_i^T X + X E_i)/2, \quad i = 1, 2, \ldots, r \tag{7.7.4}$$

and

$$X_e = (X_1, X_2, \ldots, X_r).$$

Then (7.7.1) remains stable if

$$\sum_{k=1}^{r} p_k^2 < \frac{1}{\sigma_{\max}^2(X_e)} \quad or \quad \sum_{i=1}^{r} |p_i| \sigma_{\max}(X_i) < 1$$

$$or \quad |p_i| < \frac{1}{\sigma_{\max}\left(\sum_{i=1}^{r} |X_i|\right)}, \quad i = 1, \ldots, r. \tag{7.7.5}$$

Remark

- It should be noted that Theorems 7.7.1 and 7.7.2 and others in Patel and Toda (1980) and Yedavalli (1985) all give only **sufficient conditions** for robust stability. A number of other sufficient conditions can be found in the book by Boyd *et al.* (1994).

7.8   THE STRUCTURED STABILITY RADIUS

In Section 7.6, we introduced the concept of the distance of a stable matrix from the set of unstable matrices. Here we specialize this concept to "**structured stability**," meaning that we are now interested in finding the distance from a stable matrix to the set of unstable matrices, **where the distance is measured by the size of the additive perturbations of the form** $B \triangle C$**, with B and C fixed, and $\triangle$ variable.**

Let A, B, and C be, respectively, $n \times n$, $n \times m$, and $r \times n$ matrices over the field $\mathbb{F}$ ($\mathbb{F}$ can be $\mathbb{C}$ or $\mathbb{R}$). Then the (structured) **stability radius** of the matrix triple (A, B, C) is defined as

$$r_{\mathbb{F}}(A, B, C) = \inf\{\bar{\sigma}(\triangle) : \triangle \in \mathbb{F}^{m \times r} \text{ and } A + B \triangle C \text{ is unstable }\}, \quad (7.8.1)$$

where $\bar{\sigma}(M)$ following the notation of Qiu *et al.* (1995), **denotes the largest singular value of** M (i.e., $\bar{\sigma}(M) = \sigma_{\max}(M)$). For real matrices (A, B, C), $r_{\mathbb{R}}(A, B, C)$ is called the **real stability radius** and, for complex matrices (A, B, C), $r_{\mathbb{C}}(A, B, C)$ is called the **complex stability radius. The stability radius, thus, determines the magnitude of the smallest perturbation needed to destroy the stability of the system.**

"Stability" here is referred to as either continuous-stability (with respect to the left half-plane) or discrete-stability (with respect to the unit circle).

Let $\partial \mathbb{C}_g$ denote the boundary of either the half plane or the unit circle. Let A be stable or discrete-stable.

Then,

$$r_{\mathbb{F}}(A, B, C) = \inf\{\bar{\sigma}(\triangle)|\triangle \in \mathbb{F}^{m \times r} \text{ and } A + B \triangle C \text{ has an eigenvalue on } \partial \mathbb{C}_g\}.$$

$$= \inf_{s \in \partial \mathbb{C}_g} \inf\{\bar{\sigma}(\triangle)|\triangle \in \mathbb{F}^{m \times r} \text{ and } \det(sI - A - B \triangle C) = 0\} \quad (7.8.2)$$

$$= \inf_{s \in \partial \mathbb{C}_g} \inf\{\bar{\sigma}(\triangle)|\triangle \in \mathbb{F}^{m \times r} \text{ and } \det(I - \triangle G(s)) = 0\},$$

where $G(s) = C(sI - A)^{-1}B$.

Thus, given a complex $r \times m$ matrix M, the stability radius problem reduces to the problem of computing:

$$\mu_{\mathbb{F}}(M) = [\inf\{\bar{\sigma}(\triangle) : \triangle \in \mathbb{F}^{m \times r} \text{ and } \det(I - \triangle M) = 0\}]^{-1}.$$

The Complex Stability Radius

It is easy to see that

$$\mu_{\mathbb{C}}(M) = \bar{\sigma}(M).$$

Thus, we have the following formula for the complex stability radius.

Theorem 7.8.1. *The Complex Stability Radius Formula*

$$r_{\mathbb{C}}(A, B, C) = \left\{ \sup_{s \in \partial\mathbb{C}_g} \bar{\sigma}(G(s)) \right\}^{-1}. \tag{7.8.3}$$

The Real Stability Radius

If $\mathbb{F}$ is $\mathbb{R}$, then according to the above we have

$$r_{\mathbb{R}}(A, B, C) = \left\{ \sup_{s \in \partial\mathbb{C}_g} \mu_{\mathbb{R}}[C(sI - A)^{-1}B] \right\}^{-1}. \tag{7.8.4}$$

For the real stability radius, the major problem is then is the problem of computing $\mu_{\mathbb{R}}(M)$, given M.

The following important formula for computing $\mu_{\mathbb{R}}(M)$ has been recently obtained by Qiu *et al.* (1995). We quote the formula from this paper. The proof is involved and we refer the readers to the paper for the proof. Following the notation of this paper, **we denote the second largest singular value of M by $\sigma_2(M)$**, and so on.

Denote the real and imaginary parts of a complex matrix M by $\text{Re}(M)$ and $\text{Im}(M)$, respectively. That is, $M = \text{Re}(M) + j\text{Im}(M)$.

Then the following result holds:

$$\mu_{\mathbb{R}}(M) = \inf_{\gamma \in (0,1]} \sigma_2\left(\begin{bmatrix} \text{Re}(M) & -\gamma\text{Im}(M) \\ \dfrac{1}{\gamma}\text{Im}(M) & \text{Re}(M) \end{bmatrix} \right). \tag{7.8.5}$$

The function to be minimized is a unimodular function on $(0, 1]$.

Furthermore, if $\text{rank}(\text{Im}(M)) = \lambda$, then

$$\mu_{\mathbb{R}}(M) = \max\{\bar{\sigma}(U_2^{\text{T}}\text{Re}(M)), \bar{\sigma}(\text{Re}(M)V_2)\},$$

where U_2 and V_2 are defined by the SVD of $\text{Im}(M)$, that is, they satisfy

$$\text{Im}(M) = [U_1, U_2] \begin{bmatrix} \bar{\sigma}(\text{Im}(M)) & 0 \\ 0 & 0 \end{bmatrix} [V_1, V_2]^{\text{T}}.$$

Note that since the function to be minimized is unimodular, any local minimum is also a global minimum.

Notes

(i)
$$r_{\mathbb{R}}(A, B, C) \geq r_{\mathbb{C}}(A, B, C). \qquad (7.8.6)$$

(ii)   The ratio $r_{\mathbb{R}}(A, B, C)/r_{\mathbb{C}}(A, B, C)$ can be arbitrarily large.

The following example taken from Hinrichsen and Pritchard (1990) illustrates (ii).

Example 7.8.1.   Let

$$A = \begin{pmatrix} 0 & 1 \\ -1 & -\epsilon \end{pmatrix}, \qquad B = \begin{pmatrix} 0 \\ -\epsilon \end{pmatrix},$$

and

$$C = (1, 0).$$

Then the transfer function:

$$G(s) = C(j\omega I - A)^{-1} B = \frac{-\epsilon}{1 - \omega^2 + j\omega\epsilon}.$$

By (7.8.4), the real stability radius:

$$r_{\mathbb{R}}(A, B, C) = 1/\epsilon.$$

Since

$$|G(j\omega)|^2 = \frac{\epsilon^2}{(1 - \omega^2)^2 + \epsilon^2\omega^2},$$

it is easy to see that $|G(j\omega)|^2$ is maximized when $\omega^2 = 1 - \epsilon^2/2$, if $\epsilon < \sqrt{2}$. So, by (7.8.3)

$$r_{\mathbb{C}}^2(A, B, C) = 1 - (\epsilon^2/4).$$

Thus, if ϵ is considered as a parameter, then $r_{\mathbb{C}}(A, B, C)$ is always bounded by 1 whereas $r_{\mathbb{R}}(A, B, C)$ can be made arbitrarily large by choosing ϵ small enough.

Specialization of the Stability Radius to the Distance from Unstable Matrices

From (7.8.3) we immediately have the following relation between the distance to an unstable system and the stability radius:

$$\beta = r_{\mathbb{C}}(A, I, I) = \min_{\omega \in \mathbb{R}} \sigma_{\min}(A - j\omega I) = \beta(A).$$

Also, the following formula for $\beta(A)$, when A is a real stable matrix, can be proved.

Theorem 7.8.2.   *Let A be a **real stable matrix**. Then*

$$\beta(A) \equiv r_{\mathbb{R}}(A, I, I) = \min_{s \in \partial\mathbb{C}_g} \max_{\gamma \in (0,1]} \sigma_{2n-1} \left(\begin{bmatrix} A - \text{Re}(sI) & -\gamma\text{Im}(sI) \\ \frac{1}{\gamma}\text{Im}(sI) & A - \text{Re}(sI) \end{bmatrix} \right). \qquad (7.8.7)$$

Note: For each fixed s, the function in (7.8.7) to be maximized is quasiconcave.

7.9 SOME SELECTED SOFTWARE

7.9.1 MATLAB Control System Toolbox

norm—Computes the H_2-norm of the system.
bode—Computes the magnitude and phase of the frequency response, which are used to analyze stability and robust stability.
nyquist—Calculates the Nyquist frequency response. System properties such as **gain margin, phase margin**, and **stability** can be analyzed using Nyquist plots. (The gain margin and phase margin are widely used in classical control theory as measures of robust stability).
gram controllability and observability grammrians.

7.9.2 MATCONTROL

INERTIA	Determining the inertia and stability of a matrix without solving a matrix equation or computing eigenvalues
H2NRMCG	Finding H_2-norm using the controllability Grammians
H2NRMOG	Finding H_2-norm using the observability Grammian
DISSTABC	Determining the distance to the continuous-time stability
DISSTABD	Determining the distance to the discrete-time stability

7.9.3 SLICOT

AB13BD	H_2 or L_2 norm of a system
AB13ED	Complex stability radius using bisection
AB13FD	Complex stability radius using bisection and SVD

7.10 SUMMARY AND REVIEW

The stability of the system:

$$\dot{x}(t) = Ax(t)$$

or that of

$$x(k+1) = Ax(k)$$

is essentially governed by the eigenvalues of the matrix A.

Mathematical Criteria of Stability

The continuous-time system $\dot{x}(t) = Ax(t)$ is asymptotically stable if and only if the eigenvalues of A are all in the left half plane (**Theorem 7.2.1**). Similarly, the discrete-time system $x(k+1) = Ax(k)$ is asymptotically stable if and only if all the eigenvalues of A are inside the unit circle. (**Theorem 7.3.1**). Various Lyapunov stability theorems (**Theorems 7.2.3–7.2.9**, and **Theorem 7.3.2**) have been stated and proved.

The Inertia of a Matrix

Two important inertia theorems (**Theorems 7.4.1 and 7.4.2**) and the classical **Sylvester Law of Inertia** have been stated. These inertia theorems generalize the Lyapunov stability results.

Methods for Determining Stability and Inertia

The **Characteristic Polynomial Approach** and the **Matrix Equation Approach** are two classical approaches for determining the stability of a system and the inertia of a matrix. Both these approaches have some computational drawbacks.

The zeros of a polynomial may be extremely sensitive to small perturbations. Furthermore, the numerical methods to compute the characteristic polynomial of a matrix are usually unstable.

The most numerically effective method (**The Schur method**, described in Chapter 8), for solving a Lyapunov matrix equation is based on reduction of the matrix A to RSF, and the RSF displays the eigenvalues of A or the eigenvalues can be trivially computed out of this form.

Thus, the characteristic equation approach is not numerically viable and the matrix equation approach for stability and inertia is counterproductive.

Hence, the most numerically effective approach for stability and inertia is the eigenvalue approach: compute all the eigenvalues of A.

By explicitly computing the eigenvalues, one, however, gets much more than what is needed for stability and inertia. Furthermore, since the eigenvalues of a matrix can be very sensitive to small perturbations, determining the inertia and stability by computing explicitly the eigenvalues can be misleading.

An implicit matrix equation approach (**Algorithm 7.5.1**), which does not require computation of eigenvalues nor explicit solution of any matrix equation has been described. **Algorithm 7.5.1 is about three times faster than the eigenvalue method** (According to the flop-count).

Distance to an Unstable System

Given a stable matrix A, the quantity $\beta(A)$ defined by

$$\beta(A) = \min \{\|E\|_F \text{ such that } A + E \in U\},$$

where U is the set of $n \times n$ matrices with at least one eigenvalue on the imaginary axis, is the distance to the set of unstable matrices.

A bisection algorithm (**Algorithm 7.6.1**) based on knowing if a certain Hamiltonian matrix (the matrix (7.6.3)) has a purely imaginary eigenvalue, is described. The algorithm is based on **Theorem 7.6.1**, which displays a relationship between a spectral property of the Hamiltonian matrix and the quantity $\beta(A)$.

The discrete-analog of $\beta(A)$ is defined to be

$$\gamma(A) = \min\{\|E\| \text{ for some } \theta \in \mathbb{R}; \ e^{i\theta} \in \Omega(A + E)\}.$$

An analog of Theorem 7.6.1 (**Theorem 7.6.3**) is stated and a bisection algorithm (**Algorithm 7.6.2**) based on this theorem is described.

Robust Stability

Given a stable matrix A, one naturally wonders if the matrix $A + E$ remains stable, where E is a certain perturbed matrix. Two bounds for E guaranteeing the stability of the perturbed matrix $(A + E)$ are given, in terms of solutions of certain Lyapunov equations (**Theorems 7.7.1** and **7.7.2**).

Stability Radius

Section 7.8 deals with the **structured stability radius**. If the perturbations are of the form $B \triangle C$, where $\triangle$ is an unknown perturbation matrix, then it is of interest to know the size of smallest $\triangle$ (measured using 2-norm) that will destabilize the perturbed matrix $A + B \triangle C$. In this context, the concept of **stability radius** is introduced, and formulas both for the complex stability radius (**Theorem 7.8.1**) and the real stability radius are stated.

H_2-Norm

The H_2-norm of a stable transfer, transfer function measures the steady-state covariance of the output response $y = Gv$ to the white noise inputs v. An algorithm (**Algorithm 7.2.1**) for computing the H_2-norm, based on computing the controllability or observability Grammian via Lyapunov equations is given.

7.11 CHAPTER NOTES AND FURTHER READING

A voluminous work has been published on Lyapunov stability theory since the historical monograph "**Problème de la stabilité du Mouvement**" was published by the Russian mathematician A.M. Liapunov in 1892. Some of the books that exclusively deal with Lyapunov stability are those by LaSalle and Lefschetz (1961), Lehnigk (1966), etc., and a good account of diagonal stability and diagonal-type

Lyapunov functions appears in the recent book by Kaszkurewicz and Bhaya (1999). For a good account of BIBO and BIBS stability, see the book by DeCarlo (1989).

In Section 7.2, we have just given a very brief account of the Lyapunov stability adapted to the linear case. The matrix equation version in the linear case seems to have first appeared in the book by Gantmacher (1959, Vol. II). There exist many proofs of Lyapunov stability theorem (**Theorem 7.2.3**). The proof given here is along the line of Bellman (1960). See also Hahn (1955). The proofs of the other theorems in this section can be found in most linear systems books, including the books by Chen (1984), Kailath (1980), Wonham (1986), etc.

The inertia theory has been mainly confined to the linear algebra literature. An excellent account of its control theoretic applications appear in Glover (1984) and in the book by Zhou et al. (1996).

There are also a few papers on the inertia theory with respect to more general regions in the complex plane other than the half-planes and the unit circle given in Section 7.4. Inertia theory has been applied to obtain elementary proofs of several classical root-location problems in Datta (1978a, 1978b, 1979). For an account of this work, see the recent survey paper of the author (Datta 1999). The inertia and stability algorithm is due to Carlson and Datta (1979b). The algorithm has been modified by Datta and Datta (1987) and extended to other regions in the complex plane in Datta and Datta (1986).

The concept of distance to instability was perhaps introduced by Van Loan (1985). The bisection algorithm (**Algorithm 7.6.1**) is due to Byers (1988).

There are now several good books on robust control. These include the books by Dorato and Yedavalli (1989), Hinrichsen and Martensson (1990), Barmish (1994), Bhattacharyya et al. (1995), Green and Limebeer (1995), Zhou et al. (1996). The concept of complex stability radius as robustness measures for stable matrices (in the form given here) was introduced by Hinrichsen and Pritchard (1986). There are several good papers on this subject in the book "**Control of uncertain systems**," edited by Hinrichsen and Martensson (1990). Discussion of Section 7.8 has been taken from Qiu et al. (1995).

Exercises

7.1 Verify that the spring-mass system of Example 5.2.3 is not asymptotically stable. What is the physical interpretation of the above statement?

7.2 Consider the problem of a cart with two sticks considered in Exercise 5.3 of Chapter 5. Take $M_1 = M_2 = M$.

    (a) Show that at the equilibrium states, $\bar{x}_1$ and $\bar{x}_2$ are nonzero and $\bar{x}_3 = \bar{x}_4 = 0$. What is the physical significance of this?

    (b) Show that the system is not asymptotically stable.

7.3 Consider the stick-balancing problem in Example 5.2.4. Give a mathematical explanation of the fact that without an input to the control system, if the stick is not upright with zero velocity, it will fall.

7.4 Give a proof of Theorem 7.3.2 from that of Theorem 7.2.3 using the matrix version of the Cayley transformation.

7.5 Prove that the system (7.2.2) is BIBO if and only if $G(s) = C(sI - A)^{-1}B$ has every pole with negative real part.

7.6 Prove that the discrete-time system:

$$x_{k+1} = Ax_k + Bu_k$$

is BIBO stable if and only if all the poles of the transfer functions lie inside the open unit circle of the z-plane.

7.7 Prove that the discrete-time system in Exercise 7.6 is BIBS if and only if (i) all the eigenvalues of A lie in the closed unit disc, (ii) the eigenvalues on the unit disc have multiplicity 1 in the minimal polynomial of A, and (iii) the unit circle modes are uncontrollable (consult DeCarlo (1989, p. 422)).

7.8 Let X and M be the symmetric positive definite matrices such that

$$XA + A^TX + 2\lambda X = -M,$$

then prove that all eigenvalues of A have a real part that is less than $-\lambda$.

7.9 Prove that A is a stable matrix if and only if $\|e^{At}\| \le k$, for some $k > 0$.

7.10 Prove that if M is positive definite and the discrete Lyapunov equation:

$$X - A^TXA = M$$

has a symmetric positive definite solution X, then A is discrete-stable.

7.11 Prove the following results:

(a) Suppose that A is discrete-stable. Then (A, B) is controllable if and only if the discrete Lyapunov equation:

$$X - AXA^T = BB^T$$

has a unique positive definite solution.

(b) Suppose that (A, B) is controllable. Then A is discrete-stable if and only if the discrete Lyapunov equation:

$$X - AXA^T = BB^T$$

has a unique positive definite solution

(c) Suppose that (A, C) is observable. Then A is discrete-stable if and only if there exists a unique positive definite solution X of the discrete Lyapunov equation:

$$X - A^TXA = C^TC.$$

7.12 (Glover 1984).
Let

$$X = X^T = \begin{pmatrix} X_1 & 0 \\ 0 & X_2 \end{pmatrix}$$

with $\delta(X) = 0$. Suppose that

$$AX + XA^T = -BB^T,$$
$$A^T X + XA = -C^T C.$$

Partition

$$A = \begin{pmatrix} A_{11} & A_{12} \\ A_{21} & A_{22} \end{pmatrix}, \qquad B = \begin{pmatrix} B_1 \\ B_2 \end{pmatrix},$$

and $C = (C_1, C_2)$, conformably with X. Then prove the following:

(a) If $\gamma(X_1) = 1$, then $\pi(A_{11}) = 0$.
(b) If $\delta(A) = 0$ and $\lambda_i(X_1^2) \neq \lambda_j(X_2^2)$ $\forall i, j$, then $\text{In}(A_{11}) = \text{In}(-X_1)$ and $\text{In}(A_{22}) = \text{In}(-X_2)$. (Here $\lambda_i(M)$ denotes the ith eigenvalue of M.)

7.13 Prove that in Theorem 7.4.2, the assumption that (A^T, M) is controllable implies that $\delta(X) = 0$.

7.14 Let A be a stable matrix. Prove that (i) $\beta(A) = \min_{\omega \in R} \sigma_{\min}(A - j\omega I)$, (ii) $\beta(A) \leq |\alpha(A)|$, where $\alpha(A) = \max\{Re(\lambda)|\lambda \text{ is an eigenvalue of } A\}$.

7.15 Give an example to show that the formula of $\beta(A)$ given by (7.6.2) can be arbitrary large (Consult the paper of Demmel (1987)).

7.16 Construct an example to show that a matrix A can be very near to an unstable matrix without $\alpha(A)$, defined in Exercise 7.14, being small.

7.17 Let $\text{Arg}(z)$ represent the argument of the complex number z. Let $r > 0$ and $\rho \in \mathbb{C}$, then prove that $r^{-1}\gamma(r(A + \rho I))$ is the distance from A to the nearest matrix with an eigenvalue on the circle $\{z \in C \mid |z - \rho| = r^{-1}\}$, where $\gamma(M)$ denotes the distance of a discrete-stable matrix M to instability, defined by (7.6.10).
Use the result to develop an algorithm to estimate this quantity.

7.18 Give proofs of Theorems 7.7.1 and 7.7.2 (consult the associated papers, as necessary).

7.19 Consider the perturbed system:

$$\dot{x} = (A + BKC)x,$$

where

$$A = \text{diag}(-1, -2, -3), \qquad B = \begin{pmatrix} 1 & 0 \\ 0 & 1 \\ 1 & 1 \end{pmatrix},$$

$$C = \begin{pmatrix} 1 & 0 & 1 \\ 0 & 1 & 0 \end{pmatrix}, \qquad K = \begin{pmatrix} -1 + k_1 & 0 \\ 0 & -1 + k_2 \end{pmatrix},$$

and k_1 and k_2 are two uncertain parameters varying in the intervals around zero. Use each of the Theorems 7.7.1 and 7.7.2 to calculate and compare the allowable bounds on k_1 and k_2 that guarantee the stability of $A + BKC$.

7.20 Construct an example to verify each of the followings:

(a) The real stability radius is always greater than or equal to the complex stability radius.

(b) The ratio of the real stability radius to the complex stability radius can be made arbitrarily large.

7.21 Prove that the H_2-norm of the discrete-time transfer matrix

$$G(z) = \left[\begin{array}{c|c} A & B \\ \hline C & 0 \end{array}\right]$$

can be computed as

$$\|G(z)\|_2^2 = \text{Trace}(C C_G^D C^T) = B^T O_G^D B,$$

where C_G^D and O_G^D are, respectively, the discrete-time controllability and observability Grammians given by (7.3.6) and (7.3.7), respectively. Write down a Lyapunov equation based algorithm to compute the H_2-norm of a discrete-time system based on the above formula.

7.22 Give a proof of Theorem 7.8.2.

References

Ackermann J. *Robust Control: Systems with Uncertain Physical Parameters*, Springer-Verlag, New York, 1993.

Barmish B.R. *New Tools for Robustness of Linear Systems*, McMillan Publishing Co., New York, 1994.

Bellman R. *Introduction to Matrix Analysis*, McGraw Hill, New York, 1960.

Bhattacharyya S.P., Chapellat H., and Keel L.H. In *Robust Control: the Parametric Approach*, (Thomas Kailath, ed.), Prentice Hall Information and Systems Sciences Series, Prentice Hall, Upper Saddle River, NJ, 1995.

Boyd S., El Ghaoui L., Feron E., and Balakrishnan V. "Linear Matrix Inequalities in System and Control Theory," Studies Appl Math, SIAM, Vol. 15, Philadelphia, 1994.

Bunch J.R. and Parlett B.N. "Direct methods for solving symmetric indefinite systems of linear equations," *SIAM J. Numer. Anal.*, Vol. 8, pp. 639–655, 1971.

Bunch J.R. and Kaufman L. "Some stable methods for calculating inertia and solving symmetric linear systems," *Math. Comp.*, Vol. 31, pp. 162–179, 1977.

Bunch J.R. "Analysis of the diagonal pivoting method," *SIAM J. Numer. Anal.*, Vol. 8, pp. 656–680, 1971.

Byers R. "A bisection method for measuring the distance of a stable matrix to the unstable matrices," *SIAM J. Sci. Stat. Comput.*, Vol. 9(5), pp. 875–881, 1988.

Carlson D. and Datta B.N. "The Lyapunov matrix equation $SA + A^*S = S^*B^*BS$," *Lin. Alg. Appl.*, Vol. 28, pp. 43–52, 1979a.

Carlson D. and Datta B.N. "On the effective computation of the inertia of a nonhermitian matrix," *Numer. Math.*, Vol. 33, pp. 315–322, 1979b.

Carlson D. and Schneider H. "Inertia theorems for matrices: the semidefinite case," *J. Math. Anal. Appl.*, Vol. 6, pp. 430–446, 1963.

Chen C.-T. "A generalization of the inertia theorem," *SIAM J. Appl. Math.*, Vol. 25, pp. 158–161, 1973.

Chen C.-T. *Linear Systems Theory and Design*, College Publishing, New York, 1984.

Datta B.N. "On the Routh-Hurwitz-Fujiwara and the Schur-Cohn-Fujiwara theorems for the root-separation problems," *Lin. Alg. Appl.*, Vol. 22, pp. 135–141, 1978a.

Datta B.N. "An elementary proof of the stability criterion of Liénard and Chipart," *Lin. Alg. Appl.*, Vol. 122, pp. 89–96, 1978b.

Datta B.N. "Applications of Hankel matrices of Markov parameters to the solutions of the Routh-Hurwitz and the Schur-Cohn problems," *J. Math. Anal. Appl.*, Vol. 69, pp. 276–290, 1979.

Datta B.N. "Matrix equations, matrix polynomial, and the number of zeros of a polynomial inside the unit circle," *Lin. Multilin. Alg.* Vol. 9, pp. 63–68, 1980.

Datta B.N. *Numerical Linear Algebra and Applications*, Brooks/Cole Publishing Company, Pacific Grove, CA, 1995.

Datta B.N. "Stability and Inertia," *Lin. Alg. Appl.*, Vol. 302/303, pp. 563–600, 1999.

Datta B.N. and Datta K. " On finding eigenvalue distribution of a matrix in several regions of the complex plane," *IEEE Trans. Autom. Control*, Vol. AC-31, pp. 445–447, 1986.

Datta B.N. and Datta K. "The matrix equation $XA = A^{T}X$ and an associated algorithm for inertia and stability," *Lin. Alg. Appl.*, Vol. 97, pp. 103–109, 1987.

DeCarlo R.A. *Linear Systems—A State Variable Approach with Numerical Implementation*, Prentice Hall, Englewood Cliffs, NJ, 1989.

Demmel J.W. "A Counterexample for two conjectures about stability," *IEEE Trans. Autom. Control*, Vol. AC-32, pp. 340–342, 1987.

Dorato P. and Yedavalli R.K. (eds.), *Recent Advances in Robust Control, IEEE Press*, New York, 1989.

Gantmacher F.R. *The Theory of Matrices*, Vol. 1 and Vol. II., Chelsea, New York, 1959.

Golub G.H. and Van Loan C.F. *Matrix Computations*, 3rd edn, Johns Hopkins University Press, Baltimore, MD, 1996.

Glover K. "All optimal Hankel-norm approximation of linear multivariable systems and their L_{∞} error bounds," *Int. J. Control*, Vol. 39, pp. 1115–1193, 1984.

Green M. and Limebeer D.J. *Linear Robust Control*, (Thomas Kailath, ed.), Prentice Hall Information and Systems Sciences Series, Prentice Hall, NJ, 1995.

Hahn W. "Eine Bemerkung zur zweiten methode von Lyapunov," *Math. Nachr.*, Vol. 14, pp. 349–354, 1955.

Hewer G.A. and Kenney C.S. "The sensitivity of stable Lyapunov equations," *SIAM J. Control Optimiz.*, Vol. 26, pp. 321–344, 1988.

Hinrichsen D. and Martensson B. (eds.), *Control of Uncertain Systems*, Birkhauser, Berlin, 1990.

Hinrichsen D. and Pritchard A.J. Stability radii of linear systems, *Syst. Control Lett.*, Vol. 7, pp. 1–10, 1986.

Hinrichsen D. and Pritchard A.J. Real and complex stability radii: a survey, *Control of Uncertain Systems*, (D. Hinrichsen and Martensson B., eds.), Birkhauser, Berlin, 1990.

Horn R.A. and Johnson C.R. *Matrix Analysis*, Cambridge University Press, Cambridge, UK, 1985.

Kailath T. *Linear Systems*, Prentice Hall, Englewood Cliffs, NJ, 1980.

Kaszkurewicz E. and Bhaya A. *Matrix Diagonal Stability in Systems and Computations*, Birkhauser, Boston, 1999.

Keel L.H., Bhattacharyya S.P., and Howze J.W. "*Robust control with structured perturbations*," *IEEE Trans. Automat. Control*, Vol. 33, pp. 68–78, 1988.

LaSalle J.P. and Lefschetz S. *Stability by Lyapunov's Direct Method with Applications*, Academic Press, New York, 1961.

Lehnigk S.H. *Stability Theorems for Linear Motions with an Introduction to Lyapunov's Direct Method*, Prentice Hall, Englewood Cliffs, NJ, 1966.

Lu L.Z. "A direct method for the solution of the unit circle problem," 1987 (unpublished manuscript).

Luenberger D.G. *Introduction to Dynamic Systems: Theory, Models, and Applications,* John Wiley & Sons, New York, 1979.

Liapunov A.M. "Probléme général de la stabilité du mouvement," *Comm. Math. Soc. Kharkov*, 1892; *Ann. Fac. Sci.,* Toulouse, Vol. 9, 1907; *Ann. Math. Studies*, Vol. 17, 1947; *Princeton University Press,* Princeton, NJ, 1949.

Malyshev A. and Sadkane M. "On the stability of large matrices," *J. Comput. Appl. Math.*, Vol. 102, pp. 303–313, 1999.

Marden M. *Geometry of Polynomials*, American Mathematical Society, Providence, RI, 1966.

Ostrowski A. and Schneider H. "Some theorems on the inertia of general matrices," *J. Math. Anal. Appl.*, Vol. 4, pp. 72–84, 1962.

Patel R.V. and Toda M. "Quantitative measures of robustness for multivariable systems," *Proc. Amer. Control Conf.* , San Francisco, 1980.

Petkov P., Christov N.D., and Konstantinov M.M., *Computational Methods for Linear Control Systems,* Prentice Hall, London, 1991.

Pierce D. *A Computational Comparison of Four Methods which Compute the Inertia of a General Matrix*, M. Sc. Thesis, *Northern Illinois University*, DeKalb, IL, 1983.

Qiu L., Bernhardsson B., Rantzer B., Davison E.J., Young P.M., and Doyle J.C. "A formula for computation of the real stability radius," *Automatica*, Vol. 31, pp. 879–890, 1995.

Rudin W. *Real and Complex Analysis*, McGraw Hill, New York, 1966.

Taussky O. "A generalization of a theorem of Lyapunov," *J. Soc. Ind. Appl. Math.*, Vol. 9, pp. 640–643, 1961.

Taussky O. "Matrices C with $C^n \rightarrow 0$," *J. Algebra*, Vol. 1 pp. 5–10, 1964.

Van Loan C.F. "How near is a stable matrix to an unstable matrix," in (Brualdi R., *et al.*, eds.), *Contemporary Math.*, American Mathematical Society, Providence, RI, Vol. 47, pp. 465–477, 1985.

Wilkinson J.H. *The Algebraic Eigenvalue Problem*, Clarendon Press, Oxford, 1965.

Wimmer H.K. On the Ostrowski-Schneider inertia theorem, *J. Math. Anal. Appl.*, Vol. 41, pp. 164–169, 1973.

Wimmer H.K. "Inertia theorems for matrices, controllability, and linear vibrations," *Lin. Alg. Appl.*, Vol. 8, pp. 337–343, 1974.

Wimmer H.K. and Ziebur A.D. "Remarks on inertia theorems for matrices," *Czech. Math. J.* Vol. 25, pp. 556–561, 1975.

Wonham W.M. *Linear Multivariable Systems,* Springer-Verlag, New York, 1986.

Yedavalli R.K. "Improved measures of stability robustness for linear state space models," *IEEE Trans. Autom. Control*, Vol. AC-30, pp. 577–579, 1985.

Zhou K., Doyle J.C., and Glover K. *Robust Optimal Control*, Prentice Hall, Upper Saddle River, NJ, 1996.

Zhou K. and Khargonekar P.P. "Stability robustness for linear state-space models with structured uncertainty," *IEEE Trans. Autom. Control*, Vol. AC-32, pp. 621–623, 1987.

CHAPTER 8

NUMERICAL SOLUTIONS AND CONDITIONING OF LYAPUNOV AND SYLVESTER EQUATIONS

Topics covered

- Existence and Uniqueness Results for Solutions of Lyapunov and Sylvester Equations
- Perturbation Analyses and Condition Numbers
- The Schur and the Hessenberg–Schur Methods (Both Continuous and Discrete-Time Cases)
- Backward Error Analyses of the Schur and the Hessenberg–Schur Methods
- Direct Computations of Cholesky Factors of Symmetric Positive Definite Solutions of Lyapunov Equations

---

8.1 INTRODUCTION

In **Chapter 7**, we have seen that the Lyapunov equations arise in **stability** and **robust stability** analyses, in determining **controllability** and **observability Grammians**, and in **computing H_2-norm**. The solutions of Lyapunov equations are also needed for the implementation of some **iterative methods for solving algebraic Riccati equations (AREs),** such as Newton's methods **(Chapter 13)**. The important role of Lyapunov equations in these practical applications warrants discussion of numerically viable techniques for their solutions.

The continuous-time **Lyapunov equation**:

$$XA + A^{\mathrm{T}}X = C \tag{8.1.1}$$

is a special case of another classical matrix equation, known as the **Sylvester equation**:

$$XA + BX = C. \tag{8.1.2}$$

Similarly, the discrete-time Lyapunov equation:

$$A^T X A - X = C$$

is a special case of the discrete-time Sylvester equation:

$$B X A - X = C.$$

(Note *that the matrices A, B, C in the above equations are not necessarily the system matrices.*)

The Sylvester equations also arise in a wide variety of applications. For example, we will see in **Chapter 12** that a variation of the Sylvester equation, known as the **Sylvester-observer** equation, arises in the construction of **observers** and in solutions of the **eigenvalue assignment** (EVA) (or pole-placement) problems. The Sylvester equation also arises in other areas of applied mathematics. For example, the numerical solution of elliptic boundary value problems can be formulated in terms of the solution of the Sylvester equation (Starke and Niethammer 1991). The solution of the Sylvester equation is also needed in the block diagonalization of a matrix by a similarity transformation (see Datta 1995) and Golub and Van Loan (1996). Once a matrix is transformed to a block diagonal form using a similarity transformation, the block diagonal form can then be conveniently used to compute the matrix exponential e^{At}.

In this chapter, we will first develop the basic theories on the **existence** and **uniqueness** of solutions of the Sylvester and Lyapunov equations (**Section 8.2**), next discuss **perturbation theories** (**Section 8.3**), and then finally describe **computational methods** (**Sections 8.5 and 8.6**).

The continuous-time Lyapunov equation (8.1.1) and the continuous-time Sylvester equation (8.1.2) will be referred to as just the **Lyapunov** and **Sylvester equations**, respectively.

The following methods are discussed in this chapter. They have excellent numerical properties and are recommended for use in practice:

- The **Schur methods** for the Lyapunov equations (**Sections 8.5.2 and 8.5.4**).
- The **Hessenberg–Schur Method** for the Sylvester equations (**Algorithm 8.5.1 and Section 8.5.7**).
- The **modified Schur methods** for the Cholesky factors of the Lyapunov equations (**Algorithms 8.6.1 and 8.6.2**).

Besides, a **Hessenberg method** (method based on Hessenberg decomposition only) for the Sylvester equation $AX + XB = C$ has been described in **Section 8.5.6**. The method is more efficient than the Hessenberg–Schur method, but numerical stability of this method has not been investigated yet. At present, the method is mostly of theoretical interest only.

Because of possible numerical instabilities, solving the Lyapunov and Sylvester equations via the Jordan canonical form (JCF) or a companion form of the matrix A cannot be recommended for use in practice (see discussions in Section 8.5.1).

8.2   THE EXISTENCE AND UNIQUENESS OF SOLUTIONS

In most numerical methods for solving matrix equations, it is implicitly assumed that the equation to be solved has a unique solution, and the methods then construct the unique solution. Thus, the results on the existence and uniqueness of solutions of the Sylvester and Lyapunov equations are of importance. We present some of these results in this section.

8.2.1   The Sylvester Equation: $XA + BX = C$

Assume that the matrices A, B, and C are of dimensions $n \times n$, $m \times m$, and $m \times n$, respectively. Then the following is the fundamental result on the existence and uniqueness of the Sylvester equation solution.

Theorem 8.2.1. *Uniqueness of the Sylvester Equation Solution. Let $\lambda_1, \ldots, \lambda_n$ be the eigenvalues of A, and $\mu_1, \ldots, \mu_m$, be the eigenvalues of B. Then the Sylvester equation (8.1.2) has a unique solution X if and only if $\lambda_i + \mu_j \neq 0$ for all $i = 1, \ldots, n$ and $j = 1, \ldots, m$.* **In other words, the Sylvester equation has a unique solution if and only if A and $-B$ do not have a common eigenvalue.**

Proof.   The Sylvester equation $XA + BX = C$ is equivalent to the $nm \times nm$ linear system

$$Px = c, \tag{8.2.1}$$

where $P = (I_n \otimes B) + (A^{\mathrm{T}} \otimes I_m)$,

$$x = \mathrm{vec}(X) = (x_{11}, \ldots, x_{m1}, x_{12}, x_{22}, \ldots, x_{m2}, \ldots, x_{1n}, x_{2n}, \ldots, x_{mn})^{\mathrm{T}},$$
$$c = \mathrm{vec}(C) = (c_{11}, \ldots, c_{m1}, c_{12}, c_{22}, \ldots, c_{m2}, \ldots, c_{1n}, c_{2n}, \ldots, c_{mn})^{\mathrm{T}}.$$

Thus, the Sylvester equation has a unique solution if and only if P is non-singular.

Here $W \otimes Z$ is the **Kronecker product** of W and Z. Recall from Chapter 2 that if $W = (w_{ij})$ and $Z = (z_{ij})$ are two matrices of orders $p \times p$ and $r \times r$, respectively, then their Kronecker product $W \otimes Z$ is defined by

$$W \otimes Z = \begin{pmatrix} w_{11}Z & w_{12}Z & \cdots & w_{1p}Z \\ w_{21}Z & w_{22}Z & \cdots & w_{2p}Z \\ \vdots & \vdots & & \vdots \\ w_{p1}Z & w_{p2}Z & \cdots & w_{pp}Z \end{pmatrix}. \tag{8.2.2}$$

Thus, the Sylvester equation (8.1.2) has a unique solution if and only if the matrix P of the system (8.2.1) is nonsingular.

Now, the eigenvalues of the matrix P are the nm numbers $\lambda_i + \mu_j$, where $i = 1, \ldots, n$ and $j = 1, \ldots, m$ (Horn and Johnson 1991). Since the determinant of a matrix is equal to the product of its eigenvalues, this means that P is nonsingular if and only if $\lambda_i + \mu_j \neq 0$, for $i = 1, \ldots, n$, and $j = 1, \ldots, m$. ∎

8.2.2   The Lyapunov Equation: $XA + A^T X = C$

Since the Lyapunov equation (8.1.1) is a special case of the Sylvester (8.1.2) equation, the following corollary is immediate.

Corollary 8.2.1. *Uniqueness of the Lyapunov Equation Solution. Let* λ_1, $\lambda_2, \ldots, \lambda_n$ *be the eigenvalues of A. Then the Lyapunov equation (8.1.1) has a unique solution X if and only if* $\lambda_i + \lambda_j \neq 0$, $i = 1, \ldots, n$; $j = 1, \ldots, n$.

8.2.3   The Discrete Lyapunov Equation: $A^T X A - X = C$

The following result on the uniqueness of the solution X of the discrete Lyapunov equation

$$A^T X A - X = C \tag{8.2.3}$$

can be established in the same way as in the proof of Theorem 8.2.1.

Theorem 8.2.2. *Uniqueness of the Discrete Lyapunov Equation Solution. Let* $\lambda_1, \ldots, \lambda_n$ *be the n eigenvalues of A. Then the discrete Lyapunov equation (8.2.3) has a unique solution X if and only if* $\lambda_i \lambda_j \neq 1, i = 1, \ldots, n$; $j = 1, \ldots, n$.

Remark

- In the above theorems, we have given results only for the uniqueness of solutions of the Sylvester and Lyapunov equations. However, there are certain control problems such as the **construction of Luenberger observer** and the **EVA problems**, etc., that require **nonsingular or full-rank solutions of the Sylvester equations** (see **Chapter 12**).

The nonsingularity of the unique solution of the Sylvester equation has been completely characterized recently by Datta *et al.* (1997). Also, partial results on nonsingularity of the Sylvester equation were obtained earlier by DeSouza and Bhattacharyya (1981), and Hearon (1977). **We will state these results in Chapter 12.**

8.3   PERTURBATION ANALYSIS AND THE CONDITION NUMBERS

8.3.1   Perturbation Analysis for the Sylvester Equation

In this section, we study perturbation analyses of the Sylvester and Lyapunov equations and identify the condition numbers for these problems. The results are important in assessing the accuracy of the solution obtained by a numerical algorithm. We also present an algorithm (**Algorithm 8.3.1**) for estimating the sep function that arises in computing the condition number of the Sylvester equation.

Let ΔA, ΔB, ΔC, and ΔX be the perturbations, respectively, in the matrices A, B, C, and X. Let $\hat{X}$ be the solution of the perturbed problem. That is, $\hat{X}$ satisfies

$$\hat{X}(A + \Delta A) + (B + \Delta B)\hat{X} = C + \Delta C. \qquad (8.3.1)$$

Then, proceeding as in the case of perturbation analysis for the linear system problem applied to the system (8.2.1), the following result (see Higham 1996) can be proved.

Theorem 8.3.1.   *Perturbation Theorem for the Sylvester Equation. Let the Sylvester equation $XA + BX = C$ have a unique solution X for $C \neq 0$.*
    Let

$$\epsilon = \max\left\{ \frac{\|\Delta A\|_{\mathrm{F}}}{\alpha}, \frac{\|\Delta B\|_{\mathrm{F}}}{\beta}, \frac{\|\Delta C\|_{\mathrm{F}}}{\gamma} \right\} \qquad (8.3.2)$$

where α, β, and γ are tolerances such that $\|\Delta A\|_{\mathrm{F}} \leq \epsilon\alpha$, $\|\Delta B\|_{\mathrm{F}} \leq \epsilon\beta$, and $\|\Delta C\|_{\mathrm{F}} \leq \epsilon\gamma$.
    Then,

$$\frac{\|\Delta X\|_{\mathrm{F}}}{\|X\|_{\mathrm{F}}} = \frac{\|\hat{X} - X\|_{\mathrm{F}}}{\|X\|_{\mathrm{F}}} \leq \sqrt{3}\epsilon\delta, \qquad (8.3.3)$$

where $\delta = \|P^{-1}\|_2 \dfrac{(\alpha + \beta)\|X\|_{\mathrm{F}} + \gamma}{\|X\|_{\mathrm{F}}}.$

Sep Function and its Role in Perturbation Results for the Sylvester Equation

Definition 8.3.1.   *The separation of two matrices A and B, denoted by $\mathrm{sep}(A, B)$, is defined as:*

$$\mathrm{sep}(A, B) = \min_{X \neq 0} \frac{\|AX - XB\|_{\mathrm{F}}}{\|X\|_{\mathrm{F}}}$$

Thus, in terms of the sep *function, we have*

$$\|P^{-1}\|_2 = \frac{1}{\sigma_{\min}(P)} = \frac{1}{\text{sep}(B, -A)}. \tag{8.3.4}$$

Using sep function, the inequality (8.3.3) can be re-written as:

$$\frac{\|\Delta X\|_F}{\|X\|_F} < \sqrt{3}\epsilon \frac{1}{\text{sep}(B, -A)} \frac{(\alpha + \beta)\|X\|_F + \gamma}{\|X\|_F}.$$

The perturbation result (8.3.3) clearly depends upon the norm of the solution X. However, if the relative perturbations in A, B, and C are only of the order of the machine epsilon, then the following result, independent of $\|X\|$, due to Golub *et al.* (1979), can be established.

Corollary 8.3.1. *Assume that the relative perturbations in A, B, and C are all of the order of the machine precision μ, that is, $\|\Delta A\|_F \leq \mu\|A\|_F$, $\|\Delta B\|_F \leq \mu\|B\|_F$, and $\|\Delta C\|_F \leq \mu\|C\|_F$.*

If X is a unique solution of the Sylvester equation $XA + BX = C$, C is nonzero and

$$\mu\frac{\|A\|_F + \|B\|_F}{\text{sep}(B, -A)} \leq \frac{1}{2}. \tag{8.3.5}$$

Then,

$$\frac{\|\hat{X} - X\|_F}{\|X\|_F} \leq 4\mu\frac{\|A\|_F + \|B\|_F}{\text{sep}(B, -A)}. \tag{8.3.6}$$

Example 8.3.1. Consider the Sylvester equation $XA + BX = C$ with

$$A = \begin{pmatrix} 1 & 1 & 1 \\ 0 & 1 & 1 \\ 0 & 0 & 1 \end{pmatrix}, \quad B = \begin{pmatrix} -0.9888 & 0 & 0 \\ 0 & -0.9777 & 0 \\ 0 & 0 & -0.9666 \end{pmatrix}.$$

Take $X = \begin{pmatrix} 1 & 1 & 1 \\ 1 & 1 & 1 \\ 1 & 1 & 1 \end{pmatrix}$. Then, $C = \begin{pmatrix} 0.0112 & 1.0112 & 2.0112 \\ 0.0223 & 1.0223 & 2.0223 \\ 0.0334 & 1.0334 & 2.0334 \end{pmatrix}$.

Now, change the entry $(1, 1)$ of A to 0.999999. Call this perturbed matrix $\hat{A}$. The matrices B and C remain unperturbed.

The computed solution of the perturbed problem (computed by MATLAB function **lyap**)

$$\hat{X} = \begin{pmatrix} 1.0001 & 0.9920 & 1.7039 \\ 1.0000 & 0.9980 & 1.0882 \\ 1.0000 & 0.9991 & 1.0259 \end{pmatrix}.$$

The relative error in the solution:

$$\frac{\|\hat{X} - X\|_F}{\|X\|_F} = 0.2366.$$

On the other hand, the relative error in the data:

$$\frac{\|A - \hat{A}\|_F}{\|A\|_F} = 4.0825 \times 10^{-7}.$$

The phenomenon can be easily explained by noting that $\text{sep}(B, -A)$ is small: $\text{sep}(B, -A) = 1.4207 \times 10^{-6}$.

Verification of the Bound 8.3.3
Take $\alpha = \|A\|_F$, $\beta = \|B\|_F$, and $\gamma = \|C\|_F$
    Then,

$$\epsilon = \frac{\|\hat{A} - A\|_F}{\|A\|_F} = 4.0825 \times 10^{-7} \quad \text{(Note that } \|\Delta B\| = 0 \text{ and } \|\Delta C\| = 0\text{)}.$$

The right-hand side of (8.3.3) is 2.7133.
    Since

$$\frac{\|\hat{X} - X\|_F}{\|X\|_F} = 0.2366,$$

the inequality (8.3.3) is satisfied.

8.3.2   The Condition Number of the Sylvester Equation

The perturbation bound for the Sylvester equation given in Theorem 8.3.1 does not take into account the Kronecker structure of the coefficient matrix P. The bound (8.3.3) can sometimes overestimate the effects of perturbations when A and B are only perturbed. A much sharper perturbation bound that exploits the Kronecker structure of P has been given by Higham (1996, p. 318).

    Specifically, the following result has been proved.

Theorem 8.3.2.   *Let*

$$\epsilon = \max \left\{ \frac{\|\Delta A\|_F}{\alpha}, \frac{\|\Delta B\|_F}{\beta}, \frac{\|\Delta C\|_F}{\gamma} \right\},$$

where α, β, and γ are tolerances given by $\|\Delta A\|_F \leq \epsilon\alpha$, $\|\Delta B\| \leq \epsilon\beta$, and $\|\Delta C\| \leq \epsilon\gamma$. Let ΔX denote the perturbation in the solution X of the Sylvester equation (8.1.2). Let P be defined by (8.2.1).
   Then,

$$\frac{\|\Delta X\|_F}{\|X\|_F} \leq \sqrt{3}\Psi\epsilon, \tag{8.3.7}$$

where

$$\Psi = \|P^{-1}[\beta(X^T \otimes I_m), \alpha(I_n \otimes X), -\gamma I_{mn}]\|_2/\|X\|_F. \tag{8.3.8}$$

The bound (8.3.7) can be attained for any A, B, and C and we shall call the number Ψ the **condition number** of the Sylvester equation.

Remark

- Examples can be constructed that show that the bounds (8.3.3) and (8.3.7) can differ by an arbitrary factor. For details, see Higham (1996).

MATCONTROL note: The condition number given by (8.3.7)–(8.3.8) has been implemented in MATCONTROL function **condsylvc**.

Example 8.3.2.   We verify the results of Theorem 8.3.2 with Example 8.3.1. Take $\alpha = \|A\|_F = 2.4494$. Then, $\epsilon = 4.0825 \times 10^{-7}$.
   The condition number is $\Psi = 1.0039 \times 10^6$.

$$\frac{\|\Delta X\|_F}{\|X\|_F} = 0.2366 \quad \text{and} \quad \sqrt{3}\Psi\epsilon = 0.7099.$$

Thus, the inequality (8.3.7) is verified.

8.3.3   Perturbation Analysis for the Lyapunov Equation

Since the Lyapunov equation $XA + A^T X = C$ is a special case of the Sylvester equation, we immediately have the following Corollary of Theorem 8.3.1.

Corollary 8.3.2.   *Perturbation Theorem for the Lyapunov Equation. Let X be the unique solution of the Lyapunov equation $XA + A^T X = C$; $C \neq 0$. Let $\hat{X}$ be* *the unique solution of the perturbed equation* $\hat{X}(A + \triangle A) + (A + \triangle A)^T \hat{X} = C,$

$$\text{where } \|\triangle A\|_F \le \mu \|A\|_F. \tag{8.3.9}$$

Assume that

$$\frac{\mu \|A\|_F}{\text{sep}(A^T, -A)} = \delta < \frac{1}{4}. \tag{8.3.10}$$

Then,

$$\frac{\|\hat{X} - X\|_F}{\|X\|_F} \le 8\mu \frac{\|A\|_F}{\text{sep}(A^T, -A)}.$$

8.3.4   The Condition Number of the Lyapunov Equation

For the Lyapunov equation, the **condition number** is (Higham (1996, p. 319):

$$\phi = \|(I_n \otimes A^T + A^T \otimes I_n)^{-1}[\alpha((X^T \otimes I_n) + (I_n \otimes X)\Pi^T), -\gamma I_{n^2}]\|_2 / \|X\|_F,$$

where Π is the **vec-permutation matrix** given by

$$\Pi = \sum_{i,j=1}^{n} (e_i e_j^T) \otimes (e_j e_i^T),$$

and α and γ are as defined as:

$$\|\triangle A\|_F \le \epsilon \alpha \quad \text{and} \quad \triangle C = \triangle C^T \quad \text{with} \quad \|\triangle C\|_F \le \epsilon \gamma.$$

8.3.5   Sensitivity of the Stable Lyapunov Equation

While Corollary 8.3.2 shows that the sensitivity of the Lyapunov equation under the assumptions (8.3.9) and (8.3.10) depends upon $\text{sep}(A^T, -A)$, Hewer and Kenney (1988) have shown that if A is a stable matrix, then the sensitivity can be determined by means of the 2-**norm** of the symmetric positive definite solution H of the equation

$$HA + A^T H = -I.$$

Specifically, the following result has been proved:

Theorem 8.3.3.   *Perturbation Result for the Stable Lyapunov Equation. Let A be stable and let X satisfy $XA + A^T X = -C$. Let $\triangle X$ and $\triangle C$, respectively,* *be the perturbations in X and C such that*

$$(A + \Delta A)^{\mathrm{T}}(X + \Delta X) + (X + \Delta X)(A + \Delta A) = -(C + \Delta C). \quad (8.3.11)$$

Then,

$$\frac{\|\Delta X\|}{\|X + \Delta X\|} \leq 2\|A + \Delta A\|\|H\|\left[\frac{\|\Delta A\|}{\|A + \Delta A\|} + \frac{\|\Delta C\|}{\|C + \Delta C\|}\right], \quad (8.3.12)$$

where H satisfies the following Lyapunov equation and $\|\cdot\|$ represents the 2-norm:

$$HA + A^{\mathrm{T}}H = -I.$$

Proof. Since A is stable, we may write $H = \int_0^\infty e^{A^{\mathrm{T}}t}e^{At}\,dt$. Now from $XA + A^{\mathrm{T}}X = -C$ and (8.3.11), we have

$$A^{\mathrm{T}}\Delta X + \Delta XA = -(\Delta C + \Delta A^{\mathrm{T}}(X + \Delta X) + (X + \Delta X)\Delta A). \quad (8.3.13)$$

Since (8.3.13) is a Lyapunov equation in ΔX and A is stable, we may again write

$$\Delta X = \int_0^\infty e^{A^{\mathrm{T}}t}(\Delta C + \Delta A^{\mathrm{T}}(X + \Delta X) + (X + \Delta X)\Delta A)e^{At}\,dt.$$

Let u and v be the left and right singular vectors of unit length of ΔX associated with the largest singular value. Then multiplying the above equation by u^{T} to the left and by v to the right, we have

$$\|\Delta X\| = \int_0^\infty \left| u^{\mathrm{T}}e^{A^{\mathrm{T}}t}(\Delta C + \Delta A^{\mathrm{T}}(X + \Delta X) \right.$$
$$\left. + (X + \Delta X)\Delta A)e^{At}v \right| dt,$$
$$\leq \|\Delta C + \Delta A^{\mathrm{T}}(X + \Delta X) + (X + \Delta X)\Delta A)\|$$
$$\int_0^\infty \|e^{At}u\|\|e^{At}v\|\,dt,$$
$$\leq (\|\Delta C\| + 2\|\Delta A\|\|X + \Delta X\|)\int_0^\infty \|e^{At}u\|\|e^{At}v\|\,dt. \quad (8.3.14)$$

Now, by the Cauchy–Schwarz inequality, we have

$$\int_0^\infty \|e^{At}u\|\|e^{At}v\|\,dt \leq \left[\int_0^\infty \|e^{At}u\|^2\,dt\right]^{1/2}\left[\int_0^\infty \|e^{At}v\|^2\,dt\right]^{1/2}.$$

Again

$$\int_0^\infty \|e^{At}u\|^2 \, dt = \int_0^\infty u^{\mathrm{T}} e^{A^{\mathrm{T}}t} e^{At} u \, dt,$$

$$= u^{\mathrm{T}} \left[\int_0^\infty e^{A^{\mathrm{T}}t} e^{At} \, dt \right] u,$$

$$= u^{\mathrm{T}} H u, \quad \text{where } H = \int_0^\infty e^{A^{\mathrm{T}}t} e^{At} \, dt$$

Since $\|u\|_2 = 1$ and H is symmetric positive definite (because A is stable), we have

$$u^{\mathrm{T}} H u \le \|H\|,$$

and thus

$$\int_0^\infty \|e^{At}u\|^2 \, dt \le \|H\|.$$

Similarly $\displaystyle\int_0^\infty \|e^{At}v\|^2 \, dt \le \|H\|$.
Thus, from (8.3.14), we have

$$\|\Delta X\| \le (\|\Delta C\| + 2\|\Delta A\|\|X + \Delta X\|)\|H\|. \tag{8.3.15}$$

Again from (8.3.11), we have

$$\|C + \Delta C\| \le 2\|A + \Delta A\|\|X + \Delta X\|. \tag{8.3.16}$$

Combining (8.3.15) with (8.3.16), we obtain the desired result. ∎

Remark

- The results of Theorem 8.3.3 hold for any perturbation.
  In particular, if

$$\|\Delta C\| \le \mu\|C\|, \qquad \|\Delta A\| \le \mu\|A\|,$$

and $8\mu\|A\|\|H\| \le (1 - \mu)/(1 + \mu)$, then it can be shown that

$$\frac{\|\Delta X\|}{\|X\|} \le 8\mu\|A\|\|H\|(1 - \mu) \approx 8\mu\|A\|\|H\|. \tag{8.3.17}$$

Example 8.3.3. Consider the Lyapunov equation (8.1.1) with

$$A = \begin{pmatrix} -1 & 2 & 3 \\ 0 & -0.0001 & 3 \\ 0 & 0 & -3 \end{pmatrix} \quad \text{and} \quad C = \begin{pmatrix} -2 & 0.9999 & 2 \\ 0.9999 & 3.9998 & 4.9999 \\ 2 & 4.9999 & 6 \end{pmatrix}.$$

A is stable. The exact solution X of the Lyapunov equation $XA + A^T X = C$ is

$$X = \begin{pmatrix} 1 & 1 & 1 \\ 1 & 1 & 1 \\ 1 & 1 & 1 \end{pmatrix}.$$

The solution H of the Lyapunov equation $HA + A^T H = -I$ is

$$H = 10^4 \begin{pmatrix} 0.0001 & 0.0001 & 0.0001 \\ 0.0001 & 2.4998 & 2.4999 \\ 0.0001 & 2.4999 & 2.5000 \end{pmatrix}.$$

Since $\|H\| = 4.9998 \times 10^4$ and $\|A\| = 5.3744$, according to Theorem 8.3.3, the Lyapunov equation with above A and C is expected to be ill-conditioned. We verify this as follows:

Perturb the $(1, 1)$ entry of A to -0.9999999 and keep the other entries of A and those of C unchanged. Then the computed solution $\hat{X}$ with this slightly perturbed A is

$$\hat{X} = \begin{pmatrix} 1 & 1 & 1 \\ 1 & 1.006 & 1.006 \\ 1 & 1.006 & 1.006 \end{pmatrix}.$$

Let $\hat{A}$ denote the perturbed A, then the relative perturbation in A:

$$\frac{\|\hat{A} - A\|}{\|A\|} = 1.8607 \times 10^{-8}.$$

The relative error in the solution X:

$$\frac{\|\hat{X} - X\|}{\|X\|} = 0.0040.$$

Verification of the result of Theorem 8.3.3

$$\frac{\|\Delta X\|}{\|X + \Delta X\|} = 0.0040, \qquad \Delta C = 0,$$

$$2\|A + \Delta A\|\|H\|\frac{\|\Delta A\|}{\|A + \Delta A\|} = 0.0100.$$

Thus, the inequality (8.3.12) is satisfied.

Verification of the inequality (8.3.17)

Since $\|\Delta A\|/\|A\| = 1.8607 \times 10^{-8}$, we take $\mu = 1.8607 \times 10^{-8}$.

Then $8\mu\|A\|\|H\| = 0.04 \leq (1-\mu)/(1+\mu) = 0.9999996$. Thus, the hypothesis holds.

Also, $\|\Delta X\|/\|X\| = 0.004$, $8\mu\|A\|\|H\| = 0.04$. Therefore, the inequality (8.3.17) is satisfied.

8.3.6 Sensitivity of the Discrete Lyapunov Equation

Consider now the discrete Lyapunov equation:

$$A^T X A - X = C.$$

This equation is equivalent to the linear system: $Rx = c$, where $R = A^T \otimes A^T - I_{n^2}$ (I_{n^2} is the $n^2 \times n^2$ identity matrix).

Applying the results of perturbation analysis to the linear system $Rx = c$, the following result can be proved.

Theorem 8.3.4. *Perturbation Result for the Discrete Lyapunov Equation. Let X be the unique solution of the discrete Lyapunov equation:*

$$A^T X A - X = C.$$

Let $\hat{X}$ be the unique solution of the perturbed equation where the perturbation in A is of order machine precision μ.
Assume that

$$\frac{(2\mu + \mu^2)\|A\|_F^2}{\text{sep}_d(A^T, A)} = \delta < 1,$$

where

$$\text{sep}_d(A^T, A) = \min_{x \neq 0} \frac{\|Rx\|_2}{\|x\|_2} = \min_{X \neq 0} \frac{\|A^T X A - X\|_F}{\|X\|_F} = \sigma_{\min}(A^T \otimes A^T - I_{n^2}).$$

Then,

$$\frac{\|\hat{X} - X\|_F}{\|X\|_F} \leq \frac{\mu}{1-\delta} \frac{(3+\mu)\|A\|_F^2 + 1}{\text{sep}_d(A^T, A)}. \tag{8.3.18}$$

8.3.7 Sensitivity of the Stable Discrete Lyapunov Equation

As in the case of the stable Lyapunov equation, it can be shown (Gahinet *et al.* 1990) that the sensitivity of the stable discrete Lyapunov equation can also be measured by the **2-norm** of the unique solution of the discrete Lyapunov equation: $A^T X A - X = -I$. Specifically, the following result has been proved by Gahinet *et al.* (1990).

Theorem 8.3.5.   *Sensitivity of the Stable Discrete Lyapunov Equation. Let A be discrete stable. Let H be the unique solution of*

$$A^{\mathrm{T}}HA - H = -I,$$

then $\mathrm{sep}_d(A^{\mathrm{T}}, A) \geq \sqrt{n}/\|H\|_2$.

Example 8.3.4.   Let

$$A = \begin{pmatrix} 0.9990 & 1 & 1 \\ 0 & 0.5000 & 1 \\ 0 & 0 & 0.8999 \end{pmatrix}.$$

Set

$$X = \begin{pmatrix} 1 & 1 & 1 \\ 1 & 1 & 1 \\ 1 & 1 & 1 \end{pmatrix} \quad \text{and take} \quad C = A^{\mathrm{T}}XA - X.$$

Then $\|H\|_2 = 4.4752 \times 10^5$.

By Theorem 8.3.5, the discrete Lyapunov equation $A^{\mathrm{T}}XA - X = C$ is expected to be ill-conditioned. We verify this as follows.

Let $a(2,2)$ be perturbed to 0.4990 and all other entries of A and of C remain unchanged. Let $\hat{A}$ denote the perturbed A. Let $\hat{X}$ be the solution of the perturbed problem. Then $\hat{X}$, computed by the MATLAB function **dlyap**, is

$$\hat{X} = \begin{pmatrix} 1 & 1 & 1.0010 \\ 1 & 1 & 1.0019 \\ 1.0010 & 1.0019 & 1.0304 \end{pmatrix}.$$

The relative error in X:

$$\frac{\|\hat{X} - X\|_2}{\|X\|_2} = 0.0102.$$

The relative perturbation in A:

$$\frac{\|\hat{A} - A\|_2}{\|A\|_2} = 4.8244 \times 10^{-5}.$$

8.3.8   Determining Ill-Conditioning from the Eigenvalues

Since $\|P^{-1}\|_2 = 1/\mathrm{sep}(B, -A)$ is not easily computable, and $\mathrm{sep}(B, -A) > 0$ if and only if B and $-A$ have no common eigenvalues, one may wonder if the ill-conditioning of P^{-1} (and therefore of the Sylvester or the Lyapunov equation) can be determined a priori from the eigenvalues of A and B.

The following result can be easily proved to this effect (Ghavimi and Laub 1995).

Theorem 8.3.6. *The Sylvester equation $XA + BX = C$ is ill-conditioned if both coefficient matrices A and B are ill-conditioned with respect to inversion.*

Example 8.3.5. Let

$$A = \begin{pmatrix} 1 & 1 & 1 \\ 0 & 0 & 1 \\ 1 & 1 & 0.001 \end{pmatrix}, \qquad B = \begin{pmatrix} 1 & 2 & 3 \\ 0 & 0.0001 & 1 \\ 0 & 0 & 0.0001 \end{pmatrix},$$

$$\text{and} \quad C = \begin{pmatrix} 8 & 8 & 8.001 \\ 3.0001 & 3.0001 & 3.0011 \\ 2.0001 & 2.0001 & 2.0011 \end{pmatrix}.$$

The exact solution $X = \begin{pmatrix} 1 & 1 & 1 \\ 1 & 1 & 1 \\ 1 & 1 & 1 \end{pmatrix}$.

Now change $a(3, 1)$ to 0.99999 and keep the rest of the data unchanged. Then the solution of the perturbed problem is

$$\hat{X} = \begin{pmatrix} 908.1970 & -905.2944 & -906.2015 \\ -452.6722 & 454.2208 & 454.6745 \\ 1.0476 & 0.9524 & 0.9524 \end{pmatrix},$$

which is completely different from the exact solution X.
  Note that the relative error in the solution:

$$\frac{\|X - \hat{X}\|}{\|X\|} = 585.4190.$$

However, the relative perturbation in the data:

$$\frac{\|A - \hat{A}\|}{\|A\|} = 4.5964 \times 10^{-6} \ (\hat{A} \text{ is the perturbed matrix}).$$

This drastic change in the solution X can be explained by noting that **A and B are both ill-conditioned**:

$$\text{Cond}(A) = 6.1918 \times 10^{16}, \qquad \text{Cond}(B) = 8.5602 \times 10^{8}.$$

Remark

- The converse of the above theorem is, in general, not true. To see this, consider Example 8.3.1 once more. We have seen that the Sylvester equation with the data of this example is ill-conditioned. But note that $\text{Cond}(A) = 4.0489$, $\text{Cond}(B) = 1.0230$. Thus, neither A nor B is ill-conditioned.

Near Singularity of A and the Ill-conditioning of the Lyapunov Equation

From Theorem 8.3.6, we immediately obtain the following corollary:

Corollary 8.3.3.   *If A is nearly singular, then the Lyapunov equation $XA + A^T X = C$ is ill-conditioned.*

Example 8.3.6.   Let

$$A = \begin{pmatrix} 1 & 1 & 1 \\ 0 & 0.0001 & 1 \\ 0 & 0 & 1 \end{pmatrix}, \quad C = \begin{pmatrix} 2 & 2.0001 & 4 \\ 2.0001 & 2.0002 & 4.0001 \\ 4 & 4.0001 & 6 \end{pmatrix}.$$

The exact solution

$$X = \begin{pmatrix} 1 & 1 & 1 \\ 1 & 1 & 1 \\ 1 & 1 & 1 \end{pmatrix}.$$

Now perturb the $(1, 1)$ entry of A to 0.9880. Call the perturbed matrix $\hat{A}$. The computed solution of the perturbed problem

$$\hat{X} = \begin{pmatrix} 1.0121 & 0.9999 & 1.0000 \\ 0.9999 & 2.4750 & -0.4747 \\ 1.000 & -0.4747 & 2.4747 \end{pmatrix}.$$

The relative error in X:

$$\frac{\|\hat{X} - X\|}{\|X\|} = 0.9832.$$

The relative perturbation in A:

$$\frac{\|\hat{A} - A\|}{\|A\|} = 0.0060.$$

The ill-conditioning of the Lyapunov equation with the given data can be explained from the fact that A is nearly singular. Note that $\text{Cond}(A) = 3.9999 \times 10^4$ and $\text{sep}(A^T, -A) = 5.001 \times 10^{-5}$.

8.3.9   A Condition Number Estimator for the Sylvester Equation: $A^T X - XB = C$

We have seen in **Section 8.3.1** that

$$\text{sep}(B, -A) = \frac{1}{\|P^{-1}\|_2} = \sigma_{\min}(P),$$

where P is the coefficient matrix of the linear system (8.2.1).

However, finding $\text{sep}(B, -A)$ by computing the smallest singular value of P^{-1} requires a major computational effort. Even for modest m and n, it might be computationally prohibitive from the viewpoints of both the storage and the computational cost. It will require $O(m^3 n^3)$ flops and $O(m^2 n^2)$ storage.

Byers (1984) has proposed an algorithm to estimate $\text{sep}(A, B^T)$ in the style of the LINPACK condition number estimator. The LINPACK condition number estimator for $\text{Cond}(P)$ is based on estimating $\|P^{-1}\|_2$ by $\|y\|/\|z\|$, where y, z, and w satisfy

$$P^T z = w \quad \text{and} \quad Py = z;$$

the components of the vector w are taken to be $w_i = \pm 1$, where the signs are chosen such that the growth in z is maximized.

Algorithm 8.3.1. *Estimating* $\text{sep}(A, B^T)$.
Input. $A_{m \times m}$, $B_{n \times n}$—*Both upper triangular matrices.*
Output. *Sepest—An estimate of* $\text{sep}(A, B^T)$.
  Step 1.

$$\begin{aligned}
&\textit{For } i = m, m-1, \ldots, 1 \textit{ do} \\
&\quad \textit{For } j = n, n-1, \ldots, 1 \textit{ do} \\
&\qquad p \equiv \left(\sum_{h=i+1}^{m} a_{ih} z_{hj} - \sum_{h=j+1}^{n} z_{ih} b_{jh} \right) \\
&\qquad w \equiv -\text{sign}(p) \\
&\qquad z_{ij} \equiv (w - p)/(a_{ii} - b_{jj}) \\
&\quad \textit{End} \\
&\textit{End}
\end{aligned}$$

Step 2. *Compute* $Z \equiv Z/\|Z\|$, *where* $Z = (z_{ij})$.
Step 3. *Solve for* Y: $A^T Y - YB = Z$.
Step 4. *Sepest* $= 1/\|Y\|$.

Example 8.3.7. Consider estimating $\text{sep}(B, -A)$ with

$$A = \begin{pmatrix} -1 & 2 & 3 \\ 0 & -2 & 1 \\ 0 & 0 & 0.9990 \end{pmatrix} \quad \text{and} \quad B = \begin{pmatrix} -1 & 2 & 3 \\ 0 & -2.5 & 0 \\ 0 & 0 & 1.9999 \end{pmatrix}.$$

The algorithm produces sepest $(B, -A) = O(10^{-5})$, whereas the actual value of $\text{sep}(B, -A) = 3.0263 \times 10^{-5}$.

Remarks

- If $p = 0$, $\text{sign}(p)$ can be taken arbitrarily.

- The major work in the algorithm is in the solution of the Sylvester equation in Step 3. Thus, once this equation is solved, the remaining part of the algorithm requires only a little extra work. Efficient numerical methods for solving the Sylvester and Lyapunov equations are discussed in the next section.

Flop-count: The algorithm requires $2(m^2 n + mn^2)$ flops.

Remarks

- The algorithm must be modified when A and B are in quasi-triangular forms (real Schur forms, RSFs), or one is in Hessenberg form and the other is in RSF.
- There exists a LAPACK-style (rich in Basic Linear Algebra Subroutine-Level 3 operators), estimator for $\text{sep}(B, -A)$. For details, see Kågström and Poromaa (1989, 1992, 1996).

MATCONTROL notes: The sep function can be computed using the Kronecker product in MATCONTROL function **sepkr**. Algorithm 8.3.1 has been implemented in MATCONTROL function **sepest**, which calls the function **sylvhutc** for solving the upper triangular Sylvester equation in Step 3.

8.4   ANALYTICAL METHODS FOR THE LYAPUNOV EQUATIONS: EXPLICIT EXPRESSIONS FOR SOLUTIONS

There are numerous methods for solving Lyapunov and Sylvester equations. They can be broadly classified into two classes: **Analytical** and **Numerical** Methods.

By an analytical method, we mean a method that attempts to give an explicit expression for the solution matrix (usually the unique solution).

Recall from Chapter 7 that when A is a stable matrix, a unique solution X of the continuous-time Lyapunov equation requires computations of the matrix exponential e^{At} and evaluation of a matrix integral.

Similarly, when A is discrete-stable, a unique solution X of the discrete Lyapunov equation requires computations of various powers of A and many matrix multiplications. We have already seen that there are some obvious computational difficulties with these computations.

The other analytical methods include **finite and infinite series methods** (see Barnett and Storey (1970)).

These methods again have some severe computational difficulties. For example, consider the solution of the Lyapunov equation $XA + A^T X = C$, using the **finite**

series method proposed by Jameson (1968). The method can be briefly described as follows:

Let the characteristic polynomial of A be $\det(\lambda I - A) = \lambda^n + c_1\lambda^{n-1} + \cdots + c_n$. Define the sequence of matrices $\{Q_k\}$ by

$$Q_{-1} = 0, \qquad Q_0 = C$$
$$Q_k = A^T Q_{k-1} - Q_{k-1}A + A^T Q_{k-2}A, \quad k = 1, 2, \ldots, n. \tag{8.4.1}$$

Then it has be shown that

$$X = P^{-1}(Q_n - c_1 Q_{n-1} + \cdots + (-1)^n c_n Q_0), \tag{8.4.2}$$

where $P = \left(A^T\right)^n - c_1 \left(A^T\right)^{n-1} + \cdots + (-1)^n c_n I$.

It can be seen from the description of the method that it is not numerically effective for practical computations.

Note that for computation of the matrix P, various powers of A need to be computed and the matrix P can be ill-conditioned, which will affect the accuracy of X. This, together with the fact that the sensitivity of the characteristic polynomial of a matrix A (due to the small perturbations in the coefficients) grows as the order of the matrix grows (in general), lead us to believe that such methods will give unacceptable accuracy. **Indeed, the numerical experiments show that for random matrices of size 14×14, the errors are almost as large as the solutions themselves.**

Thus, we will not pursue further with the analytic methods. However, for reader's convenience, to compare this method with other numerically viable methods, the finite series method has been implemented in MATCONTROL function **lyapfns**.

8.5 NUMERICAL METHODS FOR THE LYAPUNOV AND SYLVESTER EQUATIONS

An obvious way to solve the Sylvester equation $XA + BX = C$ is to apply Gaussian elimination with partial pivoting to the system $Px = c$ given by (8.2.1). But, unless the special structure of P can be exploited, Gaussian elimination scheme for the Sylvester equation will be **computationally prohibitive**, since $O(n^3 m^3)$ flops and $O(n^2 m^2)$ storage will be required. One way to exploit the structure of P will be to transform A and B to some **simple forms** using similarity transformations.

Thus, if U and V are nonsingular matrices such that

$$U^{-1}AU = \hat{A}, \qquad V^{-1}BV = \hat{B}, \qquad \text{and} \qquad V^{-1}CU = \hat{C},$$

then $XA + BX = C$ is transformed to

$$Y\hat{A} + \hat{B}Y = \hat{C}, \tag{8.5.1}$$

where $Y = V^{-1}XU$.

If $\hat{A}$ and $\hat{B}$ are in simple forms, then the equation $Y\hat{A} + \hat{B}Y = \hat{C}$ can be easily solved, and the solution X can then be recovered from Y. The idea, therefore, is summarized in the following steps:

Step 1. Transform A and B to "**simple**" forms (e.g., diagonal, Jordan and companion, Hessenberg, real-Schur, and Schur etc.) using similarity transformations:

$$\hat{A} = U^{-1}AU, \qquad \hat{B} = V^{-1}BV.$$

Step 2. Update the right-hand side matrix: $\hat{C} = V^{-1}CU$.
Step 3. Solve the **transformed equation** for Y: $Y\hat{A} + \hat{B}Y = \hat{C}$.
Step 4. Recover X from Y by solving the system: $XU = VY$.

8.5.1 Numerical Instability of Diagonalization, Jordan Canonical Form, and Companion Form Techniques

It is true that the rich structures of Jordan and companion matrices can be nicely exploited in solving the reduced Sylvester equation (8.5.1). However, as noted before, the companion, and JCFs, in general, cannot be obtained in a numerically stable way. (For more on numerically computing the JCF, see Golub and Wilkinson (1976).) The transforming matrices will be, in some cases, ill-conditioned and this ill-conditioning will affect the computations of $\hat{A}$, $\hat{B}$, $\hat{C}$, and X (from Y), which require computations involving inverses of the transforming matrices. Indeed, **numerical experiments performed by us show that solutions of the Sylvester equation using companion form of A with A of sizes larger than 15 have errors almost as large as the solutions themselves.** We will illustrate below by a simple example how diagonalization technique yields an inaccurate solution.

Example 8.5.1. Consider solving the Lyapunov equation: $XA + A^TX - C = 0$, with

$$A = \begin{pmatrix} 2.4618 & -1.5284 & 2.2096 & -0.3503 \\ 5.5854 & -1.2161 & 2.3825 & -1.2843 \\ 1.6935 & 2.5009 & 2.1131 & -1.2186 \\ -0.2686 & -3.2594 & 7.9205 & 0.6412 \end{pmatrix}.$$

Choose

$$X = \begin{pmatrix} 1 & 1 & 1 & 1 \\ 1 & 1 & 1 & 1 \\ 1 & 1 & 1 & 1 \\ 1 & 1 & 1 & 1 \end{pmatrix} \quad \text{and take} \quad C = XA + A^TX.$$

Let X_{Diag} be the solution obtained by the diagonalization procedure (using MATLAB Function **lyap2**).

The relative residual:

$$\frac{\|X_{\text{Diag}} A + A^T X_{\text{Diag}} - C\|}{\|X_{\text{Diag}}\|} = 1.6418 \times 10^{-7}.$$

The solution $\hat{X}$ obtained by MATLAB function **lyap** (based on the numerically viable **Schur method**):

$$\hat{X} = \begin{pmatrix} 1.0000 & 1.0000 & 1.0000 & 1.0000 \\ 1.0000 & 1.0000 & 1.0000 & 1.0000 \\ 1.0000 & 1.0000 & 1.0000 & 1.0000 \\ 1.0000 & 1.0000 & 1.0000 & 1.0000 \end{pmatrix}.$$

The relative residual:

$$\frac{\|\hat{X} A + A^T \hat{X} - C\|}{\|\hat{X}\|} = 9.5815 \times 10^{-15}.$$

Solutions via Hessenberg and Schur Forms

In view of the remarks made above, our "**simple**" forms of choice have to be **Hessenberg forms** and the (**real**) **Schur forms**, since we know that the transforming matrices U and V in these cases can be chosen to be orthogonal, which are perfectly well-conditioned. Some such methods are discussed in the following sections.

8.5.2   The Schur Method for the Lyapunov Equation: $XA + A^TX = C$

The following method, proposed by Bartels and Stewart (1972), is **now widely used as an effective computational method for the Lyapunov equation.** The method is based on reduction of A^T to RSF. It is, therefore, known as the **Schur method for the Lyapunov equation.** The method is described as follows:

Step 1. Reduction of the Problem. Let $R = U^T A^T U$ be the **RSF** of the matrix A^T. Then, employing this transformation, the Lyapunov matrix equation $XA + A^TX = C$ is reduced to

$$YR^T + RY = \hat{C}, \tag{8.5.2}$$

where $R = U^T A^T U$, $\hat{C} = U^T CU$, and $Y = U^T XU$.

Step 2. Solution of the Reduced Problem. The reduced equation to be solved is: $YR^T + RY = \hat{C}$. Let

$$Y = (y_1, \ldots, y_n), \qquad \hat{C} = (c_1, \ldots, c_n), \qquad \text{and} \qquad R = (r_{ij}).$$

Assume that the columns y_{k+1} through y_n have been computed, and consider the following two cases.

Case 1: $r_{k,k-1} = 0$. Then y_k is determined by solving the quasi-triangular system

$$(R + r_{kk}I)y_k = c_k - \sum_{j=k+1}^{n} r_{kj}y_j.$$

If, in particular, R **is upper triangular**, that is there are no "**Schur bumps**" (see **Chapter 4**) on the diagonal, then each y_i, $i = n, n-1, \ldots, 2, 1$ can be obtained by solving an $n \times n$ upper triangular system as follows:

$$\begin{aligned}
(R + r_{nn}I)y_n &= c_n, \\
(R + r_{n-1,n-1}I)y_{n-1} &= c_{n-1} - r_{n-1,n}y_n, \\
&\vdots \\
(R + r_{11}I)y_1 &= c_1 - r_{12}y_2 - \cdots - r_{1n}y_n.
\end{aligned} \qquad (8.5.3)$$

Remark

- If the complex Schur decomposition is used, that is, if $R_c = U_c^* A^T U_c$ is a complex triangular matrix, then the solution Y_c of the reduced problem is computed by solving n complex $n \times n$ linear systems (8.5.3). The MATLAB function **rsf2csf** converts an RSF to a complex triangular matrix. However, the use of complex arithmetic is more expensive and **not recommended in practice**.

Case 2: $r_{k,k-1} \neq 0$ **for some** k. This indicates that there is a "**Schur bump**" on the diagonal. This enables us to compute y_{k-1} and y_k simultaneously, by solving the following $2n \times 2n$ linear system:

$$R(y_{k-1}, y_k) + (y_{k-1}, y_k) \begin{pmatrix} r_{k-1,k-1} & r_{k,k-1} \\ r_{k-1,k} & r_{kk} \end{pmatrix}$$

$$= (c_{k-1}, c_k) - \sum_{j=k+1}^{n} (r_{k-1,j}y_j, r_{kj}y_j) = (d_{k-1}, d_k). \qquad (8.5.4)$$

Remark

- To distinguish between Case 1 and Case 2 in a computational setting, it is recommended that some threshold, for example, Tol $= \mu \|A\|_F$ be used, where μ is the machine precision.
- Thus, to see if $r_{k,k-1} = 0$, accept $r_{k,k-1} = 0$, if $|r_{k,k-1}| <$ Tol.

An Illustration

We illustrate the above procedure with $n = 3$.

Assume that $r_{21} \neq 0$, that is,

$$R = \begin{pmatrix} r_{11} & r_{12} & r_{13} \\ r_{21} & r_{22} & r_{23} \\ 0 & 0 & r_{33} \end{pmatrix}.$$

Since $r_{32} = 0$, by **Case 1**, y_3 is computed by solving the system:

$$(R + r_{33}I)\, y_3 = c_3.$$

Since $r_{21} \neq 0$, by **Case 2**, y_1 and y_2 are computed by solving

$$R(y_1, y_2) + (y_1, y_2)\begin{pmatrix} r_{11} & r_{21} \\ r_{12} & r_{22} \end{pmatrix} = (c_1 - r_{13}y_3, c_2 - r_{23}y_3). \qquad (8.5.5)$$

Step 3. Recovery of the solution of the original problem from the solution of the reduced problem. Once Y is obtained by solving the reduced problem $YR^T + RY = \hat{C}$, the solution X of the original problem $XA + A^TX = C$, is recovered as

$$X = UYU^T.$$

Example 8.5.2. Consider solving the Lyapunov equation: $XA + A^TX = C$, with

$$A = \begin{pmatrix} 0 & 2 & -1 \\ -3 & -2 & 2 \\ -2 & 1 & -1 \end{pmatrix} \quad \text{and} \quad C = \begin{pmatrix} -2 & 2 & -3 \\ -8 & -6 & -5 \\ 11 & 13 & -2 \end{pmatrix},$$

Step 1. *Reduction.* Using MATLAB function $[U, R] = \mathbf{schur}(A^T)$, we obtain

$$R = \begin{pmatrix} -1.3776 & 3.8328 & 1.3064 \\ -1.0470 & 0.8936 & -1.2166 \\ 0 & 0 & -2.5160 \end{pmatrix}, \quad U = \begin{pmatrix} 0.7052 & 0.4905 & 0.5120 \\ 0.6628 & -0.7124 & -0.2304 \\ -0.2518 & -0.5019 & 0.8275 \end{pmatrix}.$$

Then $\hat{C} = U^TCU$ is

$$\hat{C} = \begin{pmatrix} -9.3174 & 1.9816 & -7.5863 \\ -2.1855 & -1.0425 & 2.4422 \\ 16.2351 & -3.4886 & 0.3600 \end{pmatrix}.$$

Step 2. Solve $RY + YR^T = \hat{C}$. Since $r_{32} = 0$, then by **Case 1**, y_3 is computed by solving the system:

$$\begin{pmatrix} -3.8936 & 3.8328 & 1.3064 \\ -1.0470 & -1.6224 & -1.2166 \\ 0 & 0 & -5.0320 \end{pmatrix} y_3 = \begin{pmatrix} -7.5863 \\ 2.4422 \\ 0.3600 \end{pmatrix}.$$

$$y_3 = \begin{pmatrix} 0.3030 \\ -1.6472 \\ -0.0715 \end{pmatrix}.$$

Since $r_{21} \neq 0$, then by **Case 2**, y_1 and y_2 are computed by solving the system (8.5.5):

$$(y_1^T, y_2^T) = (3.4969, 0.1669, -1.2379, 0.2345, 0.5746, 3.0027)^T.$$

Step 3. *Recovery of the solution.*

$$X = UYU^T = \begin{pmatrix} 2 & 0 & -2 \\ 2 & 2 & 1 \\ 0 & -3 & 0 \end{pmatrix}.$$

Example 8.5.3. We now solve the previous example (Example 8.5.2) using the complex Schur form

Step1. $R = \begin{pmatrix} -0.2420 + 1.6503j & -0.3227 + 3.5797j & -0.0927 - 0.9538j \\ 0 & -0.2420 - 1.6503j & 1.2113 - 0.8883j \\ 0 & 0 & -2.5160 \end{pmatrix}$,

$$U = \begin{pmatrix} -0.5814 + 0.5148j & 0.0802 - 0.3581j & 0.5120 \\ -0.0030 + 0.4839j & 0.6649 + 0.5202j & -0.2304 \\ 0.3590 - 0.1838j & 0.1355 + 0.3664j & -0.8275 \end{pmatrix},$$

$$\hat{C} = \begin{pmatrix} -7.5894 + 1.4093j & 1.8383 + 4.252j & 2.6799 + 5.5388j \\ 2.1139 - 1.1953j & -2.7706 - 1.4093j & -4.7410 + 1.783j \\ -6.5401 + 11.8534j & 9.2728 + 2.5470j & 0.3600 \end{pmatrix}$$

Step 2. Since R is triangular (complex), the columns y_1, y_2, y_3 of y are successively computed by solving **complex linear systems** (8.5.3). This gives

$$y_1 = (-2.9633 + 0.000229j, -0.7772 + 0.8659j, -0.7690 - 0.9038j)^T,$$

$$y_2 = (-0.7811 - 0.8163j, 1.1082 - 0.0229j, -2.0819 - 2.923j)^T,$$

$$y_3 = (0.6108 - 0.2212j, 0.9678 - 1.2026j, -0.0715)^T.$$

Thus, with $Y = (y_1, y_2, y_3)$, we have

$$X = UYU^T = \begin{pmatrix} 2 & 0 & -2 \\ 2 & 2 & 1 \\ 0 & -3 & 0 \end{pmatrix}.$$

Note: In practice, the system (8.5.4) is solved using Gaussian elimination with partial pivoting. The LAPACK and SLICOT routines (see **Section 8.10.4**) have used Gaussian elimination with complete pivoting (see Datta (1995) and Golub and Van Loan (1996)) and the structure of the RSF has been exploited there. For details of implementations, the readers may consult the book by Sima (1996).

MATCONTROL note: The Schur method for the Lyapunov equation has been implemented in MATCONTROL function **lyaprsc**.

MATLAB note: MATLAB function **lyap** in the form

$$X = \textbf{lyap}\,(A, C)$$

solves the Lyapunov equation

$$AX + XA^T = -C$$

using the **complex Schur** triangularization of A.

Flop-count

1. Transformation of A to RSF: $26n^3$ (Assuming that the QR iteration algorithm requires about two iterations to make a subdiagonal entry negligible). (This count includes construction of U.)
2. Solution of the reduced problem: $3n^3$
3. Recovery of the solution: $3n^3$ (using the symmetry of X).
   Total flops: $32n^3$ (**Approximate**).

8.5.3 The Hessenberg–Schur Method for the Sylvester Equation

The Schur method described above for the Lyapunov equation can also be used to solve the Sylvester equation $XA + BX = C$. The matrices A and B are, respectively, transformed to the lower and upper RSFs, and then back-substitution is used to solve the reduced Schur problem. Note, that the special form of the Schur matrix S can be exploited only in the solution of the $m \times m$ linear systems with S. Some computational effort can be saved if B, the larger of the two matrices A and B, is left in Hessenberg form, while the smaller matrix A is transformed further to RSF. The reason for this is that a matrix must be transformed to a Hessenberg matrix as an initial step in the reduction to RSF (see **Chapter 4**). The important outcome here is that back-substitution for the solution of the Hessenberg–Schur problem is still possible. Noting this, Golub *et al.* (1979) developed the following Hessenberg–Schur method for the Sylvester equation problem.

Step 1. *Reduction to the Hessenberg–Schur Problem.* Assume that m is larger than n. Let $R = U^T A^T U$ and $H = V^T B V$ be, respectively, the upper RSF and the upper Hessenberg form of A and B. Then,

$$XA + BX = C \quad \text{becomes} \quad YR^T + HY = \hat{C},$$
$$\text{where} Y = V^T XU, \ \hat{C} = V^T CU. \tag{8.5.6}$$

Step 2. *Solution of the Reduced Hessenberg–Schur Problem.* In the reduced problem $HY + YR^T = \hat{C}$, let $Y = (y_1, y_2, \ldots, y_n)$ and $\hat{C} = (c_1, \ldots, c_n)$. Then, assuming that $y_{k+1}, \ldots, y_n$ have already been computed, y_k (or y_k and y_{k+1}) can be computed as in the case of the Lyapunov equation, by considering the following two cases.

Case 1. If $r_{k,k-1} = 0$, y_k is computed by solving the $m \times m$ Hessenberg system:

$$(H + r_{kk}I)y_k = c_k - \sum_{j=k+1}^{n} r_{kj}y_j.$$

Case 2. If $r_{k,k-1} \neq 0$, then equating columns $k - 1$ and k in $HY + YR^T = \hat{C}$, it is easy to see that y_{k-1} and y_k are simultaneously computed by solving the $2m \times 2m$ linear system:

$$H(y_{k-1}, y_k) + (y_{k-1}, y_k) \begin{pmatrix} r_{k-1,k-1} & r_{k,k-1} \\ r_{k-1,k} & r_{kk} \end{pmatrix}$$
$$= (c_{k-1}, c_k) - \sum_{j=k+1}^{n} (r_{k-1,j}y_j, r_{kj}y_j) = (d_{k-1}, d_k) \tag{8.5.7}$$

Note: The matrix of the system can be made upper triangular with two nonzero subdiagonals, by reordering the variables suitably. The upper triangular system can then be solved using Gaussian elimination with partial pivoting.

Step 3. *Recovery of the Original Solution.* The solution X is recovered from Y as

$$X = VYU^T.$$

Algorithm 8.5.1. *The Hessenberg–Schur Algorithm for $XA + BX = C$*
Input: *The matrices A, B, and C, respectively, of order $n \times n, m \times m$, and $m \times n; n \leq m$.*
Output: *The matrix X satisfying $XA + BX = C$.*

Step 1. *Transform A^T to a real Schur matrix R, and B to an upper Hessenberg matrix H by orthogonal similarity:*

$$U^T A^T U = R, \qquad V^T B V = H.$$

Form $\hat{C} = V^T C U$, and partition $\hat{C} = (c_1, \ldots, c_n)$ by columns.

Step 2. *Solve $HY + YR^T = \hat{C}$:*

For $k = n, \ldots, 1$ do until the columns of Y are computed
   If $r_{k,k-1} = 0$, then compute y_k by solving the Hessenberg system:

$$(H + r_{kk}I)y_k = c_k - \sum_{j=k+1}^{n} r_{kj} y_j \tag{8.5.8}$$

Else, compute y_k and y_{k-1} by solving the system:

$$\begin{pmatrix} H + r_{k-1,k-1}I & r_{k-1,k}I \\ r_{k,k-1}I & H + r_{kk}I \end{pmatrix} \begin{pmatrix} y_{k-1} \\ y_k \end{pmatrix} = \begin{pmatrix} d_{k-1} \\ d_k \end{pmatrix}, \tag{8.5.9}$$

where

$$(d_{k-1}, d_k) = (c_{k-1}, c_k) - \sum_{j=k+1}^{n} (r_{k-1,j} y_j, r_{kj} y_j). \tag{8.5.10}$$

Step 3. *Recover X: $X = VYU^T$.*

Example 8.5.4. Consider solving $XA + BX = C$ using Algorithm 8.5.1 with

$$A = \begin{pmatrix} 1 & -1 & 0 \\ 1 & 1 & 0 \\ 0 & 0 & 2 \end{pmatrix}, \qquad B = \begin{pmatrix} 1 & 2 & 3 & 4 \\ 4 & 5 & 6 & 7 \\ 7 & 8 & 9 & 1 \\ 10 & 0 & 0 & 0 \end{pmatrix}$$

and

$$C = \begin{pmatrix} 12 & 10 & 12 \\ 24 & 22 & 24 \\ 27 & 25 & 27 \\ 12 & 10 & 12 \end{pmatrix}.$$

Step 1. Reduce A^T to RSF and B to Hessenberg form H:

$$U^T A^T U = R = \begin{pmatrix} 1 & 1 & 0 \\ -1 & 1 & 0 \\ 0 & 0 & 2 \end{pmatrix},$$

$$V^T B V = H = \begin{pmatrix} 1 & -5.3766 & -0.3709 & -0.0886 \\ -12.8452 & 7.6545 & 5.3962 & -0.7695 \\ 0 & 10.6689 & 4.7871 & -0.2737 \\ 0 & 0 & -5.3340 & 1.5584 \end{pmatrix}.$$

$$U = \begin{pmatrix} 1 & 0 & 0 \\ 0 & 1 & 0 \\ 0 & 0 & 1 \end{pmatrix}, \quad V = \begin{pmatrix} 1 & 0 & 0 & 0 \\ 0 & -3.114 & -0.7398 & -0.5965 \\ 0 & -5.449 & -0.3752 & 0.7498 \\ 0 & -0.7785 & 0.5585 & -0.2863 \end{pmatrix},$$

Compute

$$\hat{C} = \begin{pmatrix} 12 & 10 & 12 \\ -31.5292 & -28.2595 & -31.5292 \\ -21.1822 & -20.0693 & -21.1822 \\ 2.4949 & 2.7608 & 2.4949 \end{pmatrix}.$$

Step 2. Solution of the reduced problem: $HY + YR^T = \hat{C}$.
Case 1. Since $r(3, 2)$ is 0, y_3 is obtained by solving: $(H + r_{33}I)y_3 = c_3$.

$$y_3 = (1, -1.6348, -0.5564, -0.1329)^T.$$

Case 2. Since $r_{21} \neq 0$, y_1 and y_2 are simultaneously computed by solving the system:

$$\begin{pmatrix} H + r_{11}I & r_{12}I \\ r_{21}I & H + r_{22}I \end{pmatrix} \begin{pmatrix} y_1 \\ y_2 \end{pmatrix} = \begin{pmatrix} d_1 \\ d_2 \end{pmatrix},$$

where $(d_1, d_2) = (c_1 - r_{13}y_3, c_2 - r_{23}y_3)$.

$$y_1 = \begin{pmatrix} 1 \\ -1.6348 \\ -0.5564 \\ -0.1329 \end{pmatrix}, \quad y_2 = \begin{pmatrix} 1 \\ -1.6348 \\ -0.5564 \\ -0.1329 \end{pmatrix}.$$

So,

$$Y = (y_1, y_2, y_3) = \begin{pmatrix} 1 & 1 & 1 \\ -1.6348 & -1.6348 & -1.6348 \\ -0.5564 & -0.5564 & -0.5564 \\ -0.1329 & -0.1329 & -0.1329 \end{pmatrix}.$$

Step 3. $X = VYU^T = \begin{pmatrix} 1 & 1 & 1 \\ 1 & 1 & 1 \\ 1 & 1 & 1 \\ 1 & 1 & 1 \end{pmatrix}.$

Flop-count:

1. Reduction to Hessenberg and RSFs: $\frac{10}{3}m^3 + 26n^3$.
2. Computation of $\hat{C}$: $2m^2n + 2mn^2$.
3. Computation of Y: $6m^2n + mn^2$.
   (To obtain Y, it was assumed that S has $n/2(2 \times 2)$ bumps, which is the worst case.)
4. Computation of X: $2m^2n + 2mn^2$.
   Total flops: **Approximately** $(10m^3/3 + 26n^3 + 10m^2n + 5mn^2)$.

Numerical Stability of the Schur and Hessenberg–Schur Methods: The round-off error analysis of the Hessenberg–Schur algorithm for the Sylvester equation $XA + BX = C$ performed by Golub *et al.* (1979) shows that "the errors no worse in magnitude than $O(\|\phi^{-1}\|\epsilon)$ will contaminate the computed $\hat{X}$, where $\|\phi^{-1}\| = 1/\text{sep}(B, -A)$, and ϵ is a small multiple of the machine precision μ."

Specifically, if

$$\frac{\epsilon(2 + \epsilon)(\|A\|_2 + \|B\|_2)}{\text{sep}(B, -A)} < \frac{1}{2}.$$

Then,

$$\frac{\|X - \hat{X}\|_F}{\|X\|_F} \leq \frac{(9\epsilon + 2\epsilon^2)(\|A\|_F + \|B\|_F)}{\text{sep}(B, -A)}. \tag{8.5.11}$$

The above result shows that the quantity $\text{sep}(B, -A)$ will indeed influence the numerical accuracy of the computed solution obtained by the Hessenberg–Schur algorithm for the Sylvester equation. (Note that $\text{sep}(B, -A)$ also appears in the perturbation bound (8.3.6).)

Similar remarks, of course, also hold for the Schur methods for the Lyapunov and Sylvester equations. We will have some more to say about the **backward error** of the computed solutions by these methods a little later in this chapter.

MATCONTROL notes: Algorithm 8.5.1 has been implemented in MATCON-TROL function **sylvhrsc**. The function **sylvhcsc** solves the Sylvester equation using **Hessenberg decomposition** of B and **complex-Schur decomposition** of A.

MATLAB note: MATLAB function **lyap** in the form:

$$X = \mathbf{lyap}\,(A, B, C)$$

solves the Sylvester equation

$$AX + XB = -C$$

using **complex-Schur decompositions** of both A and B.

8.5.4   The Schur Method for the Discrete Lyapunov Equation

We now briefly outline the Schur method for the discrete Lyapunov equation:

$$A^T X A - X = C. \tag{8.5.12}$$

The method is due to Barraud (1977).

As before, we divide the process into three steps:

Step 1. Reduction of the problem. Let $R = U^T A^T U$ be the upper RSF of the matrix A^T. Then the equation:

$$A^T X A - X = C$$

reduces to

$$R Y R^T - Y = \hat{C}, \tag{8.5.13}$$

where $Y = U^T X U$ and $\hat{C} = U^T C U$.

Step 2.   Solution of the reduced equation. Let $R = (r_{ij})$, $Y = (y_1, y_2, \ldots, y_n)$, and $\hat{C} = (c_1, c_2, \ldots, c_n)$.

Consider two cases as before.

Case 1. $r_{k,k-1} = 0$, for some k.
In this case, y_k can be determined by solving the quasi-triangular system:

$$(r_{kk} R - I)y_k = c_k - R \sum_{j=k+1}^{n} r_{kj} y_j. \tag{8.5.14}$$

In particular, if R is an upper triangular matrix, then y_n through y_1 can be computed successively by solving the triangular systems:

$$(r_{kk} R - I)y_k = c_k - R \sum_{j=k+1}^{n} r_{kj} y_j, \quad k = n, n - 1, \ldots, 2, 1. \tag{8.5.15}$$

Case 2. $r_{k,k-1} \neq 0$, for some k. In this case y_k and y_{k-1} can be simultaneously computed, as before.

For example, if $n = 3$, and $r_{2,1} \neq 0$, then y_2 and y_1 can be computed simultaneously by solving the system:

$$\begin{pmatrix} r_{11}R - I & r_{12}R \\ r_{21}R & r_{22}R - I \end{pmatrix} \begin{pmatrix} y_1 \\ y_2 \end{pmatrix} = \begin{pmatrix} c_1 - r_{13}Ry_3, \\ c_2 - r_{23}Ry_3 \end{pmatrix}. \tag{8.5.16}$$

Step 3. Recovery of X from Y. Once Y is computed, X is recovered from Y as

$$X = UYU^{\mathrm{T}}. \tag{8.5.17}$$

Example 8.5.5. Consider solving the discrete Lyapunov equation $A^{\mathrm{T}}XA - X = C$ with

$$A = \begin{pmatrix} 0 & 2 & -1 \\ -3 & -2 & 2 \\ -2 & 1 & -1 \end{pmatrix} \quad \text{and} \quad C = \begin{pmatrix} -2 & 2 & -3 \\ -8 & -6 & -5 \\ 11 & 13 & -2 \end{pmatrix}.$$

Step 1. Reduction to: $RYR^{\mathrm{T}} - Y = \hat{C}$.

$$R = \begin{pmatrix} -2.5160 & -2.7102 & -1.6565 \\ 0 & -0.2420 & 3.2825 \\ 0 & -0.8298 & -0.2420 \end{pmatrix},$$

$$U = \begin{pmatrix} -0.1972 & 0.9778 & -0.0705 \\ -0.6529 & -0.1847 & -0.7346 \\ 0.7313 & 0.0988 & -0.6749 \end{pmatrix},$$

$$\hat{C} = \begin{pmatrix} -9.4514 & 11.1896 & -12.1503 \\ -4.5736 & -0.4260 & -1.7470 \\ 7.0475 & -0.0252 & -0.1226 \end{pmatrix}.$$

Step 2. Solution of the reduced equation: $RYR^{\mathrm{T}} - Y = \hat{C}$:

$$Y = (y_1, y_2, y_3) = \begin{pmatrix} 2.2373 & -5.9557 & 2.4409 \\ 3.6415 & -0.3633 & -0.2531 \\ -5.1720 & -0.1677 & 1.5570 \end{pmatrix}.$$

Step 3. Recovery of X from Y:

$$X = UYU^{\mathrm{T}}$$
$$= \begin{pmatrix} 0.1376 & -2.1290 & 2.4409 \\ 3.6774 & 0.1419 & -1.3935 \\ -5.1721 & -0.1678 & 1.5570 \end{pmatrix}.$$

Verify: $\|A^{\mathrm{T}}XA - X - C\|_2 = O(10^{-14})$.

Flop-Count: The Schur method for the discrete Lyapunov equation requires about $34n^3$ flops ($26n^3$ for the reduction of A to the RSF).

Round-off properties: As in the case of the continuous-time Lyapunov equation, it can be shown (**Exercise 8.26**) that the computed solution $\hat{X}$ of the discrete Lyapunov equation $A^T X A - X = C$ satisfies the inequality

$$\frac{\|\hat{X} - X\|_F}{\|X\|_F} \leq \frac{cm\mu}{\text{sep}_d(A^T, A)}, \tag{8.5.18}$$

where $m = \max(1, \|A\|_F^2)$ and c is a small constant.

Thus, the accuracy of the solution obtained by the Schur method for the discrete Lyapunov equation depends upon the quantity $\text{sep}_d(A^T, A)$. (Note again that the $\text{sep}_d(A^T, A)$ appears in the perturbation bound (8.3.18).)

MATLAB note: $X = dlyap(A, C)$ solves the discrete Lyapunov equation: $A X A^T - X = -C$, using **complex-Schur** decomposition of A.

MATCONTROL notes: MATCONTROL functions **lyaprsd** and **lyapcsd** solve the discrete-time Lyapunov equation using **real-Schur** and **complex-Schur** decomposition of A, respectively.

8.5.5   Residual and Backward Error in the Schur and Hessenberg–Schur Algorithms

We consider here the following questions: How small are the relative residuals obtained by the Schur and the Hessenberg–Schur algorithms? **Does a small relative residual guarantee that the solution is accurate?**

To answer these questions, we note that there are two major computational tasks with these algorithms:

First. The reduction of the matrices to the RSF and/or to the Hessenberg form.

Second. Solutions of certain linear systems.

We know that the reduction to the RSF of a matrix by the QR iteration method, and that to the Hessenberg form by Householder's or Givens' method, are backward stable (**See Chapter 4**).

And, if the linear systems are solved using Gaussian elimination with partial pivoting, followed by the technique of iterative refinement (which is the most practical way to solve a dense linear system), then it can be shown (Golub *et al.* 1979, Higham 1996) that the relative residual norm obtained by the **Hessenberg–Schur algorithm** for the Sylvester equation satisfies

$$\frac{\|C - \left(\hat{X} A + B \hat{X}\right)\|_F}{\|\hat{X}\|_F} \leq c\mu \left(\|A\|_F + \|B\|_F\right), \tag{8.5.19}$$

where $\hat{X}$ is the computed solution and c is a small constant depending upon m and n.

This means that the relative residual is guaranteed to be small. Note that this bound does not involve sep$(B, -A)$.

Does a small relative residual imply a small backward error? We will now consider this question.

To this end, let's recall that by **backward error** we mean the amount of perturbations to be made to the data so that an approximate solution is the exact solution to the perturbed problem. If the perturbations are small, then the algorithm is backward stable.

For the Sylvester equation problem, let's define (following Higham 1996) the backward error of an approximate solution Y of the Sylvester equation $XA + BX = C$ by

$$v(Y) = \min \{\varepsilon : Y(A + \Delta A) + (B + \Delta B)Y = C + \Delta C,$$
$$\|\Delta A\|_F \leq \varepsilon\alpha, \|\Delta B\|_F \leq \varepsilon\beta, \|\Delta C\|_F \leq \varepsilon\gamma \},$$

where α, β, and γ are tolerances. The most common choice is

$$\alpha = \|A\|_F, \qquad \beta = \|B\|_F, \qquad \gamma = \|C\|_F.$$

This choice yields the **normwise relative backward error**.

As earlier, we assume that A is $n \times n$ and B is $m \times m$, and $m \geq n$.

It has been shown by Higham (1996) that

$$v(Y) \leq \delta \frac{\|\text{Res}(Y)\|_F}{(\alpha + \beta)\|Y\|_F + \gamma}, \tag{8.5.20}$$

where $\text{Res}(Y) = C - (YA + BY)$ is the residual and

$$\delta = \frac{(\alpha + \beta)\|Y\|_F + \gamma}{\sqrt{(\alpha^2\sigma_m^2 + \beta^2\sigma_n^2 + \gamma^2)}}. \tag{8.5.21}$$

Here $\sigma_1 \geq \sigma_2 \geq \cdots \geq \sigma_n \geq 0$ are the singular values of Y, and $\sigma_{n+1} = \cdots = \sigma_m = 0$.

The special case when $m = n$ is interesting. In this case

$$\delta = \frac{(\|A\|_F + \|B\|_F)\|Y\|_F + \|C\|_F}{\left(\left(\|A\|_F^2 + \|B\|_F^2\right)\sigma_{\min}^2(Y) + \|C\|_F^2\right)^{1/2}}. \tag{8.5.22}$$

Thus, δ is large only when

$$\|Y\|_F \gg \sigma_{\min}(Y) \quad \text{and} \quad \|Y\|_F \gg \frac{\|C\|_F}{\|A\|_F + \|B\|_F}. \tag{8.5.23}$$

In other words, **δ is large when Y is ill-conditioned and $\|Y\|_F$ is large.**

In the general case ($m \neq n$), δ can also be large if $\|B\|$ is large compared to the rest of the data. In these cases, the Sylvester equation is badly scaled.

Also, **note that if only A and B are perturbed, then δ is large whenever Y is ill-conditioned.**

This is because in this case,

$$\delta \geq \|Y\|_F \|Y^{\dagger}\|_2 \approx \mathrm{Cond}_2(Y)$$

(for any m and n); so, δ is large whenever Y is ill-conditioned.

From above discussions, we see that "**the backward error of an approximate solution to the Sylvester equation can be arbitrarily larger than its relative residual**" (Higham 1996). The same remark, of course, also holds for the Lyapunov equation, as we will see below.

Backward Error for the Lyapunov Equation

In case of the Lyapunov equation, $B = A^T$ (and thus $\beta = \alpha$), we have the following bound for the backward error for the Lyapunov equation.

Let Y be an approximate solution of the Lyapunov equation $XA + A^T X = C$, and let $v(Y)$ denote the backward error. Assume that C is symmetric. Then

$$v(Y) = \min\{\epsilon : Y(A + \Delta A) + (A + \Delta A)^T Y = C + \Delta C, \|\Delta A\| \leq \epsilon\alpha,$$

$$\Delta C = (\Delta C)^T, \|\Delta C\|_F \leq \epsilon\gamma\}.$$

Thus,

$$v(Y) \leq \delta \frac{\|\mathrm{Res}(Y)\|_F}{2\alpha\|Y\|_F + \gamma}. \tag{8.5.24}$$

The expression for δ in (8.5.24) can now be easily written down by specializing (8.5.22) to this case.

8.5.6   A Hessenberg Method for the Sylvester Equation: $AX + XB = C$

Though the Schur and the Hessenberg–Schur methods are numerically effective for the Lyapunov and the Sylvester equations and are widely used in practice, it would be, however nice to have methods that would require reduction of the matrices A and B to Hessenberg forms only. Note that the reduction to a Hessenberg form is preliminary to that of the RSF. Thus, such Hessenberg methods will be more efficient than the Hessenberg–Schur method. We show below how a Hessenberg method for the Sylvester equation can be developed. The method is an extension of a Hessenberg method for the Lyapunov equation by Datta and Datta (1976), and is an efficient implementation of an idea of Kreisselmeier (1972). *It answers affirmatively a question raised by Charles Van Loan (1982) as to whether a method* *can be developed to solve the Lyapunov equation just by transforming A to a Hessenberg matrix.*

Step 1. *Reduction of the problem to a Hessenberg problem.* Transform A to a lower Hessenberg matrix H_1, and B to another lower Hessenberg matrix H_2:

$$U^{T}AU = H_1, \qquad V^{T}BV = H_2.$$

(Assume that both H_1 and H_2 are **unreduced**.)

Then, $AX + XB = C$ becomes

$$H_1 Y + Y H_2 = C', \qquad \text{where } Y = U^{T}XV, \quad C' = U^{T}CV.$$

Step 2. *Solution of the reduced problem.* $H_1 Y + Y H_2 = C'$ Let $Y = (y_1, y_2, \ldots, y_n)$ and $H_2 = (h_{ij})$.

Then the equation $H_1 Y + Y H_2 = C'$ is equivalent to

$$H_1 y_n + h_{n-1,n} y_{n-1} + h_{nn} y_n = c'_n,$$
$$H_1 y_{n-1} + h_{n-2,n-1} y_{n-2} + h_{n-1,n-1} y_{n-1} + h_{nn-1} y_n = c'_{n-1},$$
$$\vdots$$
$$H_1 y_1 + h_{11} y_1 + h_{21} y_2 + \cdots + h_{n1} y_n = c'_1.$$

Eliminating y_1 through y_{n-1}, we have,

$$R y_n = d,$$

where

$$R = \frac{1}{\prod_{i=2}^{n} h_{i-1,i}} \phi(-H_1),$$

$\phi(x)$, being the characteristic polynomial of H_1 and the vector d is defined in **Step 4** below.

Thus, once y_n is obtained by solving the system $R y_n = d$, y_{n-1} through y_1 are computed recursively as follows:

$$y_{i-1} = -\frac{1}{h_{i-1,i}} \left(H_1 y_i + \sum_{j=i}^{n} h_{ji} y_j - c'_i \right), \qquad i = n, n-1, \ldots, 2.$$

Step 3. *Computing the matrix R of Step 2.* It is well known (see Datta and Datta (1976)) that by knowing only one row or a column of a polynomial matrix in an unreduced Hessenberg matrix, the other rows or columns of the matrix polynomial can be generated recursively.

Realizing that the matrix R is basically a polynomial matrix in the lower Hessenberg matrix H_1, its computation is greatly facilitated.

Thus, if $R = (r_1, \ldots, r_n)$, then, knowing r_n, r_{n-1} through r_1 can be generated recursively as follows:

$$r_{k-1} = \frac{1}{h'_{k-1,k}} \left(H_1 r_k - \sum_{i=k}^{n} h'_{ik} r_i \right),$$

where $H_1 = (h'_{ij}); k = n, n-1, \ldots, 2$

It therefore remains to know how to compute r_n. This can be done as follows.

Set $\theta_n = e_n = (0, 0, 0, \ldots, 0, 1)^{\mathrm{T}}$ and then compute θ_{n-1} through θ_0 recursively by using

$$\theta_{i-1} = -\frac{1}{h_{i-1,i}} \left(H_1 \theta_i + \sum_{j=i}^{n} h_{ji} \theta_j \right), \quad i = n, n-1, \ldots, 1.$$

Then, it can be shown (Datta and Datta 1976) that

$$r_n = \theta_0, \quad \text{setting } h_{01} = 1.$$

Step 4. *Computing the vector d of Step 2.* The vector d can also be generated from the above recursion. Thus, starting with $z_n = 0$ (a zero vector), if z_{n-1} through z_0 are generated recursively using

$$z_{i-1} = -\frac{1}{h_{i-1,i}} \left(H_1 z_i + \sum_{j=i}^{n} h_{ji} z_j - c'_i \right), \quad i = n, \cdots, 2, 1,$$

then $d = -z_0$.

Step 5. *Recovery of the original solution X from Y.*

$$X = UYV^{\mathrm{T}}.$$

Remarks

- It is to be noted that the method, as presented above, is of theoretical interest only at present. There are possible numerical difficulties. For example, if one or more of the entries of the subdiagonal of the Hessenberg matrix H_2 are small, a large round-off error can be expected in computing y_{i-1} in Step 2. A detailed study on the numerical behavior of the method is necessary, before recommending it for practical use. Probably, some modification will be necessary to make it a working numerical algorithm. The reason for including this method here is to show that *a method for the Sylvester equation can be developed just by passing through the Hessenberg transformations of the matrices A and B only; no real Schur or Schur transformation is necessary.*

Example 8.5.6. Consider solving the Sylvester equation $AX + XB = C$ with the following data

$$A = \begin{pmatrix} 1 & 2 & 3 & 4 \\ 4 & 5 & 6 & 7 \\ 7 & 8 & 9 & 1 \\ 10 & 0 & 0 & 0 \end{pmatrix}, \quad B = \begin{pmatrix} 1 & -1 & 0 \\ 1 & 1 & 0 \\ 0 & 0 & 2 \end{pmatrix}, \quad C = \begin{pmatrix} 12 & 10 & 12 \\ 24 & 22 & 24 \\ 27 & 25 & 27 \\ 12 & 10 & 12 \end{pmatrix}.$$

Step 1. Reduction of A and B to lower Hessenberg forms:

$$H_1 = \begin{pmatrix} 1.0000 & -5.3852 & 0 & 0 \\ -12.8130 & 8.7241 & 5.1151 & 0 \\ 0.8337 & 10.3127 & 4.6391 & 0.1586 \\ 0.3640 & 1.3595 & -4.8552 & 0.6368 \end{pmatrix},$$

$$U = \begin{pmatrix} 1.0000 & 0 & 0 & 0 \\ 0 & -0.3714 & -0.6009 & -0.7078 \\ 0 & -0.5571 & -0.4657 & 0.6876 \\ 0 & -0.7428 & 0.6497 & -0.1618 \end{pmatrix}.$$

$$H_2 = \begin{pmatrix} 1 & 1 & 0 \\ -1 & 1 & 0 \\ 0 & 0 & 2 \end{pmatrix}, \quad V = \begin{pmatrix} 1 & 0 & 0 \\ 0 & -1 & 0 \\ 0 & 0 & 1 \end{pmatrix},$$

$$C' = \begin{pmatrix} 12.0000 & -10.0000 & 12.0000 \\ -32.8681 & 29.5256 & -32.8681 \\ -19.1978 & 18.3641 & -19.1978 \\ -0.3640 & -0.0000 & -0.3640 \end{pmatrix}.$$

Step 2. Solution of the reduced problem: Since the matrix H_2 is **reduced** ($h_{23} = 0$), instead of an algorithm breakdown, the set of equations for y_1, y_2, y_3 decouple and we obtain:

$$H_1 y_3 + h_{33} y_3 = c'_3,$$
$$H_1 y_2 + h_{12} y_1 + h_{22} y_2 = c'_2 - h_{32} y_3 = \hat{c}_2,$$
$$H_1 y_1 + h_{11} y_1 + h_{21} y_2 = c'_1 - h_{31} y_3 = \hat{c}_1.$$

The vector y_3 is obtained as the solution of the first system, and once y_3 is known, $\hat{c}_2$ and $\hat{c}_3$ can be easily computed.

$$y_3 = \begin{pmatrix} 1.0000 \\ -1.6713 \\ -0.4169 \\ -0.1820 \end{pmatrix}, \quad \hat{c}_2 = \begin{pmatrix} -10.0000 \\ 29.5256 \\ 18.3641 \\ -0.0000 \end{pmatrix}, \quad \hat{c}_1 = \begin{pmatrix} 12.0000 \\ -32.8681 \\ -19.1978 \\ -0.3640 \end{pmatrix}.$$

We now proceed to compute y_2 and y_1 as follows:

Step 3. Computation of the vector d: starting from $z_2 = (0\ 0\ 0\ 0)^T$,

$$z_1 = -\frac{1}{h_{12}}(H_1 z_2 + h_{22} z_2 - \hat{c}_2)$$

$$= (-10.0000\ \ 29.5256\ \ 18.3641\ \ -0.0000)^T,$$

$$d = -z_0 = \frac{1}{1}(H_1 z_1 + h_{11} z_1 + h_{21} z_2 - \hat{c}_1)$$

$$= (-191.0000\ \ 542.0447\ \ 418.9050\ \ -52.2985)^T.$$

Step 4. Computation of the matrix R. Starting from $\theta_2 = (0\ 0\ 0\ 1)^T$,

$$\theta_1 = -\frac{1}{h_{12}}(H_1 \theta_2 + h_{22}\theta_2) = (0\ 0\ -0.1586\ -1.6368)^T,$$

$$\theta_0 = -\frac{1}{1}(H_1\theta_1 + h_{11}\theta_1 + h_{21}\theta_2) = (0\ 0.8112\ 1.1538\ 2.9092)^T$$

and now, starting from $r_4 = \theta_0$, we obtain

$$r_3 = \frac{1}{h'_{34}}(H_1 r_4 - h'_{44} r_4)$$

$$= (-27.5462\ \ 78.5858\ \ 84.7805\ \ -28.3717)^T,$$

$$r_2 = \frac{1}{h'_{23}}(H_1 r_3 - h'_{33} r_3 - h'_{43} r_4)$$

$$= (-63.1364\ \ 217.3103\ \ 154.1618\ -36.5856)^T,$$

$$r_1 = \frac{1}{h'_{12}}(H_1 r_2 - h'_{22} r_2 - h'_{32} r_3 - h'_{42} r_4)$$

$$= (74.0000\ \ -145.9565\ \ -125.7098\ \ -20.1428)^T,$$

which gives

$$R = \begin{pmatrix} 74.0000 & -63.1364 & -27.5462 & 0 \\ -145.9565 & 217.3103 & 78.5858 & 0.8112 \\ -125.7098 & 154.1618 & 84.7805 & 1.1538 \\ -20.1428 & -36.5856 & -28.3717 & 2.9092 \end{pmatrix}$$

and now $R y_2 = d$ gives

$$y_2 = (-1.0000\ \ 1.6713\ \ 0.4169\ \ 0.1820)^T$$

and finally we compute

$$y_1 = (1.0000\ \ -1.6713\ \ -0.4169\ \ -0.1820)^T.$$

Therefore, the solution of the reduced problem is

$$Y = \begin{pmatrix} 1.0000 & -1.0000 & 1.0000 \\ -1.6713 & 1.6713 & -1.6713 \\ -0.4169 & 0.4169 & -0.4169 \\ -0.1820 & 0.1820 & -0.1820 \end{pmatrix}.$$

The original solution X is then recovered via $X = UYV^T$:

$$X = \begin{pmatrix} 1.0000 & 1.0000 & 1.0000 \\ 1.0000 & 1.0000 & 1.0000 \\ 1.0000 & 1.0000 & 1.0000 \\ 1.0000 & 1.0000 & 1.0000 \end{pmatrix}.$$

Verification: $\|AX + XB - C\|_2 = 5.6169 \times 10^{-14}$.

MATCONTROL note: The Hessenberg methods for the Sylvester and Lyapunov equations have been implemented in MATCONTROL functions **sylvhess** and **lyaphess**, respectively. Both Hessenberg matrices are assumed to be unreduced. The above example shows that the method, however, works if one of them is reduced, but in that case the codes need to be modified.

8.5.7 The Hessenberg–Schur Method for the Discrete Sylvester Equation

In some applications, one needs to solve a general discrete Sylvester equation:

$$BXA + C = X.$$

The Schur method for the discrete Lyapunov equation described in **Section 8.5.4** can be easily extended to solve this equation.

Assume that **the order of A is smaller than that of B**. $A \in \mathbb{R}^{n \times n}$, $B \in \mathbb{R}^{m \times m}$. Let the matrices A^T and B be transformed, respectively, to an upper real Schur matrix R and an upper Hessenberg matrix H by orthogonal similarity:

$$U^T A^T U = R,$$
$$V^T B V = H.$$

Then,

$$BXA + C = X \quad \text{becomes} \quad HYR^T + \hat{C} = Y,$$

where $Y = V^T X U, \hat{C} = V^T C U$. Let $Y = (y_1, \ldots, y_n)$, and $\hat{C} = (c_1, c_2, \ldots, c_n)$.

The reduced equation can now be solved in exactly the same way as in the Hessenberg–Schur algorithm for the Sylvester equation (**Algorithm 8.5.1**). This is left as an exercise (**Exercise 8.27**) for the readers.

MATCONTROL note: MATCONTROL function **sylvhcsd** solves the discrete-time Sylvester equation, based on **complex Schur** decomposition of A.

8.6 DIRECT COMPUTATIONS OF THE CHOLESKY FACTORS OF SYMMETRIC POSITIVE DEFINITE SOLUTIONS OF LYAPUNOV EQUATIONS

In this section we describe methods for finding the Cholesky factors of the symmetric positive definite solutions of both continuous-time and discrete-time Lyapunov equations, without explicitly computing such solutions.

8.6.1 Computing the Cholesky Factor of the Positive Definite Solution of the Lyapunov Equation

Consider first the Lyapunov equation:

$$XA + A^T X = -C^T C, \tag{8.6.1}$$

where A is an $n \times n$ stable matrix (i.e., all the eigenvalues $\lambda_1, \ldots, \lambda_n$ have negative real parts), and C is an $r \times n$ matrix.

The above equation admits a unique symmetric positive semidefinite solution X. Thus, such a solution matrix X has the Cholesky factorization $X = Y^T Y$, where Y is upper triangular.

In several applications, all that is needed is the matrix Y; X is not needed as such. One such application is **model reduction problem via internal balancing and the Schur method for model reduction (Chapter 14)**, where the Cholesky factors of the controllability and observability Grammians are needed.

In these applications, it might be computationally more attractive to obtain the matrix Y directly without solving the equation for X, because X can be considerably more ill-conditioned than Y. Note that $\text{Cond}_2(X) = (\text{Cond}_2(Y))^2$. Also, it may not be computationally desirable to form the right-hand side matrix $-C^T C$ explicitly; there may be a significant loss of accuracy in this explicit formation.

We describe below a procedure due to Hammarling (1982) for finding the Cholesky factor Y without explicitly computing X and without forming the matrix product $C^T C$.

Reduction of the Problem

Substituting $X = Y^T Y$ in Eq. (8.6.1), we have

$$(Y^T Y)A + A^T(Y^T Y) = -C^T C. \tag{8.6.2}$$

The challenge is now to compute Y without explicitly forming the product $C^T C$.

Let $S = U^T AU$, where S is in upper RSF and U is orthogonal. Let

$$CU = QR$$

be the QR factorization of CU.

Then Eq. (8.6.2) becomes

$$S^T \left(\hat{Y}^T \hat{Y} \right) + \left(\hat{Y}^T \hat{Y} \right) S = -R^T R, \tag{8.6.3}$$

where $\hat{Y} = YU$ and $R^T R = (CU)^T CU$.

Solution of the Reduced Equation

To obtain $\hat{Y}$ from (8.6.3) without explicitly forming $R^T R$, we partition $\hat{Y}$, R, and S as follows:

$$\hat{Y} = \begin{pmatrix} \hat{y}_{11} & \hat{y}^T \\ 0 & Y_1 \end{pmatrix}, \qquad R = \begin{pmatrix} r_{11} & r^T \\ 0 & R_1 \end{pmatrix}, \qquad S = \begin{pmatrix} s_{11} & s^T \\ 0 & S_1 \end{pmatrix}, \tag{8.6.4}$$

where s_{11} is a scalar (a real eigenvalue in RSF S) or a 2×2 matrix ("Schur bump," corresponding to a pair of complex conjugate eigenvalues in the matrix S); and $\hat{y}$, r, and s are either column vectors or matrices with two columns.

Since $\hat{Y}$ satisfies (8.6.3) we can show, after some algebraic manipulations, that $\hat{y}_{11}$, $\hat{y}$, and Y_1 satisfy the following equations:

$$s_{11}^T \left(\hat{y}_{11}^T \hat{y}_{11} \right) + \left(\hat{y}_{11}^T \hat{y}_{11} \right) s_{11} = -r_{11}^T r_{11}, \tag{8.6.5}$$

$$S_1^T \hat{y} + \hat{y} \left(\hat{y}_{11} s_{11} \hat{y}_{11}^{-1} \right) = -r\alpha - s \hat{y}_{11}^T, \tag{8.6.6}$$

$$S_1^T \left(Y_1^T Y_1 \right) + \left(Y_1^T Y_1 \right) S_1 = -\hat{R}_1^T \hat{R}_1, \tag{8.6.7}$$

where $\alpha = r_{11} \hat{y}_{11}^{-1}$, $\hat{R}_1^T \hat{R}_1 = R_1^T R_1 + uu^T$, and $u = r - \hat{y}\alpha^T$.

Since $R_1^T R_1$ is positive definite, so is $\hat{R}_1^T \hat{R}_1$.

Note that the matrix $\hat{R}_1$ can be easily computed, once R_1 and u are known, from the QR factorization:

$$\begin{pmatrix} u^T \\ R_1 \end{pmatrix} = \hat{Q} \hat{R}_1. \tag{8.6.8}$$

Equation (8.6.7) is of the same form as the original reduced equation (8.6.2), **but is of smaller order. This is the key observation.**

The matrices S_1, Y_1, and $\hat{R}_1$ can now be partitioned further as in (8.6.4), and the whole process can be repeated. The process is continued until $\hat{Y}$ is completely determined.

Recovery of the Solution

Once $\hat{Y}$ is obtained, the "R-matrix" $\tilde{Y}$ of the QR factorization $\tilde{Q}\tilde{Y} = \hat{Y}U^T$ will be an upper triangular matrix that will satisfy Eq. (8.6.7). Let $\tilde{Y} = (y_{ij})$.

Since Y is required to have positive diagonal entries, we will take

$$Y = \text{diag}(\text{sign}(\tilde{y}_{11}), \ldots, \text{sign}(\tilde{y}_{nn})) \, \tilde{Y}.$$

Algorithm 8.6.1.   *Algorithm for the Direct Cholesky Factor of the Symmetric Positive Definite Solution of the Lyapunov Equation*
Inputs. *A—An $n \times n$ matrix*
C—An $r \times n$ matrix.

Output. *Y—The Cholesky factor of the symmetric positive definite solution of the Lyapunov equation:* $XA + A^T X = -C^T C.$

Assumption. A is stable.
Step 1. *Find the RSF S of A: $U^T A U = S.$*
Step 2. *Find the QR factorization of the $r \times n$ matrix CU : $CU = QR.$*
Step 3. *Partition* $R = \begin{pmatrix} r_{11} & r^T \\ 0 & R_1 \end{pmatrix}$, $S = \begin{pmatrix} s_{11} & s^T \\ 0 & S_1 \end{pmatrix}.$
Step 4. *Find* $\hat{Y} = \begin{pmatrix} \hat{y}_{11} & \hat{y}^T \\ 0 & Y_1 \end{pmatrix}$ *as follows:*
   4.1 *Compute $\hat{y}_{11}$ from* $s_{11}^T(\hat{y}_{11}^T \hat{y}_{11}) + (\hat{y}_{11}^T \hat{y}_{11})s_{11} = -r_{11}^T r_{11}.$
   4.2 *Compute* $\alpha = r_{11}\hat{y}_1^{-1}.$
   4.3 *Solve for $\hat{y}$:* $S_1^T \hat{y} + \hat{y}(\hat{y}_{11}s_{11}\hat{y}_{11}^{-1}) = -r\alpha - s\hat{y}_{11}^T.$
   4.4 *Compute $u = r - \hat{y}\alpha^T$ and then find the QR factorization of* $\begin{pmatrix} u^T \\ R_1 \end{pmatrix}$ *to*

find $\hat{R}_1$:

$$\hat{Q}\hat{R}_1 = \begin{pmatrix} u^T \\ R_1 \end{pmatrix}.$$

   Step 5. *Set $S = S_1, R = \hat{R}_1$ and return to Step 3 and continue until $\hat{Y}$ is completely determined.*
   Step 6. *Compute $\tilde{Y}$ from the QR factorization of $\hat{Y}U^T$:$\hat{Y}U^T = Q\tilde{Y}$. Let $\tilde{Y} = (\tilde{y}_{ij}).$*

Step 7.*Compute* $Y = \begin{pmatrix} \text{sign}(\tilde{y}_{11}) & & 0 \\ & \ddots & \\ 0 & & \text{sign}(\tilde{y}_{nn}) \end{pmatrix} \tilde{Y}.$

Example 8.6.1.   Consider solving Eq. (8.6.1) for the Cholesky factor Y with

$$A = \begin{pmatrix} -0.9501 & 0.5996 & 0.2917 \\ 0.6964 & -1.0899 & -0.6864 \\ 0 & 0.0571 & -6.6228 \end{pmatrix}, \qquad C = (1, 1, 1).$$

Step 1. Reduction of A to RSF: $[U, S] = \textbf{schur}\,(A)$ gives

$$U = \begin{pmatrix} -0.7211 & -0.6929 & 0.0013 \\ -0.6928 & 0.7210 & -0.0105 \\ -0.0063 & 0.0084 & 0.9999 \end{pmatrix},$$

$$S = \begin{pmatrix} -0.3714 & 0.0947 & 0.3040 \\ 0 & -1.6762 & -0.7388 \\ 0 & 0 & -6.6152 \end{pmatrix}.$$

Step 2. The QR factorization of CU: $[Q, R] = \textbf{qr}\,(CU)$ gives

$$R = (-1.4202 \quad 0.0366 \quad 0.9908).$$

Step 3.

$$r_{11} = -1.4202, \qquad s_{11} = -0.3714,$$

$$r = \begin{pmatrix} 0.0366 \\ 0.9908 \end{pmatrix}, \qquad S_1 = \begin{pmatrix} -1.6762 & -0.7388 \\ 0 & -6.6152 \end{pmatrix}, \qquad s = \begin{pmatrix} 0.0947 \\ 0.3040 \end{pmatrix}.$$

Step 4. Compute $\hat{y}_{11}$ and α:

$$\hat{y}_{11} = 1.6479, \qquad \alpha = r_{11}\hat{y}_{11}^{-1} = -0.8619.$$

Solve for $\hat{y}$:

$$(S_1^T + \hat{y}_{11}s_{11}\hat{y}_{11}^{-1}I)\hat{y} = -r\alpha - s\hat{y}_{11}^T$$

or

$$\begin{pmatrix} -2.0476 & 0 \\ -0.7388 & -6.9866 \end{pmatrix} \hat{y} = \begin{pmatrix} -0.1245 \\ 0.3530 \end{pmatrix},$$

$$\hat{y} = \begin{pmatrix} 0.0608 \\ -0.0569 \end{pmatrix}.$$

$$u = r - \hat{y}\alpha^T = \begin{pmatrix} 0.0890 \\ 0.9418 \end{pmatrix}, \qquad R_1 = 0$$

$$\hat{R}_1 = (0.0890, 0.9418).$$

Step 5. Solution of the reduced 2×2 problem:

$$S \equiv S_1 = \begin{pmatrix} -1.6762 & -0.7388 \\ 0 & -6.6152 \end{pmatrix}$$

$$R \equiv \hat{R}_1 = (0.0890, \ 0.9418)$$

$$\hat{y}_{11} = 0.0486, \qquad \hat{y} = (0.2036), \qquad \hat{R}_1 = 0.5689.$$

Solution of the final 1×1 problem:

$$S = -6.6152, \qquad R = 0.5689, \qquad \hat{y}_{11} = 0.1564.$$

Thus,

$$\hat{Y} = \begin{pmatrix} 1.6479 & 0.0608 & -0.0569 \\ 0 & 0.0486 & 0.02036 \\ 0 & 0 & 0.1564 \end{pmatrix}.$$

Step 6. Compute $\tilde{Y}:[Q_1, \tilde{Y}] = \mathbf{qr} \ (\hat{Y}U^T)$ (Using QR factorization):

$$\tilde{Y}_1 = \begin{pmatrix} 1.2309 & 1.0960 & 0.0613 \\ 0 & -0.0627 & -0.2011 \\ 0 & 0 & 0.1623 \end{pmatrix}.$$

Step 7. $Y = \begin{pmatrix} 1.2309 & 1.0960 & 0.0613 \\ 0 & 0.0627 & 0.2011 \\ 0 & 0 & 0.1623 \end{pmatrix}.$

MATCONTROL note: Algorithm 8.6.1 has been implemented in MATCONTROL functions **lyapchlc**.

Remark

- Note that it is possible to arrange the computation of Y with a different form of partitioning than as shown in (8.6.4). For example, let us partition matrices $\hat{Y}$, R, and S as follows:

$$\hat{Y} = \begin{pmatrix} Y_{11} & y \\ 0 & y_1 \end{pmatrix}, \qquad R = \begin{pmatrix} R_{11} & r \\ 0 & r_1 \end{pmatrix}, \qquad S = \begin{pmatrix} S_{11} & s \\ 0 & s_1 \end{pmatrix}, \quad (8.6.9)$$

where y_1, r_1, and s_1 are scalars or 2×2 matrices and y, r, and s are either column vectors or matrices with two columns.

Then, similar to Eqs. (8.6.5)–(8.6.7), one will obtain three equations. For example, the first one will be just the deflated version of the original equation.

$$S_{11}^T \left(Y_{11}^T Y_{11} \right) + \left(Y_{11}^T Y_{11} \right) S_{11} = -R_{11}^T R_{11}. \tag{8.6.10}$$

Suppose that the solution Y_{11} of this deflated equation has been computed, then the second and third equations will give us the expressions for y and y_1.

By using this new partitioning, the original algorithm of Hammarling (1982) can be slightly improved.

In the following, we will use this partitioning to solve the discrete equation.

8.6.2 Computing the Cholesky Factor of the Positive Definite Solution of the Discrete Lyapunov Equation

Consider now the discrete Lyapunov equation:

$$A^T X A + C^T C = X, \tag{8.6.11}$$

where A is an $n \times n$ discrete–stable matrix (i.e., all the eigenvalues $\lambda_1, \ldots, \lambda_n$ are inside the unit circle) and C is an $r \times n$ matrix.

Then Eq. (8.6.11) admits a unique symmetric positive semidefinite solution X. Such a solution matrix X has the Cholesky factorization: $X = Y^T Y$, where Y is upper triangular.

We would obtain the matrix Y directly without solving the equation (8.6.11) for X. Substituting $X = Y^T Y$ into the Eq. (8.6.11), we have

$$A^T (Y^T Y) A + C^T C = Y^T Y. \tag{8.6.12}$$

As in the case of the continuous-time Lyapunov equation (8.6.1), we now outline a method for finding Y of (8.6.12) without computing X and without forming the matrix $C^T C$.

Reduction of the Problem

Let $S = U^T A U$, where S is in upper RSF and U is an orthogonal matrix. Let

$$Q_1 R = C U$$

be the economy QR factorization of the matrix $C U$.

Then Eq. (8.6.12) becomes

$$S^T \left(\hat{Y}^T \hat{Y} \right) S + R^T R = \hat{Y}^T \hat{Y}, \tag{8.6.13}$$

where $\hat{Y} = Y U$ and $R^T R = (C U)^T C U$.

Solution of the Reduced Equation

To obtain $\hat{Y}$ from (8.6.13) without forming $R^T R$ explicitly, we partition $\hat{Y}$, R, and S as

$$\hat{Y} = \begin{pmatrix} Y_{11} & y \\ 0 & y_1 \end{pmatrix}, \qquad R = \begin{pmatrix} R_{11} & r \\ 0 & r_1 \end{pmatrix}, \qquad S = \begin{pmatrix} S_{11} & s \\ 0 & s_1 \end{pmatrix}.$$

From (8.6.13), we see that Y_{11}, y, and y_1 satisfy the following equations:

$$S_{11}^T \left(Y_{11}^T Y_{11} \right) S_{11} + R_{11}^T R_{11} = \left(Y_{11}^T Y_{11} \right), \tag{8.6.14}$$

$$Y_{11}^T y - (Y_{11} S_{11})^T y s_1 = S_{11}^T Y_{11}^T Y_{11} s + R_{11}^T r, \tag{8.6.15}$$

$$s_1^T y_1^T y_1 s_1 + \left(r_1^T r_1 + r^T r + (Y_{11} s + y s_1)^T (Y_{11} s + y s_1) - y^T y \right) = y_1^T y_1. \tag{8.6.16}$$

Equation (8.6.14) is of the same form as the original reduced equation (8.6.12), **but is of smaller order.**

Suppose that we have already computed the solution Y_{11} of this equation. Then y can be obtained from (8.6.15) by solving a linear system and, finally, (8.6.16) gives us y_1.

Recovery of the Solution

Once $\hat{Y}$ is obtained, the "R-matrix" $\tilde{Y}$ of the QR factorization of the matrix $\hat{Y} U^T$: $\tilde{Q} \tilde{Y} = \hat{Y} U^T$ will be the upper triangular matrix that will solve the equation (8.6.12). Let $\tilde{Y} = (\tilde{y}_{ij})$.

Since Y has to have positive diagonal entries, we take

$$Y = \text{diag}(\text{sign}(\tilde{y}_{11}), \ldots, \text{sign}(\tilde{y}_{nn})) \, \tilde{Y}.$$

Algorithm 8.6.2. *Algorithm for the Direct Cholesky Factor of the Symmetric Positive Definite Solution of the Discrete Lyapunov Equation*
Inputs. *A—An $n \times n$ matrix*
C—An $r \times n$ matrix .
Output. *Y—The Cholesky factor Y of the symmetric positive definite solution X of the discrete Lyapunov Equation: $A^T X A + C^T C = X$.*
Assumption. *A is discrete-stable, that is all its eigenvalues have moduli less than 1.*
   Step 1. *Find the RSF S of A: $U^T A U = S$.*
   Step 2. *Find the (economy size) QR factorization of the $r \times n$ matrix CU: $QR = CU$.*

Step 3. *Partition*

$$S = \begin{pmatrix} S_{11} & * \\ 0 & * \end{pmatrix}, \qquad R = \begin{pmatrix} R_{11} & * \\ 0 & * \end{pmatrix}, \qquad Y = \begin{pmatrix} Y_{11} & * \\ 0 & * \end{pmatrix},$$

where S_{11} is a scalar or 2×2 matrix (Schur bump).
Compute Y_{11} from $S_{11}^T(Y_{11}^T Y_{11})S_{11} + R_{11}^T R_{11} = Y_{11}^T Y_{11}$.
Step 4. *Do while dimension of Y_{11} < dimension of S*
   4.1. *Partition*

$$Y = \begin{pmatrix} Y_{11} & y & * \\ 0 & y_1 & * \\ 0 & 0 & * \end{pmatrix}, \qquad S = \begin{pmatrix} S_{11} & s & * \\ 0 & s_1 & * \\ 0 & 0 & * \end{pmatrix}, \qquad R = \begin{pmatrix} R_{11} & r & * \\ 0 & r_1 & * \\ 0 & 0 & * \end{pmatrix},$$

where s_1 is 1×1 scalar or 2×2 Schur bump.
   4.2. *Compute y from $Y_{11}^T y - (Y_{11}S_{11})^T ys_1 = S_{11}^T Y_{11}^T Y_{11}s + R_{11}^T r$.*
   4.3. *Compute y_1 from*

$$s_1^T y_1^T y_1 s_1 + (r_1^T r_1 + r^T r + (Y_{11}s + ys_1)^T (Y_{11}s + ys_1) - y^T y) = y_1^T y_1.$$

   4.4. *Go to Step 4 with $Y_{11} \equiv \begin{pmatrix} Y_{11} & y \\ 0 & y_1 \end{pmatrix}$.*

Step 5. *Compute $\tilde{Y}$ from the QR factorization of $Y_{11}U^T : Q\tilde{Y} = Y_{11}U^T$.*
Let $\tilde{Y} = (\tilde{y}_{ij})$.

Step 6. *Compute* $Y = \begin{pmatrix} \text{sign}(\tilde{y}_{11}) & & 0 \\ & \ddots & \\ 0 & & \text{sign}(\tilde{y}_{nn}) \end{pmatrix} \tilde{Y}.$

Example 8.6.2. Consider solving the equation (8.6.12) for the Cholesky factor Y with

$$A = \begin{pmatrix} -0.1973 & -0.0382 & 0.0675 \\ -0.1790 & -0.3042 & -0.0544 \\ 0.0794 & 0.0890 & -0.1488 \end{pmatrix}$$

and

$$C = \begin{pmatrix} 0.0651 & 0.1499 & 0.2917 \\ 0.1917 & 0.0132 & 0.4051 \end{pmatrix}.$$

Step 1. Reduction of A to the RSF: $[U, S] = \textbf{schur}(A)$ gives

$$U = \begin{pmatrix} -0.3864 & -0.7790 & 0.4938 \\ -0.7877 & 0.5572 & 0.2627 \\ 0.4798 & 0.2875 & 0.8290 \end{pmatrix},$$

$$S = \begin{pmatrix} -0.3589 & -0.0490 & 0.1589 \\ 0 & -0.1595 & -0.0963 \\ 0 & 0.0173 & -0.1319 \end{pmatrix}.$$

Step 2. The **economy size** QR factorization of CU: $[Q, R] = \mathbf{qr}\,(CU, 0)$ gives

$$R = \begin{pmatrix} 0.1100 & -0.0289 & 0.4245 \\ 0 & 0.1159 & 0.3260 \end{pmatrix}.$$

Step 3. Partitioning of R and S gives $S_{11} = (-0.3589)$, $R_{11} = (0.1100)$, which enables us to compute $Y_{11} = (0.1178)$.

Step 4. Dimension of $Y_{11} = 1 <$ dimension of $S = 3$. So we do:

    4.1.

$$s = \begin{pmatrix} -0.0490 \\ 0.1589 \end{pmatrix}^{\mathrm{T}}, \qquad s_1 = \begin{pmatrix} -0.1595 & -0.0963 \\ 0.0173 & -0.1319 \end{pmatrix},$$

$$r = \begin{pmatrix} -0.0289 \\ 0.4245 \end{pmatrix}^{\mathrm{T}}, \qquad r_1 = (0.1159\ 0.3260).$$

4.2. $y = \begin{pmatrix} -0.0291 \\ 0.4078 \end{pmatrix}^{\mathrm{T}}.$

4.3. Solve for upper triangular y_1 with positive diagonal:

$$y_1 = \begin{pmatrix} 0.1167 & 0.3242 \\ 0 & 0.1392 \end{pmatrix}.$$

4.4. Form Y_{11} :

$$Y_{11} = \begin{pmatrix} 0.1178 & -0.0291 & 0.4078 \\ 0 & 0.1167 & 0.3242 \\ 0 & 0 & 0.1392 \end{pmatrix}$$

and the loop in **Step 4** ends.

Step 5. Find the QR factorization of $Y_{11}U^{\mathrm{T}}$: $[Q_1, \tilde{Y}] = \mathbf{qr}(Y_{11}U^{\mathrm{T}})$ to obtain $\tilde{Y}$:

$$\tilde{Y} = \begin{pmatrix} -0.2034 & -0.0618 & -0.4807 \\ 0 & -0.1417 & -0.1355 \\ 0 & 0 & -0.0664 \end{pmatrix}.$$

Step 6. Compute the solution:

$$Y = \begin{pmatrix} 0.2034 & 0.0618 & 0.4807 \\ 0 & 0.1417 & 0.1355 \\ 0 & 0 & 0.0664 \end{pmatrix}.$$

MATCONTROL Note: Algorithm 8.6.2 has been implemented in MATCONTROL function **lyapchld**.

8.7 COMPARISONS OF DIFFERENT METHODS AND CONCLUSIONS

The analytical methods such as the ones based on evaluating the integral

$$X = \int_0^\alpha e^{A^\mathrm{T} t} C e^{At} \, dt$$

for the Lyapunov equation, evaluating the infinite sum $\sum (A^k)^\mathrm{T} C A^k$ for the discrete Lyapunov equation, and the finite series methods for the Sylvester and Lyapunov equations are not practical for numerical computations.

The methods, based on the reduction to Jordan and companion forms, will give inaccurate solutions when the transforming matrices are ill-conditioned. *The methods based on the reduction to Jordan and companion forms, therefore, in general should be avoided for numerical computations.*

From numerical viewpoints, the methods of choice are:

- The **Schur method (Section 8.5.2)** for the Lyapunov equation: $XA + A^\mathrm{T} X = C$.
- The **Hessenberg–Schur method (Algorithm 8.5.1)** for the Sylvester equation: $XA + BX = C$.
- The **Schur method (Section 8.5.4)** for the discrete Lyapunov equation: $A^\mathrm{T} XA - X = C$
- The **modified Schur methods (Algorithms 8.6.1 and 8.6.2)** for the Cholesky factors of the Lyapunov equation: $XA + A^\mathrm{T} X = -C^\mathrm{T} C$ and the discrete Lyapunov equation: $A^\mathrm{T} XA + C^\mathrm{T} C = X$.

8.8 SOME SELECTED SOFTWARE

8.8.1 MATLAB Control System Toolbox

Matrix equation solvers

lyap   Solve continuous Lyapunov equations
dlyap   Solve discrete Lyapunov equations.

8.8.2 MATCONTROL

CONDSYLVC   Finding the condition number of the Sylvester equation problem
LYAPCHLC   Finding the Cholesky factor of the positive definite solution of the continuous-time Lyapunov equation
LYAPCHLD   Find the Cholesky factor of the positive definite solution of the discrete-time Lyapunov equation
LYAPCSD   Solving discrete-time Lyapunov equation using complex Schur decomposition of A

LYAPFNS	Solving continuous-time Lyapunov equation via finite series method
LYAPHESS	Solving continuous-time Lyapunov equation via Hessenberg decomposition
LYAPRSC	Solving the continuous-time Lyapunov equation via real Schur decomposition
LYAPRSD	Solving discrete-time Lyapunov equation via real Schur decompostion
SEPEST	Estimating the sep function with triangular matrices
SEPKR	Computing the sep function using Kronecker product
SYLVHCSC	Solving the Sylvester equation using Hessenberg and complex Schur decompositions
SYLVHCSD	Solving the discrete-time Sylvester equation using Hessenberg and complex Schur decompositions
SYLVHESS	Solving the Sylvester equation via Hessenberg decomposition
SYLVHRSC	Solving the Sylvester equation using Hessenberg and real Schur decompositions
SYLVHUTC	Solving an upper triangular Sylvester equation.

8.8.3   CSP-ANM

Solutions of the Lyapunov and Sylvester matrix equations

- The Schur method for the Lyapunov equations is implemented as LyapunovSolve [a, b] SolveMethod → SchurDecomposition] (continuous-time case) and DiscreteLyapunovSolve [a, b, Solve-Method → SchurDecomposition] (discrete-time case).
- The Hessenberg–Schur method for the Sylvester equations is implemented as LyapunovSolve [a, b, c, SolveMethod → HessenbergSchur] (continuous-time case) and Discrete LyapunovSolve [a, b, c, SolveMethod → HessenbergSchur] (discrete-time case).
- The Cholesky factors of the controllability and observability Grammians of a stable system are computed using CholeskyFactorControllabilityGramian [system] and CholeskyFactorObservabilityGramian [system].

8.8.4   SLICOT

Lyapunov equations

SB03MD	Solution of Lyapunov equations and separation estimation
SB03OD	Solution of stable Lyapunov equations (Cholesky factor)
SB03PD	Solution of discrete Lyapunov equations and separation estimation

SB03QD   Condition and forward error for continuous Lyapunov equations
SB03RD   Solution of continuous Lyapunov equations and separation
         estimation
SB03SD   Condition and forward error for discrete Lyapunov equations
SB03TD   Solution of continuous Lyapunov equations, condition and
         forward error estimation
SB03UD   Solution of discrete Lyapunov equations, condition and forward
         error estimation

   Sylvester equations

SB04MD   Solution of continuous Sylvester equations (Hessenberg–Schur
         method)
SB04ND   Solution of continuous Sylvester equations (one matrix in Schur form)
SB04OD   Solution of generalized Sylvester equations with separation
         estimation
SB04PD   Solution of continuous or discrete Sylvester equations (Schur method)
SB04QD   Solution of discrete Sylvester equations (Hessenberg–Schur method)
SB04RD   Solution of discrete Sylvester equations (one matrix in Schur form)

   Generalized Lyapunov equations

SG03AD   Solution of generalized Lyapunov equations and separation
         estimation
SG03BD   Solution of stable generalized Lyapunov equations (Cholesky factor)

8.8.5   MATRIX$_X$

Purpose: Solve a discrete Lyapunov equation.

Syntax: P = DLYAP (A, Q)

Purpose: Solve a continuous Lyapunov equation.

Syntax: P = LYAP (A, Q)

8.8.6   LAPACK

The Schur method for the Sylvester equation, $XA + BX = C$, can be implemented in LAPACK by using the following routines in sequence: GEES to compute the Schur decomposition, GEMM to compute the transformed right-hand side, TRSY L to solve the (quasi-)triangular Sylvester equation, and GEMM to recover the solution X.

8.9   SUMMARY AND REVIEW

Applications

The applications of the Lyapunov equations include:

- Stability and robust stability analyses (**Chapter 7**).
- Computations of the controllability and observability Grammians for stable systems (needed for internal balancing and model reduction) (**Chapter 14**).
- Computations of the H_2 norm (**Chapter 7**).
- Implementation of Newton's methods for Riccati equations (**Chapter 13**).

The applications of the Sylvester equations include:

- Design of Luenberger observer (**Chapter 12**)
- Block-diagonalization of a matrix by similarity transformation.

Existence and Uniqueness Results

(1)   The Sylvester equation $XA + BX = C$ has a unique solution if and only A and $-B$ do not have an eigenvalue in common (**Theorem 8.2.1**).

(2)   The Lyapunov equation $XA + A^T X = C$ has a unique solution if and only if A and $-A$ do not have an eigenvalue in common (**Corollary 8.2.1**).

(3)   The discrete Lyapunov equation $A^T XA - X = C$ has a unique solution if and only if the product of any two eigenvalues of A is not equal to 1 or A does not have an eigenvalue of modulus 1 (**Theorem 8.2.2**).

Sensitivity Results

(1)   sep $(B, -A)$ defined by

$$\text{sep}(B, -A) = \min_{X \neq 0} \frac{\|XA + BX\|_{\text{F}}}{\|X\|_{\text{F}}} = \sigma_{\min}(P),$$

where $P = I_n \otimes B + A^T \otimes I_m$, m and n are, respectively, the orders of B and A, plays an important role in the sensitivity analysis of the Sylvester equation $XA + BX = C$ (**Theorem 8.3.1**).

(2)   sep $(A^T, -A)$ has an important role in the sensitivity analysis of the Lyapunov equation: $XA + A^T X = C$ (**Corollary 8.3.2**).

(3) $\text{sep}_d(A^T, A) = \sigma_{\min}(A^T \otimes A^T - I_{n^2})$ has an important role in the sensitivity analysis of the discrete Lyapunov equation $A^T XA - X = C$ (**Theorem 8.3.4**).

(4)   If A is stable, then the sensitivity of the Lyapunov equation can be determined by solving the Lyapunov equation $HA + A^T H = -I$. $\|H\|_2$ is an

indicator of the sensitivity of the stable Lyapunov equation $XA + A^T X = -C$ (**Theorem 8.3.3**).

(5) If A and B are ill-conditioned, then the Sylvester equation $XA + BX = C$ is ill-conditioned (**Theorem 8.3.6**). Thus, if A is ill-conditioned, then the Lyapunov equation is also ill-conditioned. **But the converse is not true in general**.

Sep-Estimation

The LINPACK style algorithm (**Algorithm 8.3.1**) gives an estimate of sep $(A, B)^T$ without computing the Kronecker product sum P, which is computationally quite sensitive.

Methods for Solving the Lyapunov and Sylvester Equations

- The analytical methods such as the finite-series method or the method based on evaluation of the integral involving the matrix exponential are not practical for numerical computations (**Section 8.4**).
- The methods based on reduction to the JCF and the companion form of a matrix should be avoided (**Section 8.5.1**).
- The Schur methods for the Lyapunov equations (**Sections 8.5.2 and 8.5.4**) and the Hessenberg–Schur method (**Algorithms 8.5.1 and Section 8.5.7**) for the Sylvester equations are by far the best for numerical computations.
- If only the Cholesky factors of stable Lyapunov equations are needed, the modified Schur methods (**Algorithms 8.6.1 and 8.6.2**) should be used. These algorithms compute the Cholesky factors of the solutions without explicitly computing the solutions themselves. The algorithms are numerically stable.

8.10 CHAPTER NOTES AND FURTHER READING

The results on the existence and uniqueness of the Lyapunov and Sylvester equations are **classical**. For proofs of these results, see Horn and Johnson (1991), Lancaster and Rodman (1995). See also Barnett and Cameron (1985), and Barnett and Storey (1970). The sensitivity issues of these equations and the perturbation results given in Section 8.3 can be found in Golub *et al.* (1979) and in Higham (1996).

The sensitivity result of the stable Lyapunov equation is due to Hewer and Kenney (1988). The sensitivity result of the stable discrete Lyapunov equation is due to Gahinet *et al.* (1990). The perturbation result of the discrete Lyapunov equation appears in Petkov *et al.* (1991). The results relating the ill-conditioning of the Sylvester equation and eigenvalues can be found in Ghavimi and Laub (1995). The LINPACK-style sep-estimation algorithm is due to Byers (1984). See Kågström and Poromaa (1996)) for LAPACK-style algorithms. For perturbation results on generalized Sylvester equation, see Kågström (1994) and Edelman *et al.* (1997, 1999). For description of LAPACK, see Anderson *et al.* (1999). A recent book by Konstantinov *et al.* (2003) Contains many results on perturbation theory for matrix equations.

The Schur method for the Lyapunov equation is due to Bartels and Stewart (1972). The Schur method for the discrete Lyapunov equation is due to Barraud (1977). Independently of Barraud, a similar algorithm was developed by Kitagawa (1977). The Hessenberg–Schur algorithms for the Sylvester and discrete Sylvester equations are due to Golub *et al.* (1979). A good account of the algorithmic descriptions and implementational details of the methods for solving the discrete Lyapunov equations appears in the recent book of Sima (1996).

The Cholesky-factor algorithms for the stable Lyapunov equations are due to Hammarling (1982). The Hessenberg algorithm for the Sylvester equation is due to Datta and Datta (1976) and Kreisselmeier (1972). For numerical solutions of the generalized Sylvester equation $AXB^{\mathrm{T}} + CXD^{\mathrm{T}} = E$, see Gardiner *et al.* (1992a). For applications of generalized Sylvester equations of the above type including the computation of stable eigendecompositions of matrix pencils see Demmel and Kågström (1987, 1993a, 1993b), Kågström and Westin (1989), etc. See Kågström and Poromaa (1989, 1992) for block algorithms for triangular Sylvester equation (with condition estimator). See Gardiner *et al.* (1992b) for a software package for solving the generalized Sylvester equation.

Exercises

8.1 Prove that the equation $A^*XB + B^*XA = -C$ has a unique solution X if and only if $\lambda_i + \bar{\lambda}_j \neq 0$, for all i and j, where λ_i is an eigenvalue of the generalized eigenvalue problem: $Ax = \lambda Bx$. (Here $A^* = (\bar{A})^{\mathrm{T}}$ and $B^* = (\bar{B})^{\mathrm{T}}$.)

8.2 Let A be a normal matrix with $\lambda_1, \ldots, \lambda_n$ as the eigenvalues. Then show that $\max_i |\lambda_i| / \min_{ij} |\bar{\lambda}_i + \lambda_j)|$ can be regarded as the condition number of the Lyapunov equation $XA + A^*X = -C$, where $A^* = (\bar{A})^{\mathrm{T}}$. Using the result, construct an example of an ill-conditioned Lyapunov equation.

8.3 If $A = (a_{ij})$ and $B = (b_{ij})$ are upper triangular matrices of order $m \times m$ and $n \times n$ respectively, then show that $X = (x_{ij})$ satisfying the Sylvester equation $AX + XB = C$ can be found from

$$x_{ij} = \frac{c_{ij} - \sum_{k=i+1}^{m} a_{ik}x_{kj} - \sum_{k=1}^{n-1} x_{ik}b_{kj}}{a_{ii} + b_{jj}}.$$

8.4 Prove Theorems 8.3.1 and 8.3.4.

8.5 Using the perturbation results in Section 8.3, construct an example to show that the Sylvester equation problem $XA + BX = C$ can be very well-conditioned even when the eigenvector matrices for A and B are ill-conditioned.

8.6 Prove or disprove that if A and $-B$ have close eigenvalues, then the Sylvester equation $XA + BX = C$ is ill-conditioned.

8.7 Construct a 2×2 example to show that the bound (8.3.7) can be much smaller than the bound (8.3.3).

8.8 Derive the expression ϕ for the **condition number of the Lyapunov equation** given in Section 8.3.4.

8.9 Using the definition of the sep function, prove that if X is a unique solution of the Sylvester equation $XA + BX = C$, then

$$\|X\|_F \leq \frac{\|C\|_F}{\text{sep}(B, -A)}.$$

8.10 Let

$$U^T AU = T = \begin{pmatrix} T_{11} & T_{12} & \cdots & T_{1p} \\ 0 & T_{22} & \cdots & T_{2p} \\ \vdots & \vdots & \ddots & \vdots \\ 0 & 0 & \cdots & T_{pp} \end{pmatrix}$$

be the RSF of A, and assume that $T_{11}, \ldots, T_{pp}$ have disjoint spectra.

(a) Develop an algorithm to transform T to the block diagonal form:

$$Y^{-1}TY = \text{diag}(T_{11}, \ldots, T_{pp}),$$

based on the solution of a Sylvester equation.

(b) Show that if the spectra of the diagonal blocks of T are not distinctly separated, then there will be a substantial loss of accuracy (consult Bavely and Stewart (1979)).

(c) Construct an example to support the statement in (b).

(d) Develop an algorithm to compute e^{At} based on the block diagonalization of A.

8.11 Construct a simple example to show that the Cholesky factor L of the solution matrix $X = L^T L$ of the Lyapunov equation: $XA + A^T X = BB^T$, where A is a stable matrix, is less sensitive (with respect to perturbations in A) than X.

8.12 Construct your own example to show that the Lyapunov equation $XA + A^T X = -C$ is always ill-conditioned if A is ill-conditioned with respect to inversion, but the converse is not true.

8.13 Repeat the last exercise with the Sylvester equation $XA + BX = C$, that is, construct an example to show that the Sylvester equation $XA + BX = C$ will be ill-conditioned if both A and B are ill-conditioned, but the converse is not true.

8.14 (a) Let A be a stable matrix. Show that the Lyapunov equation $XA + A^T X = -C$ can still be ill-conditioned if A has one or more eigenvalues close to the imaginary axes.

(b) Construct an example to illustrate the result in (a).

8.15 Give an example to show that the backward error for the Sylvester equation $XA + BX = C$, where only A and B are perturbed, is large if an approximate solution Y of the equation is ill-conditioned.

8.16 Give an example to illustrate that the backward error of an approximate solution to the Sylvester equation $XA + BX = C$ can be large, even though the relative residual is quite small.

8.17 Prove that $\text{sep}(A, -B) > 0$ if and only if A and $-B$ do not have common eigenvalues.

8.18 Let $K = I \otimes A^T + A^T \otimes I$ and $L = I \otimes S^T + S^T \otimes I$ be the Kronecker matrices, respectively, associated with the equations:

$$XA + A^TX = -C$$

and

$$\hat{X}S + S^T\hat{X} = -\hat{C},$$

where $S = U^TAU$ is the RSF of A, and

$$\hat{C} = U^TCU.$$

(a) Prove that $\|K^{-1}\|_2 = \|L^{-1}\|_2$

(b) Using the result in (a), find a bound for the error, when A is only perturbed, in terms of the norm of the matrix A and the norm of L^{-1}.

(c) Based on (a) and (b), develop an algorithm for estimating $\text{sep}(A^T, -A)$, analogous to the Byers' algorithm (Byers 1984) for estimating $\text{sep}(A, B)$.

8.19 **Relationship of the distance to instability and sep** (A) (Van Loan 1985) Define $\text{sep}(A) = \min\{\|AX + XA^*\|_F \,\big|\, X \in \mathbb{C}^{n \times n}, \|X\|_F = 1\}$
Then prove that

(a) $\text{sep}(A) = 0$, if and only if A has an eigenvalue on the imaginary axis.

(b) $\frac{1}{2}\text{sep}(A) \leq \beta(A) \leq \sigma_{\min}(A)$, where $\beta(A)$ is the distance to instability (see Chapter 7).
(**Hint:** $\text{sep}(A) = \sigma_{min}(I \otimes A + A \otimes I)$, and $\|B \otimes C\|_2 \leq \|B\|_2\|C\|_2$.)

8.20 Construct an example of an ill-conditioned discrete Lyapunov equation based on Theorem 8.3.4.

8.21 Prove that if $p(x)$ is a real polynomial of degree n having no pair of roots conjugate with respect to the unit circle, and T is the lower companion matrix of $p(x)$, then the unique solution X of the discrete-time equation: $X - T^TXT = \text{diag}(1, 0, \ldots, 0)$ can be written explicitly as: $X = (I - \phi(S)^T\phi(S))^{-1}$, where S is an unreduced lower Hessenberg matrix with 1s along the superdiagonal and zeros elsewhere, and $\phi(x) = p(x)/(x^n p(1/x))$.

Discuss the numerical difficulties of using this method for solving the discrete Lyapunov equation.

Work out an example to demonstrate the difficulties.

8.22 Develop an algorithm, analogous to Algorithm 8.6.1, to find the Cholesky factor of the symmetric positive definite solution of the Lyapunov equation $AX + XA^T = -BB^T$, where B is $n \times m$ and has full rank.

8.23 Compare the flop-count of the real Schur method and the complex Schur method for solving the Lyapunov equation: $XA + A^TX = -C$.

8.24 Work out the flop-count of the Schur method for the discrete Lyapunov equation described in Section 8.5.4.

8.25 Develop a method to solve the Lyapunov equation $A^T X A - X = -C$ based on the reduction of A to a companion form. Construct an example to show that the algorithm may not be numerically effective.

8.26 Establish the round-off error bound (8.5.18):

$$\frac{\|\hat{X} - X\|_F}{\|X\|_F} \leq \frac{cm\mu}{\text{sep}_d(A^T, A)}$$

for the Schur method to solve the discrete Lyapunov equation (8.5.12).

8.27 Develop a Hessenberg–Schur algorithm to solve the discrete Sylvester equation $BXA + C = X$.

8.28 Develop an algorithm to solve the Sylvester equation: $XA + BX = C$, based on the reductions of both A and B to RSFs.

Give a flop-count of this algorithm and compare this with that of Algorithm 8.5.1.

Research problems

8.1 Devise an algorithm for solving the equation:

$$A^T X B + B^T X A = -C$$

based on the **generalized real Schur decomposition** of the pair (A, B), described in Chapter 4.

8.2 Devise an algorithm for solving the equation:

$$AXB + LXC = D$$

using the **generalized real Schur decomposition** of the pairs (A, L) and (C^T, B^T).

8.3 Investigate if and how the norm of the solution of the discrete-stable Lyapunov equation:

$$A^T X A - X = -I$$

provides information on the sensitivity of the discrete Lyapunov equation:

$$A^T X A - X = C.$$

8.4 *Higham (1996)*. Derive conditions for the Sylvester equation: $XA + BX = C$ to have a well-conditioned solution.

References

Anderson E., Bai Z., Bischof C., Blackford S., Demmel J., Dongarra J., Du Croz J., Greenbaum A., Hammarling S., McKenney A., and Sorensen D. *LAPACK Users' Guide*, 3rd edn, SIAM, Philadelphia, 1999.

Barnett S. and Cameron R.G. *Introduction to Mathematical Control Theory*, 2nd edn, Clarendon Press, Oxford, 1985.

Barnett S. and Storey C. *Matrix Methods in Stability Theory*, Nelson, London, 1970.

Barraud A.Y. "A numerical algorithm to solve $A^T X A - X = Q$," *IEEE Trans. Autom. Control*, Vol. AC-22, pp. 883–885, 1977.

Bartels R.H. and Stewart G.W. "Algorithm 432: solution of the matrix equation $AX + XB = C$," *Comm. ACM*, Vol. 15, pp. 820–826, 1972.

Bavely C.A. and Stewart G.W. "An algorithm for computing reducing subspaces by block diagonalization," *SIAM J. Numer. Anal.*, Vol. 16, pp. 359–367, 1979.

Byers R. "A LINPACK-style condition estimator for the equation $AX - XB^T = C$," *IEEE Trans. Autom. Control*, Vol. AC-29, pp. 926–928, 1984.

Datta B.N. and Datta K. "An algorithm to compute the powers of a Hessenberg matrix and it's applications," *Lin. Alg. Appl.* Vol. 14, pp. 273–284, 1976.

Datta B.N., *Numerical Linear Algebra and Applications*, Brooks/Cole Publishing Co., Pacific Grove, CA, 1995.

Datta K., Hong Y.P., and Lee R.B. "Applications of linear transformation to matrix equations," *Lin. Alg. Appl.*, Vol. 267, pp. 221–240, 1997.

DeSouza E. and Bhattacharyya S.P. "Controllability, observability and the solution of $AX - XB = C$," *Lin. Alg. Appl.*, Vol. 39, pp. 167–188, 1981.

Demmel J. and Bo Kågström "Computing stable eigendecompositions of matrix pencils," *Lin. Alg. Appl.*, Vol. 88/89, pp. 139–186, 1987.

Demmel J. and Kågström B. "The generalized Schur decomposition of an arbitrary pencil $A - \lambda B$: Robust software with error bounds and applications, Part I: Theory and algorithms," *ACM Trans. Math. Soft.*, Vol. 19, no. 2, pp. 160–174, 1993a.

Demmel J. and Kågström B. "The generalized Schur decomposition of an arbitrary pencil $A - \lambda B$: Robust software with error bounds and algorithms, Part II: Theory and algorithms," *ACM Trans. Math. Soft.*, Vol. 19, no. 2, pp. 175–201, 1993b.

Edelman A., Elmroth E., and Kågström B. "A geometric approach to perturbation theory of matrices and matrix pencils, Part I: Versal deformations," *SIAM J. Matrix Anal. Appl.*, Vol. 18, no. 3, pp. 653–692, 1997.

Edelman A., Elmroth E., and Kågström B. "A geometric approach to perturbation theory of matrices and matrix pencils, Part II: A stratification-enhanced staircase algorithm," *SIAM J. Matrix Anal. Appl.*, Vol. 20, no. 3, pp. 667–699, 1999.

Gahinet P.M., Laub A.J., Kenney C.S., and Hewer G. "Sensitivity of the stable discrete-time Lyapunov equation," *IEEE Trans. Autom. Control*, Vol. 35, pp. 1209–1217, 1990.

Gardiner J.D., Laub A.J., Amato J.J., and Moler C.B. "Solution of the Sylvester matrix equation $AXB^T + CXD^T = E$," *ACM Trans Math. Soft.*, Vol. 8, pp. 223–231, 1992a.

Gardiner J.D., Wette M.R., Laub A.J., Amato J.J., and Moler C.B. "Algorithm 705: A FORTRAN-77 Software package for solving the Sylvester matrix equation $AXB^T + CXD^T = E$," *ACM Trans. Math. Soft.*, Vol. 18, pp. 232–238, 1992b.

Ghavimi A.R. and Laub A.J. "An implicit deflation method for ill-conditioned Sylvester and Lyapunov equations," *Num. Lin. Alg. Appl.*, Vol. 2, pp. 29–49, 1995.

Golub G.H., Nash S., and Van Loan C.F. A Hessenberg–Schur method for the problem $AX + XB = C$, *IEEE Trans. Autom. Control*, Vol. AC-24, pp. 909–913, 1979.

Golub G.H. and Van Loan C.F. *Matrix Computations*, 3rd edn, Johns Hopkins University, Baltimore, MD, 1996.

Golub G.H. and Wilkinson J.H. "Ill-conditioned eigensystems and the computation of the Jordan canonical form," *SIAM Rev.*, Vol. 18, pp. 578–619, 1976.

Hammarling S.J. "Numerical solution of the stable nonnegative definite Lyapunov equation," *IMA J. Numer. Anal.*, Vol. 2, pp. 303–323, 1982.

Hearon J.Z. "Nonsingular solutions of $TA - BT = C$," *Lin. Alg. Appl.*, Vol. 16, pp. 57–63, 1977.

Hewer G. and Kenney C. "The sensitivity of the stable Lyapunov equation," *SIAM J. Contr. Optimiz.*, Vol. 26, pp. 321–344, 1988.

Higham N.J. *Accuracy and Stability of Numerical Algorithms*, SIAM Philadelphia, 1996.

Horn R.A. and Johnson C.R. *Topics in Matrix Analysis*, Cambridge University Press, Cambridge, UK, 1991.

Jameson A. "Solution of the equation $AX + XB = C$ by the inversion of an $M \times M$ or $N \times N$ matrix," *SIAM J. Appl. Math.* Vol. 66, pp. 1020–1023, 1968.

Kaġström B. "A perturbation analysis of the generalized Sylvester equation $(AR - LB, DR - LE) = (C, F)$," *SIAM J. Matrix Anal. Appl.*, Vol. 15, no. 4, pp. 1045–1060, 1994.

Kaġström B. and Poromaa P., Distributed block algorithms for the triangular Sylvester equation with condition estimator, *Hypercube and Distributed Computers* (F. Andre and J.P. Verjus, Eds.), pp. 233–248, Elsevier Science Publishers, B.V. North Holland, 1989.

Kågström B. and Poromaa P. "Distributed and shared memory block algorithms for the triangular Sylvester equation with sep^{-1} estimators," *SIAM J. Matrix Anal. Appl.*, Vol. 13, no. 1, pp. 90–101, 1992.

Kågström B. and Poromaa P. "LAPACK-style algorithms and software for solving the generalized Sylvester equation and estimating the separation between regular matrix pairs," *ACM Trans. Math. Soft.*, Vol. 22, no. 1, pp. 78–103, 1996.

Kågström B. and Westin L. "Generalized Schur methods with condition estimators for solving the generalized Sylvester equation," *IEEE Trans. Autom. Control*, Vol. AC-34, no. 7, pp. 745–751, 1989.

Kitagawa G. "An algorithm for solving the matrix equation $X = FXF^{\mathrm{T}} + S$," *Int. J. Control*, Vol. 25, no. 5, pp. 745–753, 1977.

Konstantinov M., Gu, Da-Wei, Mehrmann Volker, Petkov Petko. Perturbation Theory for Matrix Equations, Elsevier Press, Amsterdam, 2003.

Kreisselmeier G. "A Solution of the bilinear matrix equation $AY + YB = -Q$," *SIAM J. Appl. Math.*, Vol. 23, pp. 334–338, 1972.

Lancaster P. and Rodman L., *The Algebraic Riccati Equation*, Oxford University Press, Oxford, UK, 1995.

Petkov P., Christov N.D., and Konstantinov M.M. *Computational Methods for Linear Control Systems*, Prentice Hall, London, 1991.

Sima V. *Algorithms for Linear-Quadratic Optimization*, Marcel Dekker, New York, 1996.

Starke G. and Niethammer W. "SOR for $AX - XB = C$," *Lin. Alg. Appl.*, Vol. 154–156, pp. 355–375, 1991.

Van Loan C.F. "Using the Hessenberg decomposition in control theory," in *Algorithms and Theory in Filtering and Control, Mathematical Programming Study* (Sorensen D.C. and Wets R.J., Eds.), pp. 102–111, no. 8, North Holland, Amsterdam, 1982.

Van Loan C.F. "How near is a stable matrix to an unstable matrix," *Contemporary Mathematics* (Brualdi R. *et al.*, Eds.), Vol. 47, pp. 465–477, American Mathematical Society, Providence, RI, 1985.

PART III

CONTROL SYSTEMS DESIGN

CHAPTER 9

REALIZATION AND SUBSPACE IDENTIFICATION

Topics covered

- State-Space Realization of Transfer Function
- Minimal Realization (MR)
- Subspace Identifications (Time and Frequency Domain)

9.1 INTRODUCTION

In this chapter, we consider the problems of state-space **realization** and **identification**.

The state-space realization problem is the problem to find the matrices A, B, C, and D of the transfer function $G(s)$ in the continuous-time case or $G(z)$ in the discrete-time case, given a set of large number of Markov parameters.

In case of a discrete-time system, the Markov parameters can easily be computed from the input–output sequence of the systems (see **Exercise 9.5**). Finding Markov parameters in the case of a continuous-time system is not that straightforward.

There may exist many realizations of the same transfer function matrix. Two such realizations, **controllable and observable realizations**, are obtained in **Section 9.2.1**.

A realization with the smallest possible dimension of A is called a **minimal realization** (MR). A necessary and sufficient condition for a realization to be an MR is established in **Theorem 9.2.1**, and it is shown in **Theorem 9.2.2** that two MRs are related via a nonsingular transformation.

The existing algorithms for finding MRs are all based on factoring the associated Hankel matrix (matrices) of Markov parameters. Some basic rank properties of these matrices, which are relevant to such factorizations, are established in **Section 9.3**.

Two **numerically viable algorithms (Algorithms 9.3.1 and 9.3.2)** based on the **singular value decomposition(s)** (SVD) of these matrices are then described in Section 9.3. The algorithms are valid both for continuous-time and discrete-time state-space realizations, provided the Markov parameters are known.

The identification problem is the problem of identifying system matrices A, B, C, and D from a given set of input–output data, rather than Markov parameters.

Two time-domain subspace system identification algorithms (**Algorithms 9.4.1 and 9.4.2**) are presented in **Section 9.4.** These algorithms are based on the SVD decompositions of Hankel matrices constructed directly from the input–output sequences. The algorithms are presented for discrete-time systems identification, but can be used for identifying the continuous-time systems also, provided the first and higher derivatives of the inputs and outputs can be computed. In the last section (**Section 9.4.4**), we state a **frequency-domain subspace identification** algorithm (**Algorithm 9.4.3**). A frequency-domain state-space identification is concerned with finding the system matrices, given a set of measured frequency responses. The algorithm is stated for identification of a continuous-time system; however, it can be used for discrete-time identification also, with trivial modifications.

Reader's Guide

The readers familiar with material on state-space realization can skip Sections 9.2 and 9.3.1.

9.2 STATE-SPACE REALIZATIONS OF A TRANSFER FUNCTION

In this section, we show, given a transfer matrix, how to construct state-space realizations in controllable and observable forms of this transfer matrix.

We consider here only the continuous-time case. The results are also valid for the discrete-time case by replacing the variable s by the variable z.

Definition 9.2.1. *Let $G(s)$ be the transfer matrix of order $r \times m$ which is proper. Then the quadruple (A, B, C, D) such that*

$$G(s) = C(sI - A)^{-1}B + D \qquad (9.2.1)$$

*is called a **state-space realization** of $G(s)$.*

It can be shown (**Exercise 9.1**) that given a proper rational function $G(s)$, there always exists a state-space realization of $G(s)$. However, **such a realization is not unique**, that is, there may exist many state-space realizations of the same transfer matrix.

In the following sections we show the non-uniqueness of the state-space realization of a transfer matrix (for the single-input, single-output case (SISO)), by constructing two realizations of the same transfer matrix.

9.2.1 Controllable and Observable Realizations

The transfer matrix $G(s)$ can be written in the form:

$$G(s) = D + \frac{P(s)}{d(s)}, \qquad (9.2.2)$$

where $P(s)$ is a polynomial matrix in s of degree at most $h - 1$ given by

$$P(s) = P_0 + P_1 s + \cdots + P_{h-1} s^{h-1}, \qquad (9.2.3)$$

and $d(s) = s^h + d_{h-1} s^{h-1} + \cdots + d_1 s + d_0$ is a monic polynomial of degree h (h is the least common multiple of the denominators of all the entries of $G(s)$).

Let 0_p and I_p denote, respectively, the zero and identity matrices of order p. Define now

$$A = \begin{pmatrix} 0_m & I_m & & & \\ & 0_m & I_m & & \\ \vdots & & \ddots & \ddots & \\ 0_m & & \cdots & 0_m & I_m \\ -d_0 I_m & -d_1 I_m & -d_2 I_m & \cdots & -d_{h-1} I_m \end{pmatrix}, \qquad (9.2.4)$$

$$B = \begin{pmatrix} 0_m \\ 0_m \\ \vdots \\ 0_m \\ I_m \end{pmatrix}, \qquad C = (P_0, \ldots, P_{h-1}), \qquad (9.2.5)$$

$$D = \lim_{s \to \infty} G(s), \qquad (9.2.6)$$

Then it is easily verified that

$$C(sI - A)^{-1} B + D = G(s) = D + \frac{P(s)}{d(s)}. \qquad (9.2.7)$$

Since the matrix-pair (A, B) is controllable, the above realization of $G(s)$ is called a **controllable realization.** This realization has dimension mh.

We now construct a different realization of $G(s)$.

Expand $G(s)$ in Taylor series:

$$G(s) = D' + \frac{1}{s}H_1 + \frac{1}{s^2}H_2 + \cdots \tag{9.2.8}$$

The matrices $\{D', H_k\}$ can be found as follows:

$$
\begin{aligned}
D' &= \lim_{s \to \infty} G(s) \\
H_1 &= \lim_{s \to \infty} s(G(s) - D') \\
H_2 &= \lim_{s \to \infty} s^2 \left(G(s) - D' - \frac{1}{s}H_1 \right)
\end{aligned} \tag{9.2.9}
$$

$$\vdots$$

etc.

Definition 9.2.2. *The matrices $\{H_i\}$, defined above, are called the* **Markov parameters** *of $G(s)$.*

Note: The Markov parameters $\{H_i\}$ can be expressed as:

$$H_i = CA^{i-1}B, \quad i = 1, 2, \ldots \tag{9.2.10}$$

Define now the matrices A', B', and C' as follows:

$$
A' = \begin{pmatrix}
0_r & I_r & & & \\
& 0_r & I_r & & \\
\vdots & & \ddots & \ddots & \\
0_r & & \cdots & 0_r & I_r \\
-d_0 I_r & -d_1 I_r & -d_2 I_r & \cdots & -d_{h-1} I_r
\end{pmatrix}, \tag{9.2.11}
$$

$$
B' = \begin{pmatrix}
H_1 \\
H_2 \\
H_3 \\
\vdots \\
H_h
\end{pmatrix}, \qquad C' = (I_r, 0_r, \ldots, 0_r). \tag{9.2.12}
$$

Then it can be shown that with A', B', C', and D' as defined above, we have

$$C'(sI - A')^{-1}B' + D' = G(s). \tag{9.2.13}$$

That is, we have now another realization of $G(s)$. Since (A', C') is observable, this realization is called an **observable realization** of $G(s)$. This realization has dimension rh.

9.2.2 Minimal Realization

Since there may exist more than one realization of the same transfer function $G(s)$, it is natural to look for a realization of minimal order.

Definition 9.2.3. *A state-space realization* (A, B, C, D) *of* $G(s)$ *is said to be an MR of* $G(s)$ *if the matrix* A *has the smallest possible dimension, that is, if* (A', B', C', D') *is any other realization of* $G(s)$, *then the order of* A' *is greater than or equal to the order of* A. *The dimension of an MR is called the* **McMillan degree**.

Theorem 9.2.1. *A state-space realization* (A, B, C, D) *of* $G(s)$ *is minimal if and only if* (A, B) *is controllable and* (A, C) *is observable.*

Proof.   We first prove the necessity by **contradiction.**

If (A, B) is not controllable and/or (A, C) is not observable, then from Kalman decomposition **(see Chapter 6)**, it follows that there exists a realization of smaller dimension that is both controllable and observable. This contradicts the minimality assumption.

Conversely, let (A, B, C, D) and (A', B', C', D') be two minimal realizations of $G(s)$. Assume that the order of A' is $n' < n$. Since the two realizations have the same transfer function, then they should have the same Markov parameters, that is,

$$CA^{i-1}B = C'(A')^{i-1}B'. \qquad (9.2.14)$$

This implies that

$$O_M C_M = O'_M C'_M, \qquad (9.2.15)$$

where O_M and C_M, respectively, denote the observability and controllability matrices of the realization (A, B, C, D) and, O'_M and C'_M, respectively, denote the observability and controllability matrices of the realization (A', B', C', D').

But, $\text{rank}(O_M C_M) = n$, and $\text{rank}(O'_M C'_M) = n' < n$. This is a contradiction, since $\text{rank}(O_M C_M) = \text{rank}(O'_M C'_M)$, by (9.2.15). ∎

The next question is how are two MRs of the same transfer matrices related? We answer the question in Theorem 9.2.2.

Theorem 9.2.2. *If* (A, B, C, D) *and* (A', B', C', D') *are two MRs of the same transfer function* $G(s)$, *then there exists a unique nonsingular matrix* T *such that*

$$A' = T^{-1}AT, \qquad (9.2.16)$$

$$B' = T^{-1}B, \qquad C' = CT, \qquad D' = D. \qquad (9.2.17)$$

Moreover, T is explicitly given by

$$T = \left(O_M^T O_M\right)^{-1} \cdot O_M^T O'_M \qquad (9.2.18)$$

or

$$T = C_M (C'_M)^T [C'_M (C'_M)^T]^{-1}, \qquad (9.2.19)$$

where C_M and O_M are, respectively, the controllability and observability matrices of the realization (A, B, C, D), and C_M' and O_M' are, respectively, the controllability and observability matrices of the realization (A', B', C', D').

Proof.   We just sketch a proof here and leave the details to the readers.

Let T be the matrix relating the matrices O_M and O_M', that is, T satisfies the matrix equation:

$$O_M T = O_M'. \tag{9.2.20}$$

Since O_M has full rank, such a matrix T always exists. In fact, it is unique and is given by

$$T = (O_M^T O_M)^{-1} O_M^T O_M'. \tag{9.2.21}$$

From the first block row of Eq. (9.2.20), we have $CT = C'$.

Since both the realizations have the same transfer function, and hence the same Markov parameters, we obtain

$$O_M C_M = O_M' C_M', \tag{9.2.22}$$

which gives

$$C_M = \left(O_M^T O_M\right)^{-1} O_M^T O_M' C_M' = T C_M'. \tag{9.2.23}$$

That is, T is a solution of the equation

$$T C_M' = C_M. \tag{9.2.24}$$

Since C_M' has full rank, we have

$$T = C_M (C_M')^T [C_M' (C_M')^T]^{-1}, \text{ establishing } (9.2.19).$$

Again, from the first block column of Eq. (9.2.23), we have

$$T B' = B. \tag{9.2.25}$$

All that remains to be shown is that (9.2.16) holds. To show this, first note that the Markov parameters $C A^{i-1} B$ and $C'(A')^{i-1} B'$, $i \geq 1$, are equal.

We can then write

$$O_M A C_M = O'_M A' C'_M, \tag{9.2.26}$$

which leads to

$$O_M^T O_M A C_M = O_M^T O'_M A' C'_M. \tag{9.2.27}$$

From (9.2.27) we have

$$A C_M = T A' C'_M \text{ (where } T \text{ is defined by (9.2.18)).} \tag{9.2.28}$$

But again multiplying (9.2.19) by A to the left, we have

$$A C_M (C'_M)^T (C'_M (C'_M)^T)^{-1} = AT. \tag{9.2.29}$$

From (9.2.28) and (9.2.29), we obtain

$$AT = TA'$$

That is, $A' = T^{-1} AT$. ∎

Uniqueness: Suppose that there exists another similarity transformation given by T_1 relating both the systems. Then we must have:

$$O_M(T - T_1) = 0.$$

But O_M has full rank, so, $T = T_1$.

Example 9.2.1. Let

$$G(s) = \frac{3s - 4}{s^2 - 3s + 2}.$$

Here

$$P(s) = -4 + 3s,$$
$$d(s) = s^2 - 3s + 2.$$

The Markov parameters are:

$$D' = \lim_{s \to \infty} G(s) = 0,$$
$$H_1 = \lim_{s \to \infty} s(G(s) - D') = 3,$$
$$H_2 = \lim_{s \to \infty} s^2 \left(G(s) - D' - \frac{1}{s} H_1 \right) = 5.$$

(I)   *Controllable realization*

$$A = \begin{pmatrix} 0 & 1 \\ -2 & 3 \end{pmatrix}, \qquad B = \begin{pmatrix} 0 \\ 1 \end{pmatrix}, \qquad C = (-4, 3), \qquad D = 0.$$

Verify:

$$
\begin{aligned}
C(sI - A)^{-1}B &= \frac{1}{s^2 - 3s + 2}(-4, 3) \begin{pmatrix} s-3 & 1 \\ -2 & s \end{pmatrix} \begin{pmatrix} 0 \\ 1 \end{pmatrix} \\
&= \frac{3s - 4}{s^2 - 3s + 2}.
\end{aligned}
$$

Since (A, B) is controllable and (A, C) is observable, the realization is an **MR**.

(II) *Observable realization*

$$A' = \begin{pmatrix} 0 & 1 \\ -2 & 3 \end{pmatrix}, \qquad B' = \begin{pmatrix} 3 \\ 5 \end{pmatrix}, \qquad C' = (1, 0).$$

Verify: $C'(sI - A)^{-1}B' = \dfrac{3s - 4}{s^2 - 3s + 2}.$

Since (A', B') is controllable, and (A', C') is observable, this is also an MR of $G(s)$.

(III) *Relationship.* The two realizations are related by the nonsingular transforming matrix T given by

$$T = (O_{\mathrm{M}}^{\mathrm{T}} O_{\mathrm{M}})^{-1} O_{\mathrm{M}}^{\mathrm{T}} O_{\mathrm{M}}' = \begin{pmatrix} -2.5 & 1.5 \\ -3 & 2 \end{pmatrix}.$$

Verify: $T^{-1}AT = \begin{pmatrix} 0 & 1 \\ -2 & 3 \end{pmatrix} A', \qquad T^{-1}B = \begin{pmatrix} 3 \\ 5 \end{pmatrix} = B',$

$CT = (1, 0) = C'.$

9.3   COMPUTING MINIMAL REALIZATIONS FROM MARKOV PARAMETERS

In the last section, we showed how to obtain an observable realization from a set of Markov parameters:

$$H_k = CA^{k-1}B, \qquad k = 1, 2, \ldots$$

Here we consider the problem of computing a MR using Markov parameters. Specifically, the following problem is considered.

> Given a set of large number Markov parameters $\{H_k\}$ of an unknown transfer function $G(s)$, find a minimal realization (A, B, C, D) whose transfer function $G(s) = C(sI - A)^{-1}B + D$.

Since the Markov parameters are much easier to obtain for a discrete-time system, unless otherwise stated, we assume that the given Markov parameters are of the discrete-time system:

$$x_{k+1} = Ax_k + Bu_k,$$
$$y_k = Cx_k + Du_k. \tag{9.3.1}$$

9.3.1 Some Basic Properties of the Hankel Matrix of Markov Parameters

There exist many methods for finding a minimal realization (see DeJong (1978) for a survey). Most of these methods find a minimal realization from a decomposition or a factorization of the Hankel matrix of Markov parameters of the form:

$$M_k = \begin{pmatrix} H_1 & H_2 & \cdots & H_k \\ H_2 & H_3 & \cdots & H_{k+1} \\ \vdots & \vdots & & \vdots \\ H_k & H_{k+1} & \cdots & H_{2k-1} \end{pmatrix}. \tag{9.3.2}$$

For example, a recursive method due to Rissanen (1971) obtains a minimal realization by recursively updating the decomposition of a smaller Hankel matrix to that of a larger Hankel matrix.

The following basic results due to Kalman *et al.* (see, e.g., Kalman *et al.* (1969), play an important role in the developments of Rissanen's and other methods.

Theorem 9.3.1.

(i)   Rank$(M_k) \le$ rank(M_{k+1}) *for all* k.

(ii) *If* (A, B, C, D) *is any realization of dimension* n, *then* rank$(M_k) =$ rank(M_n) *for all* $k \ge n$.

(iii) *Let* (A, B, C, D) *and* (A', B', C', D') *be two realizations of* $G(s)$ *of order* n *and* n', *respectively. Then,*

$$\text{rank}(M_n) = \text{rank}(M_n').$$

(iv) *Let* d *be the McMillan degree, then* $\max_k(\text{rank}(M_k)) = d$.

(v)   *Let* (A, B, C, D) *be any realization of dimension* n, *then*

$$d = \text{rank}(M_n) = \text{rank}(O_M C_M),$$

where O_M *and* C_M *are, respectively, the observability and controllability matrices of the realization* (A, B, C, D).

Proof.

(i)   The proof of (i) follows from the fact that M_k is a submatrix of M_{k+1}.

(ii)　The proof of (ii) follows by observing that (**Exercise 9.6**) the Hankel matrix M_k can be decomposed as:

$$M_k = \begin{pmatrix} O_M \\ \hline CA^n \\ \vdots \\ CA^{k-1} \end{pmatrix} \left(C_M | A^n B, A^{n+1} B, \ldots, A^{k-1} B) \right), \qquad (9.3.3)$$

　　　for $k \geq n$ and $M_n = O_M C_M$. 　　　　　　　　　　(9.3.4)

　　　Since the rows in $(CA^n, \ldots, CA^{k-1})$ and the columns in $(A^n B, \ldots, A^{k-1} B)$ are linear combination of the rows in O_M and the columns in C_M, respectively, we have

$$\operatorname{rank}(M_k) = \operatorname{rank}(M_n) = \operatorname{rank}(O_M C_M).$$

(iii)　Let (A', B', C', D') be another realization of $G(s)$ of order n' and let $r = \max(n, n')$. Then, since both these realization have the same Markov parameters, we must have

$$M'_r = M_r.$$

　　　Thus by (ii), $\operatorname{rank}(M_n) = \operatorname{rank}(M_r) = \operatorname{rank}(M'_r) = \operatorname{rank}(M'_n)$.

(iv)　The proof is by *contradiction*. Suppose that there exists a minimal realization (A, B, C, D) of order $d' < d$.

　　　Then by the previous two results, we should have $\max((\operatorname{rank}(M_k))) = d_1 < d'$, a contradiction.

(v)　The proof follows from (iii) and (iv). ■

Finding the McMillan Degree

The above result gives us several alternative procedures to obtain the McMillan degree of the transfer function matrix.

A simple way to do so is to find any realization of $G(s)$ and then find the rank of the product $O_M C_M$, using the SVD.

Also, if the realization is stable and C_G and O_G are, respectively, the controllability and observability Grammians obtained via solutions of respective Lyapunov equations (**see Chapter 7**), then it is well known (Glover 1984) that the McMillan degree is equal to the rank of $C_G O_G$.

9.3.2 An SVD Method for Minimal Realization

It was shown by DeJong (1978) that the **Rissanen's method is numerically unstable**.

Since the SVD provides a numerically reliable way to compute the rank of a matrix, a more numerically viable method for finding an MR should be based on the SVD of the associated Hankel matrix. We now describe below an SVD-based method for finding an MR (Ho and Kalman 1966; Zeiger and McEwen 1974; Kung 1978). **For the sake of convenience, we will assume that $D = 0$ in this section.**

Given the set $\{H_1, H_2, \ldots, H_{2N+1}\}$ of Markov parameters, consider the SVD of M_{N+1}:

$$M_{N+1} = USV^T = US^{1/2}S^{1/2}V^T = U'V',$$

where $S = \text{diag}(s_1, s_2, \ldots, s_p, 0, \ldots, 0)$, $U' = US^{1/2}$, and $V' = S^{1/2}V^T$.

Comparing this decomposition with the decomposition of M_{N+1} in the form (9.3.2) in Section 9.3.1, it is easy to see that we can take C as the first block row and the first p columns of U' and similarly B can be taken as the first p rows and the first block column of V'.

The matrix A satisfies the relations:

$$U_1 A = U_2 \quad \text{and} \quad AV_1 = V_2,$$

where
$U_1 = $ The first N block rows and the first p columns of U'
$V_1 = $ The first p rows and the first N block columns of V'.
U_2 and V_2 are similarly defined. Since U_1 and V_1 have full ranks, we immediately have from the above two equations,

$$A = U_1^\dagger U_2 \quad \text{or} \quad A = V_2 V_1^\dagger,$$

where $U_1^\dagger$ and $V_1^\dagger$ are the generalized inverses of U_1 and V_1, respectively.

This discussion leads to the following SVD algorithm for finding an **MR**:

Algorithm 9.3.1. *An SVD Algorithm for Minimal Realization*
   Inputs. *The set of Markov parameters: $\{H_1, H_2, \ldots, H_{2N+1}\}$ (N should be at least equal to the McMillan degree).*
   Outputs. *The matrices A, B, and C of a minimal realization.*
   Step 1. *Find the SVD of the block Hankel matrix*

$$M_{N+1} = \begin{pmatrix} H_1 & H_2 & \cdots & H_{N+1} \\ H_2 & H_3 & \cdots & H_{N+2} \\ \vdots & & & \\ H_{N+1} & H_{N+2} & \cdots & H_{2N+1} \end{pmatrix} = USV^T,$$

where $S = \text{diag}(s_1, s_2, \ldots, s_p, 0, \ldots, 0)$, and $s_1 \geq s_2 \geq \cdots \geq s_p > 0$

   Step 2. *Form $U' = US^{1/2}$ and $V' = S^{1/2}V^T$,*
where $S^{1/2} = \text{diag}(s_1^{1/2}, s_2^{1/2}, \ldots, s_p^{1/2}, 0, \ldots, 0)$.

Step 3. *Define*

$U_1 = $ *The first N block rows and the first p columns of U'*

$U_2 = $ *The last N block rows and the first p columns of U'*

$U^{(1)} = $ *The first block row and the first p columns of U'*

$V^{(1)} = $ *The first p rows and the first block column of V'.*
Step 4. *Compute* $A = U_1^\dagger U_2$, *Set* $B = V^{(1)}, C = U^{(1)}$.

Theorem 9.3.2 proved by Kung (1978) shows that the MR obtained by Algorithm 9.3.1 enjoys certain desirable properties.

Theorem 9.3.2. *Let E_i denote the error matrix, that is,*

$$E_i = CA^{i-1}B - H_i, \quad i \geq 1.$$

Assume that the given impulse response sequence $\{H_k\}$ is convergent. That is, $H_k \to 0$, when $k \to \infty$.

Then,

- $\sum_{i=1}^{2N+1} \|E_i\|_F^2 \leq \epsilon\sqrt{n + m + r}$, *where ϵ is a small positive number, and n, m and r are, respectively, the number of states, inputs, and outputs.*
- *The minimal realization obtained by Algorithm 9.3.1 is (i)* **discrete-stable** *and (ii)* **internally balanced**, *that is, the controllability and observability Grammians for this realization are the same and are equal to a diagonal matrix (see* **Chapter 14**).

Example 9.3.1. Let $N = 2$ and the given set of Markov parameters be:

$$\{H_1, H_2, H_3, H_4, H_5\} = \{3, 5, 9, 17, 33\}.$$

Step 1. $M_3 = \begin{pmatrix} 3 & 5 & 9 \\ 5 & 9 & 17 \\ 9 & 17 & 33 \end{pmatrix}$. Then, $U = \begin{pmatrix} 0.2414 & -0.8099 & 0.5345 \\ 0.4479 & -0.3956 & -0.8018 \\ 0.8609 & 0.4330 & 0.2673 \end{pmatrix}$,

$S = \text{diag}(44.3689 \ 0.6311 \ 0)$, and $V^{\mathrm{T}} = \begin{pmatrix} 0.2414 & 0.4479 & 0.8609 \\ -0.8099 & -0.3956 & 0.4330 \\ 0.5345 & -0.8018 & 0.2673 \end{pmatrix}$.

Step 2. $U' = \begin{pmatrix} 1.6081 & -0.64340 & 0 \\ 2.9835 & -0.31430 & 0 \\ 5.7343 & 0.34400 & 0 \end{pmatrix}$,

$$V' = \begin{pmatrix} 1.6081 & 2.9835 & 5.7343 \\ -0.6434 & -0.3143 & 0.3440 \\ 0 & 0 & 0 \end{pmatrix}.$$

Step 3.

$$U_1 = \begin{pmatrix} 1.6081 & -0.6434 \\ 2.9835 & -0.3143 \end{pmatrix},$$

$$U_2 = \begin{pmatrix} 2.9835 & -0.3143 \\ 5.7343 & 0.3440 \end{pmatrix}.$$

$$U^{(1)} = (1.6081 \quad -0.6434),$$

$$V^{(1)} = \begin{pmatrix} 1.6081 \\ -0.6434 \end{pmatrix}.$$

Step 4.

$$A = U_1^\dagger U_2 = \begin{pmatrix} 1.9458 & 0.2263 \\ 0.2263 & 1.0542 \end{pmatrix},$$

$$B = V^{(1)} = \begin{pmatrix} 1.6081 \\ -0.6434 \end{pmatrix},$$

$$C = U^{(1)} = (1.6081 \quad -0.6434).$$

Verify:

$$E_1 = CB - H_1 = -8.8818 \times 10^{-16},$$

$$E_2 = CAB - H_2 = -8.8818 \times 10^{-16},$$

$$E_3 = CA^2B - H_3 = -1.7764 \times 10^{-15},$$

$$E_4 = CA^3B - H_4 = 0,$$

$$E_5 = CA^4B - H_5 = 7.1054 \times 10^{-15},$$

$$\sum_{i=1}^{5} |E_i|^2 = 5.5220 \times 10^{-30}.$$

It is also easy to check that the realization is both controllable and observable. So, it is minimal. The controllability and observability Grammians are the same and are given by: $C_G^N = O_G^N = \text{diag}(44.3689, 0.6311)$.

Figure 9.1 shows a comparison between the graphs of the transfer functions $G_0(s) = \sum_{i=1}^{5} \frac{1}{s^i} H_i$ and $G(s) = C(sI - A)^{-1}B$. The plot shows an excellent agreement between the graphs for large values for s.

MATCONTROL notes: Algorithm 9.3.1 has been implemented in MATCONTROL function **minresvd**.

9.3.3 A Modified SVD Method for Minimal Realization

We describe now a modification of the above algorithm (see Juang 1994).

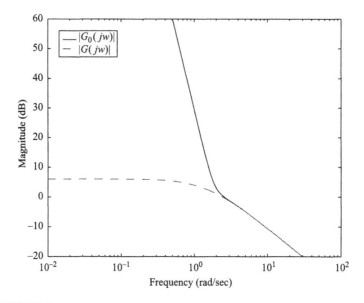

FIGURE 9.1:   Comparison of transfer functions of an SVD method.

This modified algorithm uses **lower order block Hankel matrices** in computing the system matrices A, B, and C.

Define the block Hankel matrices:

$$M_R = \begin{pmatrix} H_1 & H_2 & \cdots & H_R \\ H_2 & H_3 & \cdots & H_{R+1} \\ \vdots & \vdots & & \vdots \\ H_R & H_{R+1} & \cdots & H_{2R-1} \end{pmatrix}$$

and

$$M_{R1} = \begin{pmatrix} H_2 & H_3 & \cdots & H_{R+1} \\ H_3 & H_4 & \cdots & H_{R+2} \\ \vdots & \vdots & & \vdots \\ H_{R+1} & H_{R+2} & \cdots & H_{2R} \end{pmatrix},$$

where $R \geq n$ (n is the order of the system). Denote the controllability and observability matrices by:

$$C_M^R = (B, AB, \ldots, A^{R-1}B)$$

and

$$O_M^R = \begin{pmatrix} C \\ CA \\ \vdots \\ CA^{R-1} \end{pmatrix}.$$

Then,

$$M_{R1} = O_M^R \, A \, C_M^R,$$

and $M_R = O_M^R C_M^R$.

Consider now the SVD of M_R:

$$M_R = U \Sigma V^T = U \Sigma^{1/2} \Sigma^{1/2} V^T.$$

This means that O_M^R is related to U and C_M^R is related to V.

Define now Σ_n by:

$$\Sigma = \begin{pmatrix} \Sigma_n & 0 \\ 0 & 0 \end{pmatrix},$$

and U_n and V_n as the matrices formed by the first n columns of U and V, respectively. Also, let the matrices E_r' and E_m' be defined as:

$$E_r'^T = (I_r, 0, \ldots, 0), \qquad E_m'^T = (I_m, 0, \ldots, 0),$$

where I_s stands for identity matrix of order s, and m and r denote, respectively, the number of inputs and the number of outputs.

Then one can choose $O_M^R = U_n \Sigma_n^{1/2}$ and $C_M^R = \Sigma_n^{1/2} V_n^T$.

From the equation:

$$M_R = O_M^R C_M^R = U_n \Sigma_n^{1/2} \Sigma_n^{1/2} V_n^T$$

it follows that B and C can be chosen as:

$$B = \Sigma_n^{1/2} V_n^T E_m' \quad \text{and} \quad C = E_r'^T U_n \Sigma_n^{1/2}.$$

Also, from the equation:

$$M_{R1} = O_M^R A C_M^R = U_n \Sigma_n^{1/2} A \Sigma_n^{1/2} V_n^T$$

it follows that

$$A = \Sigma_n^{-1/2} U_n^T M_{R1} V_n \Sigma_n^{-1/2}.$$

Thus, we have the following modified algorithm using the SVD of lower order Hankel matrices.

Algorithm 9.3.2.   *A Modified SVD Algorithm for Minimal Realization*
   Inputs. *The Markov parameters* $\{H_1, H_2, \ldots, H_{2R}\}$, $R \geq n$ *(the order of the system to be identified).*
   Outputs. A, B, C *of a Minimal Realization.*
   Step 1. *Form the Hankel matrices* M_R *and* M_{R1} *as defined above.*
   Step 2. *Find the SVD of* M_R:

$$M_R = U \begin{pmatrix} \Sigma_n & 0 \\ 0 & 0 \end{pmatrix} V^T,$$

where $\Sigma_n = \mathrm{diag}(\sigma_1, \sigma_2, \ldots, \sigma_n)$; $\sigma_1 \geq \sigma_2 \geq \cdots \geq \sigma_n > 0$.
   Step 3. *Compute*

$$A = \Sigma_n^{-1/2} U_n^T M_{R1} V_n \Sigma_n^{-1/2},$$
$$B = \Sigma_n^{1/2} V_n^T E_m',$$
$$C = E_r'^T U_n \Sigma_n^{1/2},$$

where U_n *and* V_n *are the matrices of the first n columns of* U *and* V, *respectively, and* E_m' *and* E_r' *are as defined above.*

Example 9.3.2.   We consider the previous example again. Take $R = n = 2$. Let $m = 1, r = 1$.
   Then,

$$M_R = \begin{pmatrix} H_1 & H_2 \\ H_2 & H_3 \end{pmatrix} = \begin{pmatrix} 3 & 5 \\ 5 & 9 \end{pmatrix},$$

$$M_{R1} = \begin{pmatrix} H_2 & H_3 \\ H_3 & H_4 \end{pmatrix} = \begin{pmatrix} 5 & 9 \\ 9 & 17 \end{pmatrix}.$$

$$\Sigma_2 = \mathrm{diag} \begin{pmatrix} 11.8310 & 0 \\ 0 & 0.1690 \end{pmatrix}$$

$$U_2 = \begin{pmatrix} -0.4927 & -0.8702 \\ -0.8702 & 0.4927 \end{pmatrix}$$

$$V_2 = \begin{pmatrix} -0.4927 & -0.8702 \\ -0.8702 & 0.4927 \end{pmatrix}.$$

$$A = \Sigma_2^{-1/2} U_2^T M_{R1} V_2 \Sigma_2^{-1/2} = \begin{pmatrix} 1.8430 & -0.3638 \\ -0.3638 & 1.1570 \end{pmatrix},$$

$$B = \Sigma_2^{1/2} V_2^T E_1' = \begin{pmatrix} -1.6947 \\ -0.3578 \end{pmatrix},$$

$$C = E_1'^T U_2 \Sigma_2^{1/2} = (-1.6947 \quad -0.3578).$$

Verification:

$$E_1 = CB - H_1 = O(10^{-15})$$
$$E_2 = CAB - H_2 = O(10^{-15})$$
$$E_3 = CA^2B - H_3 = O(10^{-15})$$
$$E_4 = CA^3B - H_4 = O(10^{-14})$$
$$E_5 = CA^4B - H_5 = O(10^{-14}).$$

Remarks

- Algorithm 9.3.2, when extended to reconstruct the Markov parameters of a reduced-order system obtained by eliminating "noisy modes," is called **Eigensystem Realization Algorithm** (ERA) because information from the eigensystem of the realized state matrix obtained in Algorithm 9.3.2 is actually used to obtain the reduced-order model. The details can be found in Juang (1994, pp. 133–144).
- The optimal choice of the number R requires engineering intuition. The choice has to be made based on measurement data to minimize the size of the Hankel matrix M_R. See Juang (1994).

Figure 9.2 shows a comparison between the graphs of the transfer function $G_0(s) = \sum_{i=1}^{4} \frac{1}{s^i} H_i$ and $G(s) = C(sI - A)^{-1}B$. The plot shows an excellent agreement between the graphs for large values of s.

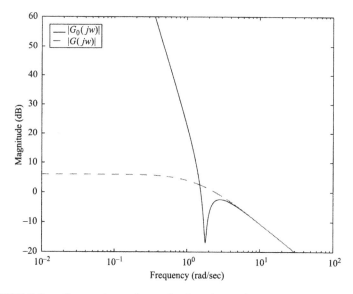

FIGURE 9.2: Comparison of transfer functions of a modified SVD method.

MATCONTROL note: Algorithm 9.3.2 has been implemented in MATCON-TROL function **minremsvd**.

9.4   SUBSPACE IDENTIFICATION ALGORITHMS

In this section we consider the problem of identifying the system matrices of an unknown system, **given a large number of input and output measurements. The problem is important in practical applications because it avoids compu- tations of Markov parameters.**

We state two SVD-based subspace algorithms, one for the deterministic case and another for the stochastic case. First, we consider the deterministic case.

Specifically, the deterministic identification problem is:

Given a large number of input and output measurements, u_k and y_k, respectively of the unknown system:

$$x_{k+1} = Ax_k + Bu_k,$$
$$y_k = Cx_k + Du_k,$$

determine the order n of the unknown system and the system matrices $\{A, B, C, D\}$ up to within a similarity transformation; $A \in \mathbb{R}^{n \times n}$, $B \in \mathbb{R}^{n \times m}$, $C \in \mathbb{R}^{r \times n}$, and $D \in \mathbb{R}^{r \times m}$.

9.4.1   A Subspace Deterministic Model Identification Algorithm

The algorithm has two major steps:

First, a state vector sequence is constructed as the intersection of the row spaces of two block Hankel matrices, constructed from the input/output data.

Second, the system matrices A, B, C, and D are obtained from the least-squares solution of a set of linear equations.

There exists different ways to compute the intersection (see Van Overschee and De Moor (1996a, 1996b for details and references). One way, presented in Moonen *et al.* (1989) is via the SVD of a concatenated Hankel matrix composed of two Hankel matrices defined by the input and output data, as follows:

$$H_{k|k+i} = \begin{pmatrix} Y_{k|k+i} \\ U_{k|k+i} \end{pmatrix}, \qquad H_{k+1|k+2i} = \begin{pmatrix} Y_{k|k+2i} \\ U_{k|k+2i} \end{pmatrix},$$

where

$$Y_{k|k+i} = \begin{pmatrix} y_k & y_{k+1} & \cdots & y_{k+j-1} \\ y_{k+1} & y_{k+2} & \cdots & y_{k+j} \\ y_{k+2} & y_{k+3} & \cdots & y_{k+j+1} \\ \vdots & \vdots & & \vdots \\ y_{k+i-1} & y_{k+i} & \cdots & y_{k+j+i-2} \end{pmatrix}$$

and

$$Y_{k|k+2i} = \begin{pmatrix} y_{k+i} & y_{k+i+1} & \cdots & y_{k+i+j-1} \\ y_{k+i+1} & y_{k+i+2} & \cdots & y_{k+i+j} \\ \vdots & \vdots & \vdots & \vdots \\ y_{k+2i-1} & y_{k+2i} & \cdots & y_{k+2i+j-2} \end{pmatrix}.$$

The matrices $U_{k|k+i}$ and $U_{k|k+2i}$ are similarly defined from the input data. Let $(X = (x_k, x_{k+1}, \ldots, x_{k+j-1}))$.

The following assumptions are made:

- $\text{rank}(X) = n$ (n is the minimal system order)
- $\text{span}_{\text{row}}(X) \cap \text{span}_{\text{row}}(U_{k|k+i}) = \emptyset$
- $\text{rank}(U_{k|k+i}) = $ Number of rows in $U_{k|k+i}$.

Theorem 9.4.1. *Let the SVD of*

$$H = \begin{pmatrix} H_{k|k+i} \\ H_{k+1|k+2i} \end{pmatrix}$$

be

$$H = \begin{pmatrix} U_{11} & U_{12} \\ U_{21} & U_{22} \end{pmatrix} \begin{pmatrix} S_{11} & 0 \\ 0 & 0 \end{pmatrix} V^{\mathrm{T}}.$$

Then the state vector sequence $X_2 = (x_{k+i}, x_{k+i+1}, \ldots, x_{k+i+j-1})$ is given by

$$X_2 = U_q^{\mathrm{T}} U_{12}^{\mathrm{T}} H_{k|k+i},$$

where U_q is defined by the SVD of $U_{12}^{\mathrm{T}} U_{11} S_{11}$:

$$U_{12}^{\mathrm{T}} U_{11} S_{11} = \begin{pmatrix} U_q, & U_q^{\perp} \end{pmatrix} \begin{pmatrix} S_q & 0 \\ 0 & 0 \end{pmatrix} \begin{pmatrix} V_q^{\mathrm{T}} \\ V_q^{\perp\mathrm{T}} \end{pmatrix}.$$

Proof. It can be shown (**Exercise 9.12**) that

$$\text{span}_{\text{row}}(X_2) = \text{span}_{\text{row}}(H_{k|k+i}) \cap \text{span}_{\text{row}}(H_{k+1|k+2i}).$$

Thus, X_2 can be realized as a basis for the row space of $U_{12}^{\mathrm{T}} H_{k|k+i}$. Then taking the SVD of $U_{12}^{\mathrm{T}} H_{k|k+i}$, we have

$$U_{12}^{\mathrm{T}} H_{k|k+i} = U_{12}^{\mathrm{T}} \begin{pmatrix} U_{11} & U_{12} \end{pmatrix} \begin{pmatrix} S_{11} & 0 \\ 0 & 0 \end{pmatrix} V^{\mathrm{T}},$$

$$= (U_{12}^{\mathrm{T}} U_{11} S_{11} \ 0) V^{\mathrm{T}},$$

$$= (U_q S_q V_q^{\mathrm{T}} \ 0) V^{\mathrm{T}},$$

$$= U_q (S_q \ 0)(V V_q)^{\mathrm{T}}.$$

Noting that $U_q^T U_q = I_{n \times n}$, we obtain from above

$$U_q^T U_{12}^T H_{k|k+i} = (S_q \ 0)(V V_q)^T,$$

which is a basis for the row space of $U_{12}^T H_{k|k+i}$ and therefore is a realization of X_2. ∎

Once X_2 is determined, the system matrices $A, B, C,$ and D are identified by solving (**in the least-squares sense**) the following overdetermined set of linear equations:

$$\begin{pmatrix} x_{k+i+1} & \cdots & x_{k+i+j-1} \\ y_{k+i} & \cdots & y_{k+i+j-2} \end{pmatrix} = \begin{pmatrix} A & B \\ C & D \end{pmatrix} \begin{pmatrix} x_{k+i} & \cdots & x_{k+i+j-2} \\ u_{k+i} & \cdots & u_{k+i+j-2} \end{pmatrix}.$$

It is, however, shown in De Moor *et al.* (1999) that the state vector sequence X_2 does not need to be explicitly computed. The system matrices $A, B, C,$ and D may be identified by making use of the already computed SVD of H. The above set of equations may then be replaced by an equivalent reduced set of equations (see **Algorithm 9.4.1**).

This way of determining $A, B, C,$ and D is computationally more efficient.

To do this, it is useful to redefine the matrices $H_{k|k+i}$ and $H_{k+i|k+2i}$ as follows:

$$H_{k|k+i} = \begin{pmatrix} u_k & u_{k+1} & \cdots & u_{k+j-1} \\ y_k & y_{k+1} & \cdots & y_{k+j-1} \\ u_{k+1} & u_{k+2} & \cdots & u_{k+j} \\ y_{k+1} & y_{k+2} & \cdots & y_{k+j} \\ \vdots & \vdots & & \vdots \\ u_{k+i-1} & u_{k+i} & \cdots & u_{k+i+j-2} \\ y_{k+i-1} & y_{k+i} & \cdots & y_{k+i+j-2} \end{pmatrix},$$

$$H_{k+i|k+2i} = \begin{pmatrix} u_{k+i} & u_{k+i+1} & \cdots & u_{k+i+j-1} \\ y_{k+i} & y_{k+i+1} & \cdots & y_{k+i+j-1} \\ u_{k+i+1} & u_{k+i+2} & \cdots & u_{k+i+j} \\ y_{k+i+1} & y_{k+i+2} & \cdots & y_{k+i+j} \\ \vdots & \vdots & & \vdots \\ \vdots & \vdots & & \vdots \\ u_{k+2i-1} & u_{k+2i} & \cdots & u_{k+2i+j-2} \\ y_{k+2i-1} & y_{k+2i} & \cdots & y_{k+2i+j-2} \end{pmatrix}.$$

The above theorem still remains valid.

The following notation will be needed to state the algorithm.

$M(p : q, l : s)$ is the submatrix of M at the intersection of rows $p, p+1, \ldots, q$ and columns $l, l+1, \ldots, s$; $M(:, l : s)$ is the submatrix of M containing columns $l, l+1, \ldots, s$ and $M(p : q, :)$ is the submatrix of M containing rows $p, p+1, \ldots, q$.

Algorithm 9.4.1. *A Deterministic Subspace Identification Algorithm.*
Inputs. *The input and output sequence $\{u_k\}$ and $\{y_k\}$, respectively. The integers $i \geq n$, where n is the order of the system to be identified and j.*
Outputs. *The identified system matrices A, B, C, and D.*
Assumptions.

1. *The system is observable.*
2. *The integers i and j are sufficiently large, and in particular $j \gg \max(mi, ri)$, where m and r are the number of inputs and outputs.*

Step 1. *Calculate U and S from the SVD of H, where*

$$H = \begin{pmatrix} H_{k|k+i} \\ H_{k+1|k+2i} \end{pmatrix}.$$

$$H = USV^{\mathrm{T}} = \begin{pmatrix} U_{11} & U_{12} \\ U_{21} & U_{22} \end{pmatrix} \begin{pmatrix} S_{11} & 0 \\ 0 & 0 \end{pmatrix} V^{\mathrm{T}}.$$

(Note that the dimensions of U_{11}, U_{12}, and S_{11} are, respectively, $(mi + ri) \times (2mi + n)$, $(mi + ri) \times (2ri - n)$, and $(2mi + n) \times (2mi + n)$).

Step 2. *Calculate the SVD of $U_{12}^{\mathrm{T}} U_{11} S_{11}$:*

$$U_{12}^{\mathrm{T}} U_{11} S_{11} = (U_q, U_q^{\perp}) \begin{pmatrix} S_q & 0 \\ 0 & 0 \end{pmatrix} \begin{pmatrix} V_q^{\mathrm{T}} \\ V_q^{\perp \mathrm{T}} \end{pmatrix}.$$

Step 3. *Solve the following set of linear equations for A, B, C, and D (in the least-squares sense):*

$$\begin{pmatrix} U_q^{\mathrm{T}} U_{12}^{\mathrm{T}} U(mi + ri + 1 : (m+r)(i+1), :)S \\ U(mi + ri + m + 1 : (m+r)(i+1), :)S \end{pmatrix}$$
$$= \begin{pmatrix} A & B \\ C & D \end{pmatrix} \begin{pmatrix} U_q^{\mathrm{T}} U_{12}^{\mathrm{T}} U(1 : mi + ri :)S \\ U(mi + ri + 1 : mi + ri + m, :)S \end{pmatrix}.$$

Remark

- It is to be noted that the system matrices are determined from U and S only; the larger matrix V is never used in the computations. Since the matrix H whose SVD to be computed could be very large in practice, computing U and S only, without computing the full SVD of H, will be certainly very useful in practice. **Also, as stated before, the state vector sequence X_2 is not explicitly computed.**

There exists also an on-line version of the above algorithm. See Moonen *et al.* (1989) for details.

Example 9.4.1.   Consider the following input–output data

$$
u = \begin{pmatrix} 0.09130 & 0.1310 & 0.6275 & 0.1301 & -0.2206 & 0.1984 & 0.4081 & -0.0175 \\ 0.2766 & 0.7047 & 0.9173 & 0.9564 & 0.6631 & 0.7419 & 0.7479 & 1.2133 \\ 1.2427 & 1.2942 & 1.3092 & 1.1574 & 1.5600 & 1.0913 & 0.7765 \end{pmatrix}^{\mathrm{T}},
$$

$$
y = \begin{pmatrix} 0.6197 & -0.4824 & 0.3221 & 0.2874 & -0.4582 & -0.1729 & 0.3162 & 0.0946 \\ -0.3497 & 0.3925 & 0.2446 & 0.2815 & 0.05621 & -0.2201 & 0.1397 & -0.0880 \\ 0.5250 & -0.1021 & 0.2294 & -0.0616 & -0.0706 & 0.3982 & -0.5695 \end{pmatrix}^{\mathrm{T}},
$$

generated from the discrete-time system:

$$
x_{k+1} = \begin{pmatrix} -0.2 & 0.3 \\ 1 & 0 \end{pmatrix} x_k + \begin{pmatrix} 1 \\ 0 \end{pmatrix} u_k,
$$

$$
y_k = (1, \ -1)\, x_k.
$$

Step 1. $S = \mathrm{diag}(9.1719, 1.9793, 1.8031, 1.6608, 1.4509, 1.3426, 1.2796, 1.0657,$
$0.5012, 0.4554, 5.1287 \times 10^{-16}, 3.5667 \times 10^{-16}, 2.2847 \times 10^{-16}, 1.3846 \times 10^{-16},$
$9.8100 \times 10^{-17}, 1.0412 \times 10^{-18}).$

Step 2.

$$
U_{12} = \begin{pmatrix}
0.4392 & -0.0372 & -0.1039 & 0.2139 & -0.0297 & -0.3324 \\
-0.1318 & 0.01116 & 0.0312 & -0.0642 & 0.0090 & 0.0997 \\
-0.3277 & 0.3783 & 0.1880 & 0.2071 & 0.3299 & 0.2790 \\
0.0544 & -0.1098 & -0.0460 & -0.0836 & -0.0960 & -0.0505 \\
-0.2282 & -0.4853 & -0.2806 & 0.1674 & -0.3377 & 0.1086 \\
0.4965 & 0.0743 & -0.0282 & 0.1215 & 0.0416 & -0.3597 \\
0.4062 & -0.0910 & 0.3218 & -0.3029 & 0.1580 & 0.3608 \\
0.0012 & 0.3828 & 0.0071 & 0.4531 & 0.2565 & -0.1672
\end{pmatrix},
$$

U_{11}

$$
= \begin{pmatrix}
0.2417 & -0.2139 & 0.2202 & 0.3049 & -0.4614 & -0.1558 & 0.0638 & 0.0142 & -0.2980 & -0.2488 \\
0.0211 & 0.3581 & 0.1578 & 0.3546 & -0.4804 & -0.0503 & -0.1130 & 0.2833 & 0.1475 & 0.5831 \\
0.2768 & -0.3634 & -0.2593 & -0.0824 & -0.3694 & 0.2078 & 0.0116 & -0.0870 & -0.1208 & 0.0008 \\
0.0383 & -0.5002 & 0.1157 & -0.0384 & -0.1351 & -0.0780 & 0.2107 & -0.2894 & 0.7057 & 0.1893 \\
0.3109 & 0.0687 & -0.5137 & 0.1861 & -0.0370 & 0.0372 & 0.2759 & -0.0140 & -0.0267 & -0.0101 \\
0.0338 & 0.0580 & -0.4635 & -0.2733 & -0.0251 & 0.3664 & -0.1282 & 0.0417 & 0.0803 & 0.3867 \\
0.3402 & 0.0787 & -0.0426 & 0.4406 & 0.2193 & 0.2340 & -0.0493 & -0.2103 & 0.0708 & -0.0626 \\
0.0389 & 0.2704 & -0.1147 & 0.3117 & 0.2969 & -0.2668 & 0.3531 & -0.0220 & 0.2897 & -0.0315
\end{pmatrix},
$$

$$
S_q = \mathrm{diag}(1.94468, \ 0.624567).
$$

Step 3.

$$
\begin{pmatrix} A & B \\ C & D \end{pmatrix} = \left(\begin{array}{cc|c} 0.2635 & 0.1752 & -0.4644 \\ 1.0153 & -0.4635 & 0.5503 \\ \hline -0.1780 & 1.6527 & 1.4416 \times 10^{-16} \end{array} \right).
$$

Verification: The first 10 Markov parameters (denoted by $H_i, i = 1, \ldots, 10$) of the original system and those of the identified system (denoted by $H'_i, i = 1, \ldots, 10$) are given below:

$$
\begin{aligned}
H_1 &= 1, \ H'_1 = 0.9922, \\
H_2 &= -1.2, \ H'_2 = -1.1962, \\
H_3 &= 0.5400, \ H'_3 = 0.5369, \\
&\vdots \\
H_{10} &= -0.0343, \ H'_{10} = -0.0341.
\end{aligned}
$$

9.4.2 A Stochastic Subspace Model Identification Algorithm

We now consider the stochastic case:

$$
\begin{aligned}
x_{k+1} &= Ax_k + Bu_k + w_k, \\
y_k &= Cx_k + Du_k + v_k,
\end{aligned}
$$

where $v_k \in \mathbb{R}^{r \times 1}$ and $w_k \in \mathbb{R}^{n \times 1}$ are unobserved vector signals; v_k is the measurement noise and w_k is the process noise. It is assumed that

$$
E\left[\begin{pmatrix} w_p \\ v_p \end{pmatrix} \begin{pmatrix} w_q^T & v_q^T \end{pmatrix} \right] = \begin{pmatrix} Q & S \\ S^T & R \end{pmatrix} \delta_{pq} \geq 0,
$$

where the matrices Q, R, and S are covariance matrices of the noise sequences w_k and v_k. The problem is now to determine the system matrices A, B, C, and D up to within a similarity transformation and also the covariance matrices Q, S, and R, given a large number of input and output data u_k and y_k, respectively.

We state a subspace algorithm for the above problem taken from the recent paper of DeMoor *et al.* (1999). The algorithm, as in the deterministic case, determines the system matrices by first finding an estimate $\tilde{X}_f$ of the state sequence $\tilde{X}$ from the measurement data.

The sequence $\tilde{X}_f$ is determined using certain oblique projections.

Define the input Hankel matrix $U_{i|l}$ from the input data as:

$$
U_{k|l} = \begin{pmatrix} u_k & u_{k+1} & \cdots & u_{k+j-1} \\ u_{k+1} & u_{k+2} & \cdots & u_{k+j} \\ \vdots & & & \\ u_l & u_{l+1} & \cdots & u_{l+j-1} \end{pmatrix}.
$$

Similarly, define the output Hankel matrix $Y_{k|l}$ from the output data.

The matrices A, B, C, D are then determined by solving the least-squares problem:

$$\min_{A,B,C,D} \left\| \begin{pmatrix} \tilde{X}_{i+1} \\ Y_{i|i} \end{pmatrix} - \begin{pmatrix} A & B \\ C & D \end{pmatrix} \begin{pmatrix} \tilde{X}_i \\ U_{i|i} \end{pmatrix} \right\|_F^2 .$$

Once the system matrices are obtained by solving above least-squares problem, the noise covariances Q, S, and R can be estimated from the residuals (**see Algorithm 9.4.2 below for details**).

The algorithm, in particular, can be used to solve the deterministic problem. Thus, it can be called a **combined deterministic stochastic identification algorithm**.

Definition 9.4.1. *The **oblique projection** of $A \in \mathbb{R}^{p \times j}$ along the row space of $B \in \mathbb{R}^{q \times j}$ on the row space of $C \in \mathbb{R}^{r \times j}$, denoted by $A/_B C$ is defined as:*

$$A/_B \, C = A(C^T \ B^T) \left[\begin{pmatrix} CC^T & CB^T \\ BC^T & BB^T \end{pmatrix}^\dagger \right]_{first \ r \ columns} C,$$

where $\dagger$ denotes the Moore–Penrose pseudo-inverse of the matrix.

For convenience, following the notations of the above paper, we write

$$U_p = U_{0|i-1}, \qquad U_f = U_{i|2i-1},$$
$$Y_p = Y_{0|i-1}, \qquad Y_f = Y_{i|2i-1},$$

where the subscript p and f denote, respectively, the past and the future. The matrix containing the past inputs U_p and outputs Y_p will be called W_p:

$$W_p = \begin{pmatrix} Y_p \\ U_p \end{pmatrix}.$$

The matrices $W_{0|i-1}$ and $W_{0|i}$ are defined in the same way as $U_{0|i-1}$ and $U_{0|i}$ from Y_p and W_p.

The following assumptions are made:

- The input u_k is uncorrelated with the noise w_k and v_k.
- The input covariance matrix $(1/j)(U_{0|2i-1}U_{0|2i-1}^T)$ is of full rank, that is, the rank is $2mi$ (the sequence u_k is then called **persistently exciting** of order $2i$).
- The number of available measurements is sufficiently large, so that $j \to \infty$.
- The noise w_k and v_k are not identically zero.

Algorithm 9.4.2. *A Subspace Stochastic Identification Algorithm.*
  Inputs. *The input and output sequences $\{u_k\}$ and $\{y_k\}$.*
  Outputs. *The order of the system and the system matrices A, B, C, D.*
  Assumptions. *As above.*

Step 1. *Find the oblique projections:*

$$\mathcal{O}_i = Y_{i|2i-1}/U_{i|2i-1} \mathbf{W}_{0|i-1}, \qquad \mathcal{O}_{i+1} = Y_{i+1|2i-1}/U_{i+1|2i-1} \mathbf{W}_{0|i}.$$

Step 2. *Compute the SVD of the oblique projection:*

$$\mathcal{O}_i = U S V^{\mathrm{T}} = \begin{pmatrix} U_1 \\ U_2 \end{pmatrix} \begin{pmatrix} S_1 & 0 \\ 0 & 0 \end{pmatrix} \begin{pmatrix} V_1^{\mathrm{T}} \\ V_2^{\mathrm{T}} \end{pmatrix},$$

(The order n of the system is equal to the order of S_1).

Step 3. *Define Γ_i and Γ_{i-1} as:*

$$\Gamma_i = U_1 S_1^{1/2}, \qquad \Gamma_{i-1} = \underline{\Gamma_i},$$

where $\underline{\Gamma_i}$ is Γ_i without the last block row.

Step 4. *Determine the state sequences:*

$$\tilde{X}_i = S_1 V_1^{\mathrm{T}}, \qquad \tilde{X}_{i+1} = \Gamma_{i-1}^{\dagger} \mathcal{O}_{i+1}.$$

Step 5. *Solve the following linear equations (**in the least-squares sense**) for A, B, C, D and the residuals ρ_w and ρ_v:*

$$\begin{pmatrix} \tilde{X}_{i+1} \\ Y_{i|i} \end{pmatrix} = \begin{pmatrix} A & B \\ C & D \end{pmatrix} \begin{pmatrix} \tilde{X}_i \\ U_{i|i} \end{pmatrix} + \begin{pmatrix} \rho_w \\ \rho_v \end{pmatrix}.$$

Step 6. *Determine the noise covariances Q, S, and R from the residuals as:*

$$\begin{pmatrix} Q & S \\ S^{\mathrm{T}} & R \end{pmatrix}_i = \frac{1}{j} \left[\begin{pmatrix} \rho_w \\ \rho_v \end{pmatrix} \begin{pmatrix} \rho_w^{\mathrm{T}} & \rho_v^{\mathrm{T}} \end{pmatrix} \right].$$

where the index i denotes a "bias" induced for finite i, which vanishes as $i \to \infty$.

Implementational remarks: In practical implementation, Step 4 should not be computed as above, because explicit computation of the latter matrix V is time consuming.

In fact, in a good software, the oblique projections in Step 1 are computed using a fast structure preserving QR factorization method and SVD in Step 2 is applied to only a part of the R-factor from the QR factorization.

For details of the proofs and practical implementations of these and other related subspace algorithms for system identification and an account of the extensive up-to-date literature (including the software on identification) on the subject, the readers are referred to the book by Van Overschee and DeMoor (1996a) and the recent review paper by DeMoor *et al.* (1999).

MATLAB note: M-files implementing Algorithm 9.4.2 (and others) come with the book by Van Overschee and DeMoor (1996b) and can also be obtained from **ftp://www.esat.kuleuven.ac.be/pub/SISTA/vanoverschee/book/subfun/**

9.4.3 Continuous-Time System Identification

Subspace system identification algorithms, analogous to Algorithms 9.4.1 and 9.4.2, can also be developed for a continuous-time system:

$$\dot{x}(t) = Ax(t) + Bu(t),$$
$$y(t) = Cx(t) + Du(t).$$

However, the input and output matrices have to be defined differently and they need computations of derivatives. Thus, define

$$
U^c_{0|i-1} = \begin{pmatrix}
u(t_0) & u(t_1) & \cdots & u(t_{j-1}) \\
u^{(1)}(t_0) & u^{(1)}(t_1) & \cdots & u^{(1)}(t_{j-1}) \\
\cdots & \cdots & \cdots & \cdots \\
u^{(i-1)}(t_0) & u^{(i-1)}(t_1) & \cdots & u^{(i-1)}(t_{j-1})
\end{pmatrix},
$$

where $u^{(p)}(t)$ denotes the pth derivative of $u(t)$, and "c" stands for "**continuous**." The matrices $Y^c_{0|i-1}$, $U^c_{i|2i-1}$, and X^c_i are similarly defined.

The continuous-time system identification problem can be stated as follows:

Given input and output measurements $u(t)$, $y(t)$, $t = t_0, t_1, \ldots, t_{j-1}$ and the estimates of the derivatives $u^{(p)}(t)$ and $y^{(p)}(t)$ up to order $2i - 1$, of the above unknown system, find the system matrices A, B, C, D, of the above continuous-time system up to within a similarity transformation.

9.4.4 Frequency-Domain Identification

The problem we consider here is the one of identifying a **continuous-time model** given a set of frequency responses. The problem can also be solved for a discrete-time system. For frequency-domain identification of discrete-time systems, see McKelvey (1994a, 1994b, 1994c). Specifically, the frequency-domain identification problem for a continuous-time system is stated as follows:

> Given N frequency domain frequency responses $G(j\omega_k)$, measured at frequencies ω_k (not necessarily distinct), $k = 1, 2, \ldots, N$, find the system matrices $A, B, C,$ and D.

One indirect approach for solving the problem is to estimate the Markov parameters via matrix-fraction descriptions of the frequency responses $G(j\omega_k)$ and then apply any of the Markov parameters based time-domain algorithms described in Section 9.3. (See **Exercise 9.10**).

We will, however, not discuss this here. For details, the readers are referred to the book by Juang (1994). Rather, we state here a direct **subspace identification algorithm** from the paper of DeMoor *et al.* (1999).

Let $\alpha > n$ be a user supplied index. Let $\text{Re}(M)$ and $\text{Im}(M)$ denote, respectively, the real and imaginary parts of a complex matrix M. Define the following matrices from the given frequency responses:

$$\mathcal{H} = \left(\text{Re}(\mathcal{H}^c), \ \text{Im}(\mathcal{H}^c) \right),$$
$$\mathcal{I} = \left(\text{Re}(\mathcal{I}^c), \ \text{Im}(\mathcal{I}^c) \right),$$

where

$$\mathcal{H}^c = \begin{pmatrix} G(j\omega_1) & G(j\omega_2) & \cdots & G(j\omega_N) \\ (j\omega_1)G(j\omega_1) & (j\omega_2)G(j\omega_2) & \cdots & (j\omega_N)G(j\omega_N) \\ \vdots & \vdots & & \vdots \\ (j\omega_1)^{\alpha-1}G(j\omega_1) & (j\omega_2)^{\alpha-1}G(j\omega_2) & \cdots & (j\omega_N)^{\alpha-1}G(j\omega_N) \end{pmatrix}$$

and

$$\mathcal{I}^c = \begin{pmatrix} I_m & I_m & \cdots & I_m \\ (j\omega_1)I_m & (j\omega_2)I_m & \cdots & (j\omega_N)I_m \\ \vdots & \vdots & & \vdots \\ (j\omega_1)^{\alpha-1}I_m & (j\omega_2)^{\alpha-1}I_m & \cdots & (j\omega_N)^{\alpha-1}I_m \end{pmatrix}.$$

Algorithm 9.4.3. *Continuous-Time Frequency-Domain Subspace Identification Algorithm.*

Inputs. *The set of measured frequencies $G(j\omega_1), G(j\omega_2), \ldots, G(j\omega_N)$, an integer α and a weighting matrix W.*

Outputs. *The system matrices A, B, C, and D.*

Step 1. *Find the orthogonal projection of the row space of $\mathcal{H}$ into the row space of $\mathcal{I}^\perp$:*

$$O_\alpha = \mathcal{H} - \mathcal{H}\mathcal{I}^\dagger\mathcal{I}.$$

Step 2. *Compute the SVD of $W O_\alpha$:*

$$W O_\alpha = U S V^{\mathsf{T}} = (U_1, U_2) \begin{pmatrix} S_1 & 0 \\ 0 & S_0 \end{pmatrix} \begin{pmatrix} V_1^{\mathsf{T}} \\ V_2^{\mathsf{T}} \end{pmatrix},$$

where W is a weighting matrix.

Step 3. *Determine $\Gamma_\alpha = W^{-1}U_1 S_1^{1/2}$.*

Step 4. *Determine A and C as follows:*
$$C = \text{the first } r \text{ rows of } \Gamma_\alpha$$
$$A = \underline{\Gamma_\alpha}^\dagger \bar{\Gamma}_\alpha,$$
where $\bar{\Gamma}_\alpha$ and $\underline{\Gamma}_\alpha$ denote Γ_α without the first and last r rows.

Step 5. *Determine B and D via the least-squares solution of the linear systems of equations:*

$$\begin{pmatrix} \text{Re}(L) \\ \text{Im}(L) \end{pmatrix} = \begin{pmatrix} \text{Re}(M) \\ \text{Im}(M) \end{pmatrix} \begin{pmatrix} B \\ D \end{pmatrix},$$

where L and M are given by:

$$L = \begin{pmatrix} G(j\omega_1) \\ \vdots \\ G(j\omega_N) \end{pmatrix} \quad and \quad M = \begin{pmatrix} C(j\omega_1 - A)^{-1} & I_r \\ \vdots \\ C(j\omega_N - A)^{-1} & I_r \end{pmatrix}.$$

(Note that L and M are, respectively, of order $rN \times m$ and $rN \times (n + r)$.)

Remarks

- The choice of the weighting matrix W is very important. If W is chosen appropriately, then the results are "unbiased"; otherwise, they will be "biased." For details of how the weighting should be chosen, the readers are referred to the paper by Van Overschee and De Moor (1996a).
- The algorithm works well when n and i are small.

    However, when i grows larger, the block Hankel matrices $\mathcal{H}$ and $\mathcal{I}$ became very highly ill-conditioned. The paper of Van Overschee and De Moore (1996a) contains a more numerically effective algorithm.

9.5   SOME SELECTED SOFTWARE

9.5.1   MATLAB Control System Toolbox

State-space models

minreal—Minimal realization and pole/zero cancellation
augstate—Augment output by appending states.

9.5.2   MATCONTROL

MINRESVD—Finding minimal realization using SVD of Hankel matrix of Markov parameters (**Algorithm 9.3.1**)
MINREMSVD—Finding minimal realization using SVD of Hankel matrix of lower order (**Algorithm 9.3.2**).

9.5.3   CSP-ANM

Model identification

- The system identification from its impulse responses is performed by
    ImpulseResponseIdentify [*response*].

- The system identification from its frequency responses is performed by FrequencyResponseIdentify [*response*].
- The system identification directly from input–output data is performed by OutputResponseIdentify [*u*, *y*].

9.5.4 SLICOT

Identification
IB—Subspace Identification

Time invariant state-space systems

IB01AD—Input–output data preprocessing and finding the system order
IB01BD—Estimating the system matrices, covariances, and Kalman gain
IB01CD—Estimating the initial state and the system matrices B and D.

TF—Time response

TF01QD   Markov parameters of a system from transfer function matrix
TF01RD   Markov parameters of a system from state-space representation

In addition to the above-mentioned software, the following toolboxes, especially designed for system identification are available.

- **MATLAB System Identification Toolbox**, developed by Prof. Lennart Ljung. (Website: http://www.mathworks.com)
- **ADAPTX**, developed by W.E. Larimore. (Website: http://adaptics.com)
- **Xmath Interactive System Identification Module**, described in the manual X-Math Interactive System Identification Module, Part 2, by P. VanOverschee, B. DeMoor, H. Aling, R. Kosut, and S. Boyd, Integrated Systems Inc., Santa Clara, California, USA, 1994 (website: http://www.isi.com/products/MATRIX$_X$/Techspec/MATRIX$_X$-Xmath/xm36.html, -/MATRIX$_X$ XMATH/inline images/pg. 37 img.html and-/MATRIX$_X$-XMath/inlineimages/pg. 38img.html).

For more details on these software packages, see the paper by DeMoor *et al.* (1999).

9.5.5 MATRIX$_X$

Purpose: Compute the minimal realization of a system.

Syntax: [SMIN, NSMIN, T]=MINIMAL (S, NS, TOL) or
[NUMMIN, DENMIN]=MINIMAL (NUM, DEN, TOL)

9.6 SUMMARY AND REVIEW

This chapter is concerned with state-space realization and model identification.

Realization

Given a transfer function matrix $G(s)$, the realization problem is the problem of finding the system matrices A, B, C, and D such that $G(s) = C(sI - A)^{-1}B + D$.

For a given proper rational function $G(s)$, there always exists a state-space realization. However, **such a realization is not unique.** In **Section 9.2.1**, the nonuniqueness of a realization is demonstrated by computing the two realizations of the same transfer function matrix $G(s)$: **controllable and observable realizations.**

Minimal Realization

A realization (A, B, C, D) of $G(s)$ is an **MR** if A has the smallest possible dimension. An important result on MR is that a **realization is minimal if and only if (A, B) is controllable and (A, C) is observable (Theorem 9.2.1).**

Two MRs are related by a nonsingular transforming matrix T (**Theorem 9.2.2**).

There are many methods for computing an MR, given a set of **Markov parameters** $H_k = CA^{k-1}B$, $k = 1, 2, 3, \ldots$, assuming that these Markov parameters are easily obtainable from a given transfer function. Most of these methods find an MR by factoring the Hankel matrix of Markov parameters:

$$
M_k = \begin{pmatrix} H_1 & H_2 & \cdots & H_k \\ H_2 & H_3 & \cdots & H_{k+1} \\ \vdots & & & \\ H_k & H_{k+1} & \cdots & H_{2k-1} \end{pmatrix}.
$$

Some basic properties of this Hankel matrix M_k that play an important role in the development of these algorithms are stated and proved in **Theorem 9.3.1**.

Two numerically viable SVD-based methods for computing an MR are given in Sections 9.3.2 and 9.3.3 (**Algorithms 9.3.1 and 9.3.2**).

Time-Domain Subspace Identification

Many times, the Markov parameters are not easily accessible. In these cases, the system matrices must be identified from a given set of input and output data.

Two subspace algorithms for system identification: **Algorithm 9.4.1** for *deterministic identification* and **Algorithm 9.4.2** for *combined deterministic and stochastic identification* are described in **Section 9.4.**

It is assumed that the number of input and output data are very large (goes to infinity) and that the data are ergodic.

Each of these two subspace algorithms comes in two steps. The first step consists of finding (implicitly or explicitly) some estimate $\tilde{X}_i$ of the state sequence, while in the second step, the system matrices A, B, C, and D are obtained by solving an overdetermined system (in the least-squares sense) using this state sequence $\hat{X}_i$.

Frequency-Domain Subspace Identification

Finally, frequency-domain subspace identification is considered in **Section 9.4.4**. The problem considered there is:

Given N frequency domain responses $G(j\omega_k)$, measured at frequencies ω_k, $k = 1, 2, \ldots, N$; find the system matrices A, B, C, and D.

A **continuous-time** frequency-domain subspace identification algorithm (**Algorithm 9.4.3**) is described in **Section 9.4.4**.

9.7   CHAPTER NOTES AND FURTHER READING

Realization theory is a classical topic in system identification. Ho and Kalman (1966) first introduced the important principles and concepts of MR theory. There are now well-known books and papers in this area such as Kung (1978), Ljung (1987, 1991a, 1991b), Silverman (1971), Zeiger and McEwen (1974), Dickinson *et al.* (1974a, 1974b), Juang (1994), Norton (1986), Aström and Eykhoff (1971), Eykhoff (1974), Rissanen (1971), DeJong (1978), Brockett (1978), Datta (1980), Gragg and Lindquist (1983). These papers and books provide a good insight into the subject of system identification from Markov parameters. The paper by Gragg and Lindquist (1983) deals with partial realization problem. The subspace system identification algorithms are the input-state-output generalizations of the realization theory and these algorithms are relatively modern.

Material on subspace algorithms in this book has been taken mostly from the recent book by Van Overschee and De Moor (1996b) and the recent review paper by De Moor *et al.* (1999). **Both references contain an up-to-date extensive list of papers and books on realization theory** and **subspace identification algorithms** (see also the papers by Lindquist and Picci (1993, 1994)). Frequency-domain identification is dealt with in some depth in the book by Juang (1994) and a Newton-type algorithm for fitting transfer functions to frequency-response measurements appears in Spanos and Mingori (1993).

There exists an intimate relation between subspace system identification and frequency weighted model reduction. The frequency weighted model reduction is discussed in Chaper 14 of this book. For details of the connection between these topics, see **Chapter 5** of the book by Van Overschee and De Moor (1996b).

Exercises

9.1   Prove that there always exists a state-space realization for a proper rational function.

9.2   Verify that the controllable realization (A, B, C, D) and the observable realization (A', B', C', D') described in Section 9.2.1 are state-space realizations of the same transfer matrix $G(s)$.

9.3   Give a complete proof of Theorem 9.2.2.

9.4   Let $G(s)$ be the transfer matrix of a SISO system and let (A, b, c, d) be a state-space realization of $G(s)$:

$$G(s) = d + c(sI - A)^{-1}b = d + \frac{b(s)}{a(s)}.$$

Prove that the realization is minimal if and only if $a(s)$ and $b(s)$ are coprime.

9.5   *Generating the Markov Parameters*

(a)   Show that for the discrete-time system (9.3.1) with initial condition $x_0 = 0$, the Markov parameters $H_0 = D$, $H_i = CA^{i-1}B$, $i = 1, 2, \ldots, l - 1$ can be determined by solving the system:

$$y = SU,$$

where $y = (y_0, y_1, y_2, \ldots, y_{l-1})_{r \times l}$

$$S = (H_0, H_1, H_2, \ldots, H_{l-1}),$$

$$U = \begin{pmatrix} u_0 & u_1 & u_2 & \cdots & u_{l-1} \\ & u_0 & u_1 & \cdots & u_{l-2} \\ & & \ddots & & \vdots \\ & & & \ddots & \vdots \\ & & & & u_0 \end{pmatrix}_{ml \times l},$$

where m is the number of inputs and r is the number of outputs; the matrix U is an $ml \times l$ block upper triangular matrix.

(b)   Assume that $A^k \approx 0$ for all time steps $k \geq p$, that is, A is discrete stable, then show that the above system can be reduced to

$$y = S'U',$$

where $y = (y_0, y_1, \ldots, y_{l-1})$,

$$S' = (H_0, H_1, H_2, \ldots, H_p)$$

and

$$U' = \begin{pmatrix} u_0 & u_1 & u_2 & \cdots & u_p & \cdots & u_{l-1} \\ & u_0 & u_1 & \cdots & u_{p-1} & \cdots & u_{l-2} \\ & & \ddots & & & & \vdots \\ & & & \ddots & & & \vdots \\ & & & & \ddots & & \vdots \\ 0 & & & & & \cdot & u_{l-p-1} \end{pmatrix}.$$

(Note that U' is of order $m(p + 1) \times l$ and S' is of order $r \times m(p + 1)$.)

(c) Discuss the numerical difficulties in solving the above system and work out an example to illustrate the difficulties.

9.6 Prove that the Hankel matrix M_k can be decomposed in the form (9.3.3).

9.7 Assuming that $H_k \to 0$ as $k \to \infty$, prove that the realization obtained by Algorithm 9.3.1 is discrete-stable. (**Hint:** Show that $\| S^{-1/2} A S^{1/2} \|_2$ is less than unity.)

9.8 (a) Construct a discrete-time system:

$$x_{k+1} = Ax_k + Bu_k$$
$$y_k = Cx_k$$

with suitable randomly generated matrices A, B, and C.

(b) Construct sufficient number of Markov parameters using the inputs $u_0 = 1, u_i = 0, i > 1$, and assuming zero initial condition.

(c) Apply Algorithms 9.3.1 and 9.3.2 to identify the system matrices A, B, and C.

(d) In each case, plot the transfer function of the original and the identified model.

9.9 A stable system is **balanced** if both controllability and observability Grammians are equal to a diagonal matrix (**Chapter 14**).

Prove that if Algorithm 9.3.2 starts with the Hankel matrix:

$$M_{\beta,\alpha} = \begin{pmatrix} H_1 & H_2 & \cdots & H_\beta \\ H_2 & H_3 & \cdots & H_{\beta+1} \\ \vdots & & & \\ H_\alpha & H_{\alpha+1} & \cdots & H_{\alpha+\beta-1} \end{pmatrix},$$

then the algorithm gives a balanced realization when the indices α and β are sufficiently large.

9.10 Frequency-Domain Realization using Markov Parameters (Juang (1994)).

Consider the frequency response function $G(z_k) = C(zI - A)^{-1}B + D$; $z_k = e^{j2\pi k/l}$, where l is the data length and $z_k, k = 0, 1, \ldots, l$ correspond to the frequency points at $2\pi k/l\Delta t$, with Δt being the sampling time interval.

Write $G(z_k) = Q^{-1}(z_k)R(z_k)$

where

$$Q(z_k) = I_r + Q_1 z_k^{-1} + \cdots + Q_p z_k^{-p}$$
$$R(z_k) = R_0 + R_1 z_k^{-1} + \cdots + R_p z_k^{-p}$$

are matrix polynomials and I_r is the identity matrix of order r.

(a) Prove that knowing $G(z_k)$, the coefficient matrices of $Q(z_k)$ and $R(z_k)$ can be found by solving a least-squares problem.

(b) How can the complex arithmetic be avoided in part (a)?

(c) Show how to obtain the Markov parameters from the coefficient matrices found in (a).

(**Hint:** $(\Sigma_{i=0}^{p} Q_i z^{-i})(\Sigma_{i=0}^{\infty} H_i z^{-i}) = \Sigma_{i=0}^{p} R_i z^{-i}$.)

(d) Derive an algorithm for frequency-domain realization similar to Algorithm 9.3.2 based on (a)–(c).

(e) (Juang 1994). Apply your algorithm to the discrete-time system model defined by the following data:

$$A = \text{diag}\left(\begin{pmatrix} 0.9859 & 1.500 \\ -1.500 & 0.9859 \end{pmatrix}, \begin{pmatrix} 0.9859 & 0.1501 \\ -0.1501 & 0.9859 \end{pmatrix}, \right.$$

$$\left. \begin{pmatrix} 0.6736 & 0.7257 \\ -0.725 & 0.6736 \end{pmatrix}, \begin{pmatrix} 0.4033 & 0.9025 \\ -0.9025 & 0.4033 \end{pmatrix} \right).$$

$$B = \begin{pmatrix} -0.0407 & -0.0454 \\ -0.5384 & -0.6001 \\ 0.0746 & -0.0669 \\ 0.9867 & -0.8850 \\ 0.0164 & 0.0373 \\ 0.0376 & 0.0860 \\ -0.0460 & -0.0421 \\ -0.0711 & -0.0650 \end{pmatrix}, \quad C^{T} = \begin{pmatrix} 0.8570 & 1.80 \\ 0.0000 & 0.00 \\ 1.5700 & -1.2 \\ 0.0000 & 0.00 \\ 1.4030 & 1.42 \\ 0.0000 & 0.00 \\ 0.9016 & 1.78 \\ 0.0000 & 0.00 \end{pmatrix},$$

$$D = \begin{pmatrix} 0. & 0. \\ 0. & 0. \end{pmatrix}$$

by calculating 200 frequency data points equally spaced in a data frequency ranging from 0 to 16.67 Hz, and assuming that the orders of $Q(z_k)$ and $R(z_k)$ are 10. Sketch the graphs of the true and estimated frequency response functions for the first input and first output and compare the results.

9.11 Consider the following discrete-time model (of a rigid body of mass m with a force f acting along the direction of the motion (Juang 1994):

$$x_{k+1} = Ax_k + Bu_k,$$
$$y_k = Cx_k,$$

where $A = \begin{pmatrix} 1 & \Delta t \\ 0 & 1 \end{pmatrix}$, $B = \begin{pmatrix} \frac{1}{2}\Delta t^2 \\ \Delta t \end{pmatrix}$, $u_k = \dfrac{f}{m}$, $C = (1, 0)$.

Δt = sampling time interval.

(a) Construct the first five Markov parameters.
(b) Apply Algorithm 9.3.2 to identify A, B, and C.
(c) Show that the original and the identified models have the identical Markov parameters.

9.12 Using the notation of Section 9.4.1, prove that $\text{span}_{\text{row}}(X_2) = \text{span}_{\text{row}}(H_{k|k+i}) \cap \text{span}_{\text{row}}(H_{k+1|k+2i})$.

9.13 Modify Algorithm 9.4.2 by incorporating weighting matrices W_1 and W_2 such that W_1 is of full rank and W_2 has the property that $\text{rank}(W_{0|i-1}) = \text{rank}(W_{0|i-1}W_2)$.

References

Åström K. and Eykhoff P. "System identification—A survey," *Automatica*, Vol. 7, pp. 123–167, 1971.

Brockett R. "The geometry of the partial realization problem," *Proc. IEEE Conf. Dec. Control*, pp. 1048–1052, 1978.

Datta K.B. "Minimal realization in companion forms," *J. Franklin Institute,* Vol. 309(2), pp. 103–123, 1980.

DeJong L.S. "Numerical aspects of recursive realization algorithms," *SIAM J. Control. Optimiz.,* Vol. 16(4), pp. 646–660, 1978.

DeMoor B., Van Overschee P., and Favoreel W. "Algorithms for subspace state-space systems identification: an overview," *Appl. Comput. Control, Signals, Circuits* (Datta B.N., et al., eds.), vol. 1, pp. 247–311, Birkhauser, Boston, MA, 1999.

Dickinson B., Morf M., and Kailath T. "A minimal realization algorithm for matrix sequences," *IEEE Trans. Autom. Control,* Vol. AC-19(1), pp. 31–38, 1974a.

Dickinson B., Kailath T., and Morf M. "Canonical matrix fraction and state space descriptions for deterministic and stochastic linear systems," *IEEE Trans. Autom. Control,* Vol. AC-19, pp. 656–667, 1974b.

Eykhoff P. *System Identification,* Wiley, London, 1974.

Glover K. "All optimal Hankel-norm approximation of linear multivariable systems and their L_∞-error bounds," *Int. J. Control,* Vol. 39, pp. 1115–1193, 1984.

Gragg W. and Lindquist A. "On the partial realization problem," *Lin. Alg. Appl.,* Vol. 50, pp. 277–319, 1983.

Ho B.L. and Kalman R.E. "Efficient construction of linear state variable models from input/output functions," *Regelungstechnik,* Vol. 14, pp. 545–548, 1966.

Jer-Nan Juang, *Applied System Identification,* Prentice Hall, Englewood Cliffs, NJ, 1994.

Kalman R.E., Falb P.L., and Arbib M.A. *Topics in Mathematical System Theory,* McGraw Hill, New York, 1969.

Kung S.Y. "A new identification method and model reduction algorithm via singular value decomposition," 12th *Asilomar Conf. Circuits, Syst. Comp.,* pp. 705–714, Asilomar, CA, 1978.

Lindquist A. and Picci G. "On subspace methods identification," *Proc. Math. Theory Networks Syst.,* Vol. 2, pp. 315–320, 1993.

Lindquist A. and Picci G. "On subspace methods identification and stochastic model reduction," *Proc. SYSID,* Vol. 2, pp. 397–404, 1994.

Ljung L. *System identification-Theory for the User,* Prentice Hall, Englewood Cliffs, NJ, 1987.

Ljung L. Issues in system identification, *IEEE Control Syst.,* Vol. 11(1), pp. 25–29, 1991a.

Ljung L. *System Identification Toolbox For Use with Matlab,* The Mathworks Inc., MA, USA, 1991b.

McKelvey T. *On State-Space Models in System Identification,* Thesis no. 447, Department of Electrical Engineering, Linköping University, Sweden, 1994a.

McKelvey T. An efficient frequency domain state-space identification algorithm, *Proc. 33rd IEEE Conf. Dec. Control,* pp. 3359–3364, 1994b.

McKelvey T. *SSID-A MATLAB Toolbox for Multivariable State-Space Model Identification,* Department of Electrical Engineering, Linköping University, Linköping, Sweden, 1994c.

Moonen M., DeMoor B., Vandenberghe L., and Vandewalle J., "On and off-line identification of linear state space models," *Int. J. Control,* Vol. 49(1), pp. 219–232, 1989.

Norton J.P. *An Introduction to Identification,* Academic Press, London, 1986.

Rissanen J. Recursive identification of linear sequences, *SIAM J. Control,* Vol. 9, pp. 420–430, 1971.

Silverman L. "Realization of linear dynamical systems," *IEEE Trans. Autom. Control*, Vol. AC-16, pp. 554–567, 1971.

Spanos J. T. and Mingori D. L. "Newton algorithm for filtering transfer functions to frequency response measurements," *J. Guidance, Control, Dynam.*, Vol. 16, pp. 34–39, 1993.

Van Overschee P. and DeMoor B. Continuous-time frequency domain subspace system identification, *Signal Processing, Special Issue on Subspace Methods, Part II: System Identification*, Vol. 52, pp. 179–194, 1996a.

Van Overschee P. and DeMoor B. *Subspace Identification for Linear Systems: Theory, Implementation and Applications*, Kluwer Academic Publishers, Boston/London/Dordrecht, 1996b.

Zeiger H. and McEwen A. "Approximate linear realizations of given dimension via Ho's algorithm," *IEEE Trans. Autom. Control*, Vol. 19, pp. 153, 1974.

CHAPTER 10

FEEDBACK STABILIZATION, EIGENVALUE ASSIGNMENT, AND OPTIMAL CONTROL

Topics covered

- State-Feedback Stabilization
- Eigenvalue Assignment (EVA)
- Linear Quadratic Regulator Problems
- H_∞ Control
- Stability Radius (Revisited)

10.1 INTRODUCTION

In this chapter, we first consider the problem of stabilizing a linear control system by choosing the control vector appropriately. Mathematically, the problem is to find a feedback matrix K such that $A - BK$ is stable in the continuous-time case or is discrete-stable in the discrete-time case. Necessary and sufficient conditions are established for the existence of stabilizing feedback matrices, and Lyapunov-style methods for constructing such matrices are described in **Section 10.2**.

A concept dual to stabilizability, called **detectability**, is then introduced and its connection with a Lyapunov matrix equation is established in **Section 10.3**.

In certain practical situations, stabilizing a system is not enough; a designer should be able to control the eigenvalues of $A - BK$ so that certain design constraints are met. This gives rise to the **eigenvalue assignment (EVA) problem** or the so-called **pole placement problem**. Mathematically, the problem is to find a feedback matrix K such that $A - BK$ has a preassigned spectrum. A well-known and a very important result on the solution of this problem is: Given a real pair of matrices (A, B) and Λ, an arbitrary set of n complex numbers, closed under complex conjugation, there exists a real matrix K such that the spectrum of $A - BK$ is the set Λ if and only if (A, B) is controllable. The matrix K is unique in the single-input case.

This important result is established in **Theorem 10.4.1**. The proof of this result is constructive, and leads to several well-known formulas, the most important of which is the **Ackermann formula**. However, these formulas do not yield numerically viable methods for pole placement. **Numerical methods for pole placement are presented in Chapter 11.**

Since there are no set guidelines as to where the poles (the eigenvalues) need to be placed, very often, in practice, a compromise is made in which a feedback matrix is constructed in such a way that not only the system is stabilized, but a certain performance criterion is satisfied. This leads to the well-known **Linear Quadratic Regulator** (LQR) problem. Both continuous-time and discrete-time LQR problems are discussed in **Section 10.5** of this chapter. The solutions of the LQR problems require the solutions of certain quadratic matrix equations, called the **algebraic Riccati equations** (AREs). **Numerical methods for the AREs are described in Chapter 13.**

The next topic in this chapter is the H_∞-control problems. Though a detailed discussion on the H_∞-control problems is beyond the scope of the book, some simplified versions of these problems are stated in **Section 10.6** in this chapter. The H_∞-control problems are concerned with stabilization of perturbed versions of a system, when certain bounds of perturbations are known. The solutions of the H_∞-control problems also require solutions of certain AREs. Two algorithms (**Algorithms 10.6.1 and 10.6.2**) are given in Section 10.6 for computing the H_∞-norm."

The concept of **stability radius** introduced in Chapter 7 is revisited in the final section of this chapter (**Section 10.7**), where a relationship between the complex stability radius and an ARE (**Theorem 10.7.3**) is established, and a bisection algorithm (**Algorithm 10.7.1**) for determining the complex stability radius is described.

Reader's Guilde for Chapter 10

The readers familiar with concepts and results of state-feedback stabilizations, pole-placement, LQR design, and H_∞ control can skip Sections 10.2–10.6. However, two algorithms for computing the H_∞-norm (**Algorithms 10.6.1 and 10.6.2**) and material on stability radius (**Section 10.7**) should be of interests to most readers.

10.2   STATE-FEEDBACK STABILIZATION

In this section, we consider the problem of stabilizing the linear system:

$$\dot{x}(t) = Ax(t) + Bu(t),$$
$$y(t) = Cx(t) + Du(t). \tag{10.2.1}$$

Suppose that the state vector $x(t)$ is known and let's choose

$$u(t) = v(t) - Kx(t), \tag{10.2.2}$$

where K is a constant matrix, and $v(t)$ is a reference input vector.

Then feeding this input vector $u(t)$ back into the system, we obtain the system:

$$\dot{x}(t) = (A - BK)x(t) + Bv(t),$$
$$y = (C - DK)x(t) + Dv(t). \tag{10.2.3}$$

The problem of stabilizing the system (10.2.1) then becomes the problem of finding K such that the system (10.2.3) becomes stable. The problem of state-feedback stabilization can, therefore, be stated as follows:

> Given a pair of matrices (A, B), find a matrix K such that $A - BK$ is stable.

Graphically, the state-feedback problem can be represented as in Figure 10.1.

In the next subsection we will investigate the conditions under which such a matrix K exists. The matrix K, when it exists, is called a **stabilizing feedback matrix**; and in this case, the pair (A, B) is called a **stabilizable pair.** The system (10.2.3) is called the **closed-loop system** and the matrix $A - BK$ is called the **closed-loop matrix.**

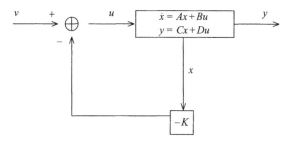

FIGURE 10.1:   State feedback configuration.

Analogously, for the discrete-time system:

$$x_{k+1} = Ax_k + Bu_k,$$
$$y_k = Cx_k + Du_k,$$

if there exists a matrix K such that $A - BK$ is discrete-stable, that is, if it has all its eigenvalues inside the unit circle, then the pair (A, B) will be called a **discrete-stabilizable** pair, and the matrix K will be called a **discrete-stabilizing feedback matrix.**

In what follows, we will present simple criteria of stabilizability and algorithms for constructing stabilizing feedback matrices via Lyapunov matrix equations.

10.2.1   Stabilizability and Controllability

In this section, we describe necessary and sufficient conditions for a given pair (A, B) to be a stabilizable pair. We start with the continuous-time case.

Theorem 10.2.1.   *Characterization of Continuous-Time Stabilizability. The following, are equivalent:*

(i)   *(A, B) is stabilizable.*

(ii)   *$Rank(A - \lambda I, B) = n$ for all $\mathrm{Re}(\lambda) \geq 0$. In other words, the unstable modes of A are controllable.*

(iii)   *For all λ and $x \neq 0$ such that $x^*A = \lambda x^*$ and $\mathrm{Re}(\lambda) \geq 0$, we have $x^*B \neq 0$.*

Proof.   We prove the equivalence of (i) and (ii) and leave the equivalence of (i) and (iii) as an exercise **(Exercise 10.1).**

Without any loss of generality we may assume (see Theorem 6.4.1) that the pair (A, B) is given in the form:

$$PAP^{-1} = \bar{A} = \begin{pmatrix} \bar{A}_{11} & \bar{A}_{12} \\ 0 & \bar{A}_{22} \end{pmatrix}, \qquad PB = \bar{B} = \begin{pmatrix} \bar{B}_1 \\ 0 \end{pmatrix},$$

where $(\bar{A}_{11}, \bar{B}_1)$ is controllable.

Since $(\bar{A}_{11}, \bar{B}_1)$ is controllable, by the eigenvalue criterion of controllability (Theorem 6.2.1 (v)), we have $\mathrm{rank}(\lambda I - \bar{A}_{11}, \bar{B}_1) = p$, where p is the order of $\bar{A}_{11}$. Therefore,

$$\mathrm{rank}(\lambda I - \bar{A}, \bar{B}) = \mathrm{rank} \begin{pmatrix} \lambda I - \bar{A}_{11} & -\bar{A}_{12} & \bar{B}_1 \\ 0 & \lambda I - \bar{A}_{22} & 0 \end{pmatrix} < n,$$

if and only if $\mathrm{rank}(\lambda I - \bar{A}_{22}) < n - p$, that is, if and only if λ is an eigenvalue of $\bar{A}_{22}$.

The proof now follows from the fact that if (A, B) is a stabilizable pair, the matrix $\bar{A}_{22}$ must be a stable matrix. This can be seen as follows:

The stabilizability of the pair (A, B) implies the stabilizability of the pair $(\bar{A}, \bar{B})$.

Since $(\bar{A}, \bar{B})$ is a stabilizable pair, there exists a matrix $\bar{K}$ such that $\bar{A} - \bar{B}\bar{K}$ is stable. This means that if $\bar{K} = (\bar{K}_1, \bar{K}_2)$, then the matrix

$$\begin{pmatrix} \bar{A}_{11} - \bar{B}_1\bar{K}_1 & \bar{A}_{12} - \bar{B}_1\bar{K}_2 \\ 0 & \bar{A}_{22} \end{pmatrix}$$

is a stable matrix, which implies that $\bar{A}_{22}$ must be stable. ■

Corollary 10.2.1. *If the pair (A, B) is controllable, then it must be stabilizable.*

Proof. If (A, B) is controllable, then again by the eigenvalue criterion of controllability, $\text{rank}(A - \lambda I, B) = n$ for every λ. In particular, $\text{rank}(A - \lambda I, B) = n$ for every λ for which $\text{Re}(\lambda) \geq 0$. Thus, (A, B) is stabilizable. ■

The above result tells us that **the controllability implies stabilizability**.

However, the converse is not true. The stabilizability is guaranteed as long as the unstable modes are controllable.

The following simple example illustrates the fact.

$$\text{Let } A = \begin{pmatrix} 1 & 1 & 1 \\ 0 & 2 & 1 \\ 0 & 0 & -3 \end{pmatrix}, \qquad b = \begin{pmatrix} 1 \\ -1 \\ 0 \end{pmatrix}.$$

(A, b) is not controllable; $\text{rank}(b, Ab, A^2b) = 2$.

However, the row vector $f^T = (-126.5, -149.5, 0)$ is such that the eigenvalues of $A - bf^T$ are $\{-10 \pm 11.4891j, -3\}$.

So, $A - bf^T$ is stable, that is, (A, b) is stabilizable.

The Discrete Case

A theorem, analogous to Theorem 10.2.1, can be proved for the discrete-time system as well. We state the result without proof. The proof is left as an exercise **(Exercise 10.2)**.

Theorem 10.2.2. *Characterization of Discrete-Stabilizability. The following conditions are equivalent:*

(i) *The pair (A, B) is discrete-stabilizable.*

(ii) $\text{Rank}(A - \lambda I, B) = n$ *for every λ such that $|\lambda| \geq 1$.*

(iii) *For all λ and $x \neq 0$ such that $x^*A = \lambda x^*$ and $|\lambda| \geq 1$, we have $x^*B \neq 0$.*

10.2.2   Stabilization via Lyapunov Equations

From the discussions of the previous section, it is clear that for finding a feedback stabilizing matrix K for a given pair (A, B), we can assume that the pair (A, B) is controllable. For, if (A, B) is not controllable but stabilizable, then we can always put it in the form:

$$TAT^{-1} = \bar{A} = \begin{pmatrix} \bar{A}_{11} & \bar{A}_{12} \\ 0 & \bar{A}_{22} \end{pmatrix}, \qquad TB = \bar{B} = \begin{pmatrix} \bar{B}_1 \\ 0 \end{pmatrix}, \qquad (10.2.4)$$

where $(\bar{A}_{11}, \bar{B}_1)$ is controllable, and $\bar{A}_{22}$ is stable.

Once a stabilizing matrix $\bar{K}_1$ for the controllable pair $(\bar{A}_{11}, \bar{B}_1)$ is obtained, the stabilizing matrix K for the pair (A, B) can be obtained as:

$$K = \bar{K}T,$$

where

$$\bar{K} = (\bar{K}_1, \bar{K}_2) \qquad (10.2.5)$$

and $\bar{K}_2$ is arbitrary.

We can therefore concentrate on stabilizing a controllable pair. Theorem 10.2.3 shows how to stabilize a controllable pair using a Lyapunov equation.

Theorem 10.2.3.   *Let (A, B) be controllable and let β be a scalar such that*

$$\beta > |\lambda_{\max}(A)|,$$

where $\lambda_{\max}(A)$ is the eigenvalue of A with the largest real part. Let K be defined by

$$K = B^{\mathrm{T}}Z^{-1}, \qquad (10.2.6)$$

where Z (necessarily symmetric positive definite) satisfies the Lyapunov equation:

$$-(A + \beta I)Z + Z[-(A + \beta I)]^{\mathrm{T}} = -2BB^{\mathrm{T}}, \qquad (10.2.7)$$

then $A - BK$ is stable, that is, (A, B) is stabilizable.

Proof.   Since $\beta > |\lambda_{\max}(A)|$, the matrix $-(A + \beta I)$ is stable.

Also, since (A, B) is controllable, the pair $(-(A+\beta I), B)$ is controllable. Thus, by Theorem 7.2.6, the Lyapunov equation (10.2.8) has a unique symmetric positive definite solution Z.

Again, Eq. (10.2.8) can be written as:

$$(A - BB^T Z^{-1})Z + Z(A - BB^T Z^{-1})^T = -2\beta Z.$$

Then, from (10.2.7) we have:

$$(A - BK)Z + Z(A - BK)^T = -2\beta Z. \tag{10.2.8}$$

Since Z is symmetric positive definite, $A - BK$ is stable by Theorem 7.2.3.

This can be seen as follows:

Let μ be an eigenvalue of $A - BK$ and y be the corresponding eigenvector.

Then multiplying both sides of Eq. (10.2.9) first by y^* to the left and then by y to the right, we have

$$2 \operatorname{Re}(\mu) y^* Z y = -2\beta y^* Z y.$$

Since Z is positive definite, $y^* Z y > 0$. Thus, $\operatorname{Re}(\mu) < 0$. So, $A - BK$ is stable. ■

The above discussion leads to the following method for finding a stabilizing feedback matrix (see Armstrong 1975).

A Lyapunov Equation Method For Stabilization

Let (A, B) be a controllable pair. Then the following method computes a stabilizing feedback matrix K.

Step 1. Choose a number β such that $\beta > |\lambda_{\max}(A)|$, where $\lambda_{\max}(A)$ denotes the eigenvalue of A with the largest real part.

Step 2. Solve the Lyapunov equation (10.2.8) for Z:

$$-(A + \beta I)Z + Z[-(A + \beta I)]^T = -2BB^T.$$

Step 3. Obtain the stabilizing feedback matrix K:

$$K = B^T Z^{-1}.$$

MATCONTROL note: The above method has been implemented in MATCONTROL function **stablyapc**.

A Remark on Numerical Effectiveness

The Lyapunov equation in Step 2 can be highly ill-conditioned, even when the pair (A, B) is robustly controllable. In this case, the entries of the stabilizing feedback matrix K are expected to be very large, giving rise to practical difficulties in implementation. See the example below.

Example 10.2.1 (*Stabilizing the Motion of the Inverted Pendulum Consider*). Example 5.2.5 (The **problem of a cart with inverted pendulum**) with the following data:

$$m = 1\,kg, \quad M = 2\,kg, \quad l = 0.5\,m, \quad \text{and } g = 9.18\,m/s^2.$$

Then,

$$A = \begin{pmatrix} 0 & 1 & 0 & 0 \\ 0 & 0 & -3.6720 & 0 \\ 0 & 0 & 0 & 1 \\ 0 & 0 & 22.0320 & 0 \end{pmatrix}.$$

The eigenvalues of A are 0, 0, ±4.6938. Thus, with no control input, there is an instability in the motion and the pendulum will fall. We will now stabilize the motion by using the Lyapunov equation method with A as given above, and

$$B = \begin{pmatrix} 0 \\ 0.4 \\ 0 \\ -0.4 \end{pmatrix}.$$

Step 1. Let's choose $\beta = 5$. This will make $-(A + \beta I)$ stable.
Step 2.

$$Z = \begin{pmatrix} 0.0009 & -0.0044 & -0.0018 & 0.0098 \\ -0.0044 & 0.0378 & 0.0079 & -0.0593 \\ -0.0018 & 0.0079 & 0.0054 & -0.0270 \\ 0.0098 & -0.0593 & -0.0270 & 0.1508 \end{pmatrix}.$$

(The computed Z is symmetric positive definite but highly ill-conditioned).
Step 3. $K = B^T Z^{-1} = 10^3(-0.5308, -0.2423, -1.2808, -0.2923)$.
Verify: The eigenvalues of $A - BK$ are $\{-5 \pm 11.2865j, -5 \pm 0.7632j\}$.

Note that the entries of K are large. The pair (A, B) is, however, robustly controllable, which is verified by the fact that the singular values of the controllability matrix are 8.9462, 8.9462, 0.3284, 0.3284.

Remark

- If the pair (A, B) is not controllable, but stabilizable, then after transforming the pair (A, B) to the form $(\bar{A}, \bar{B})$ given by

$$TAT^{-1} = \bar{A} = \begin{pmatrix} \bar{A}_{11} & \bar{A}_{12} \\ 0 & \bar{A}_{22} \end{pmatrix}, \qquad TB = \bar{B} = \begin{pmatrix} \bar{B}_1 \\ 0 \end{pmatrix},$$

we will apply the above method to the pair $(\bar{A}_{11}, \bar{B}_1)$ (which is controllable) to find a stabilizing feedback matrix $\bar{K}_1$ for the pair $(\bar{A}_{11}, B_1)$ and then obtain K that stabilizes the pair (A, B) as

$$K = (\bar{K}_1, \bar{K}_2)\, T,$$

choosing $\bar{K}_2$ arbitrarily.

Example 10.2.2. Consider the uncontrollable, but the stabilizable pair (A, B):

$$A = \begin{pmatrix} 1 & 1 & 1 \\ 0 & 2 & 1 \\ 0 & 0 & -3 \end{pmatrix}, \qquad B = \begin{pmatrix} 1 \\ -1 \\ 0 \end{pmatrix}.$$

Step 1. $A = \bar{A}$, $B = \bar{B}$. So, $T = I$.

$$\bar{A}_{11} = \begin{pmatrix} 1 & 1 \\ 0 & 2 \end{pmatrix}, \qquad \bar{A}_{22} = -3, \qquad \bar{B}_1 = \begin{pmatrix} 1 \\ -1 \end{pmatrix}.$$

Step 2. Choose $\beta_1 = 10$. The unique symmetric positive definite solution Z_1 of the Lyapunov equation:

$$-(\bar{A}_{11} + \beta_1 I)Z_1 + Z_1[-(\bar{A}_{11} + \beta_1 I)]^T = -2\bar{B}_1 \bar{B}_1^T$$

is

$$Z_1 = \begin{pmatrix} 0.0991 & -0.0906 \\ -0.0906 & 0.0833 \end{pmatrix}.$$

Step 3. $\bar{K}_1 = \bar{B}_1^T Z_1^{-1} = (-126.5, -149.5)$.
Step 4. Choose $\bar{K}_2 = 0$. Then $K = \bar{K} = (\bar{K}_1, \bar{K}_2) = (-126.5, -149.5, 0)$.
Verify: The eigenvalues of $A - BK$ are $-10 \pm 11.489 j, -3$.

Discrete-Stabilization via Lyapunov Equation

The following is a discrete-analog of Theorem 10.2.3. We state the theorem without proof. The proof is left as an exercise (**Exercise 10.3**).

Theorem 10.2.4. *Let the discrete-time system $x_{k+1} = Ax_k + Bu_k$ be controllable. Let $0 < \beta \le 1$ be such that $|\lambda| \ge \beta$ for any eigenvalue λ of A.*
Define $K = B^T(Z + BB^T)^{-1}A$, where Z satisfies the Lyapunov equation, $AZA^T - \beta^2 Z = 2BB^T$, then $A - BK$ is discrete-stable.

Theorem 10.2.4 leads to the following Lyapunov method of discrete-stabilization. The method is due to Armstrong and Rublein (1976).

A Lyapunov Equation Method for Discrete-Stabilization

Step 1. Find a number β such that $0 < \beta < \min(1, \min_i |\lambda_i|)$, where $\lambda_1, \lambda_2, \ldots, \lambda_n$ are the eigenvalues of (A).

Step 2. Solve the discrete Lyapunov equation for Z:

$$AZA^T - \beta^2 Z = 2BB^T.$$

Step 3. Compute the discrete-stabilizing feedback matrix K,

$$K = B^T(Z + BB^T)^{-1}A.$$

Example 10.2.3. Consider the **cohort population model** in Luenberger (1979, pp. 170), with $\alpha_1 = \alpha_2 = \alpha_3 = 1$, $\beta_1 = \beta_2 = \beta_3 = \beta_4 = 1$, and

$$B = \begin{pmatrix} 1 \\ 0 \\ 0 \\ 0 \end{pmatrix}.$$

Then,

$$A = \begin{pmatrix} 1 & 1 & 1 & 1 \\ 1 & 0 & 0 & 0 \\ 0 & 1 & 0 & 0 \\ 0 & 0 & 1 & 0 \end{pmatrix}.$$

The eigenvalues of A are -1.9276, -0.7748, $-0.0764 \pm 0.8147j$. The matrix A is not discrete-stable.

Step 1. Choose $\beta = 0.5$

Step 2. The solution Z to the discrete Lyapunov equation $AZA^T - \beta^2 Z = 2BB^T$ is

$$Z = - \begin{pmatrix} -0.0398 & 0.0321 & -0.0003 & 0.0161 \\ 0.0321 & -0.1594 & 0.1294 & -0.0011 \\ -0.0003 & 0.1214 & -0.6376 & 6.5135 \\ 0.0161 & -0.0011 & 6.5135 & -2.5504 \end{pmatrix}.$$

Step 3. $K = (1.2167, 1.0342, 0.9886, 0.9696)$

Verify: The eigenvalues of $A - BK$ are $-0.0742 \pm 0.4259j$, -0.4390, and 0.3708. Thus, $A - BK$ is discrete-stable.

Note: If (A, B) is not a discrete-controllable pair, but is discrete-stabilizable, then we can proceed exactly in the same way as in the continuous-time case to stabilize the pair (A, B).

The following example illustrates how to do this.

MATCONTROL note: Discrete Lyapunov stabilization method as described above has been implemented in MATCONTROL function **stablyapd**.

Example 10.2.4. Let

$$A = \begin{pmatrix} 1 & 2 & 3 \\ 1 & -1 & 1 \\ 0 & 0 & -0.9900 \end{pmatrix}, \qquad B = \begin{pmatrix} 1 \\ 0 \\ 0 \end{pmatrix}.$$

The pair (A, B) is not discrete-controllable, but is discrete-stabilizable.

Using the notations of Section *10.2.2*, we have $\bar{A} = A$, $\bar{B} = B$. The eigenvalues of $\bar{A}$ are

$$\{1.7321, -1.7321, -0.9900\}.$$

$$\bar{A}_{11} = \begin{pmatrix} 1 & 2 \\ 1 & -1 \end{pmatrix}, \quad \bar{B}_1 = \begin{pmatrix} 1 \\ 0 \end{pmatrix}.$$

The pair $(\bar{A}_{11}, \bar{B}_1)$ is controllable.

We now apply the Lyapunov method of discrete-stabilization to the pair $(\bar{A}_{11}, \bar{B}_1)$.

Step 1. Choose $\beta = 1$.

Step 2. The solution Z_1 of the discrete Lyapunov equation: $\bar{A}_{11} Z_1 \bar{A}_1^T - Z_1 = 2\bar{B}_1 \bar{B}_1^T$ is

$$Z_1 = \begin{pmatrix} 0.5 & 0.25 \\ 0.25 & 0.25 \end{pmatrix}.$$

Step 3.

$$\bar{K}_1 = (0, 2.4000).$$

Step 4. The matrix $\bar{A}_{11} - \bar{B}_1 \bar{K}_1$ is discrete-stable. To obtain $\bar{K}$ such that $\bar{A} - \bar{B}\bar{K}$ is discrete-stable, we choose $\bar{K} = (\bar{K}_1, 0)$. The eigenvalue of $\bar{A} - \bar{B}\bar{K}$ are $0.7746, -0.7746, -0.9900$, showing that $\bar{A} - \bar{B}\bar{K}$ is discrete-stable, that is, $A - B\bar{K}$ is discrete-stable.

Remark

- For an efficient implementation of the Lyapunov method for feedback stabilization using the Schur method, see Sima (1981).

10.3 DETECTABILITY

As observability is a dual concept of controllability, a concept dual to stabilizability is called **detectability.**

Definition 10.3.1. *The pair (A, C) is* **detectable** *if there exists a matrix L such that $A - LC$ is stable.*

By duality of Theorem 10.2.1, we can state the following result. The proof is left as an exercise **(Exercise 10.8)**.

Theorem 10.3.1. *Characterization of Continuous-Time Detectability. The following conditions are equivalent:*

(i) *(A, C) is detectable.*

(ii) *The matrix $\begin{pmatrix} A - \lambda I \\ C \end{pmatrix}$ has full column rank for all $\mathrm{Re}(\lambda) \geq 0$.*

(iii) *For all λ and $x \neq 0$ such that $Ax = \lambda x$ and $\mathrm{Re}(\lambda) \geq 0$, we have $Cx \neq 0$*

(iv) *(A^T, C^T) is stabilizable.*

We have seen in Chapter 7 that the controllability and observability play important role in the existence of positive definite and semidefinite solutions of Lyapunov equations.

Similar results, therefore, should be expected involving detectability. We prove one such result in the following.

Theorem 10.3.2. *Detectability and Stability. Let (A, C) be detectable and let the Lyapunov equation:*

$$XA + A^T X = -C^T C \qquad (10.3.1)$$

have a positive semidefinite solution X. Then A is a stable matrix.

Proof. The proof is by contradiction. Suppose that A is unstable. Let λ be an eigenvalue of A with $\text{Re}(\lambda) \geq 0$ and x be the corresponding eigenvector. Then premultiplying the equation (10.3.1) by x^* and postmultiplying it by x, we obtain $2\text{Re}(\lambda)(x^* X x) + x^* C^T C x = 0$. Since $X \geq 0$ and $Re(\lambda) \geq 0$, we must have that $Cx = 0$. This contradicts the fact that (A, C) is detectable. ■

Discrete-Detectability

Definition 10.3.2. *The pair (A, C) is discrete-detectable if there exists a matrix L such that $A - LC$ is discrete-stable.*

Theorems analogous to Theorems 10.3.1 and 10.3.2 also hold in the discrete case. We state the discrete counterpart of Theorem 10.3.1 in Theorem 10.3.3 and leave the proof as an exercise (**Exercise 10.10**).

Theorem 10.3.3. *The following are equivalent:*

(i)   *(A, C) is discrete-detectable.*
(ii)

$$\text{Rank} \begin{pmatrix} A - \lambda I \\ C \end{pmatrix} = n$$

for every λ such that $|\lambda| \geq 1$.
(iii)   *For all λ and $x \neq 0$ such that $Ax = \lambda x$ and $|\lambda| \geq 1$, we have $Cx \neq 0$.*
(iv)   *(A^T, C^T) is discrete-stabilizable.*

10.4   THE EIGENVALUE AND EIGENSTRUCTURE ASSIGNMENT PROBLEMS

We have just seen how an unstable system can be possibly stabilized using feedback control. However, in practical instances, stabilization alone may not be enough.

The stability of the system needs to be monitored and/or the system response needs to be altered. To meet certain design constraints, a designer should be able to choose the feedback matrix such that the closed-loop system has certain transient properties defined by the eigenvalues of the system. We illustrate this with the help of a second-order system.

Consider the second-order system:

$$\ddot{x}(t) + 2\zeta\omega_{n}\dot{x}(t) + \omega_{n}^{2}x(t) = u(t).$$

The poles of this second-order system are of the form: $\lambda_{1,2} = -\zeta\omega_{n} \pm j\omega_{n}\sqrt{1 - \zeta^{2}}$. The quantity ζ is called the **damping ratio** and ω_{n} is called the undamped **natural frequency**. The responses of the dynamical system depends upon ζ and ω_{n}. *In general, for a fixed value of ω_{n}, the larger the value of ζ ($\zeta \geq 1$), the smoother but slower the responses become; on the contrary, the smaller the value of ζ ($0 \leq \zeta < 1$), the faster but more oscillatory the response is.* Figures 10.2 and 10.3 illustrate the situations.

For Figure 10.2, $\omega_{n} = 1$ and $\zeta = 3$. It takes about eight time units to reach the steady-state value 1.

For Figure 10.3, $\omega_{n} = 1$ and $\zeta = 0.5$. The response is much faster as it reaches the steady-state value 1 in about three units time. However, it does not maintain that value; it oscillates before it settles down to 1.

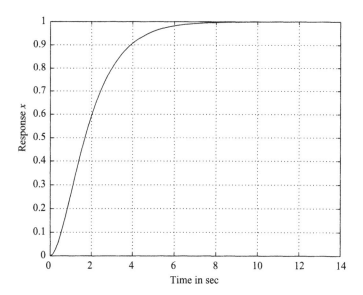

FIGURE 10.2:   Unit step response when $\zeta = 3$ and $\omega_{n} = 1$.

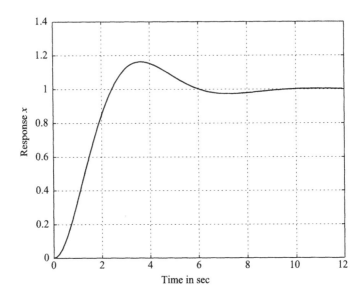

FIGURE 10.3:   Unit step response when $\zeta = 0.5$ and $\omega_n = 1$.

These quantities thus need to be chosen according to a desired transient response. If the poles are close to $j\omega$-axis in the left half s-plane, then the transient responses decay relatively slowly. On the other hand, the poles far away from the $j\omega$-axis cause rapidly decaying time responses. Normally, *"The closed-loop poles for a system can be chosen as a desired pair of dominant second-order poles, with the rest of the poles selected to have real parts corresponding to sufficiently damped modes so that the system will mimic a second-order response with a reasonable control effort"* (Franklin *et al*. 1986). The dominant poles are the poles that have dominant effects on the transient response behavior. As far as transient response is concerned, the poles with magnitudes of real parts at least five times greater than the dominant poles may be considered as insignificant. We give below some illustrative examples.

Case 1. Suppose that it is desired that the closed-loop system response have the minimum decay rate $\alpha > 0$, that is, $\text{Re}(\lambda) \le -\alpha$ for all eigenvalues λ. Then the eigenvalues should lie in the shifted half plane as shown in Figure 10.4.

Case 2. Suppose that it is desired that the system have the minimal damping ratio ζ_{min}. Then the eigenvalues should lie in the sector as shown in Figure 10.5.

Case 3. Suppose that it is desired that the closed-loop system have a minimal undamped frequency ω_{min}. Then the eigenvalues of the closed-loop matrix should lie outside of the following half of the disk: $0 < \omega_{min} \le \omega_n$, as shown in Figure 10.6.

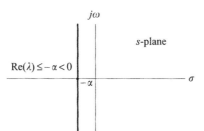

FIGURE 10.4:   The minimum decay rate α of the closed-loop system.

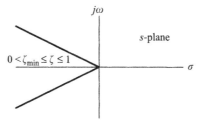

FIGURE 10.5:   Minimal damping ratio ζ of the closed-loop system: the poles lie in the sector $\{\lambda \in \mathbb{C} : |\mathrm{Im}(\lambda)| \leq -\mathrm{Re}(\lambda)\sqrt{\zeta^{-2} - 1}\}$.

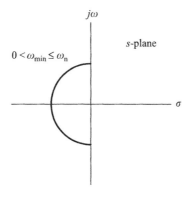

FIGURE 10.6:   The minimal undamped frequency ω_{min} of the closed-loop system: the poles lie in the region $\{\lambda \in \mathbb{C} : |\lambda| \geq \omega_{\mathrm{min}}\}$.

Knowing that to obtain certain transient responses, the eigenvalues of the closed-loop system should be placed in certain specified regions of the complex plane, the question arises: **where should these eigenvalues be placed?** An excellent discussion to this effect is given in the books by Friedland (1986, pp. 243–246) and Kailath (1980, chapter 3).

If the eigenvalues of the closed-loop system are moved far from those of the open-loop system, then from the explicit expression of the feedback vector (to be given later) in the single-input case, it is easily seen that a large feedback f will be required. From the control law,

$$u = v - f^T x(t),$$

it then follows that this would require large control inputs, and there are practical limitations on how large control inputs can be.

Thus, although the eigenvalues have to be moved to stabilize a system, "**the designer should not attempt to alter the dynamic behavior of the open-loop process more than is required**" (Friedland 1986).

10.4.1 Eigenvalue Assignment by State Feedback

The problem of assigning the eigenvalues at certain desired locations in the complex plane using the control law (10.2.2) is called the **EVA Problem by state feedback**. In control theory literature, it is more commonly known as the **pole-placement problem.**

Here is the precise mathematical statement of the problem.

Given $A \in \mathbb{R}^{n \times n}$, $B \in \mathbb{R}^{n \times m}$ $(m \leq n)$, and $\Lambda = \{\lambda_1, \ldots, \lambda_n\}$, where Λ is closed under complex conjugation, find $K \in \mathbb{R}^{m \times n}$ such that

$$\Omega(A - BK) = \Lambda.$$

Here, $\Omega(R)$ stands for the spectrum of R.

The matrix K is called the **state-feedback matrix**.

Theorem 10.4.1 gives the conditions of existence and uniqueness of K.

Theorem 10.4.1. *The State-Feedback EVA Theorem. The EVA problem is solvable for all Λ if and only if (A, B) is controllable. The solution is unique if and only if the system is a single-input system (i.e., if B is a vector). In the multi-input case, if the problem is solvable, there are infinitely many solutions.*

Proof. We first prove the **necessity**. The proof is by contradiction.

Suppose that the pair (A, B) is not controllable. Then according to the eigenvalue criteria of controllability, we have rank$(A - \lambda I, B) < n$ for some λ. Thus there exists a vector $z \neq 0$ such that $z^T(A - \lambda I) = 0, z^T B = 0$. This means that for any K, we have $z^T(A - \lambda I - BK) = 0$, which implies that λ is an eigenvalue of $A - BK$ for every K, and thus λ cannot be reassigned.

Next we prove the **sufficiency.**

Case 1. Let's consider first the **single-input case**. That is, we prove that if (A, b) is controllable, then there exists a unique vector f such that the matrix $A - bf^T$ has the desired spectrum.

Consider the (lower) controller-companion form $(C, \tilde{b})$ of the controllable pair (A, b) (see **Chapter 6**):

$$
TAT^{-1} = C = \begin{pmatrix} 0 & 1 & 0 & \cdots & 0 \\ 0 & 0 & 1 & \cdots & 0 \\ \vdots & \vdots & \vdots & & \vdots \\ 0 & 0 & 0 & \cdots & 1 \\ -a_1 & -a_2 & -a_3 & \cdots & -a_n \end{pmatrix} \tag{10.4.1}
$$

and

$$
\tilde{b} = Tb = \begin{pmatrix} 0 \\ 0 \\ 0 \\ \vdots \\ 1 \end{pmatrix}. \tag{10.4.2}
$$

We now show that there exists a row vector $\hat{f}^T$ such that the closed-loop matrix $C - \tilde{b}\hat{f}^T$ has the desired spectrum.

Let the characteristic polynomial of the desired closed-loop matrix be $d(\lambda) = \lambda^n + d_n \lambda^{n-1} + \cdots + d_1$. Let $\hat{f}^T = (\hat{f}_1, \hat{f}_2, \ldots, \hat{f}_n)$.

Then

$$
C - \tilde{b}\hat{f}^T = \begin{pmatrix} 0 & 1 & 0 & \cdots & 0 \\ 0 & 0 & 1 & \cdots & 0 \\ \cdots & \cdots & \cdots & \cdots & \cdots \\ 0 & 0 & 0 & \cdots & 1 \\ -a_1 - \hat{f}_1 & -a_2 - \hat{f}_2 & \cdots & & -a_n - \hat{f}_n \end{pmatrix}. \tag{10.4.3}
$$

The characteristic polynomial $c'(\lambda)$ of $C - \tilde{b}\hat{f}^T$, then, is $\lambda^n + (a_n + \hat{f}_n)\lambda^{n-1} + \cdots + a_1 + \hat{f}_1$. Comparing the coefficients of $c'(\lambda)$ with those of $d(\lambda)$, we immediately have

$$
\hat{f}_i = d_i - a_i, \quad i = 1, 2, \ldots, n. \tag{10.4.4}
$$

Thus, the vector $\hat{f}$ is completely determined by the coefficients of the characteristic polynomial of the matrix C and the coefficients of the characteristic polynomial of the desired closed-loop matrix. Once the vector $\hat{f}$ is known, the vector f such that the original closed-loop matrix $A - bf^T$ has the desired spectrum, can now be found from the relation:

$$f^{\mathrm{T}} = \hat{f}^{\mathrm{T}}T. \tag{10.4.5}$$

(Note that $\Omega(A - bf^{\mathrm{T}}) = \Omega(TAT^{-1} - Tbf^{\mathrm{T}}T^{-1}) = \Omega(C - \tilde{b}\hat{f}^{\mathrm{T}})$.)

Uniqueness: From the construction of $\hat{f}$, it is clear that $\hat{f}$ is unique. We now show that the uniqueness of $\hat{f}$ implies that of f. The proof is by contradiction.

Suppose there exists $g \neq f$ such that $\Omega(A - bg^{\mathrm{T}}) = \Omega(A - bf^{\mathrm{T}})$. Then $\Omega(C - \tilde{b}\hat{f}^{\mathrm{T}}) = \Omega(C - \tilde{b}\hat{g}^{\mathrm{T}})$, where $\hat{g}^{\mathrm{T}} = g^{\mathrm{T}}T^{-1} \neq \hat{f}^{\mathrm{T}}$, which contradicts the uniqueness of the vector $\hat{f}$.

Case 2. Now we turn to the **multi-input case**. Since (A, B) is controllable, there exists a matrix F and a vector g such that $(A - BF, Bg)$ is controllable (see Chen (1984, p. 344)). Thus, by Case 1, there exists a vector h such that the matrix $A - BF - Bgh^{\mathrm{T}}$ has the desired spectrum.

Then with $K = F + gh^{\mathrm{T}}$, we have that $A - BK$ has the desired spectrum.

Uniqueness: Since the choice of the pair (F, g) is not unique, there exist infinitely many feedback matrices K in the multi-input case. ∎

The Bass–Gura Formula

Note that using the expression of T from Chapter 6, the above expression for f in the single-input case can be written as (**Exercise 10.13**):

$$f = T^{\mathrm{T}}\hat{f} = [(C_{\mathrm{M}}W)^{\mathrm{T}}]^{-1}(d - a), \tag{10.4.6}$$

where d is the vector of the coefficients of the desired characteristic polynomial, a is the vector of the coefficients of the characteristic polynomial of A, C_{M} is the controllability matrix, and W is a certain Toeplitz matrix.

The above formula for f is known as the **Bass–Gura formula** (see Kailath (1980, p. 199)).

Ackermann's Formula (Ackermann 1972)

A closely related formula for the single-input feedback vector f is the well-known Ackermann formula:

$$f = e_{\mathrm{n}}^{\mathrm{T}}C_{\mathrm{M}}^{-1}d(A), \tag{10.4.7}$$

where C_{M} is the controllability matrix and $d(A)$ is the characteristic polynomial of the desired closed-loop matrix.

We also leave the derivation of Ackermann's formula as an exercise *(Exercise 10.14)*.

Notes: We remind the readers again that, since $T = C_{\mathrm{M}}^{-1}$ can be very ill-conditioned, **computing f using the constructive proof of Theorem 10.4.1 or by** the Ackermann or by the Bass–Gura formula can be highly numerically unstable. We will give some numerical examples in **Chapter 11** to demonstrate this.

The MATLAB function **acker** has implemented Ackermann's formula and **comments have been made about the numerical difficulty with this formula in the MATLAB user's manual**.

10.4.2   Eigenvalue Assignment by Output Feedback

Solving the EVA problem using the feedback law (10.2.2) requires knowledge of the full state vector $x(t)$. Unfortunately, in certain situations, the full state is not measurable or it becomes expensive to feedback each state variable when the order of the system is large. In such situations, the feedback law using the output is more practical. Thus, if we define the output feedback law by

$$u(t) = -Ky(t), \qquad y(t) = Cx(t), \tag{10.4.8}$$

we have the closed-loop system

$$\dot{x}(t) = (A - BKC)x(t).$$

The **output feedback EVA problem** then can be defined as follows.

Given the system (10.2.1), find a feedback matrix K such that the matrix $A - BKC$ has a preassigned set of eigenvalues.

The following is a well-known result by Kimura (1975) on the solution of the output feedback problem.

Theorem 10.4.2.   *The Output Feedback EVA Theorem. Let (A, B) be controllable and (A, C) be observable. Let $\mathrm{rank}(B) = m$ and $\mathrm{rank}(C) = r$. Assume that $n \leq r + m - 1$. Then an almost arbitrary set of distinct eigenvalues can be assigned by the output feedback law (10.4.8).*

10.4.3   Eigenstructure Assignment

So far, we have considered the problem of only assigning the eigenvalues. However, **if the system transient response needs to be altered, then the problem of assigning both eigenvalues and eigenvectors needs to be considered.** This can be seen as follows. We have taken the discussion here from Andry *et al.* (1983).

Suppose that the eigenvalues $\lambda_k, k = 1, \ldots, n$ of A are distinct. Let $M = (v_1, \ldots, v_n)$ be the matrix of eigenvectors, which is necessarily nonsingular.

Then every solution $x(t)$ of the system:

$$\dot{x}(t) = Ax(t), \qquad x(0) = x_0,$$

representing a free response can be written as

$$x(t) = \sum_{i=1}^{n} \alpha_i e^{\lambda_i t} v_i,$$

where $\alpha = (\alpha_1, \alpha_2, \ldots, \alpha_n)^{\mathrm{T}} = M^{-1}x_0$.

Thus, from above, we see that the **eigenvalues determine the rate at which the system response decays or grows** and the **eigenvectors determine the shape of the response.**

The problem of assigning both eigenvalues and eigenvectors is called the **eigenstructure assignment problem.**

Formally, the problem is stated as follows:

> Given the sets $S = \{\mu_1, \ldots, \mu_n\}$ and $M = \{v_1, \ldots, v_n\}$ of scalars and vectors, respectively, both closed under complex conjugation, find a feedback matrix K such that the matrix $A + BK$ has the μ_is as the eigenvalues and the v_is as the corresponding eigenvectors.

The following result, due to Moore (1976), gives a necessary and sufficient condition for a solution of the eigenstructure assignment problem by state feedback (see Andry *et al.* (1983) for details and proof).

Define

$$R_\lambda = \begin{bmatrix} N_\lambda \\ M_\lambda \end{bmatrix},$$

where the columns of R_λ form a basis for the null space of the matrix $(\lambda I - A, B)$.

Theorem 10.4.3. *The State-Feedback Eigenstructure Assignment Theorem. Assume that the numbers $\{\mu_i\}$ in the set S are distinct and self-conjugate. Then there exists a matrix K such that $(A + BK)v_i = \mu_i v_i$, $i = 1, \ldots, n$ if and only if the following conditions are satisfied:*

(i)   *The vectors $v_1, \ldots, v_n$ are linearly independent*
(ii) $v_i = v_j^*$ *whenever* $\mu_i = \mu_j^*$, $i = 1, 2, \ldots, n$
(iii) $v_i \in \mathrm{span}\{N_{\mu_i}\}$, $i = 1, 2, \ldots, n$.

If B has full rank and K exists, then it is unique. When μ_is are all real and distinct, an expression for K is

$$K = (-M_{\mu_1}z_1, -M_{\mu_2}z_2, \ldots, -M_{\mu_n}z_n)(v_1, v_2, \ldots, v_n)^{-1},$$

where the vector z_i is given by

$$v_i = N_{\mu_i}z_i, \quad i = 1, 2, \ldots, n.$$

The following result on the eigenstructure assignment by output feedback is due to Srinathkumar (1978).

Theorem 10.4.4. *The Output Feedback Eigenstructure Assignment Theorem. Let (A, B) be controllable and (A, C) be observable. Assume that $\text{rank}(B) = m$ and $\text{rank}(C) = r$. Then $\max(m, r)$ eigenvalues and $\max(m, r)$ eigenvectors with $\min(m, r)$ entries in each eigenvector can be assigned by the output feedback law* (10.4.8).

Note: Numerically effective algorithms for the output feedback problem are rare. *Perhaps, the first comprehensive work in this context is the paper by Misra and Patel (1989), where algorithms for both the single-input and the multi-output systems, using implicit shifts, have been given.* We refer the readers to the above paper for a description of this algorithm.

10.5 THE QUADRATIC OPTIMIZATION PROBLEMS

We have just seen that if a system is controllable, then the closed-loop eigenvalues can be placed at arbitrarily chosen locations of the complex plane. But, the lack of the existence of a definite guideline of where to place these eigenvalues makes the design procedure a rather difficult one in practice. *A designer has to use his or her own intuition of how to use the freedom of choosing the eigenvalues to achieve the design objective.*

It is, therefore, desirable to have a design method that can be used as an initial design process while the designer develops his or her insight.

A "compromise" is often made in practice to obtain such an initial design process. Instead of trying to place the eigenvalues at desired locations, the system is stabilized while satisfying certain performance criterion.

Specifically, the following problem, known as the **Linear Quadratic Optimization Problem**, is solved. The problem is also commonly known as the LQR problem.

10.5.1   The Continuous-Time Linear Quadratic Regulator (LQR) Problem

---

Given matrices Q and R, find a control signal $u(t)$ such that the quadratic cost function $J_C(x) = \int_0^\infty [x^T(t)Qx(t) + u^T(t)Ru(t)] \, dt$ is minimized, subject to $\dot{x} = Ax + Bu, x(0) = x_0$.

The matrices Q and R represent, respectively, weights for the states and the control vectors.

The quadratic form $x^T Q x$ represents the deviation of the state x from the initial state, and the term $u^T R u$ represents the "cost" of control. The matrices Q and R need to be chosen according to the requirements of a specific design. Note that the magnitude of the control signal u can be properly controlled by choosing R appropriately. In fact, by selecting large R, $u(t)$ can be made small (see the expression of the unique control law in Theorem 10.5.1), which is desirable. The choice of Q is related to which states are to be kept small.

Unfortunately, again it is hard to set a specific guideline of how to choose Q and R. "The choice of these quantities is again more of an art than a science" (Kailath (1980, p. 219). For a meaningful optimization problem, however, it is assumed that Q **is symmetric positive semidefinite and R is symmetric positive definite. Unless mentioned otherwise, we will make these assumptions throughout the rest of the chapter.**

The solution of the above problem can be obtained via the solution of a quadratic matrix equation called the ARE, as shown by the following result. See Anderson and Moore (1990) for details.

Theorem 10.5.1.   *The Continuous-Time LQR Theorem. Suppose the pair (A, B) is stabilizable and the pair (A, Q) is detectable. Then there exists a unique optimal control $u^0(t)$ which minimizes $J_C(x)$. The vector $u^0(t)$ is given by $u^0(t) = -Kx(t)$, where $K = R^{-1}B^T X$, and X is the unique positive semidefinite solution of the ARE:*

$$XA + A^T X + Q - XBR^{-1}B^T X = 0. \tag{10.5.1}$$

Furthermore, the closed-loop matrix $A - BK$ is stable and the minimum value of $J_C(x)$ is equal to $x_0^T X x_0$, where $x_0 = x(0)$.

The proof of the existence and uniqueness of the stabilizing solution (under the conditions that (A, B) is stabilizable and (A, Q) is detectable) will be deferred until Chapter 13. Here we give a proof of the optimal control part, assuming that such a solution exists.

Proof.   *Proof of the Optimal Control Part of Theorem 10.5.1*

$$\frac{d}{dt}\left(x^{\mathrm{T}}Xx\right) = \dot{x}^{\mathrm{T}}Xx + x^{\mathrm{T}}X\dot{x},$$

$$= (Ax + Bu)^{\mathrm{T}}Xx + x^{\mathrm{T}}X(Ax + Bu),$$
$$= (u^{\mathrm{T}}B^{\mathrm{T}} + x^{\mathrm{T}}A^{\mathrm{T}})Xx + x^{\mathrm{T}}X(Ax + Bu),$$
$$= x^{\mathrm{T}}(A^{\mathrm{T}}X + XA)x + u^{\mathrm{T}}B^{\mathrm{T}}Xx + x^{\mathrm{T}}XBu,$$
$$= x^{\mathrm{T}}(XBR^{-1}B^{\mathrm{T}}X - Q)x + u^{\mathrm{T}}B^{\mathrm{T}}Xx$$
$$+ x^{\mathrm{T}}XBu \text{ (using (10.5.1)}$$
$$= x^{\mathrm{T}}XBR^{-1}B^{\mathrm{T}}Xx + u^{\mathrm{T}}B^{\mathrm{T}}Xx + x^{\mathrm{T}}XBu + u^{\mathrm{T}}Ru$$
$$- u^{\mathrm{T}}Ru - x^{\mathrm{T}}Qx,$$
$$= (u^{\mathrm{T}} + x^{\mathrm{T}}XBR^{-1})R(u + R^{-1}B^{\mathrm{T}}Xx) - (x^{\mathrm{T}}Qx + u^{\mathrm{T}}Ru)$$

or

$$x^{\mathrm{T}}Qx + u^{\mathrm{T}}Ru = -\frac{d}{dt}(x^{\mathrm{T}}Xx) + (u^{\mathrm{T}} + x^{\mathrm{T}}XBR^{-1})R(u + R^{-1}B^{\mathrm{T}}Xx).$$

Integrating with respect to t from 0 to T, we obtain

$$\int_0^T (x^{\mathrm{T}}Qx + u^{\mathrm{T}}Ru)\,dt$$

$$= -x^{\mathrm{T}}(T)Xx(T) + x_0^{\mathrm{T}}Xx_0 + \int_0^T (u + R^{-1}B^{\mathrm{T}}Xx)^{\mathrm{T}}R(u + R^{-1}B^{\mathrm{T}}Xx)\,dt.$$

(Note that $X = X^{\mathrm{T}} \geq 0$ and $R = R^{\mathrm{T}} > 0$.)

Letting $T \to \infty$ and noting that $x(T) \to 0$ as $T \to \infty$, we obtain

$$J_{\mathrm{C}}(x) = x_0^{\mathrm{T}}Xx_0 + \int_0^\infty (u + R^{-1}B^{\mathrm{T}}Xx)^{\mathrm{T}}R(u + R^{-1}B^{\mathrm{T}}Xx)\,dt$$

Since R is symmetric and positive definite, it follows that $J_{\mathrm{C}}(x) \geq x_0^{\mathrm{T}}Xx_0$ for all x_0 and for all controls u. Since the first term $x_0^{\mathrm{T}}Xx_0$ is independent of u, the minimum value of $J_{\mathrm{C}}(x)$ occurs at

$$u^0(t) = -R^{-1}B^{\mathrm{T}}Xx(t) = -Kx(t).$$

The minimum value of $J_{\mathrm{C}}(x)$ is therefore $x_0^{\mathrm{T}}Xx_0$. ∎

Definition 10.5.1.   *The ARE:*

$$XA + A^{\mathrm{T}}X + Q - XSX = 0, \tag{10.5.2}$$

*where $S = BR^{-1}B^{\mathrm{T}}$ is called the **Continuous-Time Algebraic Riccati Equation** or in short **CARE**.*

Definition 10.5.2. *The matrix H defined by*

$$H = \begin{pmatrix} A & -S \\ -Q & -A^T \end{pmatrix} \tag{10.5.3}$$

is the Hamiltonian matrix associated with the CARE (10.5.2).

Definition 10.5.3. *A symmetric solution X of the CARE such that $A - SX$ is stable is called a* **stabilizing solution.**

Relationship between Hamiltonian Matrix and Riccati Equations

The following theorem shows that there exists a very important relationship between the Hamiltonian matrix (10.5.3) and the CARE (10.5.2). The proof will be deferred until Chapter 13.

Theorem 10.5.2. *Let (A, B) be stabilizable and (A, Q) be detectable. Then the Hamiltonian matrix H in (10.5.3) has n eigenvalues with negative real parts, no eigenvalues on the imaginary axis and n eigenvalues with positive real parts. In this case the CARE (10.5.2) has a unique stabilizing solution X. Furthermore, the closed-loop eigenvalues, that is, the eigenvalues of $A - BK$, are the stable eigenvalues of H.*

A note on the solution of the CARE: It will be shown in **Chapter 13** that the unique stabilizing solution to (10.5.2) can be obtained by constructing an invariant subspace associated with the stable eigenvalues of the Hamiltonian matrix H in (10.5.3). Specifically, if H does not have any imaginary eigenvalue and $\begin{pmatrix} X_1 \\ X_2 \end{pmatrix}$ is the matrix with columns composed of the eigenvectors corresponding to the stable eigenvalues of H, then, assuming that X_1 is nonsingular, the matrix $X = X_2 X_1^{-1}$ is a unique stabilizing solution of the CARE. For details, see **Chapter 13**.

The MATLAB function **care** solves the CARE. The matrix S in CARE is assumed to be nonnegative definite.

The Continuous-Time LQR Design Algorithm

From Theorem 10.5.1, we immediately have the following LQR design algorithm.

Algorithm 10.5.1. *The Continuous-Time LQR Design Algorithm.*
  Inputs. *The matrices A, B, Q, R, and $x(0) = x_0$.*
  Outputs. X—*The solution of the CARE.*
      K—*The LQR feedback gain matrix.*
      $J_{C\,min}$—*The minimum value of the cost function $J_C(x)$.*
  Assumptions.

1. (A, B) *is stabilizable and (A, Q) is detectable.*
2. Q *is symmetric positive semidefinite and R is symmetric positive definite.*

Step 1. *Compute the stabilizing solution X of the CARE:*

$$XA + A^T X - XSX + Q = 0, \quad S = BR^{-1}B^T.$$

Step 2. *Compute the LQR feedback gain matrix:*

$$K = R^{-1}B^T X.$$

Step 3. *Compute the minimum value of $J_C(x)$: $J_{C\min} = x_0^T X x_0$.*

Example 10.5.1 (LQR Design for the Inverted Pendulum). We consider Example 10.2.1 again, with A and B, the same as there and $Q = I_4$, $R = 1$, and

$$x_0 = \begin{pmatrix} 1 \\ 1 \\ 1 \\ 1 \end{pmatrix}.$$

Step 1. The unique positive definite solution X of the CARE (obtained by using MATLAB function **care**) is

$$X = 10^3 \begin{pmatrix} 0.0031 & 0.0042 & 0.0288 & 0.0067 \\ 0.0042 & 0.0115 & 0.0818 & 0.0191 \\ 0.0288 & 0.0818 & 1.8856 & 0.4138 \\ 0.0067 & 0.0191 & 0.4138 & 0.0911 \end{pmatrix}.$$

Step 2. The feedback gain matrix K is

$$K = (-1, -3.0766, -132.7953, -28.7861).$$

Step 3. The minimum value of $J_C(x)$ is 3100.3.

The eigenvalues of $A - BK$ are: $-4.8994, -4.5020, -0.4412 \pm 0.3718j$. Thus, X is the unique positive definite stabilizing solution of the CARE.

(Note that the entries of K in this case are smaller compared to those of K in Example 10.2.1.)

Comparison of Transient Responses with Lyapunov Stabilization

Figures 10.7a and b show the transient responses of the closed-loop systems with: (i) K from Example 10.2.1 and (ii) K as obtained above. The initial condition $x(0) = (5, 0, 0, 0)^T$.

In Figure 10.7a, the transient solutions initially have large magnitudes and then they decay rapidly. In Figure 10.7b, the solutions have smaller magnitudes but the

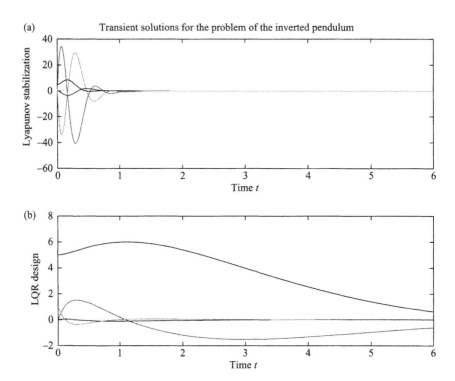

FIGURE 10.7:   Transient responses: (a) Lyapunov Method and (b) LQR design.

decay rate is much slower. The largest magnitude in transient solution in (a) is roughly six times larger than the one in (b). In some dynamical systems, strong initial oscillations in the state components must be avoided, but sometimes a faster stabilization is desired; in other cases, a slow but smooth stabilization is required.

Note that the transient solutions in (a), however, depend upon β and in (b) depend upon Q and R (discussed in the following sections).

Stability and Robustness Properties of the LQR Design

The LQR design has some very favorable stability and robustness properties. We will list some important ones here.

Guaranteed Stability Properties
Clearly, the closed-loop eigenvalues of the LQR design depend upon the matrices Q and R. We will show here how the choice of R affects the closed-loop poles.

Suppose $R = \rho I$, where ρ is a positive scalar. Then, the associated Hamiltonian matrix:

$$H = \begin{pmatrix} A & -\dfrac{1}{\rho}BB^{\mathrm{T}} \\ -Q & -A^{\mathrm{T}} \end{pmatrix}.$$

The closed-loop eigenvalues are the roots with negative real parts of the characteristic polynomial

$$d_{\mathrm{c}}(s) = \det(sI - H).$$

Let $Q = C^{\mathrm{T}}C$. It can be shown that

$$d_{\mathrm{c}}(s) = (-1)^n d(s)d(-s)\det\left[I + \dfrac{1}{\rho}G(s)G^{\mathrm{T}}(-s)\right],$$

where $d(s) = \det(sI - A)$, and $G(s) = C(sI - A)^{-1}B$.

Case 1. Low gain. When $\rho \to \infty$, $u(t) = -(1/\rho)B^{\mathrm{T}}Xx(t) \to 0$. Thus, the LQR controller has low gain. In this case, from the above expression of $d_{\mathrm{c}}(s)$, it follows that

$$(-1)^n d_{\mathrm{c}}(s) \to d(s)d(-s).$$

Since the roots with negative real parts of $d_{\mathrm{c}}(s)$; that is, the closed-loop eigenvalues, are stable, this means that **as ρ increases**:

- the stable open-loop eigenvalues remain stable.
- the unstable ones get reflected across the imaginary axis.
- if any open-loop eigenvalues are exactly on the $j\omega$-axis, the closed-loop eigenvalues start moving just left of them.

Case 2. *High gain.* If $\rho \to 0$, then $u(t)$ becomes large; thus, the LQR controller has high gain.

In this case, for finite s, the closed-loop eigenvalues approach the finite zeros of the system or their stable images.

As $s \to \infty$, the closed-loop eigenvalues will approach zeros at infinity in the so-called stable **Butterworth patterns**. (For a description of Butterworth patterns, see Friedland (1986).) An example is given in Figure 10.8.

These properties, provide good insight into the stability property of LQR controllers and, thus, can be used by a designer as a guideline of where to place the poles.

Robustness Properties of the LQR Design

As we have seen before, an important requirement of a control system design is that the closed-loop system be **robust** to uncertainties due to modeling errors, noise, and disturbances. It is shown below that the LQR design has some desirable robustness properties.

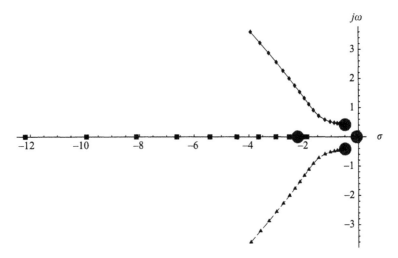

FIGURE 10.8:   Illustration of Butterworth patterns.

The classical robust stability measures are **gain** and **phase margins**, defined with respect to the **Bode plot** (see **Chapter 5**) of a single-input single-output (SISO) system.

The **gain margin** is defined to be the amount of gain increase required to make the loop gain unity where the phase angle is 180°. That is, **it is the reciprocal of the gain at the frequency where the phase angle is** 180°. Thus, the *gain margin is a measure by which the system gain needs to be increased for the closed-loop system to become just unstable.*

Similarly, the difference between the phase of the response and −180° when the loop gain is unity is called the **phase margin.** *That is, the phase margin is the minimum amount of phase lag that is needed to make the system unstable.*

The robustness properties of the LQR design for a multi-input multi-output (MIMO) system can be studied by considering the **return difference matrix:** $I + G_{LQ}(s)$, where $G_{LQ}(s)$ is the LQR loop transfer function matrix given by

$$G_{LQ}(s) = K(sI - A)^{-1}B.$$

The **optimal return difference identity** is:

$$[I+K(-sI-A)^{-1}B]^T R[I+K(sI-A)^{-1}B] = R+B^T(-sI-A)^{-T}Q(sI-A)^{-1}B.$$

or

$$(I+G_{LQ}^T(-s))R(I+G_{LQ}(s)) = R+G^T(-s)QG(s), \quad \text{where } G(s) = (sI-A)^{-1}B.$$

From the above equation, we have

$$(I + G_{LQ}^*(j\omega))R(I + G_{LQ}(j\omega)) \geq R.$$

It has been shown in Safonov and Athans (1977) that if R is a diagonal matrix so that

$$(I + G_{LQ}^*(j\omega))(I + G_{LQ}(j\omega)) \geq I,$$

then there is at least 60° of phase margin in each input channel and the gain margin is infinity. *This means that a phase shift of up to 60° can be tolerated in each of the input channels simultaneously and the gain in each channel can be increased indefinitely without losing stability.* It also follows from above (**Exercise 10.20(b)**) that for all ω,

$$\sigma_{\min}(I + G_{LQ}(j\omega)) \geq 1.$$

This means that the LQ design always results in decreased sensitivity.

See Anderson and Moore (1990, pp. 122–135), the article "Optimal Control" by F.L. Lewis in the **Control Handbook** (1996, pp. 759–778) edited by William Levine, IEEE/CRC Press, and Lewis (1986, 1992), and Maciejowski (1989) for further details.

Example 10.5.2. Consider Example 10.5.1 again. For $\omega = 1$, $G_{LQ}(j\omega) = -1.9700 + 0.5345j$,

$$\sigma_{\min}(1 + G_{LQ}(j\omega)) = 1.1076,$$

$$\sigma_{\min}(1 + G_{LQ}^{-1}(j\omega)) = 0.5426,$$

$$\text{The gain margin} = 0.4907,$$

$$\text{The phase margin} = 60.0998.$$

LQR Stability with Multiplicative Uncertainty

The inequality

$$\sigma_{\min}(I + G_{LQ}(j\omega)) \geq 1$$

also implies (**Exercise 10.20(c)**) that

$$\sigma_{\min}(I + (G_{LQ}(j\omega))^{-1}) \geq \tfrac{1}{2},$$

which means that LQR design remains stable for all unmeasured multiplicative uncertainties Δ in the system for which $\sigma_{\min}(\Delta(j\omega)) \leq \tfrac{1}{2}$.

MATLAB notes: The MATLAB command $[K, S, E] = lqr(A, B, Q, R)$ solves the LQR problem. Here, K—feedback matrix, S—steady-state solution of the CARE, E—the vector containing the closed-loop eigenvalues.

$$SA + A^T S - SBR^{-1}B^T S + Q = 0.$$

The CARE is solved using the generalized **Schur algorithm** to be described in Chapter 13.

MATLAB function **margin** can be used to compute the gain and phase margins of a system.

10.5.2   The Discrete-Time Linear Quadratic Regulator Problem

In the discrete-time case, the function to be minimized is:

$$J_D(x) = \sum_{k=0}^{\infty} (x_k^T Q x_k + u_k^T R u_k). \tag{10.5.4}$$

and the associated ARE is:

$$A^T X A - X + Q - A^T X B (R + B^T X B)^{-1} B^T X A = 0. \tag{10.5.5}$$

The above equation is called the **Discrete-time Algebraic Riccati Equation** (DARE).

A theorem on the existence and uniqueness of the optimal control u_k^0, similar to Theorem 10.5.1, is stated next. For a proof, see Sage and White (1977).

Theorem 10.5.3.   *The Discrete-Time LQR Theorem. Let (A, B) be discrete-stabilizable and (A, Q) be discrete-detectable. Then the optimal control $u_k^0, k = 0, 1, 2, \ldots$, that minimizes $J_D(x)$ is given by $u_k^0 = Kx_k$, where $K = (R + B^T X B)^{-1} B^T X A$, and X is the unique positive semidefinite solution of the DARE (10.5.5). Furthermore, the closed-loop discrete system:*

$$x_{k+1} = (A - BK) x_k$$

is discrete-stable (i.e., all the eigenvalues are strictly inside the unit circle), and the minimum value of $J_D(x)$ is $x_0^T X x_0$, where x_0 is the given initial state.

Definition 10.5.4.   *A symmetric solution X of the DARE that makes the matrix $A - BK$, where $K = (R + B^T X B)^{-1} B^T X A$, discrete-stable is called a **discrete-stabilizing solution** of the DARE.*

Example 10.5.3.

$$A = \begin{pmatrix} -1 & 1 & 1 \\ 0 & -2 & 0 \\ 0 & 0 & -3 \end{pmatrix}, \qquad B = \begin{pmatrix} 1 \\ 2 \\ 3 \end{pmatrix}, \qquad Q = \begin{pmatrix} 1 & 0 & 0 \\ 0 & 1 & 0 \\ 0 & 0 & 1 \end{pmatrix}, \qquad R = 1.$$

The solution X of the DARE (computed using MATLAB function **dlqr**) is:

$$X = 10^3 \begin{pmatrix} 0.0051 & -0.0542 & 0.0421 \\ -0.542 & 1.0954 & -0.9344 \\ 0.0421 & -0.9344 & 0.8127 \end{pmatrix}.$$

The discrete LQR gain matrix:

$$K = (-0.0437, \ 2.5872, \ -3.4543).$$

The eigenvalues of $A - BK$ are: -0.4266, -0.2186, -0.1228. Thus, X is a discrete-stabilizing solution.

MATLAB note: The MATLAB function **lqrd** computes the discrete-time feedback-gain matrix given in Theorem 10.5.3.

10.6 H_∞-CONTROL PROBLEMS

So far we have considered the stabilization of a system ignoring any effect of disturbances in the system. But, we already know that in practice a system is always acted upon by some kind of disturbances. Thus, it is desirable to stabilize perturbed versions of a system, assuming certain bounds for perturbations. This situation gives rise to the well-known "H_∞-**control problem.**"

H_∞-control theory has been the subject of intensive study for the last twenty years or so, since its introduction by Zames (1981). There are now excellent literature in the area: the books by Francis (1987), Kimura (1996), Zhou *et al.* (1996), Green and Limebeer (1995), etc., and the original important papers by Francis and Doyle (1987), Doyle *et al.* (1989), etc.

Let $\sigma_{\max}(M)$ and $\sigma_{\min}(M)$ denote, respectively, the largest and smallest singular value of M.

Definition 10.6.1.   *The H_∞-norm of the stable transfer function $G(s)$, denoted by $\|G\|_\infty$, is defined by*

$$\|G\|_\infty = \sup_{\omega \in \mathbb{R}} \sigma_{\max}(G(j\omega)).$$

In the above definition, "sup" means the supremum or the least upper bound of the function $\sigma_{\max}(G(j\omega))$.

Physical Interpretation of the H_∞-norm

Consider the system:

$$y(s) = G(s)u(s).$$

When the system is driven with a sinusoidal input of unit magnitude at a specific frequency, $\sigma_{\max}(G(j\omega))$ is the largest possible size of the output for the corresponding sinusoidal input. Thus, the H_∞-norm gives the largest possible amplification over all frequencies of a unit sinusoidal input.

A detailed discussion of H_∞ control problems is beyond the scope of this book. The original formulation was in an input/output setting. However, due to its computational attractiveness, the recent state-space formulation has become more popular. We only state two **simplified versions** of the state-space formulations of H_∞-control problems, and mention their connections with AREs. First, we prove the following well-known result that shows how the H_∞-norm of a stable transfer function matrix is related to an ARE or equivalently, to the associated Hamiltonian matrix.

Define the **Hamiltonian matrix** out of the coefficients of the matrices of the system (10.2.1)

$$M_\gamma = \begin{pmatrix} A + BR^{-1}D^T C & BR^{-1}B^T \\ -C^T(I + DR^{-1}D^T)C & -(A + BR^{-1}D^T C)^T \end{pmatrix}, \qquad (10.6.1)$$

where $R = \gamma^2 I - D^T D$.

Theorem 10.6.1.   *Let $G(s)$ be a stable transfer function and let $\gamma > 0$. Then $\|G\|_\infty < \gamma$ if and only if $\sigma_{\max}(D) < \gamma$ and M_γ defined by (10.6.1) has no imaginary eigenvalues.*

Proof.   We sketch the proof in case $D = 0$. This proof can easily be extended to the case when $D \neq 0$, and is left as an exercise (**Exercise 10.23**). Without any loss of generality, assume that $\gamma = 1$. Otherwise, we can scale G to $\gamma^{-1}G$ and B to $\gamma^{-1}B$ and work with scaled G and B (note that $\|G\|_\infty < \gamma$ if and only if $\|\gamma^{-1}G\|_\infty < 1$).

Since $\gamma = 1$ and $D = 0$, we have $R = I$. Using the notation:

$$G(s) = C(sI - A)^{-1}B \equiv \left[\begin{array}{c|c} A & B \\ \hline C & 0 \end{array} \right],$$

an easy computation shows that if

$$\Gamma(s) = I - G(-s)^T G(s), \qquad (10.6.2)$$

then

$$\Gamma^{-1}(s) = \left[\begin{array}{cc|c} A & BB^T & B \\ -C^T C & -A^T & 0 \\ \hline 0 & B^T & I \end{array}\right] = \left[\begin{array}{c|c} M_\gamma & B \\ & 0 \\ \hline 0 \quad B^T & I \end{array}\right]. \qquad (10.6.3)$$

Therefore, from above it follows that M_γ does not have an eigenvalue on the imaginary axis if and only if $\Gamma^{-1}(s)$ does not have any pole there. We now show that this is true if and only if $\|G\|_\infty$ is less than 1.

If $\|G\|_\infty < 1$, then $I - G(j\omega)^* G(j\omega) > 0$ for every ω, and hence $\Gamma^{-1}(s) = (I - G(-s)^T G(s))^{-1}$ does not have any pole on the imaginary axis. On the other hand, if $\|G\|_\infty \geq 1$, then $\sigma_{\max}(G(j\omega)) = 1$ for some ω, which means that 1 is an eigenvalue of $G(j\omega)^* G(j\omega)$, implying that $I - G(j\omega)^* G(j\omega)$ is singular. ∎

The following simple example from Kimura (1996, p. 41) illustrates Theorem 10.6.1.

Example 10.6.1. Let

$$G(s) = \frac{1}{s + \alpha}, \quad \alpha > 0.$$

The associated Hamiltonian matrix

$$H = \begin{pmatrix} -\alpha & 1 \\ -1 & \alpha \end{pmatrix}.$$

Then H does not have any imaginary eigenvalue if and only if $\alpha > 1$.
Since $\|G\|_\infty = \sup_\omega 1/\sqrt{\omega^2 + \alpha^2} = 1/\alpha$, we have, for $\alpha > 1$, $\|G\|_\infty < 1$.

10.6.1 Computing the H_∞-Norm

A straightforward way to compute the H_∞-norm is:

1. Compute the matrix $G(j\omega)$ using the Hessenberg method described in Chapter 5.
2. Compute the largest singular value of $G(j\omega)$.
3. Repeat steps 1 and 2 for many values of ω.

Certainly the above approach is impractical and computationally infeasible.

However, Theorem 10.6.1 gives us a more practically feasible method for computing the H_∞-norm. The method, then, will be as follows:

1. Choose γ.
2. Test if $\|G\|_\infty < \gamma$, by computing the eigenvalues of M_γ and seeing if the matrix M_γ has an imaginary eigenvalue.
3. Repeat, by increasing or decreasing γ, accordingly as $\|G\|_\infty < \gamma$ or $\|G\|_\infty \geq \gamma$.

The following bisection method of Boyd *et al.* (1989) is an efficient and systematic implementation of the above idea.

Algorithm 10.6.1. *The Bisection Algorithm for Computing the H_∞-**Norm***
   Inputs. *The system matrices A, B, C, and D, of dimensions $n \times n$, $n \times m$, $r \times n$, and $r \times m$, respectively.*

γ_{lb}—*A lower bound of the H_∞-norm*
γ_{ub}—*An upper bound of the H_∞-norm*
$\epsilon (> 0)$—*Error tolerance.*
   Output. *An approximation of the H_∞-norm with a relative accuracy of ϵ.*
   Assumption. *A is stable.*
   Step 1. *Set $\gamma_L \equiv \gamma_{lb}$, and $\gamma_U = \gamma_{ub}$*
   Step 2. *Repeat until $\gamma_U - \gamma_L \leq 2\epsilon\gamma_L$*
      Compute $\gamma = (\gamma_L + \gamma_U)/2$
      Test if M_γ defined by (10.6.1) has an imaginary eigenvalue
      If not, set $\gamma_U = \gamma$
      Else, set $\gamma_L = \gamma$.

Remark

- After k iterations, we have $\gamma_U - \gamma_L = 2^{-k}(\gamma_{ub} - \gamma_{lb})$. Thus, on exit, the algorithm is guaranteed to give an approximation of the H_∞-norm with a relative accuracy of ϵ.

Convergence: The convergence of the algorithm is **linear** and is independent of the data matrices A, B, C, and D.
Note: An algorithm equivalent to the above bisection algorithm was also proposed by Robel (1989).

Remark

- The condition that M_γ does not have an imaginary eigenvalue (in Step 2) can also be expressed in terms of the associated Riccati equation:

$$XA + A^TX + \gamma^{-1}XBR^{-1}B^TX + \gamma^{-1}C^TC = 0$$

(Assuming that $D = 0$.)

Example 10.6.2.

$$A = \begin{pmatrix} -1 & 2 & 3 \\ 0 & -2 & 0 \\ 0 & 0 & -4 \end{pmatrix}, \quad B = \begin{pmatrix} 1 \\ 0 \\ 0 \end{pmatrix}, \quad C = (1, \ 1, \ 1), \quad D = 0, \quad \epsilon = 0.0014.$$

Step 1. $\gamma_L = 0.2887$, $\gamma_U = 1.7321$.
Step 2.
Iteration 1

$$\gamma = 1.0104.$$

The eigenvalues of M_γ are $\{2, \quad 4, \quad -0.1429, \quad 0.1429 \; -2.0000, \quad -4,0000\}$. Since M_γ does not have purely imaginary eigenvalues, we continue.

$$\gamma_L = 0.28867, \qquad \gamma_U = 1.0103.$$

Iteration 2

$$\gamma = 0.6495.$$

The eigenvalues of M_γ are $\{2, \quad 4, \quad -2, \quad -4, \quad -0 \pm 1.1706j\}$.
Since M_γ has a purely imaginary eigenvalue, we set

$$\gamma_L = 0.6495, \qquad \gamma_U = 1.0103,$$

Iteration 3

$$\gamma = 0.8299.$$

The eigenvalues of M_γ are $\{2, \quad 4, \quad -2, \quad -4, \quad 0 \pm 0.6722j\}$. Since M_γ has a purely imaginary eigenvalue, we set

$$\gamma_L = 0.8299, \qquad \gamma_U = 1.0103.$$

The values of γ at successive iterations are found to be 0.9202, 0.9653, 0.9878, 0.9991, 0.9998, and 1; and the iterations terminated at this point satisfying the stopping criterion. Thus, we take H_∞-norm $= 1$.

Computing γ_{lb} and γ_{ub}: For practical implementation of the above algorithm, we need to know how to compute γ_{lb} and γ_{ub}. We will discuss this aspect now.

Definition 10.6.2.   *The* **Hankel singular values** *are the square roots of the eigenvalues of the matrix $C_G O_G$, where C_G and O_G are the controllability and observability Grammians, respectively.*

The bounds γ_{lb} and γ_{ub} may be computed using the *Hankel singular values* as follows:

$$\gamma_{lb} = \max\{\sigma_{max}(D), \sigma H_1\},$$

$$\gamma_{ub} = \sigma_{max}(D) + 2\sum_{i=1}^{n} \sigma H_i, \qquad (10.6.4)$$

where σH_ks are the ordered **Hankel singular values,** that is, σH_i is the ith largest Hankel singular value. These bounds are due to Enns (1984) and Glover (1984) and the formula (10.6.4) is known as the **Enns–Glover formula**.

A scheme for computing γ_{lb} and γ_{ub} will then be as follows:

1. Solve the Lyapunov equations (7.2.11) and (7.2.12) to obtain C_G and O_G, respectively.
2. Compute the eigenvalues of $C_G O_G$.
3. Obtain the Hankel singular values by taking the square roots of the eigenvalues of $C_G O_G$.
4. Compute γ_{lb} and γ_{ub} using the above Enns–Glover formula.

As an alternative to eigenvalue computations, one can also use the following formulas:

$$\gamma_{lb} = \max\{\sigma_{max}(D), \sqrt{\text{Trace}(C_G O_G)/n}\},$$
$$\gamma_{ub} = \sigma_{max}(D) + 2\sqrt{n\text{Trace}(C_G O_G)}.$$

Remark

- Numerical difficulties can be expected in forming the product $C_G O_G$ explicitly.

MATCONTROL note: Algorithm 10.6.1 has been implemented in MATCONTROL function **hinfnrm**.

The Two-Step Algorithm

Recently, Bruinsma and Steinbuch (1990) have developed a **"two-step"** algorithm to compute H_∞-norm of $G(s)$. Their algorithm is believed to be faster than the bisection algorithm just stated. **The convergence is claimed to be quadratic.**

The algorithm is called a "two-step" algorithm, because, the algorithm starts with some lower bound of $\gamma < \|G\|_\infty$, as the first step and then in the second step, this lower bound is iteratively improved and the procedure is continued until some "tolerance" is satisfied.

A New Lower Bound of the H_∞-norm

The two-step algorithm, like the bisection algorithm, requires a starting value for γ_{lb}. The Enns–Glover formula can be used for this purpose. However, the authors have proposed that the following starting value for γ_{lb} be used:

$$\gamma_{lb} = \max\{\sigma_{\max}(G(0)), \sigma_{\max}(G(j\omega_p)), \sigma_{\max}(D)\},$$

where $\omega_p = |\lambda_i|$, λ_i a pole of $G(s)$ with λ_i selected to maximize

$$\left| \frac{\text{Im}(\lambda_i)}{\text{Re}(\lambda_i)} \frac{1}{|\lambda_i|} \right|,$$

if $G(s)$ has poles with $\text{Im}(\lambda_i) \neq 0$ or to minimize $|\lambda_i|$, if $G(s)$ has only real poles.

Algorithm 10.6.2.   *The Two-Step Algorithm for Computing the H_∞-norm*
Inputs. *The system matrices A, B, C, and D, respectively, of dimensions $n \times n, n \times m, r \times n$, and $r \times m$. ϵ–error tolerance.*
Output. *An approximate value of the H_∞-norm.*
Assumption. *A is stable.*
Step 1. *Compute a starting value for γ_{lb}, using the above criterion.*
Step 2. *Repeat*
  2.1 *Compute $\gamma = (1 + 2\epsilon)\gamma_{lb}$.*
  2.2 *Compute the eigenvalues of M_γ with the value of γ computed in Step 2.1. Label the purely imaginary eigenvalues of M_γ as $\omega_1, \ldots, \omega_k$.*
  2.3 *If M_γ does not have a purely imaginary eigenvalue, set $\gamma_{ub} = \gamma$ and stop.*
    2.4 *For $i = 1, \ldots, k - 1$ do*
        (a) Compute $m_i = \frac{1}{2}(\omega_i + \omega_{i+1})$.
        (b) Compute the singular values of $G(jm_i)$.
      Update $\gamma_{lb} = \max_i(\sigma_{\max}(G(jm_i)))$.

    End
  Step 3. $\|G\|_\infty = \frac{1}{2}(\gamma_{lb} + \gamma_{ub})$.

MATLAB note: Algorithm 10.6.2 has been implemented in MATLAB Control System tool box. The usage is: **norm** (sys, inf).

In the above, "sys" stands for the linear time-invariant system in the matrices A, B, C, and D. "sys" can be generated as follows:

$A = [\ ],$ $B = [\ ],$ $C = [\ ],$ $D = [\ ],$ sys $= ss(A, B, C, D).$

Remark

- Boyd and Balakrishnan (1990) have also proposed an algorithm similar to the above "two-step" algorithm. Their algorithm converges quadratically. Algorithm 10.6.2 is also believed to converge quadratically, but no proof has been given. See also the paper by Lin *et al.* (1999) for other recent reliable and efficient methods for computing the H_∞-norms for both the state and output feedback problems.

Connection between H_∞-norm and the Distance to Unstable Matrices

Here we point out a connection between the H_∞-norm of the resolvent of A and $\beta(A)$, the distance to the set of unstable matrices. The proof is easy and left as an [**Exercise 10.27**].

Theorem 10.6.2. *Let A be a complex stable matrix, then $\beta(A) = \|(sI - A)^{-1}\|_\infty^{-1}$.*

Computing the H_∞-Norm of a Discrete-Stable Transfer Function Matrix

Let $M(z) = C(zI - A)^{-1}B$ be the transfer function matrix of the **asymptotically stable discrete-time system:**

$$x_{k+1} = Ax_k + Bu_k,$$
$$y_k = Cx_k.$$

Then

Definition 10.6.3. *The H_∞-norm of $M(z)$ is defined as*

$$\|M(z)\|_\infty = \sup_{|z| \geq 1} \sigma_{\max}(M(z)).$$

The following is a discrete-analog of Theorem 10.6.1. We state the result here without proof. For proof, we refer the readers to the book by Zhou *et al.* (1996, pp. 547–548).

Theorem 10.6.3. *Let*

$$S = \begin{pmatrix} A - BB^T(A^T)^{-1}C^TC & BB^T(A^T)^{-1} \\ -(A^T)^{-1}C^TC & (A^T)^{-1} \end{pmatrix}$$

be the symplectic matrix associated with the above stable discrete-time system. Assume that A is nonsingular and that the system does not have any uncontrollable and unobservable modes on the unit circle.

Then $\|M(z)\|_\infty < 1$ if and only if S has no eigenvalues on the unit circle and $\|C(I - A)^{-1}B\|_2 < 1$.

Computing H_∞-Norm of a Discrete-Stable System

The above theorem can now be used as a basis to develop a **bisection algorithm**, analogous to Algorithm 10.6.1, for computing the H_∞-norm of a discrete stable system. We leave this as an exercise (**Exercise 10.24**).

10.6.2 H_∞-Control Problem: A State-Feedback Case

Consider the following linear control system:

$$\dot{x}(t) = Ax(t) + Bu(t) + Ed(t), \quad x(0) = 0$$
$$z(t) = Cx(t) + Du(t). \tag{10.6.5}$$

Here $x(t)$, $u(t)$, and $z(t)$, denote the state, the control input, and controlled output vectors. The vector $d(t)$ is the disturbance vector. The matrices A, B, C, D, and E are matrices of appropriate dimensions. Suppose that a state-feedback control law

$$u(t) = Kx(t)$$

is applied to the system. Then the closed-loop system becomes:

$$\dot{x}(t) = (A + BK)x(t) + Ed(t)$$
$$z(t) = (C + DK)x(t). \tag{10.6.6}$$

The transfer function from d to z is:

$$T_{zd}(s) = (C + DK)(sI - A - BK)^{-1}E. \tag{10.6.7}$$

Suppose that the influence of the disturbance vector $d(t)$ on the output $z(t)$ is measured by the H_∞-norm of $T_{zd}(s)$. **The goal of the state feedback H_∞ control problem is to find a constant feedback matrix K such that the closed-loop system is stable and the H_∞-norm of the transfer matrix $T_{zd}(s)$ is less than a prescribed tolerance.**

Specifically, the state feedback H_∞ problem is stated as follows:

> Given a positive real number γ, find a real $m \times n$ matrix K such that the closed-loop system is stable and that $\|T_{zd}(s)\|_\infty < \gamma$.

Thus, by solving the above problem, one will stabilize perturbed versions of the original system, as long as the size of the perturbations does not exceed a certain given tolerance.

The following theorem due to Doyle *et al.* (1989) states a solution of the above problem in terms of the solution of an ARE.

Theorem 10.6.4.   *A State-Feedback H_∞ Theorem. Let the pair (A, C) be observable and the pairs (A, B), and (A, E) be stabilizable. Assume that $D^T D = I$, and $D^T C = 0$. Then the H_∞ control problem (as stated above) has a solution if and only if there exists a positive semi-definite solution X of* *the ARE:*

$$A^T X + XA - X\left(BB^T - \frac{1}{\gamma^2} EE^T\right) X + C^T C = 0, \qquad (10.6.8)$$

such that

$$A + \left(\frac{1}{\gamma^2} EE^T - BB^T\right) X$$

is stable. In this case one such state feedback matrix K is given by

$$K = -B^T X. \qquad (10.6.9)$$

Proof.   The proof follows by noting the relationship between the ARE (10.6.8) and the associated Hamilton matrix:

$$H_\gamma = \begin{pmatrix} A & \frac{1}{\gamma^2} EE^T - BB^T \\ -C^T C & -A^T \end{pmatrix}, \qquad (10.6.10)$$

as stated in Theorem 10.5.2, and then applying Theorem 10.6.1 to the transfer function matrix $T_{zd}(s)$.   ∎

Notes

1.   The application of Theorem 10.6.1 to $T_{zd}(s)$ amounts to replacing A, B, C, and R of Theorem 10.6.1 as follows:

$$A \to A + BK = A - BB^T X,$$
$$C \to C + DK = C - DB^T X,$$
$$B \to E,$$
$$R \to \gamma^2 I - I = (\gamma^2 - 1)I,$$

and using the assumptions $D^T D = I$ and $D^T C = 0$.

2.   The Riccati equation (10.6.8) is not the standard LQ Riccati equation, the CARE (Eq. (10.5.2)), because the term $(BB^T - (1/\gamma^2)EE^T)$ may be indefinite for certain values of γ.

   However, when $\gamma \to \infty$, the Riccati equation (10.6.8) becomes the CARE with $R = I$:

$$XA + A^T X - XBB^T X + C^T C = 0.$$

3.   It can be shown (**Exercise 10.26**) that if H_γ has imaginary eigenvalues, then the H_∞ problem as defined above does not have a solution.

In a more realistic situation when a dynamic measurement feedback is used, rather than the state feedback as used here, the solution of the corresponding H_∞-control problem is provided by a **pair of AREs.** Details can be found in the pioneering paper of Doyle *et al.* (1989), and in the recent books by Kimura (1996), Green and Limebeer (1995), Zhou *et al.* (1996). We only state the result from the paper of Doyle *et al.* (1989).

10.6.3   The H_∞-Control Problem: Output Feedback Case

Consider a system described by the state-space equations

$$
\begin{aligned}
\dot{x}(t) &= Ax(t) + B_1 w(t) + B_2 u(t), \\
z(t) &= C_1 x(t) + D_{12} u(t), \\
y(t) &= C_2 x(t) + D_{21} w(t),
\end{aligned}
\qquad (10.6.11)
$$

where $x(t)$—the state vector, $w(t)$—the disturbance signal, $u(t)$—the control input, $z(t)$— the controlled output, and $y(t)$—the measured output.

The transfer function from the inputs $\begin{bmatrix} w \\ u \end{bmatrix}$ to the outputs $\begin{bmatrix} z \\ y \end{bmatrix}$ is given by

$$
G(s) = \begin{pmatrix} 0 & D_{12} \\ D_{21} & 0 \end{pmatrix} + \begin{pmatrix} C_1 \\ C_2 \end{pmatrix}(sI - A)^{-1}(B_1, B_2) = \begin{bmatrix} G_{11} & G_{12} \\ G_{21} & G_{22} \end{bmatrix}.
$$

Define a feedback controller $K(s)$ by $u = K(s)y$.

Then the closed-loop transfer function matrix $T_{zw}(s)$ from the disturbance w to the output z is given by

$$
T_{zw}(s) = G_{11} + G_{12}K(I - G_{22}K)^{-1}G_{21}.
$$

Then the goal of the output feedback H_∞-control problem in this case is to find a controller $K(s)$ that $\|T_{zw}(s)\|_\infty < \gamma$, for a given positive number γ.

Figure 10.9, P is the linear system described by 10.6.11.

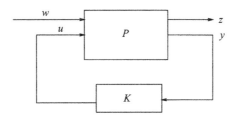

FIGURE 10.9:   Output feedback H_∞ configuration.

A solution of the above H_∞-control problem is given in the following theorem. The following **assumptions** are made:

(i) (A, B_1) is stabilizable and (A, C_1) is detectable.   (10.6.12)

(ii) (A, B_2) is stabilizable and (A, C_2) is detectable.   (10.6.13)

(iii) $D_{12}^T(C_1, D_{12}) = (0, I)$ (10.6.14)

(iv) $\begin{pmatrix} B_1 \\ D_{21} \end{pmatrix} D_{21}^T = \begin{pmatrix} 0 \\ I \end{pmatrix}$.   (10.6.15)

Here I stands for an identity matrix of appropriate dimension.

Theorem 10.6.5. *An Output Feedback H_∞ Theorem. Under the assumptions (10.6.12–10.6.15), the output feedback H_∞-control problem as stated above has a solution if and only if there exist unique symmetric positive semidefinite stabilizing solutions X and Y, respectively, to the pair of AREs*

$$XA + A^TX - X\left(B_2B_2^T - \frac{1}{\gamma^2}B_1B_1^T\right)X + C_1^TC_1 = 0, \qquad (10.6.16)$$

$$AY + YA^T - Y\left(C_2^TC_2 - \frac{1}{\gamma^2}C_1^TC_1\right)Y + B_1B_1^T = 0, \qquad (10.6.17)$$

and $\rho(XY) < \gamma^2$, where $\rho(XY)$ is the spectral radius of the matrix XY.

Furthermore, in this case, one such controller is given by the transfer function

$$K(s) = -F(sI - \hat{A})^{-1}ZL, \qquad (10.6.18)$$

where

$$\hat{A} = A + \frac{1}{\gamma^2}B_1B_1^TX + B_2F + ZLC_2 \qquad (10.6.19)$$

and

$$F = -B_2^TX, \qquad L = -YC_2^T, \qquad Z = \left(I - \frac{1}{\gamma^2}YX\right)^{-1} \qquad (10.6.20)$$

Proof. For a proof of Theorem 10.6.5, we refer the readers to the original paper of Doyle *et al.* (1989). ∎

Notes

1. The second Riccati equation is dual to the first one and is of the type that arises in the Kalman filter (**Chapter 12**).
2. A general solution without the assumptions (10.6.14) and (10.6.15) can be found in Glover and Doyle (1988).

MATLAB Note: To solve the Riccati equations (10.6.16) and (10.6.17) using **care**, these equations have to be rewritten in the usual **care** format. For example, Eq. (10.6.16) can be rewritten as:

$$A^T X + XA - X(B_1, B_2) \begin{pmatrix} -\nu^{-2}I & 0 \\ 0 & I \end{pmatrix}^{-1} \begin{pmatrix} B_1^T \\ B_2^T \end{pmatrix} X + C_1^T C_1 = 0.$$

Example 10.6.3. Zhou *et al.* (1996, pp. 440–443). Let

$$A = a, \quad B_1 = (1, 0), \quad B_2 = b_2,$$

$$C_1 = \begin{pmatrix} 1 \\ 0 \end{pmatrix}, \quad D_{12} = \begin{pmatrix} 0 \\ 1 \end{pmatrix}$$

$$C_2 = c_2, \quad D_{21} = (0, 1).$$

Then

$$D_{12}^T(C_1, D_{12}) = (0, 1) \begin{pmatrix} 1 & 0 \\ 0 & 1 \end{pmatrix} = (0, 1)$$

and

$$\begin{pmatrix} B_1 \\ D_{21} \end{pmatrix} D_{21}^T = \begin{pmatrix} 1 & 0 \\ 0 & 1 \end{pmatrix} \begin{pmatrix} 0 \\ 1 \end{pmatrix} = \begin{pmatrix} 0 \\ 1 \end{pmatrix}.$$

Thus, the conditions (10.6.14) and (10.6.15) are satisfied.

Let $a = -1, b_2 = c_2 = 1$. Let $\gamma = 2$.

Then it is easy to see that

$$\rho(XY) < \gamma^2,$$

$$T_{zw} = \begin{pmatrix} -1.7321 & 1 & -0.7321 \\ 1 & 0 & 0 \\ -0.7321 & 0 & -0.7321 \end{pmatrix},$$

and $\|T_{zw}\|_\infty = 0.7321 < \gamma = 2$.

Optimal H_∞ Control

The output H_∞-control problem in this section is usually known as the **Suboptimal H_∞-Control** problem.

Ideally, we should have considered **Optimal H_∞-Control problem:**

Find all admissible controllers $K(s)$ such that $\|T_{zw}\|_\infty$ is minimized.

Finding an optimal H_∞ controller is demanding both theoretically and computationally and, in practice, very often a suboptimal controller is enough, because suboptimal controllers are close to the optimal ones.

The behavior of the suboptimal H_∞ controller solution as γ approaches the infimal achievable norm, denoted by γ_{opt}, is discussed in the book by Zhou *et al.* (1996, pp. 438–443). It is shown there that for Example 10.6.3, $\gamma_{opt} = \|T_{zw}\|_\infty = 0.7321$.

10.7 THE COMPLEX STABILITY RADIUS AND RICCATI EQUATION

Assume in this section that A, B, and C are **complex matrices.** In Chapter 7, we introduced the concept of stability radius in the context of robust stabilization of the stable system $\dot{x} = Ax$ under structured perturbations of the form $B\Delta C$. The system:

$$\dot{x} = (A + B\Delta C)x \qquad (10.7.1)$$

may be interpreted as a closed-loop system obtained by applying the output feedback (with unknown feedback matrix Δ) to the system (10.7.3) given below. Thus, the stability radius is related to the output feedback stabilization, as well.

In this section, we will discuss the role of the complex stability radius $r_{\mathbb{C}}(A, B, C)$ in certain parametric optimization problems.

Consider the following parametric optimization problem: Minimize

$$J_\rho(x) = \int_0^\infty \left[\|u(t)\|^2 - \rho\|y(t)\|^2\right] dt \qquad (10.7.2)$$

subject to

$$\dot{x} = Ax + Bu, \qquad y = Cx. \qquad (10.7.3)$$

If $\rho \le 0$, then we have the usual LQR problem, which can be solved by solving the associated Riccati equation, as we have just seen. We will now show that for certain other values of $\rho > 0$, the above optimization problem is still solvable, by relating ρ to the stability radius. The key to that is to show the existence of a stabilizing solution of the associated Riccati equation for a given value of ρ.

Before we state the main result, we note the following, that shows that for certain values of ρ, the minimization cost is finite. **For simplicity, we will write** $r_{\mathbb{C}}(A, B, C)$ **as** r.

Theorem 10.7.1.   *Let $J_\rho(x)$ be defined by (10.7.2). Then*

(i) *$Inf\, J_\rho(0) = 0$, if and only if $\rho \le r^2$, if and only if $I - \rho G^*(i\omega)G(i\omega) \ge 0$, for all $\omega \in \mathbb{R}$.*

(ii) *Suppose A is stable and $r < \infty$. Then for all $\rho \in (-\infty, r^2)$, we have $|\inf J_\rho(x_0)| <, \infty$.*

Proof.   See Hinrichsen and Pritchard (1986a).   ∎

The ARE associated with the above minimization problem is

$$XA + A^*X - \rho C^*C - XBB^*X = 0. \tag{10.7.4}$$

Since this equation is dependent on ρ, we denote this Riccati equation by ARE_ρ. The parameterized Hamiltonian matrix associated with the ARE_ρ is

$$H_\rho = \begin{pmatrix} A & -BB^* \\ \rho C^*C & -A^* \end{pmatrix}. \tag{10.7.5}$$

The following theorems characterize $r_{\mathbb{C}} (= r)$ in terms of H_ρ.

Theorem 10.7.2. *Characterization of the Complex Stability Radius. Let H_ρ be defined by (10.7.5). Then $\rho < r$ if and only if H_{ρ^2} does not have an eigenvalue on the imaginary axis.*

Proof. See Hinrichsen and Pritchard (1986a). ■

Example 10.7.1. Consider Example 7.8.1.

$$A = \begin{pmatrix} 0 & 1 \\ -1 & -1 \end{pmatrix}, \qquad B = \begin{pmatrix} 0 \\ -1 \end{pmatrix}, \qquad C = (1, 0).$$

From Example 7.8.1, we know that $r^2 = \frac{3}{4}$.
Case 1. Let $\rho = 0.5 < r_{\mathbb{C}} = r = 0.8660$. Then,

$$H_{\rho^2} = \begin{pmatrix} 0 & 1 & 0 & 0 \\ -1 & -1 & 0 & -1 \\ 0.25 & 0 & 0 & 1 \\ 0 & 0 & -1 & 1 \end{pmatrix}.$$

The eigenvalues of H_{ρ^2} are $-0.4278 \pm 0.8264j, 0.4278 \pm 0.8264j$. Thus, H_{ρ^2} does not have a purely imaginary eigenvalue.
Case 2. Let $\rho = 1 > r_{\mathbb{C}} = r = 0.8660$. Then,

$$H_{\rho^2} = \begin{pmatrix} 0 & 1 & 0 & 0 \\ -1 & -1 & 0 & -1 \\ 1 & 0 & 0 & 1 \\ 0 & 0 & -1 & 1 \end{pmatrix}.$$

The eigenvalues of H_{ρ^2} are $0.0000 \pm 1.0000j, 0, 0$, which are purely imaginary. Therefore we obtain an upper and a lower bound for r: $0.5 < r \le 1$.

We have already mentioned the relationship between a Hamiltonian matrix and the associated ARE. In view of Theorem 10.7.2, therefore, the following result is not surprising. The proof of Theorem 10.7.3 has been taken from Hinrichsen and Pritchard (1986b).

Theorem 10.7.3.   *Stability Radius and ARE. Let A be a complex stable matrix and let $r \equiv r_{\mathbb{C}}(A, B, C) < \infty$. Let $\rho \in (-\infty, r^2)$. Then there exists a unique Hermitian stabilizing solution X of the Riccati equation (10.7.4):*

$$XA + A^*X - \rho C^*C - XBB^*X = 0.$$

Moreover, when $\rho = r^2$, there exists a unique solution X having the property that the matrix $A - BB^*X$ is unstable.

   Conversely, if A is stable and if there exists a Hermitian solution X of the above ARE, then necessarily $\rho \leq r^2$.

Remark

- Note that if the Riccati equation (10.7.4) has a stabilizing solution X, then the control law

$$u(t) = -B^*Xx(t)$$

minimizes the performance index (10.7.2), and the minimum value of the performance index is $x_0^T X x_0$.

Note: **There is no controllability assumption here on the pair** (A, B).

   Proof.   Considering the orthogonal decomposition of $\mathbb{C}^n$ into the controllability subspace and its orthogonal complement, we can find a nonsingular matrix T such that

$$TAT^{-1} = \begin{pmatrix} A_1 & A_2 \\ 0 & A_3 \end{pmatrix}, \qquad TB = \begin{pmatrix} B_1 \\ 0 \end{pmatrix},$$

and

$$CT^{-1} = (C_1, C_2),$$

where (A_1, B_1) is controllable. Multiplying the Riccati equation (10.7.4) on the left by T^{-1*} and on the right by T^{-1} and setting

$$T^{-1*}XT^{-1} = \begin{pmatrix} X_1 & X_2 \\ X_3 & X_4 \end{pmatrix},$$

we have

$$\begin{pmatrix} X_1 & X_2 \\ X_3 & X_4 \end{pmatrix} \begin{pmatrix} A_1 & A_2 \\ 0 & A_3 \end{pmatrix} + \begin{pmatrix} A_1^* & 0 \\ A_2^* & A_3^* \end{pmatrix} \begin{pmatrix} X_1 & X_2 \\ X_3 & X_4 \end{pmatrix}$$

$$- \rho \begin{pmatrix} C_1^*C_1 & C_1^*C_2 \\ C_2^*C_1 & C_2^*C_2 \end{pmatrix} - \begin{pmatrix} X_1B_1B_1^*X_1 & X_1B_1B_1^*X_2 \\ X_3B_1B_1^*X_1 & X_3B_1B_1^*X_2 \end{pmatrix} = 0 \qquad (10.7.6)$$

Eq. (10.7.6) breaks into the following four equations:

$$X_1 A_1 + A_1^* X_1 - \rho C_1^* C_1 - X_1 B_1 B_1^* X_1 = 0, \qquad (10.7.7)$$

$$X_2 A_3 + (A_1 - B_1 B_1^* X_1^*)^* X_2 + X_1 A_2 - \rho C_1^* C_2 = 0, \qquad (10.7.8)$$

$$X_3(A_1 - B_1 B_1^* X_1) + A_3^* X_3 + A_2^* X_1 - \rho C_2^* C_1 = 0, \qquad (10.7.9)$$

$$X_4 A_3 + A_3^* X_4 + X_3 A_2 + A_2^* X_2 - \rho C_2^* C_2 - X_3 B_1 B_1^* X_2 = 0. \qquad (10.7.10)$$

Since (A_1, B_1) is controllable, there is a unique solution $X_{1\rho}$ of (10.7.7) with the property that $A_1 - B_1 B_1^* X_{1\rho}$ is stable if $\rho \in (-\infty, r^2)$, and if $\rho = r^2$, then $A_1 - B_1 B_1^* X_{1\rho}$ is not stable. (In fact it has one eigenvalue on the imaginary axis). Substituting this stabilizing solution $X_{1\rho}$ into the Sylvester equations (10.7.8) and (10.7.9), it can be shown that the solutions $X_{2\rho}$ and $X_{3\rho}$ of (10.7.8) and (10.7.9) are unique and moreover $X_{3\rho} = X_{2\rho}^*$ (note that the spectrum of A_3 is in the open left-half plane). Substituting these $X_{2\rho}$ and $X_{3\rho}$ in Eq. (10.7.10), similarly, we see that $X_{4\rho}$ is also unique and $X_{4\rho}^* = X_{4\rho}$. Finally, we note that the matrix $TAT^{-1} - (TB \cdot B^* T^* X_\rho)$, where

$$X_\rho = \begin{pmatrix} X_{1\rho} & X_{2\rho} \\ X_{3\rho} & X_{4\rho} \end{pmatrix},$$

is stable. Thus, $A - BB^* X_\rho$ is stable.

To prove the converse, we note that if $X = X^*$ satisfies the Riccati equation (10.7.6), then

$$(A - j\omega I)^* X + X(A - j\omega I) - \rho C^* C - XBB^* X = 0,$$

for all $\omega \in \mathbb{R}$. Thus,

$$0 \le (B^* X(A - j\omega I)^{-1} B - I)^*(B^* X(A - j\omega I)^{-1} B - I),$$
$$= I - \rho G^*(j\omega)G(j\omega), \quad \text{for all } \omega \in \mathbb{R}.$$

Thus, $\rho \le r^2$ by the first part of Theorem 10.7.1.   ∎

Example 10.7.2.

$$A = \begin{pmatrix} 0 & 1 \\ -1 & -1 \end{pmatrix}, \qquad B = (0, -1)^\mathrm{T}, \qquad C = (1, 0).$$

Then we know from Example 7.8.1 that $r^2 = \frac{3}{4}$.

Choose $\rho = \frac{1}{2}$. Then the unique solution X_ρ to the Riccati equation:

$$XA + A^T X - XBB^T X - \rho C^T C = 0$$

is

$$X_\rho = \begin{pmatrix} -0.5449 & -0.2929 \\ -0.2929 & -0.3564 \end{pmatrix},$$

which is symmetric.

The eigenvalues of $A - BB^T X_\rho$ are $-0.3218 \pm 0.7769 j$. So, the matrix $A - BB^T X_\rho$ is stable. Thus, X_ρ is a stabilizing solution.

If ρ is chosen to be $\frac{3}{4}$, then the solution

$$X = \begin{pmatrix} -1 & -0.5 \\ -0.5 & -1 \end{pmatrix},$$

which is symmetric, but not stabilizing. The eigenvalues of $A - BB^T X$ are $0 \pm 0.7071 j$.

A Bisection Method for Computing the Complex Stability Radius

Theorem 10.7.2 suggests a bisection algorithm for computing the complex stability radius $r_{\mathbb{C}}$.

The idea is to determine $r_{\mathbb{C}}$ as that value of ρ for which the Hamiltonian matrix H_ρ given by (10.7.5), associated with the Riccati equation (10.7.4), has an eigenvalue on the imaginary axis for the first time.

If ρ_0^- and ρ_0^+ are some lower and upper estimates of $r_{\mathbb{C}}$, then the successive better estimates can be obtained by the following algorithm.

> **Algorithm 10.7.1.** *A Bisection Method for the Complex Stability Radius*
> **Inputs.**
>
> 1. *The system matrices A, B, and C.*
> 2. *Some upper and lower estimates ρ_0^+ and ρ_0^- of the complex stability radius ρ.*
>
> **Output.** *An approximate value of the complex stability radius ρ.*
> *For $k = 0, 1, 2, \ldots$, do until convergence*
> **Step 1.** *Take $\rho_k = \dfrac{\rho_k^- + \rho_k^+}{2}$ and compute $H_{\rho_k^2}$.*
> **Step 2.** *If $H_{\rho_k^2}$ has eigenvalues on the imaginary axis, set $\rho_{k+1}^- \equiv \rho_k^-$ and $\rho_{k+1}^+ \equiv \rho_k$.*

Otherwise set $\rho_{k+1}^- \equiv \rho_k$ and $\rho_{k+1}^+ \equiv \rho_k^+$.
End

Example 10.7.3. Consider Example 10.7.1. Take $\rho_0^- = 0$, $\rho_0^+ = 1$.
$k = 0$. **Step 1.** $\rho_0 = \frac{1}{2}$. $H_{\rho_0^2}$ does not have an imaginary eigenvalue.
Step 2. $\rho_1^- = \frac{1}{2}$, $\rho_1^+ = 1$.
$k = 1$. **Step 1.** $\rho_1 = \frac{3}{4}$. $H_{\rho_1^2}$ does not have an imaginary eigenvalue.
   Step 2. $\rho_2^- = \frac{3}{4}$, $\rho_2^+ = 1$.
$k = 2$. **Step 1.** $\rho_2 = \frac{7}{8}$. $H_{\rho_2^2}$ has an imaginary eigenvalue.
   Step 2. $\rho_3^- = \frac{3}{4}$, $\rho_3^+ = \frac{7}{8}$.
$k = 3$. **Step 1.** $\rho_3 = \frac{13}{16}$. $H_{\rho_3^2}$ does not have an imaginary eigenvalue.
   Step 2. $\rho_4^- = \frac{13}{16}$, $\rho_4^+ = \frac{7}{8}$.
$k = 4$. $\rho_4 = \frac{27}{32}$.
The iteration is converging toward $r = 0.8660$. The readers are asked to verify this by carrying out some more iterations.

MATCONTROL note: Algorithm 10.7.1 has been implemented in MATCONTROL function **stabradc**.

10.8 SOME SELECTED SOFTWARE

10.8.1 MATLAB Control System Toolbox

LQG design tools include:

lqr	LQ-optimal gain for continuous systems
dlqr	LQ-optimal gain for discrete systems
lqry	LQR with output weighting
lqrd	Discrete LQ regulator for continuous plant
care	Solve CARE
dare	Solve DARE
norm(sys, 2)	H_2-norm of the system
norm(sys, inf)	H_∞-norm of the system

10.8.2 MATCONTROL

STABLYAPC	Feedback stabilization of continuous-time system using Lyapunov equation
STABLYAPD	Feedback stabilization of discrete-time system using Lyupunov equation
STABRADC	Finding the complex stability radius using the bisection method
HINFNRM	Computing H_∞-norm using the bisection method.

10.8.3 CSP-ANM

Feedback stabilization

- Constrained feedback stabilization is computed by `StateFeedback Gains` [*system, region*], where the available regions are `DampingFactor Region` [α], `SettlingTimeRegion` [t_s, ϵ], `DampingRatio- Region` [ζ_{min}] and `NaturalFrequencyRegion` [$\omega_{n\,min}$], and their intersections.
- The Lyapunov algorithms for the feedback stabilization is implemented as `StateFeedbackGains` [*system, region*, `Method` $\rightarrow$ `LyapunovShift`] and `StateFeedbackGains` [*system, region*, `Method` $\rightarrow$ `PartialLyapunovShift`].

10.8.4 SLICOT

Optimal regulator problems, system norms, and complex stability radius

SB10DD	H_∞ (sub)optimal state controller for a discrete-time system
SB10FD	H_∞ (sub)optimal state controller for a continuous-time system
AB13BD	H_2- or L_2-norm of a system
AB13CD	H_∞-norm of a continuous-time stable system
AB13ED	Complex stability radius using bisection
AB13FD	Complex stability radius using bisection and SVD.

10.8.5 MATRIX$_X$

Purpose: Computing L_∞-norm of the transfer matrix of a discrete-time system.

Syntax: [SIGMA, OMEGA] = DLINFNORM (S, NS, {TOL, { MAXITER}})

Purpose: Compute optimal state-feedback gain for discrete-time system.

Syntax: [EVAL, KR] = DREGULATOR (A, B, RXX, RUU, RXU) OR
[EVAL, KR, P] = DREGULATOR (A, B, RXX, RUU, RXU)

Purpose: Computing L_∞-norm of a transfer matrix .

Syntax: [SIGMA, OMEGA] = LINFNORM (S, NS, { TOL, { MAXITER}})

Purpose: Compute optimal state-feedback gain for continuous-time system.

Syntax: [EVAL, KR]=REGULATOR (A, B, RXX, RUU, RXU) OR
[EVAL, KR, P]=REGULATOR (A, B, RXX, RUU, RXU)

Purpose: Computes and plots the Singular Values of a continuous system.

Syntax: [OMEGA, SVALS]=SVPLOT (S, NS, WMIN, WMAX, { NPTS} ,
{OPTIONS }) OR
[SVALS]=SVPLOT (S, NS, OMEGA, { OPTIONS})

10.9   SUMMARY AND REVIEW

The following topics have been discussed in this chapter.

- State-feedback stabilization
- EVA and eigenstructure assignments by state and output feedback
- The LQR design
- H_∞-control problems
- Stability radius.

Feedback Stabilization

The problem of stabilizing the continuous-time system

$$\dot{x}(t) = Ax(t) + Bu(t),$$
$$y(t) = Cx(t) + Du(t)$$

by using the state-feedback law $u(t) = -Kx(t)$ amounts to finding a feedback matrix K such that $A - BK$ is stable.

The state-feedback stabilization of a discrete-time system is similarly defined.

The **characterizations** of the continuous-time and discrete-time state-feedback stabilizations are, respectively, given in **Theorems 10.2.1** and **10.2.2**.

It is shown how a system can be stabilized using the solution of a Lyapunov equation. For the continuous-time system, the Lyapunov equation to be solved is

$$-(A + \beta I)X + X(-(A + \beta I))^{\mathrm{T}} = -2BB^{\mathrm{T}},$$

where β is chosen such that $\beta > |\lambda_{\max}(A)|$.

The stabilizing feedback matrix K is given by

$$K = B^T X^{-1}.$$

In the discrete-time case, the Lyapunov equation to be solved is

$$AXA^T - \beta^2 X = 2BB^T,$$

where β is chosen such that $0 < \beta \leq 1$ and $|\lambda| \geq \beta$ for any eigenvalue λ of A. The stabilizing feedback matrix in this case is

$$K = B^T (X + BB^T)^{-1} A.$$

Detectability

The detectability of the pair (A, C) is a dual concept of the stabilizability of the pair (A, B). Characterizations of the continuous-time and discrete-time detectability are, respectively, stated in **Theorems 10.3.1** and **10.3.3**.

The Eigenvalue Assignment

For the transient responses to meet certain designer's constraints, it is required that the eigenvalues of the closed-loop matrix lie in certain specified regions of the complex plane. This consideration gives rise to the well-known EVA problem.

The EVA problem by state feedback is defined as follows:

Given the pair (A, B) and a set Λ of the complex numbers, closed under complex conjugations, find a feedback matrix K such that $A - BK$ has the spectrum Λ.

The conditions of solvability for the EVA problem and the uniqueness of the matrix K are:

There exists a matrix K such that the matrix $A - BK$ has the spectrum Λ for every complex-conjugated set Λ if and only if (A, B) is controllable. The feedback matrix K, when it exists, is unique if and only if the system is a single-input system (**Theorem 10.4.1**).

The constructive proof of Theorem 10.4.1 and several related well-known formulas such as the **Bass–Gura formula** and the **Ackermann formula** suggest computational methods for **single-input EVA problem.** Unfortunately, however, these methods are based on the reduction of the pair (A, b) to a controller–companion pair, and **are not numerically effective.** Numerically effective algorithms for EVA are based on the reduction of the pair (A, b) or the pair (A, B) (in the multi-output case) to the controller–Hessenberg pair. **These methods will be described in Chapter 11.**

The EVA problem by output feedback is discussed in Section 10.4.2 and a well-known result on this problem is stated in **Theorem 10.4.2**.

The Eigenstructure Assignment

The eigenvalues of the state matrix A determine the rate at which the system response decays or grows. On the other hand, the eigenvectors determine the shape of the response. Thus, in certain practical applications, it is important that both the eigenvalues and the eigenvectors are assigned. The problem is known as the **eigenstructure assignment problem.** The conditions under which eigenstructure assignment is possible are stated in Theorem **10.4.3** for the state-feedback law and in Theorem **10.4.4** for the output feedback law.

The Linear Quadratic Regulator (LQR) Problem

The continuous-time LQR problem is the problem of finding a control vector $u(t)$ that minimizes the performance index

$$J_C(x) = \int_0^\infty \left[x^T(t) Q x(t) + u^T(t) R u(t) \right] dt$$

subject to

$$\dot{x}(t) = Ax(t) + Bu(t), \qquad x(0) = x_0,$$
$$y(t) = Cx(t),$$

where Q and R are, respectively, the state-weight and the control-weight matrices.

It is shown in **Theorem 10.5.1** that the continuous-time LQR problem has a solution if (A, B) is stabilizable and (A, Q) is detectable.

The solution is obtained by solving the CARE:

$$XA + A^T X - XSX + Q = 0,$$

where $S = BR^{-1}B^T$.

The optimal control vector $u^0(t)$ is given by

$$u^0(t) = -R^{-1}B^T X x(t),$$

where X is the unique symmetric positive semidefinite solution of the CARE.

The matrix $K = -R^{-1}TB^T X$ is such that $A - BK$ is stable, that is, X is a stabilizing solution.

The LQR design has the following **guaranteed stability and robustness properties:**

Stability property. The stable open-loop eigenvalues remain stable and the unstable eigenvalues get reflected across the imaginary axis (when $R = \rho I$ and $\rho \to \infty$).

Robustness properties. Using the optimal return difference identity, it can be shown that

$$\sigma_{\min}(I + G_{LQ}(j\omega)) \geq 1$$

and $\sigma_{\min}(I + G_{LQ}(j\omega)^{-1}) \geq \frac{1}{2}$, where $G_{LQ}(s) = K(sI - A)^{-1}B$.

These relations show that the upward and downward gain margins are, respectively, ∞ and 0.5. The phase margin is at least $\pm 60°$.

The discrete-time LQR problem is similarly defined. In this case, the performance index is given by

$$J_D(x) = \sum_{k=0}^{\infty} \left(x_k^T Q x_k + u_k^T R u_k \right).$$

The DARE is

$$A^T X A - X + Q - A^T X B(R + B^T X B)^{-1} B^T X A = 0.$$

If (A, B) is discrete-stabilizable and (A, Q) is discrete-detectable, then the above Riccati equation (DARE) has a unique symmetric positive semidefinite solution X and the optimal control is $u_k^0 = K x_k$, where

$$K = (R + B^T X B)^{-1} B^T X A.$$

Furthermore, X is a discrete-stabilizing solution, that is, $(A - BK)$ is discrete-stable.

H_∞-Control Problems

The H_∞-control problems are concerned with robust stabilization for unstructured perturbations in the frequency domain. The goal of a H_∞ control is to determine a controller that guarantees a closed-loop system with an H_∞-norm bounded by a certain specified quantity γ when such a controller exists. Both the state feedback and the output feedback H_∞-control problems have been discussed briefly in Sections 10.6.2 and 10.6.3, respectively. Both problems require solutions of certain Riccati equations. Under the assumptions (10.6.12)–(10.6.15), the solution of the output feedback H_∞-control problem reduces to solving a pair of Riccati equations:

$$XA + A^T X - X \left(B_2 B_2^T - \frac{1}{\gamma^2} B_1 B_1^T \right) X + C_1^T C_1 = 0,$$

$$AY + YA^T - Y \left(C_2^T C_2 - \frac{1}{\gamma^2} C_1^T C_1 \right) Y + B_1 B_1^T = 0,$$

where A, B_1, B_2, C_1, and C_2 are defined by (10.6.11). The expression for a H_∞ controller is given in (10.6.18)–(10.6.20). Also, two computational algorithms: one, the well-known **bisection algorithm** by Boyd *et al.* and the other, the two-step algorithm by Bruinsma *et al.* (1990), for computing the H_∞-norm are given in Section 10.6.1. Algorithm 10.6.2 seems to be faster than Algorithm 10.6.1. but the latter is easier to implement.

Stability Radius

The concept of stability radius has been defined in Chapter 7. Here a connection of the **complex stability radius** r is made with the ARE:

$$XA + A^*X - \rho C^*C - XBB^*X = 0$$

via the parametrized Hamiltonian matrix

$$H_\rho = \begin{pmatrix} A & -BB^* \\ \rho C^*C & -A^* \end{pmatrix}.$$

It is shown in **Theorem 10.7.3** that if $\rho \in (-\infty, r^2)$, then the above Riccati equation has a unique stabilizing solution X. Conversely, if A is stable and if there exists a Hermitian solution X of the above equation, then $\rho \leq r^2$.

In terms of the eigenvalues of the Hamiltonian matrix H_ρ, it means that $\rho < r$ if and only if H_{ρ^2} does not have an eigenvalue on the imaginary axis (**Theorem 10.7.2**).

Based on the latter result, a **bisection algorithm (Algorithm 10.7.1)** for determining the complex stability radius is described.

10.10   CHAPTER NOTES AND FURTHER READING

Feedback stabilization and EVA (pole-placement) are two central problems in control theory. For detailed treatment of these problems, see Brockett (1970), Brogan (1991), Friedland (1986), Chen (1984), Kailath (1980), Wonham (1985), Kwakernaak and Sivan (1972), etc. Most of the books in linear systems theory, however, do not discuss feedback stabilization via Lyapunov equations. Discussions on feedback stabilization via Lyapunov equations in Section 10.2 have been taken from the papers of Armstrong (1975) and Armstrong and Rublein (1976). For a Schur method for feedback stabilization, see Sima (1981). For stabilization methods of descriptor systems, see Varga (1995). For more on the output feedback problem, see Kimura (1975), Porter (1977), Sridhar and Lindorff (1973), Srinathkumar (1978), and Misra and Patel (1989).

For a discussion on eigenstructure assignment problem, see Andry *et al.* (1983).

The authoritative book by Anderson and Moore (1990) is an excellent source for a study on the LQR problem. The other excellent books on the subject include Athans and Falb (1966), Lewis (1986), Mehrmann (1991), Sima (1996), Kučěra (1979), etc. For a proof of the discrete-time LQR Theorem (**Theorem 10.5.3**), see Sage and White (1977). An excellent reference book on the theory of Riccati equations is the recent book by Lancaster and Rodman (1995). There are also several nice papers on Riccati equations in the books edited by Bittanti *et al.* (1991) and Bittanti (1989). H_∞-control problem has been dealt with in detail in the books by Kimura (1996), Zhou *et al.* (1996), Green and Limebeer (1995), Dorato *et al.* (1992, 1995). The original papers by Francis and Doyle (1987) and by Doyle *et al.* (1989) are worth reading. A recent book by Kirsten Morris (2001) contains an excellent coverage on feedback control, in particular, H_∞ feedback control. Algorithms 10.6.1 and 10.6.2 for computing the H_∞-norm have been taken, respectively, from the papers of Boyd *et al.* (1989) and Bruinsma and Steinbuch (1990). A gradient method for computing the optimal H_∞-norm has been proposed in Pandey *et al.* (1991). Recently, Lin *et al.* (2000) have proposed numerically reliable algorithms for computing H_∞-norms of the discrete-time systems, both for the state and the output feedback problems. The discussion on the complex stability radius and Riccati equation in Section 10.7 has been taken from Hinrichsen and Pritchard (1986b). For an iterative algorithm for H_∞-control by state feedback, see Scherer (1989). Theorem 10.7.3 is an extension of the work of Brockett (1970) and Willems (1971). For an application of the ARE to compute H_∞ optimization, see Zhou and Khargonekar (1988).

For more on return difference matrix, phase and gain margins of the multivariable LQR design, see Lewis (1986), Safonov *et al.* (1981), Safonov (1980), etc. For an excellent and very readable account of classical control design using H_∞ techniques, see Helton and Merino (1998).

Exercises

10.1 Prove the equivalence of (i) and (iii) in Theorem 10.2.1.

10.2 Prove Theorem 10.2.2.

10.3 Prove Theorem 10.2.4.

10.4 Construct a state-space continuous-time system that is stabilizable, but not controllable.

Apply the Lyapunov stabilization method (modify the method in the book as necessary) to stabilize this system.

10.5 Repeat Problem 10.4 with a discrete-time system.

10.6 Develop algorithms for feedback stabilization, both for the continuous-time and discrete-time systems, based on the reduction of A to the real Schur form (see Sima 1981).

Compare the efficiency of each of these Schur algorithms with the respective Lyapunov equation based algorithms given in the book.

10.7 Using the real Schur decomposition of A, develop partial stabilization algorithms, both for the continuous-time and discrete-time systems in which only the unstable eigenvalues of A are stabilized using feedback, leaving the stable eigenvalues unchanged.

10.8 Prove Theorem 10.3.1.

10.9 State and prove the discrete counterpart of Theorem 10.3.2.

10.10 Prove Theorem 10.3.3.

10.11 Give an alternative proof of the state-feedback EVA Theorem (Theorem 10.4.1).

10.12 Construct an example to verify that if the eigenvalues of the closed-loop system are moved far from those of the open-loop system, a large feedback will be required to place the closed-loop eigenvalues.

10.13 Using the expression of the transforming matrix T, which transforms the system (A, b) to a controller-companion form (10.4.1)–(10.4.2), and the expression for the feedback formula (10.4.5), derive the Bass–Gura formula (10.4.6).

10.14 Derive the Ackermann formula (10.4.7).

10.15 Work out an example to illustrate each of the following theorems: Theorems 10.5.1, 10.5.2, 10.5.3, 10.6.1, 10.6.2, 10.6.3, 10.6.4, and 10.6.5. (Use MATLAB function **care** to solve the associated Riccati equation, whenever needed.)

10.16 *Design of a regulator with prescribed degree of stability.* Consider the LQR problem of minimizing the cost function

$$J_\alpha = \int_0^\infty e^{2\alpha t} (u^T Ru + x^T Qx)\, dt.$$

(a) Show that the Riccati equation to be solved in this case is:

$$(A + \alpha I)^T X + X(A + \alpha I) + Q - XBR^{-1}B^T X = 0$$

and the optimal control is given by the same feedback matrix K as in Theorem 10.5.1.

(b) Give a physical interpretation of the problem.

10.17 *Cross-weighted LQR.* Consider the LQR problem with the quadratic cost function with a cross penalty on state and control:

$$J_{CW} = \int_0^\infty \left[x^T Qx + 2x^T Nu + u^T Ru \right] dt$$

subject to $\dot{x} = Ax + Bu$, $x(0) = x_0$, where Q, R, N are, respectively, the state-weighting matrix, the control-weighting matrix, and the cross-weighting matrix. Define $A_R = A - BR^{-1}N^T$.

(a) Show that the Riccati equation to be solved in this case is:

$$XA_R + A_R^T X + (Q - NR^{-1}N^T) - XBR^{-1}B^T X = 0,$$

and the optimal control law is given by $u(t) = -Kx(t)$, where $K = R^{-1}(N^T + B^T X)$.

(b) What are the assumptions needed for the existence of the unique, symmetric positive semidefinite solution X of the Riccati equation in (a)?

10.18 Consider the LQR problem of minimizing the cost

$$J = \int_0^\infty \left[x^2(t) + \rho^2 u^2(t) \right] dt,$$

with $\rho > 0$, subject to

$$m\ddot{q} + kq(t) = u(t).$$

(a) Find an expression for the feedback matrix K in terms of ρ, by solving an appropriate ARE.

(b) Plot the closed-loop poles as ρ varies over $0 < \rho < \infty$.

(c) Write down your observations about the behavior of the closed-loop poles with respect to stability.

10.19 Using the MATLAB function *lqr*, design an LQR experiment with a single-input system to study the behavior of the closed-loop poles and the feedback vector with varying ρ in $R = \rho I$ in the range $[1, 10^6]$, taking $\rho = 1, 10, 10^2, 10^3$, and 10^6. Plot the open loop poles, the closed loop poles, and the step responses. Make a table of the gain margins and phase margins with each feedback vector.

10.20 (a) Using the return-difference identity, show that the ith singular value σ_i of the return-difference matrix with $s = j\omega$ is:

$$\sigma_i (I + G_{LQ}(j\omega)) = \left[1 + \frac{1}{\rho} \sigma_i^2 (H(j\omega)) \right]^{1/2},$$

where $H(s) = C(sI - A)^{-1} B$, $R = \rho I$, and $Q = C^T C$.

(b) Using the result in (a), prove that

$$\sigma_{min}(I + G_{LQ}(j\omega)) \geq 1.$$

(c) Using (b), prove that

$$\sigma_{min}(I + (G_{LQ}(j\omega))^{-1}) \geq \tfrac{1}{2}.$$

10.21 In certain applications, the homogeneous ARE:

$$XA + A^T X + XWX = 0$$

is important.

Prove that the above homogeneous Riccati equation has a stabilizing solution (i.e., $A + WX$ is stable) if A has no eigenvalues on the imaginary axis.

10.22 *Computing the H_∞-norm over an interval.* Define the H_∞-norm of $G(s)$ over an interval $0 \leq \alpha < \beta$ as:

$$\|G\|_{[\alpha, \beta]} = \sup \sigma_{max}(C(j\omega)), \quad \alpha \leq \omega \leq \beta.$$

(a) Develop an algorithm to compute $\|G\|_{[\alpha, \beta]}$ by modifying the bisection algorithm (**Algorithm 10.6.1**) as follows:
   1. Take $\gamma_{lb} = \max\{\sigma_{max}(G(j\alpha)), \sigma_{max}(G(j\beta))\}$
   2. Modify the eigenvalue test in Step 2 as: if M_γ has no imaginary eigenvalues between $j\alpha$ and $j\beta$.

(b) Work out a numerical example to test your algorithm.

10.23 Give a linear algebraic proof of Theorem 10.6.1 (consult the paper by Boyd *et al.* (1989)).

10.24 Develop an algorithm to compute the H_∞-norm of a discrete-stable transfer function, based on Theorem 10.6.3.

10.25 (Kimura 1996). For the second-order system:

$$
\begin{aligned}
\dot{x}_1 &= x_2, \\
\dot{x}_2 &= w_1 + u_1, \\
z_1 &= x_1, \\
z_2 &= \delta u_1, \\
y &= c_2 x_1 + d_2 u_2,
\end{aligned}
$$

find the conditions under which the output feedback problem has a solution. Find the transfer function for H_∞ controller.

10.26 Prove that if the Hamiltonian matrix H_γ defined by (10.6.10) has an imaginary eigenvalue, then the state feedback H_∞-control problem does not have a solution.

10.27 Prove Theorem 10.6.2: If A is a complex stable matrix, then the distance to instability

$$
\beta(A) = \|(sI - A)^{-1}\|_\infty^{-1}.
$$

10.28 (a) Let $G(s) = C(sI - A)^{-1}B$. Then prove

$$
r = \begin{cases}
\dfrac{1}{\max\limits_{\omega \in \mathbb{R}} \|G(j\omega)\|}, & \text{if } G \neq 0, \\[2ex]
\infty, & \text{if } G = 0.
\end{cases}
$$

(Consult Hinrichsen and Pritchard (1986b)).

(b) Give an example to show that $r(A, I, I)$ can change substantially under similarity transformation on A.

References

Ackermann J. "Der entwurf linear regelung systeme im zustandsraum," *Regulungestechnik und prozessedatenverarbeitung*, Vol. 7, pp. 297–300, 1972.

Anderson B. D. O. and Moore J. B. *Optimal Control: Linear Quadratic Methods*, Prentice Hall, Englewood Cliffs, NJ, 1990.

Andry A. N., Jr., Shapiro E. Y., and Chung J. C. "Eigenstructure assignment for linear systems," *IEEE Trans. Aerospace and Electronic Syst.*, Vol. AES-19(5), pp. 711–729, 1983.

Armstrong E. S. "An extension of Bass' algorithm for stabilizing linear continuous constant systems," *IEEE Trans. Autom. Control*, Vol. AC-20, pp. 153–154, 1975.

Armstrong E. S. and Rublein G. T. "A stabilizing algorithm for linear discrete constant systems," *IEEE Trans. Autom. Control*, Vol. AC-21, pp. 629–631, 1976.

Athans M. and Falb P. *Optimal Control*, McGraw Hill, New York, 1966.

Bittanti S. (ed.). *The Riccati Equation in Control, Systems, and Signals*, Lecture Notes, Pitagora Editrice, Bologna, 1989.

Bittanti S., Laub A.J., and Willems J.C. (eds.), *The Riccati Equation*, Springer-Verlag, Berlin, 1991.

Boyd S., Balakrishnan V., and Kabamba P. "A bisection method for computing the H_∞-norm of a transfer function matrix and related problems," *Math. Control Signals Syst.*, Vol. 2, pp. 207–219, 1989.

Boyd S. and Balakrishnan V. "A regularity result for the singular values of a transfer function matrix and a quadratically convergenet algorithm for computing its L_∞-norm," *Syst. Control Lett.*, pp. 1–7, 1990.

Brockett R.W. *Finite Dimensional Linear Systems*, Wiley, New York, 1970.

Brogan W.L. *Modern Control Theory*, 3rd edn, Prentice Hall, Englewood Cliffs, NJ, 1991.

Bruinsma N.A. and Steinbuch M. "A fast algorithm to compute the H_∞-norm of a transfer function matrix," *Syst. Control Lett.*, Vol. 14, pp. 287–293, 1990.

Chen C.-T. *Linear System Theory and Design*, CBS College Publishing, New York, 1984.

Dorato P., Fortuna L., and Muscato G. *Robust Control for Unstructured Perturbations – An Introduction* (Thoma M. and Wyner A., eds.) Springer-Verlag, Berlin, 1992.

Dorato P., Abdallah C., and Cerone V. *Linear Quadratic Control, An Introduction*, Prentice Hall, Englewood Cliffs, NJ, 1995.

Doyle J., Glover K., Khargonekar P., and Francis B. "State-space solutions to standard H_2 and H_∞ control problems," *IEEE Trans. Autom. Control*, Vol. AC-34, pp. 831–847, 1989.

Enns D.F. "Model reduction with balanced realizations: An error bound and a frequency weighted generalization," *Proc. IEEE Conf. Dec. Control*, Las Vegas, pp. 127–132, 1984.

Francis B.A. and Doyle J.C. "Linear control theory with an H_∞ optimality criterion," *SIAM J. Control Optimiz.*, Vol. 25, pp. 815–844, 1987.

Francis B.A. *A Course in H_∞ Control Theory*, Lecture Notes in Control and Information Sciences, Vol. 88, 1987.

Franklin G.F., Powell J.D., and Emami-Naeini A. *Feedback Control of Dynamic Systems*, Addison-Wesley, New York, 1986.

Friedland B. *Control System Design: An Introduction to State-Space Methods*, McGraw-Hill, New York, 1986.

Glover K. "All optimal Hankel-norm approximations of linear multivariable systems and their L_∞-error bounds," *Int. J. Control*, Vol. 39, pp. 1115–1193, 1984.

Glover K. and Doyle J. "State-space formulae for all stabilizing controllers that satisfy an H_∞-norm bound and relations to risk sensitivity," *Syst. Control Lett.*, Vol. 11, pp. 167–172, 1988.

Green M. and Limebeer D.J.N. *Linear Robust Control*, Prentice Hall Information and System Sciences Series (Thomas Kailath, ed.), Prentice Hall, Englewood Cliffs, NJ, 1995.

Helton J. W. and Merino O., *Classical control using H^∞ Methods: Theory, optimization and Design*, SIAM, Philadelphia, 1998.

Hinrichsen D. and Pritchard A.J. "Stability radii of linear systems," *Syst. Control Lett.*, Vol. 7, pp. 1–10, 1986a.

Hinrichsen D. and Pritchard A.J. "Stability radius for structured perturbations and the algebraic Riccati equation," *Syst. Control Lett.*, Vol. 8, pp. 105–113, 1986b.

Kailath T. *Linear Systems*, Prentice Hall, Englewood Cliffs, NJ, 1980.

Kimura H. "Pole assignment by gain output feedback," *IEEE Trans. Autom. Control*, Vol. AC-20(4), pp. 509–516, 1975.

Kimura H. *Chain-Scattering to H^∞ Control*, Birkhäuser, Boston, 1996.

Kučera V. *Discrete Linear Control*, John Wiley & Sons, New York, 1979.

Kwakernaak H. and Silvan R. *Linear Optimal Control Systems*, Wiley-Interscience, New York, 1972.

Lancaster P. and Rodman L. *The Algebraic Riccati Equation*, Oxford University Press, Oxford, UK, 1995.

William S. Levine (ed.). *The Control Handbook*, CRC Press and IEEE Press, Boca Raton, Fl., 1996.

Lewis F.L. *Optimal Control*, John Wiley & Sons, New York, 1986.

Lewis F.L. *Applied Optimal Control and Estimation—Digital Design and Implementation*, Prentice Hall and Texas Instruments Digital Signal Processing Series, Prentice Hall, Englewood Cliffs, NJ, 1992.

Lin W.-W., Wang J.-S., and Xu Q.-F. "On the Computation of the optimal H_∞-norms for two feedback control problems," *Lin. Alg. Appl.*, Vol. 287, pp. 233–255, 1999.

Lin W.-W., Wang J.-S., and Xu Q.-F. "Numerical computation of the minimal H_∞-norms of the discrete-time state and output feedback control problems," *SIAM. J. Num. Anal.*, Vol. 38(2), pp. 515–547, 2000.

Luenberger D. Introduction to Dynamic Systems: Theory, Model and Applications, John Wiley, New York, 1979.

Maciejowski J.M. *Multivariable Feedback Design*, Addison-Wesley, Wokingham, 1989.

Mehrmann V. *The Autonomous Linear Quadratic Control Problem, Theory and Numerical Solution*, Lecture Notes in Control and Information Sciences, Springer-Verlag, Heidelberg, 1991.

Morris K. *Introduction to Feedback Control*, Academic Press, Boston, 2001.

Misra P. and Patel R.V. "Numerical algorithms for eigenvalue assignment by constant and dynamic output feedback," *IEEE Trans. Autom. Control*, Vol. 34, pp. 579–588, 1989.

Moore B.C. "On the flexibility offered by state feedback in multivariable systems beyond closed loop eigenvalue assignment," *IEEE Trans. Autom. Control*, Vol. AC-21, pp. 689–692, 1976.

Pandey P., Kenney C.S., Packard A., and Laub A.J. "A gradient method for computing the optimal H_∞ norm," *IEEE Trans. Autom. Control*, Vol. 36, pp. 887–890, 1991.

Porter B. "Eigenvalue assignment in linear multivariable systems by output feedback," *Int. J. Control*, Vol. 25, pp. 483–490, 1977.

Robel G. "On computing the infinity norm," *IEEE Trans. Autom. Control*, Vol. 34, pp. 882–884, 1989.

Safonov M.G. and Athans M. "Gain and phase margins of multiloop LQG regulators," *IEEE Trans. Autom. Control*, Vol. AC-22, pp. 173–179, 1977.

Safonov M.G., Laub A.J., and Hartmann G.L. "Feedback properties of multivariable systems: The role and use of the return difference matrix," *IEEE Trans. Autom. Control*, Vol. AC-26, pp. 47–65, 1981.

Safonov M.G. *Stability and Robustness of Multivariable Feedback Systems*, MIT Press, Boston, 1980.

Sage A.P. and White C.C. *Optimal Systems Control*, 2nd edn, Prentice Hall, Englewood Cliffs, NJ, 1977.

Scherer C. "H_∞-control by state feedback: An iterative algorithm and characterization of high-gain occurence," *Syst. Control Lett.*, Vol. 12, pp. 383–391, 1989.

Sima V. "An efficient Schur method to solve the stabilizing problem," *IEEE Trans. Autom. Control*, Vol. AC-26, pp. 724–725, 1981.

Sima V. *Algorithms for Linear-Quadratic Optimization*, Marcel Dekker, New York, 1996.

Sridhar B. and Lindorff D.P. "Pole placement with constant gain output feedback," *Int. J. Control*, Vol. 18, pp. 993–1003, 1973.

Srinathkumar S. "Eigenvalue/eigenvector assignment using output feedback," *IEEE Trans. Autom. Control*, Vol. AC-21, pp. 79–81, 1978.

Varga A. "On stabilization methods of descriptor systems," *Syst. Control Lett.*, Vol. 24, pp. 133–138, 1995.

Willems J.C. "Least squares stationary optimal control and the algebraic Riccati equation," *IEEE Trans. Autom. Control*, Vol. AC-16, pp. 621–634, 1971.

Wonham W.H. *Linear Multivariable Control: A Geometric Approach*, Springer-Verlag, Berlin, 1985.

Zames G. "Feedback and optimal sensitivity: Model reference transformations, multiplicative seminorms, and approximate inverses," *IEEE Trans. Autom. Control*, Vol. AC-26, pp. 301–320, 1981.

Zhou K., Doyle J.C., and Glover K. *Robust and Optimal Control*, Prentice Hall, Upper Saddle River, NJ, 1996.

Zhou K. and Khargonekar P. "An algebraic Riccati equation approach to H_∞ optimization," *Syst. Control Lett.*, Vol. 11(2), pp. 85–91, 1988.

CHAPTER 11

NUMERICAL METHODS AND CONDITIONING OF THE EIGENVALUE ASSIGNMENT PROBLEMS

Topics covered

- Numerical Methods for the Single-Input and Multi-Input Eigenvalue Assignment (EVA) Problems
- Sensitivity Analyses of the Feedback and EVA Problems and Conditioning of the Closed-Loop Eigenvalues
- Robust EVA

11.1 INTRODUCTION

We have introduced the eigenvalue assignment (EVA) problem **(pole-placement problem)** in Chapter 10 and given the results on existence and uniqueness of the solution.

In this chapter, we study numerical methods and the perturbation analysis for this problem.

There are many methods for the EVA problem. As stated in Chapter 10, some of the well-known theoretical formulas, such as the Ackermann formula, the Bass–Gura formula, etc., though important in their own rights, do not yield to computationally viable methods. The primary reason is that these methods are based on transformation of the controllable pair (A, B) to the controller-companion form, and the transforming matrix can be highly ill-conditioned in some cases. The computationally viable methods for EVA are based on transformation of the pair (A, B) to the controller-Hessenberg form or the matrix A to the real Schur form (RSF), which can be achieved using orthogonal transformations. Several methods of this type have been developed in recent years and we have described a few of them in this chapter. The methods described here include:

- A single-input recursive algorithm (**Algorithm 11.2.1**) (Datta 1987) and its RQ implementation (**Algorithm 11.2.3**) (Arnold and Datta 1998).
- A multi-input generalization (**Algorithm 11.3.1**) of the single-input recursive algorithm (Arnold and Datta 1990).
- A multi-input explicit QR algorithm (**Section 11.3.2**) (Miminis and Paige 1988).
- A multi-input Schur algorithm (**Algorithm 11.3.3**) (Varga 1981).
- A Sylvester equation algorithm for **partial eigenvalue assignment** (PEVA) (**Algorithm 11.3.4**) (Datta and Sarkissian 2002).

Algorithms 11.2.1 and 11.3.1 are the fastest algorithms, respectively, for the single-input and the multi-input EVA problems.

Unfortunately, the numerical stability of these algorithms are not guaranteed. However, many numerical experiments performed by the author and others (e.g., Varga 1996; Arnold and Datta 1998; Calvetti *et al.* 1999) show that Algorithm 11.2.1 works extremely well in practice, even with ill-conditioned problems. Furthermore, there is an RQ implementation which is numerically stable (Arnold and Datta 1998). This stable version is described in **Algorithm 11.2.3.**

The **multi-input explicit QR** algorithm in **Section 11.3.2** is also numerically stable. **However, it might give a complex feedback in some cases.**

The **Schur algorithm (Algorithm 11.3.3)**, based on the real Schur decomposition of the matrix A, is most expensive, but it has a distinguished feature that it can be used as a partial-pole placement algorithm in the sense that it lets the user reassign only a part of the spectrum leaving the rest unchanged. The algorithm is also believed to be **numerically stable.**

Besides the above-mentioned algorithms, an algorithm (**Algorithm 11.6.1**) for **robust eigenvalue assignment** (REVA), which not only assigns a desired set of eigenvalues but also a set of well-conditioned eigenvectors as well, is also included in this chapter. The REVA is important because the conditioning of the closed-loop eigenvector matrix greatly influences the sensitivity of the closed-loop eigenvalues (**see Section 11.5**). **Algorithm 11.6.1** is due to Kautsky *et al.* (1985) and is popularly known as the **KNV algorithm.** The MATLAB function **place** has implemented this algorithm.

Sections 11.4 and 11.5 are devoted, respectively, to the **conditioning of the feedback matrix and that of the closed-loop eigenvalues.** The conditioning of the feedback matrix and the conditioning of the closed-loop eigenvalues are two different matters. *It is easy to construct examples for which the feedback matrix can be computed rather very accurately by using a backward stable algorithm, but the resulting closed-loop eigenvalues might still be significantly* *different from those to be assigned.* These two problems are, therefore, treated separately.

The chapter concludes with a table of **comparison of different methods (Sections 11.7 and 11.8)** and **recommendations** are made based on this comparison (**Section 11.9**).

11.2   NUMERICAL METHODS FOR THE SINGLE-INPUT EIGENVALUE ASSIGNMENT PROBLEM

The constructive proof of Theorem 10.4.1 suggests the following method for finding the feedback vector f. Let (A, b) be controllable and S be the set of eigenvalues to be assigned.

Eigenvalue Assignment via Controller-Companion Form

Step 1. Find the coefficients $d_1, d_2, \ldots, d_n$ of the characteristic polynomial of the desired closed-loop matrix from the given set S.

Step 2. Transform (A, b) to the controller-companion form $(C, \tilde{b})$:

$$TAT^{-1} = C, \qquad Tb = \tilde{b},$$

where C is a **lower companion matrix** and $\tilde{b} = (0, 0, \ldots, 0, 1)^{\mathrm{T}}$.

Step 3. Compute $\hat{f}_i = d_i - a_i$, $i = 1, 2, \ldots, n$ where a_i, $i = 1, \ldots, n$ are the entries of the last row of C.

Step 4. Find $f^{\mathrm{T}} = \hat{f}^{\mathrm{T}}T$, where $\hat{f}^{\mathrm{T}} = (\hat{f}_1, \hat{f}_2, \ldots, \hat{f}_n)$.

Because of the difficulty of the implementation of Step 1 for large problems and of the instability of the algorithm due to possible ill-condition of T for finding the controller-canonical form in Step 2, as discussed in Chapter 6, **the above method is clearly not numerically viable.** It is of theoretical interest only.

Similar remarks hold for Ackermann's formula. The Ackermann (1972) formula, though important in its own right, is not numerically effective. Recall that the Ackermann formula for computing f to assign the spectrum $S = \{\lambda_1, \lambda_2, \ldots, \lambda_n\}$ is:

$$f^{\mathrm{T}} = e_n^{\mathrm{T}} C_{\mathrm{M}}^{-1} \phi(A),$$

where $C_{\mathrm{M}} = (b, Ab, \ldots, A^{n-1}b)$ and $\phi(x) = (x - \lambda_1)(x - \lambda_2) \cdots (x - \lambda_n)$. Thus, the implementation of Ackermann's formula requires: (i) computing $\phi(A)$ which involves computing various powers of A and (ii) computing the last row of the inverse of the controllability matrix. The **controllability matrix is usually ill-conditioned** (see the relevant comments in **Chapter 6**). The following example illustrates the point.

Example 11.2.1. Consider EVA using Ackermann's formula with

$$
A = \begin{pmatrix}
-4.0190 & 5.1200 & 0 & 0 & -2.0820 \\
-0.3460 & 0.9860 & 0 & 0 & -2.3400 \\
-7.9090 & 15.4070 & -4.0690 & 0 & -6.4500 \\
-21.8160 & 35.6060 & -0.3390 & -3.8700 & -17.8000 \\
-60.1960 & 98.1880 & -7.9070 & 0.3400 & -53.0080 \\
0 & 0 & 0 & 0 & 94.0000 \\
0 & 0 & 0 & 0 & 0 \\
0 & 0 & 0 & 0 & 0 \\
0 & 0 & 0 & 0 & 12.8000
\end{pmatrix}
$$

$$
\begin{pmatrix}
0 & 0 & 0 & 0.8700 \\
0 & 0 & 0 & 0.9700 \\
0 & 0 & 0 & 2.6800 \\
0 & 0 & 0 & 7.3900 \\
0 & 0 & 0 & 20.4000 \\
-147.2000 & 0 & 53.2000 & 0 \\
94.0000 & -147.20000 & 0 & 0 \\
12.80000 & 0 & -31.6000 & 0 \\
0 & 0 & 18.8000 & -31.6000
\end{pmatrix}
$$

$$
B = \begin{pmatrix}
-0.1510 \\
0 \\
0 \\
0 \\
0 \\
0 \\
0 \\
0 \\
0
\end{pmatrix}, \quad
S = \begin{pmatrix}
-1.0000 \\
-1.5000 \\
-2.0000 \\
-2.5000 \\
-3.0000 \\
-3.5000 \\
-4.0000 \\
-4.5000 \\
-5.0000
\end{pmatrix}
$$

The closed-loop eigenvalues assigned by the **Ackermann's formula** are:

$$
\begin{pmatrix}
-0.8824 & - & 0.4891j \\
-0.8824 & + & 0.4891j \\
-2.2850 & - & 1.0806j \\
-2.2850 & + & 1.0806j \\
-3.0575 & & \\
-3.8532 & & \\
-4.2637 & - & 0.7289j \\
-4.2637 & + & 0.7289j
\end{pmatrix}.
$$

Thus, the desired eigenvalues in S are completely different from those assigned by the Ackermann's formula. The same problem is then solved using the MAT-LAB function **place**, which uses the KNV algorithm. The spectrum assigned by MATLAB function **place** is: $\{-4.9999, -4.5001, -4.0006, -3.4999 -3.0003, -2.4984, -2.0007, -1.5004, -0.9998\}$.

Accurate results were also obtained by the recursive Algorithm (**Algorithm 11.2.1**) (see **Example 11.2.3**).

A Template of Numerical Algorithms for EVA Problem

A practical numerically effective algorithm has to be based upon the reduction of the controllable pair (A, B) to a canonical form pair that uses a well-conditioned transformation. As we have seen in Chapter 6, the controller-Hessenberg form is one such.

Indeed, several numerically effective algorithms have been proposed both for the single-input and the multi-input problems in recent years, based on the reduction of (A, B) to a controller-Hessenberg pair. We will describe some such algorithms in the sequel.

Most of these algorithms have a common basic structure which can be described as follows. In the following and elsewhere in this chapter, $\Omega(A)$ denotes the spectrum of A.

Step 1. The controllable pair (A, B) is first transformed to a controller-Hessenberg pair $(H, \tilde{B})$, that is, an orthogonal matrix P is constructed such that

$$PAP^{\mathrm{T}} = H, \quad \text{an unreduced block upper Hessenberg matrix,}$$

$$PB = \tilde{B} = \begin{pmatrix} B_1 \\ 0 \end{pmatrix}, \quad \text{where } B_1 \text{ is upper triangular.}$$

Note: In the single-input case, the controller-Hessenberg pair is $(H, \tilde{b})$, where H is an unreduced upper Hessenberg matrix and $\tilde{b} = Pb = \beta e_1, \beta \neq 0$.

Step 2. The EVA problem is now solved for the pair $(H, \tilde{B})$, by exploiting the special forms of H and $\tilde{B}$. This step involves finding a matrix F such that

$$\Omega(H - \tilde{B}F) = \{\lambda_1, \ldots, \lambda_n\}.$$

Note: In the single-input case, this step amounts to finding a row vector f^{T} such that $\Omega(H - \beta e_1 f^{\mathrm{T}}) = \{\lambda_1, \ldots, \lambda_n\}$.

Step 3. A feedback matrix K of the original problem is retrieved from the feedback matrix F of the transformed Hessenberg problem by using an orthogonal matrix multiplication: $K = FP$. (Note that $\Omega(H - \tilde{B}F) = \Omega(PAP^{\mathrm{T}} - PBFPP^{\mathrm{T}}) = \Omega(P(A - BK)P^{\mathrm{T}}) = \Omega(A - BK)$.)

The different algorithms differ in the way Step 2 is implemented. In describing the algorithms below, we will assume that Step 1 has already been implemented using the numerically stable Staircase Algorithm described in Chapter 6 (**Section 6.8**).

11.2.1   A Recursive Algorithm for the Single-Input EVA Problem

In this subsection, we present a simple recursive scheme (Datta 1987) for the single-input Hessenberg EVA problem.

Let's first remind the readers of the statement of the **single-input Hessenberg EVA problem:**

Given an unreduced upper Hessenberg matrix H, the number $\beta \neq 0$, and the set $S = \{\lambda_1, \ldots, \lambda_n\}$, closed under complex conjugation, find a row vector f^T such that

$$\Omega(H - \beta e_1 f^T) = \{\lambda_1, \ldots, \lambda_n\}.$$

We will assume temporarily, without any loss of generality, that $\beta = 1$ (recall that $Pb = \tilde{b} = \beta e_1$.)

Formulation of the Algorithm

The single-input EVA problem will have a solution if the closed-loop matrix $(H - e_1 f^T)$ can be made similar to a matrix whose eigenvalues are the same as the ones to be assigned.

Thus, the basic idea here is to construct a nonsingular matrix X such that

$$X(H - e_1 f^T)X^{-1} = \Lambda, \tag{11.2.1}$$

where $\Omega(\Lambda) = \{\lambda_1, \lambda_2, \ldots, \lambda_n\}$.

From (11.2.1), we have

$$XH - \Lambda X = X e_1 f^T. \tag{11.2.2}$$

Taking the transpose, the above equation becomes

$$H^T X^T - X^T \Lambda^T = f e_1^T X^T. \tag{11.2.3}$$

Setting $X^T = L$, Eq. (11.2.3) becomes

$$H^T L - L \Lambda^T = f e_1^T L. \tag{11.2.4}$$

The problem at hand now is to construct a nonsingular matrix L and a vector f such that Eq. (11.2.4) is satisfied. We show below how some special choices make it happen.

Let's choose

$$\Lambda^{T} = \begin{pmatrix} \lambda_1 & & & \\ * & \ddots & & 0 \\ & \ddots & \ddots & \\ 0 & & \ddots & \ddots \\ & & & * & \lambda_n \end{pmatrix}$$

and let $e_1^{T}L$ be chosen such that all but the last column of the matrix on the right-hand side of (11.2.4) are zero, that is, the matrix Eq. (11.2.4) becomes

$$H^{T}L - L\Lambda^{T} = (0, 0, \dots, \alpha f), \quad \alpha \neq 0. \tag{11.2.5}$$

The simple form of the right-hand side of (11.2.5) allows us to compute recursively the second through nth column of $L = (l_1, l_2, \dots, l_n)$, once the first column l_1 is known. The entries of the subdiagonal of Λ can be chosen as scaling factors for the computed columns of L. Once L is known, αf can be computed by equating the last column of both sides of (11.2.5):

$$\alpha f = (H^{T} - \lambda_n I)l_n. \tag{11.2.6}$$

What now remains to be shown is that how to choose l_1 such that the resulting matrix L in (11.2.5) is nonsingular.

A theorem of K. Datta (1988) tells us that if l_1 is chosen such that (H^{T}, l_1) is controllable, then L satisfying (11.2.5) will be nonsingular. Since H^{T} is an unreduced upper Hessenberg matrix, the simplest choice is $l_1 = e_n = (0, 0, \dots, 0, 1)^{T}$. It is easy to see that this choice of l_1 will yield $\alpha = l_{1n}$, the first entry of l_n. Then equating the last column of both sides of (11.2.5), we have

$$f = \frac{(H^{T} - \lambda_n I)l_n}{\alpha} = \frac{(H^{T} - \lambda_n I)l_n}{l_{1n}}.$$

The above discussion immediately leads us to the following algorithm:

Algorithm 11.2.1. *The Recursive Algorithm for the Single-input Hessenberg EVA Problem*

Inputs. *H, an unreduced upper Hessenberg matrix, and* $S = \{\lambda_1, \lambda_2, \dots, \lambda_n\}$, *a set of n numbers, closed under complex conjugation.*
Output. *The feedback vector f such that* $\Omega(H - e_1 f^{T}) = S$.

Step 1. *Set* $l_1 = e_n$, *the last column of the* $n \times n$ *identity matrix*

Step 2. *Construct a set of normalized vectors* $\{\ell_k\}$ *as follows:*

   For $i = 1, 2, \ldots, n - 1$ *do*

$$\text{Compute } \hat{\ell}_{i+1} = (H^{\mathrm{T}} - \lambda_i I)\ell_i$$

$$\ell_{i+1} = \frac{\hat{\ell}_{i+1}}{\|\hat{\ell}_{i+1}\|_2}$$

   End

Step 3. *Compute* $\ell_{n+1} = (H^{\mathrm{T}} - \lambda_n I)\ell_n$.

Step 4. *Compute* $f = \dfrac{\ell_{n+1}}{\alpha}$, *where* α *is the first entry of* ℓ_n.

Theorem 11.2.1. *The vector* f *computed by Algorithm 11.2.1 is such that the eigenvalues of the closed-loop matrix* $(H - e_1 f^{\mathrm{T}})$ *are* $\lambda_1, \ldots, \lambda_n$.

Proof. Proof follows from the above discussions. ■

Flop-count: Since l_i contains only i nonzero entries and H is an unreduced Hessenberg matrix, computations of l_2 through l_n in Algorithm 11.2.1 takes about $n^3/3$ flops. Furthermore, with these operations, one gets the transforming matrix L that transforms the closed-loop matrix to Λ by similarity. Also, it takes about $\frac{10}{3}n^3$ flops for the single-input controller-Hessenberg reduction. So, the flop-count for the EVA problem for the pair (A, b) using Algorithm 11.2.1 is about $\frac{11}{3}n^3$ flops.

Avoiding complex arithmetic: When the eigenvalues to be assigned are complex, the use of complex arithmetic in Algorithm 11.2.1 can be avoided by setting Λ as a matrix in RSF, having a 2×2 block corresponding to each pair of complex conjugate eigenvalues to be assigned. Algorithm 11.2.1 needs to be modified accordingly **[Exercise 11.1]**.

MATCONTROL note: The modified algorithm that avoids the use of complex arithmetic has been implemented in MATCONTROL function **polercs**.

Example 11.2.2. Consider EVA using Algorithm 11.2.1 with

$$H = \begin{pmatrix} 9 & 4 & 7 \\ 3 & 1 & 2 \\ 0 & 9 & 6 \end{pmatrix}, \qquad S = \{9 \; 5 \; 1\}.$$

Step 1. $l_1 = \begin{pmatrix} 0 \\ 0 \\ 1 \end{pmatrix}$

Step 2.

$i = 1$

$$\hat{l}_2 = \begin{pmatrix} 0 \\ 9 \\ -3 \end{pmatrix}, \qquad l_2 = \begin{pmatrix} 0 \\ 0.9487 \\ -0.3162 \end{pmatrix}.$$

$i = 2$

$$\hat{l}_3 = \begin{pmatrix} 2.8460 \\ -6.6408 \\ 1.5811 \end{pmatrix}, \qquad l_3 = \begin{pmatrix} 0.3848 \\ -0.8979 \\ 0.2138 \end{pmatrix}.$$

Step 3.

$$l_4 = \begin{pmatrix} 0.3848 \\ 3.4633 \\ 1.9668 \end{pmatrix}.$$

Step 4.

$$f = \begin{pmatrix} 1.0000 \\ 9.0000 \\ 5.1111 \end{pmatrix}.$$

The closed-loop matrix:

$$H - e_1 f^T = \begin{pmatrix} 8.0 & -5.0 & 1.8889 \\ 3.0 & 1.0 & 2.0 \\ 0 & 9.0 & 6.0 \end{pmatrix}.$$

Verify: The eigenvalues of the matrix $(H - e_1 f^T)$ are 9, 5, and 1.

Example 11.2.3. Let's apply Algorithm 11.2.1 to Example 11.2.1 again. The eigenvalues of the closed-loop matrix $H - e_1 f^T$ with the vector f computed by Algorithm 11.2.1 are:

$-5.0001, -4.4996, -4.0009, -3.4981, -3.0034, -2.4958, -2.0031, -1.4988, -1.0002.$

The computed closed-loop eigenvalues, thus, have the similar accuracy as those obtained by the MATLAB function **place**.

Example 11.2.4. *Eigenvalue Assignment with Ill-Conditioned Eigenvalues.* Since the matrix H and the closed-loop matrix $H - e_1 f^T$ differ only by the first row, Algorithm 11.2.1 amounts to finding a vector f such that, when the first row of the given Hessenberg matrix H is replaced by f^T, the resulting new Hessenberg matrix has the prescribed spectrum. Let H be the well-known Wilkinson bidiagonal matrix (see Datta (1995), Wilkinson (1965)) with highly ill-conditioned eigenvalues:

$$
H = \begin{pmatrix}
20 & & & & & \\
 & 19 & & & & \\
 & & 20 & \ddots & & 0 \\
 & & & 20 & \ddots & \\
 & 0 & & & \ddots & \ddots \\
 & & & & 20 & 1
\end{pmatrix}.
$$

First, the first row of H is replaced by the zero row vector and then Algorithm 11.2.1 is run on this new H with $S = \{20, 19, 18, \ldots, 1\}$, and f is computed. Since the eigenvalues of the original matrix H is the set S; in theory, f^{T} should be the same as the first row of H; namely, $f^{\mathrm{T}} = (20, 0, \ldots, 0)$. Indeed, the vector f^{T} computed by the algorithm is found to be $f^{\mathrm{T}} = (20, 0, \ldots, 0)$ and the eigenvalues of the closed-loop matrix with this computed f are $1, 2, 3, \ldots, 20$.

A closed-form solution of the feedback vector in the single-input EVA problem: We now show that Algorithm 11.2.1 yields an explicit closed-form solution for the single-input problem. The recursion in Step 2 of the algorithm yields

$$
\gamma\, l_{i+1} = (H^{\mathrm{T}} - \lambda_1 I)(H^{\mathrm{T}} - \lambda_2 I) \cdots (H^{\mathrm{T}} - \lambda_i I) l_1, \tag{11.2.7}
$$

for some (nonzero) scalar γ. Including Steps 1 and 4, (11.2.6) becomes

$$
\alpha f = (H^{\mathrm{T}} - \lambda_1 I)(H^{\mathrm{T}} - \lambda_2 I) \cdots (H^{\mathrm{T}} - \lambda_n I) e_n, \tag{11.2.8}
$$

where $\alpha = (h_{21} h_{32} \cdots h_{n,n-1})^{-1}$. If $\phi(x) = (x - \lambda_1)(x - \lambda_2) \cdots (x - \lambda_n)$, then this can be written as

$$
f^{\mathrm{T}} = \alpha e_n^{\mathrm{T}} \phi(H). \tag{11.2.9}
$$

Since this solution is unique, it **represents *the* Hessenberg representation of the Ackermann formula for the single-input EVA problem.**

11.2.2   An Error Analysis of the Recursive Single-Input Method

Knowing the above explicit expression for the feedback vector f, we can now carry out a forward error analysis of Algorithm 11.2.1. This is presented below.

By duality of (11.2.4), we see that the method computes a matrix L and a vector f such that

$$HL - L\Lambda = \alpha f e_1^T L.$$

A careful look at the iteration reveals that the forward error has a special form. Define the polynomials $\phi_{j,k}$ for $j \le k$ by

$$\phi_{j,k}(x) = (x - \lambda_j)(x - \lambda_{j+1}) \cdots (x - \lambda_k).$$

Let $\bar{l}_i$ be the computed value of the ith column of L. Define ϵ_i by

$$\bar{l}_{i+1} = (H - \lambda_i I)\bar{l}_i + \epsilon_i. \tag{11.2.10}$$

Then we have the following **forward error** formula, due to Arnold (1993) (See also Arnold and Datta (1998)).

Theorem 11.2.2. *Let $\bar{\alpha}\bar{f}$ be the* computed *feedback vector of the single-input EVA problem for (H, e_1) using Algorithm 11.2.1. If αf is the exact solution, then*

$$\bar{\alpha}\bar{f} - \alpha f = \sum_{j=1}^{n} \phi_{j,n}(H)\epsilon_j,$$

where ϵ_js are defined above.

Unfortunately, not much can be said about backward stability from a result like this. It is, however, possible to shed some light on the stability of this method by looking at ϵ_j in a different way. See Arnold and Datta (1998) for details.

Theorem 11.2.3. *Let $E = [\epsilon_1, \epsilon_2, \ldots, \epsilon_n]$ and let $\bar{L} = [\bar{\ell}_1, \bar{\ell}_2, \ldots, \bar{\ell}_n]$. Then $\bar{\alpha}\bar{f}$ solves (exactly) the single-input EVA problem for the perturbed system $(H - E\bar{L}^{-1}, \beta e_1, S)$, where the ϵ_i are the same as in the previous theorem, that is, the computed vector $\bar{f}$ is such that*

$$\Omega[(H - E\bar{L}^{-1}) - \beta e_1 \bar{f}^T] = \{\lambda_1, \ldots, \lambda_n\}.$$

Proof.   Notice that as defined, $\bar{L}$ satisfies the matrix equation:

$$H\bar{L} - \bar{L}\Lambda = E + \bar{\alpha}\bar{f}e_n^T,$$

where $\Lambda = \text{diag}(\lambda_i)$. Since $\bar{L}$ is nonsingular by construction, we can solve the perturbed equation:

$$(H + \Delta H)\bar{L} - \bar{L}\Lambda = E + \bar{\alpha}\bar{f}e_n^T$$

for ΔH. This yields $-\Delta H = E\bar{L}^{-1}$, and $\bar{\alpha}\bar{f}$ solves the EVA problem for $(H + \Delta H, \beta e_1, S)$.   ∎

Remarks on Stability and Reliability

- From the above result it cannot be said that the method is backward stable. The result simply provides an upper bound on the size of the ball around the initial data, inside which there exist $(H + \Delta H, \beta + \delta \beta)$ for which the computed solution is exact. If $\|\Delta H\|$ could be bounded above by a small quantity that was relatively independent of the initial data, then the method would be backward stable. However, Theorem 11.2.3 does allow one to say precisely when the results from the method are suspect. It is clear that $\|E\|$ is always small if the iterates are normalized every few steps, so that all of the backward error information is contained in $\bar{L}^{-1}$. **Thus, the stability of the algorithm can be monitored by monitoring the condition number of $\bar{L}$. Furthermore, since $\bar{L}$ is triangular, it is possible to estimate $\|\bar{L}^{-1}\|$ rather cheaply, even as the iteration proceeds.** (See Higham 1996.)

  The matrix L gives us even more information about the EVA problem at hand. In case the eigenvalues to be assigned are distinct, an upper bound in the condition number of the matrix of eigenvectors that diagonalizes the closed loop matrix can be obtained from the condition number of the matrix L.

  This is important because, as said in the introduction, the condition number of the matrix of eigenvectors of the closed-loop matrix is an important factor in the accuracy of the assigned eigenvalues (see Section 11.5 and the Example therein).

  Specifically, if X diagonalizes Λ, then it can be shown (Arnold and Datta 1998), that $P = (\bar{L})^{-1}X$ diagonalizes the closed-loop matrix $H - e_1 f^T$, furthermore,

  $$\text{Cond}_2(P) \leq \text{Cond}_2(X)\text{Cond}_2(L).$$

- **Computational experience has shown that if L is ill-conditioned, then so are the closed-loop eigenvalues.**

11.2.3 The QR and RQ Implementations of Algorithm 11.2.1

Algorithm 11.2.1 is an extremely efficient way to solve the Hessenberg single-input EVA problem, but as we have just seen, the backward stability of this algorithm cannot be guaranteed. It, however, turns out that there is a **numerically stable implementation** of this algorithm via QR iterations. We will discuss this below.

The QR Implementation of Algorithm 11.2.1

The idea of using QR iterations in implementing Algorithm 11.2.1 comes from the fact that the matrix $\phi(H)$ in the explicit expression of f in (11.2.9) can be written as (Stewart 1973, p. 353):

$$\phi(H) = (H - \lambda_2 I)(H - \lambda_2 I) \cdots (H - \lambda_n I) = Q_1 Q_2 \cdots Q_n R_n R_{n-1} \cdots R_1,$$

where Q_i and R_i are generated by n steps of QR iterations as follows:
$H_1 = H$
For $i = 1, 2, \ldots, n$ do
$\qquad Q_i R_i = H_i - \lambda_i I$
$\qquad H_{i+1} = R_i Q_i + \lambda_i I$
End.

Remark

- Note that since H_i is Hessenberg, so is H_{i+1}, for each i (see **Chapter 4**).

MATCONTROL note: The QR version of Algorithm 11.2.1 has been implemented in MATCONTROL function **poleqrs**.

The RQ Implementation of Algorithm 11.2.1

The difficulty of implementing the QR strategy is that the R_i need to be accumulated; the process is both expensive and unstable.

We now show how the method can be made computationally efficient by using **RQ factorizations** instead of QR factorizations, as follows:
Set $H_1 = H$
For $i = 1, 2, \ldots, n$ compute the RQ step

$$R_i Q_i = H_i - \lambda_i I$$
$$H_{i+1} = Q_i R_i + \lambda_i I$$

This time

$$\phi(H) = R_1 R_2 \cdots R_n Q_n Q_{n-1} \cdots Q_1, \qquad (11.2.11)$$

and by setting $Q = Q_n Q_{n-1} \cdots Q_1$ and $R = R_1 R_2 \cdots R_n$, we have from (11.2.9)

$$f^T = \alpha e_n^T R Q = \alpha \rho e_n^T Q. \qquad (11.2.12)$$

Here $\rho = \Pi_{i=1}^n r_{nn}^{(i)}$, where $r_{nn}^{(i)}$ denotes the (n, n)th entry of R_i. *This is a much nicer situation!* Thus, a straightforward RQ implementation of Algorithm 11.2.1 will be as follows:

Algorithm 11.2.2. *An RQ Implementation of Algorithm* 11.2.1
   Inputs. *Same as in Algorithm* 11.2.1*.*
   Output. *Same as in Algorithm* 11.2.1*.*
   Step 0. *Set $H_1 = H$.*

Step 1.   *For $i = 1, 2, \ldots, n$ do*
$$R_i Q_i = H_i - \lambda_i I$$
$$H_{i+1} = Q_i R_i + \lambda_i I$$
End

Step 2.   *Compute $f = \alpha \rho e_n^T Q_n Q_{n-1} \cdots Q_1$, where $\rho = \Pi_{i=1}^n r_{nn}^{(i)}$, $r_{nn}^{(i)}$ denotes the (n, n)th entry of R_i.*

Algorithm 11.2.2 may be made storage-efficient by observing that it is possible to **deflate** the problem at each RQ step, as follows:

$$H_{i+1} = Q_i R_i + \lambda_i I = \begin{pmatrix} * & * \\ 0 & \widetilde{H}_{i+1} \end{pmatrix}.$$

The matrix $\widetilde{H}_{i+1}$ can now be set as H_{i+1} and the iteration can be continued with $H_{i+1} \equiv \widetilde{H}_{i+1}$ after updating Q_i and ρ appropriately. Thus, algorithmically we have the **following storage-efficient version of Algorithm 11.2.2, which is recommended to be used in practice.**

Algorithm 11.2.3.   *A Storage-Efficient Version of Algorithm 11.2.2*
Inputs. *Same as in Algorithm 11.2.1.*
Output. *Same as in Algorithm 11.2.1.*

Step 0. *Set $H_1 = H$.*

Step 1. *Compute the RQ factorization of $H_1 - \lambda_1 I$; that is, compute Q_1 and R_1 such that $(H - \lambda_1 I)Q_1^T = R_1$. Compute $H_2 = Q_1 R_1 + \lambda_1 I$. Set $Q = Q_1$ and $\rho = r_{nn}^{(1)}$, where $R_1 = (r_{ij}^{(1)})$.*

Step 2. *For $i = 2, 3, \ldots, n$ do*
Compute the RQ factorization of $H_i - \lambda_i I$: $(H_i - \lambda_i I)Q_i^T = R_i$. Compute H_{i+1}, where $Q_i R_i + \lambda_i I = \begin{pmatrix} * & * \\ 0 & H_{i+1} \end{pmatrix}$. Update
$Q \equiv \begin{pmatrix} I & \\ & Q_i \end{pmatrix} Q$, *where I is a matrix consisting of the first $(i - 2)$ rows and columns of the identity matrix. Update $\rho \equiv \rho \, r_{n+2-i,n+2-i}^{(i)}$ ($r_{n+2-i,n+2-i}^{(i)}$ is the last element of R_i).*
End

Step 3.   *Compute $f^T = \alpha' \rho e_n^T Q$, where $\alpha' = 1/(h_{21} \cdots h_{n,n-1})$.*

Flop-count and numerical stability: Algorithm 11.2.3 requires about $\frac{5}{3}n^3$ flops. Since reduction to the controller-Hessenberg form requires $\frac{10}{3}n^3$ flops, the total count for EVA of the pair (A, b) using Algorithm 11.2.3 is about $5n^3$.

The algorithm is *numerically stable* (see Arnold and Datta 1998). Specifically, the method computes, given a controllable pair (H, e_1), a vector $\bar{f}$ such that it solves exactly the EVA problem for the system with the matrix $H + \Delta H$, where

$$\|\Delta H\|_F \leq \mu g(n)\|H\|_F,$$

in which μ is the machine precision and $g(n)$ is a modest function of n.

Remark

- It can be shown (Arnold 1993) that if an EVA algorithm for the single-input Hessenberg problem is backward stable, then the algorithm is also backward stable for the original problem.

Thus, the RQ implementation of Algorithm 11.2.1 is backward stable for the original problem. That is, the feedback $\bar{k}$, computed by Algorithm 11.2.3, for the problem (A, b) is exact for a nearby problem: $\bar{k}$ exactly solves the EVA problem for $(A + \Delta A, b + \delta b)$, where ΔA and δb are small. For a proof of this backward stability result, see Arnold and Datta (1998).

Example 11.2.5.   Consider Example 11.2.2 again

Step 0.

$$H_1 = H = \begin{pmatrix} 9 & 4 & 7 \\ 3 & 1 & 2 \\ 0 & 9 & 6 \end{pmatrix}, \qquad S = \{\lambda_1, \lambda_2, \lambda_3\} = \{9, 5, 1\}.$$

Step 1. Compute R_1 and Q_1 such that $H_1 - \lambda_1 I = R_1 Q_1$: $[R_1, Q_1] = rq\ (H_1 - \lambda_1 I)$.

$$\text{Compute } H_2 = Q_1 R_1 + \lambda_1 I = \begin{pmatrix} 10.5957 & -0.6123 & -6.5885 \\ -7.5693 & 7.2043 & 0.2269 \\ 0 & 2.9086 & -1.8000 \end{pmatrix}.$$

$$Q \equiv Q_1 = \begin{pmatrix} -0.2063 & 0.3094 & 0.9283 \\ 0.9785 & -0.0652 & -0.1957 \\ 0 & 0.9487 & -0.3162 \end{pmatrix}, \qquad \rho = 9.4868$$

Step 2. $i = 2$

Compute R_2 and Q_2 such that $H_2 - \lambda_2 I = R_2 Q_2$: $[R_2, Q_2] = rq\ (H_2 - \lambda_2 I)$.

$$H_3 = \begin{pmatrix} 12.9533 & 4.6561 \\ -3.0909 & -1.5411 \end{pmatrix}, \quad \text{Update } Q: Q \equiv \begin{pmatrix} -0.8109 & -0.2183 & 0.5430 \\ -0.4409 & -0.3823 & -0.8121 \\ 0.3848 & -0.8479 & 0.2138 \end{pmatrix}.$$

$$\text{Update } \rho: \rho \equiv 70.1641$$

$i = 3$

Compute R_3 and Q_3 such that $H_3 - \lambda_3 I = R_3 Q_3$: $[R_3, Q_3] = rq(H_3 - \lambda_3 I)$.

$$R_3 = \begin{pmatrix} -3.9945 & -12.1904 \\ 0 & 4.0014 \end{pmatrix}, \quad Q_3 = \begin{pmatrix} -0.6351 & 0.7725 \\ -0.7725 & -0.6351 \end{pmatrix}.$$

$H_4 = 7.8755,$

$$\text{Update } Q: Q \equiv \begin{pmatrix} -0.8109 & -0.2183 & 0.5430 \\ 0.5772 & -0.4508 & 0.6809 \\ 0.0962 & 0.8655 & 0.4915 \end{pmatrix}, \quad \text{update } \rho \equiv 280.7526.$$

Step 3.

$$f = \begin{pmatrix} 1.0000 \\ 9.0000 \\ 5.1111 \end{pmatrix}.$$

Verify:

$$H - e_1 f^T = \begin{pmatrix} 8.0000 & -5.0000 & 1.8889 \\ 3.0000 & 1.0000 & 2.0000 \\ 0 & 9.0000 & 6.0000 \end{pmatrix}.$$

The eigenvalues of $H - e_1 f^T$ are $\{5, 1, 9\}$.

MATCONTROL note: Algorithm 11.2.3 has been implemented in MATCON-TROL function **polerqs**.

11.2.4   Explicit and Implicit RQ Algorithms

We have just seen the QR and RQ versions of Algorithm 11.2.1. At least two more QR type methods were proposed earlier: An **explicit QR algorithm** by Miminis and Paige (1982) and an **implicit QR** algorithm by Patel and Misra (1984).

The explicit QR algorithm, proposed by Miminis and Paige (1982), constructs an orthogonal matrix Q such that

$$Q^T(H - \beta e_1 f^T)Q = R,$$

where R is an upper triangular matrix with the desired eigenvalues $\lambda_1, \ldots, \lambda_n$ on the diagonal. The algorithm has a forward sweep and a backward sweep. The forward sweep determines Q and the backward sweep finds f and R, if needed.

The algorithm explicitly uses the eigenvalues to be assigned as shifts and that is why it is called an **explicit QR algorithm**.

It should come as no surprise that **an implicit** RQ step is possible, and in order to handle **complex pairs of eigenvalues with real arithmetic**, an implicit double step is needed. One such method has been proposed by Patel and Misra (1984). The Patel–Misra method is similar to the Miminis–Paige method, but it includes an alternative to the "backward sweep." There now exist RQ formulations of both these algorithms (Arnold 1993; Arnold and Datta 1998). *These RQ formulations are easier to describe, understand, and implement.*

It should be mentioned here that there now exists a generalization of the implicit QR algorithm due to Varga (1996) that performs an implicit multistep in place of a double-step. The Varga algorithm is slightly more efficient than the Patel–Misra algorithm and like the latter, is believed to be numerically stable.

Methods Not Discussed

Besides the methods discussed above, there are many other methods for the single-input problem. These include the methods based on solutions of independent linear systems (Datta and Datta 1986; Bru *et al.* 1994a); the eigenvector method by Petkov *et al.* (1984), etc., parallel algorithms by Coutinho *et al.* (1995), and by Bru *et al.* (1994c), etc.; and the multishift algorithm by Varga (1996). See **Exercises 11.2–11.4 and 11.8** for statements of some of these methods.

11.3   NUMERICAL METHODS FOR THE MULTI-INPUT EIGENVALUE ASSIGNMENT PROBLEM

Some of the single-input algorithms described in the last section have been generalized in a straightforward fashion to the multi-input case or similar algorithms have been developed for the latter.

We describe here:

- A multi-input generalization of the single-input recursive algorithm (Arnold and Datta 1990).
- An explicit QR algorithm (Miminis and Paige 1988).
- A Schur method (Varga 1981).
- A Sylvester equation algorithm for PEVA algorithm (Datta and Sarkissian 2002).

There are many more algorithms for this problem that are not described here. Some of them are: a multi-input generalization of the single-input eigenvector algorithm by Petkov *et al.* (1986), a multi-input generalization of the single-input algorithm using solutions of linear systems by Bru *et al.* (1994b) (**Exercise 11.5**), a matrix equation algorithm by Bhattacharyya and DeSouza (1982) (**Exercise 11.14**), and a multi-input version of the single-input implicit QR algorithm by Patel and Misra (1984), algorithms by Tsui (1986) and Shafai and Bhattacharyya (1988), and parallel algorithms by Baksi *et al.* (1994), Datta (1989), etc.

11.3.1   A Recursive Multi-Input Eigenvalue Assignment Algorithm

The following algorithm is a generalization of the single-input recursive algorithm (**Algorithm 11.2.1**) to the multi-input case.

The development of this algorithm is along the same line as Algorithm 11.2.1.

The version of the algorithm presented here is a little different than that originally proposed in Arnold and Datta (1990).

Given a controller-Hessenberg pair $(H, \tilde{B})$ and the set $S = \{\lambda_1, \lambda_2, \ldots, \lambda_n\}$, the algorithm, like its single-input version, constructs a nonsingular matrix L recursively from where the feedback matrix F can be easily computed. Since in the multi-input case the matrix H of the controller-Hessenberg form is a block-Hessenberg matrix, by taking advantage of the block form of H, this time the matrix L can be computed in blocks. The matrix L can be computed either block column-wise or block row-wise. We compute L block row-wise here starting with the last block row.

Thus, setting

$$\Lambda = \begin{pmatrix} \Lambda_{11} & & & \\ \Lambda_{21} & \Lambda_{22} & & 0 \\ & \ddots & \ddots & \\ 0 & & \Lambda_{k,k-1} & \Lambda_{kk} \end{pmatrix},$$

where the eigenvalues $\lambda_1, \ldots, \lambda_n$ are contained in the diagonal blocks of Λ, and considering the equation:

$$LH - \Lambda L = L \begin{pmatrix} R \\ 0 \end{pmatrix} F,$$

it is easily seen that the matrices L and R can be found without knowing the matrix F. Indeed, the matrix

$$L = \begin{pmatrix} L_1 \\ L_2 \\ \vdots \\ L_k \end{pmatrix}$$

can be computed recursively block row-wise starting with L_k and if L_k is chosen as $L_k = (0, 0, \ldots 0, I_{n_k})$, then L will be nonsingular. Equating the corresponding block-rows of the equation:

$$LH - \Lambda H = \binom{R}{0} F,$$

it is easy to see that

$$\Lambda_{i+1,i} L_i = L_{i+1} H - \Lambda_{i+1,i+1} L_{i+1} = \tilde{L}_i, \quad i = k - 1, k - 2, \ldots, 2, 1,$$

from where the matrices $\Lambda_{i+1,i}$ and L_i can be computed by using the QR factorization of $\tilde{L}_i$. Once L and R are found, the matrix F can be computed from the above equation by solving a block linear system. Overall, we have the following algorithm.

Algorithm 11.3.1. *The Recursive Algorithm for the Multi-Input EVA Problem*
Inputs.
A—The $n \times n$ state matrix.
B—The $n \times m$ input matrix ($m \leq n$).
S—The set of numbers $\{\lambda_1, \lambda_2, \ldots, \lambda_n\}$, closed under complex conjugation.
Assumption. *(A, B) is controllable.*
Output. *A feedback matrix K such that $\Omega(A - BK) = \{\lambda_1, \lambda_2, \ldots, \lambda_n\}$.*

Step 1. *Using the Staircase Algorithm in Section 6.7, reduce the pair (A, B) to the controller-Hessenberg pair $(H, \tilde{B})$, that is, find an orthogonal matrix P such that $PAP^{\mathrm{T}} = H$, an unreduced block upper Hessenberg matrix with k diagonal blocks and*

$$PB = \tilde{B} = \binom{R}{0}, \quad R \text{ is upper triangular and has full rank.}$$

Step 2. *Partition S in such a way that $S = \cup\Omega(\Lambda_{ii})$, where each Λ_{ii} is an $n_i \times n_i$ diagonal matrix (Recall that n_is are defined by the dimensions of the blocks in $H = (H_{ij})$; $H_{ij} \in \mathbb{R}^{n_i \times n_j}$).*

Step 3. *Set $L_k = (0, \ldots, 0, I_{n_k})$.*
Step 4. *For $i = k - 1, \ldots, 1$ do*
    4.1. *Compute $\tilde{L}_i \equiv L_{i+1} H - \Lambda_{i+1,i+1} L_{i+1}$*
    4.2. *Compute the QR decomposition of $\tilde{L}_i^{\mathrm{T}} : \tilde{L}_i^{\mathrm{T}} = QR$*
    4.3. *Define $L_i = Q_1^{\mathrm{T}}$, where Q_1 are the first n_i columns of the matrix*
$Q = (Q_1, Q_2)$
    End

Step 5. *Solve the linear system* $(L_{11}R)F = L_1H - \Lambda_{11}L_1$ *for F, where* L_{11} *is the matrix of the first* n_1 *columns of* L_1.

Step 6. *Compute the feedback matrix K of the original problem:* $K \equiv FP$.

Theorem 11.3.1.   *The feedback matrix K constructed by the above algorithm is such that*

$$\Omega(A - BK) = \{\lambda_1, \lambda_2, \dots, \lambda_n\}.$$

Proof.   Proof follows from the discussion preceding the algorithm.   ■

Flop-count: Approximately $\frac{19}{3}n^3 + \frac{15}{2}n^2m$ flops are required to implement the algorithm.

It may be worth noting a few points regarding the complexity of this algorithm. First, the given operations count includes assigning complex eigenvalues using real arithmetic. Second, almost 95% of the total flops required for this method are in the reduction to the controller-Hessenberg form in Step 1. Finally, within the above operations count (but with some obvious additional storage requirements), the matrix L that transforms the reduced closed-loop system to the block bidiagonal matrix Δ by similarly, can be obtained.

Avoiding complex arithmetic: In order to assign a pair of complex conjugate eigenvalues using only real arithmetic, we set 2×2 "**Schur bumps**" on the otherwise diagonal Λ_{ii}. For example, if we want to assign the eigenvalues $x \pm iy$ to $A - BK$, we might set

$$\Lambda_3 = \begin{bmatrix} x & -y \\ y & x \end{bmatrix}.$$

However, the algorithm might give a complex feedback matrix if all the complex conjugate pairs cannot be distributed as above along the diagonal blocks Λ_{ii}. Some modifications of the algorithm in that case will be necessary. A *block algorithm that avoids complex feedback has been recently proposed by Carvalho and Datta* (2001).

MATCONTROL note: The modified version of Algorithm 11.3.1, proposed in Carvalho and Datta (2001), that avoids the use of complex arithmetic and is guaranteed to give a real feedback matrix has been implemented in MATCONTROL function **polercx**, while Algorithm 11.3.1 as it appears here has been implemented in MATCONTROL function **polercm**.

Example 11.3.1.   Consider EVA using Algorithm 11.3.1 with

$$A = H = \begin{pmatrix} 1 & 2 & 3 & 4 & 1 \\ 1 & 1 & 1 & 1 & 1 \\ 2 & 1 & 1 & 1 & 1 \\ 0 & 0 & 1 & 1 & 2 \\ 0 & 0 & 0 & 1 & 1 \end{pmatrix},$$

$$B = \tilde{B} = \begin{pmatrix} 1 & 1 & 1 \\ 0 & 1 & 2 \\ 0 & 0 & 3 \\ 0 & 0 & 0 \\ 0 & 0 & 0 \end{pmatrix} = \begin{pmatrix} R \\ 0 \end{pmatrix}, \qquad S = \{1, 2, 3, 4, 5\}.$$

Here $k = 3$, $n_1 = 3$, $n_2 = 1$, and $n_3 = 1$.

Step 1. The pair $(H, \tilde{B})$ is already in controller-Hessenberg form

Step 2. $\Lambda_{11} = \text{diag}(1, 2, 3)$, $\Lambda_{22} = 4$, $\Lambda_{33} = 5$.

Step 3. $L_3 = (0, 0, 0, 0, 1)$.

Step 4. $i = 2$

4.1: $\tilde{L}_2 = (0 \ 0 \ 0 \ 1 \ -4)$

4.2: (not shown)

4.3: $L_2 = (0 \ 0 \ 0 \ -0.2425 \ 0.9701)$

$i = 1$

4.1: $\tilde{L}_1 = (0 \ 0 \ -0.2425 \ 1.6977 \ -3.3955)$

4.2: (not shown)

4.3:

$$L_1 = \begin{pmatrix} 0 & 0 & 0.0638 & -0.4463 & 0.8926 \\ 0 & 1.0000 & 0 & 0 & 0 \\ 0.0638 & 0 & 0.9959 & 0.0285 & -0.0569 \end{pmatrix}.$$

Step 5.

$$F = \begin{pmatrix} -2.3333 & 3.3333 & 78.2161 & -212.4740 & 217.0333 \\ -0.3333 & -1.6667 & 5.6667 & -9.0000 & 9.6667 \\ 0.6667 & 0.3333 & -2.3333 & 5.0000 & -4.3333 \end{pmatrix}.$$

Verify: The eigenvalues of $H - \tilde{B}F$ are: $\{1.0000 \ 2.0000 \ 3.0000 \ 4.0000 \ 5.0000\}$.

11.3.2 The Explicit QR Algorithm for the Multi-Input EVA Problem

The following multi-input QR algorithm due to Miminis and Paige (1988) also follows the same "template" as that of the preceding algorithm. The algorithm consists of the following three major steps.

Step 1. The controllable pair (A, B) is transformed to the controller-Hessenberg pair $(H, \tilde{B})$:

$$PAP^\mathsf{T} = H = \begin{pmatrix} H_{11} & & & \\ H_{21} & \ddots & & H_{ij} \\ & \ddots & \ddots & \\ 0 & & H_{k,k-1} & H_{kk} \end{pmatrix} \quad \text{and} \quad PBU = \tilde{B} = \begin{pmatrix} B_{11} \\ 0 \end{pmatrix}.$$

The matrix B_{11} and the subdiagonal blocks in H are of the form $(0, R)$, where R is a nonsingular and upper triangular matrix.

Step 2. An orthogonal matrix Q and a feedback matrix F are constructed such that $\Omega(Q^{\mathrm{T}}(H - \tilde{B}F)Q) = \{\lambda_1, \ldots, \lambda_n\}$.

Step 3. A feedback matrix K of the original problem is recovered from the feedback matrix F of the Hessenberg problem in Step 2 as follows:

$$K = UFP.$$

Step 1 can be implemented using the Staircase Algorithm for the controller-Hessenberg form described in Chapter 6.

We therefore concentrate on Step 2, assuming that Step 1 has already been performed.
Let $n_1 =$ dimensions of H_{11} and $n_i = \mathrm{rank}(H_{i,i-1}), i = 2, 3, \ldots, k$. Assume also that B_{11} has n_1 columns.

We consider two cases. The algorithm comprises of implementing these two cases as the situations warrant. The feedback matrix F is obtained by accumulating feedback matrices from the individual cases.

Case 1. If $m_1 = n_1 - n_2 > 0$, that is, if $n_1 > n_2$, we can immediately allocate $m_1 = n_1 - n_2$ eigenvalues as follows:
Write

$$\begin{pmatrix} H_{11} \\ H_{21} \end{pmatrix} = \begin{pmatrix} H_{10} & * \\ 0 & R_2 \end{pmatrix}, \qquad B_{11} = \begin{pmatrix} \hat{B}_{11} & B_{12} \\ 0 & B_{22} \end{pmatrix}.$$

Then, we have

$$H - \tilde{B}F = \begin{pmatrix} H_{10} & G_1 \\ \hline 0 & H_2 \end{pmatrix} - \begin{pmatrix} \hat{B}_{11} & B_{12} \\ \hline 0 & B_{22} \\ \hline & 0 \end{pmatrix} \begin{pmatrix} F_{11} & H_1 \\ & F_2 \end{pmatrix}.$$

That is, a feedback matrix F_{11} for this allocation can be immediately found by solving

$$H_{10} - B_{11}F_{11} = \begin{pmatrix} \mathrm{diag}(\lambda_1, \ldots, \lambda_{m_1}) \\ 0 \end{pmatrix}.$$

Because of the last equation, we have

$$H - \tilde{B}F = \begin{pmatrix} \mathrm{diag}(\lambda_1, \ldots, \lambda_m) & G_1 - \hat{B}_{11}H_1 \\ & -B_{12}F_2 \\ \hline 0 & H_2 - B_2F_2 \end{pmatrix},$$

where $B_2 = \begin{pmatrix} B_{22} \\ 0 \end{pmatrix}$.

Since B_{22} is a nonsingular upper triangular matrix and H_2 is still an unreduced upper Hessenberg matrix having the same form as H, (H_2, B_2) is a controllable pair.

We then solve the problem of finding F_2 such that $H_2 - B_2 F_2$ has the remaining eigenvalues.

However, this time note that the first two blocks on the diagonal are $n_2 \times n_2$, thus, no more immediate assignment of eigenvalues is possible. The other eigenvalues have to be assigned using a different approach. If $n_2 = 1$, we then have a single-input problem to solve. Otherwise, we solve the multi-input problem with $n_1 = n_2$, using the approach below.

Case 2. Let $n_1 = n_2 = \cdots = n_r > n_{r+1} \cdots \geq n_k > 0$, for $1 < r \leq k$.
Suppose we want to assign an eigenvalue λ_1 to $H - \tilde{B} F$.
Then the idea is to find a unitary matrix Q_1 such that

$$Q_1^*(H - \tilde{B}F)Q_1 = \left(\begin{array}{c|c} \lambda_1 & * \\ \hline 0 & H_2 - B_2 F_2 \end{array} \right).$$

The unitary matrix Q_1 can be found as the product of the Givens rotations such that

$$(H - \lambda_1 I)Q_1 e_1 = \binom{a_1}{0}.$$

For example, if $n = 4$, $m = 2$, $k = 2$, and $n_1 = n_2 = 2$, then $r = 2$.

$$H - \lambda_1 I = \left(\begin{array}{cc|cc} * & * & * & * \\ * & * & * & * \\ \circledast & * & * & * \\ 0 & \circledast & * & * \end{array} \right).$$

Then Q_1 is the product of two Givens rotations Q_{11} and Q_{21}, where Q_{11} annihilates the entry h_{42} and Q_{21} annihilates the entry h_{31}. Thus,

$$(H - \lambda_1 I)Q_{11}Q_{21} = (H - \lambda I)Q_1 = \left(\begin{array}{c|ccc} * & * & * & * \\ * & * & * & * \\ \hline 0 & * & * & * \\ 0 & 0 & * & * \end{array} \right).$$

Once Q_1 is found, F can be obtained by solving $B_{11} f_1 = a_1$, where $FQ_1 = (f_1, F_2)$, and $\tilde{B} = \binom{B_{11}}{0}$. Note that B_{11} is nonsingular.

It can now be shown that (H_2, B_2) is controllable and has the original form that we started with. The process can be continued to allocate the remaining eigenvalues with the pair (H_2, B_2).

To summarize, the allocation of eigenvalues is done using unitary transformations when $n_1 = n_2$ or without unitary transformations when $n_1 > n_2$.

(Note that the Case 1 ($n_1 > n_2$) is a special case of Case 2 with $Q_1 \equiv 1$, and $r = 1$.)

Eventually, the process will end up with a single-input system which can be handled with a single-input algorithm described before.

For details of the process, see Miminis and Paige (1988).

Example 11.3.2. Let's consider Example 11.3.1 again. The eigenvalues to be assigned are: $\lambda_1 = 1, \lambda_2 = 2, \lambda_3 = 3, \lambda_4 = 4,$ and $\lambda_5 = 5$.
Then,

$$H_{11} = \begin{pmatrix} 1 & 2 & 3 \\ 1 & 1 & 1 \\ 2 & 1 & 1 \end{pmatrix}, \qquad H_{21} = (0\ 0\ 0\ 1), \qquad n_1 = 3, \qquad n_2 = 1.$$

Since $m_1 = n_1 - n_2 = 2$, the two eigenvalues 1 and 2, can be assigned immediately as in Case 1.

$$H_{10} = \begin{pmatrix} 1 & 2 \\ 1 & 1 \\ 2 & 1 \end{pmatrix}, \qquad B_{11} = \begin{pmatrix} 1 & 1 & 1 \\ 0 & 1 & 2 \\ 0 & 0 & 3 \end{pmatrix}.$$

Solving for F_{11} from

$$H_{10} - B_{11}F_{11} = \begin{pmatrix} 1 & 0 \\ 0 & 2 \\ 0 & 0 \end{pmatrix},$$

we have $F_{11} = \begin{pmatrix} -0.3333 & 3.3333 \\ -0.3333 & -1.6667 \\ 0.6667 & 0.3333 \end{pmatrix}.$

Deflation

$$H = \begin{pmatrix} 1 & 2 & 3 & 4 & 1 \\ 1 & 1 & 1 & 1 & 1 \\ 2 & 1 & 1 & 1 & 1 \\ 0 & 0 & 1 & 1 & 2 \\ 0 & 0 & 0 & 1 & 1 \end{pmatrix} = \left(\begin{array}{c|c} H_{10} & G_1 \\ \hline 0 & H_2 \end{array} \right),$$

$$\tilde{B} = \begin{pmatrix} 1 & 1 & 1 \\ 0 & 1 & 2 \\ 0 & 0 & 3 \\ 0 & 0 & 0 \\ 0 & 0 & 0 \end{pmatrix} = \left(\begin{array}{cc|c} 1 & 1 & 1 \\ 0 & 1 & 2 \\ \hline 0 & 0 & 3 \\ 0 & 0 & 0 \\ 0 & 0 & 0 \end{array} \right) = \left(\begin{array}{c|c} \hat{B}_{11} & B_{12} \\ \hline 0 & B_{22} \end{array} \right).$$

Then, $H_2 = \begin{pmatrix} 1 & 1 & 1 \\ 1 & 1 & 2 \\ 0 & 1 & 1 \end{pmatrix}$, $B_2 \equiv B_{22} = \begin{pmatrix} 3 \\ 0 \\ 0 \end{pmatrix}$.

(H_2, B_2) is controllable.

Now we find F_2 such that $H_2 - B_2 F_2$ has the eigenvalues $(3, 4, 5)$.

This is a **single-input** problem. Using any of the single-input algorithms discussed before, we obtain

$$F_2 = (-3, \; 9.6667, \; -13.6667).$$

So, the required feedback matrix F is given by

$$F = \left(\begin{array}{c|c} F_{11} & 0 \\ \hline & F_2 \end{array} \right) = \left(\begin{array}{cc|ccc} -0.3333 & 3.3333 & 0 & 0 & 0 \\ -0.3333 & -1.6667 & 0 & 0 & 0 \\ 0.6667 & 0.33333 & -3 & 9.6667 & -13.6667 \end{array} \right).$$

Verify: The eigenvalues of $H - \tilde{B}F$ are 1, 2, 5, 4, and 3.

Flop-count: The solution of the Hessenberg multi-input problem, using the above-described method requires about $\frac{23}{7}n^3$ flops.

When combined with about $6n^3$ flops required for the multi-input controller-Hessenberg reduction, the total count is about $9n^3$ flops.

Stability: The round-off error analysis performed by Miminis and Paige (1988) shows that the algorithm is **numerically backward stable.** Specifically, it can be shown that the computed feedback matrix K is such that

$$\Omega((A + \triangle A) - (B + \triangle B)K) = \Omega(L),$$

where $\| \triangle A \|$ and $\| \triangle B \|$ are small, and L is the matrix with eigenvalues $\lambda_i + \delta \lambda_i, i = 1, \cdots, n$; where $|\delta \lambda_i| \leq |\lambda_i| \mu$, μ is the machine precision.

Avoiding complex arithmetic: The method as described above might give a complex feedback matrix because it is an explicit shift algorithm. To avoid complex arithmetic to assign complex conjugate pairs, the idea of implicit shift and the double step needs to be used.

MATCONTROL note: The explicit QR algorithm described in this section has been implemented in MATCONTROL function **poleqrm.**

11.3.3   The Schur Method for the Multi-Input Eigenvalue Assignment Problem

As the title suggests, the following algorithm due to A. Varga (1981) for the multi-input EVA is based on the reduction of the matrix A to the RSF. So, unlike the other two methods just described, the Schur method does not follow the "Template."

Let

$$R = QAQ^{\mathrm{T}} = \begin{pmatrix} A_1 & A_3 \\ 0 & A_2 \end{pmatrix}$$

be the RSF of A and let $\hat{B} = QB$ be the transformed control matrix.

Let's partition $\hat{B}$ as $\hat{B} = \begin{pmatrix} B_1 \\ B_2 \end{pmatrix}$. Then, since (A, B) is controllable, so is (A_2, B_2).

Suppose that the RSF R of A has been **ordered** in such a way that A_1 contains the "**good**" eigenvalues and A_2 contains the "**bad**" ones. The "**good**" eigenvalues are the ones we want to retain and the "**bad**" ones are those we want to reassign.

It is, therefore, natural to ask how the feedback matrix F can be determined such that after the application of feedback, the eigenvalues of A_1 will remain unchanged, while those in A_2 will be changed to "desired" ones by feedback.

The answer to this question is simple. If the feedback matrix F is taken in the form $F = (0, F_2)$, then after the application of the feedback matrix to the pair $(R, \hat{B})$ we have

$$R - \hat{B}F = \begin{pmatrix} A_1 & A_3 - B_1 F_2 \\ 0 & A_2 - B_2 F_2 \end{pmatrix}.$$

This shows that the eigenvalues of $R - \hat{B}F$ are the union of the eigenvalues of A_1 and of $A_2 - B_2 F_2$.

The problem thus reduces to finding F_2 such that $A_2 - B_2 F_2$ has a desired spectrum.

The special structure of A_2 can be exploited now.

Since the diagonal blocks of A_2 are either scalars (1×1) or 2×2 matrices, all we need is a procedure to assign eigenvalues to a $p \times p$ matrix where $p = 1$ or 2.

The following is a simple procedure to do this.

Algorithm 11.3.2.   *An Algorithm to Assign p ($p = 1$ or 2) Eigenvalues*
Inputs.
M—The state matrix of order p.
G—The control matrix of order $p \times m$.
Γ_p—The set of p complex numbers, closed under complex conjugation.
r—Rank of G.

Output.
F_p—The feedback matrix such that $(M - GF_p)$ has the spectrum Γ_p.

Assumption. *(M, G) is controllable.*
Step 1. *Find the SVD of G, that is, find U and V such that $G = U(\hat{G}, 0)V^{\mathrm{T}}$, where $\hat{G}$ is $r \times r$.*
Step 2. *Update M: $\hat{M} = U^{\mathrm{T}}MU$.*
Step 3. *If $r = p$, compute $\hat{F}_p = (\hat{G})^{-1}(\hat{M} - J)$, where J is $p \times p$ and the eigenvalues of J are the set Γ_p. Go to Step 6.*

Step 4. *Let* $\Gamma_2 = \{\lambda_1, \lambda_2\}$ *and*

$$\hat{M} = \begin{pmatrix} m_{11} & m_{12} \\ m_{21} & m_{22} \end{pmatrix}, \qquad \hat{G} = \begin{pmatrix} \beta \\ 0 \end{pmatrix}.$$

Step 5. *Compute* $\hat{F}_p = (\hat{F}_{p_1}, \hat{F}_{p_2})$ *as follows:*

$$\hat{F}_{p_1} = (m_{11} + m_{22} - \lambda_1 - \lambda_2)/\beta,$$

$$\hat{F}_{p_2} = \left(\frac{m_{22}}{m_{21}}\right)\hat{F}_{p_1} - (m_{11}m_{22} - m_{12}m_{21} - \lambda_1\lambda_2)/(m_{21}\beta).$$

Step 6. *Compute* $F_p = V\begin{bmatrix} \hat{F}_p \\ 0 \end{bmatrix} U^{\mathrm{T}}.$

Algorithm 11.3.2 can now be used in an iterative fashion to assign all the eigenvalues of A_2, by shifting only 1 or 2 eigenvalues at a time.

The process starts with the last $p \times p$ block of A_2 and then after the assignment with this block is completed using the algorithm above, a new $p \times p$ diagonal block is moved, using orthogonal similarity, in the last diagonal position, and the assignment procedure is repeated on this new block.

The required feedback matrix is the sum of component feedback matrices, each of which assigns 1 or 2 eigenvalues.

The overall procedure then can be summarized as follows:

Algorithm 11.3.3. *The Schur Algorithm for the Multi-Input EVA Problem*
Inputs.
A—The $n \times n$ state matrix.
B—The $n \times m$ input matrix.
S—The set of numbers to be assigned, closed under complex conjugation.]
Output. *K—The feedback matrix such that the numbers in the set Γ belong to the spectrum of $A - BK$.*

Assumption. (A, B) *is controllable.*
Step 1. *Transform A to the* **ordered RSF***:*

$$A \equiv QAQ^{\mathrm{T}} = \begin{bmatrix} A_1 & A_3 \\ 0 & A_2 \end{bmatrix},$$

where A_1 is $r \times r$, A_2 is $(n-r) \times (n-r)$; A_1 contains the **"good"** *eigenvalues and A_2 contains the* **"bad"** *eigenvalues.*
Update $B \equiv QB$ and set $\hat{Q} = Q$.
Step 2. *Set $K \equiv 0$ (zero matrix), and $i = r + 1$.*
Step 3. *If $i > n$, stop.*
Step 4. *Set M equal to the last block in A of order p ($p = 1$ or 2) and set G equal to the last p rows of B.*

Step 5. *Compute F_p using Algorithm 11.3.2 to shift p eigenvalues from the set S.*
Step 6. *Update K and A: $K \equiv K - (0, F_p)\hat{Q}$, $A \equiv A - B(0, F_p)$.*
Step 7. *Move the last block of A in position (i, i) accumulating the transformations in Q, and update $B \equiv QB$, and $\hat{Q} = Q\hat{Q}$.*
Step 8. *Set $i \equiv i + p$ and go to Step 3.*

Remarks

- The ordering of the RSF in Step 1 has to be done according to the procedure described in Chapter 4.
- It has been tacitly assumed that "the complex numbers in S are chosen and ordered so that the ordering agrees with the diagonal structure of the matrix A_2." If this requirement is not satisfied, some interchange of the blocks of A_2 need to be done so that the required condition is satisfied, using an appropriate orthogonal similarity (**Exercise 11.9**).
- The final matrix K at the end of this algorithm is the sum of the component feedback matrices, each of them assigning 1 or 2 eigenvalues.
- The algorithm has the additional flexibility to solve a "**PEVA**," which concerns reassigning only the "bad" eigenvalues, leaving the "good" ones unchanged.

Example 11.3.3.   Let's apply Algorithm 11.3.3 with data from Example 11.3.1.
Step 1.

$$A = QAQ^T = \begin{pmatrix} -0.4543 & 1.0893 & -0.2555 & -0.7487 & -0.5053 \\ -0.7717 & -1.6068 & 0.3332 & -1.2007 & 2.6840 \\ -0.0000 & -0.0000 & 0.2805 & -0.2065 & 0.2397 \\ -0.0000 & -0.0000 & -0.0000 & 1.8369 & -3.1302 \\ 0.0000 & 0.0000 & 0.0000 & -0.0000 & 4.9437 \end{pmatrix},$$

$$Q = \begin{pmatrix} -0.2128 & 0.0287 & 0.6509 & -0.6606 & 0.3064 \\ 0.8231 & -0.1628 & -0.2533 & -0.3926 & 0.2786 \\ 0.1612 & -0.8203 & 0.4288 & 0.2094 & -0.2708 \\ -0.4129 & -0.4850 & -0.3749 & 0.0539 & 0.6714 \\ 0.2841 & 0.2539 & 0.4332 & 0.6023 & 0.5517 \end{pmatrix}.$$

Let the desired closed-loop eigenvalues be the same as in Example 11.3.1: $S = \{5\ 4\ 3\ 2\ 1\}$.

Update $B \equiv QB$:

$$B = \begin{pmatrix} -0.2128 & -0.1841 & 1.7974 \\ 0.8231 & 0.6603 & -0.2625 \\ 0.1612 & -0.6591 & -0.1929 \\ -0.4129 & -0.8979 & -2.5077 \\ 0.2841 & 0.5380 & 2.0916 \end{pmatrix}.$$

Step 2.

$$K = 0, \qquad i = 1.$$

Step 3. $i = 1 < n = 5$. Continue
Step 4.

$$p = 1,$$

$$M = (4.9437),$$

$$G = (0.2841, \ 0.5380, \ 2.0916), \qquad \Gamma_p = 5, \text{ the eigenvalue to be shifted.}$$

Step 5.

$$F_p = \begin{pmatrix} -0.0034 \\ -0.0064 \\ -0.0248 \end{pmatrix}.$$

Step 6. Updated K and A are:

$$K = \begin{pmatrix} -0.0010 & -0.0009 & -0.0015 & -0.0020 & -0.0019 \\ -0.0018 & -0.0016 & -0.0028 & -0.0038 & -0.0035 \\ -0.0071 & -0.0063 & -0.0108 & -0.0150 & -0.0137 \end{pmatrix},$$

$$A = \begin{pmatrix} -0.4543 & 1.0893 & -0.2555 & -0.7487 & -0.4626 \\ -0.7717 & -1.6068 & 0.3332 & -1.2007 & 2.6845 \\ -0.0000 & -0.0000 & 0.2805 & -0.2065 & 0.2313 \\ -0.0000 & -0.0000 & -0.0000 & 1.8369 & -3.1996 \\ 0.0000 & 0.0000 & 0.0000 & -0.0000 & 5.0000 \end{pmatrix}.$$

Step 7. Reorder A and update $\hat{Q}$ and B:

$$A = \begin{pmatrix} 5.0000 & 3.2232 & -0.3674 & -1.1307 & -2.4841 \\ 0 & 1.8369 & 0.1125 & -1.2452 & -0.4711 \\ 0 & 0 & 0.2805 & -0.0392 & -0.4337 \\ 0 & 0 & 0 & -0.5524 & -1.2822 \\ 0 & 0 & 0 & 0.5749 & -1.5087 \end{pmatrix},$$

$$\hat{Q} = \begin{pmatrix} 0.7474 & 0.3745 & 0.5264 & 0.1504 & 0.0376 \\ 0.0568 & -0.0551 & -0.2840 & 0.6816 & 0.6696 \\ 0.2416 & -0.8972 & 0.2528 & 0.1953 & -0.1859 \\ 0.3829 & 0.1005 & -0.6528 & 0.2773 & -0.5833 \\ 0.4828 & -0.2040 & -0.3902 & -0.6307 & 0.4187 \end{pmatrix},$$

$$B = \begin{pmatrix} 0.7474 & 1.1219 & 3.0756 \\ 0.0568 & 0.0017 & -0.9056 \\ 0.2416 & -0.6556 & -0.7945 \\ 0.3829 & 0.4834 & -1.3744 \\ 0.4828 & 0.2789 & -1.0956 \end{pmatrix}.$$

Step 8. $i = 2$ and return to Step 3.

$$p = 2,$$

Step 3. $i = 2 < n = 5$. Continue.
Step 4.

$$M = \begin{pmatrix} 0.5524 & -1.2822 \\ 0.5749 & -1.5087 \end{pmatrix}, \qquad G = \begin{pmatrix} 0.3829 & 0.4834 & -1.3744 \\ 0.4828 & 0.2789 & -1.0956 \end{pmatrix},$$

$$\Gamma_p = \{2, 1\}$$

Step 5.

$$F_p = \begin{pmatrix} 7.0705 & -6.3576 \\ -5.1373 & 3.4356 \\ 1.7303 & 0.7260 \end{pmatrix}.$$

Step 6.

$$K = \begin{pmatrix} -0.3630 & 2.0061 & -2.1362 & 5.9678 & -6.7883 \\ -0.3103 & -1.2184 & 2.0102 & -3.5949 & 4.4318 \\ 1.0061 & 0.0194 & -1.4235 & 0.0069 & -0.7191 \end{pmatrix},$$

$$A = \begin{pmatrix} 5.0000 & 3.2232 & -0.3674 & -5.9733 & -3.8198 \\ 0 & 1.8369 & 0.1125 & -0.0712 & 0.5417 \\ 0 & 0 & 0.2805 & -3.7408 & 3.9316 \\ 0 & 0 & 0 & 1.6016 & 0.4896 \\ 0 & 0 & 0 & 0.4896 & 1.3984 \end{pmatrix}.$$

Step 7. Reorder A and update $\hat{Q}$ and B (Recorded A and updated B are shown below):

$$A = \begin{pmatrix} 5.0000 & 3.2232 & -0.3674 & -5.9733 & -3.8198 \\ 0 & 1.0000 & 0.3087 & 0.1527 & -5.2208 \\ 0 & 0 & 2.0000 & 0.2457 & 1.2394 \\ 0 & 0.0000 & 0.0000 & 1.8369 & -0.8889 \\ 0 & 0 & 0 & 0 & 0.2805 \end{pmatrix},$$

$$B = \begin{pmatrix} 0.7474 & 1.1219 & 3.0756 \\ -0.2402 & 0.6483 & 0.5859 \\ 0.4240 & 0.1944 & -1.8404 \\ 0.4443 & 0.5319 & -0.8823 \\ 0.0804 & -0.0193 & 0.1787 \end{pmatrix}.$$

Step 8. $i = 4$ and return to Step 3.
Step 3. $i = 4 < n = 5$. Continue.

Step 4.

$$p = 1, \qquad M = 0.2805, \qquad G = (0.0804 \ - 0.0193 \ 0.1787), \qquad \Gamma_p = 4.$$

Step 5.

$$F_p = \begin{pmatrix} -7.7091 \\ 1.8532 \\ -17.1422 \end{pmatrix}.$$

Step 6.

$$K = \begin{pmatrix} -0.9827 & 2.7748 & -2.9014 & 11.1937 & -12.3165 \\ -0.1613 & -1.4032 & 2.1942 & -4.8511 & 5.7607 \\ -0.3719 & 1.7286 & -3.1250 & 11.6272 & -13.0117 \end{pmatrix},$$

$$A = \begin{pmatrix} 5.0000 & 3.2232 & -0.3674 & -5.9733 & 52.5851 \\ 0 & 1.0000 & 0.3087 & 0.1527 & 1.7698 \\ 0 & 0 & 2.0000 & 0.2457 & -27.4012 \\ 0 & 0.0000 & 0.0000 & 1.8369 & -13.5735 \\ 0 & 0 & 0 & 0 & 4.0000 \end{pmatrix}.$$

Step 7. Reorder A and update $\hat{Q}$ and B (Recorded A and updated B are shown below):

$$A = \begin{pmatrix} 5.0000 & 3.2232 & -0.3674 & -5.9733 & 52.5851 \\ 0 & 1.0000 & 0.3087 & 0.1527 & 1.7698 \\ 0 & 0 & 2.0000 & 0.2457 & -27.4012 \\ 0 & 0.0000 & 0.0000 & 4.0000 & -13.5735 \\ 0 & 0 & 0 & 0 & 1.8369 \end{pmatrix},$$

$$B = \begin{pmatrix} 0.7474 & 1.1219 & 3.0756 \\ -0.2402 & 0.6483 & 0.5859 \\ 0.4240 & 0.1944 & -1.8404 \\ 0.4261 & 0.5283 & -0.8994 \\ 0.1493 & 0.0646 & 0.0377 \end{pmatrix}.$$

Step 8. $i = 5$. Return to Step 3.
Step 3. $i = n = 5$. Continue
Step 4.

$$p = 1, \qquad M = 1.8369, \qquad G = (0.1493 \ 0.0646 \ 0.0377), \qquad \Gamma_p = 3.$$

Step 5.

$$F_p = \begin{pmatrix} -6.2271 \\ -2.6950 \\ -1.5710 \end{pmatrix}.$$

Step 6.

$$K = \begin{pmatrix} -1.9124 & 3.3022 & -3.0213 & 15.9029 & -16.2462 \\ -0.5637 & -1.1750 & 2.1423 & -2.8130 & 4.0599 \\ -0.6064 & 1.8617 & -3.1553 & 12.8153 & -14.0031 \end{pmatrix},$$

$$A = \begin{pmatrix} 5.0000 & 3.2232 & -0.3674 & -5.9733 & 65.0948 \\ 0 & 1.0000 & 0.3087 & 0.1527 & 2.9415 \\ 0 & 0 & 2.0000 & 0.2457 & -27.1285 \\ 0 & 0.0000 & 0.0000 & 4.0000 & -10.9093 \\ 0 & 0 & 0 & 0 & 3.0000 \end{pmatrix}.$$

All the eigenvalues are assigned, and the iteration terminates.

Verification: The eigenvalues of $A - BK$ are: $\{5.00000000000024,$ $3.99999999999960, 1.00000000000000, 3.00000000000018, 1.99999999999999\}$.

Flop-count and stability. The algorithm requires about $30n^3$ flops, most of which is consumed in the reduction of A to the RSF and ordering of this RSF.

The algorithm is believed to be numerically stable (note that it is based on all numerically stable operations). **However, no formal round-off error analysis of the algorithm has been performed yet.**

MATCONTROL function: Algorithm 11.3.3 has been implemented in MATCONTROL function **polesch**.

11.3.4   Partial Eigenvalue Assignment Problem

The PEVA problem is the one of reassigning by feedback only a few eigenvalues, say $\lambda_1, \ldots, \lambda_p (p < n)$, of an $n \times n$ matrix A leaving the other eigenvalues $\lambda_{p+1}, \ldots, \lambda_n$ unchanged.

Formally, PEVA is defined as follows:

Given $A \in \mathbb{R}^{n \times n}$, $B \in \mathbb{R}^{n \times m}$, a part of the spectrum $\{\lambda_1, \ldots, \lambda_p\}$ of A, and a set of self-conjugate numbers $\{\mu_1, \ldots, \mu_p\}$, find a feedback matrix F such that the spectrum of $A - BF$ is the set $\{\mu_1, \ldots, \mu_p; \lambda_{p+1}, \ldots, \lambda_n\}$.

A projection algorithm for this problem was proposed by Saad (1988). Here we describe a simple parametric algorithm via Sylvester equation. Note that Varga's algorithm described in the last section can be used for solving PEVA; however, that will require full knowledge of the eigenvalues of A and is thus not suitable for large problems. *The algorithm described below requires the knowledge of only those small number of eigenvalues (and the corresponding eigenvectors) that are required to be reassigned.* Furthermore, the algorithm is parametric in nature which can be exploited to devise a robust EVA algorithm (see **Section 11.6**). The discussion here has been taken from Datta and Sarkissian (2002). This paper also contains a result on the existence and uniqueness of the solution for PEVA in the multi-input case.

Theorem 11.3.2.   *(Parametric Solution to PEVA Problem). Assume that (i) B has full rank, (ii) the sets $\{\lambda_1, \ldots, \lambda_p\}$ and $\{\mu_1, \ldots, \mu_p\}$ are closed under complex conjunction and disjoint, and (iii) let the pair (A, B) be partially controllable with respect to $\{\lambda_1, \ldots, \lambda_p\}$. Assume further that the closed-loop matrix has a complete set of eigenvectors. Let $\Gamma = (\gamma_1, \ldots, \gamma_p)$*

be a matrix such that

$$\gamma_j = \overline{\gamma_k}, \ whenever \ \mu_j = \overline{\mu_k},$$

Let Y_1 be the matrix of left eigenvectors associated with $\{\lambda_1, \ldots, \lambda_p\}$. Set $\Lambda_1 = \text{diag}(\lambda_1, \ldots, \lambda_p)$ and $\Lambda_{c1} = \text{diag}(\mu_1, \ldots, \mu_p)$. Let Z_1 be a unique nonsingular solution of the Sylvester equation

$$\Lambda_1 Z_1 - Z_1 \Lambda_{c1} = Y_1^H B \Gamma.$$

Let Φ be defined by $\Phi Z_1 = \Gamma$, then

$$F = \Phi Y_1^H,$$

solves the partial eigenvalue assignment problem for the pair (A, B).

Conversely, if there exists a real feedback matrix F of the form that solves the PEVA problem for the pair (A, B), then the matrix Φ can be constructed satisfying Steps 2–4 of **Algorithm 11.3.4**.

Proof.   see Datta and Sarkissian (2002).   ■

Algorithm 11.3.4.   *Parametric Algorithm for PEVA Problem*
   Inputs.

   (i)   *A—The $n \times n$ state matrix.*
   (ii)   *B—The $n \times m$ control matrix.*
   (iii) *The set $\{\mu_1, \ldots, \mu_p\}$, closed under complex conjugation.*
   (iv)   *The self-conjugate subset $\{\lambda_1, \ldots, \lambda_p\}$ of the spectrum $\{\lambda_1, \ldots, \lambda_n\}$ of the matrix A and the associated right eigenvector set $\{y_1, \ldots, y_p\}$.*

   Outputs. *The real feedback matrix F such that the spectrum of the closed-loop matrix $A - BF$ is $\{\mu_1, \ldots, \mu_p; \lambda_{p+1}, \ldots, \lambda_n\}$.*
   Assumptions.

   (i)   *The matrix pair (A, B) is partially controllable with respect to the eigenvalues $\lambda_1, \ldots, \lambda_p$.*
   (ii)   *The sets $\{\lambda_1, \ldots, \lambda_p\}$, $\{\lambda_{p+1}, \ldots, \lambda_n\}$, and $\{\mu_1, \ldots, \mu_p\}$ are disjoint.*

Step 1. *Form*

$$\Lambda_1 = \text{diag}(\lambda_1, \ldots, \lambda_p), \quad Y_1 = (y_1, \ldots, y_p), \quad and \quad \Lambda_{c1} = \text{diag}(\mu_1, \ldots, \mu_p).$$

Step 2. *Choose arbitrary $m \times 1$ vectors $\gamma_1, \ldots, \gamma_p$ in such a way that $\overline{\mu_j} = \mu_k$ implies $\overline{\gamma_j} = \gamma_k$ and form $\Gamma = (\gamma_1, \ldots, \gamma_p)$.*

Step 3. *Find the unique solution Z_1 of the Sylvester equation:*

$$\Lambda_1 Z_1 - Z_1 \Lambda_{c1} = Y_1^H B\Gamma.$$

If Z_1 is ill-conditioned, then return to Step 2 and select different $\gamma_1, \ldots, \gamma_p$.

Step 4. *Solve $\Phi Z_1 = \Gamma$ for Φ.*

Step 5. *Form $F = \Phi Y_1^H$.*

A Numerical Example

In this section, we report results of our numerical experiments with Algorithm 11.3.4 on a 400×400 matrix obtained by discretization of the partial differential equation

$$\frac{\partial u}{\partial t} = \frac{\partial^2 u}{\partial x^2} + \frac{\partial^2 u}{\partial y^2} + 20\frac{\partial u}{\partial x} + 180u(x, y, t) + \sum_{i=1}^{2} F_i(x, y)g_i(t)$$

on the unit square $\Omega = (0, 1) \times (0, 1)$ with the Dirichlet boundary conditions:

$$u(x, y, t) = 0, \text{ for } (x, y) \in \partial\Omega \text{ and } t \geq 0$$

and some initial condition which is of no importance for the PEVA problem. This problem was earlier considered by Saad (1988). Using finite difference scheme of order $O(\|\Delta x\|^2, \|\Delta y\|^2)$, we discretize the equation in the region Ω with 20 interior points in both the x and y directions, thus obtaining a 400×400 matrix A. The 400×2 matrix B, whose ith column discretizes the function $F_i(x, y)$ is filled with random numbers between -1 and 1.

Using sparse MATLAB command **eigs**, the following 10 eigenvalues with the largest real parts are computed

$$\lambda_1 = 55.0660, \quad \lambda_2 = 29.2717, \quad \lambda_3 = 25.7324, \quad \lambda_4 = -0.0618,$$
$$\lambda_5 = -13.0780, \quad \lambda_6 = -22.4283, \quad \lambda_7 = -42.4115,$$
$$\lambda_8 = -48.2225, \quad \lambda_9 = -71.0371, \quad \lambda_{10} = -88.3402.$$

The residual of each eigenpair $\|y^*(A - \lambda I)\| < 4 \cdot 10^{-12}$ and each left eigenvector is normalized. Algorithm 11.3.4 was used to reassign $\lambda_1, \lambda_2, \lambda_3$, and λ_4 to $-7, -8, -9$, and -10, respectively, obtaining the 2×400 feedback matrix F with $\|F\|_2 < 127$. Note that the $\|A\|_2 = 3.3 \cdot 10^3$. The 10 eigenvalues of the closed-loop matrix $A - BF$ with the largest real parts obtained by the algorithm are the following:

$$\mu_1 = -7.00000, \quad \mu_2 = -8.0000, \quad \mu_3 = -9.0000, \quad \mu_4 = -10.0000,$$
$$\lambda_5 = -13.0780, \quad \lambda_6 = -22.4283, \quad \lambda_7 = -42.4115,$$
$$\lambda_8 = -48.2225, \quad \lambda_9 = -71.0371, \quad \lambda_{10} = -88.3402.$$

11.4 CONDITIONING OF THE FEEDBACK PROBLEM

In this section, we will discuss the sensitivity of the feedback problem, that is, **we are interested in determining a measure that describes how small perturbations in the data affect the computed feedback.** We discuss the **single-input case first.**

11.4.1 The Single-Input Case

Arnold (1993) first discussed the perturbation analysis of the single-input Hessenberg feedback problem in his Ph.D. dissertation. Based on his analysis, he derived two condition numbers for the problem and identified several condition number estimators. For details, we refer the readers to the above dissertation. We simply state here one of the condition number estimators which has worked well in several meaningful numerical experiments. Recall that the single-input Hessenberg feedback problem is defined by the data (H, β, S), where H is an unreduced upper Hessenberg matrix, $\beta = (\alpha, 0, \ldots, 0)^T$, $\alpha \neq 0$ and $S = \{\lambda_1, \ldots, \lambda_n\}$.

Estimating the Condition Numbers of the Feedback Problem

Theorem 11.4.1. *If $\nu(H, \beta)$ is the condition number of the single-input Hessenberg feedback problem, then an estimator of this number is given by*

$$\nu_{d\phi} = \frac{|\beta| \, \|\omega\| + \|H\| \, \|e_n^T \phi'(H)\|}{\|e_n^T \phi(H)\|} \qquad (11.4.1)$$

where

$$\omega = \left(\frac{1}{\beta}, \frac{1}{h_{21}}, \ldots, \frac{1}{h_{n,n-1}} \right)^T,$$

and $\phi(H) = (H - \lambda_1 I) \ldots (H - \lambda_n I)$.

Defining the quantity digits off (as in Rice (1966)) by

$$\log_{10} \left[\left(\frac{\mu \, \nu_{\text{estimate}}}{\text{err}} \right) \right], \qquad (11.4.2)$$

where μ is the machine precision ($\mu \approx 2 \times 10^{-16}$) and err stands for the error tolerance, it has be shown that the maximum digits off in estimating conditioning for 100 ill-conditioned problems are only 1.86, and minimum digits off are 0.51. **Thus, it never underestimated the error and overestimated the error by less than two digits**.

The computation of this condition estimator requires only $2n^3/3$ flops once the system is in controller-Hessenberg form. For details of these experiments, see Arnold (1993). **Note these bounds work only for the single-input Hessenberg feedback problem.**

11.4.2   The Multi-Input Case

Arnold (1993) considered a perturbation analysis of the single-input Hessenberg feedback problem by considering only the perturbation of the matrix H and the vector b.

Sun (1996) has studied perturbation analysis of both single-input and the multi-input problems by allowing perturbations of all the data, namely A, B, and $S = \{\lambda_1, \ldots, \lambda_n\}$. Below, we state his result for the multi-input problem, without proof. For more general results on perturbation analyses of feedback matrices as well as those of conditioning of the closed-loop eigenvalues, see the recent papers of Mehrmann and Xu (1996, 1997).

Let $A \in \mathbb{R}^{n \times n}$ and $B \in \mathbb{R}^{n \times m}$

Let $\lambda_i \neq \lambda_j$, for all $i \neq j$. Let $K = (k_1, \ldots, k_m)^{\mathrm{T}}$ and $X = (x_1, \ldots, x_n)$ be such that

$$A + BK = X \Lambda X^{-1}, \tag{11.4.3}$$

where $\Lambda = \mathrm{diag}(\lambda_1, \ldots, \lambda_n)$.

Also, let $y_1, \ldots, y_n$ be the normalized left eigenvectors of $A + BK$, that is, $Y = X^{-T} = (y_1, \ldots, y_n)$, which implies $y_i^{\mathrm{T}} x_j = \delta_{ij}$ for all i and j.

Suppose that the data matrices A, B, and Λ and the feedback matrix K are so perturbed that the resulting closed-loop matrix has also the distinct eigenvalues.

Let $B = (b_1, \ldots b_m)$. Define now

$$W_k = (S_1 X^{\mathrm{T}}, S_2 X^{\mathrm{T}}, \ldots, S_m X^{\mathrm{T}})_{n \times mn}, \tag{11.4.4}$$

where $S_j = \mathrm{diag}(y_1^{\mathrm{T}} b_j, \ldots, y_n^{\mathrm{T}} b_j)$, $j = 1, 2, \ldots, m$.

Also define

$$W_a = (D_1(X)X^{-1}, D_2(X)X^{-1}, \ldots, D_n(X)X^{-1})_{n \times n^2}, \tag{11.4.5}$$

$$W_b = \mathrm{diag}(T_1 X^{-1}, T_2 X^{-1}, \ldots, T_m X^{-1})_{n \times nm}, \tag{11.4.6}$$

and $W_\lambda = -I_n$, where

$$D_i(X) = \mathrm{diag}(x_{i1}, \ldots, x_{in}), \quad i = 1, \ldots, n \tag{11.4.7}$$

$$T_j = \mathrm{diag}(k_j^{\mathrm{T}} x_1, \ldots k_j^{\mathrm{T}} x_n), \quad j = 1, \ldots m, \tag{11.4.8}$$

and $x_i = (x_{i1}, \ldots, x_{in})$.

Also, let $Z = W_k^{\dagger}$, $\quad \Phi = -Z W_a$, and $\quad \Psi = -Z W_b$. $\tag{11.4.9}$

Here $W_k^{\dagger}$ denotes the generalized inverse of W_k.

Theorem 11.4.2. *Perturbation Bound for a Multi-Input Feedback Matrix*

Suppose that the controllable pair (A, B) is slightly perturbed to another controllable pair $(\tilde{A}, \tilde{B})$, and that the self-conjugate set $S = \{\lambda_1, \ldots, \lambda_n\}, \lambda_i \neq \lambda_j, i \neq j$ is slightly perturbed to the set $\tilde{S} = \{\tilde{\lambda}_1, \ldots, \tilde{\lambda}_n\}, \tilde{\lambda}_i \neq \tilde{\lambda}_j, i \neq j$.

Let K be the feedback matrix of the EVA problem with the data A, B, S. Then there is a solution $\tilde{K}$ to the problem with data $\tilde{A}, \tilde{B}$, and $\tilde{S}$ such that for any consistent norm $\| \, \|$, we have

$$
\| \tilde{K} - K \| \leq \delta_K + O \left(\left\| \begin{pmatrix} \tilde{a} \\ \tilde{b} \\ \tilde{\lambda} \end{pmatrix} - \begin{pmatrix} a \\ b \\ \lambda \end{pmatrix} \right\|^2 \right)
$$

$$
\leq \Delta_K + O \left(\left\| \begin{pmatrix} \tilde{a} \\ \tilde{b} \\ \tilde{\lambda} \end{pmatrix} - \begin{pmatrix} a \\ b \\ \lambda \end{pmatrix} \right\|^2 \right), \tag{11.4.10}
$$

where

$$
\begin{aligned}
a &= \text{vec}(A), \ \tilde{a} = \text{vec}(\tilde{A}), \\
b &= \text{vec}(B), \ \tilde{b} = \text{vec}(\tilde{B}), \\
\lambda &= (\lambda_1, \ldots, \lambda_n)^{\mathrm{T}}, \ \tilde{\lambda} = (\tilde{\lambda}_1, \tilde{\lambda}_2, \ldots, \tilde{\lambda}_n)^{\mathrm{T}}, \\
\delta_K &= \| \Phi(\tilde{a} - a) + \Psi(\tilde{b} - b) + Z(\tilde{\lambda} - \lambda) \|, \\
\Delta_K &= \| \Phi \| \| \tilde{a} - a \| + \| \Psi \| \| \tilde{b} - b \| + \| Z \| \| \tilde{\lambda} - \lambda \|,
\end{aligned} \tag{11.4.11}
$$

Z, Ψ, and Φ are defined by (11.4.9).

11.4.3 Absolute and Relative Condition Numbers

The three groups of **absolute condition numbers** that reflect the three different types of sensitivities of K with respect to the data A, B, and S, respectively, have been obtained by Sun (1996). These are:

$$
\kappa_A(K) = \| \Phi \|, \qquad \kappa_B(K) = \| \Psi \|, \qquad \text{and } \kappa_\lambda(K) = \| Z \|. \tag{11.4.12}
$$

Furthermore, the scalar $\kappa(K)$ defined by

$$
\kappa(K) = \sqrt{\kappa_A^2(K) + \kappa_B^2(K) + \kappa_\lambda^2(K)} \tag{11.4.13}
$$

can be regarded as an **absolute condition** number of K.

If $\|\cdot\|_2$ is used, and if the matrix $X = (x_1, x_2, \ldots, x_n)$ is such that $\|x_j\|_2 = 1$ for all j, then

$$\kappa_A(K) = \|\Phi\|_2 \le \|Z\|_2 \|X^{-1}\|_2, \tag{11.4.14}$$

$$\kappa_B(K) = \|\Psi\|_2 \le \max_{1 \le j \le n} \|k_j\|_2 \|Z\|_2 \|X^{-1}\|_2, \tag{11.4.15}$$

$$\kappa_\lambda(K) = \|Z\|_2. \tag{11.4.16}$$

Thus, using the Frobenius norm, the respective **relative condition numbers** are given by

$$\kappa_A^{(r)}(K) = \kappa_A(K) \frac{\|A\|_F}{\|K\|_F}, \tag{11.4.17}$$

$$\kappa_B^{(r)}(K) = \kappa_B(K) \frac{\|B\|_F}{\|K\|_F}, \tag{11.4.18}$$

and

$$\kappa_\lambda^{(r)}(K) = \kappa_\lambda(K) \frac{\|\lambda\|_2}{\|K\|_F}, \quad \lambda = (\lambda_1, \ldots, \lambda_n)^{\mathrm{T}}. \tag{11.4.19}$$

Furthermore, the **relative condition number of** K is given by

$$\kappa_K^{(r)}(K) = \sqrt{(\kappa_A^{(r)}(K))^2 + (\kappa_B^{(r)}(K))^2 + (\kappa_\lambda^{(r)}(K))^2}, \tag{11.4.20}$$

where $\kappa_A^{(r)}(K)$, $\kappa_B^{(r)}(K)$, and $\kappa_\lambda^{(r)}(K)$ are evaluated using 2-norm.

Remark

- A variation of the above results appears in Mehrmann and Xu (1996, 1997), where it has been shown that the ill-conditioning of the feedback problem is also related to the ill-conditioning of the open-loop eigenvector matrix and the distance to uncontrollability (see next section for more on this).

Example 11.4.1.   (Laub and Linnemann 1986; Sun 1996).

Consider the following single-input problem:

$$A = \begin{pmatrix} -4 & 0 & 0 & 0 & 0 \\ \alpha & -3 & 0 & 0 & 0 \\ 0 & \alpha & -2 & 0 & 0 \\ 0 & 0 & \alpha & -1 & 0 \\ 0 & 0 & 0 & \alpha & 0 \end{pmatrix}, \qquad B = \begin{pmatrix} 1 \\ 0 \\ 0 \\ 0 \\ 0 \end{pmatrix},$$

and

$$S = \{-2.9992, -0.8808, -2, -1, 7.0032 \times 10^{-14}\}.$$

Choose $\alpha = 0.0010$.

Then, $K = (3.12, -1.67, 7.45, -2.98, 0.37)$.

The feedback problem with the above data is expected to be ill-conditioned, as $\kappa_A^{(r)}(K) = 3.2969 \times 10^{12}$, $\kappa_B^{(r)}(K) = 1.01117 \times 10^{12}$, $\kappa_\lambda^{(r)}(K) = 2.3134 \times 10^{12}$, and $\kappa_K^{(r)}(K) = 4.1527 \times 10^{12}$.

Indeed, if only the 1st entry of S is changed to -3 and all other data remain unchanged, then the feedback vector for this perturbed problem becomes $\hat{K} = (3.1192, 0.0078, 7.8345, 0.0004, 0.3701)$.

11.5   CONDITIONING OF THE CLOSED-LOOP EIGENVALUES

Suppose that the feedback matrix K has been computed using a stable algorithm, that is, the computed feedback matrix $\hat{K}$ is the exact feedback matrix for a nearby EVA problem. The question now is: **How far are the eigenvalues of the computed closed-loop matrix $\hat{M}_c = A - B\hat{K}$ from the desired eigenvalues $\{\lambda_1, \ldots, \lambda_n\}$?** Unfortunately, the answer to this question is: *Even though a feedback matrix has been computed using a numerically stable algorithm, there is no guarantee that the eigenvalues of the closed-loop matrix will be near those which are to be assigned.*

The following interrelated factors, either individually, or in combination, can contribute to the conditioning of the closed-loop eigenvalues:

- The conditioning of the problem of determining the feedback matrix K from the given data.
- The condition number (with respect to a p-norm) of the eigenvector matrix of the closed-loop system.
- The distance to uncontrollability, and the distance between the open-loop and closed-loop eigenvalues.
- The norm of the feedback matrix.

Regarding the first factor, we note that if the problem of computing the feedback matrix is ill-conditioned, then even with the use of a stable numerical algorithm, the computed feedback matrix cannot be guaranteed to be accurate, and, as a result, the computed closed-loop eigenvalues might differ significantly from those to be assigned. (Note that the eigenvalue problem of a nonsymmetric matrix can be very ill-conditioned.)

The relation of the other two factors to the conditioning of the closed-loop eigenvalues can be explained by the following analysis using the **Bauer–Fike Theorem (Chapter 3)**.

Let $M_c = A - BK$ and let $E = \hat{M}_c - M_c$, where $\hat{M}_c = A - B\hat{K}$, $\hat{K}$ being the computed value of K.

Let X be the transforming matrix that diagonalizes the matrix M_c, that is, $X^{-1}M_cX = \text{diag}(\lambda_1, \ldots, \lambda_n)$. Let μ be an eigenvalue of $\hat{M}_c$. Then, by the

Bauer–Fike Theorem (**Theorem 3.3.3**), we have

$$\min_{\lambda_i} |\lambda_i - \mu| \leq \text{Cond}_2(X) \|E\|_2.$$

Again, $E = A - B\hat{K} - (A - BK) = B(K - \hat{K}) = B\Delta K$.

Thus, we see that the product of the spectral condition number of X and the $\|B\Delta K\|_2$ influences the distance between the desired poles and those obtained with a computed $\hat{K}$.

This again is related to the factors: *norm of the computed feedback matrix $\hat{K}$, distance to the uncontrollability of the pair (A, B), and the distance between closed-loop poles and the eigenvalues of A, etc.* (See Mehrmann and Xu 1997.) Note that some of these observations also follow from the explicit formula of the feedback vector (11.2.9) in the single-input case, and the one in the multi-input case derived in Arnold (1993).

Example 11.5.1. Consider EVA with the following data:

$$A = \begin{pmatrix} -4 & 0 & 0 & 0 & 0 \\ 0.001 & -3 & 0 & 0 & 0 \\ 0 & 0.001 & -2 & 0 & 0 \\ 0 & 0 & 0.001 & -1 & 0 \\ 0 & 0 & 0 & 0.001 & 0 \end{pmatrix}, \quad B = \begin{pmatrix} 1 \\ 0 \\ 0 \\ 0 \\ 0 \end{pmatrix}.$$

$$S = \{\lambda_1, \lambda_2, \lambda_3, \lambda_4, \lambda_5\} = \{10, 12, 24, 29, 30\}.$$

Then $K = (-115, 4.887 \times 10^6, -9.4578 \times 10^{10}, 8.1915 \times 10^{14}, -2.5056 \times 10^{18})$

The eigenvalue assignment problem with the above data is very ill-conditioned as the following computation shows.

Change the entry a_{51} of A to 10^{-6} and keep all other data unchanged. The eigenvalues of the closed-loop matrix then become: $\{1.5830 \times 10^6, -1.5829 \times 10^6, -3, -2, -1\}$.

The explanation of this drastic change in the closed-loop eigenvalues can be given in the light of the discussions we just had in the last section.

- **Ill-conditioning of the feedback vector:** Let $\hat{K}$ be obtained by changing the first entry of K to -114.999 and leaving the remaining entries unchanged. The eigenvalues of $(A - B\hat{K})$ are $\{29.5386 \pm 0.4856j, 23.9189, 12.0045, 9.9984\}$.

  So, **the problem of computing the feedback vector K is ill-conditioned.**
- All the subdiagonal entries of A are small, **indicating that the system is near an uncontrollable system.**
- **Distance between the open-loop and closed-loop eigenvalues:** The open-loop eigenvalues are $\{0, -1, -2, -3, -4\}$.

  Thus, **the open-loop eigenvalues are well-separated from those of the closed-loop eigenvalues.**

- **Ill-conditioning of the closed-loop eigenvector matrix:** $\text{Cond}_2(X) = 1.3511 \times 10^{24}$.

  Thus, **the spectral condition number of the closed-loop matrix is large**. The condition numbers of the individual eigenvalues are also large.

  Note: In Example 11.5.1, the feedback vector K was computed using Algorithm 11.2.3. The MATLAB function **place** cannot place the eigenvalues.

Concluding Remarks

We have identified several factors that contribute to the ill-conditioning of the closed-loop eigenvalues. In general, the problem of assigning eigenvalues is an intrinsically ill-conditioned problem. Indeed, in Mehrmann and Xu (1996), it has been shown that in the single-input case, the feedback vector K (which is unique) depends upon the solution of a linear system whose matrix is a **Cauchy matrix (Exercise 11.13)**, and a Cauchy matrix is well-known to be ill-conditioned for large order matrices. Thus, the **distribution of eigenvalues is also an important factor for conditioning of the EVA problem**, and the condition number of the problem can be reduced by choosing the eigenvalues judiciously in a prescribed compact set in the complex plane. For details, see (Mehrmann and Xu (1998)). See also Calvetti *et al.* (1999).

11.6   ROBUST EIGENVALUE ASSIGNMENT

In the last section we have discussed the aspect of the closed-loop eigenvalue sensitivity due to perturbations in the data A, B, and K.

The problem of finding a feedback matrix K such that the closed-loop eigenvalues are as insensitive as possible is called the **robust eigenvalue assignment (REVA)** problem.

Several factors affecting the closed-loop eigenvalue sensitivity were identified in the last section, the principal of those factors being the conditioning of the closed-loop eigenvector matrix.

In this section, we consider REVA with respect to minimizing the condition number of the eigenvector matrix of the closed-loop matrix. In the multi-input case, one can think of solving the problem by making use of the available freedom. One such method was proposed by Kautsky *et al.* (1985). For an excellent account of the REVA problem and discussion on this and other methods, see the paper by Byers and Nash (1989). See also Tits and Yang (1996).

11.6.1   Measures of Sensitivity

Let the matrix $M = A - BK$ be diagonalizable, that is, assume that there exists a nonsingular matrix X such that

$$X^{-1}(A - BK)X = \Lambda = \text{diag}(\lambda_1, \ldots, \lambda_n).$$

Recall from Chapter 3 (see also Wilkinson (1965), Datta (1995), etc.) that a measure of sensitivity c_j of an individual eigenvalue λ_j due to perturbations in the data A, B, and K is given by

$$c_j = \frac{1}{s_j} = \frac{\|y_j\|_2 \, \|x_j\|_2}{|y_j^T x_j|},$$

where x_j and y_j are, respectively, the right and left eigenvectors of M corresponding to λ_j. Furthermore, the overall sensitivity of all the eigenvalues of the matrix M is given by $\mathrm{Cond}_2(X) = \|X\|_2 \, \|X^{-1}\|_2$. Note also that $\max_j c_j \leq \mathrm{Cond}_2(X)$.

Thus, two natural measures of sensitivity are:

$$\nu_1 = \|C\|_\infty, \quad \text{and} \quad \nu_2 = \mathrm{Cond}_2(X),$$

where $C = (c_1, c_2, \ldots, c_n)^T$.

One could also take (see Kautsky et al. 1985)

$$\nu_3 = \|X^{-1}\|_F n^{1/2} = \|C\|_2 n^{1/2} \quad \text{and} \quad \nu_4 = \left(\sum_j \sin^2 \theta_j \right)^{1/2} \Big/ n^{1/2}$$

as other measures. Here θ_j are the angles between the eigenvectors x_j and certain corresponding orthonormal vectors $\hat{x}_j$, $j = 1, 2, \ldots, n$.

11.6.2 Statement and Existence of Solution of the Robust EigenValue Assignment Problem

In view of the above, the *REVA problem* with respect to minimizing the conditioning of the eigenvalue matrix X can be formulated as follows:

Given $A \in \mathbb{R}^{n \times n}$, $B \in \mathbb{R}^{n \times m}$ $(m \leq n)$, having full rank, and $\Lambda = \mathrm{diag}(\lambda_1, \ldots, \lambda_n)$, find a real matrix K and a nonsingular matrix X satisfying

$$(A - BK)X = X\Lambda \tag{11.6.1}$$

such that some measures ν of the sensitivity of the closed-loop eigenproblem is optimized.

The following result, due to Kautsky et al. (1985), gives conditions under which a given nonsingular X can be assigned to (11.6.1).

Theorem 11.6.1.   *Given a nonsingular X, and Λ as above, there exists a matrix K satisfying (11.6.1) if and only if*

$$U_1^{\mathrm{T}}(X\Lambda - AX) = 0, \qquad (11.6.2)$$

where U_1 is defined by

$$B = [U_0, U_1]\begin{bmatrix} Z \\ 0 \end{bmatrix}, \qquad (11.6.3)$$

with $U = [U_0, U_1]$ orthogonal and Z nonsingular.
  The matrix K is explicitly given by

$$K = Z^{-1}U_0^{\mathrm{T}}(A - X\Lambda X^{-1}). \qquad (11.6.4)$$

Proof.   Since B has full rank, the factorization of B given by (11.6.3) exists with Z nonsingular. Again, from (11.6.1), we have

$$BK = A - X\Lambda X^{-1}. \qquad (11.6.5)$$

Multiplying (11.6.5) by U^{T}, we obtain

$$\begin{aligned} ZK &= U_0^{\mathrm{T}}(A - X\Lambda X^{-1}), \\ 0 &= U_1^{\mathrm{T}}(A - X\Lambda X^{-1}). \end{aligned} \qquad (11.6.6)$$

Since X and Z are invertible, we immediately have (11.6.2) and (11.6.4).   ∎

11.6.3   A Solution Technique for the Robust Eigenvalue Assignment Problem

Theorem 11.6.1 suggests the following algorithm for a solution of the REVA problem.

Algorithm 11.6.1.   *An REVA Algorithm (The KNV Algorithm)*
  Input.
  A—The $n \times n$ state matrix.
  B–The $n \times m$ input matrix with full rank.
  Λ—The diagonal matrix containing the eigenvalues $\lambda_1, \ldots, \lambda_n$.
  Assumptions. *(A, B) is controllable and $\lambda_1, \ldots, \lambda_n$ is a self-conjugate set.*
  Output. *K—The feedback matrix such that the spectrum of $A - BK$ is the set $\{\lambda_1, \ldots, \lambda_n\}$, and the condition number of the eigenvector matrix is as small as possible.*
  Step 1. *Decompose the matrix B to determine U_0, U_1, and Z as in (11.6.3).*
  Construct orthonormal bases, comprised of the columns of matrices S_j and $\hat{S}_j$ for the space $s_j = N\{U_1^{\mathrm{T}}(A - \lambda_j I)\}$ and its complement $\hat{s}_j$, for $\lambda_j \in s_j$, $j = 1, 2, \ldots, n$.

Step 2. *Select a set of n normalized vectors* $x_1, \ldots, x_n$ *from the space* s_j *such that* $X = (x_1, \ldots, x_n)$ *is well-conditioned.*

Step 3. *Compute* $M = A - BK$ *by solving the linear systems:* $MX = X\Lambda$.

Step 4. *Compute* $K : K = Z^{-1}U_0^{T}(A - M)$.

Example 11.6.1.   Consider the REVA with the following data:

$$A = \begin{pmatrix} 1 & 2 & 3 \\ 4 & 5 & 6 \\ 7 & 8 & 9 \end{pmatrix}, \qquad B = \begin{pmatrix} 6 & 3 \\ 1 & 2 \\ 8 & 9 \end{pmatrix}, \qquad \Lambda = \text{diag}(9, 5, 1).$$

Step 1.

$$U_0 = \begin{pmatrix} -0.5970 & 0.7720 \\ -0.0995 & -0.3410 \\ -0.7960 & -0.5364 \end{pmatrix}, \qquad U_1 = \begin{pmatrix} -0.2181 \\ -0.9348 \\ 0.2804 \end{pmatrix}$$

$$Z = \begin{pmatrix} -10.0499 & -9.1543 \\ 0 & -3.1934 \end{pmatrix}.$$

Step 2. $X = \begin{pmatrix} -0.0590 & 0.9859 & -0.0111 \\ -0.7475 & 0.0215 & -0.8993 \\ -0.6617 & -0.1657 & 0.4371 \end{pmatrix}$.

Step 3. $M = \begin{pmatrix} 5.0427 & 0.0786 & 0.2640 \\ 0.9987 & 3.7999 & 5.7856 \\ 0.1399 & 2.0051 & 6.1575 \end{pmatrix}$.

Step 4. $K = \begin{pmatrix} -1.8988 & 0.0269 & 0.5365 \\ 2.4501 & 0.5866 & -0.1611 \end{pmatrix}$.

Verify: The eigenvalues of $(A - BK)$ are 5, 9, 1.
$\text{Cond}_2(K) = 6.3206$.

Some Implementational Details

Implementation of Step 1: Decomposition of B in Step 1 of the algorithm amounts to the QR factorization of B. Once this decomposition is performed, constructions of the bases can be done either by QR factorization of $(U_1^{T}(A - \lambda_j I))^{T}$ or by computing its SVD.

If QR factorization is used, then

$$(U_1^{T}(A - \lambda_j I))^{T} = (\hat{S}_j, S_j) \begin{pmatrix} R_j \\ 0 \end{pmatrix}$$

Thus, S_j and $\hat{S}_j$ are the matrices whose columns form the required orthonormal bases.

If SVD is used, then from

$$U_1^T(A - \lambda_j I) = T_j\,(\Gamma_j, 0)(\hat{S}_j, S_j)^T,$$

we see that the columns of S_j and $\hat{S}_j$ form the required orthonormal bases. Here Γ_j is the diagonal matrix containing the singular values.

Note: The QR decomposition, as we have seen, is more efficient than the SVD approach.

Implementation of Step 2: **Step 2 is the key step in the solution process.** Kautsky *et al.* (1985) have proposed four methods to implement Step 2. Each of these four methods aims at minimizing a different measure ν of the sensitivity.

We present here only one (Method 0 in their paper), which is the simplest and most natural one. This method is designed to minimize the measure $\nu_2 = \text{Cond}_2(X)$.

First, we note that $\text{Cond}_2(X)$ will be minimized if each vector $x_j \in s_j$, $j = 1, 2, \ldots, n$ is chosen such that the angle between x_j and the space

$$t_j = < x_i, i \neq j >$$

is maximized for all j. The symbol $< x_k >$ denotes the space spanned by the vectors x_k.

This can be done in an iterative fashion. Starting with an **arbitrary** set of n independent vectors $X = (x_1, \ldots, x_n)$, $x_j \in s_j$, $j = 1, \ldots, n$, we replace each vector x_j by a new vector such that the angle to the current space t_j is maximized for each j. The QR method is again used to compute the new vectors as follows:

Find $\tilde{y}_j$ by computing the QR decomposition of

$$X_j = (x_1, x_2, \ldots, x_{j-1}, x_{j+1}, \ldots, x_n),$$

$$= (\tilde{Q}_j, \tilde{y}_j) \begin{pmatrix} \tilde{R}_j \\ 0 \end{pmatrix}.$$

and then compute the new vector

$$x_j = \frac{S_j S_j^T \tilde{y}_j}{\|S_j^T \tilde{y}_j\|_2}.$$

Note that with this choice of x_j, the condition $c_j = 1/|\tilde{y}_j^T x_j|$ is minimized. The n steps of the process required to replace successively n vectors x_1 through x_n will constitute a sweep.

At the end of each sweep, $\text{Cond}_2(X)$ is measured to see if it is acceptable; if not, a new iteration is started with the current X as the starting set. The iteration is continued until $\text{Cond}_2(X)$, after a full sweep of the powers ($j = 1, 2, \ldots n$) is less than some positive tolerance.

Implementations of Step 3 and Step 4: Implementations of Step 3 and Step 4 are straightforward. M in Step 3 is computed by solving linear systems: $X^T M^T = \Lambda^T X^T$ using Gaussian elimination with partial pivoting.

K in Step 4 is computed by solving upper triangular systems:

$$ZK = U_0^T(A - M).$$

Flop-count: Step 1: $O(n^3 m)$ flops, Step 2: $O(n^3) + O(n^2 m)$ flops per *sweep*, Step 3: $O(n^3)$ flops, and Step 4: $O(mn^2)$ flops.

MATCONTROL note: Algorithm 11.6.1 has been implemented in MATCONTROL function **polerob**. It computes both the feedback matrix K and the transforming matrix X.

Remarks on convergence of Algorithm 11.6.3 and the Tits–Yang Algorithm

- Each step of the above iteration amounts to rank-one updating of the matrix X such that the sensitivity of the eigenvalue λ_j is minimized. **However, this does not necessarily mean that the overall conditioning is improved at each step**. This is because the conditioning of the other eigenvalues $(\lambda_i, i \neq j)$ will be disturbed when the old vector x_j is replaced by the new vector.

- It was, thus, stated by Kautsky *et al.* (1985) that "the process does not necessarily converge to a fixed point." It, however, turned out to be the case of "slow convergence" only. Indeed, Tits and Yang (1996) later gave a proof of the convergence of the algorithm. Tits and Yang (1996) observed that this algorithm amounts to maximize, at each iteration, the determinant of the candidate eigenvector matrix X with respect to one of its column (subject to the constraints that it is still an eigenvector matrix of the closed-loop system). Based on this observation, Tits and Yang developed a more efficient algorithm by maximizing $\det(X)$ with respect to two columns concurrently. The **Tits–Yang algorithm** can easily be extended to assign the eigenvalues with complex conjugate pairs. For details of these algorithms, we refer the readers to the paper by Tits and Yang (1996). There also exists software called **robpole** based on the Tits–Yang algorithm.

MATLAB note: The MATLAB function **place** has implemented Algorithm 11.6.1.

Given a controllable pair (A, B) and a vector p containing the eigenvalues to be assigned, $K = $ **place**(A, B, p) computes the feedback matrix K such that $(A - BK)$ has the desired eigenvalues. The software **robpole**, based on the Tits–Yang algorithm, is available in SLICOT (see Section 11.10).

Some Properties of the Closed-Loop System

The minimization of the condition number of the eigenvector matrix leads to some desirable robust properties of the closed-loop system. We state some of these properties below. The proofs can be found in Kautsky *et al.* (1985) or the readers can work out the proofs themselves.

Theorem 11.6.2.

(i) The gain matrix K obtained by Algorithm 11.6.1 satisfies the inequality

$$\|K\|_2 \leq (\|A\|_2 + \max_j |\lambda_j| \mathrm{Cond}_2(X))/(\sigma_{\min}(B)) = k',$$

where $\sigma_{\min}(B)$ denotes the smallest singular value of B.
(ii) The transient response $x(t)$ satisfies

$$\|x(t)\|_2 \leq \mathrm{Cond}_2(X) \max_j \{|e^{\lambda_j t}|\}. \|x_0\|_2,$$

where $x(0) = x_0$ or in the discrete case

$$\|x(k)\|_2 \leq \mathrm{Cond}_2(X) \cdot \max_j \{|\lambda_j|^k \cdot \|x_0\|_2\}.$$

Example 11.6.2. For Example 11.6.1, we easily see that

$$\|K\|_2 = 3.2867, \qquad k' = 12.8829.$$

Thus, the result of part (i) of Theorem 11.6.2 is verified.

Theorem 11.6.3. *If the feedback matrix K assigns a set of stable eigenvalues λ_j, then the perturbed closed-loop matrix $A - BK + \Delta$ remains stable for all perturbations Δ that satisfy*

$$\|\Delta\|_2 \leq \min_{s=j\omega} \sigma_{\min}(sI - A + BK) = \delta(K).$$

Furthermore, $\|\delta(K)\| \leq \min_j \mathrm{Re}(-\lambda_j)/\mathrm{Cond}_2(X)$.

In the discrete-case, the closed-loop system remains stable for perturbations Δ such that

$$\|\Delta\|_2 \leq \min_{s=\exp(i\omega)} \{sI - A + BK\} = \Delta(F)$$

and $\Delta(F) \geq \min_j (1 - |\lambda_j|)/\mathrm{Cond}_2(X)$.

Minimum-norm robust pole assignment: We stated in the previous section that the norm of the feedback matrix is another important factor that influences the sensitivity of closed-loop poles. Thus, it is important to consider this aspect of REVA as well.

The REVA with respect to minimizing the norm of the feedback matrix has been considered by Keel *et al.* (1985) and more recently by Varga (2000).

Both algorithms are Sylvester equation based (see **Exercise 11.11** for the statement of a Sylvester equation based EVA algorithm). The paper by Keel *et al.* (1985) addresses minimization of the performance index

$$I - \text{Trace}(K^T K),$$

whereas Varga (2000) considers the minimization of the performance index

$$J = \frac{\alpha}{2}(\|X\|_F^2 + \|X^{-1}\|_F) + \frac{1-\alpha}{2}\|K\|_F^2$$

Note that minimizing J as above for $0 < \alpha < 1$ amounts to simultaneous minimization of the norm of the feedback matrix K and of the condition number of the eigenvector matrix X (with respect to the Frobenius norm).

For space limitations, we are not able to describe these algorithms here. The readers are referred to the papers by Keel *et al.* (1985) and Varga (2000). There also exists a software, based on the Varga algorithm, called "**sylvplace**" (available from Dr. Varga (E-mail: andras.varga@dlr.de).

11.7   COMPARISON OF EFFICIENCY AND STABILITY: THE SINGLE-INPUT EVA PROBLEM

Table 11.1: Comparison of efficiency and stability of a single-input EVA problem

Method	Efficiency: Flop-count (Approximate) This count includes transformation of (A, b) to the controller-Hessenberg form	Numerical stability (backward stability and other features)
The Recursive Algorithm (**Algorithm 11.2.1**)	$\frac{11}{3}n^3$	**Stability is not guaranteed**, but the algorithm allows the users to monitor the stability. **Reliable**
The RQ implementations of the recursive algorithm (**Algorithms 11.2.2 and 11.2.3**)	$5n^3$	**Stable**
The explicit QR method (**Miminis and Paige (1982)**)	$5n^3$	**Stable**
The implicit QR method (**Patel and Misra (1984)**)	$5n^3$	**Stable**
The eigenvector method (**Petkov** *et al.* 1984; **Exercise 11.8**).	$\frac{20}{3}n^3$	**Stable**

11.8 COMPARISON OF EFFICIENCY AND STABILITY: THE MULTI-INPUT EVA PROBLEM

Table 11.2: Comparison of efficiency and stability: the multi-input EVA problem

Method	Efficiency: Flop-count (approximate). These counts include transformation of (A, B) to the controller-Hessenberg form	Numerical stability (backward stability) and other features
The recursive algorithm (**Algorithm 11.3.1**)	$\frac{19}{3}n^3$	No formal round-off error analysis available. The algorithm is believed to be **reliable**
The explicit QR algorithm (**Section 11.3.2**).	$9n^3$	**Stable**
The implicit QR algorithm (the multi-input version of the implicit single-input QR algorithm of Patel and Misra (1984))	$9n^3$	Stability not formally proven, but is **believed to be stable**
The Schur method (**Algorithm 11.3.3**)	$30n^3$	Stability not formally proven, but is **believed to be stable.** The algorithm has an attractive feature that it can also be used for partial pole placement in the sense that it allows one to reassign only the "bad" eigenvalues, leaving the "good" ones unchanged
The eigenvector method (Petkov *et al.* 1986; not described in the book)	$\frac{40}{3}n^3$	**Stable**

11.9 COMPARATIVE DISCUSSION OF VARIOUS METHODS AND RECOMMENDATION

For the single-input problem: The recursive algorithm (**Algorithm 11.2.1**) is the fastest one proposed so far. It is also extremely simple to implement. Unfortunately, the numerical stability of the algorithm cannot be guaranteed in all cases. The algorithm, however, allows the users to monitor the stability. **Algorithm 11.2.1 is thus reliable.** In a variety of test examples, this algorithm has done remarkably well, even for some ill-conditioned problems (e.g., see **Example 11.2.3**). The RQ implementation of the recursive algorithm (**Algorithm 11.2.3**), the explicit and implicit QR algorithms all have the same efficiency, and are numerically **stable.** The eigenvector algorithm (**Exercise 11.8**) if properly implemented, is also stable, but it is the most expensive one of all the single-input algorithms.

One important thing to note here is that there exist RQ implementations of all the single-input algorithms mentioned in Table 11.1 (Arnold and Datta 1998). These RQ implementations are much easier to understand and implement on computers. **We strongly recommend the use of RQ implementations of these algorithms.**

For the multi-input problem: The recursive algorithm (**Algorithm 11.3.1**) is again the fastest algorithm; however, no round-off stability analysis of this algorithm has been done yet. **The explicit QR algorithm described in Section 11.3.2 is stable.** The properly implemented eigenvector algorithm due to Petkov *et al.* (1986), is also stable but is more expensive than the explicit QR algorithm. The Schur algorithm (**Algorithm 11.3.3**) is the most expensive one. It is believed to be numerically stable. An important feature of this algorithm is that it can be used for partial pole assignment in the sense that it offers a choice to the user to place only the "bad" eigenvalues, leaving the "good" ones unchanged.

The REVA algorithm (**Algorithm 11.6.1**) exploits the freedom offered by the problem to minimize the conditioning of the eigenvector matrix which is a major factor for the sensitivity of the closed-loop poles. However, when a well-conditioned eigenvector matrix does not exist, the algorithm may give inaccurate results. When the eigenvector matrix is ill-conditioned, it may be possible to obtain more accurate results using other methods.

Based on the above observations, it is recommended that for the single-input problem, the recursive algorithm (**Algorithm 11.2.1**) *be tried first. In case of possible ill-conditioning of the matrix L, its RQ formulation* (**Algorithm 11.2.2**) *should be used.*

For the multi-input problem, the multi-input version of the recursive algorithm (*Algorithm* 11.3.1) *should be tried first.* If the algorithm appears to be unstable (as indicated by the condition number of the matrix *L*), the explicit QR algorithm (**Section 11.3.1**) is to be used.

It should, however, be noted that Algorithm 11.3.1 and the explicit QR algorithm, as stated here, might give complex feedback matrix. There now exists a modified version of Algorithm 11.3.1 (Carvalho and Datta 2001) that avoids complex arithmetic and this modified version has been implemented in MATCONTROL function **polercm**.

For REVA, the choices are either Algorithm 11.6.1 *or the Tits–Yang Algorithm.*
For PEVA, the choices are either the Schur algorithm (Algorithm 11.3.3)*, or the*
Sylvester equation algorithm (Algorithm 11.3.4)*.* Algorithm 11.6.1 does not handle
complex EVA as such, but its implementation in MATLAB function '**place**' does in
an ad hoc fashion. Numerical experimental results suggest the Tits–Yang algorithm
"typically produce more robust design" than that constructed by Algorithm 11.6.1.
For partial pole placement, Algorithm 11.3.4 seems to be very efficient and not
computationally intensive.

11.10   SOME SELECTED SOFTWARE

11.10.1   MATLAB Control System Toolbox

Classical design tools
  acker   SISO pole placement
place   MIMO pole placement.

11.10.2   MATCONTROL

POLERCS   Single-input pole placement using the recursive algorithm
POLEQRS   Single-input pole placement using the QR version of the recursive
          algorithm
POLERQS   Single-input pole placement using RQ version of the recursive
          algorithm
POLERCM   Multi-input pole placement using the recursive algorithm
POLERCX   Multi-input pole placement using the modified recursive
          algorithm that avoids complex arithmetic and complex feedback
POLEQRM   Multi-input pole placement using the explicit QR algorithm
POLESCH   Multi-input pole placement using the Schur decomposition
POLEROB   Robust pole placement.

11.10.3   CSP-ANM

Pole assignment

- The recursive algorithm is implemented as StateFeedbackGains
  [*system, poles*, Method → Recursive].
- The explicit QR algorithm is implemented as StateFeedbackGains
  [*system, poles*, Method → QRDecomposition].
- The Schur method is implemented as StateFeedbackGains [*system,
  poles*, Method → SchurDecomposition].

- The RQ implementation of the recursive single-input algorithm is implemented as StateFeedbackGains [*system, poles*, Method → RecursiveRQDecomposition].
- The implicit single-input RQ algorithm is implemented as StateFeedbackGains [*system, poles*, Method → ImplicitRQDecomposition].

11.10.4 SLICOT

Eigenvalue/eigenvector assignment

SB01BD   Pole assignment for a given matrix pair (A, B)
SB01DD   Eigenstructure assignment for a controllable matrix pair (A, B) in orthogonal canonical form
SB01MD   State feedback matrix of a time-invariant single-input system
ROBPOLE   Robust Pole Assignment (Additional function added in 2003).

11.10.5 MATRIX$_X$

Purpose: Calculate state feedback gains via pole placement for single-input continuous-time or discrete-time systems.
Syntax: KC = POLEPLACE (A, B, POLES) ...controller design
          KE = POLEPLACE (A′, B′, POLES) ...estimator design

11.10.6 POLEPACK

A collection of MATLAB programs for EVA, developed by G.S. Miminis (1991). Available on **NETLIB**.

11.11 SUMMARY AND REVIEW

Statement of the EVA Problem

Given a pair of matrices (A, B), and the set $S = \{\lambda_1, \ldots, \lambda_n\}$, closed under complex conjugation, find a matrix K such that $\Omega(A - BK) = S$.

Here $\Omega(M)$ denotes the spectrum of M.

In the single-input case, the problem reduces to that of finding a row vector f^T such that

$$\Omega(A - bf^T) = S.$$

Existence and Uniqueness

The EVA problem has a solution if and only if (A, B) is controllable. In the single-input case, the feedback vector, when it exists, is unique. In the multi-input case, when there exists a feedback matrix, there are many. Therefore, the existing freedom can be exploited to improve the conditioning of the solution and of the closed-loop eigenvectors.

Numerical Methods

There are many methods for the EVA problem. Only a few have been described here. These include:

- Recursive algorithms (**Algorithm 11.2.1** for the single-input problem and **Algorithm 11.3.1** for the multi-input problem).
- QR-type algorithms (**Algorithms 11.2.2, 11.2.3, and those described in Miminis and Paige (1982), and Patel and Misra (1984)**. For the single-input problem and the explicit QR method described in **Section 11.3.2** for the multi-input problem).
- The Schur algorithm (**Algorithm 11.3.3**) for the multi-input problem.
- PEVA (**Algorithm 11.3.4**).

Efficiency and Numerical Stability

The recursive algorithms are the most efficient algorithms. The computer implementations of these algorithms are extremely simple. The algorithms, however, do not have guaranteed numerical stability, except for the RQ version of the single-input recursive algorithm, which has been proved to be numerically stable (Arnold and Datta 1998).

In the single-input case, it has been proved (see Arnold and Datta (1998)), by forward round-off error analysis, that the stability of the recursive algorithm (**Algorithm 11.2.1**) can be monitored and it is possible for the user to know exactly when the algorithm starts becoming problematic. **It is thus reliable.** Similar results are believed to hold for the multi-input recursive algorithm as well. But no formal analysis in the multi-input case has yet been done.

The QR-type algorithms for single-input problems all have the same efficiency and are numerically stable.

For the multi-input problem, the Schur Algorithm (**Algorithm 11.3.3**) is the most expensive one. However, it has an important feature, namely, it allows one to place only the "bad" eigenvalues, leaving the "good" ones unchanged.

The explicit QR algorithm (**Section 11.3.2**) and the multi-input version of the single-input implicit QR algorithm (not described in this book) have the same efficiency. The explicit QR algorithm has been proven to be stable and the implicit QR algorithm is believed to be stable as well.

Explicit Solutions

An explicit expression for the unique feedback vector for the single-input EVA problem has been given using the recursive algorithm (**Algorithm 11.2.1**). This formula is

$$f = \frac{1}{\alpha}(H^{\mathrm{T}} - \lambda_1 I)(H^{\mathrm{T}} - \lambda_2 I) \cdots (H^{\mathrm{T}} - \lambda_n I)e_n,$$

where $H = (h_{ij})$ is the Hessenberg matrix of the controller-Hessenberg form of the pair (A, b) and $\alpha = \prod_{i=1}^{n-1} h_{i+1,i}$. In the multi-input case, the expression is rather complicated (see Arnold (1993)).

Conditioning of the Feedback Problem: From the explicit expression of the feedback vector f of the single-input EVA problem, it is clear that the Hessenberg single-input feedback problem is essentially a polynomial evaluation $\phi(H)$ at an unreduced Hessenberg matrix, where $\phi(x) = (\lambda - \lambda_1)(\lambda - \lambda_2) \cdots (\lambda - \lambda_n)$ is the characteristic polynomial of the closed-loop matrix.

A result on the Frechet derivative $D\phi(H)$ is first given in Ph.D. dissertation of Arnold (1993) and the condition numbers for the feedback problem are then defined using this derivative. Next, a condition number estimator for the problem is stated. It worked well on test examples. This estimator never underestimated the error and overestimated the error by less than 2 digits, in all 100 test examples of sizes varying from 10 to 50, both for ill-conditioned and well-conditioned problems.

In the multi-output case, **Theorem 11.4.2** gives the perturbation bound for the feedback matrix from which the absolute and relative condition numbers are defined (**Section 11.4.3**).

Conditioning of the Closed-Loop Eigenvalues

The major factors responsible for the sensitivity of the closed-loop eigenvalues have been identified in **Section 11.5**. These factors are: **the condition number of the eigenvector matrix of the closed-loop system, the distance to uncontrollability and the distance between the open-loop and the closed-loop eigenvalues, the conditioning of the feedback problem, and the norm of the feedback matrix.** The most important of them is the condition number of the eigenvector matrix.

Robust Eigenvalue Assignment

Given the pair (A, B) and the matrix $\Lambda = \mathrm{diag}(\lambda_1, \ldots, \lambda_n)$, the problem is to find a nonsingular matrix X, and a matrix K satisfying

$$(A - BK)X = X\Lambda$$

such that $\mathrm{Cond}_2(X)$ is minimum. In view of the last sentence of the preceding paragraph, the REVA problem with respect to minimizing the condition number of the eigenvector matrix is a very important practical problem. An algorithm (**Algorithm 11.6.1**) due to Kautsky *et al.* (1985) is given in Section 11.6. The algorithm requires constructions of orthonormal bases for a certain space and for its complement. The QR factorization or the SVD can be used for this purpose. An analysis of convergence and a more improved version of this algorithm can be found in Tits and Yang (1996).

Algorithm for Minimizing Feedback Norm

Our discussions on the conditioning of the closed-loop eigenvalues (**Section 11.5**) show that it is also important to have algorithms that minimize the norm of the feedback matrix.

For such algorithms, see Keel *et al.* (1985) and Varga (2000).

11.12 CHAPTER NOTES AND FURTHER READING

Many algorithms have been developed for solving the EVA by state feedback. A good account of these algorithms can be found in the recent book by Xu (1998).

The earlier algorithms, based on reduction to controller-canonical forms, turn out to be numerically unstable. For a reference of some of these earlier algorithms, see Miminis and Paige (1982, 1988).

For a comprehensive reference of the Hessenberg or controller-Hessenberg based algorithms, which are more numerically reliable, see the recent paper of Arnold and Datta (1998). For a matrix equation based algorithm for EVA, see Bhattacharyya and DeSouza (1982). For robust eigenvalue and eigenstructure assignment algorithms, see Cavin and Bhattacharyya (1983), Kautsky *et al.* (1985), Byers and Nash (1989), and Tits and Yang (1996). For REVA by output feedback, see Chu, *et al.* (1984) and references therein. The other algorithms include there is Tsui (1986), Valasek and Olgac (1995a, 1995b).

For algorithms that minimize the norm of the feedback matrix, see Keel *et al.* (1985) and Varga (2000). The EVA problem by output feedback is a difficult problem and only a few algorithms are available. See Misra and Patel (1989) for output feedback algorithms. For the EVA and eigenstructure algorithms for descriptor systems (not discussed in this chapter), see Fletcher *et al.* (1986), Chu (1988), and Kautsky *et al.* (1989), Varga (2000). For partial pole-assignment algorithms, see Varga (1981), Saad (1988), and Datta and Saad (1991), Datta and Sarkissian (2000).

For round-off error analysis of various algorithms for EVA, see Cox and Moss (1989, 1992), Arnold and Datta (1998), and Miminis and Paige (1988).

The perturbation analysis for the single-input feedback problem was considered by Arnold (1993) and our discussion in the multi-input feedback problem has been taken from Sun (1996).

For discussions on conditioning of the EVA problem, see He *et al.* (1995), Mehrmann and Xu (1996, 1997, 1998), Konstantinov and Petkov (1993), Calvetti *et al.* (1999).

For an extension of the single-input recursive algorithm to assigning Jordan canonical form (JCF), companion and Hessenberg forms, etc., see Datta and Datta (1990). For parallel algorithms for the EVA problem, see Bru *et al.* (1994c), Coutinho *et al.* (1995), Datta and Datta (1986), Datta (1991), Baksi *et al.* (1994).

Now, there also exists a block algorithm (Carvalho and Datta 2001) for the multi-input EVA. **This block algorithm, besides being suitable for high-performance computing, is guaranteed to give a real feedback matrix.**

Exercises

11.1   Modify both the single-input (**Algorithm 11.2.1**) so that the use of complex arithmetic can be avoided (consult Carvalho and Datta (2001)).

11.2   *Single-input pole placement via linear systems.* Consider the following algorithm (Datta and Datta (1986)) for the single-input Hessenberg problem $(H, \hat{b})$, where H is an unreduced upper Hessenberg matrix and $\hat{b} = (\alpha, 0, \cdots, 0)^{\mathrm{T}}, \alpha \neq 0$. Let $\{\mu_i\}_{i=1}^n$ be the eigenvalues to be assigned.
Step 1. Solve the $n \times n$ Hessenberg systems:

$$(H - \mu_i I)t_i = \hat{b}, \quad i = 1, 2, \ldots, n$$

Step 2. Solve for d:

$$T^{\mathrm{T}} d = r,$$

where $T = (t_1, t_2, \ldots, t_n)$ and $r = (\alpha, \alpha, \ldots, \alpha)^{\mathrm{T}}$.
Step 3. Compute $f^{\mathrm{T}} = \dfrac{1}{\alpha} d^{\mathrm{T}}$.

(a)   Give a proof of this algorithm, that is, prove that $\Omega(H - \hat{b}f^{\mathrm{T}}) = \{\mu_1, \ldots, \mu_n\}$; making necessary assumptions. Do an illustrative example.
(**Hint:** Take $\Lambda = \mathrm{diag}(\mu_1, \mu_2, \ldots, \mu_n)$ in the proof of Algorithm 11.2.1 and follow the lines of the proof there.)
(b)   Prove that T in Step 2 is nonsingular if the entries in the set $\{\mu_1, \mu_2, \ldots, \mu_n\}$ are pairwise distinct and none of them is an eigenvalue of H.

11.3   Consider the following modification of Algorithm in Exercise 11.2, proposed by Bru *et al.* (1994a):
   Step 1. For $i = 1, 2, \ldots, n$ do
If μ_i is not in the spectrum of H, then solve the system

$$(H - \mu_i I)t_i = \hat{b}.$$

Else solve the system $(H - \mu_i I)t_i = 0$.
   Step 2. Define the vector $u = (u_1, \ldots, u_n)^{\mathrm{T}}$ as follows:
$u_i = 1$, if μ_i is an eigenvalue of H,
$u_i = 0$, otherwise.

Step 3. Solve for f:

$$f^T T = u^T,$$

where $T = (t_1, t_2, \ldots, t_n)$.

(a) Give a proof of this algorithm assuming that the pair $(H, \hat{b})$ is controllable and that the numbers in the set $\{\mu_1, \mu_2, \ldots, \mu_n\}$ are closed under complex conjugation and pairwise distinct. Do an illustrative example.

(b) Give an example to show that the assumption that $\mu_i, i = 1, \ldots, n$ are pairwise distinct, cannot be relaxed.

Note: Bru *et al.* (1994a) have given a more general algorithm which can assign multiple eigenvalues (algorithm III in that paper.

11.4 *Assigning canonical forms (Datta and Datta 1990).* Extend Algorithm 11.2.1 to the problems of assigning the following canonical forms: a companion matrix, an unreduced upper Hessenberg matrix, a Jordan matrix with no two Jordan blocks having the same eigenvalue. (**Hint:** Follow the line of the development of Algorithm 11.2.1 replacing Λ by the appropriate canonical form to be assigned). Do illustrative examples.

11.5 *Multi-input pole-placement via Linear systems (Datta 1989).* Develop a multi-input version of the Algorithm in Exercise 11.2, making necessary assumptions. (See Tsui (1986) and Bru *et al.* (1994b).)

11.6 Give a proof of Algorithm 11.2.3 (the RQ formulation of Algorithm 11.2.1). Consult Arnold and Datta (1998), if necessary.

11.7 Show that the explicit formula for the single-input pole assignment problem (Formula 11.2.9) is a Hessenberg-form of the Ackermann's formula.

11.8 *Eigenvector method for the pole-placement (Petkov et al. 1984)*

(a) Given the single-input controller-Hessenberg pair $(H, \hat{b})$, show that it is possible to find an eigenvector $\tilde{v}$, corresponding to an eigenvalue μ to be assigned, for the closed-loop matrix $H - bf^T$, without knowing the feedback vector f.

(b) Let $\Lambda = \text{diag}(\mu_1, \mu_2, \ldots, \mu_n)$ be the matrix of the eigenvalues to be assigned, and V be the eigenvector matrix and $v^{(1)}$ be the first row of V. Assume that $\mu_i, i = 1, \ldots, n$ are all distinct. Prove that the feedback vector f can be computed from

$$f = \frac{1}{\alpha}(h_1 - v^{(1)} \Lambda V^{-1}),$$

where h_1 is the first row of H, and α is the 1st entry of $\hat{b}$.

(c) What are the possible numerical difficulties of the above method for computing f? Give an example to illustrate these difficulties.

(d) Following the same procedure as in the RQ formulation of **Algorithm 11.2.1**, work out an RQ version of the above eigenvector method. (Consult Arnold and Datta (1998), if necessary.) Compare this RQ version with the above formulation with respect to flop-count and numerical effectiveness.

11.9 Modify the Schur algorithm (**Algorithm 11.3.3**) for the multi-input problem to handle the case when the complex numbers in the matrix Γ are not so ordered that the ordering agrees with the diagonal structure of the matrix A_2. Work out a simple example with this modified Schur method.

11.10 Write MATLAB codes to implement the Algorithms 11.2.1 and 11.2.3 and those in Exercises 11.2, 11.3 and 11.8, and then using these programs, make a comparative study with respect to CPU time, flop-count, the norm of the feedback vector and the largest error-norm between the closed-loop and open-loop eigenvalues. Use randomly generated matrices.

11.11 *EVA via Sylvester matrix equation.* The following algorithm by Bhattacharyya and DeSouza (1982) solves the multi-input EVA problem:

Step 1. Pick a matrix G arbitrarily.

Step 2. Solve the Sylvester equation $AX - X\tilde{A} = -BG$, where $\tilde{A}$ is a matrix having the spectrum $\{\lambda_1, \ldots, \lambda_n\}$ to be assigned, for a full-rank solution X.

If the solution matrix X does not have full rank, return to Step 1 and pick another G.

Step 3. Compute the feedback matrix F by solving $FX = G$.

Give a proof of the algorithm and construct an example to illustrate the algorithm. (For the conditions on the existence of full-rank solution of the Sylvester equation, see Chapter 12 and the paper by DeSouza and Bhattacharyya (1981)).

11.12 Construct an example to demonstrate that even if the feedback (vector) for the single-input problem is computed reasonably accurately, the closed-loop eigenvalues may still differ from those to be assigned.

11.13 *Sensitivity analysis of the single-input pole-placement problem via Cauchy matrix (Mehmann and Xu 1998; Calvetti et al. 1999).* Let $\Lambda = \mathrm{diag}(\lambda_1, \lambda_2, \ldots, \lambda_n)$ with $\lambda_i \neq \lambda_j$ for $i \neq j$. Define $e = (1, 1, \ldots, 1)^\mathrm{T}$. Let $S = \{\mu_1, \mu_2, \ldots, \mu_n\}$, the eigenvalue set to be assigned; μ_i's are distinct and none of them is in the spectrum of Λ.

(a) Prove that the vector f_e defined by $f_e = C_h^{-T} e$ is such that $\Omega(\Lambda - e f_e^\mathrm{T}) = S$, where $C_h = (c_{ij})$ is the Cauchy matrix: $c_{ij} = 1/(\lambda_i - \mu_j)$.

(b) Show that $\mathrm{Cond}_2(C_h)$ is the spectral condition number of the closed-loop matrix $\Lambda - e f_e^\mathrm{T}$.

(c) Using the result in (a) find an expression for the feedback vector f such that $\Omega(A - b f^\mathrm{T}) = S$, assuming that A is diagonalizable.

(d) Give a bound of the condition number of the eigenvector matrix of the closed-loop matrix $A - b f^\mathrm{T}$ in terms of the condition number of C_h, the condition number of the eigenvector matrix X of A, and the minimum and maximum entries of the vector $X^{-1} b$.

(e) Give a bound of the feedback vector f in terms of the norm of f_e and the norm of the inverse of the matrix XR, where $R = \mathrm{diag}(\hat{b}_1, \hat{b}_2, \ldots, \hat{b}_n)^\mathrm{T}$, and $X^{-1} b = (\hat{b}_1, \hat{b}_2, \ldots, \hat{b}_n)^\mathrm{T}$.

(f) From the bounds obtained in (d) and (e), verify the validity of some of the factors responsible for the ill-conditioning of the single-input EVA problem, established in Section 11.5.

(g) Work out an example to illustrate (a)–(f).

11.14 From the expression (11.6.4) of the feedback matrix K, prove that if the condition of the eigenvector matrix is minimized, then a bound of the norm of the feedback matrix K is also minimized.

Give an example to show that this does not necessarily mean that the resulting feedback matrix will have the minimal norm.

11.15 (a) Perform a numerical experiment to demonstrate the slow convergence of Algorithm 11.6.1.

(b) Using MATCONTROL function **polerob** and **robpole** (from SLICOT), make a comparative study between Algorithms 11.6.1, and the Tits–Yang algorithm with respect to number of iterations, $\text{Cond}_2(X)$, and $\|K\|_2$. Use data of Example 11.6.1 and randomly generated matrices.

11.16 *Deadbeat control.* Given the discrete-system:

$$x_{i+1} = Ax_i + Bu_i,$$

the problem of "deadbeat" control is the problem of finding a state feedback $u_i = -Kx_i + v_i$ such that the resulting system:

$$x_{i+1} = (A - BK)x_i + v_i$$

has the property that $(A - BK)^p = 0$ for some $p < n$ and $\text{rank}(B) > 1$.

The solution of the homogeneous part of the closed-loop system then "dies out" after p steps; and that is why the name **deadbeat control**.

A numerically reliable algorithm for the deadbeat control has been provided by Van Dooren (1984). The basic idea of the algorithm is as follows:
If

$$\left(H = (H_{ij}), \ \tilde{B} = \begin{pmatrix} B_1 \\ 0 \\ \vdots \\ 0 \end{pmatrix} \right)$$

is the controller-Hessenberg pair of (A, B), then the solution of the problem is equivalent to finding a feedback matrix K such that

$$V^T(H - \hat{B}K)V = \begin{pmatrix} 0 & H_{12} & \cdots & \cdots & \cdots & H_{1p} \\ 0 & & H_{23} & \cdots & \cdots & H_{2p} \\ \vdots & \ddots & & & & \vdots \\ \vdots & & & \ddots & & \vdots \\ \vdots & & & & \ddots & H_{p-1,p} \\ 0 & & & & & 0 \end{pmatrix},$$

for some orthogonal matrix V. The form on the right-hand side is called the **"deadbeat"** form.

Note that in this case $(H - \hat{B}K)^p = 0$. Van Dooren's algorithm finds K recursively in p steps. At the end of the ith step, one obtains the matrices V_i^T and K_i such that

$$V_i^T(H - \hat{B}K_i)V_i = \begin{pmatrix} H_d^i & * \\ 0 & H_h^i \end{pmatrix},$$

where the matrix H_d^i is in "deadbeat" form again, and the pair $(H_h^i, \hat{B}_h^i)$; $V_i^T\hat{B} = \begin{pmatrix} \hat{B}_d^i \\ \hat{B}_h^i \end{pmatrix}$, is still in block-Hessenberg form.

Develop a scheme for obtaining V_i and hence complete the algorithm for "deadbeat" control problem. (Consult Van Dooren's paper as necessary).

Work out an illustrative example.

11.17 Using random matrices A of order $n = 5, 10, 15, 20$, and 30, and appropriate input matrices B, make a comparative study between Algorithm 11.3.1, one in Section 11.3.2, and those in Exercises 11.5 and 11.10 with respect to CPU time, flop-count, accuracy of the closed-loop eigenvalues, and norms of feedback matrices.

Research problems

11.1 Carry out a round-off error analysis of the implicit QR algorithm of Patel and Misra (1984) to establish the numerical stability of the algorithm.

11.2 Carry out a round-off error analysis of the recursive multi-input algorithm of Arnold and Datta (1990) **(Algorithm 11.3.1)**. The algorithm is believed to be reliable in practice. Prove or disprove this using the results of your analysis.

11.3 An explicit expression for the family of feedback matrices for the multi-input EVA problem has been given in Arnold (1993). Use this expression to establish the fact that the sensitivities of the closed-loop eigenvalues depend upon the nearness of the system to an uncontrollable system, the separation of the open-loop and the closed-loop eigenvalues, and the ill-conditioning of the closed-loop eigenvector matrix.

11.4 Work out an RQ version of the recursive algorithm for the multi-input EVA problem by Arnold and Datta (1990) **(Algorithm 11.3.1)**.

11.5 Carry out a round-off error analysis of the Schur algorithm of Varga **(Algorithm 11.3.3)** to establish the numerical stability of the algorithm.

11.6 In the QR algorithm of Miminis and Paige (1988) for the multi-input EVA problem, explicit shifting is used for the allocation of each eigenvalue. Work out an **implicit version** of this algorithm.

References

Ackermann J. "Der entwurf linear regelungssysteme im zustandsraum. *Regelungstechnik und prozessedatenverarbeitung*," Vol. 7, pp. 297–300, 1972.

Arnold M. *Algorithms and Conditioning for Eigenvalue Assignment*, Ph.D. dissertation, Northern Illinois University, DeKalb, May 1993.

Arnold M. and Datta B.N. "An algorithm for the multi input eigenvalue assignment problem," *IEEE Trans. Autom. Control*, Vol. 35(10), pp. 1149–1152, 1990.

Arnold M. and Datta B.N. "The single-input eigenvalue assignment algorithms: A close-look," *SIAM J. Matrix Anal. Appl.*, Vol. 19(2), pp. 444–467, 1998.

Baksi D., Datta K.B. and Roy G.D. "Parallel algorithms for pole assignment of multi-input systems," *IEEE Proc. Control Theory Appl.*, Vol. 141(6), pp. 367–372, 1994.

Bhattacharyya S.P. and DeSouza E. "Pole assignment via Sylvester's equation," *Syst. Control Letter.*, Vol. 1, pp. 261–283, 1982.

Bru R., Mas J. and Urbano A. "An Algorithm for the single input pole assignment problem," *SIAM J. Matrix Anal. Appl.*, Vol. 15, pp. 393–407, 1994a.

Bru R., Cerdan J. and Urbano A. "An algorithm for the multi-input pole assignment problem, *Lin. Alg. Appl.*," Vol. 199, pp. 427–444, 1994b.

Bru R., Cerdan J., Fernandez de Cordoba P. and Urbano A. "A parallel algorithm for the partial single-input pole assignment problem," *Appl. Math. Lett.*, Vol. 7, pp. 7–11, 1994c.

Byers R. and Nash S.G. "Approaches to robust pole assignment," *Int. J. Control*, Vol. 49(1) pp. 97–117, 1989.

Calvetti D., Lewis B. and Reichel L. "On the selection of poles in the single-input pole placement problem," *Lin. Alg. Appl.*, Vols. 302–303, pp. 331–345, 1999.

Carvalho J. and Datta B.N. "A block algorithm for the multi-input eigenvalue assignment problem," *Proc. IFAC/IEEE Sym. Syst., Struct. and Control*, Prague, 2001.

Cavin R.K. and Bhattacharyya S. P. "Robust and well conditioned eigenstructure assignment via Sylvester's equation," *Optim. Control Appl. Meth.*, Vol. 4, pp. 205–212, 1983.

Chu E.K., Nichols N.K. and Kautsky J. "Robust pole assignment for output feedback," *Proceedings of the fourth IMA Conference on Control Theory*, 1984.

Chu K.-W.E. "A controllability condensed form and a state feedback pole assignment algorithm for descriptor systems," *IEEE Trans. Autom., Control*, Vol. 33, pp. 366–370, 1988.

Coutinho M. G., Bhaya A. and Datta B. N. "Parallel algorithms for the eigenvalue assignment problem in linear systems," *Proc Int. Conf. Control and Inform. Hong Kong*, pp. 163–168, 1995.

Cox C.L. and Moss W.F. "Backward error analysis for a pole assignment algorithm," *SIAM J. Matrix Anal. Appl.*, Vol. 10(4), pp. 446–456, 1989.

Cox C.L. and Moss W.F. "Backward error analysis of a pole assignment algorithm II: The Complex Case," *SIAM J. Matrix Anal. Appl.*, Vol. 13, pp. 1159–1171, 1992.

Datta B.N. "Parallel and large-scale matrix computations in control: Some ideas," *Lin. Alg. Appl.*, Vol. 121, pp. 243–264, 1989.

Datta B.N. and Datta K. Efficient parallel algorithms for controllability and eigenvalue assignment problems, *Proc. IEEE Conf. Dec. Control*, Athens, Greece, pp. 1611–1616, 1986.

Datta B.N. "An algorithm to assign eigenvalues in a Hessenberg matrix: single input case," *IEEE Trans. Autom. Contr.*, Vol. AC-32(5), pp. 414–417, 1987.

Datta B.N. and Datta K. "On eigenvalue and canonical form assignments." *Lin. Alg. Appl.*, Vol. 131, pp. 161–182, 1990.

Datta B.N. Parallel algorithms in control theory, *Proc. IEEE Conf. on Dec. Control*, pp. 1700–1704, 1991.

Datta B.N. *Numerical Linear Algebra and Applications*, Brooks/Cole Publishing Company, Pacific Grove, CA, 1995.

Datta B.N. and Saad Y. "Arnoldi methods for large Sylvester-like matrix equations and an associated algorithm for partial spectrum assignment," *Lin. Alg. Appl.*, Vol. 154–156, pp. 225–244, 1991.

Datta B.N. and Sarkissian D.R. "Partial eigenvalue assignment in linear systems: Existence, uniqueness and numerical solution," *Proc. Math. Theory of Networks and Sys.*, (MTNS'02), Notre Dame, August, 2002.

Datta K. "The matrix equation $XA - BX = R$ and its applications," *Lin. Alg. Appl.*, Vol. 109, pp. 91–105, 1988.

DeSouza E. and Bhattacharyya S.P. "Controllability, observability and the solution of $AX - XB = C$," *Lin. Alg. Appl.*, Vol. 39, pp. 167–188, 1981.

Fletcher L.R., Kautsky J., and Nichols N.K. "Eigenstructure assignment in descriptor systems," *IEEE Trans. Autom. Control*, Vol. AC-31, pp. 1138–1141, 1986.

He C., Laub A.J. and Mehrmann V. *Placing plenty of poles is pretty preposterous, DFG-Forschergruppe Scientific Parallel Computing, Preprint 95–17*, Fak. f. Mathematik, TU Chemnitz-Zwickau, D-09107, Chemnitz, FRG, 1995.

Higham N.J. *Accuracy and Stability of Numerical Algorithms*, SIAM, Philadelphia, 1996.

Kautsky J., Nichols N.K. and Chu K.-W.E. "Robust pole assignment in singular control systems," *Lin. Alg. Appl.*, Vol. 121, pp. 9–37, 1989.

Kautsky J., Nichols N.K. and Van Dooren P. "Robust pole assignment in linear state feedback," *Int. J. Control*, Vol. 41(5), pp. 1129–1155, 1985.

Keel L.H., Fleming J.A. and Bhattacharyya S.P. Pole assignment via Sylvester's equation, in *Contemporary Mathematics*, (Brualdi R., *et al.*, eds.), Vol. 47, pp. 265, 1985, American Mathematical Society, Providence, RI.

Konstantinov M.M. and Petkov P. "Conditioning of linear state feedback," *Technical Report:* 93–61, Dept. of Engineering, Leicester University, 1993.

Laub A.J. and Linnemann A. "Hessenberg forms in linear systems theory," in *Computational and Combinatorial Methods in Systems Theory*, (Byrnes C.I. and Lindquist A., eds.), pp. 229–244, Elsevier Science publishers, North-Holland, 1986.

MathWorks, Inc., The *MATLAB User's Guide*, The MathWorks, Inc., Natick, MA, 1992.

Mehrmann V. and Xu H. "An analysis of the pole placement problem I: The single-input case," *Electron. Trans. Numer. Anal.*, Vol. 4, pp. 89–105, 1996.

Mehrmann V. and Xu H. "An analysis of the pole placement problem II: The multi-input Case," *Electron. Trams. Numer. Anal.*, Vol. 5, pp. 77–97, 1997.

Mehrmann V. and Xu H. "Choosing the poles so that the single-input pole placement problem is well-conditioned," *SIAM J. Matrix Anal. Appl.*, 1998.

Miminis G.S. and Paige C.C. "An algorithm for pole assignment of time-invariant linear systems." *Int. J. Control*, Vol. 35(2), pp. 341–354, 1982.

Miminis G.S. and Paige C.C. "A direct algorithm for pole assignment of time-invariant multi-input linear systems using state feedback," *Automatica*, Vol. 24(3), pp. 343–356, 1988.

Miminis G.S. *Polepack*, A collection of MATLAB programs for eigenvalue assignment, available on *NETLIB* (www.netlib.org), 1991.

Misra P. and Patel R.V. "Numerical algorithms for eigenvalue assignment by constant and dynamic output feedback," *IEEE Trans. Autom. Control*, Vol. 34(6), pp. 579–580, 1989.

Patel R.V. and Misra P. "Numerical algorithms for eigenvalue assignment by state feedback," *Proc. IEEE*, Vol. 72(12), pp. 1755–1764, 1984.

Petkov P., Christov N.D. and Konstantinov M.M. "A computational algorithm for pole assignment of linear multi input systems," *IEEE Trans. Autom. Control*, Vol. AC-31(11), pp. 1044–1047, 1986.

Petkov P., Christov N.D. and Konstantinov M.M. "A computational algorithm for pole assignment of linear single-input systems," *IEEE Trans. Autom. Contr.*, Vol. AC-29(11), pp. 1045–1048, 1984.

Rice J. "Theory of Conditioning," *SIAM J. Numer. Anal.*, Vol. 3(2), pp. 287–311, 1966.

Saad Y. "Projection and deflation methods for partial pole assignment in linear state feedback," *IEEE Trans. Autom. Control*, Vol. 33, pp. 290–297, 1988.

Shafai B. and Bhattacharyya S.P. "An algorithm for pole placement in high-order multivariable systems," *IEEE Trans. Autom. Control*, Vol. 33, 9, pp. 870–876, 1988.

Stewart G.W. *Introduction to Matrix Computations*, Academic Press, New York, 1973.

J.-G. Sun "Perturbation analysis of the pole assignment problem," *SIAM J. Matrix Anal. Appl.*, Vol. 17, pp. 313–331, 1996.

Szidarovszky F. and Bahill A.T. *Linear Systems Theory*, CRC Press, Boca Raton, 1991.

Tits A.L. and Yang Y. Globally convergent algorithms for robust pole assignment by state feedback, *IEEE Trans. Autom. Control*, Vol. AC-41, pp. 1432–1452, 1996.

Tsui C.C. An algorithm for computing state feedback in multi input linear systems, *IEEE Trans. Autom. Control*, Vol. AC-31(3), pp. 243–246, 1986.

Valasek M. and Olgac N. "Efficient eigenvalue assignments for general linear MIMO systems," *Automatica*, Vol. 31, pp. 1605–1617, 1995a.

Valasek M. and Olgac N. "Efficient pole placement technique for linear time-variant SISO systems," *Proc. IEEE Control Theory Appl.*, Vol. 142, 451–458, 1995b.

Van Dooren P. "Deadbeat Control: A special inverse eigenvalue problem," *BIT*, Vol. 24, pp. 681–699, 1984.

Van Dooren P.M. and Verhaegen M. "On the use of unitary state-space transformations," *Contemporary Mathematics*, (Brualdi R. *et al.* eds.) Vol. 47, pp. 447–463, 1985. American Mathematical Society, Providence, RI.

Varga A. "A multishift Hessenberg method for pole assignment of single-input systems," *IEEE Trans. Autom. Control.*, Vol. 41, pp. 1795–1799, 1996.

Varga A. "Robust pole assignment for descriptor systems," *Proc. Math. Theory of Networks and Sys.* (MTNS '2000), 2000.

Varga A. "A Schur method for pole assignment," *IEEE Trans. Autom. Control*, Vol. AC-26(2), pp. 517–519, 1981.

Xu S.-F. *An Introduction to Inverse Algebraic Eigenvalue Problems*, Peking University Press, Peking, China, 1998.

Wilkinson J.H. *The Algebraic Eigenvalue Problem*, Clarendon Press, Oxford, England, 1965.

CHAPTER **12**_____

STATE ESTIMATION: OBSERVER AND THE KALMAN FILTER

Topics covered

- State Estimation via Eigenvalue Assignment (EVA)
- State Estimation via Sylvester-Observer Equation
- Characterization of the Unique Nonsingular Solution to the Sylvester Equation
- Numerical Methods for the Sylvester-Observer Equation
- A Numerical Method for the Constrained Sylvester-Observer Equation
- Kalman Filter
- Linear Quadratic Gaussian (LQG) Design

12.1 INTRODUCTION

We have seen in Chapter 10 that all the state-feedback problems, such as feedback stabilization, eigenvalue and eigenstructure assignment, the LQR and the state-feedback H_∞-control problems, etc., require that the state vector $x(t)$ should be explicitly available. However, in most practical situations, the states are not fully accessible and but, however, the designer knows the output $y(t)$ and the input $u(t)$. The unavailable states, somehow, need to be estimated accurately from the knowledge of the matrices A, B, and C, the output vector $y(t)$, and the input vector $u(t)$.

In this chapter, we discuss how the states of a continuous-time system can be estimated. **The discussions here apply equally to the discrete-time systems, possibly with some minor changes.** So we concentrate on the continuous-time case only.

We describe two common procedures for state estimation: **one, via eigenvalue assignment** (EVA) and **the other, via solution of the Sylvester-observer equation**.

The Hessenberg–Schur method for the Sylvester equation, described in Chapter 8, can be used for numerical solution of the Sylvester-observer equation. We, however, describe two other numerical methods (**Algorithms 12.7.1** and **12.7.2**), especially designed for this equation. Both are based on the reduction of the observable pair (A, C) to the observer-Hessenberg pair, described in Chapter 6 and are recursive in nature. *Algorithm 12.7.2 is a block-generalization of Algorithm 12.7.1 and seems to be a little more efficient than the later.* Algorithm 12.7.2 is also suitable for high performance computing. Both seem to have good numerical properties.

The chapter concludes with a well-known procedure developed by Kalman (**Kalman filtering**) for optimal estimation of the states of a **stochastic system**, followed by a brief discussion on the Linear Quadratic Gaussian (**LQG**) problem that deals with optimization of a performance measure for a stochastic system.

12.2   STATE ESTIMATION VIA EIGENVALUE ASSIGNMENT

Consider the linear time-invariant continuous-time system:

$$\dot{x}(t) = Ax(t) + Bu(t),$$
$$y(t) = Cx(t), \tag{12.2.1}$$

where $A \in \mathbb{R}^{n \times n}$, $B \in \mathbb{R}^{n \times m}$, and $C \in \mathbb{R}^{r \times n}$.

Let $\hat{x}(t)$ be an estimate of the state vector $x(t)$. Obviously, we would like to construct the vector $\hat{x}(t)$ in such a way that the error $e(t) = x(t) - \hat{x}(t)$ approaches zero as fast as possible, for all initial states $x(0)$ and for every input $u(t)$). Suppose, we design a dynamical system using our available resources: the output variable $y(t)$, input variable $u(t)$, and the matrices A, B, C, satisfying

$$\dot{\hat{x}}(t) = (A - KC)\hat{x}(t) + Ky(t) + Bu(t), \tag{12.2.2}$$

where the matrix K is to be constructed. Then,

$$e(t) = \dot{x}(t) - \dot{\hat{x}}(t) = Ax(t) + Bu(t) - A\hat{x}(t) + KC\hat{x}(t) - Ky(t) - Bu(t),$$
$$= (A - KC)x(t) - (A - KC)\hat{x}(t) = (A - KC)e(t).$$

The solution of this system of differential equations is $e(t) = e^{(A-KC)t}e(0)$, which shows that the rate at which the entries of the error vector $e(t)$ approach zero can be controlled by the eigenvalues of the matrix $A - KC$. For example, if all the eigenvalues of $A - KC$ have negative real parts less than $-\alpha$, then the error $e(t)$ will approach zero faster than $e^{-\alpha t}e(0)$.

The above discussion shows that the problem of state estimation can be solved by finding a matrix K such that the matrix $A - KC$ has a suitable desired spectrum.

Note that if (A, C) is observable, then such K always exists because, the observability of (A, C) implies the controllability of (A^T, C^T). Also, if (A^T, C^T) is controllable, then by the EVA Theorem (**Theorem 10.4.1**), there always exists a matrix L such that $(A^T + C^T L)$ has an arbitrary spectrum. We can therefore choose $K = -L^T$ so that the eigenvalues of $A^T - C^T K^T$ (which are the same as those of $A - KC$) will be arbitrarily assigned.

Theorem 12.2.1. *If (A, C) is observable, then the states $x(t)$ of the system* (12.2.1) *can be estimated by*

$$\dot{\hat{x}}(t) = (A - KC)\hat{x}(t) + Ky(t) + Bu(t), \tag{12.2.3}$$

where K is constructed such that $A - KC$ is a stable matrix. The error $e(t) = x(t) - \hat{x}(t)$ is governed by

$$\dot{e}(t) = (A - KC)e(t)$$

and $e(t) \to 0$ as $t \to \infty$.

12.3 STATE ESTIMATION VIA SYLVESTER EQUATION

We now present another approach for state estimation. Knowing $A, B, C, u(t)$ and $y(t)$, let's construct the system

$$\dot{z}(t) = Fz(t) + Gy(t) + Pu(t), \tag{12.3.1}$$

where F is $n \times n$, G is $n \times r$, and P is $n \times m$, in such a way that for some constant $n \times n$ nonsingular matrix X, the error vector $e(t) = z(t) - Xx(t) \to 0$ for all $x(0), z(0)$, and for every input $u(t)$. The vector $z(t)$ will then be an estimate of $Xx(t)$. The system (12.3.1) is then said to be the **state observer** for the system (12.2.1). The idea originated with D. Luenberger (1964) and is hence referred to in control theory as the **Luenberger observer**.

We now show that the system (12.3.1) will be a state observer if the matrices X, F, G, and P satisfy certain requirements.

Theorem 12.3.1. *Observer Theorem. The system* (12.3.1) *is a state-observer of the system* (12.2.1), *that is, $z(t)$ is an estimate of $Xx(t)$ in the sense that the error $e(t) = z(t) - Xx(t) \to 0$ as $t \to \infty$ for any initial conditions $x(0), z(0)$, and $u(t)$ if*

(i) $XA - FX = GC$,
(ii) $P = XB$,
(iii) F is stable.

Proof. We need to show that if the conditions (i)–(iii) are satisfied, then $e(t) \to 0$ as $t \to 0$.

From $e(t) = z(t) - Xx(t)$, we have

$$\dot{e}(t) = \dot{z}(t) - X\dot{x}(t),$$
$$= Fz(t) + Gy(t) + Pu(t) - X(Ax(t) + Bu(t)). \tag{12.3.2}$$

Substituting $y(t) = Cx(t)$ while adding and subtracting $FXx(t)$ in Eq. (12.3.2), we get

$$\dot{e}(t) = Fe(t) + (FX - XA + GC)x(t) + (P - XB)u(t).$$

If the conditions (i) and (ii) are satisfied, then we obtain

$$\dot{e}(t) = Fe(t).$$

If, in addition, the condition (iii) is satisfied, then clearly $e(t) \to 0$ as $t \to \infty$, for any $x(0)$, $z(0)$, and $u(t)$.

Hence $z(t)$ is an estimate of $Xx(t)$.   ■

The Sylvester-Observer Equation

Definition 12.3.1.   *The matrix equation*

$$XA - FX = GC, \tag{12.3.3}$$

where A and C are given and X, F, and G are to be found will be called the **Sylvester-observer** *equation.*

The name **"Sylvester-observer equation"** is justified, because the equation arises in construction of an observer and it is a variation of the classical Sylvester equation (discussed in Chapter 8):

$$XA + TX = R,$$

where A, T, and R are given and X is the only unknown matrix.

Theorem 12.3.1 suggests the following method for the observer design.

Algorithm 12.3.1.   *Full-Order Observer Design via Sylvester-Observer Equation*

   Inputs. *The system matrices A, B, and C of order $n \times n$, $n \times m$, and $r \times n$, respectively.*

   Output. *An estimate $\hat{x}(t)$ of the state vector $x(t)$.*

   Assumptions. (A, C) *is observable.*

   Step 1. *Find a* **nonsingular solution** *X of the Sylvester-observer equation (12.3.3) by choosing F as a stable matrix and choosing G in such a way that the resulting solution X is nonsingular.*

   Step 2. *Compute $P = XB$.*

Step 3. *Construct the observer $z(t)$ by solving the system of differential equations:*

$$\dot{z}(t) = Fz(t) + Gy(t) + Pu(t), \quad z(0) = z_0.$$

Step 4. *Find an estimate $\hat{x}(t)$ of $x(t)$: $\hat{x}(t) = X^{-1}z(t)$.*

Example 12.3.1.

$$A = \begin{pmatrix} 1 & 1 \\ 1 & 1 \end{pmatrix}, \quad B = \begin{pmatrix} 1 \\ 0 \end{pmatrix}, \quad C = (1 \ 0).$$

(A, C) is observable.

Step 1. Choose $G = \begin{pmatrix} 1 \\ 3 \end{pmatrix}$, $\quad F = \text{diag}(-1, -3)$.

Then a solution X of $XA - FX = GC$ is

$$X = \begin{pmatrix} 0.6667 & -0.3333 \\ 0.8000 & -0.2000 \end{pmatrix}$$

(computed by MATLAB function **lyap**). The matrix X is nonsingular.

Step 2.

$$P = XB = \begin{pmatrix} 0.6667 \\ 0.8000 \end{pmatrix}.$$

Step 3. An estimate $\hat{x}(t)$ of $x(t)$ is

$$\hat{x}(t) = X^{-1}z(t) = \begin{pmatrix} -1.5 & 2.5 \\ -6 & 5 \end{pmatrix} \begin{pmatrix} z_1(t) \\ z_2(t) \end{pmatrix} = \begin{pmatrix} -1.5z_1 + 2.5z_2 \\ -6z_1 + 5z_2 \end{pmatrix},$$

where

$$z(t) = \begin{pmatrix} z_1(t) \\ z_2(t) \end{pmatrix}$$

is given by

$$\dot{z}(t) = \begin{pmatrix} -1 & 0 \\ 0 & -3 \end{pmatrix} z(t) + \begin{pmatrix} 1 \\ 3 \end{pmatrix} y(t) + \begin{pmatrix} 0.6667 \\ 0.8000 \end{pmatrix} u(t), \quad z(0) = z_0.$$

Comparison of the state and estimate for Example 12.3.1: In Figure 12.1, we compare the estimate $\hat{x}(t)$, obtained by Algorithm 12.3.1, with the state $x(t)$, found by directly solving Eq. (12.2.1) with $u(t)$ as the *unit step function*, and $x(0) = (6, 0)^T$. The differential equation in Step 3 was solved with $z(0) = 0$. The MATLAB function **ode23** was used to solve both the equations. The solid line corresponds to the exact states and the dotted line corresponds to the estimated state.

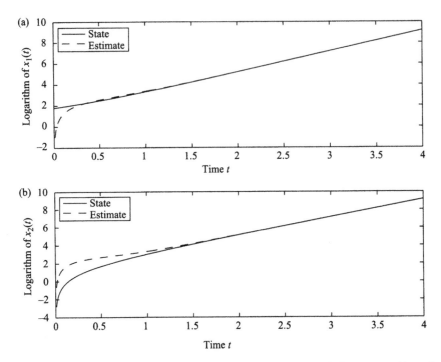

FIGURE 12.1:   The (a) first and (b) second variables of the state $x(t)$ and estimate $\hat{x}(t)$ for Example 12.3.1.

12.4   REDUCED-ORDER STATE ESTIMATION

In this section, we show that if the $r \times n$ output matrix C has full rank r, then the problem of finding a full nth order state estimator for the system (12.2.1) can be reduced to the problem of finding an $(n - r)$th order estimator.

Such an estimator is known as a **reduced-order estimator**. Once a reduced-order estimator of order $n - r$ rather than n is constructed, the full states of the original system can be obtained from the $(n - r)$ state variable of this observer together with the r variables available from measurements. As in the full-dimensional case, we will describe two approaches for finding a reduced-order estimator. We start with the **EVA approach**.

For the sake of convenience, in the next two sections, we will denote the vector $x(t)$ and its derivative $\dot{x}(t)$ just by x and $\dot{x}$. Similarly, for the other vectors.

12.4.1   Reduced-Order State Estimation via Eigenvalue Assignment

Assume as usual that A is an $n \times n$ matrix, B is an $n \times m$ matrix $(m \leq n)$, C is an $r \times n$ matrix with full rank $(r < n)$, and (A, C) is observable.

Since C has full rank, we can choose an $(n - r) \times n$ matrix R such that the matrix $S = \begin{pmatrix} C \\ R \end{pmatrix}$ is nonsingular.

Introducing the new variable $\tilde{x} = Sx$, we can then transform the system (12.2.1) to

$$\dot{\tilde{x}} = SAS^{-1}\tilde{x} + SBu,$$
$$y = CS^{-1}\tilde{x} = (I_r, 0)\tilde{x}. \tag{12.4.1}$$

Let's now partition

$$\bar{A} = SAS^{-1} = \begin{pmatrix} \bar{A}_{11} & \bar{A}_{12} \\ \bar{A}_{21} & \bar{A}_{22} \end{pmatrix}, \qquad \bar{B} = SB = \begin{pmatrix} \bar{B}_1 \\ \bar{B}_2 \end{pmatrix}, \tag{12.4.2}$$

and $\tilde{x} = \begin{pmatrix} \tilde{x}_1 \\ \tilde{x}_2 \end{pmatrix}$, where $\bar{A}_{11}$ and $\tilde{x}_1$ are, respectively, $r \times r$ and $r \times 1$. Then we have

$$\begin{pmatrix} \dot{\tilde{x}}_1 \\ \dot{\tilde{x}}_2 \end{pmatrix} = \begin{pmatrix} \bar{A}_{11} & \bar{A}_{12} \\ \bar{A}_{21} & \bar{A}_{22} \end{pmatrix} \begin{pmatrix} \tilde{x}_1 \\ \tilde{x}_2 \end{pmatrix} + \begin{pmatrix} \bar{B}_1 \\ \bar{B}_2 \end{pmatrix} u,$$
$$y = (I_r, 0) \begin{pmatrix} \tilde{x}_1 \\ \tilde{x}_2 \end{pmatrix} = \tilde{x}_1.$$

That is,

$$\dot{y} = \dot{\tilde{x}}_1 = \bar{A}_{11}\tilde{x}_1 + \bar{A}_{12}\tilde{x}_2 + \bar{B}_1 u, \tag{12.4.3}$$

$$\dot{\tilde{x}}_2 = \bar{A}_{21}\tilde{x}_1 + \bar{A}_{22}\tilde{x}_2 + \bar{B}_2 u, \tag{12.4.4}$$

$$y = \tilde{x}_1. \tag{12.4.5}$$

Since $y = \tilde{x}_1$, we only need to find an estimator for the vector $\tilde{x}_2$ of this transformed system.

The transformed system is not in standard state-space form. However, the system can be easily put in standard form by introducing the new variables

$$\bar{u} = \bar{A}_{21}\tilde{x}_1 + \bar{B}_2 u = \bar{A}_{21} y + \bar{B}_2 u \tag{12.4.6}$$

and

$$v = \dot{y} - \bar{A}_{11} y - \bar{B}_1 u. \tag{12.4.7}$$

From (12.4.4)–(12.4.7), we then have

$$\dot{\tilde{x}}_2 = \bar{A}_{22}\tilde{x}_2 + \bar{u}, \qquad v = \bar{A}_{12}\tilde{x}_2, \tag{12.4.8}$$

which is in standard form.

Since (A, C) is observable, it can be shown (**Exercise 12.3**) that $(\bar{A}_{22}, \bar{A}_{12})$ is also observable. Since $\tilde{x}_2$ has $(n - r)$ elements, we have thus reduced the full n-dimensional estimation problem to an $(n - r)$-dimensional problem. We, therefore, now concentrate on finding an estimate of $\tilde{x}_2$.

By (12.2.2) an $(n - r)$ dimensional estimate $\hat{\tilde{x}}_2$ of $\tilde{x}_2$ defined by (12.4.8) is of the form:

$$\dot{\hat{\tilde{x}}}_2 = (\bar{A}_{22} - L\bar{A}_{12})\hat{\tilde{x}}_2 + Lv + \bar{u},$$

for any matrix L chosen such that $\bar{A}_{22} - L\bar{A}_{12}$ is stable.

Substituting the expressions for $\bar{u}$ and v from (12.4.6) and (12.4.7) into the last equation, we have

$$\dot{\hat{\tilde{x}}}_2 = (\bar{A}_{22} - L\bar{A}_{12})\hat{\tilde{x}}_2 + L(\dot{y} - \bar{A}_{11}y - \bar{B}_1u) + (\bar{A}_{21}y + \bar{B}_2u).$$

Defining another new variable

$$z = \hat{\tilde{x}}_2 - Ly,$$

we can then write

$$\dot{z} = (\bar{A}_{22} - L\bar{A}_{12})(z + Ly) + (\bar{A}_{21} - L\bar{A}_{11})y + (\bar{B}_2 - L\bar{B}_1)u$$
$$= (\bar{A}_{22} - L\bar{A}_{12})z + [(\bar{A}_{22} - L\bar{A}_{12})L + (\bar{A}_{21} - L\bar{A}_{11})]y + (\bar{B}_2 - L\bar{B}_1)u \tag{12.4.9}$$

Comparing Eq. (12.4.9) with (12.2.2) and noting that $\bar{A}_{22} - L\bar{A}_{12}$ is a stable matrix, we see that $z + Ly$ is also an estimate of $\tilde{x}_2$.

Once an estimate of $\tilde{x}_2$ is found, an estimate of the original n-dimensional state vector x from the estimate of $\tilde{x}_2$ can be easily constructed, as shown below.

Since $y = \tilde{x}_1$ and $\hat{\tilde{x}}_2 = z + Ly$, we immediately have

$$\hat{\tilde{x}} = \begin{pmatrix} \hat{\tilde{x}}_1 \\ \hat{\tilde{x}}_2 \end{pmatrix} = \begin{pmatrix} y \\ Ly + z \end{pmatrix} \tag{12.4.10}$$

as an estimate of $\tilde{x}$.

Finally, since $\tilde{x} = Sx$, an estimate $\hat{x}$ of x can be constructed from an estimate of $\tilde{x}$ as:

$$\hat{x} = S^{-1}\hat{\tilde{x}} = \begin{pmatrix} C \\ R \end{pmatrix}^{-1} \begin{pmatrix} y \\ Ly + z \end{pmatrix}.$$

The above discussion can be summarized in the following algorithm:

Algorithm 12.4.1. *Reduced-Order Observer Design via EVA*
    Inputs. *The system matrices A, B, C, respectively, of order $n \times n$, $n \times m$, and $r \times n$.*
    Output. *An estimate $\hat{x}$ of the state vector x.*

Assumptions. (i) (A, C) *is observable.* (ii) C *is of full rank.*

Step 1. *Find an* $(n - r) \times n$ *matrix* R *such that* $S = \begin{pmatrix} C \\ R \end{pmatrix}$ *is nonsingular.*

Step 2. *Compute* $\bar{A} = SAS^{-1}$, $\bar{B} = SB$, *and partition them as*

$$\bar{A} = \begin{pmatrix} \bar{A}_{11} & \bar{A}_{12} \\ \bar{A}_{21} & \bar{A}_{22} \end{pmatrix}, \qquad \bar{B} = \begin{pmatrix} \bar{B}_1 \\ \bar{B}_2 \end{pmatrix}, \qquad (12.4.11)$$

where $\bar{A}_{11}, \bar{A}_{12}, \bar{A}_{21}, \bar{A}_{22}$ *are, respectively,* $r \times r, r \times (n - r), (n - r) \times r,$ *and* $(n - r) \times (n - r)$ *matrices.*

Step 3. *Find a matrix* L *such that* $\bar{A}_{22} - L\bar{A}_{12}$ *is stable.*

Step 4. *Construct a reduced-order observer by solving the systems of differential equations:*

$$\dot{z} = (\bar{A}_{22} - L\bar{A}_{12})z + [(\bar{A}_{22} - L\bar{A}_{12})L + (\bar{A}_{21} - L\bar{A}_{11})]y$$
$$+ (\bar{B}_2 - L\bar{B}_1)u, z(0) = z_0. \qquad (12.4.12)$$

Step 5. *Find* $\hat{x}$, *an estimate of* x:

$$\hat{x} = \begin{pmatrix} C \\ R \end{pmatrix}^{-1} \begin{pmatrix} y \\ Ly + z \end{pmatrix}. \qquad (12.4.13)$$

Example 12.4.1. Consider the design of a reduced-order observer for the linearalized model of the Helicopter problem discussed in Doyle and Stein (1981), and also considered in Dorato *et al.* (1995), with the following data:

$$A = \begin{pmatrix} -0.02 & 0.005 & 2.4 & -32 \\ -0.14 & 0.44 & -1.3 & -30 \\ 0 & 0.018 & -1.6 & 1.2 \\ 0 & 0 & 1 & 0 \end{pmatrix}, \qquad B = \begin{pmatrix} 0.14 & -0.12 \\ 0.36 & -8.6 \\ 0.35 & 0.009 \\ 0 & 0 \end{pmatrix}$$

and

$$C = \begin{pmatrix} 0 & 1 & 0 & 0 \\ 0 & 0 & 0 & 57.3 \end{pmatrix}.$$

Since $\text{rank}(C) = 2, r = 2$.

Step 1. Choose $R = \begin{pmatrix} 1 & 1 & 1 & 1 \\ 0 & 1 & 1 & 1 \end{pmatrix}$. The matrix $S = \begin{pmatrix} C \\ R \end{pmatrix}$ is nonsingular.

Step 2. $\bar{A} = SAS^{-1} = \begin{pmatrix} 1.7400 & -0.5009 & -0.1400 & -1.1600 \\ -57.3006 & -1 & 0 & 57.3000 \\ -0.0370 & -1.0698 & -0.1600 & 0.6600 \\ 2.3580 & -0.4695 & -0.1400 & -1.7600 \end{pmatrix}$,

$\bar{B} = SB = \begin{pmatrix} 0.3600 & -8.6000 \\ 0 & 0 \\ 0.8500 & -8.7191 \\ 0.7100 & -8.5991 \end{pmatrix}$.

$\bar{A}_{11} = \begin{pmatrix} 1.7400 & -0.5009 \\ -57.3006 & -1 \end{pmatrix}$, $\bar{A}_{12} = \begin{pmatrix} -0.14 & -1.16 \\ 0 & 57.3 \end{pmatrix}$,

$\bar{A}_{21} = \begin{pmatrix} -0.0370 & -1.0698 \\ 2.3580 & -0.4695 \end{pmatrix}$, $\bar{A}_{22} = \begin{pmatrix} -0.1600 & 0.6600 \\ -0.1400 & -1.7600 \end{pmatrix}$,

$\bar{B}_1 = \begin{pmatrix} 0.3600 & -8.6000 \\ 0 & 0 \end{pmatrix}$, $\bar{B}_2 = \begin{pmatrix} 0.85 & -8.7191 \\ 0.7100 & -8.5991 \end{pmatrix}$.

Step 3. The matrix

$$L = \begin{pmatrix} -6 & -0.1099 \\ 1 & 0.0244 \end{pmatrix}$$

is such that the eigenvalues of $\bar{A}_{22} - L\bar{A}_{12}$ (the observer eigenvalues) are $\{-1, -2\}$ (L is obtained using MATLAB function **place**).

Step 4. The reduced-order 2-dimensional observer is given by:

$$\dot{z} = (\bar{A}_{22} - L\bar{A}_{12})z + [(\bar{A}_{22} - L\bar{A}_{12})L + (\bar{A}_{21} - L\bar{A}_{11})]y + (\bar{B}_2 - L\bar{B}_1)u, \ Z(0) = Z_0.$$

with $\bar{A}_{11}, \bar{A}_{12}, \bar{A}_{21}, \bar{A}_{22}, \bar{B}_1$, and $\bar{B}_2$ as computed above.

An estimate $\hat{x}$ of the state vector x is then

$$\hat{x} = \begin{pmatrix} C \\ R \end{pmatrix}^{-1} \begin{pmatrix} y \\ Ly+z \end{pmatrix} = \begin{pmatrix} 0 & 0 & 1 & -1 \\ 1 & 0 & 0 & 0 \\ -1 & -0.0175 & 0 & 1 \\ 0 & 0.0175 & 0 & 0 \end{pmatrix} \begin{pmatrix} y \\ Ly+z \end{pmatrix},$$

where z is determined from (12.4.12).

Remark

- The explicit inversion of the matrix $S = \begin{pmatrix} C \\ R \end{pmatrix}$, which could be a source of large round-off errors in case this matrix is ill-conditioned, can be avoided by taking the QR decomposition of the matrix C: $C = RQ_1$ and then choosing an orthogonal matrix Q_2 such that the matrix $Q = \begin{pmatrix} Q_1 \\ Q_2 \end{pmatrix}$ is orthogonal. The matrix Q can then be used in place of S. We leave the details for the readers as an exercise (**Exercise 12.18**).

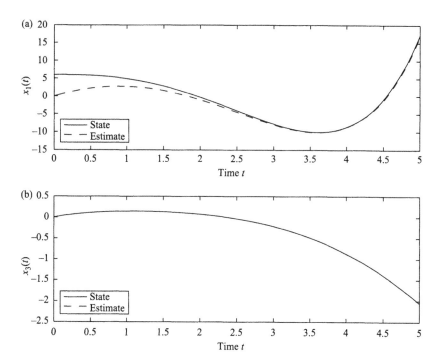

FIGURE 12.2: The (a) first and (b) third variables of the state $x(t)$ and estimate $\hat{x}(t)$, for Example 12.4.1.

Comparison of the state and estimate for example 12.4.1: In Figure 12.2, we compare the estimate $\hat{x}(t)$, obtained by Algorithm 12.4.1 with the state $x(t)$, found by directly solving Eq. (12.2.1) with $u(t) = H(t)[1\ 1]^{\mathrm{T}}$, $H(t)$ is the unit step function and $x(0) = (6, 0, 0, 0)^{\mathrm{T}}$. To solve Eqs. (12.2.1) and (12.4.12), **MATLAB** function **ode23** was used. For Eq. (12.4.12), the initial condition was $z(0) = 0$. The first and the third components of the solutions are compared. The solid line corresponds to the exact state and the dotted line corresponds to the estimated state.

12.4.2 Reduced-Order State Estimation via Sylvester-Observer Equation

As in the case of a full-dimensional observer, a reduced-order observer can also be constructed via solution of a Sylvester-observer equation. The procedure is as follows:

Algorithm 12.4.2. *Reduced-order Observer Design via Sylvester-Observer Equation*
    Inputs. *The matrices A, B, and C of order $n \times n$, $n \times m$, and $r \times n$, respectively.*
    Output. *An estimate $\hat{x}$ of the state vector x.*

Assumptions. *(i)* (A, C) *is observable. (ii)* C *is of full rank.*
Step 1. *Choose an* $(n - r) \times (n - r)$ **stable** *matrix* F.
Step 2. *Solve the reduced-order Sylvester-observer equation for a full rank* $(n - r) \times n$ *solution matrix* X:

$$XA - FX = GC,$$

choosing the $(n - r) \times r$ *matrix* G *appropriately.* **(Numerical methods for solving the Sylvester-observer equation will be described in Section 12.7).**
Step 3. *Compute* $P = XB$.
Step 4. *Find the* $(n - r)$ *dimensional reduced-order observer* z *by solving the system of differential equations:*

$$\dot{z} = Fz + Gy + Pu, \qquad z(0) = z_0. \tag{12.4.14}$$

Step 5. *Find an estimate* $\hat{x}$ *of* x:

$$\hat{x} = \begin{pmatrix} C \\ X \end{pmatrix}^{-1} \begin{pmatrix} y \\ z \end{pmatrix}.$$

Note: If we write $\begin{pmatrix} C \\ X \end{pmatrix}^{-1} = (\bar{S}_1, \bar{S}_2)$, *then* $\hat{x}$ *can be written in the compact form:*

$$\hat{x} = \bar{S}_1 y + \bar{S}_2 z. \tag{12.4.15}$$

Example 12.4.2.   Consider Example 12.4.1 again.
Step 1. Choose $F = \begin{pmatrix} -1 & 0 \\ 0 & -2 \end{pmatrix}$.

Step 2. Choose $G = \begin{pmatrix} 1 & 2 \\ 3 & 4 \end{pmatrix}$.
The solution X of the Sylvester-observer equation $XA - FX = GC$ is

$$X = \begin{pmatrix} -0.117 & -0.0822 & 62.1322 & 37.2007 \\ -0.1364 & -1.9296 & 428.2711 & -173.4895 \end{pmatrix}.$$

Step 3. $P = XB = \begin{pmatrix} 21.7151 & 1.2672 \\ 149.1811 & 20.4653 \end{pmatrix}$.

Step 4. The two-dimensional reduced-order observer is given by $\dot{z} = Fz + Gy + Pu$, where F, G, and P are the matrices found in Step 1, Step 2, and Step 3, respectively.

An estimate $\hat{x}$ of x is

$$\hat{x} = \begin{pmatrix} C \\ X \end{pmatrix}^{-1} \begin{pmatrix} y \\ z \end{pmatrix} = \begin{pmatrix} -24.5513 & -135.1240 & 124.1400 & -18.0098 \\ 1 & 0 & 0 & 0 \\ -0.0033 & -0.0360 & 0.0395 & -0.0034 \\ 0 & 0.0175 & 0 & 0 \end{pmatrix} \begin{pmatrix} y \\ z \end{pmatrix}.$$

(Note that if

$$\hat{x} = \begin{pmatrix} \hat{x}_1 \\ \hat{x}_2 \\ \hat{x}_3 \\ \hat{x}_4 \end{pmatrix} \quad \text{and} \quad y = \begin{pmatrix} y_1 \\ y_2 \end{pmatrix},$$

then $\hat{x}_2 = y_1$, $\hat{x}_4 = 0.0175y_2$, same as was obtained in Example 12.4.1 using the EVA method).

Comparison of the states and estimates for Example 12.4.2: In Figure 12.3, we compare the actual state vector with the estimated one obtained by Algorithm 12.4.2 on the data of Example 12.4.1. The solid line corresponds to

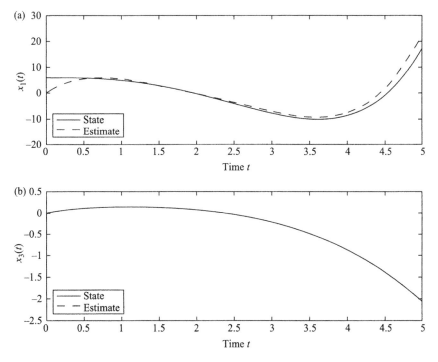

FIGURE 12.3: The (a) first and (b) third variables of the state $x(t)$ and the estimate $\hat{x}(t)$, for Example 12.4.2.

the actual state and the dotted line corresponds to the estimated state. MATLAB function **ode23** was used to solve the underlying differential equations with the same initial conditions as in Example 12.4.1. *The third components are indistinguishable.*

12.5   COMBINED STATE FEEDBACK AND OBSERVER DESIGN

When an estimate $\hat{x}$ of x is used in the feedback control law

$$u = s - K\hat{x} \tag{12.5.1}$$

in place of x, one naturally wonders: **what effect will there be on the EVA?** We consider only the reduced-order case, here. The same conclusion, of course, is true for a full-order observer.

Using (12.5.1) in (12.2.1), we obtain

$$\dot{x} = Ax + B(s - K\hat{x}),$$
$$= Ax + B(s - K\bar{S}_1 y - K\bar{S}_2 z), \quad \text{(using (12.4.15))}$$
$$= Ax + B(s - K\bar{S}_1 Cx - K\bar{S}_2 z),$$
$$= (A - BK\bar{S}_1 C)x - BK\bar{S}_2 z + Bs.$$

Also, Eq. (12.4.14), can be written as

$$\dot{z} = Fz + Gy + Pu = Fz + GCx + P(s - K\bar{S}_1 y - K\bar{S}_2 z),$$
$$= (GC - PK\bar{S}_1 C)x + (F - PK\bar{S}_2)z + Ps$$

(using (12.5.1) and (12.4.15)).

Thus, the combined (feedback and observer) system (Figure 12.4) is given by

$$\begin{pmatrix} \dot{x} \\ \dot{z} \end{pmatrix} = \begin{pmatrix} A - BK\bar{S}_1 C & -BK\bar{S}_2 \\ GC - PK\bar{S}_1 C & F - PK\bar{S}_2 \end{pmatrix} \begin{pmatrix} x \\ z \end{pmatrix} + \begin{pmatrix} B \\ P \end{pmatrix} s,$$
$$y = (C, 0) \begin{pmatrix} x \\ z \end{pmatrix}. \tag{12.5.2}$$

Applying to this system the equivalence transformation, given by the nonsingular matrix

$$\begin{pmatrix} I & 0 \\ -X & I \end{pmatrix}$$

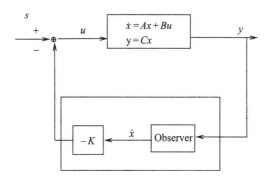

FIGURE 12.4:   Observer-based state feedback.

and noting that $e = z - Xx$, $XA - FX = GC$, and $P = XB$, we have, after some algebraic manipulations,

$$\begin{pmatrix} \dot{x} \\ \dot{e} \end{pmatrix} = \begin{pmatrix} A - BK & -BK\bar{S}_2 \\ 0 & F \end{pmatrix} \begin{pmatrix} x \\ e \end{pmatrix} + \begin{pmatrix} B \\ 0 \end{pmatrix} s,$$

$$y = (C, 0) \begin{pmatrix} x \\ e \end{pmatrix}.$$

(12.5.3)

Thus, the eigenvalues of the combined system are the union of the eigenvalues of the closed-loop matrix $A - BK$ and of the observer matrix F.

Therefore, the observer design and feedback design can be carried out independently, and the calculation of the feedback gain is not affected whether the true state x or the estimated state $\hat{x}$ is used.

This property is known as the **separation property**.

12.6   CHARACTERIZATION OF NONSINGULAR SOLUTIONS OF THE SYLVESTER EQUATION

We have just seen that the design of an observer via the Sylvester-observer equation requires a nonsingular solution X for the full-order design (**Algorithm 12.3.1**) or a full rank solution X for the reduced-order design (**Algorithm 12.4.2**). In this section, we describe some necessary conditions for a unique solution of the Sylvester equation to have such properties. For the sake of convenience, we consider the full-order case (i.e., A and F are $n \times n$) only. The results, however, hold for the reduced-order case also and the proofs given here can be easily modified to deal with the latter and are left as an exercise (**Exercise 12.7**).

The following theorem was proved by Bhattacharyya and DeSouza (1981). The proof here has been taken from Chen (1984).

Theorem 12.6.1.   *Necessary Conditions for Nonsingularity of the Sylvester Equation Solution. Let A, F, G, and C, respectively, be of order $n \times n$, $n \times n$, $n \times r$, and $r \times n$. Let X be a unique solution of the Sylvester-observer equation*

$$XA - FX = GC. \tag{12.6.1}$$

Then, **necessary conditions** *for X to be nonsingular are that (A, C) is observable and (F, G) is controllable.*

Proof.   From the given Eq. (12.6.1), we have

$$XA^0 - F^0 X = 0, \quad \text{(Noting that } A^0 = I_{n \times n} \text{ and } F^0 = I_{n \times n}.)$$
$$XA - FX = GC,$$
$$XA^2 - F^2 X = GCA + FGC,$$
$$\vdots$$
$$XA^n - F^n X = GCA^{n-1} + FGCA^{n-2} + \cdots + F^{n-1}GC \ .$$

Let $a(\lambda) = \lambda^n + a_1 \lambda^{n-1} + \cdots + a_n$ be the characteristic polynomial of A, and let's denote the controllability matrix of the pair (F, G) by C_{FG}, and the observability matrix of the pair (A, C) by O_{AC}.

First of all, we note that the uniqueness of X implies that the matrix $a(F)$ is nonsingular and vice versa. This is seen as follows: By Theorem 8.2.1, X is a unique solution of (12.6.1) if and only if A and F do not have a common eigenvalue. Again, A and F do not have a common eigenvalue if and only if the matrix $a(F)$ is non-singular because the eigenvalues of $a(F)$ are the n numbers $\prod_{j=1}^{n}(\mu_i - \lambda_j)$, $i = 1, \ldots, n$; where, λ_is are the eigenvalues of A and μ_is are the eigenvalues of F. Thus, $a(F)$ is nonsingular if and only if X is a unique solution of (12.6.1).

Now, multiplying the above equations, respectively, by $a_n, a_{n-1}, \ldots, 1$, and using the Cayley–Hamilton theorem, we obtain after some algebraic manipulations:

$$X = -[a(F)]^{-1} C_{FG} \, R O_{AC}, \tag{12.6.2}$$

where

$$R = \begin{pmatrix} a_{n-1}I & a_{n-2}I & \cdots & a_1 I & I \\ a_{n-2}I & a_{n-3}I & \cdots & I & 0 \\ \vdots & \vdots & & \vdots & \vdots \\ a_1 I & I & \cdots & 0 & 0 \\ I & 0 & \cdots & 0 & 0 \end{pmatrix}.$$

From (12.6.2), it then immediately follows that for X to be nonsingular, the rectangular matrices C_{FG} and O_{AC} must have full rank; or, in other words, the pair (F, G) must be controllable and the pair (A, C) must be observable. ∎

Corollary 12.6.1. *If G is $n \times 1$ and C is $1 \times n$, then* **necessary and sufficient conditions** *for the unique solution X of* (12.6.1) *to be nonsingular are that* (F, G) *is controllable and* (A, C) *is observable.*

Proof. In this case, both the matrices C_{FG} and O_{AC} are square matrices. Thus, from (12.6.2), it immediately follows that X is nonsingular if and only if (F, G) is controllable and (A, C) is observable. ∎

Theorem 12.6.1 has recently been generalized by Datta *et al.* (1997) giving a necessary and sufficient condition for nonsingularity of X. We state the result below and refer the readers to the paper for the proof.

Theorem 12.6.2. *Characterization of the Nonsingularity of the Sylvester Equation Solution. Let A, F, and R be $n \times n$ matrices. Let $a(\lambda) = \lambda^n + a_1\lambda^{n-1} + \cdots + a_n$ be the characteristic polynomial of A.*
   Define

$$S = (F^{n-1} + a_1 F^{n-2} + \cdots + a_{n-1}I)R + (F^{n-2} + a_1 F^{n-3} + \cdots + a_{n-2}I)$$
$$\times RA + \cdots + (F + a_1 I)RA^{n-2} + RA^{n-1}.$$

Then a unique solution X of the Sylvester equation

$$FX - XA = R$$

is nonsingular if and only if S is nonsingular. Furthermore, the unique solution X is given by

$$X = (a(F))^{-1}S.$$

(Note again that the uniqueness of X implies that $a(F)$ is nonsingular).

Remark

- The results of Theorems 12.6.1 and 12.6.2 also hold in case the matrix X is not necessarily a square matrix. In fact, this general case has been dealt with in the papers by Bhattacharyya and DeSouza (1981), and Datta *et al.* (1997), and conditions for the unique solution to have full rank have been derived there.

12.7 NUMERICAL SOLUTIONS OF THE SYLVESTER-OBSERVER EQUATION

In this section, we discuss numerical methods for solving the Sylvester-observer equation. These methods are based on the reduction of the observable pair (A, C) to the observer-Hessenberg form $(H, \bar{C})$, described in Chapter 6.
   The methods use the following template.

Step 1. *Reduction of the problem.* The pair (A, C) is transformed to observer-Hessenberg form by orthogonal similarity, that is, an orthogonal matrix O is constructed such that

$$OAO^{\mathrm{T}} = H, \text{ an unreduced block upper-Hessenberg matrix,}$$
$$CO^{\mathrm{T}} = \bar{C} = (0, C_1).$$

The equation $XA - FX = GC$ is then transformed to $XO^{\mathrm{T}}OAO^{\mathrm{T}} - FXO^{\mathrm{T}} = GCO^{\mathrm{T}}$

or

$$YH - FY = G\bar{C}, \tag{12.7.1}$$

where $Y = XO^{\mathrm{T}}$.

Step 2. *Solution of the reduced problem.* The reduced Hessenberg Sylvester-observer equation (12.7.1) is solved.

Step 3. *Recovery of the Solution X of the Original Problem.* The solution X of the original problem is recovered from the solution of the reduced problem:

$$X = YO. \tag{12.7.2}$$

We now discuss the implementation of Step 2. Step 3 is straightforward. Implementation of Step 1 has been described in Chapter 6.

The simplest way to solve Eq. (12.7.1) is to choose the matrices F and G completely satisfying the controllability requirement of the pair (F, G). In that case, the Sylvester-observer equation reduces to an ordinary Sylvester equation, and, therefore, can be solved using the **Hessenberg–Schur method**, described in Chapter 8.

Indeed, F can be chosen in the lower real Schur form (RSF), as required by the method. Therefore, computations will be greatly reduced. We will not repeat the procedure here. Instead, we will present below two simple recursive procedures, designed specifically for solution of the reduced-order Sylvester-observer equation (12.7.1).

12.7.1   A Recursive Method for the Hessenberg Sylvester-Observer Equation

In the following, we describe a recursive procedure for solving the reduced multi-output Sylvester-observer equation

$$YH - FY = G\bar{C}. \tag{12.7.3}$$

The procedure is due to Van Dooren (1984). The procedure computes simultaneously the matrices F, Y, and G, assuming that $(H, \bar{C})$ is observable.

Set $q = n - r$ and assume that Y has the form:

$$Y = \begin{pmatrix} 1 & y_{12} & \cdots & & \cdots & y_{1,n} \\ & \ddots & \ddots & & & \vdots \\ 0 & & 1 & y_{q,q+1} & \cdots & y_{q,n} \end{pmatrix} \tag{12.7.4}$$

and choose F in lower triangular form (for simplicity):

$$F = \begin{pmatrix} f_{11} & 0 & \cdots & \cdots & 0 \\ f_{21} & f_{22} & 0 & \cdots & 0 \\ \vdots & & \ddots & & \vdots \\ f_{q1} & \cdots & \cdots & \cdots & f_{qq} \end{pmatrix}, \qquad G = \begin{pmatrix} g_1^T \\ g_2^T \\ \cdots \\ g_q^T \end{pmatrix}, \tag{12.7.5}$$

where the diagonal entries f_{ii}, $i = 1, \ldots, q$ are known and the remaining entries of F are to be found. It has been shown in (Van Dooren 1984) that a solution Y in the above form always exists. The reduced Sylvester-observer equation can now be solved recursively for Y, F, and G, as follows.

Let g_i^T denote the ith row of G. Comparing the first row of Eq. (12.7.3), we obtain

$$(1, y_1)H - f_{11}(1, y_1) = g_1^T \bar{C}. \tag{12.7.6}$$

Similarly, comparing the ith row of that equation, we have

$$(0, 0, \ldots, 0, 1, y_i)H - (f_i, f_{ii}, 0, \ldots, 0)Y = g_i^T \bar{C}, \quad i = 2, 3, \ldots, q \tag{12.7.7}$$

In the above, $y_i = (y_{i,i+1}, \ldots, y_{i,n})$ and $f_i = (f_{i1}, f_{i2}, \ldots, f_{i,i-1})$.

The Eqs. (12.7.6) and (12.7.7) can be, respectively, written as

$$(y_1, g_1^T) \begin{bmatrix} (H - f_{11}I)_{\text{bottom}(n-1)} \\ -\bar{C} \end{bmatrix} = -[1\text{st row } of \ (H - f_{11}I)], \tag{12.7.8}$$

and

$$(f_i, y_i, g_i^T) \begin{bmatrix} -Y_{\text{top}(i-1)} \\ (H - f_{ii}I)_{\text{bottom}(n-i)} \\ -\bar{C} \end{bmatrix} = -[i\text{th row of } (H - f_{ii}I)], \tag{12.7.9}$$

where $Y_{\text{top}(i-1)}$ and $(H - f_{ii}I)_{\text{bottom}(n-i)}$ denote, respectively, the top $i - 1$ rows of Y and the bottom $n - i$ rows of $H - f_{ii}I$. Because of the structure of the observer-Hessenberg form $(H, \bar{C})$, the above systems are consistent and these systems can be solved recursively to compute the unknown entries of the matrices Y, F, and G.

We illustrate how to solve these equations in the special case with $n = 3, r = 1$. The reduced equation to be solved in this case is:

$$\begin{pmatrix} 1 & y_{12} & y_{13} \\ 0 & 1 & y_{23} \end{pmatrix} \begin{pmatrix} h_{11} & h_{12} & h_{13} \\ h_{21} & h_{22} & h_{23} \\ 0 & h_{32} & h_{33} \end{pmatrix} - \begin{pmatrix} f_{11} & 0 \\ f_{21} & f_{22} \end{pmatrix} \begin{pmatrix} 1 & y_{12} & y_{13} \\ 0 & 1 & y_{23} \end{pmatrix}$$

$$= \underbrace{\begin{pmatrix} g_{11} \\ g_{21} \end{pmatrix}}_{G} \underbrace{\begin{pmatrix} 0 & 0 & c_1 \end{pmatrix}}_{\bar{C}} .$$

Comparing the first row of the last equation, we have

$$\begin{cases} y_{12}h_{21} = f_{11} - h_{11}, \\ y_{12}(h_{22} - f_{11}) + y_{13}h_{32} = -h_{12}, \\ y_{13}(h_{33} - f_{11}) + y_{12}h_{23} - g_{11}c_1 = -h_{13}. \end{cases} \tag{12.7.10}$$

Similarly, comparing the second row, we have

$$\begin{cases} -f_{21} = -h_{21}, \\ y_{23}h_{32} - f_{21}y_{12} - f_{22} = -h_{22}, \\ y_{23}h_{33} - f_{21}y_{13} - f_{22}y_{23} - g_{21}c_1 = -h_{23}. \end{cases} \tag{12.7.11}$$

The system (12.7.10) can be written as

$$(y_{12}, y_{13}, g_{11}) \begin{pmatrix} h_{21} & h_{22} - f_{11} & h_{23} \\ 0 & h_{32} & h_{33} - f_{11} \\ 0 & 0 & -c_1 \end{pmatrix} = \begin{pmatrix} f_{11} - h_{11} \\ -h_{12} \\ -h_{13} \end{pmatrix}^T$$

Similarly, the system (12.7.11) can be written as

$$(f_{21}, y_{23}, g_{21}) \begin{pmatrix} -1 & -y_{12} & -y_{13} \\ 0 & h_{32} & h_{33} - f_{22} \\ 0 & 0 & -c_1 \end{pmatrix} = \begin{pmatrix} -h_{21} \\ f_{22} - h_{22} \\ -h_{23} \end{pmatrix}^T .$$

Note that since the pair (A, C) is observable, h_{21}, h_{32}, and c_1 are different from zero and, therefore, the matrices of the above two systems are nonsingular.

Algorithm 12.7.1. *A Recursive Algorithm for the Multi-Output Sylvester-Observer Equation*

    Inputs. *The matrices* $A_{n \times n}$, *and* $C_{r \times n}$.

    Output. *A full-rank solution* X *of the reduced-order Sylvester-observer equation:*

$$XA - FX = GC.$$

Assumption. (A, C) *is observable.*

Step 0. *Set* $n - r = q$.

Step 1. *Transform the pair (A, C) to the observer-Hessenberg pair $(H, \bar{C})$:*

$$OAO^{\mathrm{T}} = H, \ CO^{\mathrm{T}} = \bar{C}.$$

Step 2. *Choose $F = (f_{ij})$ as a $q \times q$ lower triangular matrix, where the diagonal entries f_{ii}, $i = 1, \ldots, q$ are arbitrarily given numbers, and the off-diagonal entries are to be computed.*

Step 3. *Solve for Y satisfying*

$$YH - FY = G\bar{C},$$

where Y has the form (12.7.4), as follows:

Compute the first row of Y and the first row of G by solving the system (12.7.8). Compute the second through qth rows of Y, the second through qth rows of F, and the second through qth rows of G simultaneously, by solving the system (12.7.9).

Step 4. *Recover X from Y:*

$$X = YO.$$

Example 12.7.1. Consider Example 12.4.1 again.

Here $n = 4, r = 2$.

Step 1. The observer-Hessenberg pair of (A, C) is given by:

$$H = \begin{pmatrix} -0.0200 & 2.4000 & 0.0050 & -32.0000 \\ 0 & -1.6000 & 0.0180 & 1.2000 \\ -0.1400 & -1.3000 & 0.4400 & -30.0000 \\ 0 & 1.000 & 0 & 0 \end{pmatrix},$$

$$\bar{C} = \begin{pmatrix} 0 & 0 & 1 & 0 \\ 0 & 0 & 0 & 57.3 \end{pmatrix}.$$

The transforming matrix

$$O = \begin{pmatrix} 1 & 0 & 0 & 0 \\ 0 & 0 & 1 & 0 \\ 0 & 1 & 0 & 0 \\ 0 & 0 & 0 & 1 \end{pmatrix}.$$

Step 2. Let's choose $f_{11} = -1$, $f_{22} = -2$

Step 3. The solution of the system (12.7.8) is $(0, 7, 6.7, 10.085, -4.1065)$. Thus, $y_1 = (0, 7, 6.7)$, $g_1 = (10.085, -4.1065)$. The first row of $Y = (1, 0, 7, 6.7)$.

The solution of the system (12.7.9) is $(0.0007, -0.0053, -0.4068, 0, 0.0094)$. Thus, $f_{21} = 0.0007$ and $y_2 = (-0.0053, -0.4068)$, $g_2 = (0, 0.0094)$. So,

$$F = \begin{pmatrix} -1 & 0 \\ 0.0007 & -2 \end{pmatrix}, G = \begin{pmatrix} 10.085 & -4.1065 \\ 0 & 0.0094 \end{pmatrix}.$$

The second row of $Y = (0, 1, -0.0053, -0.4068)$.

Therefore,

$$Y = \begin{pmatrix} 1 & 0 & 7 & 6.7 \\ 0 & 1 & -0.0053 & -0.4068 \end{pmatrix}.$$

Step 4. Recover X from Y:

$$X = YO = \begin{pmatrix} 1 & 7 & 0 & 6.7 \\ 0 & -0.0053 & 1 & -0.4068 \end{pmatrix}.$$

Flop-count: Solving for F, G, and Y (using the special structures of these matrices): $2(n - r)rn^2$ flops.

Obtaining the observer-Hessenberg form: $2(3n + r)n^2$ flops (including the construction of O).

Recovering X from Y: $2(n - r)n$

Total: (**About**) $(6 + 2r)n^3$.

MATCONTROL note: Algorithm 12.7.1 has been implemented in MATCONTROL function **sylvobsm**.

12.7.2   A Recursive Block-Triangular Algorithm for the Hessenberg Sylvester-Observer Equation

A block version of Algorithm 12.7.1 has recently been obtained by Carvalho and Datta (2001). This block algorithm seems to be computationally slightly more efficient than Algorithm 12.7.1 and is suitable for high-performance computing. We describe this new block algorithm below.

As in Algorithm 12.7.1, assume that the observable pair (A, C) has been transformed to an observer-Hessenberg pair $(H, \bar{C})$, that is, an orthogonal matrix O has been computed such that

$$OAO^T = H \quad \text{and} \quad \bar{C} = C\,O^T = \begin{bmatrix} 0 & \cdots & 0, & C_1 \end{bmatrix},$$

where $H = (H_{ij})$ is block upper Hessenberg with diagonal blocks $H_{ii} \in \mathbb{R}^{n_i \times n_i}$, $i = 1, 2, \ldots, p$ and $n_1 + \cdots + n_p = n$.

Given the Observer-Hessenberg pair $(H, \bar{C})$, we now show how to compute the matrices Y, F, and G in **blocks** such that

$$YH - FY = G\bar{C}. \tag{12.7.12}$$

Partitioning the matrices F, Y, and G conformably with H allows us to write the above equation as

$$
\begin{bmatrix} Y_{11} & Y_{12} & \cdots & Y_{1p} \\ & Y_{22} & \cdots & Y_{2p} \\ & & Y_{qq} & Y_{qp} \end{bmatrix}
\begin{bmatrix} H_{11} & H_{12} & \cdots & H_{1p} \\ H_{21} & H_{22} & \cdots & H_{2p} \\ & H_{32} & \cdots & H_{3p} \\ & & H_{p-1,p} & H_{pp} \end{bmatrix}
$$

$$
- \begin{bmatrix} F_{11} & & \\ F_{21} & F_{22} & \\ F_{q1} & \cdots & F_{qq} \end{bmatrix}
\begin{bmatrix} Y_{11} & Y_{12} & \cdots & Y_{1p} \\ & Y_{22} & \cdots & Y_{2p} \\ & & Y_{qq} & Y_{qp} \end{bmatrix}
$$

$$
= \begin{bmatrix} G_1 \\ \cdots \\ G_q \end{bmatrix} \begin{bmatrix} 0 & 0 & \cdots & 0 & C_1 \end{bmatrix}.
$$

We set $Y_{ii} = I_{r \times r}, i = 1, 2, \ldots, q$ for simplicity. Since matrix F is required to have a preassigned spectrum $\mathcal{S}$, we distribute the elements of $\mathcal{S}$ among the diagonal blocks of F in such a way that $\Omega(F) = \mathcal{S}$, where $\Omega(M)$ denotes the spectrum of M. A complex conjugate pair is distributed as a 2×2 matrix and a real one as a 1×1 scalar on the diagonal of F. Note that, some compatibility between the structure of $\mathcal{S}$ and the parameters $n_i, i = 1, \ldots, p$ is required to exist for this to be possible.

Equating now the corresponding blocks on left- and right-hand sides, we obtain:

$$
\sum_{k=i}^{j+1} Y_{ik} H_{kj} - \sum_{k=1}^{\min(i,j)} F_{ik} Y_{kj} = 0, \quad j = 1, 2, \ldots, p-1. \tag{12.7.13}
$$

$$
\sum_{k=i}^{p} Y_{ik} H_{kp} - \sum_{k=1}^{i} F_{ik} Y_{kp} = G_i C_1. \tag{12.7.14}
$$

From (12.7.13) and (12.7.14), we conclude $F_{ij} = 0$ for $j = 1, 2, \ldots, i-2$, and $F_{ij} = H_{ij}$ for $j = i - 1$.

Thus, Eqs. (12.7.13) and (12.7.14) are reduced to

$$
\sum_{k=i}^{j+1} Y_{ik} H_{kj} - \sum_{k=\max(i-1,1)}^{i} F_{ik} Y_{kj} = 0, \quad j = i, i+1, \ldots, p-1. \tag{12.7.15}
$$

$$
\sum_{k=i}^{p} Y_{ik} H_{kp} - \sum_{k=\max(i-1,1)}^{i} F_{ik} Y_{kp} = G_i C_1, \quad \text{for } i = 1, 2, \ldots, q. \tag{12.7.16}
$$

For a computational purpose we rewrite Eq. (12.7.15) as

$$\sum_{k=i}^{j} Y_{ik} H_{kj} + Y_{i,j+1} H_{j+1,j} - \sum_{k=\max(i-1,1)}^{i} F_{ik} Y_{kj} = 0, \qquad j = i, i+1, \ldots, p-1,$$

that is, for $j = i, i+1, \ldots, p-1$,

$$Y_{i,j+1} H_{j+1,j} = -\sum_{k=i}^{j} Y_{ik} H_{kj} + \sum_{k=\max(i-1,1)}^{i} F_{ik} Y_{kj}. \qquad (12.7.17)$$

Equations (12.7.16) and (12.7.17) allow us to compute the off-diagonal blocks Y_{ij} of Y and the blocks G_i of G recursively.

This is illustrated in the following, in the special case when $p = 4, q = 3$:

First row: $i = 1$

$H_{11} + Y_{12} H_{21} - F_{11} = 0$ (solve for Y_{12})
$H_{12} + Y_{12} H_{22} + Y_{13} H_{32} - F_{11} Y_{12} = 0$ (solve for Y_{13})
$H_{13} + Y_{12} H_{23} + Y_{13} H_{33} + Y_{14} H_{43} - F_{11} Y_{13} = 0$ (solve for Y_{14})
$H_{14} + Y_{12} H_{24} + Y_{13} H_{34} + Y_{14} H_{44} - F_{11} Y_{14} = G_1 C_1$ (solve for G_1).

Second row: $i = 2$

$H_{22} + Y_{23} H_{32} - F_{21} Y_{12} - F_{22} = 0$ (solve for Y_{23})
$H_{23} + Y_{23} H_{33} + Y_{24} H_{43} - F_{21} Y_{13} - F_{22} Y_{23} = 0$ (solve for Y_{24})
$H_{24} + Y_{23} H_{34} + Y_{24} H_{44} - F_{21} Y_{14} - F_{22} Y_{24} = G_2 C_1$ (solve for G_2)

Third row: $i = 3$

$H_{33} + Y_{34} H_{43} - F_{32} Y_{23} - F_{33} = 0$ (solve for Y_{34})
$H_{34} + Y_{34} H_{44} - F_{32} Y_{24} - F_{33} Y_{34} = G_3 C_1$ (solve for G_3)

The above discussion leads to the following algorithm:

Algorithm 12.7.2. *A Recursive Block Triangular Algorithm for the Multi-Output Sylvester Observer Equation*

   Input. *Matrices $A \in \mathbb{R}^{n \times n}$, $C \in \mathbb{R}^{r \times n}$ of full-rank and the self-conjugate set $S \in \mathbb{C}^{n-r}$.*

   Output. *Block matrices X, F, and G, such that $\Omega(F) = S$ and $XA - FX = GC$.*

   Step 1. Reduce (A, C) to observer-Hessenberg form $(H, \bar{C})$. Let $n_i, i = 1, \ldots, p$ be the dimension of the diagonal blocks H_{ii} of the matrix H.

   Step 2. Partition matrices Y, F, and G in blocks according to the block structure of H. Let $q = p - 1$.

   Step 3. Distribute the elements of S along the diagonal blocks $F_{ii}, i = 1, 2, \ldots, q$ such that $\Omega(F) = S$; the complex conjugate pairs as 2×2 blocks and the real ones as 1×1 scalars along the diagonal of the matrix F.

Step 4. Set $Y_{11} = I_{n_1 \times n_1}$.
Step 5. For $i = 2, 3, \ldots, q$, set

$$F_{i,i-1} = H_{i,i-1}, \qquad Y_{ii} = I_{n_i \times n_i}.$$

Step 6. For $i = 1, 2, \ldots, q$ do
  6.1. For $j = i, i+1, \ldots, p - 1$, solve the upper triangular system for $Y_{i,j+1}$:

$$Y_{i,j+1} H_{j+1,j} = -\sum_{k=i}^{j} Y_{ik} H_{kj} + \sum_{k=\max(i-1,1)}^{i} F_{ik} Y_{kj}.$$

  6.2. Solve the triangular system for G_i:

$$G_i C_1 = \sum_{k=i}^{p} Y_{ik} H_{kp} - \sum_{k=\max(i-1,1)}^{i} F_{ik} Y_{kp}.$$

Step 7. Form the matrices Y, F, and G from their computed blocks.
Step 8. Recover $X = Y\, O$.

Return

Remark

- Recall that once the matrix X is obtained, the estimated state-vector $\hat{x}(t)$ can be computed from

$$\begin{bmatrix} C \\ X \end{bmatrix} \hat{x}(t) = \begin{bmatrix} y(t) \\ z(t) \end{bmatrix}.$$

It is interesting to note that the matrix X does not need to be computed explicitly for this purpose because the above system is equivalent to:

$$\begin{bmatrix} \bar{C} \\ Y \end{bmatrix} \hat{x}(t) = \begin{bmatrix} y(t) \\ z(t) \end{bmatrix} O^{\mathrm{T}}.$$

The matrix $\begin{pmatrix} \bar{C} \\ Y \end{pmatrix}$ is a nonsingular block upper Hessenberg by the construction of Y. This structure is very important from the computational point of view since it can possibly be exploited in high-performance computations.

Flop-count and comparison of efficiency

Flop-count of Algorithm 12.7.2

1. Reduction to observer-Hessenberg form using the staircase algorithm:

$$6n^3 + 2rn^2 \text{ flops}$$

2. Computation of Y using Steps 4–8 of the algorithm:

$$\sum_{i=1}^{p-1} \sum_{j=i}^{p} \left[(j - i + 1)(2r^3) + 2(2r^3) + r^2 \right]$$

$$= \sum_{i=1}^{p-1} \sum_{j=i}^{p} \left\{ [2(j - i) + 7]r^3 + r^2 \right\} \approx \sum_{i=1}^{p-1} \left[(p - i)^2 + (p - i) \right] r^3$$

$$\approx \left[\frac{(p - 1)p(2p - 1)}{6} + \frac{(p - 1)p}{2} \right] r^3 \approx \frac{n^3}{3} + \frac{rn^2}{2} \text{flops.}$$

3. Computation of X from Y: n^3 flops (note that the matrix Y is a unit block triangular matrix).

Thus, total count is $(19n^3/3) + (5r/2)n^2$ flops.

Comparison of Efficiency. Algorithm 12.7.1 requires about $(6 + 2r)n^3$ flops. [*Note*: the flop count given in Van Dooren (1984) is nearly one half of that given here; this is because a "flop" is counted there as a multiplication/division coupled with an addition/subtraction.]

Also, it can be shown that a recent block algorithm of Datta and Sarkissian (2000) requires about $52n^3/3$ flops.

*Thus Algorithm 12.7.2 is much faster than both Van Dooren's (***Algorithm** 12.7.1) and the Datta–Sarkissian algorithms.*

Besides, this algorithm is suitable for implementations using the recently developed and widely used scientific computing software package LAPACK (Anderson et al. (1999)), since it is composed of BLAS-3 (Basic Linear Algebra Subroutines Level 3) operations such as matrix–matrix multiplications, QR factorizations, and solutions of triangular systems with multiple right hand sides.

Example 12.7.2. We consider Example 12.7.1 again,
Step 1. The matrices H, $\bar{C}$, and O are given by:

$$H = \begin{bmatrix} -0.0200 & 2.4000 & 0.0050 & -32.0000 \\ 0 & -1.6000 & 0.0180 & 1.2000 \\ -0.1400 & -1.3000 & 0.4400 & -30.0000 \\ 0 & 1.0000 & 0 & 0 \end{bmatrix},$$

$$C_1 = \begin{bmatrix} 1.0000 & 0 \\ 0 & 57.3000 \end{bmatrix}, \quad O = \begin{bmatrix} 1 & 0 & 0 & 0 \\ 0 & 0 & 1 & 0 \\ 0 & 1 & 0 & 0 \\ 0 & 0 & 0 & 1 \end{bmatrix}, \quad \text{and} \quad \bar{C} = (0, C_1).$$

Step 2. $q = 1$.

Steps 3 and 4.

$$F = F_{11} = \begin{bmatrix} -1.00 & 0 \\ 0 & -2.00 \end{bmatrix}, \qquad Y_{11} = \begin{bmatrix} 1 & 0 \\ 0 & 1 \end{bmatrix}.$$

Step 5. Skipped ($q = 1$).
Step 6. $i = 1$.
   6.1. $j = 1$. Solve the triangular system $Y_{12}H_{21} = -Y_{11}H_{11} + F_{11}Y_{11}$ for Y_{12}:

$$Y_{12} = \begin{bmatrix} 7.0000 & 6.7000 \\ 0 & -0.4000 \end{bmatrix}.$$

   6.2: Solve triangular system $G_1C_1 = Y_{11}H_{12} + Y_{12}H_{22} - F_{11}Y_{12}$ for G_1:

$$G_1 = \begin{bmatrix} 10.0850 & -4.1065 \\ 0.0180 & 0.0070 \end{bmatrix}.$$

Step 7. Form matrices Y, F, and G from the computed blocks:

$$Y = \begin{bmatrix} 1 & 0 & 7.0000 & 6.7000 \\ 0 & 1 & 0 & -0.4000 \end{bmatrix},$$

$$F = \begin{bmatrix} -1.000 & 0 \\ 0 & -2.000 \end{bmatrix}, \qquad G = \begin{bmatrix} 10.0850 & -4.1065 \\ 0.0180 & 0.0070 \end{bmatrix}$$

Step 8. Recover $X = YO$:

$$X = \begin{bmatrix} 1 & 7.0000 & 0 & 6.7000 \\ 0 & 0 & 1 & -0.4000 \end{bmatrix}.$$

Verify: $\|XA - FX - GC\|_2 = O(10^{-13})$ and $\Omega(F) = \{-2.0000, -1.0000\}$.
Thus, the residue is small and the spectrum of F has been assigned accurately.
   MATCONTROL note: Algorithm 12.7.2 has been implemented in MATCON-
TROL function **sylvobsmb**.
   Comparison of the state and estimate for Example 12.7.2: Figure 12.5 shows
the relative error between the exact state $x(t)$ and the estimate $\hat{x}(t)$ satisfying

$$\begin{bmatrix} \overline{C} \\ Y \end{bmatrix} \hat{x}(t) = \begin{bmatrix} y(t) \\ z(t) \end{bmatrix} O^{\mathrm{T}}$$

with the data above and $u(t)$ as the unit step function. The underlying systems of
ordinary differential equations were solved by using **MATLAB** procedure **ode45**
with zero initial conditions. The relative error is defined by

$$\frac{\|x(t) - \hat{x}(t)\|_2}{\|x(t)\|_2}.$$

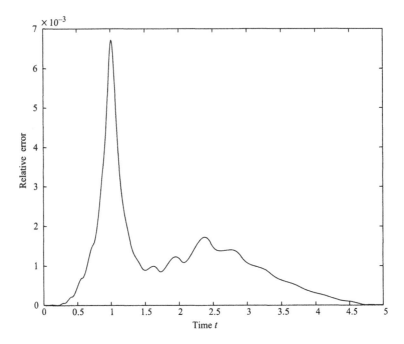

FIGURE 12.5:   Relative error between the state and estimate.

12.8   NUMERICAL SOLUTION OF A CONSTRAINED SYLVESTER-OBSERVER EQUATION

In this section, we consider the problem of solving a constrained reduced-order Sylvester-observer equation. Specifically, the following problem is considered:

Solve the reduced-order Sylvester-observer equation

$$XA - FX = GC \tag{12.8.1}$$

such that

$$XB = 0 \tag{12.8.2}$$

and

$$\begin{bmatrix} X \\ C \end{bmatrix} \tag{12.8.3}$$

has full rank.

The importance of solving the constrained Sylvester equation lies in the fact that if the constraint (12.8.2) is satisfied, then the feedback system with the reduced-order observer has the same robustness properties as that of the direct feedback system (see Tsui (1988)).

We state a recent method of Barlow *et al.* (1992) to solve the above problem.

A basic idea behind the method is to transform the given equation to a reduced-order unconstrained equation and then recover the solution of the constrained equation from that of the reduced unconstrained equation. We skip the details and present below just the algorithm. For details of the development of the algorithm, see the above paper by Barlow *et al.* (1992).

Algorithm 12.8.1. *An Algorithm for Constrained Sylvester-observer Equation*
  Inputs.

(i) *The system matrices A, B, and C of order $n \times n$, $n \times m$, and $r \times n$, respectively.*
(ii) *A matrix F of order $(n - r)$.*

Output. *An $(n - r) \times n$ matrix X and an $(n - r) \times r$ matrix G satisfying*
(12.8.1) *such that $\begin{pmatrix} X \\ C \end{pmatrix}$ is nonsingular and $XB = 0$.*

Assumptions. *(A, C) is observable, $n > r > m$, and $\operatorname{rank}(CB) = \operatorname{rank}(B) = m$.*

Step 1. *Find the QR factorization of B:*

$$B = W \begin{pmatrix} S \\ 0 \end{pmatrix},$$

where S is $m \times m$, upper triangular and has full rank, and W is $n \times n$ and orthogonal.

Partition $W = (W_1, W_2)$, where W_1 is $n \times m$ and W_2 is $n \times (n - m)$.

Step 2. *Set*

$$A_1 = W_2^T A W_1, \qquad A_2 = W_2^T A W_2, \qquad C_1 = C W_1, \qquad C_2 = C W_2$$

Step 3. *Find a QR factorization of C_1:*

$$C_1 = (Q_1, Q_2) \begin{pmatrix} R \\ 0 \end{pmatrix},$$

where Q_1 is $r \times m$, Q_2 is $r \times (r - m)$, and R is an $m \times m$ upper triangular matrix with full rank.

Step 4. *Define E by*

$$E = \begin{pmatrix} E_1 \\ E_2 \end{pmatrix} = Q^T C_2,$$

where E_1 is $m \times (n - m)$, E_2 is $(r - m) \times (n - m)$, and $Q = (Q_1, Q_2)$.

Step 5. *Form* $\hat{A} = A_2 - A_1 R^{-1} E_1$. *Solve the Sylvester equation:*

$$Z\hat{A} - FZ = G_2 E_2,$$

choosing G_2 *randomly.* (*Use* **Algorithm 8.5.1.**)

Step 6. Set $G_1 = ZA_1 R^{-1} = ZJ$, $G = (G_1, G_2)Q^{\mathrm{T}}$, and $X = ZW_2^{\mathrm{T}}$.

(*Note that* Z *is of order* $(n - r) \times (n - m)$ *and* $J = A_1 R^{-1}$ *is computed by solving the upper triangular system* $JR = A_1$).

MATHCONTROL note: Algorithm 12.8.1 has been implemented in MATCON-TROL functions **sylvobsc**.

Example 12.8.1. Consider solving the Eq. (12.8.1) using Algorithm 12.8.1 with

$$A = \begin{pmatrix} -0.02 & 0.005 & 2.4 & -3.2 \\ -0.14 & 0.44 & -1.3 & -3 \\ 0 & 0.018 & -1.6 & 1.2 \\ 0 & 0 & 1 & 0 \end{pmatrix}, \quad B = \begin{pmatrix} 1 \\ 1 \\ 1 \\ 1 \end{pmatrix},$$

$$C = \begin{pmatrix} 0 & 1 & 0 & 0 \\ 0 & 0 & 0 & 57.3 \end{pmatrix}, \quad \text{and} \quad F = \begin{pmatrix} -1 & 0 \\ 1 & -2 \end{pmatrix}.$$

Then, $n = 4$, $r = 2$, $m = 1$.

Step 1. $W_1 = (-0.5, -0.5, -0.5, -0.5)^{\mathrm{T}}$.

$$W_2 = \begin{pmatrix} -0.5 & -0.5 & -0.5 \\ 0.8333 & -0.1667 & -0.1667 \\ -0.1667 & 0.8333 & -0.1667 \\ -0.1667 & -0.1667 & 0.8333 \end{pmatrix}, \quad S = -2.$$

Step 2. $A_1 = W_2^{\mathrm{T}} A W_1 = \begin{pmatrix} 1.5144 \\ -0.2946 \\ -0.9856 \end{pmatrix}$,

$$A_2 = W_2^{\mathrm{T}} A W_2 = \begin{pmatrix} 0.9015 & -1.6430 & 0.5596 \\ 0.1701 & -2.5976 & 2.9907 \\ -0.4185 & -0.2230 & 1.5604 \end{pmatrix},$$

$$C_1 = CW_1 = \begin{pmatrix} -0.5 \\ -28.65 \end{pmatrix},$$

$$C_2 = CW_2 = \begin{pmatrix} 0.8333 & -0.1667 & -0.1667 \\ -9.55 & -9.55 & 47.75 \end{pmatrix}.$$

Step 3. $Q = \begin{pmatrix} -0.0174 & -0.9998 \\ -0.9998 & 0.0174 \end{pmatrix}$, $R = 28.6544$.

Step 4. $E = Q^T C_2 = \begin{pmatrix} 9.5340 & 9.5515 & -47.7398 \\ -0.9998 & 0 & 0.9998 \end{pmatrix}$,

$$E_2 = \begin{pmatrix} -0.9998 & 0 & 0.9998 \end{pmatrix}$$

Step 5. $\hat{A} = A_2 - A_1 R^{-1} E_1 = \begin{pmatrix} 0.3976 & -2.1478 & 1.96 \\ -0.0721 & -2.4944 & 2.4999 \\ -0.0905 & 0.1056 & -0.0817 \end{pmatrix}$.

Choose

$$G_2 = \begin{pmatrix} 1 \\ 0 \end{pmatrix}.$$

The solution Z of the Sylvester equation: $Z\hat{A} - FZ = G_2 E_2$

$$Z = \begin{pmatrix} -0.6715 & 0.9589 & -0.0860 \\ -0.2603 & -0.5957 & 0.9979 \end{pmatrix}.$$

Step 6. $G_1 = ZA_1 R^{-1} = \begin{pmatrix} -0.0424 \\ -0.0420 \end{pmatrix}.$

$$G = (G_1, G_2)Q^T = \begin{pmatrix} -0.9991 & 0.0598 \\ 0.0007 & 0.0419 \end{pmatrix},$$

$$X = \begin{pmatrix} -0.1007 & -0.7050 & 0.9254 & -0.1196 \\ -0.0709 & -0.2839 & -0.6199 & 0.9743 \end{pmatrix}.$$

Verify:

(i) $\|XA - FX - GC\| = O(10^{-3})$,

(ii) $XB = 10^{-3} \begin{pmatrix} 0.1000 \\ -0.4000 \end{pmatrix}$, and

(iii) $\text{rank} \begin{pmatrix} X \\ C \end{pmatrix} = 4.$

Note: If G_2 were chosen as $G_2 = \begin{pmatrix} 1 \\ 1 \end{pmatrix}$, then the solution X would be rank-deficient and consequently $\begin{pmatrix} X \\ C \end{pmatrix}$ would be also rank-deficient. Indeed, in this case,

$$X = \begin{pmatrix} -0.1006 & -0.7044 & 0.9246 & -0.1195 \\ -0.1006 & -0.7044 & 0.9246 & -0.1195 \end{pmatrix},$$

which has rank 1.

12.9   OPTIMAL STATE ESTIMATION: THE KALMAN FILTER

So far we have discussed the design of an observer ignoring the "noise" in the system, that is, we assumed that all the inputs were given exactly and all the outputs were measured exactly without any errors. But in a practical situation, the measurements are always corrupted with noise. Therefore, it is more practical to consider a system with noise. In this section, we consider the problem of finding the optimal steady-state estimation of the states of a **stochastic system**. Specifically, the following problem is addressed.

Consider the stochastic system:

$$\dot{x}(t) = Ax(t) + Bu(t) + Fw(t)$$
$$y(t) = Cx(t) + v(t), \qquad (12.9.1)$$

where $w(t)$ and $v(t)$ represent "noise" in the input and the output, respectively. The problem is to find the linear estimate $\hat{x}(t)$ of $x(t)$ from all past and current output $\{y(s), s \leq t\}$ that minimizes the mean square error:

$$E[\|x(t) - \hat{x}(t)\|^2], \text{ as } t \rightarrow \infty, \qquad (12.9.2)$$

where $E[z]$ is the expected value of a vector z.

The following assumptions are made:

1. The system is **controllable** and **observable**.   (12.9.3)

Note that the controllability assumption implies that the noise $w(t)$ excites all modes of the system and the observability implies that the noiseless output $y(t) = Cx(t)$ contains information about all states.

2. Both w and v are white noise, **zero-mean** stochastic processes. That is, for all t and s,

$$E[w(t)] = 0, \qquad E[v(t)] = 0, \qquad (12.9.4)$$
$$E[w(t)w^T(s)] = W\delta(t - s), \qquad (12.9.5)$$
$$E[v(t)v^T(s)] = V\delta(t - s), \qquad (12.9.6)$$

where W and V are **symmetric and positive semidefinite and positive definite covariance matrices**, respectively, and $\delta(t - s)$ is the Dirac delta function.

3. The noise processes w and v are **uncorrelated** with one another, that is,

$$E[w(t)v^T(s)] = 0. \qquad (12.9.7)$$

4. The initial state x_0 is a **Gaussian zero-mean** random variable with known covariance matrix, and uncorrelated with w and v. That is,

$$E[x_0] = 0,$$
$$E[x_0 x_0^T] = S, \qquad E[x_0 w^T(t)] = 0, \qquad E[x_0 v^T(t)] = 0, \qquad (12.9.8)$$

where S is the positive semidefinite covariance matrix.

The following is a well-known (almost classical) result on the solution of the above problem using an algebraic Riccati equation (ARE). For a proof, see Kalman and Bucy (1961). For more details on this topic, see Kailath *et al.* (2000).

Theorem 12.9.1.   *Under the assumptions* (12.9.3)–(12.9.8), *the best estimate* $\hat{x}(t)$ (*in the linear least-mean-square sense*) *can be generated by the* **Kalman filter** (*also known as the* **Kalman-Bucy filter**).

$$\dot{\hat{x}}(t) = (A - K_{\mathrm{f}}C)\hat{x}(t) + Bu(t) + K_{\mathrm{f}}y(t), \qquad (12.9.9)$$

where $K_{\mathrm{f}} = X_{\mathrm{f}}C^{\mathrm{T}}V^{-1}$, *and* X_{f} *is the symmetric positive definite solution of the ARE:*

$$AX + XA^{\mathrm{T}} - XC^{\mathrm{T}}V^{-1}CX + FWF^{\mathrm{T}} = 0. \qquad (12.9.10)$$

Definition 12.9.1.   *The matrix* $K_{\mathrm{f}} = X_{\mathrm{f}}C^{\mathrm{T}}V^{-1}$ *is called the* **filter gain matrix.**

Note: The output estimate $\hat{y}(t)$ is given by $\hat{y}(t) = C\hat{x}(t)$.
The error between the measured output $y(t)$ and the predicted output $C\hat{x}(t)$ is given by the residual $r(t)$:

$$r(t) = y(t) - C\hat{x}(t).$$

where $\hat{x}$ is generated by (12.9.9).

Algorithm 12.9.1.   *The State Estimation of the Stochastic System Using Kalman Filter*
   Inputs.
      1.   *The matrices* A, B, C, *and* F *defining the system* (12.9.1)
      2.   *The covariance matrices* V *and* W (*both symmetric and positive definite*).
   Output. *An estimate* $\hat{x}(t)$ *of* $x(t)$ *such that* $E[\|x(t) - \hat{x}(t)\|^2]$ *is minimized, as* $t \to \infty$.
   Assumptions. (12.9.3)–(12.9.8).
   Step 1. *Obtain the unique symmetric positive definite solution* X_{f} *of the ARE:*

$$AX_{\mathrm{f}} + X_{\mathrm{f}}A^{\mathrm{T}} - X_{\mathrm{f}}C^{\mathrm{T}}V^{-1}CX_{\mathrm{f}} + FWF^{\mathrm{T}} = 0.$$

   Step 2. *Form the filter gain matrix* $K_{\mathrm{f}} = X_{\mathrm{f}}C^{\mathrm{T}}V^{-1}$.
   Step 3. *Obtain the estimate* $\hat{x}(t)$ *by solving* (12.9.9).

Duality Between Kalman Filter and the LQR Problems

The ARE (12.9.10) in Theorem 12.9.1 is dual to the Continuous-time Algebraic Riccati Equation (CARE) that arises in the solution of the LQR problem. To distinguish it from the CARE, it will be referred to as the **Continuous-time Filter Algebraic Riccati Equation** (CFARE).

Using this duality, the following important properties of the Kalman filter, *dual to those of the LQR problem described in Chapter 10*, can be established (**Exercise 12.15**).

1. *Guaranteed stability.* The filter matrix $A - K_f C$ is stable, that is, $\operatorname{Re}\lambda_i (A - K_f C) < 0$; $i = 1, 2, \ldots, n$, where $\lambda_i, i = 1, \ldots, n$, are the eigenvalues of $A - K_f C$.

2. *Guaranteed robustness.* Let V be a diagonal matrix and let $W = I$. Let $G_{KF}(s)$ and $G_{FOL}(s)$ denote, respectively, the Kalman-filter loop-transfer matrix and the filter open-loop transfer matrix (from $w(t)$ to $y(t)$), that is,

$$G_{KF}(s) \equiv C(sI - A)^{-1}K_f \qquad (12.9.11)$$

and

$$G_{FOL}(s) \equiv C(sI - A)^{-1}F. \qquad (12.9.12)$$

Then the following equality holds:

$$(I + G_{KF}(s))V(I + G_{KF}(s))^* = V + G_{FOL}(s)G_{FOL}^*(s). \qquad (12.9.13)$$

Using the above equality, one obtains

$$(I + G_{KF}(s))(I + G_{KF}(s))^* \geq I. \qquad (12.9.14)$$

In terms of singular values, one can then deduce that

$$\sigma_{\min}(I + G_{KF}(s)) \geq 1 \qquad (12.9.15)$$

or

$$\sigma_{\max}(I + G_{KF}(s))^{-1} \leq 1$$

and

$$\sigma_{\min}(I + G_{KF}^{-1}(s)) \geq \tfrac{1}{2}. \qquad (12.9.16)$$

See the article by Athans on "Kalman filtering" in the *Control Handbook* (1996, pp. 589–594), edited by W.S. Levine, IEEE Press/CRC Press.

Example 12.9.1. Consider the stochastic system:

$$\dot{x}(t) = Ax(t) + Bu(t) + w(t),$$
$$y(t) = Cx(t) + v(t)$$

with A, B, and C as in Example 12.4.1.

Take

$$W = BB^T, \qquad V = \begin{pmatrix} 1 & 0 \\ 0 & 1 \end{pmatrix}, \qquad F = I_{4\times4}.$$

Step 1. The symmetric positive definite solution X_f of the CFARE

$$AX + XA^T - XC^T V^{-1} CX + FWF^T = 0$$

is

$$X_f = \begin{pmatrix} 8.3615 & 0.0158 & 0.0187 & -0.0042 \\ 0.0158 & 9.0660 & 0.0091 & -0.0031 \\ 0.0187 & 0.0091 & 0.0250 & 0.0040 \\ -0.0042 & -0.0031 & 0.0040 & 0.0016 \end{pmatrix}.$$

Step 2. The filter gain matrix $K_f = X_f C^T V^{-1}$ is

$$K_f = \begin{pmatrix} 0.0158 & -0.2405 \\ 9.0660 & -0.1761 \\ 0.0091 & 0.2289 \\ -0.0031 & 0.0893 \end{pmatrix}.$$

The optimal state estimator of $\hat{x}(t)$ is given by

$$\dot{\hat{x}}(t) = (A - K_f C)\hat{x}(t) + Bu(t) + K_f y(t).$$

The filter eigenvalues, that is, the eigenvalues of $A - K_f C$, are $\{-0.0196,$ $-8.6168, -3.3643 \pm j2.9742\}$.

MATLAB note: The MATLAB function **kalman** designs a Kalman state estimator given the state-space model and the process and noise covariance data. **kalman** is available in MATLAB **Control System Toolbox**.

Comparison of the state and the estimate for Example 12.9.1: In Figure 12.6 we compare the actual state with the estimated state obtained in Example 12.9.1 with $x(0) = \hat{x}(0) = (-6 \quad -1 \quad 1 \quad 2)^T$ and $u(t) = H(t)(1 \quad 1 \quad 1 \quad 1)^T$, where $H(t)$ is the unit step function. Only the first and second variables are compared. The solid line corresponds to the exact state and the dotted line corresponds to the estimated state. *The graphs of the second variables are indistinguishable.*

The Kalman Filter for the Discrete-Time System

Consider now the discrete stochastic system:

$$\begin{aligned} x_{k+1} &= Ax_k + Bu_k + Fw_k, \\ y_k &= Cx_k + v_k, \end{aligned} \tag{12.9.17}$$

where w and v are the process and measurement noise. Then, under the same assumptions as was made in the continuous-time case, it can be shown that the

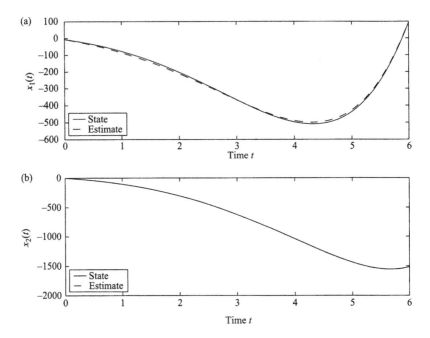

FIGURE 12.6: The (a) first and (b) second variables of the state $x(t)$ and estimate $\hat{x}(t)$, obtained by Kalman filter.

state error covariance is minimized in steady-state when the filter gain is given by

$$K_{\mathrm{d}} = A X_{\mathrm{d}} C^{\mathrm{T}} (C X_{\mathrm{d}} C^{\mathrm{T}} + V)^{-1}, \tag{12.9.18}$$

where X_{d} is the symmetric positive semidefinite solution of the Riccati equation:

$$X = A(X - X C^{\mathrm{T}} (C X C^{\mathrm{T}} + V)^{-1} C X) A^{\mathrm{T}} + F W F^{\mathrm{T}}, \tag{12.9.19}$$

and V and W are the symmetric positive definite and positive semidefinite covariance matrices, that is,

$$E[v_k v_j^{\mathrm{T}}] = V \delta_{kj}, \quad E[w_k w_j^{\mathrm{T}}] = W \delta_{kj}; \tag{12.9.20}$$

and

$$\delta_{kj} = \begin{cases} 0 & \text{if } k \neq j, \\ 1 & \text{if } k = j. \end{cases} \tag{12.9.21}$$

For details, see Glad and Ljung (2000, pp. 137–138).

Definition 12.9.2.   *In analogy with the continuous-time case, the discrete-time algebraic Riccati equation (DARE) (12.9.19), arising in discrete Kalman filter will be called the discrete filter algebraic Riccati equation or DFARE, for short.*

12.10   THE LINEAR QUADRATIC GAUSSIAN PROBLEM

The linear quadratic regulator (LQR) problems deal with optimization of a performance measure for a deterministic system. The **Linear Quadratic Gaussian** (LQG) problems deal with optimization of a performance measure for a stochastic system.

Specifically, the **continuous-time** LQG problem is defined as follows:

Consider the controllable and observable stochastic system (12.9.1) and the quadratic objective function

$$J_{QG} = \lim_{T \to \infty} \frac{1}{2T} E\left[\int_{-T}^{T} (x^T Q x + u^T R u) dt \right],$$

where the weighting matrices Q and R are, respectively, symmetric positive semidefinite and positive definite. Suppose that the noise $w(t)$ and $v(t)$ are both Gaussian, white, zero-mean, and stationary processes with positive semidefinite and positive definite covariance matrices W and V. The problem is to find the optimal control $u(t)$ that minimizes the average cost.

Solution of the LQG Problem via Kalman Filter

The solution of the LQG problem is obtained by combining the solutions of the deterministic LQR problem and the optimal state estimation problem using the Kalman filter (see the next subsection on the separation property of the LQG design).

The control vector $u(t)$ for the LQG problem is given by

$$u(t) = -K_c \hat{x}(t), \tag{12.10.1}$$

where

(i)   the matrix K_c is the feedback matrix of the associated LQR problem, that is,

$$K_c = R^{-1} B^T X_c, \tag{12.10.2}$$

X_c satisfying the CARE: $X_c A + A^T X_c + Q - X_c B R^{-1} B^T X_c = 0.$
$$\tag{12.10.3}$$

(ii)   the vector $\hat{x}(t)$ is generated by the Kalman filter:

$$\dot{\hat{x}}(t) = (A - K_f C)\hat{x}(t) + Bu(t) + K_f y(t). \tag{12.10.4}$$

The filter gain matrix $K_f = X_f C^T V^{-1}$ and X_f satisfies the CFARE

$$A X_f + X_f A^T - X_f C^T V^{-1} C X_f + F W F^T = 0. \tag{12.10.5}$$

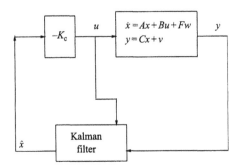

FIGURE 12.7:   The LQG design via Kalman filter.

For a proof of the above, see Dorato *et al.* (1995).

The LQG design via Kalman filter is illustrated in Figure 12.7.

The LQG Separation Property

In this section, we establish the LQG separation property. For the sake of convenience, we assume that $F = I$. By substituting (12.10.1) into (12.10.3), we obtain the compensator:

$$\dot{\hat{x}}(t) = (A - BK_c - K_fC)\hat{x}(t) + K_fy(t),$$
$$u(t) = -K_c\hat{x}(t).$$
$$(12.10.6)$$

The transfer function $M(s)$ of this compensator (from $y(t)$ to $u(t)$) can be easily written down:

$$M(s) = -K_c(sI - A + BK_c + K_fC)^{-1}K_f. \qquad (12.10.7)$$

From (12.10.6) and (12.9.1), it is easy to see that the closed-loop matrix satisfies the differential equation

$$\begin{pmatrix} \dot{x}(t) \\ \dot{\hat{x}}(t) \end{pmatrix} = \begin{pmatrix} A & -BK_c \\ K_fC & A - BK_c - K_fC \end{pmatrix} \begin{pmatrix} x(t) \\ \hat{x}(t) \end{pmatrix} + \begin{pmatrix} I & O \\ O & K_f \end{pmatrix} \begin{pmatrix} w(t) \\ v(t) \end{pmatrix}.$$
$$(12.10.8)$$

Define the error vector

$$e(t) = x(t) - \hat{x}(t). \qquad (12.10.9)$$

Then from (12.10.8) and (12.10.9), we obtain

$$\begin{pmatrix} \dot{x}(t) \\ \dot{e}(t) \end{pmatrix} = \begin{pmatrix} A - BK_c & BK_c \\ O & A - K_fC \end{pmatrix} \begin{pmatrix} x(t) \\ e(t) \end{pmatrix} + \begin{pmatrix} I & O \\ I & -K_f \end{pmatrix} \begin{pmatrix} w(t) \\ v(t) \end{pmatrix}.$$

Thus, the $2n$ closed-loop eigenvalues are the union of the n eigenvalues of $A - BK_c$ and the n eigenvalues of $A - K_fC$.

Furthermore, if (A, B) is controllable and (A, C) is observable, then both the matrices $A - BK_c$ and $A - K_f C$ are stable. However, the matrix $A - BK_c - K_f C$ is not necessarily stable.

Algorithm 12.10.1. *The Continuous-time LQG Design Method*
Inputs.

(i)   *The matrices A, B, C, and F defining the system (12.9.1).*
(ii) *The covariance matrices V and W.*

Output. *The control vector $u(t)$ generated by the LQG regulator.*
Assumptions. (12.9.3)–(12.9.8).
Step 1. *Obtain the symmetric positive definite stabilizing solution X_c of the CARE:*

$$XA + A^{\mathrm{T}}X - XBR^{-1}B^{\mathrm{T}}X + Q = 0. \tag{12.10.10}$$

Step 2. *Compute $K_c = R^{-1}B^{\mathrm{T}}X_c$*
Step 3.
   3.1. *Solve the CFARE:*

$$AX + XA^{\mathrm{T}} - XC^{\mathrm{T}}V^{-1}CX + FWF^{\mathrm{T}} = 0 \tag{12.10.11}$$

to obtain the symmetric positive definite stabilizing solution X_f.
   3.2. *Compute filter gain matrix*

$$K_f = X_f C^{\mathrm{T}}V^{-1}. \tag{12.10.12}$$

Step 4. *Solve for $\hat{x}(t)$:*

$$\dot{\hat{x}}(t) = (A - BK_c - K_f C)\hat{x}(t) + K_f y(t), \quad \hat{x}(0) = \hat{x}_0. \tag{12.10.13}$$

Step 5. *Determine the control law:*

$$u(t) = -K_c \hat{x}(t). \tag{12.10.14}$$

Remarks

- Though the optimal closed-loop system will be asymptotically stable, the LQG design method described above does not have the same properties as the LQR design method; in fact, most of the nice properties of the LQR design are lost by the introduction of the Kalman filter. See Doyle (1978) and Zhou *et al.* (1996, pp. 398–399).

- Overall, the LQG design has lower stability margins than the LQR design and its sensitivity properties are not as good as those of the LQR design.
- It might be possible to recover some of the desirable properties of the LQR design by choosing the weights appropriately. This is known as the **Loop Transfer Recovery** (LTR). The details are beyond the scope of this book. See Doyle and Stein (1979, 1981) and the book by Anderson and Moore (1990).

Example 12.10.1. We consider the LQG design for the helicopter problem of Example 12.9.1, with

$$Q = C^T C \quad \text{and} \quad R = I_{2\times 2},$$

and the same W and V.

Step 1. The stabilizing solution X_c of the CARE (computed by MATLAB function **care**) is

$$X_c = \begin{pmatrix} 0.0071 & -0.0021 & -0.0102 & -0.0788 \\ -0.0021 & 0.1223 & 0.0099 & -0.1941 \\ -0.0102 & 0.0099 & 41.8284 & 174.2 \\ -0.0788 & -0.1941 & 174.2 & 1120.9 \end{pmatrix}.$$

Step 2. The control gain matrix K_c is

$$K_c = R^{-1} B^T X_c = \begin{pmatrix} -0.0033 & 0.0472 & 14.6421 & 60.8894 \\ 0.0171 & -1.0515 & 0.2927 & 3.2469 \end{pmatrix}.$$

Step 3. The filter gain matrix K_f computed in Example 12.9.1 is

$$K_f = \begin{pmatrix} 0.0158 & -0.2405 \\ 9.0660 & -0.1761 \\ 0.0091 & 0.2289 \\ -0.0031 & 0.0893 \end{pmatrix}.$$

The closed-loop eigenvalues: The closed-loop eigenvalues are the union of the eigenvalues of $A - BK_c$ (the controller eigenvalues) and those of $A - K_f C$ (the filter eigenvalues):

$$\{-3.3643 \pm 2.9742j, -0.0196, -8.6168\}$$
$$\cup \{-0.0196, -8.6168, -3.3643 \pm 2.9742j\}.$$

MATLAB note: The MATLAB function (from the **control system toolbox**) **lqgreg** forms the LQG regulator by combining the Kalman estimator designed with **Kalman** and the optimal state feedback gain designed with **lqr**. In case of a discrete-time system, the command **dlqr** is used in place of **lqr**.

12.11 SOME SELECTED SOFTWARE

12.11.1 MATLAB Control System Toolbox

LQG design tools

kalman	Kalman estimator
kalmd	Discrete Kalman estimator for continuous plant
lqgreg	Form LQG regulator given LQ gain and Kalman estimator.

12.11.2 MATCONTROL

SYLVOBSC	Solving the constrained multi-output Sylvester-observer equation
SYLVOBSM	Solving the multi-output Sylvester-observer equation
SYLVOBSMB	Block triangular algorithm for the multi-output Sylvester-observer equation

12.11.3 CSP-ANM

Design of reduced-order state estimator (observer)

- The reduced-order state estimator using pole assignment approach is computed by ReducedOrderEstimator [*system, poles*].
- The reduced-order state estimator via solution of the Sylvester-observer equation using recursive bidiagonal scheme (a variation of the triangular scheme of van Dooren (1984)) is computed by ReducedOrderEstimator [*system, poles*, Method → RecursiveBidiagonal] and ReducedOrderEstimator [*system, poles*, Method → RecursiveBlockBidiagonal] (block version of the recursive bidiagonal scheme).
- The reduced-order state estimator via solution of the Sylvester-observer equation using recursive triangular scheme is computed by ReducedOrderEstimator [*system, poles*, Method → RecursiveTriangular] and ReducedOrderEstimator [*system, poles*, Method → RecursiveBlockTriangular] (block version of the recursive triangular scheme).

12.11.4 SLICOT

FB01RD	Time-invariant square root covariance filter (Hessenberg form)
FB01TD	Time-invariant square root information filter (Hessenberg form)
FB01VD	One recursion of the conventional Kalman filter
FD01AD	Fast recursive least-squares filter.

12.11.5   MATRIX$_X$

Purpose: Calculate optimal state estimator gain matrix for a discrete time system.

Syntax: [EVAL, KE]=DESTIMATOR (A, C, QXX, QYY, QXY) OR
[EVAL, KE, P]=DESTIMATOR (A, C, QXX, QYY, QXY)

Purpose: Calculate optimal state estimator gain matrix for a continuous time system.

Syntax: [EVAL, KE]=ESTIMATOR (A, C, QXX, QYY, QXY)
[EVAL, KE, P]=ESTIMATOR (A, C, QXX, QYY, QXY)

Purpose: Given a plant and optimal regulator, this function designs an estimator which recovers loop transfer robustness via the design parameter RHO. Plots of singular value loop transfer response are made for the (regulator) and (estimator + regulator) systems.

Syntax:
[SC, NSC, EVE, KE, SLTF, NSLTF]=LQELTR (S, NS, QXX, QYY, KR, RHO, WMIN, WMAX,
{NPTS} , {OPTION}); OR
[SC, NSC, EVE, KR, SLTF, NSLTF]=LQRLTR (S, NS, RXX, RUU, KE, RHO, OMEGA,
{OPTION});

Purpose: Given a plant and optimal estimator, this function designs a regulator which recovers loop transfer robustness via the design parameter RHO. Plots of singular value loop transfer response are made for the (estimator) and (regulator + estimator) systems.

Syntax:
[SC, NSC, EVR, KR, SLTF, NSLTF]=LQRLTR (S, NS, RXX, RUU, KE, RHO, WMIN, WMAX,
{NPTS} , {OPTION}); OR
[SC, NSC, EVR, KR, SLTF, NSLTF]=LQRLTR (S, NS, RXX, RUU, KE, RHO, OMEGA,
{OPTION});

12.12   SUMMARY AND REVIEW

In Chapters 10 and 11 we have discussed feedback stabilization, EVA and related problems. Solutions of these problems require that the states are available for measurements. Unfortunately, in many practical situations, all the states are not accessible. One therefore needs to estimate the states by knowing only input and output. This gives rise to **state estimation** problem, which is the subject matter of this chapter.

Full State Estimation

The states can be estimated using

- EVA approach **(Theorem 12.2.1)**
- Solving the associated Sylvester-like matrix equation, called the **Sylvester-observer equation (Algorithm 12.3.1)**.

In "**the eigenvalue assignment approach**," the states x can be estimated by constructing the observer

$$\dot{\hat{x}}(t) = (A - KC)\hat{x}(t) + Ky(t) + Bu(t),$$

where the matrix K is constructed such that $A - KC$ is a stable matrix, so that the error $e(t) = x(t) - \hat{x}(t) \to 0$ as $t \to \infty$.

Using "**the Sylvester equation approach**," the states are estimated by solving the Sylvester-observer equation

$$XA - FX = GC,$$

where the matrix F is chosen to be a stable matrix and G is chosen such that the solution X is nonsingular. The estimate $\hat{x}(t)$ is given by $\hat{x}(t) = X^{-1}z(t)$, where $z(t)$ satisfies $\dot{z}(t) = Fz(t) + Gy(t) + XBu(t)$.

Reduced-Order State Estimation

If the matrix C has full rank r, then the full state estimation problem can be reduced to the problem of estimating only the $n - r$ states.

Again, two approaches: **the EVA approach** and **the Sylvester-observer matrix equation** can be used for reduced-order state estimation.

Reduced-order state estimation via EVA (**Algorithm 12.4.1**) is discussed in **Section 12.4.1**. Here the EVA problem to be solved is of order $n - r$.

In the Sylvester equation approach for reduced-order state estimation, one solves a reduced-order equation

$$XA - FX = GC$$

by choosing F as an $(n-r) \times (n-r)$ stable matrix and choosing G as an $(n-r) \times r$ matrix such that the solution matrix X has full rank. The procedure is described in **Algorithm 12.4.2**.

Two numerical methods for the multi-output equation, both based on reduction of the pair (A, C) to the observer-Hessenberg pair $(H, \bar{C})$, are proposed to solve the above reduced-order Sylvester-observer equation. These methods are described in Section 12.7 (**Algorithms 12.7.1** and **12.7.2**).

Optimal State Estimation: The Kalman Filter

If there is "noise" in the system, then one has to consider the state estimation problem for a **stochastic system.** The optimal steady-state estimation of a stochastic system is traditionally done by constructing the **Kalman filter.**

For the continuous-time stochastic system (12.9.1), the Kalman filter is given by

$$\dot{\hat{x}}(t) = (A - K_f C)\hat{x}(t) + Bu(t) + K_f y(t),$$

where $K_f = X_f C^T V^{-1}$, and X_f the symmetric positive definite solution of the CFARE: $AX + XA^T - XC^T V^{-1} CX + FWF^T = 0$.

The matrices V and W are the covariance matrices associated with "noise" in the output and input, respectively. The matrix K_f is called the **Kalman filter gain.**

It can be shown that under the assumptions (12.9.3)–(12.9.8), the above Riccati equation has a symmetric positive definite solution and the estimate $\hat{x}(t)$ is such that

$$E[\|x(t) - \hat{x}(t)\|^2]$$

is minimized as $t \to \infty$.

Like the LQR design, the Kalman filter also possesses **guaranteed stability** and **robustness properties**:

- The matrix $A - K_f C$ is stable.
- $\sigma_{\min}(I + G_{KF}(s)) \geq 1$
- $\sigma_{\min}(I + G_{KF}^{-1}(s)) \geq \frac{1}{2}$,

where $G_{KF}(s) = C(sI - A)^{-1} K$.

For the discrete-time system, the DFARE to be solved is

$$X = A(X - XC^T(CXC^T + V)^{-1}CX)A^T + FWF^T$$

and the discrete **Kalman filter** gain is given by

$$K_d = X_d C^T (CX_d C^T + V)^{-1},$$

where X_d is the stabilizing solution of the above discrete Riccati equation (DFARE).

The Linear Quadratic Gaussian (LQG) Problem

The LQG problem is the problem of finding an optimal control that minimizes a performance measure given a **stochastic** system. **Thus, it is the counterpart of the deterministic LQR problem for a stochastic system.**

Given the stochastic system (12.9.1) and the performance measure J_{QG} defined in Section 12.10, the optimal control $u(t)$ for the LQG problem can be computed as

$$u(t) = -K_c \hat{x}(t),$$

where $K_c = R^{-1} B^T X_c$, X_c being the solution of the CARE arising in the solution of the deterministic LQR problem. The estimate $\hat{x}(t)$ is determined by using the Kalman filter. Specifically, $\hat{x}(t)$ satisfies

$$\dot{\hat{x}}(t) = (A - K_f C)\hat{x}(t) + Bu(t) + K_f y(t),$$

where K_f is the Kalman filter gain computed using the stabilizing solution of the CFARE.

Thus, the LQG problem is solved by first solving the LQR problems followed by constructing a Kalman filter.

Unfortunately, the LQG design described as above does not have some of the nice properties of the LQR problem that we have seen before in Chapter 10. They are lost by the introduction of the Kalman filter.

12.13   CHAPTER NOTES AND FURTHER READING

State estimation is one of the central topics in control systems design and has been discussed in many books (Kailath 1980; Chen 1984; Anderson and Moore 1990; etc.). The idea of reduced-order observers is well-known (Luenberger (1964, 1966, 1971, 1979)). The treatment of Section 12.4 on the reduced-order estimation has been taken from Chen (1984).

The term "**Sylvester-observer equation**" was first introduced by the author (Datta 1994). Algorithm 12.7.1 was developed by Van Dooren (1984), while Algorithm 12.7.2 was by Carvalho and Datta (2001). For large-scale solution of this equation, see Datta and Saad (1991); for computing an orthogonal solution to the Sylvester-observer equation, see Datta and Hetti (1997). For a discussion of the numerical properties of the method in Datta and Saad (1991), see Calvetti *et al.* (2001). A parallel algorithm for the multi-output Sylvester-observer equation appears in Bischof *et al.* (1996). For numerical solution of the Sylvester-observer equation with F as the JCF see Tsui (1993) and the references therein. For other algorithms for this problem see Datta (1989) and Datta and Sarkissian (2000). The last paper contains an algorithm for designing a "**functional observer**," which can be used to compute the feedback control law $y = K\hat{x}(t)$ without any matrix inversion.

The method for the constrained Sylvester-observer equation presented in Section 12.8 was developed by Barlow *et al.* (1992). For numerical methods dealing with nearly singular constrained Sylvester-observer equation, see Ghavimi and Laub (1996).

The topic of Kalman filter is now a classical topic. Since the appearance of the pioneering papers by Kalman (1960), Kalman and Bucy (1961), and Kalman (1964), many books and papers have been written on the subject (see, e.g., Kwakernaak and Sivan 1972; Anderson and Moore 1979; Maybeck 1979; Lewis 1986, 1992; etc.).

A special issue of *IEEE Transactions on Automatic Control*, edited by Athans (1971b) was published on the topic of LQG design, which contains many important earlier papers in this area and an extensive bibliography on this subject until 1971. See Dorato *et al.* (1995) for up-to-date references. For aerospace applications of LQG design see McLean (1990). Gangsaas (1986), Bernstein and Haddad (1989) have discussed LQG control with H_∞ performance bound.

We have not discussed in detail the stability and robustness properties of the LQG design. See the papers of Safonov and Athans (1977) and Doyle (1978) in this context.

For discussions on the LQG loop transfer recovery, see the original paper of Doyle and Stein (1979) and the survey of Stein and Athans (1987), and Section 7.2 of the recent book by Dorato *et al.* (1995).

Exercises

12.1   Consider Example 5.2.5 with the following data: $M = 2, m = 1, g = 0.18$, and $l = 1$. Take $C = (1, 1, 1, 1)$.
    (i)   Find a feedback matrix K such that the closed-loop matrix $A - BK$ has the eigenvalues $-1, -2, -3, -4$.
    (ii) Assuming now that the state x is not available for feedback, construct a full-dimensional observer using (a) the eigenvalue assignment method and (b) the Sylvester-observer equation. Compare the results by plotting the error between the true and observed states.
    (iii) Construct a three-dimensional reduced-order observer using (a) the eigenvalue assignment method and (b) the Sylvester-observer equation. Compare the results by plotting the error between the true and observed states.
    In each case (ii) and (iii), choose the observer eigenvalues to be three times as those of the matrix $A - BK$.

12.2   Are the conditions of Theorem 12.3.1 also necessary? Give reasons for your answer.

12.3   Prove that the pair (A, C) is observable if and only if the pair $(\bar{A}_{22}, \bar{A}_{12})$ is observable, where $\bar{A}_{12}$ and $\bar{A}_{22}$ are given by (12.4.2).

12.4   Establish the "**separation property**" stated in Section 12.5 for a full-dimensional observer.

12.5   Prove that the transfer function matrix of the combined system (12.5.2) of the state feedback and observer can be computed from

$$\dot{x} = (A - BK)x + Br, \qquad y = Cx$$

and the transfer function matrix is

$$\hat{G}(s) = C(sI - A + BK)^{-1}B.$$

How do you interpret this result?

12.6   Construct an example to show that the necessary conditions for the unique solution X of the Sylvester equation stated in Theorem 12.6.1 are not sufficient, unless $r = 1$.

12.7   Using the ideas from the proof of Theorem 12.6.1 prove that necessary conditions for the existence of a unique full rank solution X in $XA - FX = GC$ such that $T = \begin{pmatrix} C \\ X \end{pmatrix}$ is nonsingular are that (A, C) is observable and (F, G) is controllable. Prove further that for the single-output case ($r = 1$), the conditions are sufficient as well.

12.8   Deduce Theorem 12.6.1 from Theorem 12.6.2.

12.9   Workout a proof of Algorithm 12.8.1 (consult Barlow *et al.* (1992)).

12.10   Prove that the EVA approach and the Sylvester-observer equation approach, both for full-dimensional and reduced-order state-estimation, are mathematically equivalent.

12.11   Compare flop-count of Algorithm 12.4.1 with that of Algorithm 12.4.2. (To implement Step 3 of Algorithm 12.4.1, assume that Algorithm 11.3.1 has been used, and to implement Step 2 of Algorithm 12.4.2, assume that Algorithm 12.7.2 has been used).

12.12   *Functional estimator* (Chen (1984, p. 369)). Consider the problem of finding an estimator of the form:

$$\dot{z}(t) = Fz(t) + Gy(t) + Hu(t),$$
$$w(t) = Mz(t) + Ny(t),$$

where M and N are row vectors, so that $w(t)$ will approach $kx(t)$ for a constant row vector k, as $t \to \infty$.

(a)   Show that if M and N are chosen so as to satisfy the equation:

$$MT + N\bar{C} = \bar{k},$$

with T given by

$$T\bar{A} - FT = G\bar{C},$$
$$H = T\bar{B},$$

where $\bar{A}$ and $\bar{B}$ are the same as in (12.4.2), and $\bar{C} = CS^{-1}$, $\bar{k} = kS^{-1}$, and F is a stable matrix, then $w(t)$ will approach $kx(t)$ as $t \to \infty$.

(b)   Based on the result in (a), formulate an algorithm for designing such an estimator and apply your algorithm to Example 12.4.1.

12.13   Prove that if (A, C) is observable, then a state-estimator for the discrete-time system

$$x_{k+1} = Ax_k + Bu_k,$$
$$y_k = Cx_k$$

may be constructed as

$$\hat{x}_{k+1} = A\hat{x}_k + Bu_k + L(y_k - C\hat{x}_k),$$

where L is such that the eigenvalues of $A - LC$ have moduli less than 1.

12.14   Show that for a "deadbeat" observer, that is, for an observer with the "observer eigenvalues" equal to zero, the observer state equals the original state.

12.15   Establish the **"Guaranteed stability"** and **"Guaranteed robustness"** properties of the Kalman Filter, stated in Section 12.9.

12.16   Design an experiment for the Kalman filter estimation of the linearized state-space model of the motion of a satelite in Example 5.2.6, choosing the initial values of the variables and the covariance matrices appropriately. Show the error behavior by means of a graph.

12.17   Design an experiment to show that the LQG design has lower stability margins than the LQR design.

12.18   Rework Algorithm 12.4.1 using the QR decomposition of the matrix C, so that the explicit inversion of the matrix S can be avoided.

References

Alfriend K.T. "Special section on robust control design for a benchmark problem," *AIAA J. Guidance, Control and Dynam.*, Vol. 15, pp. 1060–1149, 1992.

Anderson B.D.O. and Moore J.B. *Optimal Control: Linear Quadratic Methods*, Prentice Hall, Englewood Cliffs, NJ, 1990.

Anderson B.D.O. and Moore J.B. *Optimal Filtering*, Prentice Hall, Englewood Cliffs, NJ, 1979.

Anderson E., Bai Z., Bischof C., Blackford S., Demmel J., Dongarra J., Du Croz J., Greenbaum, A. Hammarling S., McKenney A., and Sorensen D. *LAPACK Users' Guide*, 3rd edn, SIAM, Philadelphia, 1999.

Athans M. "The role and use of the stochastic linear quadratic Gaussian problem in control system design," *IEEE Trans. Autom. Control*, Vol. AC-16, pp. 529–552, 1971a.

Athans M. (ed.), "Special issue on linear-quadratic-Gaussian problem," *IEEE Trans. Autom. Control*, Vol. AC-16, 1971b.

Barlow J.B., Monahemi M.M., and O'Leary D.P. "Constrained matrix Sylvester equations," *SIAM J. Matrix Anal. Appl.*, Vol. 13, pp. 1–9, 1992.

Bernstein D.S. and Haddad W.M. "LQG Control with $\mathcal{H}^\infty$ performances bound: A Riccati equation approach," *IEEE Trans. Autom. Control*, Vol. AC-34, pp. 293–305, 1989.

Bhattacharyya S.P. and DeSouza E. "Controllability, observability and the solution of $AX - XB = C$," *Lin. Alg. Appl.*, Vol. 39, pp. 167–188, 1981.

Birdwell J.P. and Laub A.J. "Balanced singular values for LQG/LTR design," *Int. J. Control*, Vol. 45, pp. 939–950, 1986.

Bischof C., Datta B.N., and Purkayastha A. "A parallel algorithm for the Sylvester-observer equation," *SIAM J. Sci. Comput.*, Vol. 17, no. 3, pp. 686–698, 1996.

Calvetti D., Lewis B., and Reichel L. "On the solution of large Sylvester-observer equation," *Num. Lin. Alg. Appl.*, Vol. 8, pp. 435–451, 2001.

Carvalho J. and Datta B.N. "A block algorithm for the Sylvester-Observer equation arising in state-estimation," *Proc. IEEE Conf. Dec. Control*, Orlando, Florida, 2001.

Chen C.-T. *Linear System Theory and Design*, CBS College Publishing, New York, 1984.

The Control Handbook, edited by William S. Levine, CRC Press and IEEE Press, Boca Raton, Florida, 1995.

Chiang R.Y. and Safonov M.G. *Robust-Control Toolbox for Use with MATLAB*, Math Works, Natick, MA, 1988.

Datta B.N. "Parallel and large-scale matrix computations in control; some ideas," *Lin. Alg. Appl.*, Vol. 12, pp. 243–264, 1989.

Datta B.N. "Linear and numerical linear algebra in control theory: some research problems," *Lin. Alg. Appl.*, Vol. 197/198, pp. 755–790, 1994.

Datta B.N. and Saad Y. "Arnoldi methods for large Sylvester-like matrix equations and an associated algorithm for partial spectrum assignment," *Lin. Alg. Appl.*, Vol. 156, pp. 225–244, 1991.

Datta B.N. and Sarkissian D. Block algorithms for state estimation and functional observers, Proc. *IEEE Joint Conf. Control Appl. Comput. aided Control Syst. Des.*, pp. 19–23, 2000.

Datta K., Hong Y.P., and Lee R.B. "Applications of linear transformations to matrix equations," *Lin. Alg. Appl.*, Vol. 267, pp. 221–240, 1997.

Datta B.N. and Hetti C. "Generalized Arnoldi methods for the Sylvester-observer equation and the multi-input pole placement problem," *Proc. IEEE Conf. Dec. Control*, pp. 4379–4383, 1997.

Dorato P., Abdallah C., and Cerone V. *Linear Quadratic Control: An Introduction*, Prentice Hall, Englewood Cliffs, NJ, 1995.

Doyle J.C. and Stein G. "Multivariable feedback design: Concepts for a classical/modern synthesis," *IEEE Trans. Autom. Control*, Vol. AC-26, pp. 4–16, 1981.

Doyle J.C. "Guaranteed margins for LQG regulators," *IEEE Trans. Autom. Control*, Vol. AC-23, pp. 756–757, 1978.

Doyle J.C. and Stein G. "Robustness with observers," *IEEE Trans. Autom. Control*, Vol. AC-24, pp. 607–611, 1979.

Friedland B. *Control System Design*, McGraw-Hill, New York, 1986.

Gangsaas D. "Application of modern synthesis to aircraft control: three case studies," *IEEE Trans. Autom. Control*, Vol. AC-31, pp. 995–1104, 1986.

Ghavimi A. and Laub A.J. "Numerical methods for nearly constrained matrix Sylvester equations," *SIAM J. Matrix Anal. Appl.*, Vol. 17, no. 1, pp. 212–221, 1996.

Glad T. and Ljung L. *Control Theory: Multivariable and Nonlinear Methods,* Taylor and Francis, London, 2000.

Kailath T., Sayed A.H., and Hassibi B. *Linear Estimation*, Prentice Hall, Englewood Cliffs, NJ, 2000.

Kailath T. *Linear Systems*, Prentice Hall, Englewood Cliffs, NJ, 1980.

Kalman R.E. "Contribution to the theory of optimal control," *Bol. Soc. Matem. Mex.*, Vol. 5, pp. 102–119, 1960.

Kalman R.E. and Bucy R.S. "New results in linear filtering and prediction theory," *ASME Trans. Ser. D: J. Basic Engr.*, Vol. 83, pp. 95–107, 1961.

Kalman R.E. "When is a linear control system optimal?" *ASME Trans. Ser. D: J. Basic Engr.*, Vol. 86, pp. 51–60, 1964.

Kučera V. *Discrete Linear Control*, John Wiley & Sons, New York, 1979.

Kushner H.J. *Introduction to Stochastic Control*, Holt, Rinehart, and Winston, New York, 1971.

Kwakernaak H. and Sivan R. *Linear Optimal Control Systems*, Wiley-Interscience, New York, 1972.

Lewis F.L. *Optimal Control*, John Wiley & Sons, New York, 1986.

Lewis F.L. *Applied Optimal Control and Estimation*, Prentice Hall, Englewood Cliffs, NJ, 1992.

Levine W.S. (ed.) *The Control Handbook*, CRC Press and IEEE Press, Boca Raton, Florida, 1996.

Luenberger D.G. "Observing the state of a linear system," *IEEE Trans. Mil. Electr.*, Vol. 8, 74–80, 1964.

Luenberger, D.G. "Observers for multivariable systems," *IEEE Trans. Autom. Control*, Vol. 11, pp. 190–197, 1966.

Luenberger D.G. "An introduction to observers," *IEEE Trans. Autom. Control*, Vol. AC-16, pp. 596–602, 1971.

Luenberger D.G. *Introduction to Dynamic Systems; Theory, Models, and Applications*, John Wiley & Sons, New York, 1979.

Mahmoud M.S. "Structural properties of discrete systems with slow and fast modes," *Large Scale Systems*, Vol. 3, pp. 227–336, 1982.

MATLAB User's Guide, The Math Works, Inc., Natick, MA, 1992.

Maybeck P.S. *Stochastic Models, Estimation, and Control*, Vols. 1–3, Academic Press, New York, 1979.

McLean D. *Automatic Flight Control Systems*, International Series in Systems and Control Engineering, Prentice Hall, London, 1990.

Safonov M.G. and Athans M. "Gain and phase margins of multiloop LQG regulators," *IEEE Trans. Autom. Control*, Vol. AC-22, pp. 173–179, 1977.

DeSouza C.E. and Fragoso M.D. "On the existence of maximal solution for generalized algebraic Riccati equations arising in stochastic control," *Syst. Contr. Lett.* Vol. 14, pp. 233–239, (1990).

Stein G. and Athans M. "The LQG/LTR procedure for multi-variable feedback control design," *IEEE Trans. Autom. Control*, Vol. AC-32, pp. 105–114, 1987.

Tsui C.-C. "An algorithm for the design of multi-functional observers," *IEEE Trans. Autom. Control*, Vol. AC-30, pp. 89–93, 1985.

Tsui C.-C. "A new approach to robust observer design," *Int. J. Control*, Vol. 47, pp. 745–751, 1988.

Tsui C.-C. "On the solution to matrix equation $TA - FT = LC$," *SIAM J. Matrix Anal. Appl.*, Vol. 14, pp. 33–44, 1993.

Van Dooren P. "Reduced order observers: a new algorithm and proof," *Syst. Contr. Lett.*, Vol. 4, pp. 243–251, 1984.

Zhou K., Doyle J.C. and Glover K. *Robust and Optimal Control*, Prentice Hall, Upper Saddle River, NJ, 1996.

CHAPTER 13

NUMERICAL SOLUTIONS AND CONDITIONING OF ALGEBRAIC RICCATI EQUATIONS

> ### Topics covered
>
> - Results on Existence and Uniqueness of Solutions of the CARE and DARE
> - Perturbation Analyses and Condition Numbers
> - The Schur Methods, Newton's Methods, and the Matrix Sign Function Methods
> - Convergence Results for Newton's Methods
> - The Generalized Eigenvector and the Generalized Schur Methods
> - Inverse-Free Generalized Schur Methods
> - The Schur and Inverse-Free Schur Methods for the Descriptor Riccati Equations
> - Comparative Study and Recommendations

13.1 INTRODUCTION

This chapter is devoted to the study of numerical solutions of the continuous-time algebraic Riccati equation (CARE):

$$XA + A^{\mathrm{T}}X + Q - XBR^{-1}B^{\mathrm{T}}X = 0 \qquad (13.1.1)$$

and of its discrete counterpart (DARE)

$$A^{\mathrm{T}}XA - X + Q - A^{\mathrm{T}}XB(R + B^{\mathrm{T}}XB)^{-1}B^{\mathrm{T}}XA = 0. \qquad (13.1.2)$$

Equation (13.1.1) is very often written in the following compact form:

$$XA + A^{\mathrm{T}}X + Q - XSX = 0, \qquad (13.1.3)$$

519

where
$$S = BR^{-1}B^{\mathrm{T}}. \tag{13.1.4}$$

Equation (13.1.2) can also be written in the compact form:

$$A^{\mathrm{T}}X(I + SX)^{-1}A - X + Q = 0, \tag{13.1.5}$$

where S is again as given by (13.1.4).

These equations have long been subject of research in mathematics, physics, and engineering. They play major roles in many design problems in control and filter theory. As we have seen in Chapter 10, historically, AREs started as an important tool in the solution of *Linear Quadratic Optimization problems.* In recent years, they became a subject of intensive study, both from theoretical and computational viewpoints, because of their important roles in **state-space solutions of H_∞ and robust control problems.** For a brief history of the importance, applications, and historical developments of the AREs, see Bittanti *et al.* (1991).

The following computational methods for the CARE and DARE are widely known in the literature and most of them are discussed in **Section 13.5** of this chapter.

1. **The Eigenvector Methods** (McFarlane 1963; Potter 1966).
2. **The Schur Methods and the Structure-Preserving Schur Methods** (Laub 1979; Byers 1983, 1986a, 1990; Bunse-Gerstner and Mehrmann 1986; Mehrmann 1988; Benner *et al.* 1997c).
3. **The Generalized Eigenvector, the Generalized Schur, and Inverse-Free Generalized Methods** (Pappas *et al.* 1980; Van Dooren 1981; Arnold and Laub 1984; Mehrmann 1991).
4. **The Matrix Sign Function Methods** (Roberts 1980 [1971]; Denman and Beavers 1976; Bierman 1984; Gardiner and Laub 1986; Byers 1987; Kenney and Laub 1995).
5. **Newton's Methods** (Kleinman 1968; Hewer 1971; Benner and Byers 1998; Guo and Lancaster 1998; Guo 1998).

The eigenvector methods are well known to have numerical difficulties in case the Hamiltonian matrix associated with the CARE or the symplectic matrix associated with the DARE has some multiple or near-multiple eigenvalues (the corresponding eigenvectors will be ill-conditioned).

In these cases, the Schur methods, based on the real Schur decompositions of the Hamiltonian matrix for the CARE and of the symplectic matrix for the DARE, should be preferred over the eigenvector methods. The Schur method is widely used in practice for the CARE. Unfortunately, it cannot be applied to the DARE when A is singular. Indeed, even if A is theoretically nonsingular, but is computationally close to a singular matrix, the Schur method for the DARE should be avoided. An alternative for the DARE then is to use the generalized Schur method which is based on the Schur decomposition of a matrix pencil and does not involve computation of the inverse of A. Having said this, it should be noted that the Schur methods and the generalized Schur methods require explicit computation of the inverse of the matrix R both for the CARE and the DARE. So, when R is close to a singular matrix, the methods of choice are the inverse-free generalized Schur methods.

Newton's methods are iterative in nature and are usually used as iterative refinement techniques for solutions obtained by the Schur methods or the matrix sign function methods. **Table 13.1 presents a comparison of the different methods and recommendation based on this comparison.**

Sections **13.2 and 13.3** deal, respectively, with the results on the **existence** and **uniqueness of the stabilizing solutions** of the CARE and the DARE. **The condition numbers and bounds of the condition numbers** of the CARE and DARE are identified in Section **13.4.**

13.2   THE EXISTENCE AND UNIQUENESS OF THE STABILIZING SOLUTION OF THE CARE

The goal of this section is to derive conditions under which the CARE admits a unique symmetric positive semidefinite stabilizing solution.

For this we first need to develop an important relationship between the CARE and the associated Hamiltonian matrix and some spectral properties of this matrix.

Recall from Chapter 10 that associated with the CARE is the $2n \times 2n$ Hamiltonian matrix:

$$H = \begin{pmatrix} A & -S \\ -Q & -A^{\mathrm{T}} \end{pmatrix}. \tag{13.2.1}$$

The Hamiltonian matrix H has the following interesting spectral property.

Theorem 13.2.1.   *For each eigenvalue λ of H, $-\bar{\lambda}$ is also an eigenvalue of H (with the same geometric and algebraic multiplicity as λ).*

Proof.   Define the $2n \times 2n$ matrix:

$$J = \begin{pmatrix} 0 & I \\ -I & 0 \end{pmatrix}, \tag{13.2.2}$$

where I is the $n \times n$ identity matrix. Then it is easy to see that $J^{-1}HJ = -JHJ = -H^{\mathrm{T}}$, which shows that H and $-H^{\mathrm{T}}$ are similar. Hence, λ is also an eigenvalue of $-H^{\mathrm{T}}$. Since the eigenvalues of $-H^{\mathrm{T}}$ are the negatives of the eigenvalues of H, and the complex eigenvalues occur in conjugate pairs, the theorem is proved.   ∎

The following theorems show that a solution X of the CARE is determined by the associated Hamiltonian matrix.

Theorem 13.2.2. *A matrix X is a solution of the CARE if and only if the columns of $\begin{pmatrix} I \\ X \end{pmatrix}$ span an n-dimensional invariant subspace of the Hamiltonian matrix H defined by (13.2.1).*

Proof. We first prove that if the columns of $\begin{pmatrix} I \\ X \end{pmatrix}$ span an n-dimensional invariant subspace of H, then X is a solution of the CARE.

So, assume there exists an $n \times n$ matrix L such that:

$$H \begin{pmatrix} I \\ X \end{pmatrix} = \begin{pmatrix} I \\ X \end{pmatrix} L. \tag{13.2.3}$$

Multiplying both sides of (13.2.3) by J^{-1}, where J is defined by (13.2.2), we have

$$J^{-1} H \begin{pmatrix} I \\ X \end{pmatrix} = J^{-1} \begin{pmatrix} I \\ X \end{pmatrix} L. \tag{13.2.4}$$

Noting that $J^{-1} = \begin{pmatrix} 0 & -I \\ I & 0 \end{pmatrix}$, we obtain from (13.2.4)

$$\begin{pmatrix} Q & A^T \\ A & -S \end{pmatrix} \begin{pmatrix} I \\ X \end{pmatrix} = \begin{pmatrix} -X \\ I \end{pmatrix} L. \tag{13.2.5}$$

Premultiplying both sides of (13.2.5) by (I, X), we get

$$XA + A^T X + Q - XSX = 0,$$

showing that X satisfies the CARE.

To prove the converse, we note that if X is a solution of the CARE, then

$$H \begin{pmatrix} I \\ X \end{pmatrix} = \begin{pmatrix} A - SX \\ -Q - A^T X \end{pmatrix} = \begin{pmatrix} A - SX \\ X(A - SX) \end{pmatrix} = \begin{pmatrix} I \\ X \end{pmatrix} (A - SX), \tag{13.2.6}$$

that is, the columns of $\begin{pmatrix} I \\ X \end{pmatrix}$ span an invariant subspace of H. ∎

Corollary 13.2.1. *If the columns of $\begin{pmatrix} X_1 \\ X_2 \end{pmatrix}$ span an n-dimensional invariant subspace of the Hamiltonian matrix H associated with the CARE and X_1 is invertible, then $X = X_2 X_1^{-1}$ is a solution of the CARE.*

Proof.

$$\text{The span of the columns of } \begin{pmatrix} X_1 \\ X_2 \end{pmatrix}$$

$$= \text{the span of the columns of } \begin{pmatrix} X_1 \\ X_2 \end{pmatrix} X_1^{-1}$$

$$= \text{the span of the columns of } \begin{pmatrix} I \\ X_2 X_1^{-1} \end{pmatrix}.$$

Therefore, by Theorem 13.2.2, we see that $X = X_2 X_1^{-1}$ is a solution of the CARE. ∎

The next theorem shows how the eigenvalues of the Hamiltonian matrix H are related to those of the optimal closed-loop matrix.

Theorem 13.2.3. *Let X be a symmetric solution of the CARE. Then the eigenvalues of the Hamiltonian matrix H are the eigenvalues of $A - BK$ together with those of $-(A - BK)^T$, where $K = R^{-1} B^T X$.*

Proof. Define $T = \begin{pmatrix} I & 0 \\ X & I \end{pmatrix}$, where I and X are $n \times n$. Then,

$$T^{-1}HT = \begin{pmatrix} I & 0 \\ -X & I \end{pmatrix} \begin{pmatrix} A & -S \\ -Q & -A^T \end{pmatrix} \begin{pmatrix} I & 0 \\ X & I \end{pmatrix},$$

$$= \begin{pmatrix} A - SX & -S \\ -(A^T X + XA + Q - XSX) & -(A - SX)^T \end{pmatrix},$$

$$= \begin{pmatrix} A - SX & -S \\ 0 & -(A - SX)^T \end{pmatrix}. \tag{13.2.7}$$

Thus, the eigenvalues of H are the eigenvalues of $A - SX$ together with those of $-(A - SX)^T$.

The result now follows by noting that:

$$A - SX = A - BR^{-1}B^T X = A - BK.$$

(Recall that $S = BR^{-1}B^T$.) ∎

Symmetric Positive Semidefinite Stabilizing Solutions of the CARE

As we have seen in Chapter 10, several applications require a symmetric positive semidefinite stabilizing solution of the associated Riccati equation. We derive in this subsection a necessary and sufficient condition for the existence of such a solution.

Recall that a symmetric solution X of (13.1.1) is a **stabilizing solution** if $A - BK = A - BR^{-1}B^T X = A - SX$ is stable.

Proof of Theorem 13.2.4 below has been taken from Kimura (1997).

Theorem 13.2.4.   *Existence and Uniqueness of the Stabilizing Solution.*
Assume that $R > 0$ and $Q \geq 0$, $Q \neq 0$.
Then the following conditions are equivalent:

1.   *The CARE:*

$$XA + A^T X - XBR^{-1}B^T X + Q = 0 \qquad (13.2.8)$$

has a unique symmetric positive semidefinite stabilizing solution X.
2.   *(A, B) is stabilizable and the associated Hamiltonian matrix H has no pure imaginary eigenvalues.*

Proof of necessity.   First suppose that X is a stabilizing solution of the CARE. We then show that H does not have an imaginary eigenvalue.

Since X is a stabilizing solution, $A - SX$ is stable, that is $A - BK$ is stable. From Theorem 13.2.3, we then have that n eigenvalues of H are stable and the other n have positive real parts. Thus, H does not have a purely imaginary eigenvalue.

Proof of sufficiency.   Next assume that H given in (13.2.1), with $S = BR^{-1}B^T$, has no eigenvalues on the imaginary axis. We shall then show that under the assumption of the stabilizability of (A, B), there exists a unique stabilizing solution of the CARE.

The proof will be divided in several parts.

First of all we note that the stabilizability of (A, B) implies the stabilizability of (A, S).

Since H has no pure imaginary eigenvalues, there are n stable eigenvalues of H (by Theorem 13.2.1).

Then,

$$H \begin{pmatrix} X_1 \\ X_2 \end{pmatrix} = \begin{pmatrix} X_1 \\ X_2 \end{pmatrix} E, \qquad (13.2.9)$$

where E is a stable matrix and the columns of $\begin{pmatrix} X_1 \\ X_2 \end{pmatrix}$ form the eigenspace of H corresponding to these stable eigenvalues.

A. $X_2^T X_1$ is symmetric.

The relation (13.2.9) can be expressed as

$$AX_1 - SX_2 = X_1 E \qquad (13.2.10)$$

and

$$-QX_1 - A^T X_2 = X_2 E. \qquad (13.2.11)$$

Multiplying (13.2.10) by X_2^T on the left, we have

$$X_2^T A X_1 - X_2^T S X_2 = X_2^T X_1 E. \qquad (13.2.12)$$

Now taking the transpose of (13.2.11), we have

$$X_2^T A = -X_1^T Q - E^T X_2^T.$$

Multiplying the last equation by X_1 to the right, we get

$$E^T X_2^T X_1 = -X_2^T A X_1 - X_1^T Q X_1. \qquad (13.2.13)$$

Using (13.2.12) in (13.2.13), we then have

$$E^T X_2^T X_1 + X_2^T X_1 E = -X_2^T S X_2 - X_1^T Q X_1. \qquad (13.2.14)$$

Since S and Q are symmetric, the right-hand side matrix is symmetric, and therefore the left-hand side matrix is also symmetric. This means that

$$E^T X_2^T X_1 + X_2^T X_1 E = X_1^T X_2 E + E^T X_1^T X_2$$

or

$$E^T (X_2^T X_1 - X_1^T X_2) + (X_2^T X_1 - X_1^T X_2) E = 0.$$

Since E is stable, this Lyapunov equation has a unique solution which implies that $X_2^T X_1 - X_1^T X_2 = 0$. That is, $X_2^T X_1 = X_1^T X_2$, proving that $X_2^T X_1$ is symmetric.

B.   **X_1 is invertible.**

Suppose that X_1 is not invertible. Then there exists a vector $d \neq 0$ such that

$$X_1 d = 0. \tag{13.2.15}$$

Now multiplying the transpose of (13.2.10) by d^T to the left and by $X_2 d$ to the right, we have

$$
\begin{aligned}
d^T X_2^T S X_2 d &= -d^T E^T X_1^T X_2 d + d^T X_1^T A^T X_2 d, \\
&= -d^T E^T X_2^T X_1 d + d^T X_1^T A^T X_2 d = 0
\end{aligned}
$$

$$\text{(because } X_1^T X_2 = X_2^T X_1 \text{ and } X_1 d = 0\text{).}$$

Again, since $S \geq 0$, we must have

$$S X_2 d = 0.$$

The Eq. (13.2.10) therefore yields

$$X_1 E d = 0.$$

As this holds for all $d \in K_{er}(X_1)$, this means that $K_{er}(X_1)$ is E-invariant, that is, there exists an eigenvalue μ of E such that

$$Ed' = \mu d', \qquad X_1 d' = 0, \quad d' \neq 0. \tag{13.2.16}$$

Again, multiplying (13.2.11) by d' and using the relation (13.2.16), we obtain

$$(\mu I + A^T) X_2 d' = 0. \tag{13.2.17}$$

Also, from (13.2.10) and (13.2.16), we have

$$S X_2 d' = 0. \tag{13.2.18}$$

Since $\text{Re}(\mu) < 0$ and (A, S) is stabilizable, we conclude from (13.2.18) that

$$X_2 d' = 0. \tag{13.2.19}$$

Finally, $X_2 d' = 0$ and $X_1 d' = 0$ imply that $\begin{pmatrix} X_1 \\ X_2 \end{pmatrix}$ does not have the full rank which contradicts (13.2.9).

Thus, X_1 is nonsingular.

C. **X is symmetric.**
Since X_1 is nonsingular, we have from Corollary 13.2.1 that $X = X_2 X_1^{-1}$ is a solution of the CARE and, since $X_2^T X_1$ is symmetric, so is X. This is seen as follows:

$$
\begin{aligned}
X^T - X &= X_1^{-T} X_2^T - X_2 X_1^{-1} \\
&= X_1^{-T}(X_2^T X_1) X_1^{-1} - X_1^{-T}(X_1^T X_2) X_1^{-1} \\
&= X_1^{-T}(X_2^T X_1 - X_1^T X_2) X_1^{-1} = 0.
\end{aligned}
$$

D. **X is a stabilizing solution.**
Multiplying (13.2.10) by X_1^{-1} to the right, we obtain

$$
A - SX_2 X_1^{-1} = X_1 E X_1^{-1}.
$$

Since E is stable, so is $A - SX_2 X_1^{-1} = A - SX$. Thus, X is a stabilizing solution.

E. **X is unique.**
Let X_1 and X_2 be two stabilizing solutions. Then,

$$
\begin{aligned}
A^T X_1 + X_1 A - X_1 S X_1 + Q &= 0 \\
A^T X_2 + X_2 A - X_2 S X_2 + Q &= 0
\end{aligned}
$$

Subtracting these two equations, we have

$$
A^T(X_1 - X_2) + (X_1 - X_2)A + X_2 S X_2 - X_1 S X_1 = 0
$$

or

$$
(A - SX_1)^T(X_1 - X_2) + (X_1 - X_2)(A - SX_2) = 0.
$$

Since the last equation is a homogeneous Sylvester equation and the coefficient matrices $A - SX_1$ and $A - SX_2$ are both stable, it follows that $X_1 - X_2 = 0$, that is, $X_1 = X_2$.

F. **X is positive semidefinite.**
Since X is symmetric and satisfies (13.2.8), Eq. (13.2.8) can be written in the form of the following Lyapunov equation:

$$
(A - BK)^T X + X(A - BK) = -Q - XSX,
$$

where $K = R^{-1} B^T X$. Furthermore, $A - BK = A - BR^{-1}B^T X = A - SX$ is stable. Thus, X can be expressed in the form (**see Chapter 7**): $X = \int_0^\infty e^{(A-BK)^T t}(Q + XSX)e^{(A-BK)t} dt$. Since Q and S are positive semidefinite, it follows that X is positive semidefinite. ∎

Theorem 13.2.5. *Let* (A, B) *be stabilizable and* (A, Q) *be detectable. Assume that* $Q \geq 0$, $S \geq 0$. *Then the Hamiltonian matrix:*

$$H = \begin{pmatrix} A & -S \\ -Q & -A^T \end{pmatrix}$$

associated with the CARE does not have a purely imaginary eigenvalue.

Proof. The proof is by contradiction.

Suppose that H has a purely imaginary eigenvalue $j\alpha$, where α is a nonnegative real number, and let $\begin{pmatrix} r \\ s \end{pmatrix}$ be the corresponding eigenvector. Then,

$$H\begin{pmatrix} r \\ s \end{pmatrix} = j\alpha \begin{pmatrix} r \\ s \end{pmatrix}, \quad \begin{pmatrix} r \\ s \end{pmatrix} \neq \begin{pmatrix} 0 \\ 0 \end{pmatrix}. \tag{13.2.20}$$

Multiplying both sides of (13.2.20) by (s^*, r^*) to the left, we obtain

$$s^*Ar - r^*Qr - s^*Ss - r^*A^Ts = j\alpha(s^*r + r^*s)$$

or

$$(s^*Ar - r^*A^Ts) - r^*Qr - s^*Ss = j\alpha(s^*r + r^*s).$$

Considering the real part of this equation, we get

$$-r^*Qr - s^*Ss = 0.$$

Since $S \geq 0$ and $Q \geq 0$, we conclude that

$$Ss = 0 \tag{13.2.21}$$

and

$$Qr = 0. \tag{13.2.22}$$

So, from (13.2.20), we have

$$Ar = j\alpha r \tag{13.2.23}$$

and

$$-A^Ts = j\alpha s. \tag{13.2.24}$$

Thus, combining (13.2.23) and (13.2.22), we have $\begin{pmatrix} A - j\alpha I \\ Q \end{pmatrix} r = 0$. Since (A, Q) is detectable, we have $r = 0$. Similarly, using (13.2.24) and (13.2.21), one can show that $s = 0$. This gives us a contradiction that $\begin{pmatrix} r \\ s \end{pmatrix}$ is an eigenvector. Thus, H cannot have a purly imaginary eigenvalue. ∎

An Expression for the Stabilizing Solution

Combining Theorem 13.2.4, Corollary 13.2.1, and Theorem 13.2.5, we arrive at the following result:

Theorem 13.2.6. *An Expression for the Unique Stabilizing Solution of the CARE. Suppose that (A, B) is stabilizable and (A, Q) is detectable. Assume that $Q \geq 0$ and $R > 0$. Then there exists a unique positive semidefinite stabilizing solution X of the CARE: $XA + A^T X - XBR^{-1}B^T X + Q = 0$. This solution is given by $X = X_2 X_1^{-1}$, where the columns of the matrix $\begin{pmatrix} X_1 \\ X_2 \end{pmatrix}$ span the invariant subspace of the Hamiltonian matrix (13.2.1) associated with its stable eigenvalues.*

Remark

- The following simple example shows that the detectability of (A, Q) is not necessary for the existence of a symmetric positive semidefinite stabilizing solution of the CARE.

$$A = \begin{pmatrix} -1 & 0 \\ 0 & 2 \end{pmatrix}, \qquad B = \begin{pmatrix} 1 \\ 1 \end{pmatrix}, \qquad Q = \begin{pmatrix} 0 & 0 \\ 0 & 0 \end{pmatrix}, \qquad R = 1.$$

  Then (A, B) is stabilizable, but (A, Q) is not detectable. The matrix $X = \begin{pmatrix} 0 & 0 \\ 0 & 4 \end{pmatrix}$ is the stabilizing solution of the CARE and is positive semidefinite.

13.3   THE EXISTENCE AND UNIQUENESS OF THE STABILIZING SOLUTION OF THE DARE

The existence and uniqueness of the stabilizing solution of the DARE can be studied via a symplectic matrix which takes the role of the Hamiltonian matrix of the CARE.

Definition 13.3.1. *A matrix M is **symplectic** if*

$$J^{-1}M^T J = J^T M^T J = M^{-1},$$

where J is defined by (13.2.2).

Assume A is invertible and consider the matrix:

$$M = \begin{pmatrix} A + S(A^{-1})^T Q & -S(A^{-1})^T \\ -(A^{-1})^T Q & (A^{-1})^T \end{pmatrix}, \tag{13.3.1}$$

where $S = BR^{-1}B^T$, $Q = Q^T$, and $S = S^T$.

Then, it can be shown (**Exercise 13.3**) that

1. M is symplectic.
2. If λ is a nonzero eigenvalue of M, so is $1/\bar{\lambda}$.

We now state the discrete counterparts of Theorems 13.2.5 and 13.2.6. The proofs can be found in Lancaster and Rodman (1995).

Theorem 13.3.1. *Let (A, B) be discrete-stabilizable and let (A, Q) be discrete-detectable. Assume that $Q \geq 0$ and $S \geq 0$. Then the symplectic matrix (13.3.1) has no eigenvalues on the unit circle.*

Suppose that the symplectic matrix M has no eigenvalues on the unit circle. Then it must have n eigenvalues inside the unit circle and n outside it. As in the continuous-time case, it can then be shown that if the columns of the matrix $\begin{pmatrix} X_1 \\ X_2 \end{pmatrix}$ form a basis for the invariant subspace associated with the eigenvalues inside the unit circle, then X_1 is nonsingular and $X = X_2 X_1^{-1}$ is a unique symmetric positive semidefinite stabilizing solution of the DARE.

Thus, we have the following theorem as the discrete counterpart of Theorem 13.2.6.

Theorem 13.3.2. *An Expression for the Unique Stabilizing Solution of the DARE. Suppose that (A, B) is discrete-stabilizable and (A, Q) is discrete-detectable. Assume that $Q \geq 0$, $R > 0$. Then the DARE:*

$$A^T XA - X + Q - A^T XB(R + B^T XB)^{-1} B^T XA = 0$$

has a unique symmetric positive semidefinite discrete-stabilizing solution X.

Furthermore, X is given by $X = X_2 X_1^{-1}$, where the columns of $\begin{pmatrix} X_1 \\ X_2 \end{pmatrix}$ span the n-dimensional invariant subspace of the symplectic matrix M associated with the eigenvalues inside the unit circle.

13.4 CONDITIONING OF THE RICCATI EQUATIONS

Before we describe the solution methods for Riccati equations, we state some results on the perturbation theory of such equations that will help us identify the *condition numbers* of the equations. These condition numbers, as usual, will help us understand the sensitivity of the solutions of the Riccati equations when the entries of the data matrices are slightly perturbed.

13.4.1 Conditioning of the CARE

Consider first the CARE:

$$A^T X + XA + Q - XSX = 0, \tag{13.4.1}$$

where

$$S = BR^{-1}B^T. \tag{13.4.2}$$

Let ΔA, ΔX, ΔQ, and ΔS be small perturbations in A, X, Q, and S, respectively. Suppose that X is the unique stabilizing solution of the CARE and that $X + \Delta X$ is the unique stabilizing solution of the perturbed Riccati equation:

$$(A + \Delta A)^T (X + \Delta X) + (X + \Delta X)(A + \Delta A)$$
$$+ (Q + \Delta Q) - (X + \Delta X)(S + \Delta S)(X + \Delta X) = 0. \tag{13.4.3}$$

We are interested in finding an upper bound for the relative error $\|\Delta X\|/\|X\|$.

Several results exist in literature. Byers (1985) and Kenney and Hewer (1990) obtained the first-order perturbation bounds and Chen (1988), Konstantinov *et al.* (1990) gave global perturbation bound. Xu (1996) has improved Chen's result and Konstantinov *et al.* (1995) have sharpened the results of Konstantinov *et al.* (1990). The most recent result in this area is due to Sun (1998), who has improved Xu's result. We present below Sun's result and the condition numbers derived on the basis of this result.

Following the notations in Byers (1985), we define three operators:

$$\Omega(Z) = (A - SX)^T Z + Z(A - SX), \tag{13.4.4}$$
$$\Theta(Z) = \Omega^{-1}(Z^T X + XZ) \tag{13.4.5}$$
$$\Pi(Z) = \Omega^{-1}(XZX). \tag{13.4.6}$$

Note: Since the closed-loop matrix $A_C = A - SX$ is stable, Ω^{-1} exists. In fact, if $\Omega(Z) = W$, then

$$Z = \Omega^{-1}(W) = -\int_0^\infty e^{A_C^T t} W e^{A_C t} dt,$$

and $\|\Omega^{-1}\|_F = 1/\text{sep}(A_C^T, -A_C)$.

Define $l = \|\Omega^{-1}\|^{-1}$, $p = \|\Theta\|$, and $q = \|\Pi\|$, where $\|\cdot\|$ is any unitarily invariant norm.

Then the following perturbation result due to Sun (1998) holds:

Theorem 13.4.1. *A Perturbation Bound for the CARE. Let X and $X + \Delta X$ be, respectively, the symmetric positive semidefinite stabilizing solutions of the CARE (13.4.1) and the perturbed CARE (13.4.3).*

Then, for sufficiently small $[\Delta Q, \Delta A, \Delta S]$,

$$\frac{\|\Delta X\|}{\|X\|} \lesssim \frac{\|Q\|}{l\|X\|} \cdot \frac{\|\Delta Q\|}{\|Q\|} + p\frac{\|A\|}{\|X\|} \cdot \frac{\|\Delta A\|}{\|A\|} + \frac{q\|S\|}{\|X\|} \cdot \frac{\|\Delta S\|}{\|S\|}. \tag{13.4.7}$$

Using the results of Theorem 13.4.1, Sun has defined a set of condition numbers of the CARE.
The numbers:

$$\kappa_{CARE}^{AB}(Q) = \frac{1}{l}, \qquad \kappa_{CARE}^{AB}(A) = p, \qquad \text{and} \qquad \kappa_{CARE}^{AB}(S) = q$$

are the **absolute condition numbers** of X with respect to Q, A, S, respectively.
The numbers:

$$\kappa_{CARE}^{REL}(Q) = \frac{\|Q\|}{l\|X\|}, \qquad \kappa_{CARE}^{REL}(A) = \frac{p\|A\|}{\|X\|}, \qquad \text{and} \qquad \kappa_{CARE}^{REL}(S) = \frac{q\|S\|}{\|X\|}$$

are then the **relative condition numbers**.
Moreover, the scalar:

$$\kappa_{CARE}^{REL}(X) = \frac{1}{\|X\|}\sqrt{\left(\frac{\|Q\|}{l}\right)^2 + (p\|A\|)^2 + (q\|S\|)^2} \tag{13.4.8}$$

can be regarded as the **relative condition number of X**.
Using a local linear estimate, Byers (1985) has obtained an approximate condition number given by

$$\kappa_{CARE}^{B} = \frac{1}{\|X\|_F}\left(\frac{\|Q\|_F}{l} + p\|A\|_F + q\|S\|_F\right),$$

in which the operator norm $\|\cdot\|$ for defining l, p, and q is induced by the Frobenius norm $\|\cdot\|_F$.

The above is known as **Byers' approximate condition number.** Indeed, taking the Frobenius norm in (13.4.8), and comparing (13.4.8) with Byers' condition number, one obtains:

Theorem 13.4.2.

$$\frac{1}{\sqrt{3}} \kappa_{CARE}^{B} \leq \kappa_{CARE}^{REL}(X) \leq \kappa_{CARE}^{B}$$

Expressions for l, p, and q: If the operator norm $\| \cdot \|$ for defining l, p, and q is induced by the Frobenius norm $\| \cdot \|_F$, then it can be shown (Sun 1998) that

$$l = \|T^{-1}\|_2^{-1}, \qquad p = \|T^{-1}(I_n \otimes X + (X^T \otimes I_n)E)\|_2$$

and

$$q = \|T^{-1}(X^T \otimes X)\|_2,$$

where

$$T = I_n \otimes (A - SX)^T + (A - SX)^T \otimes I_n, \qquad \text{and } A - SX \text{ is stable.}$$

E is the vec-permutation matrix:

$$E = \sum_{i,j=1}^{n} (e_i e_j^T) \otimes (e_j e_i^T).$$

Remark

- A recent paper of Petkov *et al.* (1998) contains results on estimating the quantities l, p, and q.

Estimating Conditioning of the CARE using Lyapunov Equations

Computing the quantities l, p, and q using the Kronecker products is computationally intensive. On the other hand (*using the 2-norm in the definition of Byers' condition number*) Kenney and Hewer (1990) have obtained an upper and a lower bound of κ_{CARE}^{B} by means of solutions of certain Lyapunov equations, which are certainly computationally much less demanding than computing Kronecker products. Using these results, ill-conditioning of κ_{CARE} can be more easily detected.

Assume that $A - SX$ is stable and let H_k be the solution to the Lyapunov equation:

$$(A - SX)^T H_k + H_k(A - SX) = -X^k, \quad k = 0, 1, 2. \qquad (13.4.9)$$

Furthermore, let's define $H_1^{(1)}$ as follows:

Set $\tilde{Q} = 2X$ and solve the successive Lyapunov equations for $\tilde{H}$ and H, respectively:

$$(A - SX)^T \tilde{H} + \tilde{H}(A - SX) = \tilde{Q} \tag{13.4.10}$$

and

$$(A - SX)H + H(A - SX)^T = \tilde{H}. \tag{13.4.11}$$

Let

$$W = 2XH \quad \text{and} \quad H_1^{(1)} = \Theta\left(\frac{W}{\|W\|}\right).$$

Define

$$U = \frac{\|H_0\|\|Q\| + 2\|H_0\|^{1/2}\|H_2\|^{1/2}\|A\| + \|H_2\|\|S\|}{\|X\|} \tag{13.4.12}$$

and

$$L = \frac{\|H_0\|\|Q\| + \|H_1^{(1)}\|\|A\| + \|H_2\|\|S\|}{\|X\|}, \tag{13.4.13}$$

Then it has been shown that:

$$L \leq \kappa_{CARE}^B \leq U \tag{13.4.14}$$

From the relations (13.4.12)–(13.4.14), we see that κ_{CARE}^B will be large (and consequently the CARE will be ill-conditioned) if H_0, $H_1^{(1)}$, and H_2 have large norms (relative to that of X). Conversely, if the norms of H_0, $H_1^{(1)}$, and H_2 are not large, then the CARE will be well-conditioned.

If the norms vary widely in the sense that there is a mixture of large and small norms, then there will be **selective sensitivity.** More specifically, the ratios:

$$r_1 = \frac{\|H_0\|\|Q\|}{\|X\|}, \qquad r_2 = \frac{\|H_1^{(1)}\|\|A\|}{\|X\|}, \qquad \text{and } r_3 = \frac{\|H_2\|\|S\|}{\|X\|}$$

measure, respectively, the sensitivity of X with respect to perturbations in the matrix Q, the matrix A, and the matrix S.

Example 13.4.1.   An Ill-Conditioned CARE.

$$A = \begin{pmatrix} 1 & 2 & 3 \\ 0.0010 & 4 & 5 \\ 0 & 7 & 8 \end{pmatrix}, \quad B = \begin{pmatrix} 1 \\ 0 \\ 0 \end{pmatrix}, \quad R = 1.$$

$$Q = \begin{pmatrix} 1 & 1 & 1 \\ 1 & 5 & 3 \\ 1 & 3 & 5 \end{pmatrix}.$$

$$X = 10^9 \begin{pmatrix} 0 & 0.0003 & 0.0004 \\ 0.0003 & 4.5689 & 5.3815 \\ 0.0004 & 5.3815 & 6.3387 \end{pmatrix}.$$

The residual norm of the solution X: $\|XA + A^T X - XSX + Q\| = O(10^{-5})$.

$$\|H_0\| = 5.6491 \times 10^8, \quad \|H_1\| = 1.8085 \times 10^9, \quad \text{and} \quad \|H_2\| = 4.8581 \times 10^{18}.$$

U and L **are both of order 10^8.**

Thus, the Riccati equation is expected to be ill-conditioned with the given data.

Indeed, this is an example of mixed sensitivity. Note that the ratios r_2 and r_3 are large, but r_1 is quite small. Thus, X **should be sensitive with respect to perturbation in A and S.** This is verified as follows.

Let M_{new} stand for a perturbed version of the matrix M and X_{new} stands for the new solution of the ARE with the perturbed data.

Case 1. *Perturbation in A.*   Let A be perturbed to $A + \Delta A$, where

$$\Delta A = 10^{-8} \begin{pmatrix} 3.169 & 2.668 & 3.044 \\ -1.259 & -0.5211 & -2.364 \\ 2.798 & 3.791 & -3.179 \end{pmatrix}.$$

The matrices B, Q, and R remain unperturbed.

Then,

Relative error in X: $\dfrac{\|X_{\text{new}} - X\|}{\|X\|} = 4.1063 \times 10^{-5}$,

Relative perturbation in A: $\dfrac{\|A_{\text{new}} - A\|}{\|A\|} = 4.9198 \times 10^{-9}$.

Case 2. *Perturbation in B.* Let B be perturbed to $B + \Delta B$, where

$$\Delta B = 10^{-8} \begin{pmatrix} -4.939 \\ 0.7715 \\ -0.9411 \end{pmatrix}.$$

A, Q, R remain unperturbed.

$$\text{Relative error in} X: \frac{\|X_{\text{new}} - X\|}{\|X\|} = 7.5943 \times 10^{-5},$$

$$\text{Relative perturbation in} B: \frac{\|B_{\text{new}} - B\|}{\|B\|} = 5.086 \times 10^{-8}.$$

Case 3. *Perturbation in Q.* The matrix Q is perturbed such that the relative perturbation in Q

$$\frac{\|Q_{\text{new}} - Q\|}{\|Q\|} = 4.048 \times 10^{-9}.$$

The matrices A, B, and R remain unperturbed.

$$\text{Then the relative error in } X: \frac{\|X_{\text{new}} - X\|}{\|X\|} = 4.048 \times 10^{-9}.$$

Note: All the solutions to the CARE in this example were computed using the Schur method (**Algorithm 13.5.1**) followed by Newton's iterative refinement procedure (**Algorithm 13.5.8**). The residual norms of the solutions obtained by the Schur method alone were of order 10^5. On the other hand, the residual norm of the solution with the Schur method followed by Newton's iterative procedure was, in each case, of order 10^{-5}.

Example 13.4.2. A Well-Conditioned CARE.

$$A = \begin{pmatrix} -1 & 1 & 1 \\ 0 & -2 & 0 \\ 0 & 0 & -3 \end{pmatrix}, \qquad B = \begin{pmatrix} 1 \\ 1 \\ 1 \end{pmatrix},$$

$$Q = \begin{pmatrix} 1 & 0 & 0 \\ 0 & 1 & 0 \\ 0 & 0 & 1 \end{pmatrix}, \quad \text{and} \quad R = 1.$$

In this case $\|H_0\| = 0.3247$, $\|H_1\| = 0.1251$, $\|H_2\| = 0.0510$, and $U = 3.1095$.

The CARE is, therefore, expected to be well-conditioned.

Indeed, if $(1, 1)$ entry of A is perturbed to -0.9999999, and the other data remain unchanged, then we find

Relative error in X: $\|X_{\text{new}} - X\|/\|X\| = 5.6482 \times 10^{-8}$.

Relative perturbation in A: $\|A_{\text{new}} - A\|/\|A\| = 3.1097 \times 10^{-8}$, where A_{new} and X_{new}, respectively, denote the perturbed A and the solution of the CARE with the perturbed data.

Conditioning and Accuracy

Suppose that $\hat{X}$ is an approximate stabilizing solution of the CARE:

$$XA + A^T X - XSX + Q = 0,$$

where $S = BR^{-1}B^T$ and let $\text{Res}(\hat{X}) = \hat{X}A + A^T\hat{X} - \hat{X}S\hat{X} + Q$.

Then the question arises: If $\text{Res}(\hat{X})$ **is small, does it guarantee that the error in the solution is also small (Exercise 13.8).** In the case of linear system problem, it is well-known that the smallness of the residual does not guarantee that the error in the solution is small, if the linear system problem is ill-conditioned. Similar result can be proved in the case of the Riccati equations (see Kenney *et al.* 1990). **The result basically says that even if the residual is small, the computed solution may be inaccurate, if the CARE is ill-conditioned.** On the other hand, if $\text{Res}(\hat{X})$ is small and the CARE is well-conditioned, then the solution is guaranteed to be accurate. Below, we quote a recent result of Sun (1997a) which is an improvement of the result of Kenney *et al.* (1990).

Theorem 13.4.3. *Residual Bound of an Approximate Stabilizing Solution. Let $\hat{X} \geq 0$ approximate the positive semidefinite stabilizing solution X to the CARE. Define the linear operator $T : \mathbb{R}^{n \times n} \to \mathbb{R}^{n \times n}$ by*

$$T(Z) = (A - S\hat{X})^T Z + Z(A - S\hat{X}), \qquad Z = Z^T \in \mathbb{R}^{n \times n}.$$

Assuming that $4\|T^{-1}\|\|T^{-1}(\text{Res}(\hat{X}))\|\|S\| < 1$ for any unitarily invariant norm $\|\cdot\|$, then

$$\frac{\|\hat{X} - X\|}{\|\hat{X}\|} \leq \frac{2}{1 + \sqrt{1 - 4\|T^{-1}\|\|T^{-1}\text{Res}(\hat{X})\|\|S\|}} \frac{\|T^{-1}\text{Res}(\hat{X})\|}{\|\hat{X}\|}.$$

13.4.2 Conditioning of the DARE

Consider now the DARE:

$$A^T XA - X + Q - A^T XB(R + B^T XB)^{-1}B^T XA = 0.$$

The condition number of the DARE, denoted by κ_{DARE}, may be obtained by means of the **Frechet derivative** of the DARE (Gudmundsson *et al.* 1992).

Define $A_d = A - B(R + B^T XB)^{-1}B^T XA$, $\qquad S = BR^{-1}B^T$. (13.4.15)

Assume that X is the stabilizing solution of the DARE. Then the condition number of the DARE is given by:

$$\kappa_{\text{DARE}} = \frac{\|[Z_1, Z_2, Z_3]\|_2}{\|X\|_F}, \tag{13.4.16}$$

where

$$Z_1 = \|A\|_F P^{-1}(I \otimes A_d^T X + (A_d^T X \otimes I)E), \qquad (13.4.17)$$

$$Z_2 = -\|S\|_F P^{-1}(A^T X(I + SX)^{-1} \otimes A^T X(I + SX)^{-1}), \qquad (13.4.18)$$

and

$$Z_3 = \|Q\|_F P^{-1}. \qquad (13.4.19)$$

In the above, E is the vec-permutation matrix:

$$E = \sum_{i,j=1}^{n} e_i e_j^T \otimes e_j e_i^T, \qquad (13.4.20)$$

and P is a matrix representation of the Stein operator:

$$\Omega(Z) = Z - A_d^T Z A_d. \qquad (13.4.21)$$

Note that, since A_d is discrete-stable, P^{-1} exists.

The condition number (13.4.16) measures the sensitivity of the stabilizing solution X of the DARE with respect to first-order perturbations.

Assume that the bounds for ΔA, ΔS, and ΔQ are **sufficiently small**. Then, using first-order perturbation only, it can be shown **(Exercise 13.7)** that the following quantity is an **approximate condition number** of the DARE:

$$\frac{2\|A\|_F^2\|Q\|_F/\|X\|_F + \|A\|_F^2\|S\|_F\|X\|_F}{sep_d(A_d^T, A_d)}, \qquad (13.4.22)$$

where

$$sep_d(A_d^T, A_d) = \min_{X \neq 0} \frac{\|A_d^T X A_d - X\|_F}{\|X\|_F}. \qquad (13.4.23)$$

Note: The quantity $sep(A_d^T, A_d)$ can be computed as the minimum singular value of the matrix:

$$A_d^T \otimes A_d^T - I_{n^2}.$$

Remark

- A perturbation theorem for the DARE, analogous to Theorem 13.4.1 (for the CARE), and the absolute and relative condition numbers using the results of that theorem can be obtained. For details see Sun (1998).

Also, a recent paper of Sima *et al.* (2000) contains efficient and reliable condition number estimators both for the CARE and DARE.

Example 13.4.3. (An Ill-Conditioned DARE.)   Let's take A, B, Q, and R the same as in Example 13.4.1.

Let $A_{\text{new}} = \begin{pmatrix} 0.9980 & 2 & 3 \\ 0.0010 & 4 & 5 \\ 10^{-8} & 7 & 8 \end{pmatrix}$. Let B, Q, and R remain unchanged.

The solution X of the DARE (computed by MATLAB function **dare**) is

$$X = 10^{10} \begin{pmatrix} 0.0000 & 0.0005 & 0.0005 \\ 0.0005 & 5.4866 & 6.4624 \\ 0.0005 & 6.4624 & 7.6118 \end{pmatrix}.$$

The solution X_{new} of the perturbed version of the DARE is

$$X_{\text{new}} = 10^{10} \begin{pmatrix} 0.0000 & 0.0005 & 0.0005 \\ 0.0005 & 5.4806 & 6.4554 \\ 0.0005 & 6.4554 & 7.6036 \end{pmatrix}.$$

Relative error in X: $\|X - X_{\text{new}}\|/\|X\| = 0.0010$, while the perturbations in A were of order $O(10^{-4})$.

Example 13.4.4. (A Well-Conditioned DARE.)   Let

$$A = \begin{pmatrix} 1 & 2 & 3 \\ 2 & 3 & 4 \\ 3.999 & 6 & 7 \end{pmatrix}.$$

Take B, Q, and R the same as in Example 13.4.1.
   Let

$$A_{\text{new}} = \begin{pmatrix} 0.9990 & 2 & 3 \\ 2 & 3 & 4 \\ 4 & 6 & 7 \end{pmatrix}, \quad B_{\text{new}} = B, \quad Q_{\text{new}} = Q, \quad \text{and} \quad R_{\text{new}} = R.$$

Then both the relative error in X and the relative perturbation in A are of $O(10^{-4})$. In this case, $\text{sep}(A_{\text{d}}^{\text{T}}, A_{\text{d}}) = 0.0011$.

13.5   COMPUTATIONAL METHODS FOR RICCATI EQUATIONS

The computational methods (**listed in the Introduction**) for the AREs can be broadly classified into three classes:

- The Invariant Subspace Methods
- The Deflating Subspace Methods
- Newton's Methods.

The eigenvector, Schur vector, and matrix sign function methods are examples of the invariant subspace methods. The generalized eigenvector and generalized Schur vector methods are examples of the deflating subspace methods.

The following methods have been included in our discussions here. For the CARE:

- **The eigenvector method** (Section 13.5.1)
- The **Schur method** (Algorithm 13.5.1)
- The **Hamiltonian Schur method** (Section 13.5.1)
- **The inverse-free generalized Schur method** (Algorithm 13.5.3)
- **The matrix sign function method** (Algorithm 13.5.6)
- **Newton's method** (Algorithm 13.5.8)
- **Newton's method with line search** (Algorithm 13.5.9).

For the DARE:

- **The Schur method** (Section 13.5.1)
- **The generalized Schur method** (Algorithm 13.5.2)
- The **inverse-free generalized Schur method** (Algorithm 13.5.4).
- The **matrix sign function method** (Algorithm 13.5.7)
- **Newton's method** (Algorithm 13.5.10)
- **Newton's method with line search** (Algorithm 13.5.11).

13.5.1   The Eigenvector and Schur Vector Methods

An invariant subspace methods for solving the CARE (DARE) is based on computing a stable invariant subspace of the associated Hamiltonian (symplectic) matrix; that is the subspace corresponding to the eigenvalues with the negative real parts (inside the unit circle). If this subspace is spanned by $\binom{X_1}{X_2}$ and X_1 is invertible, then $X = X_2 X_1^{-1}$ is a stabilizing solution.

To guarantee the existence of such a solution, it will be assumed throughout this section that (A, B) *is stabilizable (discrete-stabilizable) and the Hamiltonian matrix H (symplectic matrix M) does not have an imaginary eigenvalue (an eigenvalue on the unit circle).* Note that a sufficient condition for the existence of a unique positive *semidefinite* stabilizing solution of the CARE(DARE) was given Theorem 13.2.6 (Theorem 13.3.2).

The Eigenvector Method for the CARE

Let H be diagonalizble and have the eigendecomposition:

$$V^{-1} H V = \begin{pmatrix} -\bar{\Lambda} & 0 \\ 0 & \Lambda \end{pmatrix},$$

where $\Lambda = \text{diag}(\lambda_1, \dots, \lambda_n)$ and $\lambda_1, \dots, \lambda_n$ are the n eigenvalues of H with positive real parts.

Let V be partitioned conformably:

$$V = \begin{pmatrix} V_{11} & V_{12} \\ V_{21} & V_{22} \end{pmatrix}$$

such that $\begin{pmatrix} V_{11} \\ V_{21} \end{pmatrix}$ is the matrix of eigenvectors corresponding to the stable eigenvalues. Then it is easy to see that

$$H \begin{pmatrix} V_{11} \\ V_{21} \end{pmatrix} = \begin{pmatrix} V_{11} \\ V_{21} \end{pmatrix} (-\bar{\Lambda}).$$

Thus, $X = V_{21} V_{11}^{-1}$ is the unique stabilizing solution.

Remark

- **The eigenvector method, in general, cannot be recommended for practical use.** The method becomes highly unstable if the Hamiltonian matrix H is defective or nearly defective, that is, if there are some multiple or near multiple eigenvalues of H. In these cases, the matrix V_{11} will be poorly conditioned, making $X = V_{21} V_{11}^{-1}$ inaccurate; and this might happen even if the CARE itself is not ill-conditioned.

The eigenvector method, in principle, is applicable even when H is not diagonalizable by computing the principal vectors, **but again is not recommended in practice.**

MATCONTROL note: The eigenvector method for the CARE has been implemented in MATCONTROL function **riceigc.**

The Eigenvector Method for the DARE

An analogous method for the DARE can be developed by taking the eigendecomposition of the associated symplectic matrix M. However, since forming the matrix M requires computation of A^{-1}, **the eigenvector method for the DARE works only when A is nonsingular. But even in this case, the results will be inaccurate if A is ill-conditioned.** Moreover, the method will have the same sort of difficulties as those mentioned above for the CARE. We, thus, skip the description of the eigenvector method for the DARE.

The Schur Vector Method for the CARE

The numerical difficulties of the eigenvector method for the CARE may somehow be reduced or eliminated if the Hamiltonian matrix H is transformed to **an ordered Real Schur form** (RSF) by using the QR iteration algorithm, rather than using its eigendecomposition.

Let $U^T H U$ be an **ordered Real Schur matrix:**

$$U^T H U = \begin{pmatrix} T_{11} & T_{12} \\ 0 & T_{22} \end{pmatrix},$$

where the eigenvalues of H with negative real parts have been stacked in T_{11} and those with positive real parts are stacked in T_{22}.

Let

$$U = \begin{pmatrix} U_{11} & U_{12} \\ U_{21} & U_{22} \end{pmatrix}$$

be a conformable partitioning of U. Then,

$$H \begin{pmatrix} U_{11} \\ U_{21} \end{pmatrix} = \begin{pmatrix} U_{11} \\ U_{21} \end{pmatrix} T_{11}.$$

Thus, the matrix $X = U_{21} U_{11}^{-1}$ is then the unique stabilizing solution of the CARE.

The above discussion leads to the following algorithm, called the *Schur algorithm*, due to Laub (1979).

Algorithm 13.5.1. *The Schur Algorithm for the CARE*
  Inputs.
A—An n × n matrix
B—An n × m (m ≤ n) matrix
Q—An n × n symmetric matrix
R—An m × m symmetric matrix.
  Output.
X—The unique stabilizing solution of the CARE.
  Step 1. *Form the Hamiltonian matrix*

$$H = \begin{pmatrix} A & -BR^{-1}B^T \\ -Q & -A^T \end{pmatrix}.$$

Step 2. *Transform H to the ordered RSF:*

$$U^T H U = \begin{pmatrix} T_{11} & T_{12} \\ 0 & T_{22} \end{pmatrix},$$

where the n eigenvalues of H with negative real parts are contained in T_{11}.
Step 3. *Partition U conformably:*

$$U = \begin{pmatrix} U_{11} & U_{12} \\ U_{21} & U_{22} \end{pmatrix}.$$

Step 4. *Compute the solution X by solving the linear systems:*

$$X U_{11} = U_{21}.$$

Software for the ordered RSF

The ordered RSF of H can be obtained by transforming H first to the RSF by orthogonal similarity, followed by another orthogonal similarity applied to the RSF to achieve the desired ordering of the eigenvalues (**See Chapter 4**).

There exists an efficient algorithm and an associated software developed by Stewart (1976) for this purpose: Algorithm 506 of the Association for Computing Machinery Trans. Math Software (1976), pp. 275–280. See also the LAPACK routine STRSEN.

The MATLAB program **ordersch** from MATCONTROL can also be used for this purpose.

Flop-count: The Schur method is based on reduction to RSF, which is done by QR iterations algorithm; so, an exact flop-count cannot be given. However, assuming that the average number of iterations per eigenvalue is 2, about $200n^3$ flops will be necessary to execute the algorithm. (This count also takes into account of the ordering of RSF).

Example 13.5.1. Consider solving the CARE with:

$$A = \begin{pmatrix} -1 & 1 & 1 \\ 0 & -2 & 0 \\ 0 & 0 & -3 \end{pmatrix}, \qquad Q = I_{3\times3}, \qquad S = \begin{pmatrix} 1 & 1 & 1 \\ 1 & 1 & 1 \\ 1 & 1 & 1 \end{pmatrix},$$

$$B = \begin{pmatrix} 1 \\ 1 \\ 1 \end{pmatrix}, \qquad R = 1.$$

Step 1. Form the Hamiltonian matrix

$$H = \begin{pmatrix} A & -S \\ -Q & -A^T \end{pmatrix} = \left(\begin{array}{ccc|ccc} -1 & 1 & 1 & -1 & -1 & -1 \\ 0 & -2 & 0 & -1 & -1 & -1 \\ 0 & 0 & -3 & -1 & -1 & -1 \\ \hline -1 & 0 & 0 & 1 & 0 & 0 \\ 0 & -1 & 0 & -1 & 2 & 0 \\ 0 & 0 & -1 & -1 & 0 & 3 \end{array} \right).$$

Step 2. Transform the Hamiltonian matrix to the ordered RSF:

$$U^T H U = \begin{pmatrix} T_{11} & T_{12} \\ 0 & T_{22} \end{pmatrix},$$

$$= \left(\begin{array}{ccc|ccc} -2.9940 & -0.0216 & 1.3275 & & & \\ 0 & -2.1867 & 0.7312 & & * & \\ 0 & -0.2573 & -1.9055 & & & \\ \hline & & & 2.9940 & 1.3285 & 0.2134 \\ & 0 & & 0 & 1.9623 & 0.2434 \\ & & & 0 & -0.7207 & 2.1298 \end{array} \right).$$

The eigenvalues of T_{11} are: -2.9940, $-2.0461 + 0.4104j$, $-2.0461 - 0.4104j$. Thus, all the stable eigenvalues are contained in T_{11}.

Step 3. Extract U_{11} and U_{21} from U:

$$U_{11} = \begin{pmatrix} 0.4417 & 0.3716 & 0.7350 \\ 0.0053 & -0.8829 & 0.3951 \\ -0.8807 & 0.1802 & 0.3986 \end{pmatrix}, \quad U_{21} = \begin{pmatrix} 0.1106 & 0.0895 & 0.3260 \\ 0.0232 & -0.1992 & 0.1552 \\ -0.1285 & 0.0466 & 0.1199 \end{pmatrix}.$$

Step 4. Compute the stabilizing solution:

$$X = U_{21} U_{11}^{-1} = \begin{pmatrix} 0.3732 & 0.0683 & 0.0620 \\ 0.0683 & 0.2563 & 0.0095 \\ 0.0620 & 0.0095 & 0.1770 \end{pmatrix}.$$

The eigenvalues of $A - SX$ are: $-2.0461 + 0.4104j$, $-2.0461 - 0.4104j$, -2.9940. Thus, $A - SX$ is stable, that is, X is a unique stabilizing solution.

MATCONTROL note: The Schur method for the CARE (using ordered RSF) has been implemented in MATCONTROL function **ricschc**.

Stability Analysis of the Schur Method and Scaling

The round-off properties of the Schur method are quite involved. It can be shown (Petkov *et al.* 1991) that the relative error in the computed solution is proportional to $\|U_{11}^{-1}\|/\mathrm{sep}(T_{11}, T_{22})$.

This means that the Schur method can be numerically unstable even if the CARE is not ill-conditioned. For example, the Schur method can be unstable if the Hamiltonian matrix H is nearly defective.

However, the difficulty can be overcome by proper scaling (Kenney *et al.* 1989). **Thus, for all practical purposes, the Schur method, when combined with an appropriate scaling, is numerically stable.** For a discussion on scaling procedure, see Kenney *et al.* (1989), and Benner (1997). See also Pandey (1993).

Benner (1997) has given an extensive discussion on scaling. Based on several existing scaling strategies and considering the practical difficulties with these strategies, he has proposed a mixture of these procedures for scaling the CARE. Benner's strategy is as follows:

Write the CARE:

$$XA + A^{\mathrm{T}}X - XSX + Q = 0$$

in the form:

$$X_\rho A_\rho + A_\rho^{\mathrm{T}} X_\rho - X_\rho S_\rho X_\rho + Q = 0,$$

where $A_\rho = \rho A$, $A_\rho^{\mathrm{T}} = (\rho A)^{\mathrm{T}}$, $X_\rho = X/\rho$, and $S_\rho = \rho^2 S$, ρ being a positive scalar.

Choose ρ as

$$\rho = \begin{cases} \dfrac{\|S\|_2}{\|Q\|_2}, & \text{if } \|Q\|_2 > \|S\|_2 \\[2mm] \dfrac{\|A\|_2}{\|S\|_2}, & \text{if } \|Q\|_2 \leq \|S\|_2 \text{ and } \|Q\|_2\|S\|_2 < \|A\|_2^2 \\[2mm] 1, & \text{otherwise.} \end{cases}$$

For a rationale of choosing ρ this way, see Benner (1997).

Note: Note that the relative condition number of the CARE remains invariant under the above scaling.

The Schur Method for the DARE

The Schur method for the DARE:

$$A^T X A - X - A^T X B (R + B^T X B)^{-1} B^T X A + Q = 0$$

is analogous. Form the symplectic matrix:

$$M = \begin{pmatrix} A + S(A^{-1})^T Q & -S(A^{-1})^T \\ -(A^{-1})^T Q & (A^{-1})^T \end{pmatrix},$$

where $S = BR^{-1}B^T$.

Let M be transformed to an ordered RSF such that the eigenvalues with moduli less than 1 appear in the first block, that is, an orthogonal matrix U can be constructed such that

$$U^T M U = \begin{pmatrix} S_{11} & S_{12} \\ 0 & S_{22} \end{pmatrix},$$

where each eigenvalue of S_{11} is inside the unit circle. Partition U conformably:

$$U = \begin{pmatrix} U_{11} & U_{12} \\ U_{21} & U_{22} \end{pmatrix}.$$

Then $X = U_{21}U_{11}^{-1}$ is the unique stabilizing solution of the DARE.

Remarks

- Since one needs to form A^{-1} explicitly to compute M, the Schur method for the DARE is not applicable if A is singular. Even if A is theoretically nonsingular, **the method is expected to give an inaccurate answer in case A is ill-conditioned with respect to inversion.**
- A slightly faster method (Sima (1996, p. 244)) forms the matrix M^{-1} and orders the RSF so that the eigenvalues with moduli less than 1 appear in the first block.

MATCONTROL note: The Schur method for the DARE has been implemented in MATCONTROL function **ricschd**.

The Hamiltonian–Schur Methods for the CARE

The Schur methods for the AREs are based on orthogonal similarity transformations of the associated Hamiltonian and symplectic matrices to RSFs. The rich structures of these matrices are, however, not exploited in these methods. The Hamiltonian and the symplectic matrices are treated just as $2n \times 2n$ general matrices in these methods. **It would be useful if methods could be developed that could take advantage of Hamiltonian and Symplectic structures.** Such structure-preserving methods, besides reflecting physical structures, are often faster.

Theorem 13.5.1 below shows that developments of such structure-preserving methods are possible.

Definition 13.5.1. *If a matrix U is both symplectic and unitary, it is called a* **symplectic–unitary** *matrix. A* **symplectic–orthogonal** *matrix can be similarly defined.*

From the above definition, it follows that a $2n \times 2n$ symplectic–unitary matrix U can be written as:

$$U = \begin{pmatrix} U_{11} & U_{12} \\ -U_{12} & U_{11} \end{pmatrix},$$

where U_{11} and U_{12} are $n \times n$. If $\hat{U}$ is $n \times n$ unitary, then

$$U = \begin{pmatrix} \hat{U} & 0_{n \times n} \\ 0_{n \times n} & \hat{U} \end{pmatrix}$$

is symplectic–unitary.

Theorem 13.5.1. *The Hamiltonian–Schur Decomposition (HSD) Theorem. (Paige and Van Loan 1981). If the real parts of all the eigenvalues of a Hamiltonian matrix H are nonzero, then there exists a symplectic–orthogonal matrix U and a Hamiltonian matrix T such that*

$$U^{\mathrm{T}} H U = T = \begin{pmatrix} T_1 & T_2 \\ 0_{n \times n} & -T_1^{\mathrm{T}} \end{pmatrix},$$

where T_1 is an $n \times n$ upper triangular, and T_2 is an $n \times n$ symmetric matrix. Furthermore, U and T can be chosen so that the eigenvalues of T_1 have negative real parts.

Definition 13.5.2. *The Hamiltonian matrix T in Theorem 13.5.1 is called a* **Hamiltonian–Schur matrix** *and the decomposition itself is called the* **HSD.**

Note: The first n columns of U in the above HSD span the invariant subspace corresponding to the stabilizing solution of the CARE.

Symplectic-Schur Decomposition (SSD)

For a symplectic matrix, we have the following theorem.

Theorem 13.5.2.   *The SSD Theorem. If M is symplectic and has no eigenvalues on the unit circle, then there exists a symplectic–orthogonal matrix U such that*

$$U^{\mathrm{T}}MU = R = \begin{pmatrix} R_1 & R_2 \\ 0_{n \times n} & R_1^{-\mathrm{T}} \end{pmatrix},$$

where R_1 is $n \times n$ upper triangular. Moreover, $R_2 R_1$ is symmetric.

Definition 13.5.3.   *The above decomposition is called an* **SSD**.

The existence of the HSD and the SSD naturally lead to the following problem: **How to obtain these decompositions in a numerically effective way by exploiting the structures of the Hamiltonian and the symplectic matrices?**

Byers (1983, 1986a) first developed such a structure-preserving method for the HSD in the case the matrix S in the Hamiltonian matrix:

$$H = \begin{pmatrix} A & -S \\ -Q & -A^{\mathrm{T}} \end{pmatrix},$$

has rank 1. (For example, a single-input problem).

Definition 13.5.4.   *A Hamiltonian matrix H has* **Hamiltonian-Hessenberg** *form, if it has the zero structure of a $2n \times 2n$ upper Hessenberg matrix with the order of the last n rows and columns reversed.*

As in the standard QR iteration algorithm for the RSF of a matrix A, Byers' method also comes in two stages:

Stage I. The matrix H is reduced to a **Hamiltonian-Hessenberg matrix** H_{H} by an orthogonal–symplectic transformation.

Stage II. The Hamiltonian-Hessenberg matrix H_{H} is further reduced to Hamiltonian–Schur form using Hamiltonian QR iterations.

Of course, once such a reduction is done, this can immediately be used to solve the CARE.

For a complete description of the method and details of numerical implementations, see Byers (1986a).

Unfortunately, in spite of several attempts, such a reduction in the general case of a Hamiltonian matrix remained a difficult problem, until the recent paper of Benner *et al.* (1997c).

A Hamiltonian–Schur Method for the CARE (rank $S \geq 1$)

We next outline briefly the Hamiltonian–Schur method of Benner *et al.* (1997c) for solving the CARE in the multi-input case. The method also uses symplectic–orthogonal transformations in the reduction to the Hamiltonian–Schur form of the matrix H_E defined below.

The method is based on an interesting relationship between the invariant subspaces of the Hamiltonian matrix H and the extended matrix

$$\begin{pmatrix} 0 & H \\ H & 0 \end{pmatrix}.$$

It makes use of the **symplectic URV-like decomposition** that was also introduced by the authors (Benner *et al.* 1999c).

Theorem 13.5.3. *Symplectic-URV Decomposition. Given a $2n \times 2n$ Hamiltonian matrix H, there exist symplectic–orthogonal matrices U_1 and U_2 such that*

$$H = U_2 \begin{pmatrix} H_t & H_r \\ 0 & -H_b^T \end{pmatrix} U_1^T,$$

where H_r is an $n \times n$ matrix, H_t is an $n \times n$ upper triangular matrix and H_b is an $n \times n$ real Schur matrix.

Furthermore, the positive and negative square roots of the eigenvalues of $H_t H_b$ are the eigenvalues of H.

The basis of the Hamiltonian–Schur method is the following result.

Theorem 13.5.4. *Extended HSD Theorem. Suppose that the Hamiltonian matrix*

$$H = \begin{pmatrix} A & -S \\ -Q & -A^T \end{pmatrix}$$

has no purely imaginary eigenvalues. Define

$$H_E = \begin{pmatrix} 0 & H \\ H & 0 \end{pmatrix}.$$

Then there exists an orthogonal matrix U of order $4n$ such that

$$U^T H_E U = T = \begin{pmatrix} T_1 & T_2 \\ 0 & -T_1^T \end{pmatrix}$$

is in Hamiltonian–Schur form and no eigenvalues of T_1 have negative real parts.

Remark

- Note that the transforming matrix U in Theorem 13.5.4 is not symplectic–orthogonal. But this non-symplectic transformation can be computed without rounding errors!

Solution of the CARE using the Extended HSD

Let H have no eigenvalue on the imaginary axis. Let the matrix U in Theorem 13.5.4 be partitioned as

$$U = \begin{pmatrix} U_{11} & U_{12} \\ U_{21} & U_{22} \end{pmatrix},$$

where each U_{ij} is of order $2n \times 2n$. Define the matrix $\hat{Y}$ as

$$\hat{Y} = \frac{\sqrt{2}}{2}(U_{11} - U_{21}).$$

Let Y be an orthogonal basis of Range($\hat{Y}$). Then it has been shown (Benner *et al.* 1997c) that

$$\text{Range}(Y) = \text{Inv}(H),$$

where Inv(H) is the invariant subspace associated with the eigenvalues of H with negative real parts.

Furthermore, if

$$\hat{Y} = \begin{pmatrix} \hat{Y}_1 \\ \hat{Y}_2 \end{pmatrix},$$

where $\hat{Y}_1$ and $\hat{Y}_2$ are of order $n \times 2n$, then the stabilizing solution X of the CARE is given by

$$X\hat{Y}_1 = -\hat{Y}_2.$$

Note that the above equations represent an overdetermined consistent set of linear equations.

The symplectic-URV decomposition is used to compute the matrix U to achieve the Hamiltonian–Schur matrix T. Note also that it is not necessary to explicitly compute Y, if only the stabilizing solution of the CARE is sought.

The details are rather involved and we refer the readers to the paper of Benner *et al.* 1997c).

Efficiency and stability: The method based on the above discussion is more efficient than the Schur method. It has also been shown that the method computes the Hamiltonian–Schur form of a Hamiltonian matrix close to $\tilde{H}_{\mathrm{E}}$, where $\tilde{H}_{\mathrm{E}}$ is permutationally similar to H_{E}, that is, there exists a permutation matrix P such that $PH_{\mathrm{E}}P^{\mathrm{T}} = \tilde{H}_{\mathrm{E}}$.

13.5.2   The Generalized Eigenvector and Schur Vector Methods

The deflating subspace methods are generalizations of the invariant subspace methods in the sense that the solutions of the Riccati equations are now computed by finding the bases for the stable deflating subspaces of certain matrix pencils rather than finding those of the Hamiltonian and the symplectic matrices. As of the invariant subspace methods, it will be assumed that for solving the CARE (DARE) with deflating subspace methods, the pair (A, B) is stabilizable (discrete stabilizable) and the associated Hamiltonian (symplectic) matrix pencil does not have an imaginary eigenvalue (an eigenvalue on the unit circle).

For the CARE, the pencil is $P_{\text{CARE}} - \lambda N_{\text{CARE}}$, where

$$P_{\text{CARE}} = \begin{pmatrix} A & -S \\ -Q & -A^{\text{T}} \end{pmatrix}, \qquad N_{\text{CARE}} = \begin{pmatrix} I & 0 \\ 0 & I \end{pmatrix}. \tag{13.5.1}$$

For the DARE, the pencil is $P_{\text{DARE}} - \lambda N_{\text{DARE}}$, where

$$P_{\text{DARE}} = \begin{pmatrix} A & 0 \\ -Q & I \end{pmatrix}, \qquad N_{\text{DARE}} = \begin{pmatrix} I & S \\ 0 & A^{\text{T}} \end{pmatrix}. \tag{13.5.2}$$

Since no inversion of A is required to form the above pencils, **this generalization is significant for the DARE,** because, as we have seen, the eigenvector and the Schur methods cannot be applied to the DARE when A is singular.

As in the case of an invariant subspace method, a basis for a deflating subspace of a pencil can be constructed either by using the generalized eigendecomposition or the generalized Schur decomposition of the pencil. As before, an eigenvector method will have numerical difficulties in case the pencil has a multiple or near-multiple eigenvalue. **We will thus skip the descriptions of the generalized eigenvector methods and describe here only the generalized Schur method for the DARE.** We leave the description of the generalized Schur method for the CARE as an exercise (**Exercise 13.18**).

The following results form a mathematical foundation for a deflating subspace method for the DARE. The results are due to Pappas *et al.* (1980).

Theorem 13.5.5.   *Suppose that (A, B) is discrete-stabilizable and (A, Q) is discrete-detectable. Then the symplectic pencil $P_{\text{DARE}} - \lambda N_{\text{DARE}}$ does not have any eigenvalue λ with $|\lambda| = 1$.*

Proof.   The proof is by contradiction.
Let $|\lambda| = 1$ be an eigenvalue of the pencil $P_{\text{DARE}} - \lambda N_{\text{DARE}}$ with the eigenvector

$$z = \begin{pmatrix} z_1 \\ z_2 \end{pmatrix} \neq 0.$$

Then we can write:

$$\begin{pmatrix} A & 0 \\ -Q & I \end{pmatrix} \begin{pmatrix} z_1 \\ z_2 \end{pmatrix} = \lambda \begin{pmatrix} I & S \\ 0 & A^T \end{pmatrix} \begin{pmatrix} z_1 \\ z_2 \end{pmatrix}.$$

This means that

$$Az_1 = \lambda z_1 + \lambda S z_2; \tag{13.5.3}$$

$$-Qz_1 + z_2 = \lambda A^T z_2. \tag{13.5.4}$$

Premultiplying the first equation by $\bar{\lambda} z_2^*$ and postmultiplying the conjugate transpose of the second by z_1, we have

$$\bar{\lambda} z_2^* A z_1 = |\lambda|^2 z_2^* z_1 + |\lambda|^2 z_2^* S z_2 \tag{13.5.5}$$

and

$$z_2^* z_1 = z_1^* Q z_1 + \bar{\lambda} z_2^* A z_1. \tag{13.5.6}$$

Substituting (13.5.5) into (13.5.6), we obtain

$$z_2^* z_1 = z_1^* Q z_1 + |\lambda|^2 z_2^* z_1 + |\lambda|^2 z_2^* S z_2 \tag{13.5.7}$$

or

$$z_2^* S z_2 + z_1^* Q z_1 = 0 \quad \text{(since } |\lambda|^2 = 1\text{)}. \tag{13.5.8}$$

Since $S = BR^{-1}B^T$, Eq. (13.5.8) can be written as:

$$(z_2^* B) R^{-1} (B^T z_2) + z_1^* Q z_1 = 0. \tag{13.5.9}$$

Since R is positive definite, this implies that

$$B^T z_2 = 0 \quad \text{and} \quad Q z_1 = 0. \tag{13.5.10}$$

Therefore, from (13.5.3) and (13.5.4), we have $Az_1 = \lambda z_1$ and $A^T z_2 = (1/\lambda)z_2$. (Note that since $|\lambda| = 1$, $\lambda \neq 0$).

Thus, from (13.5.10) and from the last equation, we have $z_2^* B = 0$ and $z_2^* A = (1/\bar{\lambda})z_2^*$.

This means that for any F, $z_2^*(A - BF) = (1/\bar{\lambda})z_2^*$, that is, $(1/\bar{\lambda})$ is an eigenvalue of $A - BF$ for every F. Since (A, B) is discrete-stabilizable, this means that $z_2 = 0$. Similarly, since (A, Q) is detectable, it can be shown (**Exercise 13.17**) that $z_1 = 0$. Therefore,

$$z = \begin{pmatrix} z_1 \\ z_2 \end{pmatrix}$$

is a zero vector, which is a contradiction. ∎

Theorem 13.5.5, together with the fact that if $\lambda \neq 0$ is an eigenvalue with multiplicity r of the pencil $P_{DARE} - \lambda N_{DARE}$, so is $1/\lambda$ with the same multiplicity, allows us to state the following theorem:

Theorem 13.5.6. *Suppose that (A, B) is discrete-stabilizable and (A, Q) is discrete-detectable. Let $\lambda = 0$ be an eigenvalue of multiplicity r. Then the eigenvalues of the pencil $P_{DARE} - \lambda N_{DARE}$ can be arranged as follows (adopting the convention that the reciprocal of a zero is infinity):*

$$\underbrace{0, \ldots, 0;}_{r} \quad \underbrace{\lambda_{r+1}, \ldots, \lambda_n;}_{n-r} \quad \underbrace{\frac{1}{\lambda_n}, \ldots, \frac{1}{\lambda_{r+1}};}_{n-r} \quad \underbrace{\infty, \infty, \ldots, \infty}_{r}.$$

with $0 < |\lambda_i| < 1, i = r + 1, \ldots, n$.

MATCONTROL note: The generalized eigenvector method for the DARE has been implemented in MATCONTROL function **ricgeigd**.

The Generalized Schur-Vector Method for the DARE

Assume that the generalized Schur form of the pencil $P_{DARE} - \lambda N_{DARE}$ has been ordered such that the generalized eigenvalues of the pencil with moduli less than 1 can be obtained from the first quarters of the matrices, that is, the orthogonal matrices Q' and Z have been computed such that:

$$Q'(P_{DARE} - \lambda N_{DARE})Z = P_1 = \begin{pmatrix} P_{11} & P_{12} \\ 0 & P_{22} \end{pmatrix}$$

and

$$Q'(P_{DARE} - \lambda N_{DARE})Z = N_1 = \begin{pmatrix} N_{11} & N_{12} \\ 0 & N_{22} \end{pmatrix}$$

and the generalized eigenvalues of the pencil $P_{11} - \lambda N_{11}$ have modulii less than 1 (see below for details of how to do this).

Let

$$Z = \begin{pmatrix} Z_{11} & Z_{12} \\ Z_{21} & Z_{22} \end{pmatrix}.$$

Then the columns of

$$\begin{pmatrix} Z_{11} \\ Z_{21} \end{pmatrix}$$

form a basis for the discrete stable (deflating) subspace and the matrix $X = Z_{21}Z_{11}^{-1}$ is a unique symmetric positive semidefinite stabilizing solution of the DARE. We leave the details as an exercise (**Exercise 13.18**).

Algorithm 13.5.2.   *The Generalized Schur Algorithm for the DARE*
Inputs.
A—*An $n \times n$ matrix*
B—*An $n \times m$ matrix*
Q—*An $n \times n$ symmetric matrix*
R—*An $m \times m$ symmetric matrix*

Output. *X*—*The unique stabilizing solution of the DARE:* $A^T X A + Q - X - A^T X B (R + B^T X B)^{-1} B^T X A = 0$. **Step 1.** *Form* $P_{\text{DARE}} = \begin{pmatrix} A & 0 \\ -Q & I \end{pmatrix}$, $N_{\text{DARE}} = \begin{pmatrix} I & S \\ 0 & A^T \end{pmatrix}$.

Step 2. *Transform the pencil* $P_{\text{DARE}} - \lambda N_{\text{DARE}}$ *to the* **generalized RSF** *using the QZ algorithm, that is, find orthogonal matrices Q_1 and Z_1 such that:*

$$Q_1 P_{\text{DARE}} Z_1 = P_1 = \begin{pmatrix} P_{11} & P_{12} \\ 0 & P_{22} \end{pmatrix}$$

and

$$Q_1 N_{\text{DARE}} Z_1 = N_1 = \begin{pmatrix} N_{11} & N_{12} \\ 0 & N_{22} \end{pmatrix},$$

where P_1 is quasi-upper triangular and N_1 is upper triangular.

Step 3. *Reorder the above generalized RSF by using an orthogonal transformation, so that the pencil $P_{11} - \lambda N_{11}$ has all its eigenvalues with moduli less than 1. That is, find orthogonal matrices Q_2 and Z_2 such that $Q_2 Q_1 P_{\text{DARE}} Z_1 Z_2$ is quasi-upper triangular and $Q_2 Q_1 N_{\text{DARE}} Z_1 Z_2$ is upper triangular, and moreover, the diagonal blocks corresponding to the eigenvalues with moduli less than 1 are in the upper left quarter of these matrices.*

Step 4. *Form*

$$Z = Z_1 Z_2 = \begin{pmatrix} Z_{11} & Z_{12} \\ Z_{21} & Z_{22} \end{pmatrix}.$$

Step 5. *Compute* $X = Z_{21} Z_{11}^{-1}$, *that is, solve for X:* $X Z_{11} = Z_{21}$.

Example 13.5.2.   Consider solving the DARE with

$$A = \begin{pmatrix} 1 & 2 \\ 3 & 4 \end{pmatrix}, \qquad B = \begin{pmatrix} 1 \\ 0 \end{pmatrix}, \qquad Q = \begin{pmatrix} 1 & 0 \\ 0 & 1 \end{pmatrix}, \qquad R = 1.$$

Step 1. $P_{\text{DARE}} = \begin{pmatrix} 1 & 2 & 0 & 0 \\ 3 & 4 & 0 & 0 \\ -1 & 0 & 1 & 0 \\ 0 & -1 & 0 & 1 \end{pmatrix}$, $N_{\text{DARE}} = \begin{pmatrix} 1 & 0 & 1 & 0 \\ 0 & 1 & 0 & 0 \\ 0 & 0 & 1 & 3 \\ 0 & 0 & 2 & 4 \end{pmatrix}$.

Step 2. The generalized RSF of the pencil $P_{DARE} - \lambda N_{DARE}$ is given by: $Q_1(P_{DARE} - \lambda N_{DARE})Z_1 = P_1 - \lambda N_1$, where

$$P_1 = \begin{pmatrix} -5.5038 & 0.3093 & 0.7060 & 0.0488 \\ 0 & 1.4308 & 0.1222 & 0.0903 \\ 0 & 0 & 0.2665 & 0.2493 \\ 0 & 0 & 0 & 0.9530 \end{pmatrix},$$

$$N_1 = \begin{pmatrix} -0.9912 & -0.3540 & 0.2965 & -0.8012 \\ 0 & -0.2842 & 0.8565 & -0.5442 \\ 0 & 0 & -1.3416 & 0.9885 \\ 0 & 0 & 0 & 5.2920 \end{pmatrix}.$$

Step 3. The eigenvalues with moduli less than 1 are:

$$\frac{P_1(3, 3)}{N_1(3, 3)} = -0.1986 \quad \text{and} \quad \frac{P_1(4, 4)}{N_1(4, 4)} = 0.1801.$$

Step 4. The matrix

$$\begin{pmatrix} Z_{11} \\ Z_{21} \end{pmatrix}$$

is given by

$$\begin{pmatrix} Z_{11} \\ Z_{21} \end{pmatrix} = \begin{pmatrix} 0.5518 & -0.1074 \\ -0.3942 & 0.0847 \\ 0.6400 & 0.4499 \\ -0.3614 & 0.8825 \end{pmatrix}.$$

Step 5.

$$X = Z_{22}Z_{11}^{-1} = \begin{pmatrix} 54.9092 & 75.2247 \\ 75.2247 & 106.1970 \end{pmatrix}$$

is the stabilizing solution.

Implementational Details

The reduction to the generalized RSF can be achieved using the QZ algorithm, as described in Chapter 4.

Unfortunately, however, the eigenvalues might appear in any arbitrary order. Some reordering needs to be done. A systematic way to do this is as follows:

First, check if the last eigenvalue in the upper left quarter has modulus less than 1, if not, move it to the last position in the lower right quarter. Check the next eigenvalue now in the upper left quarter, if it does not have modulus less than 1, move it to the next position in the lower right quarter.

Note that each move is equivalent to finding a pair of orthogonal matrices such that pre- and postmultiplications by these matrices perform the necessary change.

The process can be continued until all the n eigenvalues with moduli greater than 1 have been moved to the lower right quarter and the upper left quarter contains only the eigenvalues with moduli less than 1.

There is also a slightly more efficient algorithm (Sima 1996, pp. 262–264) for ordering the eigenvalues of the pencil $P_{\text{DARE}} - \lambda \, N_{\text{DARE}}$.

There exists FORTRAN routines, developed by Van Dooren (1982) to compute deflating subspaces with specified spectrum. These subroutines are available as Algorithm 590–DSUBSP and EXCHQZ in ACM software library. Also, the LAPACK package (Anderson et al. 1999) includes the routine STGSEN, which performs a specified reordering of the eigenvalues of the generalized RSF.

Numerical stability and scaling: It can be shown (see Petkov et al. 1989) that the generalized Schur method may yield inaccurate results if the DARE is not properly scaled. For a scaling strategy that can be used to overcome this problem, see Gudmundsson et al. (1992) and Benner (1997).

The Generalized Schur Methods Without Explicit Computation of the Inverse of the Control Weighting Matrix R

All the methods we have considered so far require the explicit computation of the inverse of the control weighting matrix R. **These methods, therefore, may not yield accurate solutions when R is severely ill-conditioned.**

For example, consider the following example from Arnold and Laub (1984):

$$A = \begin{pmatrix} -0.1 & 0 \\ 0 & -0.02 \end{pmatrix}, \qquad B = \begin{pmatrix} 0.1 & 0 \\ 0.001 & 0.01 \end{pmatrix},$$
$$Q = \begin{pmatrix} 100 & 1000 \\ 1000 & 10{,}000 \end{pmatrix}, \qquad R = \begin{pmatrix} 1+\epsilon & 1 \\ 1 & 1 \end{pmatrix}.$$

The pair (A, B) is controllable. The matrix R becomes progressively ill-conditioned as $\epsilon \to 0$. The CARE with the above data was solved by Arnold and Laub, using RICPACK, **a software package especially designed for solving Riccati equations.** It was shown that the accuracy of the solution deteriorated as R became more and more ill-conditioned. For $\epsilon = 10^{-16}$, the relative accuracy was of order 10^{-1} only.

In this case, an **inverse-free generalized Schur method**, that avoids computations of R^{-1} is useful.

The Continuous-Time Case

First, we observe that the Hamiltonian eigenvalue problem $Hx = \lambda x$ associated with the CARE, can be replaced by the eigenvalue problem for the extended

$(2n + m) \times (2n + m)$ pencil:

$$P^{\mathrm{E}}_{\mathrm{CARE}} - \lambda N^{\mathrm{E}}_{\mathrm{CARE}},$$

where $P^{\mathrm{E}}_{\mathrm{CARE}} = \begin{pmatrix} A & 0 & B \\ -Q & -A^{\mathrm{T}} & 0 \\ 0 & B^{\mathrm{T}} & R \end{pmatrix}$, and $N^{\mathrm{E}}_{\mathrm{CARE}} = \begin{pmatrix} I & 0 & 0 \\ 0 & I & 0 \\ 0 & 0 & 0 \end{pmatrix}$.

(Note that this pencil does not involve R^{-1}.) The solution of the CARE can now be obtained by constructing a basis of the stable deflating subspace of this pencil. It was further observed by Van Dooren (1981) that this $(2n + m) \times (2n + m)$ pencil can be compressed, using an orthogonal factorization of the matrix

$$\begin{pmatrix} R \\ B \end{pmatrix},$$

into a $2n \times 2n$ pencil, without affecting the deflating subspaces. Thus, if

$$\begin{pmatrix} W_{11} & W_{12} \\ W_{21} & W_{22} \end{pmatrix} \begin{pmatrix} R \\ B \end{pmatrix} = \begin{pmatrix} \hat{R} \\ 0 \end{pmatrix},$$

then instead of considering the $(2n + m) \times (2n + m)$ pencil $P^{\mathrm{E}}_{\mathrm{CARE}} - \lambda N^{\mathrm{E}}_{\mathrm{CARE}}$, we consider the $2n \times 2n$ compressed pencil $P^{\mathrm{EC}}_{\mathrm{CARE}} - \lambda N^{\mathrm{EC}}_{\mathrm{CARE}}$, where

$$P^{\mathrm{EC}}_{\mathrm{CARE}} = \begin{pmatrix} W_{22}A & W_{21}B^{\mathrm{T}} \\ -Q & -A^{\mathrm{T}} \end{pmatrix} \quad \text{and} \quad N^{\mathrm{EC}}_{\mathrm{CARE}} = \begin{pmatrix} W_{22} & 0 \\ 0 & I \end{pmatrix}.$$

This leads to the following algorithm:

Algorithm 13.5.3. *Inverse-Free Generalized Schur Algorithm for the CARE.*
   Inputs.
A—An $n \times n$ matrix
B—An $n \times m$ matrix ($m \leq n$)
Q—An $n \times n$ symmetric matrix
R—An $m \times m$ symmetric matrix
   Output.
X—The unique stabilizing solution of the CARE
   Step 1. *Find the QR factorization of the matrix*

$$\begin{pmatrix} R \\ B \end{pmatrix} : W \begin{pmatrix} R \\ B \end{pmatrix} = \begin{pmatrix} \hat{R} \\ 0 \end{pmatrix}.$$

Partition

$$W = \begin{pmatrix} W_{11} & W_{12} \\ W_{21} & W_{22} \end{pmatrix},$$

where W_{22} is an $n \times n$ matrix.
   Step 2. *Form $P^{\mathrm{EC}}_{\mathrm{CARE}}$ and $N^{\mathrm{EC}}_{\mathrm{CARE}}$ as shown above.*
   Step 3. *Find the* **ordered generalized Schur form** *of the pencil $P^{\mathrm{EC}}_{\mathrm{CARE}} - \lambda N^{\mathrm{EC}}_{\mathrm{CARE}}$ using the QZ algorithm, that is, find orthogonal matrices Q_1 and Z*

such that $Q_1(P_{CARE}^{EC} - \lambda N_{CARE}^{EC})Z = \tilde{M} - \lambda \tilde{N}$; *where $\tilde{M}$ and $\tilde{N}$ are, respectively, quasi-upper and upper triangular matrices, and the n eigenvalues with negative real parts appear first.*

Step 4. *Compute* $X = Z_{21}Z_{11}^{-1}$, *where*

$$Z = \begin{pmatrix} Z_{11} & Z_{12} \\ Z_{21} & Z_{22} \end{pmatrix}.$$

Remark

- In his paper, Van Dooren (1981) described the compression technique by using an orthogonal factorization of the matrix

$$\begin{pmatrix} B \\ 0 \\ R \end{pmatrix}.$$

Instead, here we have used (an equivalent) factorization of $\begin{pmatrix} R \\ B \end{pmatrix}$ in the form $\begin{pmatrix} \hat{R} \\ 0 \end{pmatrix}$, so that a standard QR factorization algorithm can be used to achieve this factorization.

Example 13.5.3.

$$A = \begin{pmatrix} 2 & -1 \\ 1 & 0 \end{pmatrix}, \qquad B = \begin{pmatrix} 1 \\ 0 \end{pmatrix}, \qquad Q = \begin{pmatrix} 1 & 0 \\ 0 & 1 \end{pmatrix}, \qquad R = 10^{-10}.$$

Step 1.

$$W = \begin{pmatrix} -0.0000 & -1.0000 & 0 \\ -1.0000 & 0.0000 & 0 \\ 0 & 0 & 1.0000 \end{pmatrix} = \begin{pmatrix} W_{11} & W_{12} \\ W_{21} & W_{22} \end{pmatrix}.$$

Step 2.

$$P_{CARE}^{EC} = \begin{pmatrix} 0 & 0 & -1 & 0 \\ 1 & 0 & 0 & 0 \\ -1 & 0 & -2 & -1 \\ 0 & -1 & 1 & 0 \end{pmatrix}, \qquad N_{CARE}^{EC} = \begin{pmatrix} 0 & 0 & 0 & 0 \\ 0 & 1 & 0 & 0 \\ 0 & 0 & 1 & 0 \\ 0 & 0 & 0 & 1 \end{pmatrix}.$$

Step 3.

$$Z = \begin{pmatrix} -1.0000 - 0.0028i & 0.0000 + 0.0000i & -0.0000 + 0.0000i & -0.0000 + 0.0000i \\ 0.0000 + 0.0000i & 0.7071 + 0.0025i & -0.7071 + 0.0044i & 0.0000 - 0.0000i \\ -0.0000 - 0.0000i & 0.0000 + 0.0000i & 0.0000 - 0.0000i & 1.0000 - 0.0000i \\ 0.0000 + 0.0000i & 0.7071 + 0.0025i & 0.7071 - 0.0044i & -0.0000 + 0.0000i \end{pmatrix}.$$

Step 4.

$$X\,(\text{in Long Format}) = \begin{pmatrix} 0.00001000030018 & 0.00000999990018 \\ 0.00000999990018 & 1.00001000029721 \end{pmatrix}.$$

Verify: The residual norm$= 7.357 \times 10^{-8}$.

The Discrete-Time Case

The discrete problem is analogous. Here we consider the $(2n + m) \times (2n + m)$ pencil $P_{\text{DARE}}^{\text{E}} - \lambda N_{\text{DARE}}^{\text{E}}$, where

$$P_{\text{DARE}}^{\text{E}} = \begin{pmatrix} A & 0 & -B \\ -Q & -I & 0 \\ 0 & 0 & R \end{pmatrix}, \qquad \text{and} \qquad N_{\text{DARE}}^{\text{E}} = \begin{pmatrix} I & 0 & 0 \\ 0 & A^{\text{T}} & 0 \\ 0 & B^{\text{T}} & 0 \end{pmatrix}.$$

This pencil is then compressed into the $2n \times 2n$ pencil $P_{\text{DARE}}^{\text{EC}} - \lambda N_{\text{DARE}}^{\text{EC}}$, where

$$P_{\text{DARE}}^{\text{EC}} = \begin{pmatrix} W_{22}A & 0 \\ -Q & -I \end{pmatrix}, \qquad \text{and} \qquad N_{\text{DARE}}^{\text{EC}} = \begin{pmatrix} W_{22} & W_{21}B^{\text{T}} \\ 0 & A^{\text{T}} \end{pmatrix},$$

by taking the QR factorization of the matrix $\begin{pmatrix} R \\ -B \end{pmatrix}$:

$$W\begin{pmatrix} R \\ -B \end{pmatrix} = \begin{pmatrix} \tilde{R} \\ 0 \end{pmatrix}, \qquad \text{where } W = \begin{pmatrix} W_{11} & W_{12} \\ W_{21} & W_{22} \end{pmatrix}.$$

This leads to the following algorithm:

Algorithm 13.5.4. *Inverse-free Generalized Schur Method for the DARE.*
Inputs.
A—An $n \times n$ matrix
B—An $n \times m$ matrix $(m \leq n)$
Q—An $n \times n$ symmetric matrix
R—An $m \times m$ symmetric matrix.
Output.
X—The unique stabilizing solution of the DARE.

Step 1. *Find the QR factorization of* $\begin{pmatrix} R \\ -B \end{pmatrix}$, *that is, find an orthogonal matrix W such that*

$$W\begin{pmatrix} R \\ -B \end{pmatrix} = \begin{pmatrix} \tilde{R} \\ 0 \end{pmatrix}.$$

Partition

$$W = \begin{pmatrix} W_{11} & W_{12} \\ W_{21} & W_{22} \end{pmatrix}.$$

Step 2. *Form* $P_{\text{DARE}}^{\text{EC}} = \begin{pmatrix} W_{22}A & 0 \\ -Q & -I \end{pmatrix}$, $\qquad N_{\text{DARE}}^{\text{EC}} = \begin{pmatrix} W_{22} & W_{21}B^{\text{T}} \\ 0 & A^{\text{T}} \end{pmatrix}.$

Step 3. *Compute the* **ordered generalized Schur form** *of the pencil* $P_{DARE}^{EC} - \lambda N_{DARE}^{EC}$, *using the QZ algorithm followed by some ordering procedure so that the eigenvalues of moduli less than 1 appear in the first quarter, that is, find orthogonal matrices Q_1 and Z such that $Q_1(P_{DARE}^{EC} - \lambda N_{DARE}^{EC})Z = \tilde{P} - \lambda \tilde{N}$ and the n eigenvalues with moduli less than 1 appear first.*
Step 4. *Form $X = Z_{21}Z_{11}^{-1}$, where*

$$Z = \begin{pmatrix} Z_{11} & Z_{12} \\ Z_{21} & Z_{22} \end{pmatrix}.$$

Example 13.5.4.   Consider solving the DARE with **a singular matrix A**:

$$A = \begin{pmatrix} 0 & 1 \\ 0 & 0 \end{pmatrix}, \quad B = \begin{pmatrix} 0 \\ 1 \end{pmatrix}, \quad Q = \begin{pmatrix} 1 & 2 \\ 2 & 4 \end{pmatrix}, \quad R = 1.$$

Step 1.

$$W = \left(\begin{array}{c|c|c} -0.7071 & 0 & 0.7071 \\ \hline 0 & 1.0000 & 0 \\ \hline 0.7071 & 0 & 0.7071 \end{array} \right) = \left(\begin{array}{c|c} W_{11} & W_{12} \\ \hline W_{21} & W_{22} \end{array} \right).$$

Step 2.

$$P_{DARE}^{EC} = \begin{pmatrix} 0 & 1 & 0 & 0 \\ 0 & 0 & 0 & 0 \\ -1 & -2 & -1 & 0 \\ -2 & -4 & 0 & -1 \end{pmatrix},$$

$$N_{DARE}^{EC} = \begin{pmatrix} 1 & 0 & 0 & 0 \\ 0 & 0.7071 & 0 & 0.7071 \\ 0 & 0 & 0 & 0 \\ 0 & 0 & 1 & 0 \end{pmatrix}.$$

Step 3.

$$Z = \begin{pmatrix} 0.8615 & -0.2781 & 0.3731 & -0.2034 \\ -0.3290 & 0.3256 & 0.8231 & -0.3290 \\ 0.2034 & 0.3731 & 0.2781 & 0.8615 \\ 0.3290 & 0.8231 & -0.3256 & 0.9329 \end{pmatrix}.$$

Step 4.

$$X = \begin{pmatrix} 1.0000 & 2.0000 \\ 2.0000 & 4.5000 \end{pmatrix}.$$

Verify: The residual norm$= 7.772 \times 10^{-16}$.

MATLAB note: MATLAB functions **care** and **dare** solve the CARE and DARE, respectively, using generalized Schur methods, when R is well-conditioned and inverse free methods when R is ill-conditioned or singular.

13.5.3 The Matrix Sign Function Methods

Let A be an $n \times n$ matrix with **no zero** or **purely imaginary eigenvalues**. Let

$$J = X^{-1} A X = D + N,$$

be the Jordan canonical form (JCF) of A, where $D = \text{diag}(d_1, \dots, d_n)$ and N is nilpotent and commutes with D. Then the matrix sign function of A is defined as: $\text{Sign}(A) = X \text{ diag } (\text{sign}(d_1), \text{sign}(d_2), \dots, \text{sign}(d_n)) X^{-1}$, where

$$\text{sign}(d_i) = \begin{cases} 1 & \text{if Re}(d_i) > 0, \\ -1 & \text{if Re}(d_i) < 0. \end{cases}$$

Some important properties of $\text{Sign}(A)$ **are (Exercise 13.16):**

1. $\text{Sign}(A)$ has the same stable invariant subspace as A.
2. The eigenvalues of $\text{Sign}(A)$ are ± 1, depending upon the sign of the corresponding eigenvalues of A.
3. The range of $\text{Sign}(A) - I$ is the stable invariant subspace of A.
4. The eigenvectors of $\text{Sign}(A)$ are the eigenvectors and principal vectors of A.
5. $\text{Sign}(TAT^{-1}) = T \text{Sign}(A) T^{-1}$.

We will now show how sign function can be used to solve the CARE and DARE. Before doing so, let's first describe an algorithm for computing $\text{Sign}(A)$.

The basic sign function algorithm is:

$$Z_0 = A,$$
$$Z_{k+1} = \tfrac{1}{2} \left(Z_k + Z_k^{-1} \right), \quad k = 0, 1, \dots$$

It can be shown that the sequence $\{Z_k\}$ converges to $\text{Sign}(A)$ quadratically.

The initial convergence can, however, be very slow. Byers (1987) has shown that the convergence can be accelerated if Z_k is scaled by $|\det(Z_k)|^{1/n}$. For a discussion of scaling, see Kenney and Laub (1992).

Thus, a **practical algorithm** for computing $\text{Sign}(A)$ is:

Algorithm 13.5.5. *Computing $Sign(A)$*
  Input. *An $n \times n$ matrix A.*
  Output. *$Sign(A)$, the matrix sign function of A.*
  Step 1. *Set $Z_0 = A$.*
  Step 2. *For $k = 0, 1, 2, \dots$, do until convergence*
  Compute $c = |\det Z_k|^{1/n}$.
  Compute $Z_{k+1} = (1/2c) \left(Z_k + c^2 Z_k^{-1} \right)$.
  End

Stopping criteria: The algorithm can be terminated if

- the norm of the difference between two successive iterates is small enough or
- the number of iterations exceeds the maximum number prescribed.

The Matrix Sign Function Method for the CARE

The mathematical basis for the matrix sign function method for the CARE is the following theorem.

Theorem 13.5.7. *Roberts (1971). Let H be the Hamiltonian matrix (13.2.1) associated with the CARE: $XA + A^T X + Q - XSX = 0$.*
Let (A, B) be stabilizable and let (A, Q) be detectable.
Let

$$\text{Sign}(H) = \begin{pmatrix} W_{11} & W_{12} \\ W_{21} & W_{22} \end{pmatrix},$$

where W_{ij} are $n \times n$ real matrices.
  Then a stabilizing solution X of the CARE is a solution of the following **overdetermined** *consistent linear systems:*

$$\begin{pmatrix} W_{12} \\ W_{22} + I \end{pmatrix} X = - \begin{pmatrix} W_{11} + I \\ W_{21} \end{pmatrix}.$$

Proof.   Define

$$T = \begin{pmatrix} I & Y \\ 0 & I \end{pmatrix} \begin{pmatrix} I & 0 \\ -X & I \end{pmatrix} = \begin{pmatrix} I - YX & Y \\ -X & I \end{pmatrix},$$

where Y satisfies

$$(A - SX)Y + Y(A - SX)^T = -S.$$

An easy computation then shows that

$$THT^{-1} = \begin{pmatrix} A - SX & 0 \\ 0 & -(A - SX)^T \end{pmatrix}.$$

Note that $T^{-1} = \begin{pmatrix} I & -Y \\ X & I - XY \end{pmatrix}.$

Then, using Property 5 of the sign function matrix, we obtain

$$\text{Sign}(H) = T^{-1} \text{Sign} \begin{pmatrix} A - SX & 0 \\ 0 & -(A - SX)^T \end{pmatrix} T,$$

$$= T^{-1} \begin{pmatrix} -I & 0 \\ 0 & I \end{pmatrix} T \text{ (since } A - SX \text{ is asymptotically stable),}$$

$$= \begin{pmatrix} 2YX - I & -2Y \\ 2XYX - 2X & I - 2XY \end{pmatrix}.$$

Thus, $\text{Sign}(H) + I_{2n} = \begin{pmatrix} 2YX & -2Y \\ 2XYX - 2X & 2I - 2XY \end{pmatrix}$

or $\begin{pmatrix} W_{11} + I & W_{12} \\ W_{21} & W_{22} + I \end{pmatrix} = \left(\begin{pmatrix} 2Y \\ 2(XY - I) \end{pmatrix} X, - \begin{pmatrix} 2Y \\ 2(XY - I) \end{pmatrix} \right).$

Now comparing both sides of the equation, we see that X must satisfy:

$$\begin{pmatrix} W_{12} \\ W_{22} + I \end{pmatrix} X = - \begin{pmatrix} W_{11} + I \\ W_{21} \end{pmatrix}. \quad \blacksquare$$

Symmetric Version of the Matrix Sign Function Algorithm

Theorem 13.5.7 yields a computational method to solve the CARE. However, the convergence can be painfully slow. The method can be made more efficient by using the following trick (Bierman 1984; Byers 1987) in which one works only with symmetric matrices.

Define

$$W_0 = JH = \begin{pmatrix} 0 & I \\ -I & 0 \end{pmatrix} \begin{pmatrix} A & -S \\ -Q & -A^T \end{pmatrix} = \begin{pmatrix} -Q & -A^T \\ -A & S \end{pmatrix}.$$

The matrix W_0 is symmetric.

Now compute $\text{Sign}(H)$ by performing the following iterations:

$$W_{k+1} = \frac{1}{2c_k} \left(W_k + c_k^2 J W_k^{-1} J \right), \quad k = 0, 1, 2, \ldots$$

Then each W_k is symmetric and $\lim_{k \to \infty} W_k = J \text{sign}(H)$.

The parameter c_k is chosen to enhance the rate of convergence, as before.

Let

$$J \, \text{Sign}(H) = Y = \begin{pmatrix} Y_{11} & Y_{12} \\ Y_{21} & Y_{22} \end{pmatrix}.$$

Then

$$\text{Sign}(H) = \begin{pmatrix} W_{11} & W_{12} \\ W_{21} & W_{22} \end{pmatrix} = J^T Y,$$

The equation:

$$\begin{pmatrix} W_{12} \\ W_{22} + I \end{pmatrix} X = - \begin{pmatrix} W_{11} + I \\ W_{21} \end{pmatrix}$$

then becomes

$$\begin{pmatrix} Y_{22} \\ Y_{12} + I \end{pmatrix} X = \begin{pmatrix} I - Y_{21} \\ -Y_{11} \end{pmatrix}.$$

This leads to the following symmetric version of the matrix sign function algorithm for the CARE:

Algorithm 13.5.6.   *The Matrix Sign Function Algorithm for the CARE.*
   Inputs.
A—An $n \times n$ matrix
B—An $n \times m$ matrix
Q—An $n \times n$ symmetric matrix
R—An $m \times m$ symmetric matrix
ϵ—Error tolerance.
   Output.
X—The unique stabilizing solution of the CARE:
$A^T X + XA - XBR^{-1}B^T X + Q = 0$

   Step 1.
   1.1 *Form $S = BR^{-1}B^T$*

   1.2 *Define $J = \begin{pmatrix} 0 & I \\ -I & 0 \end{pmatrix}$. Form $W = JH = \begin{pmatrix} -Q & -A^T \\ -A & S \end{pmatrix}$.*
   Step 2. *For $k = 1, 2, \ldots$ do until convergence with the given tolerance ϵ*

$$c = |\det W|^{1/2n}$$

$$W = \frac{1}{2c}(W + c^2 J W^{-1} J),$$

   Step 3. *Partition $W = \begin{pmatrix} W_{11} & W_{12} \\ W_{21} & W_{22} \end{pmatrix}$, where each W_{ij} is of order n.*
   Step 4. *Form $M = \begin{pmatrix} W_{22} \\ W_{12} + I_n \end{pmatrix}$, $N = \begin{pmatrix} I - W_{21} \\ -W_{11} \end{pmatrix}$.*
   Step 5. *Solve for $X : MX = N$.*

Example 13.5.5. Consider solving the CARE using Algorithm 13.5.6 with

$$A = \begin{pmatrix} 0 & 1 \\ 0 & 0 \end{pmatrix}, \quad B = \begin{pmatrix} 0 \\ 1 \end{pmatrix}, \quad Q = \begin{pmatrix} 1 & 0 \\ 0 & 1 \end{pmatrix}, \quad R = 1.$$

Step 1. $S = \begin{pmatrix} 0 & 0 \\ 0 & 1 \end{pmatrix}, \quad H = \begin{pmatrix} 0 & 1 & 0 & 0 \\ 0 & 0 & 0 & -1 \\ -1 & 0 & 0 & 0 \\ 0 & -1 & -1 & 0 \end{pmatrix},$

$$W_0 = JH = \begin{pmatrix} -1 & 0 & 0 & 0 \\ 0 & -1 & -1 & 0 \\ 0 & -1 & 0 & 0 \\ 0 & 0 & 0 & 1 \end{pmatrix}.$$

Step 2. $W_1 = \frac{1}{2}\left(W_0 + JW_0^{-1}J\right) = \begin{pmatrix} -1 & 0 & 0 & -0.5 \\ 0 & -1 & -0.5 & 0 \\ 0 & -0.5 & 0.5 & 0 \\ -0.5 & 0 & 0 & 0.5 \end{pmatrix},$

$$c = |\det(W_1)|^{1/4} = 0.8660.$$

$$W_2 = \frac{1}{2c}(W_1 + c^2 JW_1^{-1}J) = \begin{pmatrix} -1.1547 & 0 & 0 & -0.5774 \\ 0 & -1.1548 & -0.5774 & 0 \\ 0 & -0.5774 & 0.5774 & 0 \\ -0.5774 & 0 & 0 & 0.5774 \end{pmatrix}.$$

(Note that each W_i, $i = 0, 1, 2$ is symmetric.)

Step 3. $J \mathrm{Sign}(H) = W_2 = W = \begin{pmatrix} W_{11} & W_{12} \\ W_{21} & W_{22} \end{pmatrix}.$

Step 5. $X = \begin{pmatrix} 1.7321 & 1 \\ 1 & 1.7321 \end{pmatrix}.$

Verify: The residual norm $= 9.9301 \times 10^{-16}$.

Example 13.5.6. Now consider solving the CARE using Algorithm 13.5.6 with the following data:

$$A = \begin{pmatrix} -1 & 1 & 1 \\ 0 & -2 & 0 \\ 0 & 0 & -3 \end{pmatrix}, \quad B = \begin{pmatrix} 1 \\ 1 \\ 1 \end{pmatrix}, \quad Q = \begin{pmatrix} 1 & 0 & 0 \\ 0 & 1 & 0 \\ 0 & 0 & 1 \end{pmatrix}, \quad R = 1.$$

Step 1.

$$W_0 = \begin{pmatrix} -1 & 0 & 0 & 1 & 0 & 0 \\ 0 & -1 & 0 & -1 & 2 & 0 \\ 0 & 0 & -1 & -1 & 0 & 3 \\ 1 & -1 & -1 & 1 & 1 & 1 \\ 0 & 2 & 0 & 1 & 1 & 1 \\ 0 & 0 & 3 & 1 & 1 & 1 \end{pmatrix}.$$

Step 2. After five iterations, $\|W_5 - W_4\|/\|W_4\| = 6.4200 \times 10^{-15}$ (The readers are asked to verify this by carrying out 5 iterations).

Step 3.

$$W \equiv W_5.$$

Step 5.

$$X = \begin{pmatrix} 0.3732 & 0.0683 & 0.0620 \\ 0.0683 & 0.2563 & 0.0095 \\ 0.0620 & 0.0095 & 0.1770 \end{pmatrix}.$$

Verify: The residual norm $= 3.1602 \times 10^{-16}$.

Flop-count and stability: It can be shown that Algorithm 13.5.6 requires about $4n^3$ flops per iteration. **The algorithm is not stable in general (Byers 1986b)**, unless used with an iterative refinement technique such as Newton's method (see **Section 13.5.4**).

MATCONTROL note: Algorithm 13.5.6 has been implemented in MATCONTROL function **ricsgnc**.

The Matrix Sign Function Method for the DARE

The matrix sign function method for the CARE described in the previous section can now be applied to solve the DARE by converting the symplectic matrix M to the Hamiltonian matrix H using the bilinear transformation:

$$H = (M + I)^{-1}(M - I).$$

Because A needs to be nonsingular, the method is not applicable if A is singular, and is not numerically effective when A is ill-conditioned.

Avoiding Explicit Inversion of A

The explicit inversion of A, however, may be avoided, by using the following simple trick (Gardiner and Laub 1986).

Write

$$M = N^{-1}P,$$

where

$$N = \begin{pmatrix} I & S \\ 0 & A^T \end{pmatrix} \quad \text{and} \quad P = \begin{pmatrix} A & 0 \\ -Q & I \end{pmatrix}.$$

Then it can be shown that even if A is singular, the matrix $(P + N)$ is invertible and the matrix H can be expressed as $H = (P + N)^{-1}(P - N)$.

Algorithm 13.5.7.   *The Matrix Sign Function Algorithm for the DARE.*

Inputs.
A—An $n \times n$ matrix
B—An $n \times m$ matrix
Q—An $n \times n$ symmetric matrix
R—An $m \times m$ symmetric matrix
   Output.
X—The unique stabilizing solution X of the DARE:
$A^T X A - X + Q - A^T X B (R + B^T X B)^{-1} B^T X A = 0.$
   Step 1. *Form $S = B R^{-1} B^T$,*

$$N = \begin{pmatrix} I & S \\ 0 & A^T \end{pmatrix}, \qquad P = \begin{pmatrix} A & 0 \\ -Q & I \end{pmatrix}.$$

   Step 2. *Form $H = (P + N)^{-1}(P - N)$.*
   Step 3. *Apply the matrix sign function algorithm for the CARE (Algorithm 13.5.6) with H in Step 2.*

Example 13.5.7.   Consider solving the DARE using Algorithm 13.5.7 with

$$A = \begin{pmatrix} 0 & 1 \\ 0 & 0 \end{pmatrix}, \qquad B = \begin{pmatrix} 0 \\ 1 \end{pmatrix}, \qquad Q = \begin{pmatrix} 1 & 0 \\ 0 & 1 \end{pmatrix}, \qquad R = 1.$$

Step 1. $S = \begin{pmatrix} 0 & 0 \\ 0 & 1 \end{pmatrix}, \qquad N = \begin{pmatrix} 1 & 0 & 0 & 0 \\ 0 & 1 & 0 & 1 \\ 0 & 0 & 0 & 0 \\ 0 & 0 & 1 & 0 \end{pmatrix}, \qquad P = \begin{pmatrix} 0 & 1 & 0 & 0 \\ 0 & 0 & 0 & 0 \\ -1 & 0 & 1 & 0 \\ 0 & -1 & 0 & 1 \end{pmatrix}.$

Step 2. $H = \begin{pmatrix} -0.3333 & 0.6667 & -0.6667 & 0.6667 \\ -0.6667 & 0.3333 & 0.6667 & -0.6667 \\ -1.3333 & 0.6667 & 0.3333 & 0.6667 \\ 0.6667 & -1.3333 & -0.6667 & -0.3333 \end{pmatrix}.$

Step 3. $X = \begin{pmatrix} 1 & 0 \\ 0 & 2 \end{pmatrix}.$

Verify: The residual norm $= 6.7195 \times 10^{-16}$.

MATCONTROL note: Algorithm 13.5.7 has been implemented in MATCONTROL function **ricsgnd**.

13.5.4   Newton's Methods

Recall that the classical Newton's method for finding a root x of $f(x) = 0$ can be stated as follows:

- Choose x_0, an initial approximation to x.

- Generate a sequence of approximations $\{x_i\}$ defined by

$$x_{i+1} = x_i - \frac{f(x_i)}{f'(x_i)}, \quad i = 0, 1, 2, \ldots \tag{13.5.11}$$

Then, *whenever x_0 is chosen close enough to x*, the sequence $\{x_i\}$ converges to the root x and the convergence is *quadratic* if $f'(x) \neq 0$. Newton's methods for the CARE and DARE can similarly be developed.

Newton's Method for the CARE

Consider first the CARE: $XA + A^T X - XBR^{-1}B^T X + Q = 0$.

Starting from an initial approximate solution X_0, the computed solutions are iteratively refined until convergence occurs; this is done by solving a Lyapunov equation at each iteration. The way how the Lyapunov equations arise can be explained as follows. Write $X = X_0 + (X - X_0)$. Substituting this into the CARE, we have

$$(A - BR^{-1}B^T X_0)^T X + X(A - BR^{-1}B^T X_0)$$
$$= -X_0 BR^{-1}B^T X_0 - Q + (X - X_0)BR^{-1}B^T(X - X_0).$$

Assuming that $X - X_0$ is small (i.e., the initial approximate solution is good), we can neglect the last term on the right-hand side of the above equation. Thus we obtain the following Lyapunov equation for the next approximation X_1:

$$(A - BR^{-1}B^T X_0)^T X_1 + X_1(A - BR^{-1}B^T X_0) = -X_0 BR^{-1}B^T X_0 - Q.$$

Assuming that X_1 is a better approximation than X_0 (i.e., $\| X - X_1 \| < \| X - X_0 \|$), the process can be continued until the convergence occurs, if there is convergence.

The above discussion immediately suggests the following **Newton method for the CARE**: (Kleinman 1968):

Step 1. Choose an initial approximation X_0.
Step 2. Compute $\{X_k\}$ iteratively by solving the Lyapunov equation:

$$(A - SX_k)^T X_{k+1} + X_{k+1}(A - SX_k) = -X_k SX_k - Q, \quad k = 0, 1, 2, \ldots,$$

where $S = BR^{-1}B^T$.
Step 3. Continue until and if convergence occurs.

Newton's method, as stated above, is not in the familiar form. However, the above steps can be easily reorganized to obtain Newton's method in the familiar form (see Benner (1997), Hammarling (1982) and Lancaster and Rodman (1995) for details).

To do this, let's define

$$R_C(X) = XA + A^T X - XSX + Q,$$

where $S = BR^{-1}B^T$.

Now, the Fréchet derivative of $R_C(X)$ is given by

$$R'_X(Z) :\equiv (A - SX)^T Z + Z(A - SX).$$

Thus, Newton's method for $R_C(X) = 0$ is

$$R'_{X_i}(\Delta_i) + R_C(X_i) = 0, \quad i = 0, 1, 2, \ldots$$

$$X_{i+1} = X_i + \Delta_i.$$

The above observation leads to the following Newton algorithm for the CARE.

Algorithm 13.5.8. *Newton's Method for the CARE*
Inputs.
A—An $n \times n$ matrix
B—An $n \times m$ matrix
Q—An $n \times n$ symmetric matrix
R—An $m \times m$ symmetric matrix
Output. *The set $\{X_k\}$ converging to an approximate stabilizing solution matrix X of the CARE.*
Assumptions. (A, B) *is stabilizable, $R > 0$ and the CARE has a stabilizing solution X, and is unique.*
Step 1. *Set $S = BR^{-1}B^T$.*
Step 2. *Choose an initial approximate solution $X_0 = X_0^T$ such that $A - SX_0$ is stable.*
Step 3. *Construct the sequence of solutions $\{X_i\}$ as follows:*
For $i = 0, 1, 2, \ldots$ do until convergence occurs
  3.1. *Compute $A_i = A - SX_i$*
  3.2. *Compute $R_C(X_i) = A^T X_i + X_i A + Q - X_i S X_i$*
  3.3. *Solve the Lyapunov equation for Δ_i: $A_i^T \Delta_i + \Delta_i A_i + R_C(X_i) = 0$.*
  3.4. *Compute $X_{i+1} = X_i + \Delta_i$.*
End

Remark

- The above form of Newton's method is usually known as **Newton's Method in incremental form.** This form has some computational advantages over that presented in the beginning of this section in the sense that, in general, more accurate answers can be expected. This is because, in the incremental form algorithm, we solve the Lyapunov equation for the increment Δ_i and not for the solution directly and therefore, the solution X_i will have more correct digits.

The proof of the following theorem can be found in Lancaster and Rodman (1995, pp. 232–233). It gives conditions under which the above iterates converge.

Theorem 13.5.8. *Convergence of Newton's Method for the CARE. Let the assumptions for Algorithm 13.5.8 hold. Let X_0 be an approximate stabilizing solution and let X be a unique stabilizing solution X of the CARE. Then the matrices A_i and X_i, $i = 0, 1, \ldots$, constructed by the above algorithm are such that*

(i) *All A_i are stable; that is, all iterates X_i are stabilizing.*
(ii) $X \leq \cdots \leq X_{i+1} \leq X_i \leq \cdots \leq X_1$.
(iii) $\text{Lim}_{i \to \infty} X_i = X$, *where X is the unique symmetric positive-semidefinite stabilizing solution of the CARE.*
(iv) *There exists a constant $c > 0$ such that $\|X_{i+1} - X\| \leq c\|X_i - X\|^2$, for $i \geq 1$; that is, the sequence $\{X_i\}$ converges quadratically.*

Stopping criterion: The following can be used as a stopping criterion. Stop the iteration if

I. for a certain value of k and the prescribed *tolerance* ϵ

$$\frac{\|X_{k+1} - X_k\|_F}{\|X_k\|_F} \leq \epsilon,$$

or

II. the number of iterations k exceeds a prescribed number N.

If a condition-number estimator for the CARE is available, then Criterion I can be replaced by the following more appropriate stopping criterion: Stop the iteration if

$$\frac{\|X_{k+1} - X_k\|_F}{\|X_k\|_F} \leq \mu \kappa_{\text{CARE}}^{\text{E}},$$

where $\kappa_{\text{CARE}}^{\text{E}}$ denotes an estimate of the κ_{CARE} and μ is the machine precision.

Example 13.5.8. Consider solving the CARE using Newton's method (**Algorithm 13.5.8**) with

$$A = \begin{pmatrix} -1 & 1 & 1 \\ 0 & -2 & 0 \\ 0 & 0 & -3 \end{pmatrix}, \qquad B = \begin{pmatrix} 1 \\ 1 \\ 1 \end{pmatrix}, \qquad Q = \begin{pmatrix} 1 & 0 & 0 \\ 0 & 1 & 0 \\ 0 & 0 & 1 \end{pmatrix}, \qquad R = 1.$$

Step 1. $S = \begin{pmatrix} 1 & 1 & 1 \\ 1 & 1 & 1 \\ 1 & 1 & 1 \end{pmatrix}$.

Step 2. $X_0 = \begin{pmatrix} 0.4 & 0.1 & 0.1 \\ 0.1 & 0.3 & 0.0 \\ 0.1 & 0 & 0.2 \end{pmatrix}$

Step 3.

$$i = 0$$

$$\Delta_0 = \begin{pmatrix} -0.0248 & -0.0302 & -0.0369 \\ -0.0302 & -0.0426 & 0.0103 \\ -0.0369 & 0.0103 & -0.0224 \end{pmatrix},$$

$$X_1 = X_0 + \Delta_0 = \begin{pmatrix} 0.3752 & 0.0698 & 0.0631 \\ 0.0698 & 0.2574 & 0.0103 \\ 0.0631 & 0.0103 & 0.1776 \end{pmatrix}.$$

Relative Change: $\dfrac{\|X_1 - X_0\|}{\|X_0\|} = 0.1465.$

$$i = 1.$$

$$\Delta_1 = \begin{pmatrix} -0.0020 & -0.0015 & -0.0010 \\ -0.0015 & -0.0011 & -0.0008 \\ -0.0010 & -0.0008 & -0.0005 \end{pmatrix},$$

$$X_2 = X_1 + \Delta_1 = \begin{pmatrix} 0.3732 & 0.0683 & 0.0620 \\ 0.0683 & 0.2563 & 0.0095 \\ 0.0620 & 0.0095 & 0.1770 \end{pmatrix}.$$

Relative Change: $\dfrac{\|X_2 - X_1\|}{\|X_1\|} = 0.0086.$

$$i = 2$$

$$\Delta_2 = 10^{-5} \begin{pmatrix} -0.4561 & -0.3864 & -0.2402 \\ -0.3864 & -0.3311 & -0.2034 \\ -0.2402 & -0.2034 & -0.1265 \end{pmatrix},$$

$$X_3 = X_2 + \Delta_2 = \begin{pmatrix} 0.3732 & 0.0683 & 0.0620 \\ 0.0683 & 0.2563 & 0.0095 \\ 0.0620 & 0.0095 & 0.1770 \end{pmatrix}.$$

Relative Change: $\dfrac{\|X_3 - X_2\|}{\|X_2\|} = 2.1709 \times 10^{-5}.$

MATHCONTROL note: Algorithm 13.5.8 has been implemented in MATCON-TROL function **ricnwtnc**.

Convergence: We know that there exist infinitely many X_0 for which $A - SX_0$ is stable. *The choice of proper X_0 is crucial.* If the initial solution matrix X_0 is not close enough to the exact solution X, then, as in the case of scalar Newton's method, the convergence can be painfully slow. *The method might even converge* *to a nonstabilizing solution in the presence of round-off errors.* Things might go wrong even at the first step. To see this, let's consider the following example from Kenney *et al.* (1990):

$$A = 0, \qquad B = Q = I, \qquad R = I.$$

The exact solution is $X = I$. Let $X_0 = \epsilon I$, where $\epsilon > 0$ is a small positive number. Then,

$$A - BB^T X_0 = -\epsilon I$$

is stable for all $\epsilon > 0$ and the initial error is $\|X - X_0\| = 1 - \epsilon \cong 1$ for small ϵ. However,

$$X_1 = \frac{1 + \epsilon^2}{2\epsilon} I \quad \text{and} \quad \|X - X_1\| \simeq \frac{1}{2\epsilon},$$

which is quite large. Thus, even though the errors at subsequent steps decrease, a large number of steps will be needed for the error made at the first step to damp out.

Some conditions guaranteeing convergence from the first step on have been given by Kenney *et al.* (1990). This is stated in the following Theorem (**assuming that $R = I_{m \times m}$**).

Theorem 13.5.9. *Let X_0 be an initial approximation such that $A - BB^T X_0$ is stable and assume that $\|X - X_0\| < 1/(3\|B\|^2 \|\Omega^{-1}\|)$, where $\Omega(Z) = (A - BB^T X)^T Z + Z(A - BB^T X)$, then $\|X - X_1\| \leq \|X - X_0\|$, with equality only when $X_0 = X$.*

Flop-count: Newton's method is iterative; therefore, an exact flop count cannot be given. However, if the Schur method is used to solve the Lyapunov equations at each iteration, then about $40n^3$ flops are needed per iteration.

Stability: Since the principal computational task in Newton's method is the solution of a Lyapunov matrix equation at each iteration, **the method can be shown to be stable if a numerically stable method such as the Schur method is used to solve the Lyapunov equation.** Specifically, if $\hat{X}$ is the computed solution obtained by Newton's method, then it can be shown (Petkov *et al.* 1991) that

$$\frac{\|\hat{X} - X\|_F}{\|X\|_F} \leq \mu \kappa_{\text{CARE}},$$

where κ_{care} is the condition number of the CARE. **That is, the method does not introduce more errors than what is already inherent in the problem.**

Modified Newton's Methods

Several modifications of Newton's methods for the AREs have been obtained in recent years (Benner 1990; Benner and Byers 1998; Guo 1998; Guo and Lancaster

1998; Guo and Laub 2000; etc.). We just state in the following the line search modification of Newton's method by Benner and Byers (1998).

Newton's Method with Line Search

The performance of Newton's method can be improved by using an optimization technique called **line search.**

The idea is to take a Newton step at each iteration in the direction so that $\|R_C(X_{i+1})\|_F^2$ is minimized. Thus the iteration:

$$X_{i+1} = X_i + \Delta_i$$

in Step 3 of Newton's method will be replaced by

$$X_{i+1} = X_i + t_i \Delta_i,$$

where t_i is a real scalar to be chosen so that $\|R_C(X_i + t_i \Delta_i)\|_F^2$ will be minimized.

This is equivalent to minimizing

$$f_i(t) = \text{Trace}(R_C(X_i + t\Delta_i)^T R_C(X_i + t\Delta_i)) = \text{Trace}(R_C(X_i + t\Delta_i)^2),$$
$$= \alpha_i(1 - t)^2 - 2\beta_i(1 - t)t^2 + v_i t^4,$$

where

$$\alpha_i = \text{Trace}(R_C(X_i)^2), \qquad \beta_i = \text{Trace}(R_C(X_i)V_i),$$
$$v_i = \text{Trace}(V_i^2), \quad V_i = \Delta_i S \Delta_i.$$

It can be shown (see Benner 1997; Benner and Byers 1998) that the function $f_i(t)$ has a local minimum at some value $t_i \in [0, 2]$.

We thus have the following modified Newton's algorithm.

Algorithm 13.5.9. *Newton's Method with Line Search for the CARE*
Inputs. *Same as in Algorithm* 13.5.8.
Output. *Same as in Algorithm* 13.5.8.
Assumptions. *Same as in Algorithm* 13.5.8.
Step 1. *Same as in Algorithm* 13.5.8.
Step 2. *Same as in Algorithm* 13.5.8.
Step 3. *For $i = 0, 1, 2, \ldots$ do until convergence occurs*
   3.1 *Same as in Algorithm* 13.5.8.
   3.2 *Same as in Algorithm* 13.5.8.
   3.3 *Same as in Algorithm* 13.5.8.
   3.4 *Compute $V_i = \Delta_i S \Delta_i$*
   3.5 *Compute α_i, β_i, and v_i of f_i as given above.*
   Step 3.6 *Compute $t_i \in [0, 2]$ such that $f_i(t_i) = \min_{t \in [0,2]} f_i(t)$.*
   Step 3.7 *Compute $X_{i+1} = X_i + t_i \Delta_i$.*
      End.

Example 13.5.9. The input matrices A, B, Q, and R are the same as in Example 13.5.8.

Step 1. $S = \begin{pmatrix} 1 & 1 & 1 \\ 1 & 1 & 1 \\ 1 & 1 & 1 \end{pmatrix}$.

Step 2. $X_0 = \begin{pmatrix} 0.4 & 0.1 & 0.1 \\ 0.1 & 0.3 & 0 \\ 0.1 & 0 & 0.2 \end{pmatrix}$.

Step 3. $i = 0$: $\Delta_0 = \begin{pmatrix} -0.0248 & -0.0302 & -0.0369 \\ -0.0302 & -0.0426 & 0.0103 \\ -0.0369 & 0.0103 & -0.0224 \end{pmatrix}$.

$\alpha_0 = 0.1761$, $\qquad \beta_0 = -0.0049$, $\qquad \gamma_0 = 2.1827 \times 10^{-4}$, $\qquad t_0 = 1.0286$.

$$X_1 = X_0 + t_0 \Delta_0 = \begin{pmatrix} 0.3745 & 0.0690 & 0.0620 \\ 0.0690 & 0.2562 & 0.0105 \\ 0.0620 & 0.0105 & 0.1770 \end{pmatrix}.$$

Relative change: $\|X_1 - X_0\|/\|X_0\| = 0.1507$.

$$i = 1: \Delta_1 = \begin{pmatrix} -0.0012 & -0.0006 & 0.0000 \\ -0.0006 & 0.0001 & -0.0011 \\ 0.0000 & -0.0011 & 0.0001 \end{pmatrix}$$

$\alpha_1 = 8.9482 \times 10^{-5}$, $\quad \beta_1 = -4.2495 \times 10^{-8}$, $\quad \gamma_1 = 4.9519 \times 10^{-11}$, $t_1 = 1.0005$.

$$X_2 = X_1 + t_1 \Delta_1 = \begin{pmatrix} 0.3732 & 0.0683 & 0.0620 \\ 0.0683 & 0.2563 & 0.0095 \\ 0.0620 & 0.0095 & 0.1770 \end{pmatrix}.$$

Relative change: $\|X_2 - X_1\|/\|X_1\| = 0.0038587$.

$$i = 2: \Delta_2 = 10^{-6} \begin{pmatrix} -0.1677 & -0.4428 & -0.4062 \\ -0.4428 & -0.7620 & 0.1277 \\ -0.4062 & 0.1277 & -0.2505 \end{pmatrix}.$$

$\alpha_2 = -2.9393 \times 10^{-10}$, $\quad \beta_2 = -1.0425 \times 10^{-17}$, $\quad \gamma_2 = 6.1179 \times 10^{-24}$, $t_2 = 1.0000$.

$$X_3 = X_2 + t_2 \Delta_2 = \begin{pmatrix} 0.3732 & 0.0683 & 0.0620 \\ 0.0683 & 0.2563 & 0.0095 \\ 0.0620 & 0.0095 & 0.1770 \end{pmatrix}.$$

Relative change: $\|X_3 - X_2\|/\|X_2\| = 2.4025 \times 10^{-6}$.

$$i = 3: \quad \Delta_3 = 10^{-12} \begin{pmatrix} -0.1593 & -0.0972 & 0.0319 \\ -0.0972 & -0.0286 & -0.1791 \\ 0.00319 & -0.1791 & 0.0308 \end{pmatrix}.$$

$\alpha_3 = 2.4210 \times 10^{-24}$, $\quad \beta_3 = -1.4550 \times 10^{-37}$, $\quad \gamma_3 = 2.4612 \times 10^{-50}$, $t_3 = 1.0000$.

$$X_4 = X_3 + t_3 \Delta_3 = \begin{pmatrix} 0.3732 & 0.0683 & 0.0620 \\ 0.0683 & 0.2563 & 0.0095 \\ 0.0620 & 0.0095 & 0.1770 \end{pmatrix}.$$

Relative change: $\|X_4 - X_3\|/\|X_3\| = 5.5392 \times 10^{-13}$.

Theorem 13.5.10. *Convergence of Newton's Method with Line Search for the CARE. If (A, B) is a controllable pair, and if the step sizes t_i are bounded away from zero, then Newton's method with the line search (**Algorithm 13.5.9**) converges to the stabilizing solution.*

Proof. See Benner and Byers (1998), Guo and Laub (2000). ■

Flop-count: Algorithm 13.5.9 is slightly more expensive (about 8% to the cost of one Newton step) than Algorithm 13.5.8. **However, one saves about one iteration step out of 15;** often much more, but seldom less.

MATCONTROL note: Algorithm 13.5.9 has been implemented in MATCONTROL function **ricnwlsc**.

Newton's method for the DARE

Newton's method for the DARE:

$$A^T X A - X + Q - A^T X B (R + B^T X B)^{-1} B^T X A = 0$$

is analogous. It is based on successive solutions of **Stein equations (discrete-time Lyapunov equations)** associated with the discrete-time system. We state the algorithm below without detailed discussions. The algorithm was originally developed by Hewer (1971). See also Kleinman (1974).

Algorithm 13.5.10. *Newton's Method for the DARE*
    Inputs. *A—An $n \times n$ matrix*
           B—An $n \times m$ matrix
           Q—An $n \times n$ symmetric matrix
           R—An $m \times m$ symmetric matrix

Output. *The set $\{X_k\}$ converging to the unique stabilizing solution X of the DARE:*

$$R_D(X) = A^T X A - X + Q - A^T X B (R + B^T X B)^{-1} B^T X A = 0.$$

Assumptions. *(i) (A, B) is discrete-stabilizable (ii) $R \geq 0$, (iii) A stabilizing solution X exists and is unique, and (iv) $R + B^T X B > 0$.*

Step 1. *Choose $X_0 = X_0^T$ such that $A - B(R + B^T X_0 B)^{-1} B^T X_0 A$ is a discrete-stable matrix, that is, it has all its eigenvalues inside the unit circle.*

Step 2. *For $i = 0, 1, 2, \ldots$ do until convergence.*

    2.1 *Compute $K_i = (R + B^T X_i B)^{-1} B^T X_i A$*

    2.2 *Compute $A_i = A - B K_i$*

    2.3 *Compute $R_D(X_i) = A^T X_i A - X_i + Q - A^T X_i B (R + B^T X_i B)^{-1} B^T X_i A$*

    2.4 *Solve the discrete-time Lyapunov equation (Stein equation) for Δ_i:*
$$A_i^T \Delta_i A_i - \Delta_i + R_D(X_i) = 0$$

    2.5 *Compute $X_{i+1} = X_i + \Delta_i$.*

  End

The following theorem gives conditions under which the sequence $\{X_i\}$ converges. The proof of this theorem can be found in Lancaster and Rodman (1995, pp. 308–310), in case R is nonsingular. See also Benner (1997), Mehrmann (1991).

Theorem 13.5.11. *Convergence of Newton's Method for the DARE. Let the assumptions of Algorithm 13.5.10 hold. Let X_0 be a stabilizing approximate solution of the DARE. Then the matrices A_i and X_i, constructed by the above algorithm, are such that*

  (i)   All A_i are discrete-stable,

  (ii) $\lim_{i \to \infty} X_i = X$, where X is the unique symmetric positive semidefinite discrete-stabilizing solution of the DARE.

  (iii) $X \leq \cdots \leq X_{i+1} \leq X_i \leq \cdots \leq X_1$

  (iv) There exists a constant $c > 0$ such that $\|X_{i+1} - X\| \leq c\|X_i - X\|^2$, $i \geq 1$, that is, the sequence $\{X_i\}$ converges quadratically.

Stopping criterion: The same stopping criteria as in the case of Newton's method for the CARE can be used.

Example 13.5.10. Consider solving the DARE using Algorithm 13.5.10 with

$$A = \begin{pmatrix} -1 & 1 & 1 \\ 0 & -2 & 0 \\ 0 & 0 & -3 \end{pmatrix}, \quad B = \begin{pmatrix} 1 \\ 1 \\ 1 \end{pmatrix}, \quad R = 1, \quad Q = \begin{pmatrix} 1 & 0 & 0 \\ 0 & 1 & 0 \\ 0 & 0 & 1 \end{pmatrix}.$$

Step 1. $\quad X_0 = \begin{pmatrix} 1 & -5 & 10 \\ -5 & 1600 & -2000 \\ 10 & -2000 & 2700 \end{pmatrix}.$

Step 2. $i = 0$

The eigenvalues of $A - B(R + B^T X_0 B)^{-1} B^T X_0 A$ are $-0.8831 \pm j0.2910$, -0.0222.
Then X_0 is a discrete-stabilizing approximate solution of the DARE.

$$K_0 = \begin{pmatrix} -0.0192 & 2.6154 & -6.8077 \end{pmatrix},$$

$$A_0 = \begin{pmatrix} -0.9808 & -1.6154 & 7.8077 \\ 0.0192 & -4.6154 & 6.8077 \\ 0.0192 & -2.6154 & 3.8077 \end{pmatrix},$$

$$X_1 = 10^4 \begin{pmatrix} 0.0008 & -0.0137 & 0.0167 \\ -0.0137 & 0.6808 & -0.9486 \\ 0.0165 & -0.9486 & 1.3364 \end{pmatrix}.$$

Relative change: $\dfrac{\|X_1 - X_0\|}{\|X_0\|} = 3.7654.$

$$i = 1.$$

$$K_1 = \begin{pmatrix} -0.0301 & 4.4699 & -9.5368 \end{pmatrix},$$

$$A_1 = \begin{pmatrix} -0.9699 & -3.4699 & 10.5368 \\ 0.0301 & -6.4699 & 9.5368 \\ 0.0301 & -4.4699 & 6.5368 \end{pmatrix},$$

$$X_2 = 10^3 \begin{pmatrix} 0.0067 & -0.0893 & 0.1029 \\ -0.0893 & 2.0297 & -2.5658 \\ 0.1029 & -2.5658 & 3.3125 \end{pmatrix}.$$

Relative change: $\dfrac{\|X_2 - X_1\|}{\|X_1\|} = 0.7364.$

$$i = 2.$$

$$K_2 = \begin{pmatrix} -0.0826 & 5.1737 & -10.2938 \end{pmatrix},$$

$$A_2 = \begin{pmatrix} -0.9174 & -4.1737 & 11.2938 \\ 0.0826 & -7.1737 & 10.2938 \\ 0.0826 & -5.1737 & 7.2938 \end{pmatrix},$$

$$X_3 = 10^3 \begin{pmatrix} 0.0054 & -0.0670 & 0.0767 \\ -0.0670 & 1.6234 & -2.0796 \\ 0.0767 & -2.0796 & 2.7283 \end{pmatrix}.$$

Relative change: $\dfrac{\|X_3 - X_2\|}{\|X_2\|} = 0.1862.$

The relative changes continue to decrease from this step onwards.

$$X_7 = 10^3 \begin{bmatrix} 0.0053 & -0.0658 & 0.0751 \\ -0.0658 & 1.5943 & -2.0428 \\ 0.0751 & -2.0428 & 2.6817 \end{bmatrix}.$$

For $i = 6$, relative change: $\|X_7 - X_6\|/\|X_6\|$ is 2.3723×10^{-15}.

MATCONTROL note: Algorithm 13.5.10 has been implemented in MATCON-TROL function **ricnwtnd**.

Newton's Method with Line Search for the DARE

Algorithm 13.5.10 can be modified in a similar way as in case of the CARE to include the line search.

The function $f_i(t)$ to be minimized in this case is given by:

$$f_i(t) = \alpha_i(1-t)^2 - 2\beta_i(1-t)t^2 + \gamma_i t^4,$$

where $\alpha_i = \text{Trace}(R_d(X_i)^2)$, $\beta_i = \text{Trace}(R_d(X_i)V_i)$, $\gamma_i = \text{Trace}(V_i^2)$, and $V_i = A_i^T \Delta_i B(R + B^T X_i B)^{-1} B^T \Delta_i A_i$

For details, see Benner (1997).

> **Algorithm 13.5.11.** *Newton's Method with Line Search for the DARE*
> **Inputs.** *Same as in Algorithm* 13.5.10.
> **Output.** *Same as in Algorithm* 13.5.10.
> **Assumptions.** *Same as in Algorithm* 13.5.10.
> **Step 1.** *Same as in Algorithm* 13.5.10.
> **Step 2.** *For* $k = 0, 1, 2, \ldots$ *do until convergence*
>   **2.1** *Same as in Algorithm* 13.5.10
>   **2.2** *Same as in Algorithm* 13.5.10
>   **2.3** *Same as in Algorithm* 13.5.10
>   **2.4** *Same as in Algorithm* 13.5.10
>   **2.5** *Compute* $S_i = B(R + B^T X_i B)^{-1}B^T$
>   **2.6** *Compute* $V_i = A_i^T \Delta_i S_i \Delta_i A_i$
>   **2.7** *Compute the coefficients* $\alpha_i, \beta_i,$ *and* γ_i *of* $f_i(t)$ *as above*
>   **2.8** *Compute* $t_i \in [0, 2]$ *such that* $f_i(t_i) = \min_{t\in[0,2]} f_i(t)$
>   **2.9** $X_{i+1} = X_i + t_i\Delta_i.$
> *End*

Flop-count: The algorithm is again just slightly more expensive than Algorithm 13.5.10. The additional cost of forming V_i, the coefficients of f_i, a local minimizer t_i of f_i and scaling Δ_i by t_i is cheap as compared to $O(n^3)$ flops required for other computations.

Convergence: The line search procedure can sometimes significantly improve the convergence behavior of Newton's method. For details, see Benner (1997).

Example 13.5.11. Consider solving the DARE using Algorithm 13.5.11 with

$$A = \begin{pmatrix} -1 & 1 & 1 \\ 0 & -2 & 0 \\ 0 & 0 & -3 \end{pmatrix}, \qquad B = \begin{pmatrix} 1 \\ 1 \\ 1 \end{pmatrix}, \qquad Q = \begin{pmatrix} 1 & 0 & 0 \\ 0 & 1 & 0 \\ 0 & 0 & 1 \end{pmatrix},$$

$$R = 1.$$

Step 1. $X_0 = \begin{pmatrix} 1 & -5 & 10 \\ -5 & 1600 & -2000 \\ 10 & -2000 & 2700 \end{pmatrix}.$

Step 2. $i = 0,$ $\Delta_0 = 10^4 \begin{pmatrix} 0.0007 & -0.0132 & 0.0157 \\ -0.0132 & 0.5208 & -0.7486 \\ 0.0157 & -0.7486 & 1.0664 \end{pmatrix},$

$\alpha_0 = 9.7240 \times 10^7,$ $\beta_0 = 5.5267 \times 10^8,$ $\gamma_0 = 3.1518 \times 10^9,$
$t_0 = 0.3402.$

$X_1 = X_0 + t_0 D_0 = 10^3 \begin{pmatrix} 0.0034 & -0.0500 & 0.0635 \\ -0.0500 & 3.3718 & -4.5471 \\ 0.0635 & -4.5471 & 6.3283 \end{pmatrix}.$

Relative change: $\|X_1 - X_0\|/\|X_0\| = 1.2812.$

Step 3. $i = 1,$ $\Delta_1 = 10^3 \begin{pmatrix} 0.0029 & -0.0405 & 0.0431 \\ -0.0405 & -1.1655 & 1.7233 \\ 0.0431 & 1.7233 & -2.6498 \end{pmatrix},$

$\alpha_1 = 1.1123 \times 10^7,$ $\beta_1 = 1.7963 \times 10^6,$ $\gamma_1 = 3.0428 \times 10^5,$
$t_1 = 0.8750.$

$X_2 = X_1 + t_1 \Delta_1 = 10^3 \begin{pmatrix} 0.0059 & -0.0854 & 0.1012 \\ -0.0854 & 2.3520 & -3.0392 \\ 0.1012 & -3.0392 & 4.0097 \end{pmatrix}.$

Relative change: $\|X_2 - X_1\|/\|X_1\| = 0.3438.$

$i = 2,$ $\Delta_2 = 10^{-3} \begin{pmatrix} -0.0006 & 0.0196 & -0.0261 \\ 0.0196 & -0.7570 & 0.9955 \\ -0.0261 & 0.9955 & -1.3267 \end{pmatrix},$

$\alpha_2 = 1.9251 \times 10^5,$ $\beta_2 = -157.2798,$ $\gamma_2 = 0.1551,$
$t_2 = 1.0008.$

$X_3 = X_2 + t_2 \Delta_2 = 10^3 \begin{pmatrix} 0.0053 & -0.0658 & 0.0751 \\ -0.0658 & 1.5944 & -2.0429 \\ 0.0751 & -2.0429 & 2.6819 \end{pmatrix}.$

Relative change: $\|X_3 - X_2\|/\|X_2\| = 0.3283$.

$$i = 3, \qquad \Delta_3 = \begin{pmatrix} -0.0003 & 0.0024 & -0.0011 \\ 0.0024 & -0.0481 & 0.1094 \\ -0.0011 & 0.1094 & -0.2202 \end{pmatrix},$$

$$\alpha_3 = 0.0912, \quad \beta_3 = -2.8785 \times 10^{-5}, \quad \gamma_3 = 1.6525 \times 10^{-8}, \quad t_3 = 1.0003.$$

$$X_4 = X_3 + t_3\Delta_3 = 10^3 \begin{pmatrix} 0.0053 & -0.0658 & 0.0751 \\ -0.0658 & 1.5943 & -2.0428 \\ 0.0751 & -2.0428 & 2.6817 \end{pmatrix}.$$

Relative change: $\|X_4 - X_3\|/\|X_3\| = 6.4273 \times 10^{-5}$.

The relative changes continue to decrease after each iteration. For example, for $i = 5$, we have

Relative change: $\|X_6 - X_5\|/\|X_5\| = 1.0906 \times 10^{-13}$, and **Relative Residual** = 3.2312×10^{-11}.

MATCONTROL note: Algorithm 13.5.11 has been implemented in MATCON-TROL function **ricnwlsd**.

Newton's Method as an Iterative Refinement Technique

Newton's method is often used as an **iterative refinement technique.** First, a direct robust method such as the Schur method or the matrix sign function method is applied to obtain an approximate solution and this approximate solution is then refined by using a few iterative steps of Newton's method. **For higher efficiency, Newton's method with the line search (Algorithm 13.5.9 for the CARE and Algorithm 13.5.11 for the DARE) should be preferred over Newton's method.**

13.6 THE SCHUR AND INVERSE-FREE GENERALIZED SCHUR METHODS FOR THE DESCRIPTOR RICCATI EQUATIONS

We have seen in Chapter 5 that several practical applications give rise to the descriptor systems:

$$E\dot{x}(t) = Ax(t) + Bu(t) \quad \text{(Continuous-time)}, \tag{13.6.1}$$

$$Ex_{k+1} = Ax_k + Bu_k \quad \text{(Discrete-time)}. \tag{13.6.2}$$

The AREs **associated** with the above systems, respectively, are:

$$A^\mathrm{T} X E + E^\mathrm{T} X A - E^\mathrm{T} X B R^{-1} B^\mathrm{T} X E + Q = 0, \qquad (13.6.3)$$

and

$$E^\mathrm{T} X E = A^\mathrm{T} X A - A^\mathrm{T} X B (B^\mathrm{T} X B + R)^{-1} B^\mathrm{T} X A + Q. \qquad (13.6.4)$$

The Riccati equations (13.6.3) and (13.6.4) will be, respectively, called as the **descriptor continuous-time algebraic Riccati equation** (DCARE) and the **descriptor discrete-time algebraic Riccati equation** (DDARE).

Most of the methods, such as the Schur method, the matrix sign function method, and Newton's method, can be easily extended to solve DCARE and DDARE.

Below we state how the generalized Schur methods and the inverse-free generalized Schur methods can be extended to solve these equations. The derivations of the others are left as **Exercises.** See Bender and Laub (1985, 1987), Benner (1997), Laub (1991), Mehrmann (1988), Benner *et al.* (1999a) etc. in this context. For descriptor discrete-time Lyapunov and Riccati equations, see Zhang *et al.* (1999).

13.6.1 The Generalized Schur Method for the DCARE

The matrix pencil associated with the DCARE is

$$P_{\mathrm{DCARE}} - \lambda N_{\mathrm{DCARE}} = \begin{pmatrix} A & -S \\ -Q & -A^\mathrm{T} \end{pmatrix} - \lambda \begin{pmatrix} E & O \\ O & E^\mathrm{T} \end{pmatrix},$$

where $S = BR^{-1}B^\mathrm{T}$.

The Schur method for the DCARE, then, can be easily developed by transforming the above pencil to the **Ordered** RSF using the QZ iteration algorithm (Chapter 4). Thus, if Q_1 and Z_1 are orthogonal matrices such that

$$Q_1 P_{\mathrm{DCARE}} Z_1 = \begin{pmatrix} L_{11} & L_{12} \\ O & L_{22} \end{pmatrix}, \qquad Q_1 N_{\mathrm{DCARE}} Z_1 = \begin{pmatrix} N_{11} & N_{12} \\ O & N_{22} \end{pmatrix},$$

where $Q_1 P_{\mathrm{DCARE}} Z_1$ is upper quasi-triangular, $Q_1 N_{\mathrm{DCARE}} Z_1$ is upper triangular, and $L_{11} - \lambda N_{11}$ is stable, then the columns of $\begin{pmatrix} Z_{11} \\ Z_{21} \end{pmatrix}$, where

$$Z_1 = \begin{pmatrix} Z_{11} & Z_{12} \\ Z_{21} & Z_{22} \end{pmatrix},$$

span the stable deflating subspace. So, the matrix $X = Z_{21} Z_{11}^{-1}$ is a solution of the DCARE.

MATLAB note: MATLAB function **care** in the form:

$$[X, L, G, rr] = \text{care}(A, B, Q, R, E)$$

solves the DCARE.

Here $G = R^{-1}(B^T X E)$; the gain matrix, $L = \text{eig}(A - BG, E)$, and $rr =$ the Frobenius norm of the relative residual matrix.

13.6.2   The Inverse-Free Generalized Schur Method for the DCARE

In case R is singular or nearly singular, one needs to use the inverse-free generalized Schur method. The extended pencil to be considered in this case is

$$\begin{pmatrix} A & 0 & B \\ -Q & -A^T & 0 \\ 0 & B^T & R \end{pmatrix} - \lambda \begin{pmatrix} E & 0 & 0 \\ 0 & E^T & 0 \\ 0 & 0 & 0 \end{pmatrix}.$$

This extended pencil is then compressed into a $2n \times 2n$ pencil in the same way as in Algorithm 13.5.3 and the rest of the procedure is the same as that algorithm.

13.6.3   The Inverse-Free Generalized Schur Method for the DDARE

The matrix pencil associated with the DDARE is

$$\begin{pmatrix} A & 0 \\ -Q & E^T \end{pmatrix} - \lambda \begin{pmatrix} E & S \\ 0 & A^T \end{pmatrix}, \quad \text{where } S = BR^{-1}B^T.$$

The extended pencil for the **Inverse-free generalized Schur method for the DDARE** is

$$\begin{pmatrix} A & 0 & -B \\ -Q & E^T & 0 \\ 0 & 0 & R \end{pmatrix} - \lambda \begin{pmatrix} E & 0 & 0 \\ 0 & A^T & 0 \\ 0 & B^T & 0 \end{pmatrix}.$$

The pencil is now compressed into a $2n \times 2n$ pencil as in Algorithm 13.5.4 and the rest of the steps of Algorithm 13.5.4 is then followed.

MATLAB note: The MATLAB function **dare** in the form $[X, L, G, rr] = \text{dare}(A, B, Q, R, E)$ solves the DDARE. Here $G = (B^T X B + R)^{-1} B^T X A$, $L = \text{eig}(A - BG, E)$, and $rr =$ the Frobenius norm of the relative residual matrix.

13.7   CONCLUSIONS AND TABLE OF COMPARISONS

In this section, we present Table 13.1 which compares the different methods discussed in this chapter and gives a guideline for practical uses of these methods, based on this comparative study. We only present data for the CARE. A similar table can be set up for the DARE as well. However, the comments made about the Schur method for the CARE are not valid for the DARE, because *the Schur method*

Table 13.1: A table of comparisons of different methods for the CARE

Method	Efficiency, stability, and convergence properties	Remarks
The Eigenvector and the Generalized Eigenvector Methods	The methods are in general **not numerically stable** (they become unstable when the Hamiltonian matrix has nearly multiple eigenvalues).	**Not recommended to be used in practice.**
The Schur Method	**Stable in practice.**	**Widely used.**
The Symplectic Hamiltonian– Schur Method	**Stable and structure-preserving.** Requires less computations and storage for problems of size greater than 20.	**Works in the single-input.**
The Extended Hamiltonian– Schur Method	**Stable and structure-preserving. More-efficient** than the **Schur-method.**	**Works in the multi-input case.**
Newton's Method	Convergence is ultimately quadratic if the initial approximation is close to the solution slow initial convergence can be improved by using Newton's methods with line search.	Usually used as an **iterative refinement procedure.**
The Matrix Sign Function Method	**Not stable** in general. Though iterative in nature; unlike Newton's method, it does not require the knowledge of a stabilizing initial guess.	Simple to use and is structure preserving. **Recommended to be used in conjunction with Newton's method, with line search.**
The Generalized Schur Method	**Stable in practice.**	Does not work if the control weighting matrix R is singular. Even if R is theoretically nonsingular, the method should not be used if it is ill-conditioned.
The Inverse-Free Generalized Schur Method	**Stable in practice**	**The best way to solve the CARE is when R is nearly singular.**

for the DARE does not work when A is singular and is expected to give inaccurate results when A is theoretically nonsingular, but is computationally nearly singular.

Conclusions and Recommendations

In conclusion, the following recommendations are made: **For the CARE:** *The Schur method (**Algorithm** 13.5.1), the extended Hamiltonian–Schur method or the matrix sign function (**Algorithm** 13.5.6) method followed by Newton's iteration with line search (**Algorithm** 13.5.9) is recommended. If R is singular or nearly singular, then the inverse-free generalized Schur method (**Algorithm** 13.5.3) should be used in place of the Schur method or the matrix sign function method.*

For the DARE: *The inverse-free generalized Schur method (**Algorithm** 13.5.4) or the matrix sign function method (**Algorithm** 13.5.7) followed by Newton's method with line search (**Algorithm** 13.5.11) is recommended. However, the matrix sign function method should be avoided if R is nearly singular.*

13.8   SOME SELECTED SOFTWARE

13.8.1   MATLAB Control System Toolbox

Matrix equation solvers.

care   Solve continuous algebraic Riccati equations
dare   Solve discrete algebraic Riccati equations.

13.8.2   MATCONTROL

RICEIGC	The eigenvector method for the continuous-time Riccati equation
RICSCHC	The Schur method for the continuous-time Riccati equation
RICSCHD	The Schur method for the discrete-time Riccati equation
RICGEIGD	The generalized eigenvector method for the discrete-time Riccati equation
RICNWTNC	Newton's method for the continuous-time Riccati equation
RICNWTND	Newton's method for the discrete-time Riccati equation
RICSGNC	The matrix sign function method for the continuous-time Riccati equation
RICSGND	The matrix sign function method for the discrete-time Riccati equation
RICNWLSC	Newton's method with line search for the continuous-time Riccati equation
RICNWLSD	Newton's method with line search for the discrete-time Riccati equation.

13.8.3 CSP-ANM

Solutions of the AREs:

- The Schur method is implemented as `RiccatiSolve` $[a, b, q, r,$ `SolveMethod` $\rightarrow$ `SchurDecomposition`] (continuous-time case) and `DiscreteRiccatiSolve` $[a, b, q, r,$ `SolveMethod` $\rightarrow$ `SchurDecomposition`] (**discrete-time case**).
- Newton's method is implemented as `RiccatiSolve` $[a, b, q, r,$ `SolveMethod` $\rightarrow$ `Newton`, `InitialGuess` $\rightarrow w_0$] (discrete-time case).
- The matrix sign function method is implemented as `RiccatiSolve` $[a, b, q, r,$ `SolveMethod` $\rightarrow$ `MatrixSign`] (continuous-time case) and `DiscreteRiccatiSolve` $[a, b, q, r,$ `SolveMethod` $\rightarrow$ `MatrixSign`] (discrete-time case).
- The inverse-free method based on generalized eigenvectors is implemented as `RiccatiSolve` $[a, b, q, r,$ `SolveMethod` $\rightarrow$ `Generalized Eigendecomposition`] (continuous-time case) and `DiscreteRiccatiSolve` $[a, b, q, r,$ `SolveMethod` $\rightarrow$ `GeneralizedEigendecomposition`] (discrete-time case).
- The inverse-free method based on generalized Schur decomposition is implemented as `RiccatiSolve` $[a, b, q, r,$ `SolveMethod` $\rightarrow$ `GeneralizedSchurDecomposition`] (continuous-time case) and `DiscreteRiccatiSolve` $[a, b, q, r,$ `Solvemethod` $\rightarrow$ `GeneralizedSchurDecomposition`] (discrete-time case).

13.8.4 SLICOT

Riccati equations

SB02MD	Solution of AREs (Schur vectors method)
SB02MT	Conversion of problems with coupling terms to standard problems
SB02ND	Optimal state feedback matrix for an optimal control problem
SB02OD	Solution of AREs (generalized Schur method)
SB02PD	Solution of continuous algebraic Riccati equations (matrix sign function method) with condition and forward error bound estimates
SB02QD	Condition and forward error for continuous Riccati equation solution
SB02RD	Solution of AREs (refined Schur vectors method) with condition and forward error bound estimates
SB02SD	Condition and forward error for discrete Riccati equation solution

13.8.5   MATRIX$_X$

Purpose: Solve Riccati equation. Using the option "DISC" solves the discrete Riccati equation.

Syntax: [EV, KC] = RICCATI (S, Q, NS, 'DISC')
[EV, KC, P] = RICCATI (S, Q, NS, 'DISC')

Purpose: Solves the indefinite ARE: $A'P + PA - PRP + Q = 0$

Syntax: [P, SOLSTAT] = SINGRICCATI (A, Q, R { ,TYPE})

13.9   SUMMARY AND REVIEW

As we have seen in Chapters 10 and 12 that the AREs (13.1.1) and (13.1.2.) and their variations arise in many areas of control systems design and analysis, such as:

- The LQR and LQG designs
- Optimal state estimation (Kalman filter)
- H_∞-control
- Spectral factorizations (not described in this book, see Van Dooren 1981).

Existence and Uniqueness of Stabilizing Solution

Let $Q \geq 0$ and $R > 0$. If (A, B) is stabilizable and (A, Q) is detectable, then the CARE admits a unique symmetric positive semidefinite stabilizing solution (**Theorem 13.2.6**).

Conditioning of the Riccati Equations

The absolute and the relative condition numbers of the CARE have been identified using a perturbation result (**Theorem 13.4.1**).

An approximate condition number of the CARE, using a first-order estimate is Byers' condition number (in **Frobenius norm**):

$$\kappa_{\text{CARE}}^{\text{B}} = \frac{\|\Omega^{-1}\| \|Q\| + \|\Theta\| \|A\| + \|\Pi\| \|S\|}{\|X\|},$$

where X is the stabilizing solution of the CARE and Ω, Π, and Θ are defined by (13.4.4)–(13.4.6), and $\|\Omega^{-1}\|_{\text{F}} = 1/\text{sep}(A_{\text{C}}^{\text{T}}, -A_{\text{C}})$, where $A_{\text{C}} = A - SX$, $S = BR^{-1}B^{\text{T}}$.

The quantities $\|\Omega^{-1}\|$, $\|\Theta\|$, and $\|\Pi\|$ are computationally intensive. Upper bounds of κ^{B}_{CARE} can be obtained by solving the following Lyapunov equations:

$$(A - SX)^{T} H_k + H_k(A - SX) = -X^k, \quad k = 0, 1, 2.$$

The large norms of these matrices (relative to the stabilizing solution X), in general, indicate that the CARE is ill-conditioned.
The condition number of the DARE is given by (13.4.16).
A first-order estimator for the condition number of the DARE is

$$\kappa^{E}_{DARE} = \frac{2\|A\|^2_F\|Q\|_F/\|X\|_F + \|A\|^2_F\|S\|_F\|X\|_F}{\text{sep}(A^T_d, A_d)},$$

where $A_d = A - B(R + B^T X B)^{-1} B^T X A$, $S = B R^{-1} B^T$. The quantity $\text{sep}(A^T_d, A_d)$ can be determined as the minimum singular value of the matrix $A^T_d \otimes A^T_d - I^2_n$.

Numerical Methods for the Riccati Equations

The numerical methods for the Riccati equations discussed here can be broadly classified into the following three classes:

- Invariant subspace methods
- Deflating subspace methods
- Newton's methods.

A basic idea of finding a stabilizing solution of the CARE (DARE), using eigenvector and Schur methods is to construct a basis for the stable invariant subspace of the Hamiltonian matrix H (symplectic matrix M). Such a basis can be constructed using the eigenvectors or the Schur vectors of the Hamiltonian matrix H (the symplectic matrix M). The eigenvector matrix can be ill-conditioned if the matrix H (the matrix M) is nearly defective and, therefore, **the eigenvector approach is not recommended to be used in practice**. The Schur method is preferable to the eigenvector method. If

$$U^T H U = \begin{pmatrix} T_{11} & T_{12} \\ 0 & T_{22} \end{pmatrix}$$

is the **ordered** RSF of H, and the eigenvalues with negative real parts are contained in T_{11}, then $X = U_{21} U^{-1}_{11}$ is the unique stabilizing solution of the CARE, where

$$U = \begin{pmatrix} U_{11} & U_{12} \\ U_{21} & U_{22} \end{pmatrix}.$$

The Schur method for the DARE can be similarly developed by finding an ordered RSF of the symplectic matrix M. However, since computation of the matrix M

requires the explicit inversion of A, **the Schur method for the DARE does not work if A is singular or can be problematic if A is theoretically nonsingular, but is computationally singular.** In such cases, a deflating subspace method should be used.

The idea behind the generalized eigenvector and Schur vector methods is basically the same as that of an invariant subspace method except that the solution of the Riccati equation is now found by computing a basis for the deflating subspace of a matrix pencil. For the CARE, the pencil is $P_{\text{CARE}} - \lambda N_{\text{CARE}}$, where

$$P_{\text{CARE}} = \begin{pmatrix} A & -S \\ -Q & A^{\mathrm{T}} \end{pmatrix}, \qquad N_{\text{CARE}} = \begin{pmatrix} I & 0 \\ 0 & I \end{pmatrix}.$$

For the DARE, the matrices of the pencil are

$$P_{\text{DARE}} = \begin{pmatrix} A & 0 \\ -Q & I \end{pmatrix}, \qquad N_{\text{DARE}} = \begin{pmatrix} I & S \\ 0 & A^{\mathrm{T}} \end{pmatrix}.$$

Again, for reasons stated above, the **generalized Schur decomposition using the QZ algorithm should be used to compute such a basis. See Section 13.5.2 for details. The eigenvector approach should be avoided**.

Both the Schur methods and the generalized Schur methods require an explicit inversion of the matrix R. In case R is ill-conditioned with respect to matrix inversion, these methods may not give accurate solutions. The difficulties can be overcome by using an extended $(2n + m) \times (2n + m)$ pencil.

For the CARE, the extended pencil is $P_{\text{CARE}}^{\text{E}} - \lambda N_{\text{CARE}}^{\text{E}}$, where

$$P_{\text{CARE}}^{\text{E}} = \begin{pmatrix} A & 0 & B \\ -Q & -A^{\mathrm{T}} & 0 \\ 0 & B^{\mathrm{T}} & R \end{pmatrix}, \qquad N_{\text{CARE}}^{\text{E}} = \begin{pmatrix} I & 0 & 0 \\ 0 & I & 0 \\ 0 & 0 & 0 \end{pmatrix}.$$

This extended $(2n + m) \times (2n + m)$ pencil can then be compressed into a $2n \times 2n$ pencil by finding the QR factorization of $\begin{pmatrix} R \\ B \end{pmatrix}$, without affecting the deflating subspace. The solution of the CARE can then be obtained by finding the ordered generalized Schur form of the compressed pencil.

For the DARE, the extended pencil is $P_{\text{DARE}}^{\text{E}} - \lambda N_{\text{DARE}}^{\text{E}}$, where

$$P_{\text{DARE}}^{\text{E}} = \begin{pmatrix} A & 0 & -B \\ -Q & -I & 0 \\ 0 & 0 & R \end{pmatrix}, \qquad N_{\text{DARE}}^{\text{E}} = \begin{pmatrix} I & 0 & 0 \\ 0 & A^{\mathrm{T}} & 0 \\ 0 & B^{\mathrm{T}} & 0 \end{pmatrix}.$$

This $(2n + m) \times (2n + m)$ can be compressed into a $2n \times 2n$ pencil by using the QR factorization of $\begin{pmatrix} R \\ -B \end{pmatrix}$. For details, see **Section 13.5.2.**

Again, the required basis should be constructed by finding the generalized RSF of the pencil using the QZ algorithm.

13.10   CHAPTER NOTES AND FURTHER READING

The AREs have been very well studied in the literatures of mathematics and control and filter theory.

For an excellent account of up-to-date theoretical developments, see the recent book of Lancaster and Rodman (1995). Some of the earlier theoretical developments are contained in Kučera (1972, 1979), Coppel (1974), and Singer and Hammarling (1983), Willems (1971), Wimmer (1984, 1994), Lancaster and Rodman (1980). The books by Anderson and Moore (1990), Ando (1988), Kwakernaak and Sivan (1972), Kimura (1997), Zhou *et al.* (1996) also contain a fair amount of theory of AREs. The existence of maximal solutions for generalized AREs arising in stochastic control has been discussed in DeSouza and Fragoso (1990). The paper by DeSouza *et al.* (1986) deals with Riccati equations arising in optimal filtering of nonstabilizable systems having singular state transition matrices. For some application of Riccati equations in general forms to dynamic games, see Basar (1991).

Important numerical methods have been dealt with in details in the books by Sima (1996) and Mehrmann (1991). Benner (1999) has given an up-to-date review with special attention to structure-preserving methods. An extensive bibliography on numerical methods appear in Laub (1991) and Benner (1997). See Jamshidi (1980) for an earlier review.

For a review of periodic Riccati equations see the article of Bittanti *et al.* and the references therein in the book "**The Riccati Equation**" edited by Bittanti *et al.* (1991). The latter contains several important papers on Riccati equations and the paper by Bittanti gives a brief life history of Count Jacopo Riccati (1676–1754), which is certainly worth reading.

The sensitivity of the continuous-time Riccati equations has been studied by several people: Byers (1985), Kenney and Hewer (1990), Chen (1988), Konstantinov *et al.* (1990), Xu (1996), Sun (1998), and Ghavimi and Laub (1995), etc. **Theorem 13.4.1** is due to Sun (1998). The bound (13.4.14) is due to Kenney and Hewer (1990). The residual of an approximate stabilizing solution (**Theorem 13.4.3**) is due to Sun (1997a). The sensitivity of the DARE has been studied in Gudmundsson *et al.* (1992), Konstantinov *et al.* (1993), Sun (1998), and Gahinet *et al.* (1990). The paper by Ghavimi and Laub (1995) relates backward error and sensitivity to accuracy and discusses techniques for refinement of computed solutions of the AREs. For recent results, see Petkov *et al.* (2000). For results on the upper and lower bounds of the solutions of CARE and DARE, see Lee (1997a, 1997b).

The eigenvector methods for the Riccati equations were proposed by McFarlane (1963) and Potter (1966). The Schur method for the Riccati equations originally appeared in the celebrated paper by Laub (1979). Petkov *et al.* (1987) studied the numerical properties of the Schur method and concluded that the Schur method can be unstable in some cases and the solutions may be inaccurate. A further analysis by Kenney *et al.* (1989) attributed such inaccuracy to poor scaling. For an excellent account of scaling of the Schur methods, see Benner (1997). The structure-preserving Hamiltonian–Schur method was first proposed by Byers in his Householder-prize winning Ph.D. thesis (1983) in the case of a single-input system (rank(B) = 1). See Byers (1986a) for details of the method. The theoretical foundation of this method is contained in the well-known paper by Paige and Van Loan (1981). Their result was later extended to the case when the Hamiltonian matrix has eigenvalues on the imaginary axis by Lin and Ho (1990). Patel *et al.* (1994) have discussed computation of stable invariant subspaces of Hamiltonian matrices. Another method, called the multishift method to compute the invariant subspace of the Hamiltonian matrix corresponding to the stable eigenvalues was developed by Ammar and Mehrmann (1993). The algorithm is called the multishift algorithm because n stable eigenvalues of the Hamiltonian matrix are used as shifts to isolate the desired invariant subspace. The multishift method sometimes has convergence problems, particularly for large n. The Hamiltonian–Schur algorithm in the multi-input case is due to Benner *et al.* (1997c). For structure-preserving eigenvalue methods see Benner *et al.* (1999c) and Bunse-Gerstner *et al.* (1992). Mehrmann (1988) has given a structure-preserving method for the discrete-time Riccati equation with single-input and single-output. The non-orthogonal symplectic methods have been discussed by Bunse-Gerstner and Mehrmann (1986) and Bunse-Gerstner *et al.* (1989) for the CARE, and by Benner *et al.* (1999b), Fassbender and Benner (2001) for the DARE. The details of these methods and other references can be found in the recent book by Fassbender (2000). Interesting connection between structure-preserving HR and SR algorithms appears in Benner *et al.* (1997a).

The generalized eigenvalue problem approach leading to deflating subspace method for the discrete-time Riccati equation was proposed by Pappas *et al.* (1980). See also Arnold and Laub (1984), Emami-Naeini and Franklin (1979, 1982). The inverse-free methods (the extended pencil approach (**Algorithm 13.5.3** and **Algorithm 13.5.4**)) and the associated compressed techniques were proposed by Van Dooren (1981).

The idea of using matrix sign function to solve the CARE was first introduced by Roberts (1980, [1971]). Byers (1986b, 1987) discussed numerical stability of the method and studied the computational aspects in details. See also Bierman (1984) and Bai and Demmel (1998). A generalization of the matrix sign function method to a matrix pencil and its application to the solutions of DCARE and DDARE was proposed by Gardiner and Laub (1986). For an account of the matrix sign function, see the recent paper of Kenney and Laub (1995). For a perturbation analysis of the matrix sign function, see Sun (1997c). Howland (1983) relates matrix sign function to separation of matrix eigenvalues.

For details of Newton's algorithm for the CARE (**Algorithm 13.5.8**) and that of the DARE (**Algorithm 13.5.10**), as presented here, see Benner (1997), Lancaster and Rodman (1995). The correct proof of convergence of Newton's method (**Theorem 13.5.8**) seemed to appear for the first time in Lancaster and Rodman (1995).

Kenney *et al.* (1990) gave results on error bounds for Newton's method, where it was first pointed out that if the initial solution X_0 is not chosen carefully, the error on the first step can be disastrous. They also gave conditions which guarantee monotone convergence from the first step on (**Theorem 13.5.9**). Several modifications of Newton's methods have appeared in recent years (Guo 1998; Guo and Lancaster 1998; Guo and Laub 2000; etc.). The line search modification proposed by Benner and Byers (1998) is extremely useful in practice. In general, it improves the convergence behavior of Newton's method and avoids the problem of a disastrously large first step. For acceleration techniques of the DARE, see Benner (1998).

Ghavimi *et al.* (1992) have discussed the local convergence analysis of conjugate gradient methods for solving the AREs.

For an account of parallel algorithms for AREs, see Bai and Qian (1994), Gardiner and Laub (1991), and Laub (1991) and references therein, Quintana and Hernández (1996a, 1996b, 1996c), etc.

For large-scale solutions of the AREs see Ferng *et al.* (1997), Lu and Lin (1993), Jaimoukha and Kasenally (1994) and Benner and Fassbender (1997). The recent book by Ionescu *et al.* (1999) gives a nice treatment of AREs for the indefinite sign and singular cases. See also Campbell (1980). For least-squares solutions of stationary optimal control using the AREs, see Willems (1971).

Some discussions on finding the Cholesky factor of the solution to an ARE without first computing the solution itself appears in Singer and Hammarling (1983). Lin (1987) has given a numerical method for computing the closed-loop eigenvalues of a discrete-time Riccati equation. Patel (1993) has given a numerical method for computing the eigenvalues of a simplectic matrix. For numerical algorithms for descriptor Riccati equations, see Benner (1999), Mehrmann (1991), Bender and Laub (1985, 1987), Benner *et al.* (1999a), etc. A description of discrete-time descriptor Riccati equations also appears in Zhang *et al.* (1999). A comparative study with respect to efficiency and accuracy of most of the methods described in this chapter for the CARE (the **eigenvector, Schur, inverse-free generalized Schur, Hamiltonian–Schur and Newton's Methods**) has been made in the recent M.Sc. Thesis of Ho (2000), using MATLAB and FORTRAN-77 codes. (In particular, this thesis contains MATLAB codes for ordered **Real Schur** and **Generalized Real Schur** decompositions). Numerical experiments were performed on 12 benchmark examples taken from the collection of Benner *et al.* (1995a, 1997b). The conclusions drawn in this thesis are almost identical to those mentioned in **Section 13.7**. For a recent collection of benchmark examples for Riccati equations, see Abels and Benner (1999a, 1999b).

Exercises

13.1 Derive necessary and sufficient conditions for the CARE (13.1.1) to have a unique symmetric positive definite stabilizing solution X.

13.2 Construct an example to show that the observability of (A, Q) is not necessary for the solution X of the CARE (13.1.1) to be positive definite.

13.3 Prove that the matrix defined in (13.3.1) is symplectic, and that if λ is a nonzero eigenvalue of M, so is $1/\bar{\lambda}$.

13.4 Establish the relation (13.2.16).

13.5 (a) Prove the discrete counterpart of Theorem 13.2.4, that is, prove an analogous theorem for the DARE.

(b) Using the results of Problem 13.3, and those of 13.5(a), prove Theorem 13.3.2.

13.6 Prove that the homogeneous CARE: $XA + A^T X + XSX = 0$ has a stabilizing solution if A has no eigenvalues on the imaginary axis. Prove or disprove a discrete-counterpart of this result.

13.7 Prove that the quantity (13.4.22) serves as an approximate condition number of the DARE (13.1.2). Construct an example of an ill-conditioned DARE using this quantity.

13.8 Find an example to illustrate that a small relative residual in a computed solution of the CARE does not guarantee a small error in the solution.

13.9 Prove that if Ω is singular, then $\text{sep}((A - SX), -(A - SX)^T)$ is zero.

13.10 Give a proof of Algorithm 13.5.1, making necessary assumptions.

13.11 Construct an example to show that the solution of the CARE, obtained by Algorithm 13.5.1, might be inaccurate, even though the problem is not ill-conditioned. (**Hint:** Construct an example for which U_{11} is ill-conditioned, but the CARE is well-conditioned.)

13.12 Give an example to demonstrate the superiority of the Schur algorithm for the CARE over the eigenvector algorithm, in case the associated Hamiltonian matrix is nearly defective.

13.13 Using Theorem 13.5.1 and the transformation

$$H = (M + I_{2n})(M - I_{2n})^{-1},$$

prove Theorem 13.5.2.

13.14 Construct an example to demonstrate the numerical difficulties of the Schur algorithm for the DARE in case the matrix A is nearly singular.

13.15 Write down an algorithm for solving the discrete algebraic Riccati equation, using the eigenvectors of the symplectic matrix. Discuss the computational drawbacks of the algorithm. Construct an example to illustrate the computational drawbacks.

13.16 Prove the properties 1 through 5 of the matrix sign function $\text{Sign}(A)$ stated in Section 13.5.3.

13.17 Prove that if $|\lambda| = 1$ is an eigenvalue of the pencil $P_{\text{DARE}} - \lambda N_{\text{DARE}}$ with the eigenvector $z = \begin{pmatrix} z_1 \\ z_2 \end{pmatrix}$, where P_{DARE} and N_{DARE} are the same as given in Theorem 13.5.5, then the detectability of (A, Q) implies that $z_1 = 0$.

13.18 Formulate the generalized Schur method for the CARE and develop that for the DARE in details.

13.19   Why is the generalized Schur method not preferable over the Schur method for the CARE if R is not nearly singular?

13.20   Construct an example to demonstrate the poor accuracy of the generalized eigenvector method for the DARE in case the pencil $P_{DARE} - \lambda N_{DARE}$ has near multiple eigenvalues.

Apply the generalized Schur algorithm (Algorithm 13.5.2) to the same example and verify the improvement in the accuracy of the solution.

13.21   Work out the details of how the pencil $P^E_{CARE} - \lambda N^E_{CARE}$ can be transformed to the compressed pencil $P^{EC}_{CARE} - \lambda N^{EC}_{CARE}$ using the QR factorization of the matrix $\begin{pmatrix} R \\ B \end{pmatrix}$.

13.22   Repeat the previous exercise for the DARE, that is, work out the details of the transformation to the pencil $P^{EC}_{DARE} - \lambda N^{EC}_{DARE}$ using the QR factorization of the matrix $\begin{pmatrix} R \\ -B \end{pmatrix}$.

13.23   Prove that the pencil $P^E_{CARE} - \lambda N^E_{CARE}$ and the pencil $P^{EC}_{CARE} - \lambda N^{EC}_{CARE}$ as defined in Section 13.5.2 for the CARE have the same deflating subspaces, and similarly for the DARE.

13.24   Develop the following algorithms in detail for both the DCARE and DDARE (consult Laub (1991) and Benner (1997)):

The Schur algorithms, the generalized Schur algorithms, the inverse-free generalized Schur algorithms, the matrix sign function algorithms, and Newton's algorithms.

Construct a simple example to illustrate each of the above algorithms.

13.25   Perform numerical experiments to compare Newton's methods with Newton's methods with line search, both for the CARE and DARE, by using several examples from the Benchmark collections in Benner *et al.* (1995a, 1995b, 1997b). Display your results on number of iterations and norms of the residual matrices using tables and graphs.

13.26   Construct an example to demonstrate the superiority of the inverse-free generalized Schur algorithm over the Schur algorithm for the CARE, in case the control weighting matrix R is positive definite but nearly singular.

13.27   Carry out a numerical experiment with a 150×150 randomly generated problem to make a comparative study with respect to computer-time and accuracy of the solution to the CARE with the following methods: *the eigenvector method, the Schur method, inverse-free generalized Schur method, the matrix sign function method,* and *the Hamiltonian structure-preserving Schur method.* Write down your observations and conclusions with tables and graph.

13.28   Repeat the previous exercise with the DARE using the following methods: The *eigenvector method, the generalized eigenvector method, the Schur method, the generalized Schur method, inverse-free generalized Schur method, and the matrix sign function method.*

13.29   (Kenney and Hewer 1990). Study the sensitivity of the solution of the CARE with the following data, for $\epsilon = 10^0, 10^{-1}, 10^{-2}, 10^{-3}$. Present your results with tables and graphs.

$$A = \begin{pmatrix} -\epsilon & 1 & 0 & 0 \\ -1 & -\epsilon & 0 & 0 \\ 0 & 0 & \epsilon & 1 \\ 0 & 0 & -1 & \epsilon \end{pmatrix}, \qquad B = \begin{pmatrix} 1 \\ 1 \\ 1 \\ 1 \end{pmatrix}, \qquad R = 1, \qquad Q = BB^{\mathrm{T}}.$$

Research Problems

13.1 Develop a structure-preserving method to compute the symplectic Schur decomposition and apply the method to solve the DARE, thus obtaining a symplectic structure-preserving method for the DARE.

References

Abels J. and Benner P. "CAREX – a collection of benchmark examples for continuous-time algebraic Riccati equations (version 2.0)". *SLICOT Working Note* 1999-14, November 1999a. (Available at the NICONET Website: http://www.win.tue.ne/niconet/niconet.html).

Abels J. and Benner P. "DAREX–a collection of benchmark examples for discrete-time algebraic Riccati equations (version 2.0)". *SLICOT Working Note* 1999-15, November 1999b. (Available at the NICONET Website: http://www.win.tue.ne/niconet/niconet.html).

Ammar G. and Mehrmann V. "A multishift algorithm for the numerical solution of algebraic Riccati equations," *Electr. Trans. Num. Anal.*, Vol. 1, pp. 33–48, 1993.

Anderson B.D.O. and Moore J.B. *Optimal Control: Linear Quadratic Methods*, Prentice Hall, Englewood Cliffs, NJ, 1990.

Anderson E., Bai Z., Bischof C., Blockford S., Demmel J., Dongarra J., DuCroz J., Greenbaum A., Hammarling S., McKenney A., and Sorensen D. *LAPACK Users' Guide*, 3rd edn, SIAM, Philadelphia, 1999.

Ando T. *Matrix Quadratic Equations*, Hokkaido University, Research Institute of Applied Electricity, Division of Applied Mathematics, Sapporo, Japan, 1988.

Arnold W. III and Laub A. "Generalized eigenproblem algorithms and software for algebraic Riccati equations," *Proc. IEEE*, Vol. 72, pp. 1746–1754, 1984.

Bai Z. and Demmel J. "Using the matrix sign function to compute invariant subspaces," *SIAM J. Matrix Anal. Appl.*, Vol. 19, pp. 205–225, 1998.

Bai Z. and Qian Q. "Inverse free parallel method for the numerical solution of algebraic Riccati equations," in *Proc. Fifth SIAM Conf. Appl. Lin. Alg.* (Lewis J. ed.) June, pp. 167–171, Snowbird, UT, 1994.

Basar T. "Generalized Riccati equations in dynamic game," in *The Riccati Equation*, (Bittanti S., Alan Laub and Williams J.C. eds.), Springer-Verlag, Berlin, 1991.

Bender D. and Laub A. "The linear-quadratic optimal regulator problem for descriptor systems," *Proc.* 24th *IEEE Conf. Dec. Control*, Ft. Lauderdale, Florida, December, pp. 957–962, 1985.

Bender D. and Laub A. "The linear-quadratic optimal regular for descriptor systems: discrete-time case," *Automatica*, Vol. 23, pp. 71–85, 1987.

Benner P. *Contributions to the Numerical Solution of Algebraic Riccati Equations and Related Eigenvalue Problems*, Dissertation for Dipl.-Math., Technischen Universität Chemnitz-Zwickau, Germany 1997.

Benner P. "Acclerating Newton's method for discrete-time Riccati equations," in *Proc. MTNS'98*, (Beghi A., Finesso L., and Picci G. eds.) pp. 569–572, Padova, Italy, 1998.

Benner P. "Computational methods for linear-quadratic optimization," *Rendiconti del Circulo Matematico di Palermo*, Supplemento, Serie II, no. 58, pp. 21–56, 1999.

Benner P. and Byers R. "An exact line search method for solving generalized continuous-time algebraic Riccati equations," *IEEE Trans. Autom. Control*, Vol. 43, pp. 101–107, 1998.

Benner P., Byers R., Mehrmann V., and Xu H. "Numerical solution of linear–quadratic control problems for descriptor systems," in *Proc. 1999 IEEE Intl. Symp. CACSD*, (Gonzalez O. ed.) Kohala Coast-Island of Hawaii, Hawaii, USA, pp. 46–51, August 22–27, 1999a.

Benner P. and Fassbender H. "An implicitly restarted symplectic Lanczos method for the Hamiltonian eigenvalue problem," *Lin. Alg. Appl.*, Vol. 263, pp. 75–111, 1997.

Benner P., Fassbender H., and Watkins D., "Two connections between the SR and HR eigenvalue algorithms," *Lin. Alg. Appl.*, Vol. 272, pp. 17–32, 1997a.

Benner P., Fassbender H. and Watkins D. "SR and SZ algorithms for the symplectic (butterfly) eigenproblem," *Lin. Alg. Appl.*, Vol. 287, pp. 41–76, 1999b.

Benner P., Laub A., and Mehrmann V. "A collection of benchmark examples for the numerical solution of algebraic Riccati equations I: continuous-time case," *Tech. Report SPC 95-22, Fak. f. Mathematik, TU Chemnitz-Zwickau, 09107 Chemnitz, FRG, 1995a*.

Benner P., Laub A., and Mehrmann V. "A collection of benchmark examples for the numerical solution of algebraic Riccati equations II: discrete-time case," *Tech. Report SPC 95-23, Fak. f. Mathematik, TU Chemnitz-Zwickau, 09107 Chemnitz, FRG, 1995b*.

Benner P., Laub A., and Mehrmann V. "Benchmarks for the numerical solution of algebraic Riccati equations," *IEEE Control Syst. Mag.*, Vol. 7(5), pp. 18–28, 1997b.

Benner P., Mehrmann V., and Xu H. "A new method for computing the stable invariant subspace of a real Hamiltonian matrix," *J. Comput. Appl. Math.*, Vol. 86, pp. 17–43, 1997c.

Benner P., Mehrmann V., and Xu H. "A numerically stable, structure preserving method for computing the eigenvalues of real Hamiltonian or symplectic pencils," *Numer. Math.*, Vol. 78, pp. 329–358, 1999c.

Bierman G.J. "Computational aspects of the matrix sign function solution to the ARE," *Proc. 23rd IEEE Conf. Dec. Contr.*, Las Vegas, Nevada, pp. 514–519, 1984.

Bittanti S., Laub A., and Willems J.C. (eds.), *The Riccati Equation*, Springer-Verlag, Berlin, 1991.

Bunse-Gerstner A., Mehrmann V., and Watkins D. "An SR algorithm for Hamiltonian matrices based on Gaussian elimination," *Meth. Oper. Res.*, Vol. 58, pp. 339–358, 1989.

Bunse-Gerstner A., Byers R., and Mehrmann V. "A chart of numerical methods for structured eigenvalue problems," *SIAM J. Matrix Anal. Appl.*, Vol. 13(2), pp. 419–453, 1992.

Bunse-Gerstner A. and Mehrmann V. "A symplectic QR-like algorithm for the solution of the real algebraic Riccati equation," *IEEE Trans. Autom. Control*, Vol. AC-31, pp. 1104–1113, 1986.

Byers R. *Hamiltonian and Symplectic Algorithms for the Algebraic Riccati Equation*, Ph.D. thesis, Cornell University, Department of Computer Science, Ithaca, NY, 1983.

Byers R. "Numerical condition of the algebraic Riccati equation," in *Contemporary Mathematics*, (Brualdi R. *et al.*, ed.) Vol. 47, pp. 35–49, American Mathematical Society, Providence, RI, 1985.

Byers R. "A Hamiltonian QR-algorithm," *SIAM J. Sci. Statist. Comput.*, Vol. 7, pp. 212–229, 1986a.

Byers R. "Numerical stability and instability in matrix sign function based algorithms," in *Computational and Combinatorial Methods in Systems Theory* (Byrnes C.I. and Lindquist A. eds.) pp. 185–200, North Holland, New York, 1986b.

Byers R. "Solving the algebraic Riccati equation with the matrix sign function," *Lin. Alg. Appl.*, Vol. 85, pp. 267–279, 1987.

Byers R. "A Hamiltonian Jacobi Algorithm," *IEEE Trans. Autom. Control*, Vol. 35(5), pp. 566–570, 1990.

Campbell S.L. *Singular Systems of Differential Equations*, Pitman, Marshfield, MA, 1980.

Chen C.-H. "Perturbation analysis for solutions of algebraic Riccati equations," *J. Comput. Math.*, Vol. 6, pp. 336–347, 1988.

Coppel W.A. "Matrix quadratic equations," *Bull. Australian Math. Soc.*, Vol. 10, pp. 327–401, 1974.

Denman E.D. and Beavers A.N. "The matrix sign function and computations in systems," *Appl. Math. Comput.*, Vol. 2, pp. 63–94, 1976.

DeSouza C.E. and Fragoso M.D. "On the existence of maximal solution for generalized algebraic Riccati equations arising in stochastic control," *Syst. Control Lett.*, Vol. 14, pp. 223–239, 1990.

DeSouza C.E., Gevers M.R., and Goodwin G.C. "Riccati equations in optimal filtering of nonstabilizable systems having singular state transition matrices," *IEEE Trans. Autom. Control*, Vol. AC-31, pp. 831–838, 1986.

Emami-Naeini A. and Franklin G.F. "Design of steady-state quadratic-loss optimal digital controls for systems with a singular system matrix, Proc. 13th Astimor Conf. Circuits, Systems and Computers, pp. 370–374, 1979.

Emami-Naeini A. and Franklin G.F. "Deadbeat control and tracking of discrete-time systems," *IEEE Trans. Autom. Control*, Vol. AC-27, pp. 176–181, 1982.

Fassbender H. and Benner P. "A hybrid method for the numerical solution of discrete-time algebraic Riccati equations," in *Contemporary Mathematics on Structured Matrices in Mathematics, Computer Science, and Engineering* (Olshevsky V. ed.) Vol. 280, pp. 255–269, American Mathematical Society, Providence, RI, 2001.

Fassbender H. *Symplectic Method for the Symplectic Eigenproblem,* Kluwer Academic/Plenum Publishers, New York, 2000.

Ferng W.R., Lin W.-W., and Wang C.-S. "The shift-inverted J-Lanczos algorithm for the numerical solutions of large sparse algebraic Riccati equations," *Comput. Math. Appl.*, Vol. 33(10), pp. 23–40, 1997.

Gahinet P.M., Laub A.J., Kenney C.S., and Hewer G.A. "Sensitivity of the stable discrete-time Lyapunov equation," *IEEE Trans. Autom. Control*, Vol. 35, pp. 1209–1217, 1990.

Gardiner J.D. and Laub A.J. "A generalization of the matrix-sign-function solution for algebraic Riccati equations," *Int. J. Control*, Vol. 44, pp. 823–832, 1986.

Gardiner J.D. and Laub A.J. "Parallel algorithms for algebraic Riccati equations," *Int. J. Control*, Vol. 54, pp. 1317–1333, 1991.

Ghavimi A., Kenney C., and Laub A.J. "Local convergence analysis of conjugate gradient methods for solving algebraic Riccati equations," *IEEE Trans. Autom. Control*, Vol. AC-37, pp. 1062–1067, 1992.

Ghavimi A.R. and Laub A.J. "Backward error, sensitivity, and refinement of computed solutions of algebraic Riccati equations," *Num. Lin. Alg. Appl.*, Vol. 2, pp. 29–49, 1995.

Gudmundsson T., Kenney C., and Laub A.J. "Scaling of the discrete-time algebraic Riccati equation to enhance stability of the Schur method," *IEEE Trans. Autom. Control*, Vol. AC-37, pp. 513–518, 1992.

Guo C.-H. "Newton's method for discrete algebraic Riccati equations when the closed-loop matrix has eigenvalues on the unit circle," *SIAM J. Matrix Anal. Appl.*, Vol. 20, pp. 279–294, 1998.

Guo C.-H. and Lancaster P. "Analysis and modification of Newton's method for algebraic Riccati equations," *Math. Comp.*, Vol. 67, pp. 1089–1105, 1998.

Guo C.-H. and Laub A.J. "On a Newton-like method for solving algebraic Riccati equations," *SIAM J. Matrix Anal. Appl.*, Vol. 21, pp. 694–698, 2000.

Hammarling S.J. *Newton's Method for Solving the Algebraic Riccati Equation*, NPL Report DITC 12/82, National Physical Laboratory, Teddington, Middlesex TW11 OLW, U.K., 1982.

Hewer G.A. "An iterative technique for the computation of steady state gains for the discrete optimal controller," *IEEE Trans. Autom. Control*, Vol. AC-16, pp. 382–384, 1971.

Ho T. *A Study of Computational Methods for the Continuous-time Algebraic Riccati Equation*, M.Sc. Thesis, Northern Illinois University, DeKalb, Illinois, 2000.

Howland J.L. "The sign matrix and the separation of matrix eigenvalues," *Lin. Alg. Appl.*, Vol. 49, pp. 221–232, 1983.

Ionescu V., Oara C., and Weiss M. *Generalized Riccati Theory and Robust Control*, John Wiley, New York, 1999.

Jaimoukha I.M. and Kasenally E.M. "Krylov subspace methods for solving large Lyapunov equations," *SIAM J. Numer. Anal.* Vol. 31, 227–251, 1994.

Jamshidi M. "An overview on the solutions of the algebraic Riccati equation and related problems," *Large-Scale Syst.*, Vol. 1, pp. 167–192, 1980.

Kenney C.S. and Hewer G. "The sensitivity of the algebraic and differential Riccati equations," *SIAM J. Control Optimiz.*, Vol. 28, pp. 50–69, 1990.

Kenney C.S. and Laub A.J. "On scaling Newton's method for polar decomposition and the matrix sign function," *SIAM J. Matrix Anal. Appl.*, Vol. 13, pp. 688–706, 1992.

Kenney C.S. and Laub A.J. "The matrix sign function," *IEEE Trans. Autom. Control*, Vol. 40, pp. 1330–1348, 1995.

Kenney C.S., Laub A.J., and Wette M. "A stability-enhancing scaling procedure for Schur-Riccati solvers," *Syst. Control Lett.*, Vol. 12, pp. 241–250, 1989.

Kenney C.S., Laub A.J., and Wette M. "Error bounds for Newton refinement of solutions to algebraic Riccati equations," *Math. Control, Signals, Syst.*, Vol. 3, pp. 211–224, 1990.

Kimura, H. *Chain-Scattering Approach to H∞-Control*, Birkhäuser, Boston, 1997.

Kleinman D.L. "On an iterative technique for Riccati equation computations," *IEEE Trans. Autom. Control*, Vol. AC-13, pp. 114–115, 1968.

Kleinman D.L. "Stabilizing a discrete, constant linear system with application to iterative methods for solving the Riccati equation, *IEEE Trans. Autom. Control*, Vol. AC-19, pp. 252–254, 1974.

Konstantinov M.M., Petkov P., and Christov N.D. "Perturbation analysis of matrix quadratic equations," *SIAM J. Sci. Stat. Comput.*, Vol. 11, pp. 1159–1163, 1990.

Konstantinov M.M., Petkov P., and Christov N.D. "Perturbation analysis of the discrete Riccati equation," *Kybernetika*, Vol. 29, pp. 18–29, 1993.

Konstantinov M.M., Petkov P., Gu D.W., and Postlethwaite I. *Perturbation Techniques for Linear Control Problems*, Report 95-7, Control Systems Research, Department of Engineering, Leicester University, UK, 1995.

Kučera V. "A contribution to matrix quadratic equations," *IEEE Trans. Autom. Control*, Vol. 17, pp. 344–347, 1972.

Kučera, V. *Discrete Linear Control*, John Wiley & Sons, New York, 1979.

Kwakernaak H. and Sivan R. *Linear Optimal Control Systems*, Wiley-Interscience, New York, 1972.

Lancaster P. and Rodman L. "Existence and uniqueness theorems for algebraic Riccati equations," *Int. J. Control*, Vol. 32, pp. 285–309, 1980.

Lancaster P. and Rodman L. *Algebraic Riccati Equations*, Oxford University Press, Oxford, 1995.

Lee C.-H. "On the upper and lower bounds of the solution of the continuous algebraic Riccati matrix equation," *Int. J. Control*, Vol. 66, pp. 105–118, 1997a.

Lee C.-H. "Upper and lower bounds of the solutions of the discrete algebraic Riccati and Lyupunov matrix equations," *Int. J. Control*, Vol. 68, pp. 579–598, 1997b.

Laub A.J. "A Schur method for solving algebraic Riccati equations," *IEEE Trans. Autom. Control*, Vol. AC-24, pp. 913–921, 1979.

Laub, A.J. "Invariant subspace methods for the numerical solution of Riccati equations," in *The Riccati Equation* (Bittanti S. *et al.*, eds.) pp. 163–196, Springer-Verlag, Berlin 1991.

Lin W.-W. "A new method for computing the closed-loop eigenvalues of a discrete-time algebraic Riccati equation," *Lin. Alg. Appl.*, Vol. 6, pp. 157–180, 1987.

Lin W.-W. and Ho T.-C. *On Schur type decompositions of Hamiltonian and Symplectic Pencils*, Tech. Report, Institute of Applied Mathematics, National Tsing Hua University Taiwan, 1990.

Lu L. and Lin W.-W. "An iterative algorithm for the solution of the discrete-time algebraic Riccati equation," *Lin. Alg. Appl.*, Vol. 188/189, pp. 465–488, 1993.

McFarlane A. "An eigenvector solution of the optimal linear regulator problem," *J. Electron. Control*, Vol. 14, pp. 643–654, 1963.

Mehrmann V. *The Autonomous Linear Quadratic Control Problem*. Lecture Notes in Control and Information Sciences, Vol. 163, Springer-Verlag, Berlin, 1991.

Mehrmann V. "A symplectic orthogonal method for single-input single-output discrete-time optimal linear quadratic control problems," *SIAM J. Matrix Anal. Appl.*, Vol. 9, pp. 221–247, 1988.

Paige C. and Van Loan C. "A Schur decomposition for Hamiltonian matrices," *Lin. Alg. Appl.*, pp. 11–32, 1981.

Pandey P. "On scaling an algebraic Riccati equation," *Proc. Amer. Control Conf.*, Vol. 9, pp. 1583–1587, June, 1993.

Pappas T., Laub A.J., and Sandell N. "On the numerical solution of the discrete-time algebraic Riccati equation," *IEEE Trans. Autom. Control*, Vol. AC-25, pp. 631–641, 1980.

Patel R.V. "On computing the eigenvalues of a symplectic pencil," *Lin. Alg. Appl.*, Vol. 188/189, pp. 591–611, 1993.

Patel R.V., Lin Z., and Misra P. "Computation of stable invariant subspaces of Hamiltonian matrices," *SIAM J. Matrix Anal. Appl.*, Vol. 15, pp. 284–298, 1994.

Petkov P., Christov N.D., and Konstantinov M.M. "Numerical properties of the generalized Schur approach for solving the discrete matrix Riccati equation," *Proc. 18th Spring Conference of the Union of Bulgarian Mathematicians*, Albena, pp. 452–457, 1989.

Petkov P., Christov N.D., and Konstantinov M.M. "On the numerical properties of the Schur approach for solving the matrix Riccati equation," *Syst. Control Lett.*, Vol. 9, pp. 197–201, 1987.

Petkov P., Christov N.D., and Konstantinov M.M. *Computational Methods for Linear Control Systems*, Prentice Hall, London, 1991.

Petkov P., Konstantinov M.M., and Mehrmann V. DGRSVX and DMSRIC: *Fortran 77 Subroutines for Solving Continuous-time Matrix Algebraic Riccati Equations with Condition and Accuracy Estimates*. Tech. Report SFB393/98-116, Fakultät für Mathematik, TU Chemnitz, 09107 Chemnitz, FRG, 1998.

Petkov P., Gu D., Konstantinou M.M., and Mehrmann V. "Condition and Error Estimates in the Solution of Lyapunov and Riccati equations, *SLICOT Working Note*, 2000-1.

Potter J.E. "Matrix quadratic solutions," *SIAM J. Appl. Math.*, Vol. 14, pp. 496–501, 1966.

Quintana E. and Hernández V. *Algoritmos por bloques y paralelos para resolver ecuaciones matriciales de Riccati mediante el método de Newton*, Tech. Report DSIC-II/6/96, Dpto. de Sistemas Informáticos y Computación, Universidad Politécnica de Valencia, Valencia, Spain, 1996a.

Quintana E. and Hernández V. *Algoritmos por bloques y paralelos para resolver ecuaciones matriciales de Riccati mediante el método de Schur*, Tech. Report DSIC-II/7/96, Dpto. de Sistemas Informáticos y Computación, Universidad Politécnica de Valencia, Valencia, Spain, 1996b.

Quintana E. and Hernández V. *Algoritmos por bloques y paralelos para resolver ecuaciones matriciales de Riccati mediante la división espectral*, Tech. Report DSIC-II/6/96, Dpto de Sistemas Informáticos y Computación, Universidad Politécnica de Valencia, Valencia, Spain, 1996c.

Roberts J. "Linear model reduction and solution of the algebraic Riccati equation by use of the sign function," *Int. J. Control*, Vol. 32, pp. 677–687, 1980 (reprint of a technical report form Cambridge University in 1971).

Sima V. *Algorithms for Linear-Quadratic Optimization*, Marcel Dekker, New York, 1996.

Sima V., Petkov P. and VanHuffel S. "Efficient and reliable algorithms for condition estimation of Lyapunov and Riccati equations," *Proc. Mathematical Theory of Networks and Systems* (MTNS – 2000), 2000.

Singer M.A. and Hammarling S.J. *The Algebraic Riccati Equation*, National Physical Laboratory Report, DITC 23/83, January, 1983.

Stewart G.W. "Algorithm 506-HQR3 and EXCHNG: Fortran subroutines for calculating and ordering the eigenvalues of a real upper Hessenberg matrix," *ACM Trans. Math. Soft.*, Vol. 2, pp. 275–280, 1976.

Sun J.-G. "Residual bounds of approximate solutions of the algebraic Riccati equations," *Numer. Math.*, Vol. 76, pp. 249–263, 1997a.

Sun J.-G. "Backward error for the discrete-time algebraic Riccati equation," *Lin. Alg. Appl.*, Vol. 25, pp. 183–208, 1997b.

Sun J.-G. "Perturbation analysis of the matrix sign function," *Lin. Alg. Appl.*, Vol. 250, pp. 177–206, 1997c.

Sun J.-G. "Perturbation theory for algebraic Riccati equation," *SIAM J. Matrix Anal. Appl.*, Vol. 19(1), pp. 39–65, 1998.

Van Dooren P. "A generalized eigenvalue approach for solving Riccati equations," *SIAM J. Sci. Stat. Comput.*, Vol. 2, pp. 121–135, 1981.

Van Dooren P. "Algorithm 590-DSUBSP and EXCHQZ: Fortran subroutines for computing deflating subspaces with specified spectrum," *ACM Trans. Math. Soft.*, Vol. 8, pp. 376–382, 1982.

Van Loan C.F. "A symplectic method for approximating all the eigenvalues of a Hamiltonian matrix," *Lin. Alg. Appl.*, Vol. 16, pp. 233–251, 1984.

Willems J.C. "Least squares stationary optimal control and the algebraic Riccati equation," *IEEE Trans. Autom. Control*, Vol. AC-16, pp. 621–634, 1971.

Wimmer H.K. "The algebraic Riccati equation: Conditions for the existence and uniqueness of solutions," *Lin. Alg. Appl.*, Vol. 58, pp. 441–452, 1984.

Wimmer H.K. "Existence of positive-definite and semidefinite solutions of discrete-time algebraic Riccati equations," *Int. J. Control*, Vol. 59, pp. 463–471, 1994.

Xu S.-F. "Sensitivity analysis of the algebraic Riccati equations," *Numer. Math.*, Vol. 75, pp. 121–134, 1996.

Zhang L.Q., Lam J., and Zhang Q.L. "Lyapunov and Riccati equations of discrete-time descriptor systems," *IEEE Trans. Autom. Control*, Vol. 44(1), pp. 2134–2139, 1999.

Zhou K. (with Doyle J.C. and Glover K.), *Robust and Optimal Control*, Prentice Hall, Upper Saddle River, NJ, 1996.

CHAPTER 14

INTERNAL BALANCING AND MODEL REDUCTION

Topics covered

- Internal Balancing
- Model Reduction via Internal Balancing
- Model Reduction via Schur Decomposition
- Hankel Norm Approximation

14.1 INTRODUCTION

Several practical situations such as the design of large space structures (LSS), control of power systems, and others, give rise to very large-scale control problems. Typically, these come from the discretization of distributed parameter problems and have thousands of states in practice. Enormous computational complexities hinder the computationally feasible solutions of these problems.

As a result, control theorists have always sought ways to construct *reduced-order models* of appropriate dimensions (depending upon the problem to be solved) which can then be used in practice in the design of control systems. This process is known as *model reduction*. The idea of model reduction is to construct a reduced-order model from the original full-order model such that the reduced-order model is close, in some sense, to the full-order model. The closeness is normally measured by the smallness of $\|G(s) - G_R(s)\|$, where $G(s)$ and $G_R(s)$ are, respectively, the transfer function matrices of the original and the reduced-order model. Two norms, $\|\cdot\|_\infty$ norm and the Hankel-norm are considered here. The problem of constructing a reduced-order model such that the Hankel-norm error is minimized is called an **Optimal Hankel-norm approximation problem**. A widely used practice of model reduction is to first find a balanced realization (i.e., a realization with controllability and observability Grammians equal to a diagonal matrix) and then to

601

truncate the balanced realization in an appropriate manner to obtain a reduced-order model. The process is known as **balanced truncation method**. Balanced truncation does not minimize the H_∞ model reduction error, it only gives an upper bound. Balancing of a continuous-time system is discussed in **Section 14.2**, where two algorithms are described. **Algorithm 14.2.1** (Laub 1980; Glover 1984) constructs internal balancing of a stable, controllable and observable system, whereas **Algorithm 14.2.2** (Tombs and Postlethwaite 1987) is designed to extract a balanced realization, if the original system is not minimal. **Internal balancing** of a discrete-time system is described in **Section 14.3**.

In **Section 14.4**, it is shown (**Theorem 14.4.1**) that a reduced-order model constructed by **truncating a balanced realization (Algorithm 14.4.1) remains stable** and the H_∞-norm error is bounded.

A Schur method (**Algorithm 14.4.2**) for model reduction is then described. The Schur method due to Safonov and Chiang (1989) is designed to overcome the numerical difficulties in **Algorithm 14.4.1** due to the possible ill-conditioning of the balancing transforming matrices. In **Theorem 14.4.2**, it is shown that the transfer function matrix of the reduced-order model obtained by the Schur method is the same as that of the model reduction procedure via internal balancing using **Algorithm 14.2.1**. The Schur method, however, has its own computational problem. It requires computation of the product of the controllability and observability Grammians, which might be a source of round-off errors. The method, can be modified by using Cholesky factors of the Grammians which then leads to the **square-root algorithm (Algorithm 14.2.2)**.

The advantages of the Schur and the square-root methods can be combined into a **balancing-free square-root algorithm** (Varga 1991). This algorithm is briefly sketched in **Section 14.4.3**.

Section 14.5 deals with **Hankel-norm approximation.** A state-space characterization of all solutions to optimal Hankel-norm approximation due to Glover (1984) is stated (**Theorem 14.5.2**) and then an algorithm to compute an **optimal Hankel-norm approximation (Algorithm 14.5.1)** is described.

Section 14.6 shows how to obtain a **reduced-order model of an unstable system.**

The **frequency-weighted model reduction** problem due to Enns(1984) is considered in **Section 14.7**. The errors at the high frequencies can sometimes possibly be reduced by using suitable weights on the frequencies.

Finally, in **Section 14.8**, a numerical **comparison of different model reduction procedures** is given.

14.2   INTERNAL BALANCING OF CONTINUOUS-TIME SYSTEMS

Let (A, B, C, D) be an n-th order stable system that is both controllable and observable. Then it is known (Glover 1984) that there exists a transformation such that the transformed controllability and observability Grammians are equal to a diagonal matrix Σ. Such a realization is called a **balanced realization** (or **internally balanced realization**).

Internal balancing of a given realization is a preliminary step to a class of methods for model reduction, called **Balance Truncation Methods**. In this section, we describe two algorithms for internal balancing of a **continuous-time system**. The matrix D of the system (A, B, C, D) remains unchanged during the transformation of the system to a balanced system. **We, therefore, drop the matrix D from our discussions in this chapter.**

14.2.1 Internal Balancing of a Minimal Realization (MR)

Suppose that the n-th order system (A, B, C) is stable and minimal. Thus, it is both controllable and observable. Therefore, the controllability Grammian C_G and the observability Grammian O_G are symmetric and positive definite (**see Chapter 7**) and hence admit the Cholesky factorizations.

Let $C_G = L_c L_c^T$ and $O_G = L_o L_o^T$ be the respective Cholesky factorizations.
Let

$$L_o^T L_c = U \Sigma V^T \tag{14.2.1}$$

be the singular value decomposition (SVD) of $L_o^T L_c$.
Define now

$$T = L_c V \Sigma^{-1/2}, \tag{14.2.2}$$

where $\Sigma^{1/2}$ denotes the square root of Σ.

Then T is nonsingular, and furthermore using the expressions for C_G and Eq. (14.2.2), we see that the transformed controllability Grammian $\tilde{C}_G$ is

$$\tilde{C}_G = T^{-1} C_G T^{-T} = \Sigma^{1/2} V^T L_c^{-1} L_c L_c^T L_c^{-T} V \Sigma^{1/2} = \Sigma.$$

Similarly, using the expression for O_G and the Eqs. (14.2.1) and (14.2.2), we see that the transformed observability Grammian $\tilde{O}_G$ is

$$\tilde{O}_G = T^T O_G T = \Sigma^{-1/2} V^T L_c^T L_o L_o^T L_c V \Sigma^{-1/2}$$
$$= \Sigma^{-1/2} V^T V \Sigma U^T U \Sigma V^T V \Sigma^{-1/2} = \Sigma^{1/2} \cdot \Sigma^{1/2} = \Sigma.$$

Thus, the particular choice of

$$T = L_c V \Sigma^{-1/2} \tag{14.2.3}$$

reduces both the controllability and observability Grammians to the same diagonal matrix Σ. The system $(\tilde{A}, \tilde{B}, \tilde{C})$, where the system matrices are defined by

$$\tilde{A} = T^{-1}AT, \qquad \tilde{B} = T^{-1}B, \qquad \tilde{C} = CT \qquad (14.2.4)$$

is then a **balanced realization** of the system (A, B, C). The decreasing positive numbers $\sigma_1 \geq \sigma_2 \geq \cdots \geq \sigma_n$ in $\Sigma = \text{diag}(\sigma_1, \sigma_2, \ldots, \sigma_n)$, are the **Hankel singular values.**

The above discussion leads to the following algorithm for internal balancing.

Algorithm 14.2.1.   *An Algorithm for Internal Balancing of a Continuous-Time MR*

   Inputs.
A—The $n \times n$ state matrix.
B—The $n \times m$ input matrix.
C—The $r \times n$ output matrix.
   Outputs.
T—An $n \times n$ nonsingular balancing transforming matrix.
$\tilde{A}, \tilde{B}, \tilde{C}$—The matrices of internally balanced realization:

$$\tilde{A} = T^{-1}AT, \qquad \tilde{B} = T^{-1}B, \qquad \tilde{C} = CT.$$

   Assumptions. (A, B) *is controllable,* (A, C) *is observable, and* A *is stable.*
   Result. $T^{-1}C_GT^{-T} = T^TO_GT = \Sigma$, *a diagonal matrix with positive diagonal entries.*

Step 1. Compute the controllability and observability Grammians, C_G and O_G, by solving, respectively, the Lyapunov equations:

$$AC_G + C_GA^T + BB^T = 0, \qquad (14.2.5)$$

$$A^TO_G + O_GA + C^TC = 0. \qquad (14.2.6)$$

(Note that since A is a stable matrix, the matrices C_G and O_G can be obtained by solving the respective Lyapunov equations above (see **Chapter 7**).)

Step 2. Find the Cholesky factors L_c and L_o of C_G and O_G:

$$C_G = L_cL_c^T \quad \text{and} \quad O_G = L_oL_o^T \qquad (14.2.7)$$

Step 3. Find the SVD of the matrix $L_o^TL_c$: $L_o^TL_c = U\Sigma V^T$.

Step 4. Compute $\Sigma^{-1/2} = \text{diag}\left(\dfrac{1}{\sqrt{\sigma_1}}, \dfrac{1}{\sqrt{\sigma_2}}, \ldots, \dfrac{1}{\sqrt{\sigma_n}}\right)$, where $\Sigma = \text{diag}(\sigma_1, \sigma_2, \ldots, \sigma_n)$. (Note that $\sigma_i, i = 1, 2, \ldots, n$ are positive).

Step 5. Form $T = L_cV \Sigma^{-1/2}$

Step 6. Compute the matrices of the balanced realization:

$$\tilde{A} = T^{-1}AT, \qquad \tilde{B} = T^{-1}B, \quad \text{and} \quad \tilde{C} = CT.$$

Remark

- The original method of Laub (1980) consisted in finding the transforming matrix T by diagonalizing the product $L_c^T O_G L_c$ or $L_o^T C_G L_o$, which is symmetric and positive definite. The method described here is mathematically equivalent to Laub's method and is numerically more effective.

Example 14.2.1. Consider finding the balanced realization using Algorithm 14.2.1 of the system (A, B, C) given by:

$$A = \begin{pmatrix} -1 & 2 & 3 \\ 0 & -2 & 1 \\ 0 & 0 & -3 \end{pmatrix}, \qquad B = (1, 1, 1)^T, \qquad C = (1, 1, 1).$$

Step 1. By solving the Lyapunov equation (14.2.5), we obtain

$$C_G = \begin{pmatrix} 3.9250 & 0.9750 & 0.4917 \\ 0.9750 & 0.3667 & 0.2333 \\ 0.4917 & 0.2333 & 0.1667 \end{pmatrix}.$$

Similarly, by solving the Lyapunov equation (14.2.6), we obtain

$$O_G = \begin{pmatrix} 0.5000 & 0.6667 & 0.7917 \\ 0.6667 & 0.9167 & 1.1000 \\ 0.7917 & 1.1000 & 1.3250 \end{pmatrix}.$$

Step 2. The Cholesky factors of C_G and O_G are:

$$L_c = \begin{pmatrix} 1.9812 & 0 & 0 \\ 0.4912 & 0.3528 & 0 \\ 0.2482 & 0.3152 & 0.0757 \end{pmatrix}, \qquad L_o = \begin{pmatrix} 0.7071 & 0 & 0 \\ 0.9428 & 0.1667 & 0 \\ 1.1196 & 0.2667 & 0.0204 \end{pmatrix}.$$

Step 3. From the SVD of $L_o^T L_c$ (using MATLAB function **svd**):

$$[U, \Sigma, V] = \text{svd}(L_o^T L_c),$$

we have

$$\Sigma = \text{diag} \begin{pmatrix} 2.2589 & 0.0917 & 0.0006 \end{pmatrix},$$

$$V = \begin{pmatrix} -0.9507 & 0.3099 & 0.0085 \\ -0.3076 & -0.9398 & -0.1488 \\ -0.0381 & -0.1441 & 0.9888 \end{pmatrix}.$$

Step 4.
$$\Sigma^{1/2} = \text{diag}(1.5030, 0.3028, 0.0248).$$

Step 5. The transforming matrix T is:

$$T = L_c V \Sigma^{-1/2} = \begin{pmatrix} -1.2532 & 2.0277 & 0.6775 \\ -0.3835 & -0.5914 & -1.9487 \\ -0.2234 & -0.7604 & 1.2131 \end{pmatrix}.$$

Step 6. The balanced matrices are:

$$\tilde{A} = T^{-1}AT = \begin{pmatrix} -0.7659 & 0.5801 & -0.0478 \\ -0.5801 & -2.4919 & 0.4253 \\ 0.0478 & 0.4253 & -2.7422 \end{pmatrix}.$$

$$\tilde{B} = T^{-1}B = \begin{pmatrix} -1.8602 \\ -0.6759 \\ 0.0581 \end{pmatrix}, \qquad \tilde{C} = CT = \begin{pmatrix} -1.8602 & 0.6759 & -0.0581 \end{pmatrix}.$$

Verify:

$$T^{-1}C_G T^{-T} = T^T O_G T = \Sigma = \text{diag}(2.2589, 0.0917, 0.0006).$$

Computational Remarks

- **The explicit computation of the product $L_o^T L_c$ can be a source of round-off errors.** The small singular values might be almost wiped out by the rounding errors in forming the explicit product $L_o^T L_c$. It is suggested that the algorithm of Heath *et al.* (1986), which computes the singular values of a product of two matrices without explicitly forming the product, be used in practical implementation of this algorithm.

MATLAB notes: The MATLAB function in the form:

$$\text{SYSB} = \textbf{balreal} \ (\text{sys})$$

returns a balanced realization of the system (A, B, C). The use of the function **balreal** in the following format:

$$[\text{SYSB, G, T, TI}] = \textbf{balreal} \ (\text{sys})$$

returns, in addition to $\tilde{A}, \tilde{B}, \tilde{C}$ of the balanced system, a vector G containing the diagonal of the Grammian of the balanced realization. The matrix T is the matrix of the similarity transformation that transforms (A, B, C) to $(\tilde{A}, \tilde{B}, \tilde{C})$ and TI is its inverse.

MATCONTROL notes: Algorithm 14.2.1 has been implemented in MATCONTROL function **balsvd**.

14.2.2   Internal Balancing of a Nonminimal Realization

In the previous section we showed how to compute the balanced realization of a stable minimal realization. Now we show how to obtain a balanced realization given a stable **nonminimal continuous-time realization.** The method is due to Tombs and Postlethwaite (1987) and is known as the **square-root method.** The algorithm is based on a partitioning of the SVD of the product $L_o^T L_c$ and the balanced matrices $\tilde{A}$, $\tilde{B}$, $\tilde{C}$ are found by applying two transforming matrices L and Z to the matrices A, B, and C. The matrices L_o and L_c are, respectively, the Cholesky factors of the positive semidefinite observability and controllability matrices O_G and C_G.

The balanced realization in this case is of order $k(k < n)$ in contrast with the previous one where the balanced matrices are of the same orders as of the original model.

Let the SVD of $L_o^T L_c$ be represented as

$$L_o^T L_c = (U_1, U_2) \operatorname{diag}(\Sigma_1, \Sigma_2)(V_1, V_2)^T$$

where $\Sigma_1 = \operatorname{diag}(\sigma_1, \ldots, \sigma_k) > 0$, $\Sigma_2 = 0_{n-k \times n-k}$.

The matrices U_1, V_1^T, and Σ_1 are of order $n \times k$, $k \times n$, and $k \times k$, respectively. Define now

$$L = L_o U_1 \Sigma_1^{-1/2}, \qquad Z = L_c V_1 \Sigma_1^{-1/2}$$

Then it has been shown in Tombs and Postlethwaite (1987) that the realization $(\tilde{A}, \tilde{B}, \tilde{C})$, where the matrices $\tilde{A}$, $\tilde{B}$, and $\tilde{C}$ are defined by $\tilde{A} = L^T A Z$, $\tilde{B} = L^T B$, and $\tilde{C} = C Z$ is balanced, truncated to k states, of the system (A, B, C).

Remark

- Note that no assumption on the controllability of (A, B) or the observability of (A, C) is made.

Algorithm 14.2.2. *The Square-Root Algorithm for Balanced Realization of a Continuous-Time Nonminimal Realization.*

　　Inputs. *The system matrices A, B, and C of a nonminimal realization.*

　　Outputs. *The transforming matrices L, Z, and the balanced matrices $\tilde{A}$, $\tilde{B}$, and $\tilde{C}$.*

　　Assumption. *A is stable.*

　　Step 1. *Compute L_o and L_c, using the LDL^T decomposition of O_G and C_G, respectively. (Note that L_o and L_c may be symmetric positive semidefinite, rather than positive definite.)*

　　Step 2. *Compute the SVD of $L_o^T L_c$ and partition it in the form:*

$$L_o^T L_c = (U_1, U_2) \operatorname{diag}(\Sigma_1, \Sigma_2)(V_1, V_2)^T$$

where $\Sigma_1 = \operatorname{diag}(\sigma_1, \sigma_2, \ldots \sigma_k) > 0$.

　　Step 3. *Define*

$$L = L_o U_1 \Sigma_1^{-1/2} \quad and \quad Z = L_c V_1 \Sigma_1^{-1/2}.$$

　　Step 4. *Compute the balanced matrices $\tilde{A}$, $\tilde{B}$, and $\tilde{C}$ as:*

$$\tilde{A} = L^T A Z, \quad \tilde{B} = L^T B, \quad and \quad \tilde{C} = C Z.$$

Example 14.2.2. Let A, C be the same as in Example 14.2.1, and let $B = (1, 0, 0)^{\mathrm{T}}$. Thus, (A, B) is **not controllable.**

Step 1.

$$L_0 = \begin{pmatrix} 0.7071 & 0 & 0 \\ 0.9428 & 0.1667 & 0 \\ 1.1196 & 0.2667 & 0.0204 \end{pmatrix}, \qquad L_c = \mathrm{diag}(0.7071, 0, 0).$$

Step 2. $U_1 = \begin{pmatrix} 1 \\ 0 \\ 0 \end{pmatrix}$, $\qquad V_1 = \begin{pmatrix} 1 \\ 0 \\ 0 \end{pmatrix}$, $\qquad \Sigma_1 = 0.5000$, $\qquad \kappa = 1$.

Step 3.

$$L = L_0 U_1 \Sigma_1^{-1/2} = \begin{pmatrix} 1 \\ 1.3333 \\ 1.5833 \end{pmatrix}, \qquad Z = L_c V_1 \Sigma_1^{-1/2} = \begin{pmatrix} 1 \\ 0 \\ 0 \end{pmatrix}.$$

Step 4. $\tilde{A} = -1$, $\tilde{B} = 1$, $\tilde{C} = 1$.

Thus, $(\tilde{A}, \tilde{B}, \tilde{C})$ is a **balanced realization** of order 1, since the realized system is both controllable and observable. Indeed both the controllability and observability Grammians for this realization are equal to 0.5.

MATCONTROL note: Algorithm 14.2.2 has been implemented in MATCONTROL function **balsqt.**

Numerical difficulties of Algorithm 14.2.1 and 14.2.2: Algorithm 14.2.1 of the last section can be numerically unstable in the case when the matrix T is ill-conditioned.

To see this, we borrow the following simple example from Safonov and Chiang (1989):

$$\mathrm{Let} \left(\begin{array}{c|c} A & B \\ \hline C & 0 \end{array} \right) = \left(\begin{array}{cc|c} -\frac{1}{2} & -\epsilon & \epsilon \\ 0 & -\frac{1}{2} & 1 \\ \hline 1 & \epsilon & 0 \end{array} \right).$$

The transforming matrix T of Algorithm 14.2.1 in this case is given by

$$T = \begin{pmatrix} \sqrt{\frac{1}{\epsilon}} & 0 \\ 0 & \sqrt{\epsilon} \end{pmatrix}.$$

Thus, as ϵ becomes smaller and smaller, T becomes more and more ill-conditioned. Indeed, when $\epsilon \to 0$, Cond(T) becomes infinite.

In such cases, the model reduction procedure via internal balancing becomes unstable.

Similarly, the **square-root algorithm (Algorithm 14.2.2)** can be unstable if the matrices L and T are ill-conditioned.

14.3   INTERNAL BALANCING OF DISCRETE-TIME SYSTEMS

In this section, we consider internal balancing of the stable discrete-time system:

$$x_{k+1} = Ax_k + Bu_k,$$
$$y_k = Cx_k. \tag{14.3.1}$$

We assume that the system is controllable and observable, and give here a discrete analog of Algorithm 14.2.1.

The **discrete analog of Algorithm 14.2.2** can be similarly developed and is left as an (Exercise 14.11(b)).

The controllability Grammian C_G^D and the observability Grammian O_G^D, defined by (Chapter 7):

$$C_G^D = \sum_{i=0}^{\infty} A^i B B^T (A^T)^i$$

and

$$O_G^D = \sum_{i=0}^{\infty} (A^T)^i C^T C A^i$$

satisfy, in this case, respectively, the discrete Lyapunov equations:

$$A C_G^D A^T - C_G^D + B B^T = 0 \tag{14.3.2}$$

and

$$A^T O_G^D A - O_G^D + C^T C = 0. \tag{14.3.3}$$

It can then be shown that the transforming matrix T defined by

$$T = L_c V \Sigma^{-1/2}, \tag{14.3.4}$$

where L_c, V, and Σ are defined in the same way as in the continuous-time case, will transform the system (14.3.1) to the internally balanced system:

$$\tilde{x}_{k+1} = \tilde{A}\tilde{x}_k + \tilde{B}u_k,$$
$$\tilde{y}_k = \tilde{C}\tilde{x}_k. \tag{14.3.5}$$

The Grammians again are transformed to the same diagonal matrix Σ, the matrices $\tilde{A}$, $\tilde{B}$, and $\tilde{C}$ are defined in the same way as in the continuous-time case.

Example 14.3.1.

$$A = \begin{pmatrix} 0.0010 & 1 & 1 \\ 0 & 0.1200 & 1 \\ 0 & 0 & -0.1000 \end{pmatrix}, \qquad B = \begin{pmatrix} 1 \\ 1 \\ 1 \end{pmatrix}, \qquad C = (1, 1, 1).$$

(Note that the eigenvalues of A have moduli less than 1, so it is discrete-stable.)

Step 1. The discrete controllability and observability Grammians obtained, respectively, by solving (14.3.2) and (14.3.3) are:

$$C_G^D = \begin{pmatrix} 6.0507 & 3.2769 & 0.8101 \\ 3.2769 & 2.2558 & 0.8883 \\ 0.8101 & 0.8883 & 1.0101 \end{pmatrix},$$

and

$$O_G^D = \begin{pmatrix} 1 & 1.0011 & 1.0019 \\ 1.0011 & 2.2730 & 3.2548 \\ 1.0019 & 3.2548 & 5.4787 \end{pmatrix}.$$

Step 2. The Cholesky factors of C_G^D and O_G^D are:

$$L_c^D = \begin{pmatrix} 2.4598 & 0 & 0 \\ 1.3322 & 0.6936 & 0 \\ 0.3293 & 0.6482 & 0.6939 \end{pmatrix},$$

and

$$L_o^D = \begin{pmatrix} 1 & 0 & 0 \\ 1.0011 & 1.1273 & 0 \\ 1.0019 & 1.9975 & 0.6963 \end{pmatrix}.$$

Step 3. The SVD of $(L_o^D)^T L_c^D$:

$$[U, \Sigma, V] = \mathbf{svd}(L_o^D)^T L_c^D$$

gives

$$\Sigma = \mathrm{diag}(\ 5.3574, \quad 1.4007, \quad 0.1238\ ),$$

$$V = \begin{pmatrix} 0.8598 & -0.5055 & 0.0725 \\ 0.4368 & 0.6545 & -0.6171 \\ 0.2645 & 0.5623 & 0.7835 \end{pmatrix}.$$

Step 4.
$$\Sigma^{1/2} = \mathrm{diag}(2.3146, 1.1835, 0.3519).$$

Step 5. The transforming matrix T is:

$$T = L_c^D V \Sigma^{-1/2},$$

$$= \begin{pmatrix} 0.9137 & -1.0506 & 0.5068 \\ 0.6257 & -0.1854 & -0.9419 \\ 0.3240 & 0.5475 & 0.4759 \end{pmatrix}.$$

Step 6. The balanced matrices are:

$$\tilde{A} = T^{-1} A T = \begin{pmatrix} 0.5549 & 0.4098 & 0.0257 \\ -0.4098 & -0.1140 & 0.2629 \\ 0.0257 & -0.2629 & -0.4199 \end{pmatrix},$$

$$\tilde{B} = T^{-1}B = \begin{pmatrix} 1.8634 \\ 0.6885 \\ 0.0408 \end{pmatrix},$$

$$\tilde{C} = CT = (1.8634, \ -0.6885, \ 0.0408).$$

Verify:

$$T^{-1}C_G^D T^{-T} = T^T O_G^D T = \Sigma = \text{diag}(5.3574, \ 1.4007, \ 0.1238).$$

14.4 MODEL REDUCTION

Given an nth order realization (A, B, C) with the transfer function matrix $G(\lambda) = C(\lambda I - A)^{-1}B$, where "$\lambda$" is complex variable "$s$" for the continuous-time case and is the complex variable $z = (1 + s)/(1 - s)$ in the discrete time, the ideal **model reduction problem** aims at finding a state-space system of order $q < n$ such that the H_∞ error-norm

$$E = \|G(\lambda) - G_R(\lambda)\|_\infty$$

is minimized over all state-space systems of order q, where $G_R(\lambda)$ is the transfer function of the reduced-order model.

The exact minimization is a difficult computational task, and, in practice, a less strict requirement, such a guaranteed upper bound on E is sought to be achieved. We will discuss two such methods in this chapter:

- **Balanced Truncation Method**
- **The Schur Method**

We shall also describe briefly an **optimal Hankel-Norm Approximation** (HNA) method in **Section 14.5**. This optimal HNA method minimizes the error in Hankel norm (defined in **Section 14.5**). Furthermore, we will state another model reduction approach, called **Singular Perturbation** (SP) **Method** in **Exercise 14.23**. For properties of SP method, see Anderson and Liu (1989). Finally, an idea of **Frequency-Weighted Model Reduction** due to Enns (1984) will be discussed in **Section 14.7**.

14.4.1 Model Reduction via Balanced Truncation

As the title suggests, the idea behind model reduction via balanced truncation is to obtain a reduced-order model by deleting those states that are least controllable and observable (as measured by the size of **Hankel singular values**). Thus, if $\Sigma_R = \text{diag}(\sigma_1 I_{s_1}, \ldots, \sigma_N I_{s_N})$, is the matrix of Hankel singular values (which are arranged in decreasing order), obtained by a balanced realization, where s_i is the multiplicity of σ_i, and $\sigma_d \gg \sigma_{d+1}$, then the balanced realization implies that the states corresponding to the Hankel singular values $\sigma_{d+1} \ldots, \sigma_N$ are less controllable and less observable than those corresponding to $\sigma_1, \ldots, \sigma_d$. Thus, the reduced-order model obtained by eliminating these less controllable and less observable states are likely to retain some desirable information about the original system. Indeed, to this end, the following result (**Theorem 14.4.1**) holds. The idea of obtaining such a reduced-order model is due to Moore (1981). Part (a) of the theorem is due to Pernebo and Silverman (1982), and Part (b) was proved by Enns (1984) and independently by Glover (1984). We shall discuss here only the continuous-time case; the discrete-time case in analogous.

Theorem 14.4.1.   *Stability and Error Bound of the Truncated Subsystem. Let*

$$G(s) = \left[\begin{array}{cc|c} A_R & A_{12} & B_R \\ A_{21} & A_{22} & B_2 \\ \hline C_R & C_2 & 0 \end{array} \right] \tag{14.4.1}$$

be the transfer function matrix of an nth order internally balanced stable system with the Grammian $\Sigma = \mathrm{diag}(\Sigma_R, \Sigma_2)$, *where*

$$\begin{aligned} \Sigma_R &= \mathrm{diag}(\sigma_1 I_{s_1}, \ldots, \sigma_d I_{s_d}), \quad d < N \\ \Sigma_2 &= \mathrm{diag}(\sigma_{d+1} I_{s_{d+1}}, \ldots, \sigma_N I_{s_N}) \end{aligned} \tag{14.4.2}$$

and

$$\sigma_1 > \sigma_2 > \cdots > \sigma_d > \sigma_{d+1} > \sigma_{d+2} > \cdots > \sigma_N.$$

The multiplicity of σ_i *is* s_i, $i = 1, 2, \ldots, N$ *and* $s_1 + s_2 + \cdots + s_N = n$.

(a)   *Then the truncated system* (A_R, B_R, C_R) *with the transfer function:*

$$G_R(s) = \left[\begin{array}{c|c} A_R & B_R \\ \hline C_R & 0 \end{array} \right] \tag{14.4.3}$$

is balanced and stable.

(b)   *Furthermore, the error:*

$$\|G(s) - G_R(s)\|_\infty \le 2(\sigma_{d+1} + \cdots + \sigma_N). \tag{14.4.4}$$

In particular, if $d = N - 1$, *then* $\|G(s) - G_{N-1}(s)\|_\infty = 2\sigma_N$.

Proof.   We leave the proof part (a) as an exercise (**Exercise 14.1**). We assume that the **singular values** σ_i **are distinct** and prove part (b) only in the case $n = N$. The proof in the general case can be easily done using the proof of this special case and is also left as an exercise (**Exercise 14.1**). The proof has been taken from Zhou *et al.* (1996, pp. 158–160).

Let

$$\phi(s) = (sI - A_R)^{-1}, \tag{14.4.5}$$

$$\psi(s) = sI - A_{22} - A_{21}\phi(s)A_{12}, \tag{14.4.6}$$

$$\bar{B}(s) = A_{21}\phi(s)B_R + B_2, \tag{14.4.7}$$

$$\bar{C}(s) = C_R\phi(s)A_{12} + C_2. \tag{14.4.8}$$

Then,

$$G(s) - G_R(s) = C(sI - A)^{-1}B - C_R\phi(s)B_R,$$

$$= (C_R, C_2) \begin{pmatrix} sI - A_R & -A_{12} \\ -A_{21} & sI - A_{22} \end{pmatrix}^{-1} \begin{pmatrix} B_R \\ B_2 \end{pmatrix} - C_R\phi(s)B_R,$$

$$= \bar{C}(s)\psi^{-1}(s)\bar{B}(s). \tag{14.4.9}$$

For $s = j\omega$, we have

$$\sigma_{max}[G(j\omega) - G_R(j\omega)] = \lambda_{max}^{1/2}[\psi^{-1}(j\omega)\bar{B}(j\omega)\bar{B}^*(j\omega)\psi^{-*}(j\omega)\bar{C}^*(j\omega)\bar{C}(j\omega)], \tag{14.4.10}$$

where $\lambda_{max}(M)$ denotes the largest eigenvalue of the matrix M.

Since the singular values are distinct, we have $\Sigma_2 = \text{diag}(\sigma_{r+1}, \ldots, \sigma_n)$, and since Σ_2 satisfies

$$A_{22}\Sigma_2 + \Sigma_2 A_{22}^T + B_2 B_2^T = 0,$$

we obtain

$$\bar{B}(j\omega)\bar{B}^*(j\omega) = \psi(j\omega)\Sigma_2 + \Sigma_2\psi^*(j\omega).$$

Similarly, since Σ_2 also satisfies

$$\Sigma_2 A_{22} + A_{22}^T\Sigma_2 + C_2^T C_2 = 0,$$

we obtain

$$\bar{C}^*(j\omega)\bar{C}(j\omega) = \Sigma_2\psi(j\omega) + \psi^*(j\omega)\Sigma_2.$$

Substituting these expressions of $\bar{B}(j\omega)\bar{B}^*(j\omega)$ and $\bar{C}^*(j\omega)\bar{C}(j\omega)$ into (14.4.10), we obtain after some algebraic manipulations

$$\sigma_{max}[G(j\omega) - G_R(j\omega)] = \lambda_{max}^{1/2}\{[\Sigma_2 + \psi^{-1}(j\omega)\Sigma_2\psi^*(j\omega)]$$
$$\times [\Sigma_2 + \psi^{-*}(j\omega)\Sigma_2\psi(j\omega)]\}.$$

If $d = n - 1$, then $\Sigma_2 = \sigma_n$, and we immediately have

$$\sigma_{max}[G(j\omega) - G_R(j\omega)] = \sigma_n\lambda_{max}^{1/2}\{[1 + \Theta^{-1}(j\omega)][1 + \Theta(j\omega)]\}$$

where $\Theta = \psi^{-*}(j\omega)\psi(j\omega)$.

Note that $\Theta^{-*} = \Theta$ is a scalar function. So, $|\Theta(j\omega)| = 1$.
Using the triangle inequality, we then have

$$\sigma_{\max}[G(j\omega) - G_{\mathrm{R}}(j\omega)] \leq \sigma_n[1 + |\Theta(j\omega)|] = 2\sigma_n$$

Thus, we have proved the result for one-step, that is, we have proved the result assuming that the order of the truncated model is just one less than the original model. Using this "one-step" result, Theorem 14.4.1 can be proved for any order of the truncated model (**Exercise 14.1**). ■

The above theorem shows that once a system is internally balanced, the balanced system can be truncated by eliminating the states corresponding to the less controllable and observable modes (as measured by the sizes of the Hankel singular values) to obtain a reduced-order model that still preserves certain desirable properties of the original model (see Zhou *et al.* 1996). However, the reduced-order model obtained this way does not minimize the H_∞ error.

Choosing the order of the Reduced Model

If the reduced-order model is obtained by truncating the states corresponding to the smaller Hankel singular values $\sigma_{d+1}, \ldots, \sigma_N$, then the order q of the reduced-order model is

$$q = \sum_{i=1}^{d} s_{\mathrm{i}},$$

where s_i is the multiplicity of σ_{i}.

Computationally, the decision on choosing which Heinkel singular values are to be ignored, has to be made in a judicious way so that the matrix $\sum$ which needs to be inverted to compute the balanced matrices does not become too ill-conditioned. Thus, the ratios of the largest Hankel singular value to the consecutive ones need to be monitored. See discussion in Tombs and Postlethwaite (1987).

Algorithm 14.4.1. *Model Reduction via Balanced Truncation*
Inputs. *The system matrices A, B, and C.*
Outputs. *The reduced-order model with the matrices A_{R}, B_{R}, and C_{R}.*
Assumption. *A is stable.*
Step 1. *Find a balanced realization.*
Step 2. *Choose q, the order of model reduction.*
Step 3. *Partition the balanced realization in the form*

$$A = \begin{pmatrix} A_{\mathrm{R}} & A_{12} \\ A_{21} & A_{22} \end{pmatrix}, \qquad B = \begin{pmatrix} B_{\mathrm{R}} \\ B_2 \end{pmatrix}, \qquad C = (C_{\mathrm{R}}, C_2),$$

where A_{R} is of order q, and B_{R} and C_{R} are similarly defined.

The MATLAB function **modred** in the form

$$RSYS = modred(SYS, ELIM)$$

reduces the order of the model sys, by eliminating the states specified in the vector ELIM.

Example 14.4.1. Consider Example 14.2.1 once more. Choose $q = 2$. Then $A_R =$ The 2×2 leading principal submatrix of $\tilde{A}$ is

$$\begin{pmatrix} -0.7659 & 0.5801 \\ -0.5801 & -2.4919 \end{pmatrix}.$$

The eigenvalues of A_R are: -0.9900 and -2.2678.
  Therefore, A_R is stable.
  The matrices B_R and C_R are:

$$B_R = \begin{pmatrix} -1.8602 \\ -0.6759 \end{pmatrix}, \qquad C_R = (-1.8602, 0.6759).$$

Let $G_R(s) = C_R(sI - A_R)^{-1}B_R$.

  Verification of the Error Bound: $\|G(s) - G_R(s)\|_\infty = 0.0012$. Since $2\sigma_3 = 0.0012$, the error bound given by (14.4.4) is satisfied.

14.4.2 The Schur Method for Model Reduction

The numerical difficulties of model reduction via balanced truncation using Algorithm 14.2.1 or Algorithm 14.2.2 (because of possible ill-conditioning of the transforming matrices) can be overcome if orthogonal matrices are used to transform the system to another **equivalent system** from which the reduced-order model is extracted. Safonov and Chiang (1989) have proposed a Schur method for this purpose.
  A key observation here is that in Algorithm 14.2.1, the rows $\{1, \ldots, d\}$ and rows $\{d + 1, \ldots, n\}$ of T^{-1} form bases for the left eigenspaces of the matrix $C_G O_G$ associated with the eigenvalues $\{\sigma_1^2, \ldots, \sigma_d^2\}$ and $\{\sigma_{d+1}^2, \ldots, \sigma_n^2\}$, respectively (**Exercise 14.7**).
  Thus the idea will be to replace the matrices T and T^{-1} (which can be very ill-conditioned) by the orthogonal matrices (which are perfectly conditioned) sharing the same properties.
  The Schur method, described below, constructs such matrices, using the RSF of the matrix $C_G O_G$.
  Specifically, the orthonormal bases for the right and left invariant subspaces corresponding to the "**large**" eigenvalues of the matrix $C_G O_G$ will be computed by finding the **ordered** real Schur form of $C_G O_G$.

Once the "large" eigenvalues are isolated from the "small" ones, the reduced order model preserving the desired properties, can be easily extracted.

We will now state the algorithm.

Algorithm 14.4.2. *The Schur Method for Model Reduction (Continuous-time System).*

Inputs.

A—The $n \times n$ state matrix.

B—The $n \times m$ control matrix.

C—The $r \times n$ output matrix.

q—The dimension of the desired reduced-order model.

Outputs.

$\hat{A}_R$*—The $q \times q$ reduced state matrix*

$\hat{B}_R$*—The $q \times m$ reduced control matrix*

$\hat{C}_R$*—The $r \times q$ reduced output matrix.*

S_1 *and* S_2*—Orthogonal transforming matrices such that* $\hat{A}_R = S_1^T A S_2$,

$\hat{B}_R = S_1^T B$, *and* $\hat{C}_R = C S_2$

Assumption. *A is stable.*

Step 1. *Compute the controllability Grammian C_G and the observability Grammian O_G by solving, respectively, the Lyapunov equations (14.2.5) and (14.2.6).*

Step 2. *Transform the matrix $C_G O_G$ to the RSF Y, that is, find an orthogonal matrix X such that $X^T C_G O_G X = Y$.*

Note: The matrix $C_G O_G$ does not have any complex eigenvalues. Thus, Y is actually an upper triangular matrix. Furthermore, the real eigenvalues are nonnegative.

Step 3. *Reorder the eigenvalues of Y in ascending and descending order, that is, find orthogonal matrices U and V such that*

$$U^T Y U = U^T X^T C_G O_G X U = U_S^T C_G O_G U_S = \begin{pmatrix} \lambda_1 & & * \\ & \ddots & \\ 0 & & \lambda_n \end{pmatrix}, \quad (14.4.11)$$

$$V^T Y V = V^T X^T C_G O_G X V = V_S^T C_G O_G V_S = \begin{pmatrix} \lambda_n & & * \\ & \ddots & \\ 0 & & \lambda_1 \end{pmatrix}, \quad (14.4.12)$$

where $\lambda_1 \le \lambda_2 \le \cdots \le \lambda_n$.

(Note that $\lambda_n = \sigma_1^2, \lambda_{n-1} = \sigma_2^2, \ldots$, and so on; where $\sigma_1 \ge \sigma_2 \ge \cdots \ge \sigma_n$ are the Hankel singular values.)

Step 4. *Partition the matrices* U_S, V_S *as follows:*

$$U_S = (U_{1S}, U_{2S}), \qquad V_S = (V_{1S}, V_{2S}).$$

Here U_{1S} contains the first $n - q$ columns of U_S and U_{2S} contains the remaining q columns. On the other hand, V_{1S} contains the first q columns of V_S and V_{2S} contains the remaining $n - q$ columns.

Note: Note that the columns of the V_{1S} and those of the matrix U_{1S} form, respectively, orthonormal bases for the right invariant subspace of $C_G O_G$ associated with the large eigenvalues $\{\sigma_1^2, \ldots, \sigma_q^2\}$ and the small eigenvalues $\{\sigma_{q+1}^2, \ldots, \sigma_n^2\}$. The columns of U_{2S} and V_{2S}, similarly, form orthonormal bases for the left invariant subspace of $C_G O_G$, with the large and the small eigenvalues, respectively.

Step 5. *Find the SVD of $U_{2S}^T V_{1S}$: $Q \Sigma R^T = U_{2S}^T V_{1S}$.*

Step 6. Compute *the transforming matrices:* $\quad S_1 = U_{2S} Q \Sigma^{-1/2}$, $S_2 = V_{1S} R \Sigma^{-1/2}$

Step 7. *Form the reduced-order matrices:*

$$\hat{A}_R = S_1^T A S_2, \qquad \hat{B}_R = S_1^T B, \qquad and \ \hat{C}_R = C S_2. \qquad (14.4.13)$$

Flop-count: Since the reduction to the Schur form using the QR iteration is an iterative process, an exact count cannot be given. The method just outlined requires approximately $100n^3$ flops.

Properties of the Reduced-Order Model by the Schur Method

The Schur method for model reduction does not give balanced realization. But the essential properties of the original model are preserved in the reduced-order model, as shown in the following theorem.

Theorem 14.4.2. *The transfer function matrix $\hat{G}_R(s) = \hat{C}_R(sI - \hat{A}_R)^{-1}\hat{B}_R$ obtained by the Schur method (Algorithm 14.4.2) is exactly the same as that of the one obtained via balanced truncation (Algorithm 14.4.1). Furthermore, the controllability and observability Grammians of the reduced-order model are, respectively, given by:*

$$\hat{C}_G^R = S_1^T C_G S_1, \qquad \hat{O}_G^R = S_2 O_G S_2.$$

Proof. We prove the first part and leave the second part as an exercise **(Exercise 14.9).**

Let the transforming matrix T of the internal balancing algorithm and its inverse be partitioned as:

$$T = (T_1, T_2), \qquad (14.4.14)$$

and

$$T^{-1} = \begin{pmatrix} T_I \\ * \end{pmatrix}, \qquad (14.4.15)$$

where T_1 is $n \times q$ and T_I is of order $q \times n$.

Then the transfer function $G_R(s)$ of the reduced-order model obtained by Algorithm 14.2.1 is given by

$$G_R(s) = C_R(sI - A_R)^{-1} B_R = CT_1(sI - T_I A T_1)^{-1} T_I B. \qquad (14.4.16)$$

Again, the transfer function $\hat{G}_R(s)$ of the reduced-order model obtained by the Schur algorithm (Algorithm 14.4.2) is given by:

$$\hat{G}_R(s) = \hat{C}_R(sI - \hat{A}_R)^{-1} \hat{B}_R = CS_2(sI - S_1^T A S_2)^{-1} S_1^T B. \qquad (14.4.17)$$

The proof now amounts to establishing a relationship between S_1, S_2 and T_1 and T_I.

Let's define

$$V_R = (V_{1S} \ U_{1S}), \qquad \text{and} \qquad V_L = \begin{pmatrix} U_{2S}^T \\ V_{2S}^T \end{pmatrix}. \qquad (14.4.18)$$

Then, since the first q and the last $(n - q)$ columns of T^{-1}, V_R, and V_L^{-1} span, respectively, the right eigenspaces associated with $\sigma_1^2, \ldots, \sigma_q^2$ and $\sigma_{q+1}^2, \ldots, \sigma_n^2$, it follows that there exist nonsingular matrices X_1, and X_2, such that

$$V_R = T \begin{pmatrix} X_1 & 0 \\ 0 & X_2 \end{pmatrix} = V_L^{-1} \begin{pmatrix} E_1 & 0 \\ 0 & E_2 \end{pmatrix}. \qquad (14.4.19)$$

From (14.4.18) and (14.4.19) we have

$$V_{1S} = T_1 X_1. \qquad (14.4.20)$$

Thus,

$$S_2 = V_{1S} R \Sigma^{-1/2} = T_1 X_1 R \Sigma^{-1/2} \ \text{(using (14.4.20))}.$$

Similarly,

$$\begin{aligned} S_1^T &= \Sigma^{-1/2} Q^T U_{2S}^T = \Sigma^{1/2} R^T (R \Sigma^{-1} Q^T) U_{2S}^T, \\ &= \Sigma^{1/2} R^T (U_{2S}^T V_{1S})^{-1} U_{2S}^T \\ &= \Sigma^{1/2} R^T (U_{2S}^T T_1 X_1)^{-1} U_{2S}^T \ \text{(using (14.4.20))} = \Sigma^{1/2} R^T X_1^{-1} T_I. \end{aligned}$$

Thus,

$$
\begin{aligned}
\hat{G}_R(s) &= C S_2 (sI - S_1^T A S_2)^{-1} S_1^T B \\
&= C T_1 X_1 R \Sigma^{-1/2} \left(sI - \Sigma^{1/2} R^T X_1^{-1} T_I A T_1 X_1 R \Sigma^{-1/2} \right)^{-1} \\
&\quad \times \left(\Sigma^{1/2} R^T X_1^{-1} T_I B \right) \\
&= C T_1 (sI - T_I A T_1)^{-1} T_I B = G_R(s). \quad \blacksquare
\end{aligned}
$$

Note: Since $G_R(s)$ of Theorem 14.4.1 and $\hat{G}_R(s)$ of Theorem 14.4.2 are the same, from Theorem 14.4.1, we conclude that $\hat{A}_R$ is stable and $\|G(s) - \hat{G}_R(s)\|_\infty \le 2 \sum_{i=q+1}^n \sigma_i$, where σ_{q+1} through σ_n are the $(q + 1)$th through nth entries of the diagonal matrix Σ of the balancing algorithm, that is, σs are the Hankel singular values.

Relation to the square-root method: There could be large round-off errors in explicit computation of the matrix product $C_G O_G$. The formation of the explicit product, however, can be avoided by computing the Cholesky factors L_c and L_o of the matrices C_G and O_G, using **Algorithm 8.6.1** described in Chapter 8. This then leads to a **square-root** method for model reduction. We leave the derivation of the modified algorithm to the readers **(Exercise 14.10)**. For details, see Safonov and Chiang (1989).

Example 14.4.2.

$$
A = \begin{pmatrix} -1 & 2 & 3 \\ 0 & -2 & 1 \\ 0 & 0 & -3 \end{pmatrix}, \qquad b = \begin{pmatrix} 1 \\ 1 \\ 1 \end{pmatrix}, \qquad C = (1, 1, 1), \qquad q = 2.
$$

The system (A, B, C) is **stable, controllable**, and **observable**.

Step 1. Solving the Lyapunov equations (14.2.5) and (14.2.6) we obtain

$$
C_G = \begin{pmatrix} 3.9250 & 0.9750 & 0.4917 \\ 0.9750 & 0.3667 & 0.2333 \\ 0.4917 & 0.2333 & 0.1667 \end{pmatrix},
$$

and

$$
O_G = \begin{pmatrix} 0.5000 & 0.6667 & 0.7917 \\ 0.6667 & 0.9167 & 1.1000 \\ 0.7917 & 1.1000 & 1.3250 \end{pmatrix}.
$$

Step 2. The Schur decomposition Y and the transforming matrix X obtained using the MATLAB function **schur**:

$$
[X, Y] = \text{schur}(C_G O_G)
$$

are

$$X = \begin{pmatrix} -0.9426 & -0.3249 & 0.0768 \\ -0.2885 & 0.6768 & -0.6773 \\ -0.1680 & 0.6606 & 0.7317 \end{pmatrix},$$

$$Y = \begin{pmatrix} 5.1028 & -5.2629 & -1.0848 \\ 0 & 0.0084 & 0.0027 \\ 0 & 0 & 0 \end{pmatrix}.$$

Step 3. Since the eigenvalues of Y are in decreasing order of magnitude, we take

$$V_S = X.$$

Next, we compute U_S such that the eigenvalues of $U_S^T C_G O_G U_S$ appear in increasing order of magnitude:

$$U_S = \begin{pmatrix} 0.2831 & -0.8579 & 0.4289 \\ -0.8142 & 0.0214 & 0.5802 \\ 0.5069 & 0.5134 & 0.6924 \end{pmatrix},$$

$$U_S^T C_G O_G U_S = \begin{pmatrix} 0 & -0.0026 & 0.8663 \\ 0 & 0.0084 & -5.3035 \\ 0 & 0 & 5.1030 \end{pmatrix}.$$

Step 4. Partitioning U_S and $V_S = X$, we obtain

$$U_{1S} = \begin{pmatrix} 0.2831 \\ -0.8142 \\ 0.5069 \end{pmatrix}, \qquad U_{2S} = \begin{pmatrix} -0.8579 & 0.4289 \\ 0.0214 & 0.5802 \\ 0.5134 & 0.6924 \end{pmatrix}.$$

$$V_{1S} = \begin{pmatrix} -0.9426 & -0.3249 \\ -0.2885 & 0.6768 \\ -0.1680 & 0.6606 \end{pmatrix}, \qquad V_{2S} = \begin{pmatrix} 0.0768 \\ -0.6773 \\ 0.7317 \end{pmatrix}.$$

Step 5. The SVD of the product $U_{2S}^T V_{1S}$ is given by:

$$[Q, \Sigma, R] = \text{svd}(U_{2S}^T V_{1S})$$

$$Q = \begin{pmatrix} -0.4451 & 0.8955 \\ 0.8955 & 0.4451 \end{pmatrix}, \qquad R = \begin{pmatrix} -0.9348 & 0.3550 \\ 0.3550 & 0.9349 \end{pmatrix} \qquad \text{and}$$

$$\Sigma = \text{diag}(1, 0.9441).$$

Step 6. The transforming matrices are:

$$S_1 = \begin{pmatrix} 0.7659 & -0.5942 \\ 0.5100 & 0.2855 \\ 0.3915 & 0.7903 \end{pmatrix} \quad \text{and} \quad S_2 = \begin{pmatrix} 0.7659 & -0.6570 \\ 0.5099 & 0.5458 \\ 0.3916 & 0.5742 \end{pmatrix}.$$

Step 7. The matrices of the reduced order model are:

$$\hat{A}_R = S_1^T A S_2 = \begin{pmatrix} 0.3139 & 1.7204 \\ -1.9567 & -3.5717 \end{pmatrix}, \quad \hat{B}_R = S_1^T B = \begin{pmatrix} 1.6674 \\ 0.4817 \end{pmatrix}, \quad \text{and}$$

$$\hat{C}_R = C S_2 = (1.6674, 0.4631).$$

Verification of the properties of the reduced-order model: We next verify that the reduced-order model has desirable properties such as stability and the error bound (14.4.4) is satisfied.

1. The eigenvalues of $\hat{A}_R$ are: $\{-0.9900, -2.2678\}$. Thus, $\hat{A}_R$ is **stable.**
   (**Note that these eigenvalues are the same as those of A_R of order 2 obtained by Algorithm 14.4.1**).
2. The controllability Grammian $\hat{C}_G^R$ of the reduced order model is given by:

$$\hat{C}_G^R = S_1^T C_G S_1 = \begin{pmatrix} 3.5732 & -1.4601 \\ -1.4601 & 0.8324 \end{pmatrix}.$$

The eigenvalues of $\hat{C}_G^R$ are 4.2053, and 0.2003. Thus, $\hat{C}_G^R$ is positive definite.
   It is easily verified by solving the Lyapunov equation $\hat{A}_R \hat{C}_G^R + \hat{C}_G^R \hat{A}_R^T = -\hat{B}_R \hat{B}_R^T$, that the $\hat{C}_G^R$ given above is indeed the controllability Grammian of the reduced order model.

Similar results hold for the observability Grammian $\hat{O}_{GR}$.

Verification of the error bound: $\|G(s) - \hat{G}_R(s)\|_\infty = 0.0012$.
Since $2\sigma_3 = 0.0012$, the error bound (14.4.4) is verified.

Comparison of the reduced order models obtained by balanced truncation and the schur method with the original Model: Figure 14.1 compares the errors of the reduced-order models with the theoretical error bound given by (14.4.4). Figure 14.2 compares the step response of the original model with the step responses of the reduced-order models obtained by balanced truncation and the Schur method.

MATCONTROL note: The MATCONTROL function **modreds** implements the Schur Algorithm (**Algorithm 14.4.2**) in the following format:

$$[A_R, B_R, C_R, S, T] = \text{modreds}(A, B, C, d).$$

The matrices A_R, B_R, C_R are the matrices of the reduced-order model of dimension d. The matrices S and T are the transforming matrices.

14.4.3 A Balancing-Free Square-Root Method for Model Reduction

By computing the matrices L and T a little differently than in the square-root method (**Algorithm 14.2.2**), the main advantages of the Schur method and the square-root method can be combined. The idea is due to Varga (1991).

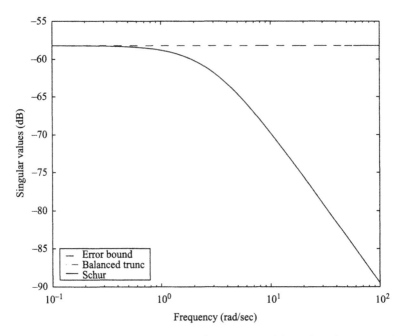

FIGURE 14.1: Theoretical error bound and errors of the reduced-order models.

Consider the **economy** QR factorizations of the matrices $L_c V_1$ and $L_o^T U_1$:

$$L_c V_1 = XW, \qquad L_o^T U_1 = YZ,$$

where W and Z are nonsingular upper triangular and X and Y are orthonormal matrices.

Then L and T defined by

$$L = (Y^T X)^{-1} Y^T, \qquad Z = X$$

are such that the system $(\tilde{A}, \tilde{B}, \tilde{C})$ with $\tilde{A} = LAZ$, $\tilde{B} = LB$, and $\tilde{C} = CZ$ form a **minimal realization** and therefore can be used to obtain a reduced-order model.

Example 14.4.3. Let's consider Example 14.2.2 once more.
  Then,

$$X = \begin{pmatrix} -1 \\ 0 \\ 0 \end{pmatrix}, \qquad Y = \begin{pmatrix} -1 \\ 0 \\ 0 \end{pmatrix}, \qquad W = -0.7071, \qquad Z = -0.7071.$$

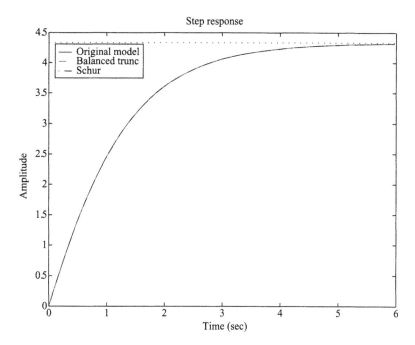

FIGURE 14.2:   Step responses of the original and the reduced-order models.

The matrices L and Z in this case are

$$L = (-1 \ \ 0 \ \ 0), \qquad Z_2 = \begin{pmatrix} -1 \\ 0 \\ 0 \end{pmatrix}.$$

The matrices $\tilde{A}$, $\tilde{B}$ and $\tilde{C}$ are: $\tilde{A} = -1$, $\tilde{B} = -1$, $\tilde{C} = -1$.

14.5   HANKEL-NORM APPROXIMATIONS

Let (A, B, C) be a stable realization of $G(s) = C(sI - A)^{-1}B$. Then the **Hankel-norm** of $G(s)$ is defined as

$$\|G(s)\|_H = \lambda_{\max}^{1/2}(C_G O_G), \tag{14.5.1}$$

where C_G and O_G are the controllability and observability Grammians, respectively, and $\lambda_{\max}(M)$ stands for the largest eigenvalue of M.

The **optimal Hankel-norm approximation problem** is the problem of finding an approximation $\hat{G}(s)$ of McMillan degree $k < n$ such that the norm of the error $\|G(s) - \hat{G}(s)\|_H$ is minimized.

The following theorem gives an achievable lower bound of the error of an approximation in Hankel-norm. Proof can be found in Glover (1984) or in Zhou *et al.* (1996, pp. 189–190).

Theorem 14.5.1. *Let $G(s)$ be a stable rational transfer function with Hankel singular values $\sigma_1 \geq \sigma_2 \geq \cdots \geq \sigma_k \geq \sigma_{k+1} \cdots \geq \sigma_n > 0$. Then for all stable $\hat{G}(s)$ of McMillan degree $\leq k$*

$$\| G(s) - \hat{G}(s) \|_{\bar{H}} \geq \sigma_{k+1}.$$

We now give a result on characterization of all solutions to optimal Hankel-norm approximations and then state an algorithm to find an optimal Hankel-norm approximation.

The presentation here is based on Glover (1984). For proofs and other details, the readers are referred to the paper of Glover or the book by Zhou *et al.* (1996).

14.5.1   A Characterization of All Solutions to the optional Hankel-Norm Approximation

The following theorem gives necessary and sufficient conditions for $\hat{G}(s)$ to be an optimal Hankel-norm approximation to $G(s)$. (For proof, see Glover (1984, lemma 8.1).)

Theorem 14.5.2. *Let $G(s) = C(sI - A)^{-1}B$ be a stable, rational, $m \times m$ transfer function with singular values*

$$\sigma_1 \geq \sigma_2 \geq \sigma_3 \geq \cdots \geq \sigma_k > \sigma_{k+1} = \sigma_{k+2} = \cdots = \sigma_{k+p} > \sigma_{k+p+1} \geq \cdots$$
$$\geq \sigma_n > 0.$$

Let $\hat{G}(s)$ be of McMillan degree $k \leq n - p$. Then $\hat{G}(s)$ is an optimal Hankel-norm approximation to $G(s)$ if and only if there exists $(\hat{A}, \hat{B}, \hat{C})$, P_e, Q_e such that

(a)   *$\hat{G}(s)$ is the stable part of*

$$\hat{C}(sI - \hat{A})^{-1}\hat{B}. \tag{14.5.2}$$

(b)   *The matrices P_e and Q_e satisfy*

(*i*) $$A_e P_e + P_e A_e^{\mathrm{T}} + B_e B_e^{\mathrm{T}} = 0, \tag{14.5.3}$$

(*ii*) $$A_e^{\mathrm{T}} Q_e + Q_e A_e + C_e^{\mathrm{T}} C_e = 0 \tag{14.5.4}$$

(*iii*) $$P_e Q_e = \sigma_{k+1}^2 I, \tag{14.5.5}$$

where A_e, B_e and C_e are defined by

$$A_e = \begin{pmatrix} A & 0 \\ 0 & \hat{A} \end{pmatrix}, \qquad B_e = \begin{pmatrix} B \\ \hat{B} \end{pmatrix}, \qquad C_e = (C, \; -\hat{C}). \quad (14.5.6)$$

and

(c) *If P_e and Q_e are partitioned conformally with A_e in (14.5.6) as:*

$$P_e = \begin{pmatrix} P_{11} & P_{12} \\ P_{12}^T & P_{22} \end{pmatrix}, \qquad Q_e = \begin{pmatrix} Q_{11} & Q_{12} \\ Q_{12}^T & Q_{22} \end{pmatrix},$$

then

$$\mathrm{In}(P_{22}) = \mathrm{In}(Q_{22}) = (k, l, 0). \quad (14.5.7)$$

Further, $\dim(\hat{A}) = k + l$ can be chosen $\leq n + 2k - 1$.

We now give a construction of $\hat{A}$, $\hat{B}$, $\hat{C}$ that satisfy the Eqs. (14.5.3)–(14.5.7) for a **balanced realization** (A, B, C) of $G(s)$, which will be the basis of an algorithm for Hankel-norm approximation. The construction, however, remains valid for a more general class of realization, and the details can be found in Glover (1984).

Theorem 14.5.3. *Let (A, B, C) be a balanced realization of $G(s)$ with the matrix of the singular values*

$$\Sigma = \mathrm{diag}(\sigma_1, \sigma_2, \ldots, \sigma_n),$$

$$\sigma_1 \geq \sigma_2 \geq \cdots \geq \sigma_k > \sigma_{k+1} = \sigma_{k+2} = \cdots = \sigma_{k+p} > \sigma_{k+p+1} \geq \cdots \geq \sigma_n > 0.$$

Partition $\Sigma = (\Sigma_1, \sigma_{k+1}I_p)$, $\sigma_{k+1} \neq 0$, and then partition A, B, C conformally:

$$A = \begin{pmatrix} A_{11} & A_{12} \\ A_{21} & A_{22} \end{pmatrix}, \qquad B = \begin{pmatrix} B_1 \\ B_2 \end{pmatrix}, \qquad C = (C_1, C_2). \quad (14.5.8)$$

Define now

$$\hat{A} = \Gamma^{-1}(\sigma_{k+1}^2 A_{11}^T + \Sigma_1 A_{11} \Sigma_1 - \sigma_{k+1} C_1^T U B_1^T), \quad (14.5.9)$$

$$\hat{B} = \Gamma^{-1}(\Sigma_1 B_1 + \sigma_{k+1} C_1^T U), \quad (14.5.10)$$

$$\hat{C} = C_1 \Sigma_1 + \sigma_{k+1} U B_1^T, \quad (14.5.11)$$

where

$$\Gamma = \Sigma_1^2 - \sigma_{k+1}^2 I \quad (14.5.12)$$

and U is such that

$$U = -C_2 \left(B_2^T \right)^{\dagger}, \quad (14.5.13)$$

where '$\dagger$' denotes generalized inverse.

Then A_e, B_e, and C_e defined by (14.5.6) satisfy (14.5.3)–(14.5.5), with

$$P_e = \begin{pmatrix} \Sigma_1 & 0 & I \\ 0 & \sigma_{k+1}I & 0 \\ I & 0 & \Sigma_1\Gamma^{-1} \end{pmatrix}, \qquad (14.5.14)$$

$$Q_e = \begin{pmatrix} \Sigma_1 & 0 & -\Gamma \\ 0 & \sigma_{k+1}I & 0 \\ -\Gamma & 0 & \Sigma_1\Gamma \end{pmatrix}. \qquad (14.5.15)$$

Based on Theorems 14.5.2 and 14.5.3, the following algorithm can be written down for finding a Hankel-norm approximation of a balanced realization (A, B, C) of $G(s)$.

Algorithm 14.5.1. *An Algorithm for optimal Hankel-Norm Approximation of a Continuous-Time System*
   Inputs.
   1.   *The matrices A, B, and C of a stable realization $G(s)$.*
   2.   *k—McMillan degree of Hankel-norm approximation*
   Outputs. *The matrices $\hat{A}_{11}$, $\hat{B}_1$, and $\hat{C}_1$ of a Hankel-norm approximation $\hat{G}(s)$ of McMillan degree k.*
   Assumptions.
   1.   *A is stable.*
   2.   *The Hankel singular values σ_i are such that $\sigma_1 \geq \sigma_2 \geq \cdots \geq \sigma_k > \sigma_{k+1} > \sigma_{k+2} \geq \cdots \geq \sigma_n > 0$.*

   Step 1. *Find a balanced realization $(\tilde{A}, \tilde{B}, \tilde{C})$ of $G(s)$ using* **Algorithm 14.2.1** *or* **Algorithm 14.2.2**, *whichever is appropriate.*
   Step 2. *Partition $\Sigma = \mathrm{diag}(\Sigma_1, \sigma_{k+1})$ and then order the balanced realization $(\tilde{A}, \tilde{B}, \tilde{C})$ conformally so that*

$$\tilde{A} = \begin{pmatrix} A_{11} & A_{12} \\ A_{21} & A_{22} \end{pmatrix}, \qquad \tilde{B} = \begin{pmatrix} B_1 \\ B_2 \end{pmatrix}, \qquad \tilde{C} = (C_1, C_2).$$

(Note that A_{11} is $(n-1) \times (n-1)$).
   Step 3. *Compute the matrix U satisfying (14.5.13) and form the matrices Γ, $\hat{A}$, $\hat{B}$, and $\hat{C}$ using Eqs. (14.5.9)–(14.5.12).*
   Step 4. Block diagonalize *$\hat{A}$ to obtain $\hat{A}_{11}$:*

   (a)   *Transform $\hat{A}$ to an upper real Schur form and then order the real Schur form so that the eigenvalues with negative real parts appear first; that is, find an orthogonal matrix V_1 such that $V_1^T \hat{A} V_1$ is in upper real Schur* *form and then find another orthogonal matrix V_2 such that*

$$V_2^T V_1^T \hat{A} V_1 V_2 = \begin{pmatrix} \hat{A}_{11} & \hat{A}_{12} \\ 0 & \hat{A}_{22} \end{pmatrix}, \tag{14.5.16}$$

where the eigenvalues of $\hat{A}_{11}$ have negative real parts and those of $\hat{A}_{22}$ have positive real parts. (Note that $\hat{A}_{11}$ is $k \times k$).

(b) *Solve the Sylvester equation for $X \in \mathbb{R}^{k \times (n-k-1)}$ (using **Algorithm 8.5.1**):*

$$\hat{A}_{11} X - X \hat{A}_{22} + \hat{A}_{12} = 0 \tag{14.5.17}$$

(c) *Let*

$$T = V_1 V_2 \begin{pmatrix} I & X \\ 0 & I \end{pmatrix} = (T_1, T_2), \tag{14.5.18}$$

$$S = \begin{pmatrix} I & -X \\ 0 & I \end{pmatrix} V_2^T V_1^T = \begin{pmatrix} S_1 \\ S_2 \end{pmatrix}. \tag{14.5.19}$$

Step 5. *Form*

$$\hat{B}_1 = S_1 \hat{B}, \tag{14.5.20}$$

$$\hat{C}_1 = \hat{C} T_1. \tag{14.5.21}$$

Example 14.5.1. Consider Example 14.2.1 once more. Then
Step 1. The balanced matrices $\tilde{A}$, $\tilde{B}$, and $\tilde{C}$ are the same as of Example 14.2.1.
Step 2.

$$\Sigma = \text{diag}(2.2589, 0.0917, 0.0006), \qquad \Sigma_1 = \text{diag}(2.2589, 0.0917).$$

$k = 2$ and $\sigma_3 = 0.0006$.

$$A_{11} = \begin{pmatrix} -0.7659 & 0.5801 \\ -0.5801 & -2.4919 \end{pmatrix}, \qquad B_1 = \begin{pmatrix} -1.8602 \\ -0.6759 \end{pmatrix}, \qquad B_2 = (0.0581)$$

$$C_1 = (-1.8602, 0.6759), \qquad C_2 = (-0.0581).$$

Step 3.

$$U = 1, \qquad \Gamma = \text{diag}(5.1026, 0.0084),$$

$$\hat{A} = \begin{pmatrix} -0.7659 & 0.0235 \\ -14.2961 & -2.4919 \end{pmatrix}, \qquad \hat{B} = \begin{pmatrix} -0.8235 \\ -7.3735 \end{pmatrix}, \qquad \hat{C} = (-4.2020, -0.0620).$$

Step 4.

(a) $\hat{A}$ is already in upper Schur form. Thus, $V_1 = I$.
    The eigenvalues of $\hat{A}$ are $-0.9900, -2.2678$. Since both have negative real parts, no reordering is needed.
    Thus, $V_2 = I$, $\hat{A}_{11} = \hat{A}$, $\hat{A}_{12} = 0$, $\hat{A}_{22} = 0$.

(b) $X = 0$.
(c)

$$T = \begin{pmatrix} I & 0 \\ 0 & I \end{pmatrix}, \qquad T_1 = I, \qquad T_2 = I.$$

$$S = \begin{pmatrix} I & 0 \\ 0 & I \end{pmatrix}, \qquad S_1 = I, \qquad S_2 = I.$$

Step 5.

$$\hat{B}_1 = \hat{B}, \qquad \hat{C}_1 = \hat{C}.$$

Obtaining an error bound: Next, we show how to construct a matrix $\hat{D}_1$ such that with

$$\hat{G}(s) = \hat{D}_1 + \hat{C}_1(sI - \hat{A}_{11})^{-1}\hat{B}_1,$$

an error bound for the approximation $\|G(s) - \hat{G}(s)\|_\infty$ can be obtained.
Define

$$\hat{B}_2 = S_2\hat{B}, \qquad \hat{C}_2 = \hat{C}T_2, \qquad \hat{D}_1 = -\sigma_{k+1}U. \qquad (14.5.22)$$

Step 6. Update now $\hat{D}_1$ as follows:
   6.1. Find a balanced realization of the system $(-\hat{A}_{22}, \hat{B}_2, \hat{C}_2, \hat{D}_1)$, say (A_3, B_3, C_3, D_3). Compute the Hankel singular values of this balanced system and call them $\mu_1, \mu_2, \ldots, \mu_{n-k-1}$.
   6.2. Let q be an integer greater than or equal to $r + m$, where r and m are the number of outputs and inputs, respectively. Define $Z, Y \in \mathbb{R}^{q \times (n-k-1)}$ by

$$Z = \begin{pmatrix} B_3^T \\ 0 \end{pmatrix}, \qquad Y = \begin{pmatrix} C_3 \\ 0 \end{pmatrix}.$$

Denote the ith columns of Z and Y by z_i and y_i, respectively.
   6.3. For $i = 1, 2, \ldots, n - k - 1$ do

(i) Find Householder matrices H_1 and H_2 such that

$$H_1 y_i = - \begin{pmatrix} \alpha & 0 & \cdots & 0 \end{pmatrix}^T$$

and

$$H_2 z_i = - \begin{pmatrix} \beta & 0 & \cdots & 0 \end{pmatrix}^T.$$

(ii) Define

$$U_i = H_1 \begin{pmatrix} -\alpha/\beta & 0 & 0 & 0 \\ 0 & 0 & I_{r-1} & 0 \\ 0 & I_{m-1} & 0 & 0 \\ 0 & 0 & 0 & I_{q-r-m+1} \end{pmatrix} H_2.$$

(iii) If $i < n - k + 1$, then for $j = i + 1$ to $(n - k + 1)$ do

$$y \equiv -(y_j \mu_j + U z_j \mu_i)(\mu_i^2 - \mu_j^2)^{-1/2},$$
$$z_j \equiv (z_j \mu_j + U^T y_j \mu_i)(\mu_i^2 - \mu_j^2)^{-1/2},$$
$$y_j = y$$

(iv) Compute $\hat{D}_1 = \hat{D}_1 + (-1)^i \mu_i \begin{pmatrix} I_r & 0 \end{pmatrix} U \begin{pmatrix} I_m \\ 0 \end{pmatrix}$

Theorem 14.5.4. *(An Error Bound).* $\|G(s) - \hat{G}(s)\|_\infty \le \sigma_{k+1} + \mu_1 + \mu_2 + \cdots + \mu_{n-k-1}.$

Example 14.5.2. Consider $k = 2$ and

$$A = \begin{pmatrix} -1 & 2 & -1 & 3 \\ 0 & -2 & 2 & 0 \\ 0 & 0 & -3 & -2 \\ 0 & 0 & 0 & -4 \end{pmatrix}, \quad B = \begin{pmatrix} 1 & -2 \\ 2 & 0 \\ -1 & 5 \\ 2 & 3 \end{pmatrix}, \quad C = \begin{pmatrix} -1 & 0 & 2 & -3 \\ 1 & 1 & -2 & 1 \end{pmatrix}$$

Step 1. The Hankel singular values:

$$\{4.7619 \quad 1.3650 \quad 0.3614 \quad 0.0575\}$$

Step 2.

$$\tilde{A} = \begin{pmatrix} -1.1663 & 1.7891 & -0.2132 & 0.3266 \\ -0.0919 & -2.5711 & 0.7863 & -1.0326 \\ 0.3114 & 2.6349 & -3.7984 & 1.5960 \\ -0.3641 & 0.5281 & -0.2582 & -2.4642 \end{pmatrix},$$

$$\tilde{B} = \begin{pmatrix} 3.3091 & -0.3963 \\ -0.2903 & 2.6334 \\ -0.6094 & -1.5409 \\ 0.4947 & -0.1967 \end{pmatrix}$$

$$\tilde{C} = \begin{pmatrix} -2.4630 & 1.3077 & 0.9011 & 0.1600 \\ 2.2452 & -2.3041 & 1.3905 & -0.5078 \end{pmatrix}$$

The permutation matrix that does the reordering is:

$$P = \begin{pmatrix} 1 & 0 & 0 & 0 \\ 0 & 1 & 0 & 0 \\ 0 & 0 & 0 & 1 \\ 0 & 0 & 1 & 0 \end{pmatrix}.$$

The reordered balanced realization gives

$$\Sigma_1 = \begin{pmatrix} 4.7619 & 0 & 0 \\ 0 & 1.3650 & 0 \\ 0 & 0 & 0.0575 \end{pmatrix}, \qquad \Sigma_2 = 0.3614 = \sigma_3.$$

$$A_{11} = \begin{pmatrix} -1.1663 & 1.7891 & 0.3266 \\ -0.0919 & -2.5711 & -1.0326 \\ -0.3641 & 0.5281 & -2.4642 \end{pmatrix}, \qquad A_{12} = \begin{pmatrix} -0.2132 \\ 0.7863 \\ -0.2582 \end{pmatrix},$$

$$A_{21} = \begin{pmatrix} 0.3114 & 2.6349 & 1.5960 \end{pmatrix}, \qquad A_{22} = -3.7984.$$

$$B_1 = \begin{pmatrix} 3.3091 & -0.3963 \\ -0.2903 & 2.6334 \\ 0.4947 & -0.1967 \end{pmatrix}, \qquad B_2 = \begin{pmatrix} -0.6094 & -1.5409 \end{pmatrix},$$

$$C_1 = \begin{pmatrix} -2.4630 & 1.3077 & 0.1600 \\ 2.2452 & -2.3041 & -0.5078 \end{pmatrix}, \qquad C_2 = \begin{pmatrix} 0.9011 \\ 1.3905 \end{pmatrix}.$$

Step 3.

$$U = \begin{pmatrix} 0.2000 & 0.5057 \\ 0.3086 & 0.7804 \end{pmatrix}.$$

$$\Gamma = \begin{pmatrix} 22.5447 & 0.0000 & 0.0000 \\ 0.0000 & 1.7326 & 0.0000 \\ 0.0000 & 0.0000 & -0.1273 \end{pmatrix}, \qquad \hat{A} = \begin{pmatrix} -1.1872 & 0.4948 & -0.0019 \\ 0.0063 & -2.3615 & 0.0072 \\ 0.3687 & 1.5211 & 2.5932 \end{pmatrix}$$

$$\hat{B} = \begin{pmatrix} -0.7022 & 0.0756 \\ 0.3225 & -1.8376 \\ 0.1306 & 0.9841 \end{pmatrix}, \qquad \hat{C} = \begin{pmatrix} 11.5617 & -2.2454 & 0.0090 \\ -10.9487 & 2.4347 & -0.0295 \end{pmatrix}.$$

Step 4.

$$V_1 = \begin{pmatrix} -0.3754 & -0.9222 & -0.0930 \\ 0.8936 & -0.3335 & -0.3006 \\ -0.2462 & 0.1960 & -0.9492 \end{pmatrix}, \qquad V_2 = \begin{pmatrix} 1 & 0 & 0 \\ 0 & 1 & 0 \\ 0 & 0 & 1 \end{pmatrix},$$

$$\hat{A}_{11} = \begin{pmatrix} -2.3661 & 0.4343 \\ 0 & -1.1847 \end{pmatrix}, \qquad \hat{A}_{12} = \begin{pmatrix} 1.3682 \\ -0.7792 \end{pmatrix}, \qquad \hat{A}_{22} = 2.5953.$$

Solution of (14.5.17) gives

$$X = \begin{pmatrix} 0.2577 \\ -0.2061 \end{pmatrix}$$

and then

$$T_1 = \begin{pmatrix} -0.3754 & -0.9222 \\ 0.8936 & -0.3335 \\ -0.2462 & 0.1960 \end{pmatrix}, \qquad T_2 = \begin{pmatrix} 0.0003 \\ -0.0015 \\ -1.0530 \end{pmatrix},$$

$$S_1 = \begin{pmatrix} -0.3515 & 0.9710 & -0.0015 \\ -0.9413 & -0.3954 & 0.0003 \end{pmatrix},$$

$$S_2 = \begin{pmatrix} -0.0930 & -0.3006 & -0.9492 \end{pmatrix}.$$

Step 5. Using (14.5.20)–(14.5.22), we obtain

$$\hat{B}_1 = \begin{pmatrix} 0.5597 & -1.8124 \\ 0.5335 & 0.6558 \end{pmatrix}, \qquad \hat{C}_1 = \begin{pmatrix} -6.3494 & -9.9113 \\ 6.2934 & 9.2788 \end{pmatrix},$$

$$\hat{B}_2 = \begin{pmatrix} -0.1556 & -0.3889 \end{pmatrix}, \qquad \hat{C}_2 = 10^{-1} \begin{pmatrix} -0.0237 \\ 0.2383 \end{pmatrix},$$

$$\hat{D}_1 = \begin{pmatrix} -0.0723 & -0.1828 \\ -0.1115 & -0.2820 \end{pmatrix}.$$

Step 6.

6.1. The matrices of the balanced realization of the system $(-\hat{A}_{22}, \hat{B}_2, \hat{C}_2, \hat{D}_1)$ are:

$$A_3 = \begin{pmatrix} -2.5953 \end{pmatrix}, \qquad B_3 = 10^{-1} \begin{pmatrix} -0.3720 & -0.9298 \end{pmatrix},$$

$$C_3 = 10^{-1} \begin{pmatrix} -0.0991 \\ 0.9966 \end{pmatrix}, \qquad D_3 = \begin{pmatrix} -0.0723 & -0.1828 \\ -0.1115 & -0.2820 \end{pmatrix}.$$

The system (A_3, B_3, C_3, D_3) has only one Hankel Singular value $\mu_1 = 0.0019$.

6.2. Taking $q = r + m = 4$, we obtain

$$Z = 10^{-1} \begin{pmatrix} -0.3720 \\ -0.9298 \\ 0.0000 \\ 0.0000 \end{pmatrix}, \qquad Y = 10^{-1} \begin{pmatrix} -0.0991 \\ 0.9966 \\ 0.0000 \\ 0.0000 \end{pmatrix}.$$

6.3. $i = 1, \alpha = \beta = 0.1001$

$$H_1 = \begin{pmatrix} -0.0989 & 0.9951 & 0.0000 & 0.0000 \\ 0.9951 & 0.0989 & 0.0000 & 0.0000 \\ 0.0000 & 0.0000 & 1. & 0.0000 \\ 0.0000 & 0.0000 & 0.0000 & 1.0000 \end{pmatrix},$$

$$H_2 = \begin{pmatrix} 0.3714 & 0.9285 & 0.0000 & 0.0000 \\ 0.9285 & -0.3714 & 0.0000 & 0.0000 \\ 0.0000 & 0.0000 & -1.0000 & 0.0000 \\ 0.0000 & 0.0000 & 0.0000 & -1.0000 \end{pmatrix}.$$

$$U_1 = \begin{pmatrix} 0.0368 & 0.0919 & -0.9951 & 0.0000 \\ -0.3696 & -0.9239 & -0.0989 & 0.0000 \\ 0.9285 & -0.3714 & 0.0000 & 0.0000 \\ 0.0000 & 0.0000 & 0.0000 & -1.0000 \end{pmatrix},$$

$$\hat{D}_1 = \begin{pmatrix} -0.0723 & -0.1829 \\ -0.1108 & -0.2803 \end{pmatrix}.$$

Verification: Let $\hat{G}(s) = \hat{C}_1(sI - \hat{A}_{11})^{-1}\hat{B}_1 + \hat{D}_1$. Then,

$$0.3627 = \|G(s) - \hat{G}(s)\|_\infty \le \sigma_3 + \mu_1 = 0.3633.$$

Also, $\|G(s) - \hat{G}(s)\|_H = 0.3614 = \sigma_3$.

Figure 14.3 compares the step response of the original model with that of the Hankel-norm approximation model.

MATCONTROL note: Algorithm 14.5.1 has been implemented in Matcontrol function **hnaprx**.

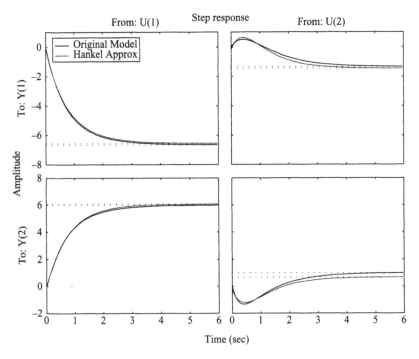

FIGURE 14.3: Step responses of the original and Hankel-norm approximation models.

14.6   MODEL REDUCTION OF AN UNSTABLE SYSTEM

We have so far considered model reduction of a stable system.

However, model reduction of an unstable system can also be performed. Varga (2001) has proposed two approaches. The first approach consists of finding only the reduced-order model of the stable part and then including the unstable part in the resulting reduced model. The second approach is based on computing a stable rational coprime factorization of the transfer function matrix and then reducing the stable system. We describe just the first approach here. For details of the second approach, see Varga (2001).

Step 1. Decompose the transfer function matrix $G(\lambda)$ additively as:

$$G(\lambda) = G_S(\lambda) + G_U(\lambda)$$

such that $G_S(\lambda)$ is the stable part and $G_U(\lambda)$ is the unstable part.

Step 2. Find a reduced-order model $G_{RS}(\lambda)$ of the stable part $G_S(\lambda)$.

Step 3. The reduced-order model $G_R(\lambda)$ of $G(\lambda)$ is then given by

$$G_R(\lambda) = G_{RS}(\lambda) + G_U(\lambda).$$

Computational remarks. The decomposition in Step 1 can be performed by block-diagonalizing the matrix A using the procedure of Step 4 of Algorithm 14.5.1.

14.7   FREQUENCY-WEIGHTED MODEL REDUCTION

In this section, we consider the frequency-weighted model reduction, proposed by Enns (1984). Specifically, the following problem is considered.

Given a stable transfer function matrix $G(s) = C(sI - A)^{-1}B$ and the two input and output weighting transfer function matrices $W_i = C_i(sI - A_i)^{-1}B_i$, and $W_o = C_o(sI - A_o)^{-1}B_o$, find a reduced-order model (A_R, B_R, C_R) with

$$G_R(s) = C_R(sI - A_R)^{-1}B_R$$

such that $\| W_o(G - G_R)W_i \|_\infty$ is minimized and $G(s)$ and $G_R(s)$ have the same number of unstable poles.

The effect of weighting on the model reduction is the possible reduction of the errors at the high frequencies.

The weighting model reduction problem can be solved in a similar way as the model reduction procedure by balanced truncation described in Section 14.4.

First, we note that the state space realization for the weighted transfer matrix is given by

$$\bar{W_o}GW_i = \bar{C}(sI - \bar{A})^{-1}\bar{B},$$

where

$$\bar{A} = \begin{pmatrix} A & 0 & BC_i \\ B_oC & A_o & 0 \\ 0 & 0 & A_i \end{pmatrix}, \qquad \bar{B} = \begin{pmatrix} 0 \\ 0 \\ B_i \end{pmatrix}, \qquad \bar{C} = (0, C_o, 0). \quad (14.7.1)$$

Let $\hat{C}_G$ and $\hat{O}_G$ be the solutions to the Lyapunov equations:

$$\bar{A}\hat{C}_G + \hat{C}_G(\bar{A})^T + \bar{B}\bar{B}^T = 0,$$

$$\hat{O}_G\bar{A} + (\bar{A})^T\hat{O}_G + (\bar{C})^T\bar{C} = 0.$$

Then the input weighted Grammian $\bar{C}_G$ and the output weighted Grammian $\bar{O}_G$ are defined by

$$\bar{C}_G \equiv (I_n, 0)\hat{C}_G \begin{pmatrix} I_n \\ 0 \end{pmatrix} \quad \text{and} \quad \bar{O}_G \equiv (I_n, 0)\hat{O}_G \begin{pmatrix} I_n \\ 0 \end{pmatrix}.$$

It can be shown (**Exercise 14.21**) that $\bar{C}_G$ and $\bar{O}_G$ satisfy:

$$\begin{pmatrix} A & BC_i \\ 0 & A_i \end{pmatrix} \begin{pmatrix} \bar{C}_G & \bar{C}_{G12} \\ \bar{C}_{G12}^T & \bar{C}_{G22} \end{pmatrix} + \begin{pmatrix} \bar{C}_G & \bar{C}_{G12} \\ \bar{C}_{G12}^T & \bar{C}_{G22} \end{pmatrix} \begin{pmatrix} A^T & 0 \\ C_i^T B^T & A_i^T \end{pmatrix}$$
$$+ \begin{pmatrix} 0 \\ B_i \end{pmatrix} \begin{pmatrix} 0 & B_i^T \end{pmatrix} = \begin{pmatrix} 0 & 0 \\ 0 & 0 \end{pmatrix} \qquad (14.7.2)$$

$$\begin{pmatrix} \bar{O}_G & \bar{O}_{G12} \\ \bar{O}_{G12}^T & \bar{O}_{G22} \end{pmatrix} \begin{pmatrix} A & 0 \\ B_oC & A_o \end{pmatrix} + \begin{pmatrix} A^T & C^T B_o^T \\ 0 & A_o^T \end{pmatrix} \begin{pmatrix} \bar{O}_G & \bar{O}_{G12} \\ \bar{O}_{G12}^T & \bar{O}_{G22} \end{pmatrix}$$
$$+ \begin{pmatrix} 0 \\ C_o^T \end{pmatrix} \begin{pmatrix} 0 & C_o \end{pmatrix} = \begin{pmatrix} 0 & 0 \\ 0 & 0 \end{pmatrix}. \qquad (14.7.3)$$

Consider now two special cases.
Case 1. $W_i = I$. Then $\bar{C}_G$ can be obtained from

$$\bar{C}_G A^T + A\bar{C}_G + BB^T = 0.$$

Case 2. $W_o = I$. Then $\bar{O}_G$ can be obtained from

$$\bar{O}_G A + A^T\bar{O}_G + C^TC = 0.$$

Now, let T be a nonsingular matrix such that

$$T\bar{C}_G T^{\mathrm{T}} = (T^{-1})^{\mathrm{T}} \bar{O}_G T^{-1} = \mathrm{diag}(\bar{\Sigma}_1, \bar{\Sigma}_2); \qquad (14.7.4)$$

that is, the matrix T makes the realization balanced.

Let $\bar{\Sigma}_1 = \mathrm{diag}(\sigma_1 I_{s_1}, \ldots, \sigma_r I_{s_r})$ and $\bar{\Sigma}_2 = \mathrm{diag}(\sigma_{r+1} I_{s_{r+1}} \cdots \sigma_n I_{s_n})$.
Partition the system (TAT^{-1}, TB, CT^{-1}) accordingly; that is

$$TAT^{-1} = \begin{pmatrix} \bar{A}_R & \bar{A}_{12} \\ \bar{A}_{21} & \bar{A}_{22} \end{pmatrix}, \qquad TB = \begin{pmatrix} \bar{B}_R \\ \bar{B}_2 \end{pmatrix},$$

and

$$CT^{-1} = (\bar{C}_R, \bar{C}_2).$$

Then $(\bar{A}_R, \bar{B}_R, \bar{C}_R)$ is a weighted reduced-order model.

If the full-order original model is minimal, then $\bar{\Sigma}_1 > 0$.

Unfortunately, the stability of the reduced-order model here cannot, in general, be guaranteed.

However, there are some special cases of weightings for which the reduced-order models are stable (**Exercise 14.22**). Also, no a priori error found for the approximate is known.

14.8 SUMMARY AND COMPARISONS OF MODEL REDUCTION PROCEDURES

We have described the following techniques for model reduction of a stable system:

(i)   The balanced truncation procedure (**Algorithm 14.4.1**)
(ii) The Schur method (Algorithm **14.4.2**)
(iii) The Hankel-norm approximation algorithm (**Algorithm 14.5.1**)
(iv) Frequency-weighted model reduction (**Section 14.7**).

For the first two methods (i)-(ii), the error satisfies

$$\|G(s) - G_R(s)\|_\infty \le 2(\sigma_{d+1} + \sigma_{d+2} + \cdots + \sigma_N),$$

In the method (i), $G_R(s)$ is obtained by truncating the balanced realization of $G(s)$ to the first $(s_1 + s_2 \ldots + s_d)$ states, where s_i is the multiplicity of σ_i. For the method (ii), $G_R(s)$ is obtained by Algorithm 14.4.2. For a similar error bound for the method (iii), see Theorem 14.5.4. Furthermore, for this method, the reduced-order model $G_R(s)$ has the property: $\inf \| G - G_R(s) \|_H = \sigma_{k+1}$, where $G_R(s)$ is of McMillan degree k.

The weighted model reduction procedure in Section 14.7 does not enjoy any of the above properties. Even the stability in general cannot be guaranteed. Stability, however, in some special cases can be proved. See Enns (1984) for details. Discussion of this section has been taken from Zhou *et al.* (1996).

If the system is not stable, model reduction is still possible using the three simple steps of Section 14.6.

The balanced truncation procedure for model reduction (Algorithm 14.4.1) and Algorithm 14.5.1 need computation of a balanced realization. Two algorithms (**Algorithms 14.2.1** and **14.2.2**) have been described for this purpose. Both these algorithms suffer from the danger of possible ill-conditioning of the transforming matrices. **However, the methods usually work well in practice for well-equilibrated systems.**

The Schur method has been designed to avoid such possible ill-conditioning.

Unfortunately, because of the requirement of explicitly computing the product of the controllability and observability Grammians, *the Schur method is usually less accurate for moderately ill-conditioned systems than the square-root method* (see Varga 2001). The main advantages of the balanced truncation procedure and the Schur method have been combined in the **balanced-free square-root method** by Varga (1991). Numerical experiments performed by Varga (2001) show that the accuracy of this method is usually better than either of the Schur methods or the balanced truncation method using the square-root algorithm for balancing.

Finally, we remark that it is **very important that the system be scaled properly for the application of the balanced-truncation or the Hankel-norm approximation method.** One way to do this is to attempt to reduce the 1-norm of the scaled system matrix

$$S = \begin{pmatrix} Z^{-1}AZ & Z^{-1}B \\ CZ & 0 \end{pmatrix}, \text{ where } Z \text{ is a positive definite matrix.}$$

Note that the Hankel singular values are not affected by such a coordinate transformation; in particular, by coordinate scaling of diagonal matrices.

For a comparative study of different model reduction algorithms and detailed description of available software, see Varga (2001). See also Varga (1994).

14.9   SOME SELECTED SOFTWARE

14.9.1   MATLAB Control System Toolbox

State-space models
  balreal   Grammian-based Balancing of state-space realization.
  modred   Model state reduction.
  ssbal   Balancing of state-space model using diagonal similarity.

14.9.2   MATCONTROL

BALSVD	Internal balancing using the SVD
BALSQT	Internal balancing using the square-root algorithm
MODREDS	Model reduction using the Schur method
HNAPRX	Hankel-norm approximation.

14.9.3   CSP-ANM

Model reduction

- The Schur method for model reduction is implemented as `DominantSub-system` [*system*, `Method`→`SchurDecomposition`].
- The square-root method for model reduction is implemented as `DominantSubsystem` [*system*, `Method`→`SquareRoot`].

14.9.4   SLICOT

Model reduction

AB09AD	Balance and truncate model reduction
AB09BD	Singular perturbation approximation based model reduction
AB09CD	Hankel-norm approximation based model reduction
AB09DD	Singular perturbation approximation formulas
AB09ED	Hankel-norm approximation based model reduction of unstable systems
AB09FD	Balance and truncate model reduction of coprime factors
AB09GD	Singular perturbation approximation of coprime factors
AB09ID	Frequency-weighted model reduction based on balanced truncations
AB09KD	Frequency-weighted Hankel-norm approximation
AB09MD	Balance and truncate model reduction for the stable part
AB09ND	Singular perturbation approximation based model reduction for the stable part.

State-space transformations

TB01ID	Balancing a system matrix for a given triplet.

14.9.5   MATRIX$_X$

Purpose: Convert a discrete dynamic system into an internally balanced dynamic form.

Syntax: [SB, SIGMASQ, T] = DBALANCE (SD, NS)

Purpose: Compute a reduced order form of a discrete-time system.
Syntax: [SR, NSR] = DMREDUCE (SD, NS, KEEP)

Purpose: Compute a reduced-order form of a continuous system.
Syntax: [SR, NSR] = MREDUCE (S, NS, KEEP)

Purpose: Perform model structure determination.
Syntax: [THETA, COR, COV] = MSD (X, Y)
The other software packages dealing with model reduction include:

- **MATRIX$_X$ Model Reduction Module** (1998) by B.D.O. Anderson and B. James.
- **μ-Analysis and Synthesis Toolbox 1.0** by G. Balas, J. Doyle, K. Glover, A. Packard and R. Smith (1998).
- **Robust Control Toolbox** 2.0 by R.Y. Chiang and M.G. Safonov.

14.10   SUMMARY AND REVIEW

The chapter covers the topics:

- Internal balancing
- Model reduction
- Hankel-norm approximation.

Internal Balancing

Given an $n \times n$ stable minimal realization (A, B, C), there always exists a transformation T that simultaneously diagonalizes both the controllability and observability Grammians to the same diagonal matrix $\Sigma = \mathrm{diag}(\sigma_1, \ldots, \sigma_n)$, where $\sigma_1 \geq \sigma_2 \geq \cdots \geq \sigma_r > \sigma_{r+1} > \sigma_{r+2} > \cdots > \sigma_n$. The numbers $\sigma_i, i = 1, \ldots, n$ are the **Hankel singular values.**

In this case, the transformed system $(\tilde{A}, \tilde{B}, \tilde{C})$, is called **internally balanced.** **Algorithms 14.2.1** and **14.2.2** compute balanced realization of a **continuous-time system.** The internal balancing of a **discrete-time system is discussed in Section 14.3.**

Model Reduction

The problem of model reduction is the problem of constructing a qth order model from a given nth order model $(n > q)$ in such a way that the reduced qth order model is close to the original system in some sense. The precise mathematical definition of model reduction appears in **Section 14.4.**

Model reduction via internal balancing: Once a system is internally balanced, a desired reduced-order model can be obtained by eliminating the states corresponding to the less controllable and observable modes (**Algorithm 14.4.1**).

Theorem 14.4.1 shows that a truncated model is also balanced and stable, and furthermore, if $G(s)$ and $G_R(s)$ are the respective transfer functions of the original and the truncated model, then

$$\|G(s) - G_R(s)\|_\infty \leq 2(\sigma_{d+1}, \ldots, \sigma_N),$$

where the states corresponding to $\sigma_{d+1}, \ldots, \sigma_N$ are eliminated.

The Schur method for model reduction: There are some numerical difficulties associated with the procedure of finding a reduced order model via internal balancing using Algorithms 14.2.1 and 14.2.2. The transforming matrix T in Algorithm 14.2.1 and the matrices L and Z in Algorithm 14.2.2 can be, in some cases, highly ill-conditioned. An alternative method (**Algorithm 14.4.2**) for model reduction based on the real Schur decomposition of the product of the controllability and observability Grammians, is described in **Section 14.4. The transforming matrix T in this case is orthogonal, and, therefore, well-conditioned.** The Schur method does not give an internally balanced system; however, the essential properties of the original system are preserved. In fact, **Theorem 14.4.2** shows that the transfer function matrix obtained by the Schur method is exactly the same as that of the one obtained via **Algorithm 14.4.1**.

A possible numerically difficulty with Algorithm 14.4.2 is the explicit computation of the product of the controllability and observability Grammians. In this case, instead of explicitly computing the controllability and observability Grammians, their Cholesky factors can be computed using the Hammarling algorithm (**Algorithm 8.6.1**) in Chapter 8.

Combining the advantages of the Schur method and the square-root algorithm, a **balancing-free square root method** has been developed. This is described in **Section 14.4.3**.

Hankel-norm approximation: Given a stable $G(s)$, the problem of finding a $\hat{G}(s)$ of McMillan degree k such that $\|G(s) - \hat{G}(s)\|_H$ is minimized is called an optimal Hankel-norm approximation.

A characterization of all solutions to Hankel-norm approximation is given in Section 14.5.1 (**Theorem 14.5.2**). An algorithm (**Algorithm 14.5.1**) for computing an optimal Hankel-norm approximation is then presented.

Model Reduction of an Unstable System

The model reduction of an unstable system can be achieved by decomposing the model into its stable and unstable part, followed by finding a model reduction of the stable part and finally adding the reduced-order model of the stable part with the unstable part. This is descried in **Section 14.6**. For this and another approach, based on stable rational coprime factorization, see Varga (2001).

Weighted Model Reduction

Sometimes the errors at high frequencies in a reduced-order model can be reduced using weights on the model. This is discussed in **Section 14.7**.

Comparison of the Model Reduction Procedures

The model reduction procedures are summarized and a brief comparative discussion of different procedures is presented in **Section 14.8.**

14.11 CHAPTER NOTES AND FURTHER READING

The internal balancing algorithms, **Algorithms 14.2.1** and **14.2.2** are due to Laub (1980) and Tombs and Postlethwaite (1987), respectively. The idea of model reduction via balanced truncation was first introduced by Moore (1981).

The stability property of the truncated subsystem (part (a) of Theorem 14.4.1) was obtained by Pernebo and Silverman (1982) and the error bound (part (b) of Theorem 14.4.1) is due to Glover (1984) and Enns (1984).

The Schur algorithm for model reduction and Theorem 14.4.2 is due to Safonov and Chiang (1989). The balancing-free square-root method for model reduction is due to Varga (1991).

The Hankel-norm approximation problem was introduced and solved by Glover in a celebrated paper (Glover 1984). Besides the topic of Hankel-norm approximation of a transfer function, the paper contains many other beautiful results on systems theory and linear algebra. A good discussion of this topic can also be found in the book by Zhou *et al.* (1996). See also Glover (1989).

For results on discrete-time balanced model reduction, see Al-Saggaf and Franklin (1987), and Hinrichsen and Pritchard (1990).

The idea of frequency weighted model reduction is due to Enns (1984). Other subsequent results on this and related topics can be found in Al-Saggaf and Franklin (1988), Glover (1986, 1989), Glover *et al.* (1992), Hung and Glover (1986), Liu and Anderson (1990), Zhou (1993), etc.

For a discussion on Balanced Stochastic Truncation (BST) method, see Zhou *et al.* (1996).

The idea of singular perturbation approximation is due to Liu and Anderson (1989).

For an optimal Hankel norm approximation procedure with stable weighting functions, see Hung and Glover (1986).

The other papers on Hankel norm approximation include Kung and Lin (1981) and Latham and Anderson (1986).

A recent book by Obinata and Anderson (2000) deals exclusively with the topic of model reduction.

The paper by Green (1988) deals with stochastic balanced realization. For more on this topic, see Zhou *et al.* (1996).

Exercises

14.1   Prove part (a) of Theorem 14.4.1 and fill in the missing details of part (b), whenever indicated in the book.

14.2   Let

$$G(s) = \left[\begin{array}{c|c} A & B \\ \hline C & D \end{array}\right].$$

Suppose that there exists a symmetric matrix $P = \text{diag}(P_1, 0)$, with P_1 nonsingular, such that

$$AP + PA^{\mathrm{T}} + BB^{\mathrm{T}} = 0.$$

Partition $G(s)$ conformably with P as

$$\left[\begin{array}{cc} A & B \\ C & D \end{array}\right] = \left[\begin{array}{cc|c} \hat{A}_{\mathrm{R}} & A_{12} & \hat{B}_{\mathrm{R}} \\ A_{21} & A_{22} & B_2 \\ \hline \hat{C}_{\mathrm{R}} & C_2 & D \end{array}\right].$$

Then prove that $\left[\begin{array}{c|c} \hat{A}_{\mathrm{R}} & \hat{B}_{\mathrm{R}} \\ \hline \hat{C}_{\mathrm{R}} & D \end{array}\right]$ is also a realization of $G(s)$. Moreover, if $\hat{A}_{\mathrm{R}}$ is stable, then $(\hat{A}_{\mathrm{R}}, \hat{B}_{\mathrm{R}})$ is controllable.

14.3   Based on the result of Exercise 14.2 develop a method for extracting a controllable subsystem from a stable noncontrollable system.

14.4   (Zhou *et al.* (1996)) Let $G(s)$ be the same as in Exercise 14.2. Suppose that there exists a symmetric matrix $Q = \text{diag}(Q_1, 0)$, with Q_1 is nonsingular, such that $QA + A^{\mathrm{T}}Q + C^{\mathrm{T}}C = 0$. Partition the realization (A, B, C, D) conformably with Q as in Exercise 14.2. Then prove that

$$\left[\begin{array}{c|c} \hat{A}_{\mathrm{R}} & \hat{B}_{\mathrm{R}} \\ \hline \hat{C}_{\mathrm{R}} & D \end{array}\right]$$

is also a realization of $G(s)$. Prove further that $(\hat{A}_{\mathrm{R}}, \hat{C}_{\mathrm{R}})$ is observable if $\hat{A}_{\mathrm{R}}$ is stable.

14.5   Based on Exercise 14.4, develop a method for extracting an observable subsystem from a stable nonobservable system.

14.6   Construct your own example to illustrate the numerical difficulties of Algorithm 14.2.1.

14.7 Prove that the rows $\{1, \ldots, d\}$ and the rows $\{d + 1, \ldots, n\}$ of T^{-1} in Algorithm 14.2.1, form bases for the left eigenspaces of the matrix $C_G O_G$ associated with the eigenvalues $\{\sigma_1^2, \ldots, \sigma_d^2\}$, and $\{\sigma_{d+1}^2, \ldots, \sigma_n^2\}$, respectively.

14.8 Prove that the columns of the matrix V_{1S} and those of the matrix U_{2S} in the Schur algorithm (Algorithm 14.4.2) for model reduction, form orthonormal bases for the right and left invariant subspace of $C_G O_G$ associated with the large eigenvalues $\sigma_1^2, \ldots, \sigma_d^2$.

14.9 Prove that the controllability and observability Grammians of the reduced-order model obtained by the Schur algorithm (Algorithm 14.4.2) are, respectively, given by $\hat{C}_G^R = S_1^T C_G S_1$ and $\hat{O}_G^R = S_2 O_G S_2$, where C_G and O_G are the controllability and observability Grammians of the original model.

14.10 (a) Modify the Schur algorithm for model reduction by making use of Hammarling's algorithm (**Algorithm 8.6.1**) so that the explicit formation of the product $C_G O_G$ is avoided, and only the Cholesky factors L_c and L_o are computed. (Consult Safonov and Chiang (1989)).

   (b) Work out an example to demonstrate the superiority of this modified Schur algorithm over the Schur algorithm.

14.11 (a) Prove that the matrix T defined by (14.3.4) transforms the discrete-time system (14.3.1) to the balanced system (14.3.5).

   (b) Work out a discrete analog of Algorithm 14.2.2.

14.12 (Zhou *et al.* (1996)). Let

$$G(s) = \left[\begin{array}{c|c} A & B \\ \hline C & O \end{array}\right]$$

be the transfer function of a balanced realization. Then prove that

$$\sigma_1 \leq \|G\|_\infty \leq \int_0^\infty \|Ce^{At}B\|dt \leq 2\sum_{i=1}^N \sigma_i.$$

14.13 Construct an example to show that if the diagonal entries of the matrix Σ of the balanced Grammian are all distinct, then every subsystem of the balanced system is asymptotically stable. Construct another example to show that this condition is only sufficient.

14.14 Construct an example to show that the bound of Theorem 14.4.1 can be loose if the quantities $\sigma_i, i = 1, \ldots, n$ are close to each other.

(**Hint:** Construct a stable realization $G(s)$ such that $G^T(-s)G(s) = I$ and then construct a balanced realization of $G(s)$. Now make a small perturbation to this balanced realization and work with this perturbed system.)

14.15 (a) Develop a Schur method for model reduction of the discrete-time system.

   (b) Give a simple example to illustrate the method.

   (c) Give a flop-count of the method.

14.16 *Minimal realization using block diagonalization (Varga 1991).* Consider the following algorithm:

   Step 1. Reduce A to block diagonal form and update B and C, that is, find a nonsingular matrix T such that $T^{-1}AT = \text{diag}(\bar{A}_1, \bar{A}_2, \ldots, \bar{A}_r)$, $T^{-1}B = (\bar{B}_1, \bar{B}_2, \ldots, \bar{B}_r)$, $CT = (\bar{C}_1, \bar{C}_2, \ldots, \bar{C}_r)$. (see Exercise 8.10).

Assume that the diagonal blocks in A have disjoint spectra.

Step 2. Find an MR of each of the system $(\bar{A}_i, \bar{B}_i, \bar{C}_i), i = 1, \ldots, r$ using Algorithm 14.2.1 or 14.2.2, as appropriate.

Let $(\hat{A}_i, \hat{B}_i, \hat{C}_i), i = 1, \ldots, r$ be the computed MR of $(\bar{A}_i, \bar{B}_i, \bar{C}_i)$ in Step 2. Then show that the system $(\hat{A}, \hat{B}, \hat{C})$ defined by:

$$\hat{A} = \operatorname{diag}(\hat{A}_1, \hat{A}_2, \ldots, \hat{A}_r), \qquad \hat{B} = \begin{pmatrix} \hat{B}_1 \\ \hat{B}_2 \\ \vdots \\ \hat{B}_r \end{pmatrix}, \qquad \hat{C} = (\hat{C}_1, \hat{C}_2, \ldots, \hat{C}_r),$$

is an MR of (A, B, C).

14.17 Using the matrix version of bilinear transformation:

$$s = \frac{z - 1}{z + 1},$$

prove that the Hankel singular values of the discrete and continuous systems are identical.

(**Hint:** Obtain the system matrices (A, B, C) for the continuous-time system from the system matrices $(\bar{A}, \bar{B}, \bar{C})$ of the discrete-time system and then show that the controllability Grammian (observability Grammian) of (A, B, C) is the same as the controllability Grammian (observability Grammian) of $(\bar{A}, \bar{B}, \bar{C})$.)

14.18 Using the bilinear transformation of Exercise 14.17 and Algorithm 14.5.1, find an optimal Hankel-norm approximation for the discrete-time system defined by the matrices in Example 14.3.1.

14.19 Write down a discrete analog of Algorithm 14.2.2 and apply the algorithm to the system (A, B, C) defined by

$$A = \begin{pmatrix} 0.0001 & 1 & 0 \\ 0 & 0.1200 & 1 \\ 0 & 0 & 0 \end{pmatrix}, \qquad B = (1, 1, 0)^{\mathrm{T}}, \qquad C = (1, 1, 1).$$

14.20 (Safonov and Chiang 1989). Consider the system (A, B, C) given by

$$A = \begin{pmatrix} -6 & -1 & 0 & 0 & 0 & 0 & 0 & 0 & 0 & 0 \\ 1 & -8 & 0 & 0 & 0 & 0 & 0 & 0 & 0 & 0 \\ 0 & 0 & -10 & 3 & 0 & 0 & 0 & 0 & 0 & 0 \\ 0 & 0 & 1 & -8 & 0 & 0 & 0 & 0 & 0 & 0 \\ 0 & 0 & 0 & 0 & -13 & -3 & 9 & 0 & 0 & 0 \\ 0 & 0 & 0 & 0 & 1 & -8 & 0 & 0 & 0 & 0 \\ 0 & 0 & 0 & 0 & 0 & 1 & -8 & 0 & 0 & 0 \\ 0 & 0 & 0 & 0 & 0 & 0 & 0 & -14 & -9 & 0 \\ 0 & 0 & 0 & 0 & 0 & 0 & 0 & 1 & -8 & 0 \\ 0 & 0 & 0 & 0 & 0 & 0 & 0 & 0 & 0 & -2 \end{pmatrix},$$

$$B = \begin{pmatrix} 1 & 0 & 0 & 0 & 1 & 0 & 0 & 0 & 0 & 10^{-3} \\ 0 & 0 & 1 & 0 & 0 & 0 & 0 & 1 & 0 & 10^{-3} \end{pmatrix}^{\mathrm{T}},$$

$$C = \begin{pmatrix} 0 & 1 & 0 & 1 & 0 & 0 & 0 & 0 & 0 & 5 \times 10^5 \\ 0 & 0 & 0 & 0 & 0 & 0 & -6 & 1 & -2 & 5 \times 10^5 \end{pmatrix}.$$

Find a reduced-order model of order 4 using
(a) Balanced truncation via Algorithms 14.2.1 and 14.2.2.
(b) The Schur method (Algorithm 14.4.2).
Compare the results with respect to the condition numbers of the transforming matrices and the $\| \cdot \|_\infty$ norm errors.

14.21 Prove that the weighting Grammians $\bar{C}_G$ and $\bar{O}_G$ are given by the equations (14.7.2) and (14.7.3).

14.22 Consider the two special cases of the frequency-weighted model reduction:
Case 1. $W_i(s) = I$ and $W_o(s) \neq I$,
Case 2. $W_i(s) \neq I$ and $W_o(s) = I$.
Prove that the reduced-order model $(\bar{A}_R, \bar{B}_R, \bar{C}_R)$ is stable provided that it is controllable in Case 1 and is observable in Case 2.
(**Hint:** Write the balanced Grammian $\Sigma = \mathrm{diag}(\Sigma_1, \Sigma_2)$. Then show that

$$\bar{A}_R \Sigma_1 + \Sigma_1 \bar{A}_R^{\mathrm{T}} + \bar{B}_R \bar{B}_R^{\mathrm{T}} = 0, \quad \text{and} \quad \bar{A}_R^{\mathrm{T}} \Sigma_1 + \Sigma_1 \bar{A}_R + \bar{C}_R^{\mathrm{T}} \bar{C}_R = 0).$$

Work out an example to illustrate the result.

14.23 *Singular perturbation approximations.* Let $(\bar{A}, \bar{B}, \bar{C})$ be a balanced realization of (A, B, C). Partition the matrices $\bar{A}, \bar{B}, \bar{C}$ as:

$$\bar{A} = \begin{pmatrix} \bar{A}_{11} & \bar{A}_{12} \\ \bar{A}_{21} & \bar{A}_{22} \end{pmatrix}, \qquad \bar{B} = \begin{pmatrix} \bar{B}_1 \\ \bar{B}_2 \end{pmatrix}, \qquad \bar{C} = (\bar{C}_1, \bar{C}_2).$$

Then the system $(\hat{A}, \hat{B}, \hat{C})$ defined by

$$\hat{A} = \bar{A}_{11} + \bar{A}_{12}(\gamma I - \bar{A}_{22})^{-1}\bar{A}_{21}, \qquad \hat{B} = \bar{B}_1 + \bar{A}_{12}(\gamma I - \bar{A}_{22})^{-1}\bar{B}_2,$$
$$\hat{C} = \bar{C}_1 + \bar{C}_2(\gamma I - \bar{A}_{22})^{-1}\bar{A}_{21}$$

is called **the balanced singular perturbation approximation** of $(\bar{A}, \bar{B}, \bar{C})$ (Liu and Anderson 1989). ($\gamma = 0$ for a continuous-time system and $\gamma = 1$ for a discrete-time system).
(a) Compute singular perturbation approximations of the system in Example 14.2.1
     using Algorithms 14.2.1 and 14.2.2.
(b) Show how the balancing-free square root method in Section 14.4.3 can be modified to compute singular perturbation approximation (**Hint.** Find the SVD of $Y^{\mathrm{T}}X$ and then compute L and Z from the matrices of the SVD). See Varga (1991).
(c) Apply the modified balancing-free square-root method in (b) to the system in Example 14.2.1 and compare the results.

References

Al-Saggaf U.M. and Franklin G.F., "An error bound for a discrete reduced order model of a linear multivariable sytem", *IEEE Trans. Autom. Control*, Vol. AC-32, pp. 815–819, 1987.

Al-Saggaf U.M. and Franklin G.F. "Model reduction via balanced realizations: an extension and frequency weighting techniques," *IEEE Trans. Autom. Control*, Vol. AC-33(7), pp. 687–692, 1988.

Balas G., Doyle J., Glover K., Packard A. and Smith R. *μ-Analysis and Synthesis Toolbox 3.0.4*, The MathWorks Inc., Natick, MA, 1998.

Chiang R.Y. and Safonov M.G. *Robust Control Toolbox 2.0.6.*, The MathWorks Inc., Natick, MA, 1997.

Enns D.F. "Model reduction with balanced realizations: An error bound and a frequency weighted generalization", *Proc. 23rd IEEE Conf. Dec. Control*, pp. 127–132, 1984.

Glover K. "Robust stabilization of linear multivariable systems: relations to approximation", *Int. J. Control*, vol. 43(3), pp. 741–766, 1986.

Glover K. *A Tutorial on Hankel-norm Approximation, Data to Model*, Springer-Verlag, New York, 1989. (Willems J.C. ed.)

Glover K., Limebeer D.J.N., and Hung Y.S. "A structured approximation problem with applications to frequency weighted model reduction," *IEEE Trans. Autom. Control*, Vol. AC-37(4), pp. 447–465, 1992.

Glover K. "All optimal Hankel-norm approximations of linear multivariable systems and their L^∞-error bounds," *Int. J. Cont.*, 39, pp. 1115–1193, 1984.

Green M. "Balanced stochastic realizations," *Lin. Alg. Appl.*, Vol. 98, pp. 211–247, 1988.

Heath M.T., Laub A.J., Paige C.C., and Ward R.C. "Computing the singular value decomposition of the product of two matrices," *SIAM J. Sci. Stat. Comput.*, Vol. 7, pp. 1147–1159, 1986.

Hinrichsen D. and Pritchard A.J. "An improved error estimate for reduced-order models of discrete time system," *IEEE Trans. Autom. Control*, Vol. AC-35, pp. 317–320, 1990.

Hung Y.S. and Glover K. "Optimal Hankel-norm approximation of stable systems with first order stable weighting functions," *Syst. Control Lett.*, Vol. 7, pp. 165–172, 1986.

Kung S.K. and Lin D.W. "Optimal Hankel norm model reduction: multivariable systems," *IEEE Trans. Autom. Control*, Vol. AC-26, pp. 832–852, 1981.

Latham G.A. and Anderson B.D.O. "Frequency weighted optimal Hankel-norm approximation of stable transfer function," *Syst. Control Lett.*, Vol. 5, pp. 229–236, 1986.

Laub A.J. "Computation of 'balancing' transformations," *Proc. 1980 JACC*, Session FA8-E, 1980.

Liu Y. and Anderson B.D.O. "Singular perturbation approximation of balanced systems," *Int. J. Control*, Vol. 50, pp. 1379–1405, 1989.

Liu Y. and Anderson B.D.O. "Frequency weighted controller reduction methods and loop transfer recovery," *Automatica*, Vol. 26(3), pp. 487–497, 1990.

MATLAB, *Control System Toolbox* 4.2, The MathWorks Inc., Natick, MA, 1998.

MATRIX$_X$, *Xmath Model Reduction module*, ISI, Santa Clara, CA, January 1998.

Moore B.C. "Principal component analysis in linear systems: controllability, observability and model reduction," *IEEE Trans. Autom. Control*, Vol. AC-26, pp. 17–31, 1981.

Obinata G. and Anderson B.D.O. *Model Reduction for Control System Design*, Springer-Verlag, New York, 2000.

Pernebo L. and Silverman L.M. "Model reduction via balanced state space representation," *IEEE Trans. Autom. Control*, Vol. AC-27(2), pp. 382–387, 1982.

Safonov M.G. and Chiang R.Y. "A Schur method for balanced-truncation model reduction," *IEEE Trans. Autom. Control*, Vol. 34, pp. 729–733, 1989.

Tombs M.S. and Postlethwaite I. "Truncated balanced realization of a stable non-minimal state-space system," *Int. J. Control*, Vol. 46, pp. 1319–1330, 1987.

Varga A. "Balancing-free square-root algorithm for computing singular perturbation approximations," *Proc. 30th IEEE Conf. Dec. Control*, pp. 1062–1065, 1991.

Varga A. "Numerical methods and software tools for model reduction," *Proc. 1st MATH-MOD Conf., Vienna*, Troch I. and Breitenecker F., eds. Vol. 2, pp. 226–230, 1994.

Varga A. "Model reduction software in the SLICOT library," *Applied and Computational Control, Signals and Circuits: Recent Developments*, Datta B.N. *et al.*, eds. pp. 239–282, Kluwer Academic Publisher, Boston, 2001.

Zhou K (with Doyle J.C. and Glover K.), *Robust and Optimal Control,* Prentice Hall, Upper Saddle River, NJ, 1996.

Zhou K. "Frequency weighted model reduction with $\mathcal{L}_\infty$ error bounds," *Syst. Control Lett.*, Vol. 21, pp. 115–125, 1993.

SPECIAL TOPICS

CHAPTER 15

LARGE-SCALE MATRIX COMPUTATIONS IN CONTROL: KRYLOV SUBSPACE METHODS

15.1 INTRODUCTION

Numerically effective computational methods for various control problems discussed in preceding chapters are viable only for dense computations. Unfortunately, these methods are not suitable for solutions of many large practical problems such as those arising in the design of large sparse structures, power systems, etc. There are two main reasons for this. First, they destroy the sparsity, inherited in most large practical problems and second, they are $O(n^3)$ methods and thus, are computationally prohibitive for large problems. The sparsity is lost by the use of canonical forms such as, triangular Hessenberg and real-Schur, which are obtained by using Gaussian eliminations, Householder and Givens transformations, and those techniques are well-known to destroy the sparsity.

On the other hand, there have been some fine recent developments in the area of large-scale matrix computations. A class of classical methods known as the **Krylov subspace methods** (Lanczos 1950; Arnoldi 1951) have been found to be suitable for sparse matrix computations. The reason is that these methods can be implemented using matrix-vector multiplications only; therefore, the sparsity in the original problem can be preserved. The examples are the **Generalized Minimal Residual** (GMRES) and the **Quasi-Minimal Residual** (QMR) methods for linear systems problem; the **Arnoldi, Lanczos**, and the **Jacobi–Davidson** methods, and several variants of them such as the **restarted** and **block Arnoldi methods and band Lanczos method** for eigenvalue problems.

It is only natural to develop algorithms for large-scale control problems using these effective large-scale techniques of matrix computations. Some work to this effect has been done in the last few years.

In this chapter, we will briefly review some of these methods. In Section 15.2, we give a brief description of the basic Arnoldi and Lanczos methods to facilitate the understanding of how these methods are applied to solve large-scale control problems. We stress that *the descriptions of our Krylov subspace methods are basic.* For practically implementable versions of these methods and associated software, we refer the readers to the books by Bai *et al.* (2000), and Saad (1992a, 1996). *In particular, the homepage, ETHOME of the book by Bai et al.* (2000) *contains valuable information of available software.* Our only goal of this chapter is to show the readers how these modern iterative numerical methods can be gainfully employed to solve some of the large and sparse matrix problems arising in control.

15.2　THE ARNOLDI AND BLOCK ARNOLDI METHODS

In this section, we summarize the essentials of the scalar Arnoldi and block Arnoldi methods.

15.2.1　The Scalar Arnoldi Method

Given an $n \times n$ matrix A, a vector v, and an integer $m \leq n$, the scalar Arnoldi method computes simultaneously a set of orthonormal vectors $\{v_1, \ldots, v_{m+1}\}$, an $(m + 1) \times m$ matrix $\tilde{H}_m$ such that

$$A V_m = V_{m+1} \tilde{H}_m, \tag{15.2.1}$$

where $V_m = (v_1, \ldots, v_m)$ and $V_{m+1} = (V_m, v_{m+1})$. *The vectors* $\{v_1, \ldots, v_m\}$ *form an orthonormal basis of the Krylov subspace* $K_m(A, v_1) = \mathrm{span}\{v_1, Av_1, \ldots, A^{m-1}v_1\}$. Furthermore, it is easy to establish that

$$V_m^{\mathrm{T}} A V_m^{\mathrm{T}} = H_m, \tag{15.2.2}$$

where H_m is an $m \times m$ upper Hessenberg matrix obtained from $\tilde{H}_m$ by deleting its last row. **The algorithm breaks down at step** j, **i.e.,** $v_{j+1} = 0$, **if and only if the degree of the minimal polynomial of** v_1 **is exactly** j, that is, it is a combination of j eigenvectors.

15.2.2　The Block Arnoldi Method

The block Arnoldi method is a generalization of the scalar Arnoldi method. Starting with a block vector V_1 of order $n \times p$ and norm unity, the block Arnoldi method constructs a set of block vectors $\{V_1, \ldots, V_{m+1}\}$ such that if $U_m = (V_1, \ldots, V_m)$, then $U_m^{\mathrm{T}} U_m = I_{mp \times mp}$, and $U_m^{\mathrm{T}} A U_m$ is an upper block Hessenberg matrix $H_m = (H_{ij})$. Furthermore, $A U_m - U_m H_m = V_{m+1} H_{m+1,m} E_m^{\mathrm{T}}$, where E_m is the last p columns of the $mp \times mp$ identity matrix. **The block Arnoldi algorithm is particularly suitable for handling multivariable control problems.**

Algorithm 15.2.1.   *The Block Arnoldi Algorithm (Modified Gram–Schmidt Version).*

Let V be an $n \times p$ matrix.

Step 0. *Compute the $n \times p$ orthogonal matrix V_1 by finding the QR factorization of V: $V = V_1 R$ (Note that R is here $p \times p$). (Use column pivoting if V does not have full rank).*

Step 1. *For $k = 1, 2, \ldots, m$ do*
   Compute $\hat{V} = AV_k$.
   For $j = 1, 2, \ldots, k$ do
   $H_{j,k} = V_j^T \hat{V}$
   $\hat{V} = \hat{V} - V_j H_{j,k}$
   End
Compute $H_{k+1,k}$ by finding the QR factorization of $\hat{V}$: $\hat{V} = V_{k+1} H_{k+1,k}$
End

   The block Arnoldi algorithm clearly breaks down if $H_{k+1,k}$ becomes zero for some k. **Such a breakdown has positive consequences in some applications.** *(See Section 4.1.1.)*

Remarks

- Define the block $mp \times mp$ upper Hessenberg matrix $H_m = (H_{ij})$,

$$U_m = (V_1, V_2, \ldots, V_m)$$

  and

$$U_{m+1} = (U_m, V_{m+1}).$$

  Then relations analogous to (15.2.1) and (15.2.2) hold:

$$AU_m = U_{m+1} \tilde{H}_m,$$

  where

$$\tilde{H}_m = \begin{pmatrix} H_m \\ 0 \ldots 0 H_{m+1,m} \end{pmatrix}_{(m+1)p \times mp},$$

  and

$$U_m^T AU_m = H_m.$$

15.2.3   The Lanczos and Block Lanczos Methods

For a nonsymmetric matrix A, the Lanczos algorithm constructs, starting with two vectors v_1, and w_1, a **pair** of **biorthogonal bases** $\{v_1, \ldots, v_m\}$ and $\{w_1, \ldots, w_m\}$ for the two Krylov subspaces: $K_m(A, v_1) = \text{span}\{v_1, Av_1, \ldots, A^{m-1}v_1\}$ and $K_m(A^T, w_1) = \text{span}\{w_1, A^T w_1, \ldots, (A^T)^{m-1}w_1\}$.

Algorithm 15.2.2.   *The Nonsymmetric Lanczos Algorithm*
   Step 0. *Scale the vectors v and w to get the vectors v_1 and w_1 such that*
$w_1^T v_1 = 1$. *Set $\beta_1 \equiv 0$, $\delta_1 \equiv 0$, $w_0 = v_0 \equiv 0$.*
   Step 1. *For $j = 1, 2, \ldots, m$ do*

$$\alpha_j = w_j^T A v_j$$
$$\hat{v}_{j+1} = A v_j - \alpha_j v_j - \beta_j v_{j-1}$$
$$\hat{w}_{j+1} = A^T w_j - \alpha_j w_j - \delta_j w_{j-1}$$
$$\delta_{j+1} = \sqrt{|\hat{w}_{j+1}^T \hat{v}_{j+1}|}$$
$$\beta_{j+1} = \hat{w}_{j+1}^T \hat{v}_{j+1} / \delta_{j+1}$$
$$w_{j+1} = \hat{w}_{j+1} / \beta_{j+1}$$
$$v_{j+1} = \hat{v}_{j+1} / \delta_{j+1}$$

   End.

If the algorithm does not break down before completion of m steps, then, defining $V_m = (v_1, \ldots, v_m)$ and $W_m = (w_1, \ldots, w_m)$, we obtain (i) $W_m^T A V_m = T_m$, (ii) $A V_m = V_m T_m + \delta_{m+1} v_{m+1} e_m^T$, and $A^T W_m = W_m T_m^T + \beta_{m+1} w_{m+1} e_m^T$, where T_m is tridiagonal $(\alpha_1, \ldots, \alpha_m; \beta_2, \ldots, \beta_m; \delta_2, \ldots, \delta_m)$.
   Breakdown of the Lanczos method: If neither v_j nor w_j is zero, but $w_j^T v_j = 0$, then we have a breakdown (see Wilkinson (1965, p. 389)). In that case, the **look-ahead Lanczos** idea has to be applied (see Bai *et al.* 2000), Parlett *et al.* (1985), and Freund *et al.* (1993).

The Block Lanczos Method

Starting with $n \times p$ block vectors P_1 and Q_1 such that $P_1^T Q_1 = I$, the block Lanczos method generates right and left Lanczos block vectors $\{Q_j\}$ and $\{P_i\}$ of dimension $n \times p$, and a block tridiagonal matrix $T_B = $ Tridiagonal $(T_1, \ldots, T_m;$ $L_2, \ldots, L_m; M_2, \ldots, M_m)$ such that defining

$$P_{[m]} = (P_1, P_2, \ldots, P_m) \quad \text{and} \quad Q_{[m]} = (Q_1, Q_2, \ldots, Q_m),$$

we have (i) $Q_{[m]}^T A Q_{[m]} = T_B$, (ii) $A Q_{[m]} = Q_{[m]} T_B + Q_{m+1} M_{m+1} E_m^T$, and (iii) $A^T P_{[m]} = P_{[m]} T_B^T + P_{m+1} L_{m+1}^T E_m^T$, where E_m is an $mp \times m$ matrix of which bottom square is an identity matrix and zeros elsewhere.
   For details of the algorithm, see Bai *et al.* (2000) and Golub and Van Loan (1996). *The block Lanczos method breaks down if $P_{j+1}^T Q_{j+1}$ is singular.* In such a situation, an *adaptively blocked Lanczos method* (Bai *et al.* 1999) can be used

to deal with the situation of breakdown. ABLE adaptively changes the block size and maintains the full or semi biorthogonality of the block Lanczos vectors.

15.3 SCOPES OF USING THE KRYLOV SUBSPACE METHODS IN CONTROL

Since the Krylov subspace methods (such as, the Arnoldi and Lanczos methods) are the projection methods onto K_m, it is only natural to use these methods as the projection techniques to solve large-scale control problems, as has been done in numerical linear algebra for matrix problems.

A **template** is then as follows: First, the original large control problem is projected onto an m-dimensional Krylov subspace by constructing a basis of the subspace. The projected smaller problem is then solved using a standard well-established technique. Finally, an approximate solution of the original problem is obtained form the solution of the projected problem. The solution of the projected problem is constructed such that either a **Galerkin property** is satisfied, that is, the residual is orthogonal to the associated Krylov subspace, or the norm of the residual error is minimized (**GMRES type**). These projected methods usually give cheaply computed residual error norms, which, in turn can be used as a stopping criteria in case the methods need to be restarted. For a description of the GMRES method, see Saad and Schultz (1986).

15.4 ARNOLDI METHODS FOR LYAPUNOV, SYLVESTER, AND ALGEBRAIC RICCATI EQUATIONS

Numerical methods for solving the **Lyapunov equations** $AX + XA^T + BB^T = 0$ (**Continuous-time**), and $AXA^T - X + BB^T = 0$ (**Discrete-time**) have been discussed in Chapter 8. The standard **Schur-method** (**Section 8.5.2, Section 8.5.4**), and **Algorithms 8.6.1** and **8.6.2**, based on the Schur decomposition of A, is not suitable for sparse problems. In the following subsections, we show the use of scalar Arnoldi to solve the *single-input continuous-time* and that of the block Arnoldi to solve the *multi-input discrete-time problem*.

The matrix A is assumed to be *stable* in each case; that is, in the continuous-time case, A is assumed to have all eigenvalue negative real parts and in the discrete case, A is assumed to have all its eigenvalues within the unit circle.

Algorithm 15.4.1. *An Arnoldi Method for the Single-Input Stable Lyapunov Equation*

   Step 1. *Run m steps of the Arnoldi algorithm with* $v_1 = b/\|b\|_2 = b/\beta$. *Obtain* V_m *and* H_m.

   Step 2. *Solve the projected* $m \times m$ *Lyapunov matrix equation:* $H_m G_m + G_m H_m^T + \beta^2 e_1 e_1^T = 0$, *using the Schur-method (see* **Section 8.5.2***).*

   Step 3. *Compute* X_m, *an approximation to* X: $X_m = V_m G_m V_m^T$.

Galerkin condition, residual error and re-start: (i) It is shown (Saad 1990; Jaimoukha and Kasenally 1994) that the residual $\text{Res}(X_m) = AX_m + X_m A^\mathsf{T} + bb^\mathsf{T}$ satisfies: $V_m^\mathsf{T} \text{Res}(X_m) V_m = 0$ and (ii) the residual error-norm for the projected solution: $\|\text{Res}(G_m)\|_\mathsf{F} = \sqrt{2}\|h_{m+1,m} e_m^\mathsf{T} G_m\|_\mathsf{F}$. Using this cheaply computed residual error-norm as a stopping criterion, Algorithm 15.4.1 can be restarted at every fixed number (say m_1) of iterations, wherever needed.

Algorithm 15.4.2. *A Block Arnoldi Algorithm for Stable Discrete-Time Lyapunov Equation*
  Step 1. *Find the QR factorization of B to compute V_1 of order $n \times p$:*

$$B = V_1 R.$$

  Step 2. *Run m steps of the block Arnoldi algorithm to obtain H_m, U_m, and $H_{m+1,m}$ with V_1 as obtained in Step 1.*
  Step 3. *Obtain an $mp \times mp$ matrix G_m by solving the projected discrete Lyapunov equation using the Schur-method* (**Section 8.5.4**):

$$H_m G_m H_m^\mathsf{T} + \begin{pmatrix} R \\ 0 \end{pmatrix} (R^\mathsf{T}\ 0) = G_m.$$

  Step 4. *Compute the approximate solution $X_m = U_m G_m U_m^\mathsf{T}$.*

Galerkin condition, residual error norm, and Restart

  1. The residual $\text{Res}(X_m) = AX_m A^\mathsf{T} - X_m + BB^\mathsf{T}$ satisfies the **Galerkin property**: $U_m^\mathsf{T}\text{Res}(X_m)U_m = 0$.
     Furthermore, the **residual error norm** for the solution of the projected problem is given by (Jaimoukha and Kasenally 1994):

$$\|\text{Res}(G_m)\|_\mathsf{F} = \left\| H_{m+1,m} E_m^\mathsf{T} G_m \left(\sqrt{2} H_m^\mathsf{T}\ \ E_m H_{m+1,m}^\mathsf{T} \right) \right\|_\mathsf{F}.$$

  2. If H_m is also discrete-stable, then the error bound $\|X - X_m\|_2$ converges to zero as m increases (Boley 1994).
  3. As in the continuous-time case, the cheaply computed residual can be used to restart the process if necessary.

Arnoldi Methods for Sylvester Equation

Let A, B, and C be the matrices of order n. (**Note that the matrix B here is not the usual control matrix.**) We have seen in **Section 8.2.1** that the Sylvester equation: $AX - XB = C$ can be written as the linear systems of equations: $(I \otimes A - B^\mathsf{T} \otimes I)x = c$, where $\otimes$ denotes the Kronecker product, and x and c are vectors with

n^2 components. Solving the Sylvester equation this way will require a formidable amount of storage and time for large and sparse problems. Hu and Reichel (1992) have proposed a method to solve this system requiring a considerable reduced amount of storage space. Their idea is to replace the Krylov subspace $K_m(I \otimes A - B^T \otimes I, r_0)$ with a subspace of the form $K_m(B^T, g) \otimes K_m(A, f)$ for certain vectors f and g. The vectors f and g are chosen so that the initial residual vector $r_0 = b - Ax_0$, where x_0 is the initial approximate solution, lies in the Krylov subspace $K_m(B^T, g) \otimes K_m(A, f)$.

Algorithm 15.4.3. *A Restarted Arnoldi Algorithm for the Sylvester Equation* $AX - XB = C$ *(Galerkin type)*

    Step 1. *Choose x_0 and compute $r_0 \equiv c - (I \otimes A - B^T \otimes I)x_0$.*

    Step 2. *If $\|r_0\|_2 \le \epsilon$, then compute the approximate solution matrix X_0 of the equation $AX - XB = C$ from the entries of x_0.*

    Step 3. *Choose f and g using the following scheme:*

    Let R_0 be defined by: $e_j^T R_0 e_k = e_{j+n(k-1)}^T r_0$, $1 \le j, k \le n$. Then, if $\|R_0\|_1 \ge \|R_0\|_\infty$, determine $g = R_0^T f / \|f\|^2$, taking f as a column of R_0 of the largest norm. Else, determine $f = R_0 g / \|g\|^2$, taking g as a row of R_0 of the largest norm.

    Using the Arnoldi algorithm, compute the orthonormal bases of $K_{m+1}(A, f)$ and $K_{m+1}(B^T, g)$; that is, obtain $H_A, H_B, \tilde{H}_A, \tilde{H}_B, V_m, V_{m+1}, W_m, W_{m+1}$.

    Step 4. *Compute $\tilde{r}_0 = (W_m \otimes V_m)^T r_0$.*

    Step 5. *Determine Q_A and R_A from H_A, and Q_B and R_B from H_B by Schur factorizations. That is, find Q_A, R_A; Q_B, R_B such that $H_A = Q_A R_A Q_A^*$ and $H_B = Q_B R_B Q_B^*$. Compute $r_0' = (Q_B \otimes Q_A)^* \tilde{r}_0$.*

    Step 6. *Solve the triangular or the quasi-triangular system: $(I \otimes R_A - R_B \otimes I)y_0' = r_0'$ and compute $y_0 = (Q_B \otimes Q_A)y_0'$.*

    Step 7. *Compute the correction vector: $z_0 = (W_m \otimes V_m)y_0$ and update the solution: $x_0 \equiv x_0 + z_0$.*

    Step 8. *Compute the updated residual vector: $r_0 \equiv r_0 - (W_m \otimes V_{m+1}\tilde{H}_A)y_0 + (W_{m+1}\tilde{H}_B \otimes V_m)y_0$ and go to Step 2.*

A **breakdown of the algorithm** occurs **when the matrix of the linear system in Step 6 becomes singular**. In this case, one can either reduce m or restart the algorithm with different f and g, for example, random vectors. The same action should be taken when dim $K_m(A, f) < m$ or dim $K_m(B^T, g) < m$.

Block Arnoldi Methods for Sylvester Equation

While the Hu–Reichel algorithm is a projection algorithm on the linear algebraic system associated with the Sylvester equation, projection algorithms on the actual

Sylvester equation have recently been developed (El Guennouni *et al.* 2003; Robbé and Sadkane 2002). Furthermore, the Hu–Reichel algorithm has been extended to the block form by Simoncini (1996). Let $A \in \mathbb{R}^{n \times n}$, $B \in \mathbb{R}^{p \times p}$.

Algorithm 15.4.4. *Block Arnoldi Methods for the Sylvester Equation $AX + XB = C$.*

Step 1. *Choose X_0 and compute the residual matrix $R_0 = C - (AX_0 - X_0 B)$. Assume that* rank$(R_0) = q$.

Step 2. *Obtain an upper triangular matrix Λ_1 by computing the full rank QR factorization of R_0: $R_0 = V_1 \Lambda_1$ and run m steps of the block Arnoldi algorithm with V_1 to obtain U_m, H_m, and $\tilde{H}_m$* **(Algorithm 15.2.1)**.

Step 3. **(Galerkin-type)**: *Compute the approximate solution: $X_m^G = X_0 + U_m Z_m$, obtaining Z_m by solving the Sylvester equation using the Hessenberg–Schur method* **(Algorithm 8.5.1)**:

$$H_m Z_m - Z_m B = \Lambda, \text{ where } \Lambda = \begin{pmatrix} \Lambda_1 \\ 0 \end{pmatrix} \in \mathbb{R}^{mq \times p}.$$

(GMRES-type): *Compute the approximate solution: $X_m^{GM} = X_0 + U_m Z_m$, obtaining Z_m by solving the minimization problem:*

$$\min_{Z \in R^{mq \times p}} \|\bar{\Lambda} - \bar{S}_m(Z)\|_F, \text{ where } \bar{\Lambda} = \begin{pmatrix} \Lambda_1 \\ 0 \end{pmatrix} \in \mathbb{R}^{(m+1)q \times p}$$

and

$$\bar{S}_m(Z) = \tilde{H}_m Z - \begin{pmatrix} Z \\ 0 \end{pmatrix} B,$$

Residuals and restart: It can be shown (Robbé and Sadkane (2002)) that the residuals $R_m^{GM} = S(X_m^{GM}) - C$ and $R_m^G = S(X_m^G) - C$ satisfy, respectively:

$$\|R_m^{GM}\|_F = \|\bar{\Lambda} - \bar{S}_m(Z_m)\|_F \quad \text{and} \quad \|R_m^G\|_F = \|H_{m+1,m} Z_m^L\|_F,$$

where Z_m^L is the last $q \times p$ block of Z_m. Using these easily computed residuals, the method should be periodically restarted with $X_0 = X_m^G$ or $X_0 = X_m^{GM}$, where $X_m^G | X_m^{GM}$ is the last computed approximate solution with Galerkin/GMRES method.

Convergence analysis (Robbé and Sadkane 2002)

1. The GMRES algorithm converges if the field of values of A and B are disjoint. If the Galerkin algorithm converges, then the GMRES algorithm also converges. However, if **GMRES** stagnates (i.e., $\|R_m^{GM})\|_F = \|R_0\|_F$), then the Galerkin algorithm fails.
   Note: It is assumed that R_0 and the parameter m are the same in both these algorithms.

2. (*Breakdown*). If the block Arnoldi algorithm breaks down at iteration m; that is, if $H_{m+1,m} = 0$, then the approximate solutions computed by GMRES and the Galerkin algorithm are exact; that is, $X_m^G = X_m^{GM} = X$.

Arnoldi Method for Sylvester-Observer Equation (Single-Output Case)

The Sylvester-observer equation $A^T X - X H = C^T G$ arises in the construction of Luenberger observer (see Chapter 12; Datta 1994). For a full-rank solution X, it is necessary that (A, C) is observable and (H^T, G) is controllable. If H is an upper Hessenberg matrix (as in the case of the scalar Arnoldi), then in the single-output case g can be chosen to be $e_n = (0, 0, \ldots, 0, 1)$ and the Sylvester-observer equation in this case reduces to $A^T X - X H = (0, 0, \ldots, 0, c^T)$. An Arnoldi method was developed by Datta and Saad (1991) to solve this equation by observing the striking resemblance of this equation with the Arnoldi equation: $A V_m - V_m H_m = (0, 0, \ldots, 0, h_{m+1,m} v_{m+1})$. Naturally, the Arnoldi vector v_1 should be chosen so that the last vector v_{m+1} becomes the vector c, given a priori. This is done by observing that, apart from a multiplicative scalar, the polynomial $p_m(x)$ such that $v_{m+1} = p_m(A) v_1$, is the characteristic polynomial of H_m (see Saad 1992a). The matrix H_m is constructed to have a pre-assigned spectrum $\{\mu_1, \ldots, \mu_m\}$ for which an eigenvalue assignment algorithm (e.g., Datta 1987) is invoked at the end of $(m - 1)$ steps of the Arnoldi algorithm with the chosen vector v_1.

Algorithm 15.4.5. *An Arnoldi Algorithm for Single-output Sylvester-Observer Equation*
   Step 1. *Solve the linear system:* $q(A^T)x = c^T$, *and compute* $v_1 = x/\|x\|$; *where* $q(t) = (t - \mu_1)(t - \mu_2) \cdots (t - \mu_m)$.
   Step 2. *Run $m - 1$ steps of the Arnoldi method on A^T with v_1 as the initial vector to generate V_m and the first $m - 1$ columns of H_m. Let $\tilde{H}_{m-1}$ denote the matrix of the first $m - 1$ columns of H_m.*
   Step 3. *Find a column vector y such that $\Omega([\tilde{H}_{m-1}, y]) = \Omega(H_m) = \{\mu_1, \ldots, \mu_m\}$, where $\Omega(K)$ denotes the spectrum of the matrix K.*
   Step 4. *Compute $\alpha = (c')^T c' / \|c\|^2$, where c' is the last column of $A^T V_m - V_m H_m$.*
   Step 5. *Set $X_m = (1/\alpha) V_m$.*

Solving the equation $q(A^T)x = c^T$ using the partial fraction approach: A partial fraction approach suggested in Datta and Saad (1991) to solve the above polynomial system of equations consists in decomposing the system into m linearly independent systems: $(A^T - \mu_i I)x_i = c^T$, $i = 1, \ldots, m$ and then obtaining the solution x as the linear combination: $x = \sum_{i=1}^{n} \dfrac{1}{q'(\mu_i)} x_i$, where $q'(\mu_j) = \Pi_{i=1,\ldots,m, i \neq j}$. Each of these systems can be solved by applying k steps of the Arnoldi method, constructing an orthonomial basis V_k of the span $\{c, A, c, \ldots, A^{k-1}c\}$ and then solving k independent small $m \times m$ Hessenberg linear systems. The bulk of the work is in constructing V_k, and this is done only once. A detailed stability (numerical) property of the approach studied in Calvetti *et al.* (1995), Calvetti and Reichel (1997), and Calvetti *et al.* (2001) shows that *the performance of the scheme can be improved by choosing μ_is as the equidistant points on a circle or on the zeros of a certain Chebyshev polynomial.*

Remarks

- Observe that the solution obtained by this algorithm has the nice additional property of being **orthonormal**.
- A full Arnoldi-type of method ($m = n$) for the construction of an orthogonal solution to the multi-output Sylvester-observer equation has been developed by Datta and Hetti (1997). Also, there now exists a singular value decomposition (SVD)-based algorithm (Datta and Sarkissian 2000) for solving the multi-output Sylvester-observer equation, which might be suitable for large-scale computing.

Arnoldi Method for Continuous-Time Algebraic Riccati Equation (CARE)

In **Chapter 13**, we have described numerical solutions of the algebraic Riccati equations. *The Schur method, the generalized Schur method, or similar methods based on matrix decompositions are not practical for large problems.* An idea to solve the CARE using the block Arnoldi method developed by Jaimoukha and Kasenally (1994) is as follows. For simplicity, we write the CARE as: $XA + A^T X - XBB^T X + LL^T = 0$ (i.e., $R = I$ and $Q = LL^T$). **Assume that the associated Hamiltonian matrix does not have a purely imaginary eigenvalue.**

Algorithm 15.4.6. *An Arnoldi Algorithm for CARE (Galerkin-type)*
    Step 1. *Compute U_m, H_m, $H_{m+1,m}$ by running m steps of the block Arnoldi method starting with V_1 given by: $L = V_1 R$ (**QR factorization of L**). Define B_m by $U_m^T B = B_m$ and L_m by $U_m L_m = L$.*
    Step 2. *Solve the projected equation for G_m:*

$$G_m H_m + H_m^T G_m - G_m B_m B_m^T G_m + L_m L_m^T = 0$$

    Step 3. *Compute approximation X_m of X: $X_m = U_m G_m U_m^T$*

Galerkin condition and restart

1. **The residual norm satisfies $\text{Res}(X_m)$ the Galerkin property: V_m^T $\text{Res}(X_m) V_m = 0$.**

2. Algorithm 15.4.6 can be **restarted** by using the cheaply computed residual error norm: $\|\text{Res}(G_m)\|_F = \sqrt{2}\|H_{m+1,m}E_m^T G_m\|_F$, as a stopping criterion.

15.5   ARNOLDI METHOD FOR PARTIAL EIGENVALUE ASSIGNMENT

Let the spectrum of an $n \times n$ matrix A be denoted by $\Omega(A) = \{\lambda_1, \ldots, \lambda_p, \lambda_{p+1}, \ldots, \lambda_n\}$. Recall from Chapter 11 that the Partial Eigenvalue Assignment (PEVA) is defined as follows: Given an $n \times n$ large and sparse matrix A, with partial spectrum $\{\lambda_1, \ldots, \lambda_p\}$, an $n \times m$ control matrix B, and a set of self-conjugate scalars $\{\mu_1, \ldots, \mu_p\}$, the problem is the one of finding a feedback matrix K such that $\Omega(A - BK) = \{\mu_1, \ldots, \mu_p; \lambda_{p+1}, \ldots, \lambda_n\}$. The problem naturally arises in feedback stabilization of large systems such as large space structures, power plants.

We have described a Sylvester equation approach due to Datta and Sarkissian (2002) in Chapter 11. Here we describe a projection method due to Saad (1988), which can be implemented using the Arnoldi method. It is based on computing an orthonormal basis for the left invariant subspace associated with the p eigenvalues that are to be reassigned.

Algorithm 15.5.1.   *A Projection Algorithm for Partial Pole-Placement*
   Step 1. *Compute the partial Schur decomposition:* $A^T Q = QR$ *associated with the eigenvalues* $\lambda_1, \lambda_2, \ldots, \lambda_p$.
   Step 2. *Compute* $S_0 = Q^T B$ *and solve the projected* $p \times p$ *eigenvalue assignment problem. That is, find a matrix* G *such that* $\Omega(R^T - S_0 G^T) = \{\mu_1, \mu_2, \ldots, \mu_p\}$, *using a* **standard multi-input EVA method** **(Algorithm 11.3.1)**.
   Step 3. *Form the feedback matrix:* $K = (QG)^T$.

15.6   LANCZOS AND ARNOLDI METHODS FOR MODEL REDUCTION

In **Chapter 14**, we have described several techniques for model reduction. These include model reduction via **balancing** and **the Schur method**. Since these methods require reduction of the state-matrix A to real-Schur form, they are not suitable for large and sparse computations. Here we describe some Krylov-subspace ideas. These Krylov methods are designed to construct a reduced-order model (ROM) such that the first few Markov parameters (see **Chapter 9**) of this model match with those of the original model.

Several Krylov subspace methods for model reduction have been developed in recent years. These include the *Padé via Lanczos (PVL)* approach, the *interpolation approach*, based on the rational Krylov method of Ruhe, *implicitly restarted Lanczos method*, and *Arnoldi and implicitly restarted dual Arnoldi methods*. The *PVL technique has been proven to be effective in circuit simulation and the multipoint rational interpolation approach has been successful in moment matching of the transfer function at selected frequencies*. The machinery needed to describe these techniques has not been developed here and, therefore, we have to skip the descriptions of these techniques. For state-of-the-art survey on this topic, see Antoulas (2003) and Van Dooren (2000), and Datta (2003).

We will describe here only a basic Lanczos and an Arnoldi method for model reduction in the single-input, single-output (SISO) case and just mention the existence of the block Lanczos and band Lanczos methods in the multi-input, multi-output (MIMO) case.

15.6.1 Lanczos Methods for Model Reduction

Algorithm 15.6.1.   *A Lanczos Algorithm for SISO Model Reduction*
    Step 0. *Scale the vectors b and c to obtain the vectors v_1 and w_1 such that* $w_1^T v_1 = 1$.
    Step 1. *Run k steps of the Lanczos algorithm* (**Algorithm 15.2.2**) *to generate the matrices W_k and V_k and then compute $A_k = W_k^T A V_k$, $b_k = W_k^T b$, $c_k = c V_k$.*
    Step 2. *Form the reduced-order model (A_k, b_k, c_k).*

It can be shown that the reduced-order model defined by (A_k, b_k, c_k) preserves the first $2k$ Markov parameters of the original system. (See Gragg 1974; Gragg and Lindquist 1983). That is, $c A^{i-1} b = c_k A_k^{i-1} b_k$, $i = 1, 2, \ldots, 2k$.

Numerical Disadvantages and Possible Cures

There are several numerical difficulties with the above algorithm: first, there can be serious "breakdowns" in the Lanczos process due to the ill-conditioning of the submatrices in the system's Hankel matrix; second, the steady-state error can be large; third, the stability of the ROM is not guaranteed even though the original model is stable. An **implicit restated Lanczos scheme** due to Grimme *et al.* (1996), to stabilize the ROM is as follows: Suppose that the matrix A_k is not stable and assume that there are q unstable modes: $\mu_1, \ldots, \mu_q$. Then the idea is to restart Algorithm 15.6.1 with the new starting vectors $\bar{v}_1 = \bar{p}_\vartheta (A - \mu_q I) \cdots (A - \mu_1 I) v_1$, and $\bar{w}_1 = \bar{p}_w (A^T - \mu_q I) \cdots (A^T - \mu_1 I) w_1$, where $\bar{p}_\vartheta$ and $\bar{p}_w$ are certain scalars. The scheme is implemented implicitly using a technique similar to the one proposed in Sorensen (1992). There also exist relations between the modified

Markov parameters of the original system and the above restarted Lanczos model (see Grimme *et al.* 1996).

15.6.2   Block Lanczos and Band Lanczos Methods for MIMO Model Reduction

In the MIMO case, when $m = r$, the **block Lanczos** method can be used. Specifically, the following result (see Boley 1994) can be proved.

Theorem 15.6.1. *Let j steps of the block Lanczos method be applied to the MIMO system (A, B, C), starting with block vectors generated from the QR decompositions of B and C, obtaining the matrices $P_{[j]}$ and $Q_{[j]}$. Define $\hat{A} = Q_{[j]}^T A P_{[j]}$, $\hat{B} = Q_{[j]}^T B$, $\hat{C} = C P_{[j]}$. Then the ROM $(\hat{A}, \hat{B}, \hat{C})$ has the following properties: $\hat{C} \hat{A}^i \hat{B} = C A^i B$ for $i = 0, 1, \ldots, 2(j - 1)$.*

The *band Lanczos method* is an extension of the standard nonsymmetric Lanczos method for single vectors to blocks of starting vectors of different sizes. This method is thus ideal for the MIMO case when $m \neq r$. For space limitations, the detailed description of the algorithm cannot be given here. For description of the algorithm, we refer the readers to the paper by Aliga *et al.* (2000). For application of the band Lanczos algorithm to the MIMO model reduction, see Freund (1999) and the paper by Freund in Bai *et al.* (2000, pp. 205–216). See also, Bai *et al.* (1997), Bai and Freund (1999), and Freund (1997).

15.6.3   An Arnoldi Method for SISO Model Reduction

The idea is to use the Arnoldi method simultaneously on (A, b) and (A^T, c^T) and then combine the results to obtain ROMs. The ROMs have been shown to satisfy the Galerkin conditions (Jaimoukha and Kasenally 1997).

Algorithm 15.6.2.   *An Arnoldi Algorithm for SISO Model Reduction*
   Step 1. *Perform m steps of the Arnoldi method with (A, b) to obtain H_m, $\tilde{H}_m$, V_m, $\tilde{V}_m$ and l_m, with $v_1 = b/\|b\|_2$. ($\tilde{V}_m = v_{m+1}$, $\tilde{H}_m = h_{m+1,m} e_m^T$ and $l_m = \|b\|_2 e_1$).*
   Step 2. *Perform m steps of the Arnoldi method with (A^T, c^T) to produce G_m, $\tilde{G}_m$, W_m, $\tilde{W}_m$ and k_m, with $w_1 = c^T/\|c\|_2$, $((\tilde{G}_m)^T = g_{m,m+1} e_m^T$, $\tilde{W}_m = w_{m+1}$ and $k_m = \|c\|_2 e_1^T$).*
   Step 3. *Form $T_m = W_m^T V_m$, $\hat{H}_m = T_m^{-1} W_m^T A V_m = H_m + T_m^{-1} W_m^T \tilde{V}_m \tilde{H}_m$ and $\hat{G}_m = W_m^T A V_m T_m^{-1} = G_m + \tilde{G}_m \tilde{W}_m^T V_m T_m^{-1}$*
   Step 4. *Form the ROM $(\hat{H}_m, l_m, k_m T_m)$ or $(\hat{G}_m, T_m l_m, k_m)$.*

Galerkin conditions and residual errors: Let $h_m(s) = (sI - \hat{H}_m)^{-1} l_m$ and $g_m(s) = k_m(sI - \hat{G}_m)^{-1}$. Then the Galerkin conditions $W_m^T((sI - A)V_m h_m(s) - b) = 0$, and $(g_m(s)W_m^T(sI - A) - c)V_m = 0$, $\forall s$ are satisfied.

Remarks

- *Jaimoukha and Kasenally (1997) have described a restarted Arnoldi frame-work which may be employed to make the ROMs stable and to remove redundant modes in the models.* For space limitation, we skip the description of this implicit method here.
- Antoulas *et al.* (2001) have recently proposed a restarted Arnoldi method, closely related to the one described above, based on the concept of the **Cross Grammian**. For space limitation, we skip the description here and refer the readers to the above paper.

15.7   CHAPTER NOTES AND FURTHER READING

In this chapter, we have provided a very brief review of some of the existing Krylov subspace methods for a few large problems arising in design and analysis of control problems. These include Arnoldi methods for Lyapunov and Sylvester equations by Saad (1990), Hu and Reichel (1992), Jaimoukha and Kasenally (1994); Arnoldi method for the single-output Sylvester-observer equation by Datta and Saad (1991); a projection algorithm (which can be implemented using Arnoldi method) for PEVA problem by Saad (1988); and Lanczos and Arnoldi methods for model reduction by Boley (1994), Grimme *et al.* (1996), and Jaimoukha and Kasenally (1995, 1997). See Boley and Golub (1984, 1991) for Krylov subspace methods for determining controllability.

The Hu–Reichel algorithm was extended by Simoncini (1996) to block form. There have also been some recent developments on the Krylov subspace methods for Sylvester equation. El Guennouni *et al.* (2001) have developed block Arnoldi and nonsymmetric block Lanczos algorithms for Sylvester equation. Robbé and Sadkane (2002) have proposed new block Arnoldi and block GMRES methods for Sylvester equation and analyzed their convergence properties in details.

In the context of model reduction, it is noted that there are other important methods, such as the PVL, the interpolation methods, etc., which have not been included here. For details of these methods, the readers are referred to the associated papers cited in the reference section of this Chapter. In particular, for Lanczos methods of model reduction see, Feldman and Freund (1995a, 1995b, 1995c), Jaimoukha and Kasenally (1997), Grimme *et al.* (1996), Papakos and Jaimoukha (2001), Papakos (2001), Papakos and Jaimoukha (2002), Gallivan *et al.* (1996), etc. The paper by Papakos and Jaimoukha (2002) contains a procedure for model reduction combining nonsymmetric Lanczos algorithm and Linear Fractional Transformations (LFT). The delightful recent surveys by Freund (1999), the recent research monograph by Antoulas (2003), Ph.D. thesis by Grimme (1994), and short course lecture notes by Van Dooren (1995, 2000) and Feldman and Freund (1995b) are good sources of knowledge for model reduction. The paper by Freund (1999) includes 123 references on large-scale matrix computations using Krylov methods and their applications to model reduction. The earlier general surveys on Krylov subspace methods in control include the papers by Boley (1994), Datta (1997), Boley and Datta (1996), Van Dooren (1992), Bultheel and Van Barel (1986), and Fortuna *et al.* (1992). Some other papers of interest on Krylov subspace methods for model reduction include the papers by Villemagne and Skelton (1987), and Su and Craig, Jr. (1991).

For recent algorithms on partial eigenvalue and eigenstructure assignments which are not Krylov subspace methods, but suitable for large-scale computations, see Sarkissian (2001) and Datta and Sarkissian (2002). See also Calvetti *et al.* (2001).

Research Problems

1. Develop a block Arnoldi type algorithm to solve the multi-output Sylvester-observer equation $AX - XB + GC$, analogous to single-output algorithm (**Algorithm 15.4.5**).
2. Develop a block Arnoldi algorithm for the discrete-time Algebraic Riccati equation (DARE): $A^T X A - X + Q - A^T X B (R + B^T X B)^{-1} B^T X A = 0$, analogous to Algorithm 15.4.6 in the continuous-time case.
3. Develop a block Arnoldi algorithm for the generalized Sylvester equation: $AXB - X = C$.
4. Develop a block Arnoldi algorithm for MIMO model reduction that preserves stability of the original system.

References

Aliaga J.I., Boley D.L., Freund R.W., and Hernández V. "A Lanczos-type method for multiple starting vectors," *Math. Comp.*, Vol. 69, pp. 1577–1601, 2000.

Antoulas A.C., Sorensen D.C., and Gugercin S. "A survey of model reduction methods for large-scale systems," in *Contemporary Math.* (Olshevsky V., ed.), American Mathematical Society, Providence, Vol. 280, pp. 193–219, 2001.

Antoulas A.C. *Lectures on Approximations of Large-Scale Dynamical Systems,* SIAM, Philadelphia, 2003 (To appear).

Arnoldi W.E. "The principle of minimized iterations in the solution of the matrix eigenvalue problem," *Quart. Appl. Math.,* Vol. 9, pp. 17–29, 1951.

Bai Z., Feldmann P., and Freund R.W. "Stable and passive reduced-order models based on partial Padé approximation via the Lanczos process," *Numerical Analysis Manuscript*, No. 97-3-10, Bell Laboratories, Murray Hill, NJ, 1997.

Bai Z., Day D., and Ye Q. "ABLE: an adaptive block Lanczos method for nonhermitian eigenvalue problem," *SIAM J. Matrix Anal. Appl.*, pp. 1060–1082, 1999.

Bai Z. and Freund R. "A band symmetric Lanczos process based on coupled recurrences with applications in model-order reduction," *Numerical Analysis Manuscript*, Bell Laboratories, Murray Hill, NJ, 1999.

Bai Z., Demmel J., Dongarra J., Ruhe A., and Van der Vorst H. *Templates for the Solution of Algebraic Eigenvalue Problems, A Practical Guide*, SIAM, Philadelphia, 2000.

Boley D.L. "Krylov space methods on state-space control models," *Proc. Circuit Syst. Signal*, Vol. 13, pp. 733–758, 1994.

Boley D. and Datta B.N. "Numerical methods for linear control systems," in *Systems and Control in the Twenty-First Century* (Byrnes C., Datta B., Gilliam D., and Martin C., eds.), pp. 51–74, Birkhauser, Boston, 1996.

Boley D. and Golub G. "The Lanczos–Arnoldi algorithm and controllability," *Syst. Control Lett.*, Vol. 4, pp. 317–327, 1984.

Boley D. and Golub G. "The nonsymmetric Lanczos algorithm and controllability," *Syst. Control Lett.*, Vol. 16, pp. 97–105, 1991.

Brezinski C., Redivo Zaglia M., and Sadok H. "Avoiding breakdown and near-breakdown in Lanczos type algorithms," *Numer. Algorithms*, Vol. 1, pp. 26–284, 1991.

Bultheel A. and Van Barel M. "Padé techniques for model reduction in linear systems theory: a survey," *J. Comput. Appl. Math.*, Vol. 14, pp. 401–438, 1986.

Calvetti D. and Reichel L. "Numerical aspects of solution methods for large Sylvester-like observer-equations," *Proc. IEEE Conf. Decision Control*, pp. 4389–4397, 1997.

Calvetti D., Gallopoulas E., and Reichel L. "Incomplete partial fractions for parallel evaluations of rational matrix functions," *J. Comp. Appl. Math.*, Vol. 59, pp. 349–380, 1995.

Calvetti D., Lewis B., and Reichel L. "On the solution of large Sylvester-observer equation," *Num. Lin. Alg. Appl.*, Vol. 8, pp. 435–451, 2001.

Calvetti D., Lewis B., and Reichel L. "Partial eigenvalue assignment for large linear control systems," in *Contemporary Mathematics* (Olshevsky V. ed.), American Mathematical Society, Providence, RI, Vol. 28, pp. 24–254, 2001.

Datta B.N. "An algorithm to assign eigenvalues in a Hessenberg matrix: single-input case," *IEEE Trans. Autom. Control*, AC-32, pp. 414–417, 1987.

Datta B.N. "Linear and numerical linear algebra in control theory: some research problems," *Lin. Alg. Appl.*, Vol. 197/198, pp. 755–790, 1994.

Datta B.N. *Numerical Linear Algebra and Applications*, Brooks/Cole Publishing Company, Pacific Grove, CA, 1995.

Datta B.N. "Krylov-subspace methods for control: an overview," *Proc. IEEE Conf. Decision Control*, 1997.

Datta B.N. "Krylov-subspace methods for large-scale matrix problems in control," *Future Generation of Computer Systems*, Vol. 19, pp. 125–126, 2003.

Datta B.N. and Hetti C. "Generalized Arnoldi methods for the Sylvester-observer matrix equation and the multi-input eigenvalue assignment problems," *Proc. IEEE Conf. Decision Control*, pp. 4379–4383, 1997.

Datta B.N. and Saad Y. "Arnoldi methods for large Sylvester-like observer matrix equations, and an associated algorithm for partial spectrum assignment," *Lin. Alg. Appl.*, Vol. 154–156, pp. 225–244, 1991.

Datta B.N. and Sarkissian D. "Block algorithms for state estimation, and functional observers," *Proc. IEEE Joint Conf. on Control Appl.*, pp. 19–23, 2000.

Datta B.N. and Sarkissian D. Partial eigenvalue assignment: Existence, uniqueness, and numerical solutions, Proc. Mathematical Theory of Networks and Systems, Notre Dame, August, 2002.

El Guennouni A., Jbilou K., and Riquet J. "Block Krylov subspace methods for solving large Sylvester equations," preprint, LMPA, No. 132, Université du Littoral, 2001 (To appear in *Numer. Algorithms*), 2003.

Feldmann P. and Freund R.W. "Efficient linear circuit analysis by Padé approximation via the Lanczos process," *IEEE Trans. Comput.-Aided Design*, Vol. 14, pp. 639–649, 1995a.

Feldman P. and Freund R.W. *Numerical Simulation of Electronic Circuits: State-of-the-Art Techniques and Challenges*, Course Notes, 1995b (*Available on-line from http:/cm.bell-labs.com/who/Freund*).

Feldman P. and Freund R.W. "Reduced-order modeling of large linear subcircuits via a block Lanczos algorithm," *Proc. 32nd Design Autom. Conf., Assoc. Comp. Mach.*, pp. 474–479, 1995c.

Fortuna L., Nunnari G., and Gallo A. *Model Order Reduction Techniques with Applications in Electric Engineering*, Springer-Verlag, London, UK, 1992.

Freund R.W. "Computing minimal partial realization via a Lanczos-type algorithm for multiple starting vectors," *Proc. IEEE Conf. Decision Control*, pp. 4394–4399, 1997.

Freund R.W. "Reduced-order modeling techniques based on Krylov subspace methods and their uses in circuit simulation," in *Applied and Computational Control, Signals, and Circuits* (Datta B.N. *et al.*, eds.), Vol. 1, pp. 435–498, Birkhauser, Boston, 1999.

Freund R.W., Gutknecht M.H., and Nachtigal N.M. "An implementation of the look-ahead Lanczos algorithm for non-hermitian matrices," *SIAM J. Sci. Comput.*, Vol. 14, pp. 137–158, 1993.

Gallivan K., Grimme E.J., and Van Dooren P. "A rational Lanczos algorithm for model reduction," *Numer. Algorithms*, Vol. 12, pp. 33–63, 1996.

Golub G.H. and Van Loan C.F. *Matrix Computations*, 3rd edn, Johns Hopkins University, Baltimore, MD, 1996.

Gragg W.B. "Matrix interpolations and applications of the continued fraction algorithm," *Rocky Mountain J. Math.*, Vol. 4, pp. 213–225, 1974.

Gragg W.B. and Lindquist A. "On the partial realization problem," *Lin. Alg. Appl.*, Vol. 50, pp. 277–319, 1983.

Grimme E.J., Sorensen D.C., and Van Dooren P. "Model reduction of state space systems via an implicitly restarted Lanczos method," *Numer. Algorithms*, Vol. 1, pp. 1–32, 1996.

Grimme E.J. *Krylov Projection Methods for Model Reduction*, Ph.D. Thesis, University of Illinois at Urbana-Champaign, Urbana, Illinois, 1994.

Hu D.Y. and Reichel L. "Krylov subspace methods for the Sylvester equation," *Lin. Alg. Appl. Appl.*, Vol. 172, pp. 283–313, 1992.

Jaimoukha I.M. "A general minimal residual Krylov subspace method for large-scale model reduction," *IEEE Trans. Autom. Control,* Vol. 42, pp. 1422–1427, 1997.

Jaimoukha I.M. and Kasenally E.M. "Krylov subspace methods for solving large Lyapunov equations," *SIAM J. Numer. Anal.,* Vol. 31, pp. 227–251, 1994.

Jaimoukha I.M. and Kasenally E.M. "Oblique projection methods for large scale model reduction," *SIAM J. Matrix Anal. Appl.,* Vol. 16, pp. 602–627, 1995.

Jaimoukha I.M. and Kasenally E.M. "Implicitly restarted Krylov Subspace methods for stable partial realizations," *SIAM J. Matrix Anal. Appl.,* Vol. 18(3), pp. 633–652, 1997.

Lanczos C. "An iteration method for the solution of the eigenvalue problem of linear differential and integral operators," *J. Res. Nat. Bur. Standards,* Vol. 45, pp. 255–282, 1950.

Papakos V. and Jaimoukha I.M. "Implicitly restarted Lanczos algorithm for model reduction," *Proc. 40th IEEE Conf. Decision Control,* pp. 3671–3672, 2001.

Papakos V. *An Implicitly Restarted Nonsymmetric Lanczos Algorithm for Model Reduction,* M.Phil. to Ph.D. Transfer Report, Imperial College, London, UK, 2001.

Papakos V. and Jaimoukha I.M. "Model reduction via an LFT-based explicitly restarted Lanczos algorithm (preprint)," *Proc. Math Theory Networks Syst.* (MTNS' 2002), Notre Dame, 2002.

Parlett B.N., Taylor D.R., and Liu Z.A. "A look-ahead Lanczos algorithm for unsymmetric matrices," *Math. Comp.,* Vol. 44, pp. 105–124, 1985.

M. Robbé and Sadkane M. "A convergence analysis of GMRES and FOM methods for Sylvester equations," *Numer. Algorithms,* Vol. 30, pp. 71–84, 2002.

Saad Y. "Projection and deflation methods for partial pole assignment in linear state feedback," *IEEE Trans. Autom. Control,* Vol. 33, pp. 290–297, 1988.

Saad Y. "Numerical solutions of large Lyapunov equations," in *Signal Processing, Scattering, Operator Theory, and Numerical Methods* (Kaashoek M.A., Van Schuppen J.H., and Ran A.C. eds.), pp. 503–511, Birkhauser, 1990.

Saad Y. *Numerical Methods for Large Eigenvalue Problems,* John Wiley, New York, 1992a.

Saad Y. *Iterative Methods for Sparse Linear Systems,* PWS, Boston, MA, 1996.

Saad Y. and Schultz M.H. "GMRES: a generalized minimal residual algorithm for solving nonsymmetric linear systems," *SIAM J. Sci. Statist. Comput.,* Vol. 7, pp. 856–869, 1986.

Sarkissian D. *Theory and Computations of Partial Eigenvalue and Eigenstructure Assignment Problems in Matrix Second-order and Distributed Parameter Systems,* Ph.D. Dissertation. Northern Illinois University, DeKalb, Illinois, 2001.

Simoncini V. "On the numerical solution of $AX - XB = C$," *BIT,* Vol. 36, pp. 814–830, 1996.

Sorensen D.C. "Implicit application of polynomial filters in a k-step Arnoldi method," *SIAM J. Matrix Anal. Appl.,* Vol. 13, pp. 357–385, 1992.

Sorensen D.C. and Antoulas A.C. *Projection Methods for Balanced Model Reduction,* Unpublished manuscript, 2001 (available from the webpage: *http://www.ece.rice.edu/ aca*).

Su T.-J. and Craig R.R., Jr. "Model reduction and control of flexible structures using Krylov vectors," *J. Guidance Control Dynam.,* Vol. 14, pp. 260–267, 1991.

Van Dooren P. *The Lanczos Algorithm and Padé Approximations,* Short Course, Benelux Meeting on Systems and Control, 1995.

Van Dooren P. *Gramian-Based Model Reduction of Large-scale Dynamical Systems,* Short Course SIAM Annual Meeting, San Juan, Puerto Rico, July 2000.

Van Dooren P. "Numerical linear algebra techniques for large-scale matrix problems in systems and control," *Proc. IEEE Conf. Decision Control,* 1992.

de Villemagne C. and Skelton R.E. "Model reductions using a projection formulation," *Int. J. Control*, Vol. 46, pp. 2141–2169, 1987.

Wilkinson J.H. *The Algebraic Eigenvalue Problem,* Clarendon Press, Oxford, UK, 1965.

SOME EXISTING SOFTWARE FOR CONTROL SYSTEMS DESIGN AND ANALYSIS

In this appendix, we will give a brief description of some of the existing software for control systems design and analysis.

A.1 MATLAB CONTROL SYSTEM TOOLBOX

As the title suggests, MATLAB *Control System Toolbox* is based on the well-known matrix computations software "MATLAB." It is a collection of M-files which implement some of the numerically viable algorithms for control system design, analysis, and modeling.

The control systems can be modeled either as transfer functions or in state-space form. Both continuous-time and discrete-time systems can be handled. The toolbox has excellent graphic capabilities and various time and frequency responses can be viewed on the screen and analyzed.

The software can be obtained from The MathWorks, Inc., 24 Prime Park Way, Natick, MA 01760-1500
Tel: (508) 647-7000, Fax: (508) 647-7001, URL: http://www.mathworks.com
Newsgroup: Comp. soft. sys. matlab.
See MATLAB Control System Toolbox: Users Guide (1996) for details.

A.2 MATCONTROL

MATCONTROL is also a collection of **M-files implementing major algorithms of this book.** MATCONTROL is primarily designed for classroom use—by using this toolbox, the students (and the instructors) will be able to compare different algorithms for the same problem with respect to efficiency, stability, accuracy, easiness-to-use, and specific design and analysis requirements.

A.3   CONTROL SYSTEM PROFESSIONAL—ADVANCED NUMERICAL METHODS (CSP-ANM)

Control System Professional (*CSP*) based on *"Mathematica"* is a collection of *Mathematica* programs (1996) to solve control systems problems. *CSP-ANM* extends the scope of CSP by adding new numerical methods for a wide class of control problems as well as for a number of matrix computations problems that have extensive uses in control systems design and analysis.

ANM is compatible with, and requires, *Control System Professional* 2.0 or later. The software has been developed by Biswa Nath Datta and Daniil Sarkissian (with the help of Igor Bakshee from Wolfram Research Incorporation).

"Typically, *Advanced Numerical Methods* provides several numerical methods to solve each problem enabling the user to choose from most appropriate tool for a particular task based on computational efficiency and accuracy." Users can select the most appropriate tool for a given task or have the package choose a suitable method automatically based on the size of data and the required accuracy. Thus, the package, though oriented mostly for professional users, is also an important tool for students, researchers, and educators alike.

The algorithms implemented in the package have been taken mostly form the current book by the author. More details can be found from *http://www.wolfram.com/products/applications/ann*

Software and manual: There is a User's Manual written by Biswa Nath Datta and Daniil Sarkissian (2003) with help from Igor Bakshee and published by *Wolfram Research, Inc.*. Both the software and the manual can be obtained from:

Wolfram Research, Inc., 100 Trade Center Drive, Champaign, Illinois 61820-7237, USA
Tel.: (217) 398-0700, Fax: (217) 398-0747
E-mail: Info@wolfram.com, URL: www.wolfram.com

A.4   SLICOT

SLICOT is a Fortran 77 Subroutine Library in Control Theory. It is built on the well-established matrix software packages, the **B**asic **L**inear **A**lgebra **S**ubroutines (BLAS) and the **L**inear **A**lgebra **P**ackage (LAPACK). The library also contains other mathematical tools such as discrete sine/cosine and Fourier transformations. The routines can be embedded in MATLAB by an appropriate interface thus enhancing the applicability of the library.

For a brief description of the library, see the paper **"SLICOT—A Subroutine Library in Systems and Control Theory"** by Peter Benner, Volker Mehrmann, Vasile Sima, Sabine Van Huffel, and Andras Varga in *Applied and Computational Control, Signals, and Circuits* (Biswa Nath Datta, Editor), Birkhauser, 2001. The official website for SLICOT is: *http://www.win.tuc.nc/niconet*

A.5 MATRIX$_X$

MATRIX$_X$, as the title suggests, is built on functions that are most commonly used for matrix computations. It is broken into several modules. The principal ones are *MATRIX*$_X$ Core, Control, and System Build, Optimization, and Robust Control. The core module contains the core MATLAB commands with some modifications and extensions. The control module contains both classical and modern control commands.

The *MATRIX*$_X$ core and control modules are command driven, while the system build module is menu driven. This module allows the users to simulate the systems by building the block diagrams of the systems on the screen. MATRIX$_X$ is a product of *Integrated Systems, Inc.* There exist a *MATRIX*$_X$ User's Guide (1991) and a book by Shahian and Hassul (1992) describing the functional details of the software.

A.6 SYSTEM IDENTIFICATION SOFTWARE

Each of the software packages **MATLAB Control System Toolbox, Control System Professional, Control System Professions—Advanced Numerical Methods, SLICOT, MATRIX**$_X$, etc., has its own software module for system identification. See Chapter 9 of this book for details.

There now also exist a few software packages, especially designed for system identification. We describe three of them in the following.

A.6.1 MATLAB System Identification Toolbox

This toolbox has been developed by Prof. Lennart Ljung of Linköping University, Sweden. The toolbox can be used either in command mode or via a Graphical User Interface (GUI). The details can be found in the Users' manual (Ljung 1991) and MathWorks website: **http://www.mathworks.com**

A.6.2 *X*math Interactive System Identification Module, Part-2

This is a product of Integrated System Inc., Santa Clara, USA, 1994. It is a GUI-based software for multivariable system identification. The details can be found in User's Manual (VanOverschee *et al.* 1994).
Website: http://www.isi.com/products/MATRIX$_X$/Techspec/MATRIX$_X$-*X*math/ *xm*36.html.

A.6.3 ADAPT$_X$

This software package has been developed by W.E. Larimore. For details, see the Users Manual (Larimore 1997). **Website:** *http://adaptics.com*

Some further details on these softwares and subspace state-space system identification software can be found in the recent paper by DeMoor *et al.* (1999).

References

MATLAB Control System Toolbox: User's Guide, The MathWorks, Inc. Natick, MA, 1996.

MATRIX$_X$ User's Guide, Integrated Systems, Inc., Santa Clara, CA, 1991.

MATHEMATICA Control System Professional, Wolfram Research Inc., Champaign, Illinois, 1996.

Datta B.N. and Sarkissian D. (with Bakshee I.) *Advanced Numerical Methods. Control system professional suite component*, Software and Manual, Wolfram Research Inc., Champaign, Il., 2003.

DeMoor B., VanOverschee P., and Favoreel W. "Subspace state-space system identification," in *Applied and Computational Control, Signals, and Circuits*, (Datta B.N. *et al.*, eds.), pp. 247–311, Birkhauser, Boston, 1999.

Larimore W.E. *ADAPT$_X$ Automatic System Identification Software, Users Manual*, Adaptics Inc., Reading, MA 01867, USA, 1997.

Ljung L. *System Identification Toolbox for Use with MATLAB*, The MathWorks Inc., MA, USA, 1991.

Shahian B. and Hassul M. *Control System Design Using MATRIX$_X$*, Prentice Hall, Englewood Cliffs, NJ, 1992.

VanOverschee P., DeMoor B., Aling H., Kosut R., and Boyd S. *Xmath Interactive System Identification Module, Part* 2, Integrated Systems, Inc., Santa Clara, CA, 1994.

APPENDIX **B**

MATCONTROL AND LISTING OF MATCONTROL FILES

B.1 ABOUT MATCONTROL

What is the MATCONTROL library?
The MATCONTROL library is a set of M-files implementing the majority of algorithms of the book:
Numerical Methods for Linear Control Systems Design and Analysis by B.N. Datta.

Who wrote the MATCONTROL library?
The MATCONTROL library was written by several graduate students of Professor Datta. The most contributions were made by Joao Carvalho and Daniil Sarkissian.

How can I get the MATCONTROL library?
The MATCONTROL library is distributed with the book mentioned above.

What to do if a routine is suspected to give wrong answers?
Please let us know immediately. Send an email to dattab@math.niu.edu and, if possible, include a MATLAB diary file that calls the routine and produces the wrong answer.

How to install MATCONTROL
The MATCONTROL library is distributed in a subdirectory called "Matcontrol." This directory must be copied from the media that accompanies the book into anywhere in your system.

After the "Matcontrol" directory has been copied, you just have to let MATLAB know where MATCONTROL is located. In order to do that, you must include it in MATLAB's path.

The easiest way to do so is by including the proper MATLAB commands in your MATLAB startup file (startup.m). If you do not have this file already, please create it.

Using your preferred text editor, open (or create) startup.m and add the following line:

Unix/Linux systems:
matlabpath([matlabpath,'path_of_Matcontrol']);
MS-Windows* systems:
path(path,'path_of_Matcontrol');

Examples: Here, "Mfiles" is the working directory of MATLAB.
On Linux PC:
matlabpath([matlabpath,':/home/carvalho/Mfiles/Matcontrol']);
On Unix-Solaris Workstation:
matlabpath([matlabpath,':/export/home/grad/carvalho/Mfiles/Matcontrol']);
On MS-Windows PC:
path(path,'C:\Mfiles\Matcontrol');

Once you have done that, you can use MATCONTROL in the next MATLAB session. Please issue the command **"help Matcontrol"** to see if MATCONTROL was properly included in MATLAB's path. You should see a list of all MATCONTROL M-files.

*Disclaimer: MATLAB and Windows are trademarks of their respective owners.

B.2   CHAPTERWISE LISTING OF MATCONTROL FILES

Here is the chapterwise listing of MATCONTROL files.

Reference: *Numerical Methods for Linear Control Systems Design and Analysis* by **B.N. Datta**.

Chapter 5:   **Linear State-Space Models and Solutions of the State Equations**

EXPMPADE   The Padé approximation to the exponential of a matrix
EXPMSCHR   Computing the exponential of a matrix using Schur
            decomposition
FREQRESH   Computing the frequency response matrix using Hessenberg
            decomposition
INTMEXP   Computing an integral involving matrix exponentials

Chapter 6:   **Controllability,   Observability,   and   Distance   to Uncontrollability**

CNTRLHS   Finding the controller-Hessenberg form
CNTRLHST   Finding the controller-Hessenberg form with triangular sub-
            diagonal blocks.
OBSERHS   Finding the observer-Hessenberg form

CNTRLC   Finding the controller canonical form (Lower Companion)
DISCNTRL   Distance to controllability using the Wicks–DeCarlo
            algorithm

Chapter 7:   Stability, Inertia, and Robust Stability

INERTIA   Determining the inertia and stability of a matrix without
            solving a matrix equation or computing eigenvalues
H2NRMCG   Finding H_2-norm using the controllability Grammians
H2NRMOG   Finding H_2-norm using the observability Grammian
DISSTABC   Determining the distance to the continuous-time stability
DISSTABD   Determining the distance to the discrete-time stability

Chapter 8:   Numerical Solutions and Conditioning of Lyapunov and
            Sylvester Equations

CONDSYLVC   Finding the condition number of the Sylvester equation
            problem
LYAPCHLC   Finding the Cholesky factor of the positive definite
            solution of the continuous-time Lyapunov equation
LYAPCHLD   Finding the Cholesky factor of the positive definite
            solution of the discrete-time Lyapunov equation
LYAPCSD   Solving the discrete-time Lyapunov equation using
            complex Schur decomposition of A
LYAPFNS   Solving the continuous-time Lyapunov equation via finite
            series method
LYAPHESS   Solving the continuous-time Lyapunov equation via
            Hessenberg decomposition
LYAPRSC   Solving the continuous-time Lyapunov equation via real
            Schur decomposition
LYAPRSD   Solving the discrete-time Lyapunov equation via real
            Schur decomposition
SEPEST   Estimating the *sep* function with triangular matrices
SEPKR   Computing the *sep* function using Kronecker product
SYLVHCSC   Solving the Sylvester equation using Hessenberg
            and complex Schur decompositions
SYLVHCSD   Solving the discrete-time Sylvester equation using
            Hessenberg and complex Schur decompositions
SYLVHESS   Solving the Sylvester equation via Hessenberg
            decomposition

SYLVHRSC   Solving the Sylvester equation using Hessenberg and real
            Schur decompositions
SYLVHUTC   Solving an upper triangular Sylvester equation

Chapter 9:   **Realization and Subspace Identification**

MINRESVD   Finding minimal realization using singular value decom-
            position of the Hankel matrix of Markov parameters
MINREMSVD   Finding minimal realization using singular value
            decomposition of a Hankel matrix of lower order

Chapter 10:   **Feedback Stabilization, Eigenvalue Assignment, and
                   Optimal Control**

STABLYAPC   Feedback stabilization of continuous-time system using
            Lyapunov equation
STABLYAPD   Feedback stabilization of discrete-time system using
            Lyupunov equation
STABRADC   Finding the complex stability radius using the bisection
            method
HINFNRM   Computing H_∞-norm using the bisection method

Chapter 11:   **Numerical Methods and Conditioning of the Eigen-
                   value Assignment Problems**

POLERCS   Single-input pole placement using the recursive algorithm
POLEQRS   Single-input pole placement using the QR version of the
            recursive algorithm
POLERQS   Single-input pole placement using RQ version of the
            recursive algorithm
POLERCM   Multi-input pole placement using the recursive algorithm
POLERCX   Multi-input pole placement using the modified recursive
            algorithm that avoids complex arithmetic and complex
            feedback
POLEQRM   Multi-input pole placement using the explicit QR
            algorithm
POLESCH   Multi-input pole placement using the Schur
            decomposition
POLEROB   Robust pole placement

Chapter 12:   State Estimation: Observer and the Kalman Filter

SYLVOBSC	Solving the constrained multi-output Sylvester-observer equation
SYLVOBSM	Solving the multi-output Sylvester-observer equation
SYLVOBSMB	Block triangular algorithm for the multi-output Sylvester-observer equation

Chapter 13:   Numerical Solutions and Conditioning of the Algebraic Riccati Equations

RICEIGC	The eigenvector method for the continuous-time Riccati equation
RICSCHC	The Schur method for the continuous-time Riccati equation
RICSCHD	The Schur method for the discrete-time Riccati equation
RICGEIGD	The generalized eigenvector method for the discrete-time Riccati equation
RICNWTNC	Newton's method for the continuous-time Riccati equation
RICNWTND	Newton's method for the discrete-time Riccati equatioin
RICSGNC	The matrix sign-function method for the continuous-time Riccati equation
RICSGND	The matrix sign-function method for the discrete-time Riccati equation
RICNWLSC	Newton's method with line search for the continuous-time Riccati equation
RICNWLSD	Newton's method with line search for the discrete-time Riccati equation

Chapter 14:   Internal Balancing and Model Reduction

BALSVD	Internal balancing using the singular value decomposition
BALSQT	Internal balancing using the square-root algorithm
MODREDS	Model reduction using the Schur method
HNAPRX	Hankel-norm approximation

CASE STUDY: CONTROL OF A 9-STATE AMMONIA REACTOR

C.1 INTRODUCTION

In this section, we present the results of a case study on the control of a 9-state Ammonia Reactor taken from the Benchmark Collection (Benner *et al.* 1995; see also Patnaik *et al.* 1980). The dynamics of the system is described by:

$$\dot{x}(t) = Ax(t) + Bu(t)$$
$$y(t) = Cx(t) + Du(t),$$

with system matrices as:

$$
A = \begin{bmatrix}
-4.019 & 5.120 & 0. & 0. & -2.082 & 0. & 0. & 0. & 0.870 \\
-0.346 & 0.986 & 0. & 0. & -2.340 & 0. & 0. & 0. & 0.970 \\
-7.909 & 15.407 & -4.069 & 0. & -6.450 & 0. & 0. & 0. & 2.680 \\
-21.816 & 35.606 & -0.339 & -3.870 & -17.800 & 0. & 0. & 0. & 7.390 \\
-60.196 & 98.188 & -7.907 & 0.340 & -53.008 & 0. & 0. & 0. & 20.400 \\
0. & 0. & 0. & 0. & 94. & -147.200 & 0. & 53.200 & 0. \\
0. & 0. & 0. & 0. & 0. & 94. & -147.200 & 0. & 0. \\
0. & 0. & 0. & 0. & 0. & 12.800 & 0. & -31.600 & 0. \\
0. & 0. & 0. & 0. & 12.800 & 0. & 0. & 18.800 & -31.600
\end{bmatrix},
$$

$$
B = \begin{bmatrix}
0.010 & 0.003 & 0.009 & 0.024 & 0.068 & 0. & 0. & 0. & 0. \\
-0.011 & -0.021 & -0.059 & -0.162 & -0.445 & 0. & 0. & 0. & 0. \\
-0.151 & 0. & 0. & 0. & 0. & 0. & 0. & 0. & 0.
\end{bmatrix}^{\mathrm{T}},
$$

$$
C = \begin{bmatrix}
0. & 0. & 0. & 0. & 0. & 0. & 1. & 0. & 0. \\
0. & 0. & 0. & 0. & 0. & 0. & 0. & 1. & 0. \\
0. & 0. & 0. & 0. & 0. & 0. & 0. & 0. & 1.
\end{bmatrix}, \quad
D = \begin{bmatrix}
0. & 0. & 0. \\
0. & 0. & 0. \\
0. & 0. & 0.
\end{bmatrix}.
$$

The controllability, observability, and the asymptotic stability of the system are determined first using, respectively, the MATCONTROL functions **cntrlhs**, **obserhs**, and the MATLAB function **eig**.

The feedback controller is then designed via **Lyapunov stabilization, pole-placement;** and **LQR** and **LQG techniques.** The MATCONTROL functions **stablyapc** and **polercm**, and the MATLAB functions **lqr** and **lqgreg**, are, respectively, used for this purpose.

The **impulse responses** of the system are then compared in each case.

The states of the system are estimated using the MATLAB function **Kalman** and the MATCONTROL function **sylvobsmb**. The relative errors between the estimated and actual states are then plotted in each case.

Finally, the system is identified using the MATCONTROL functions **minresvd** and **minremsvd** and then the identified model is reduced further by the MATCON-TROL function **modreds**. The frequency response in each case is compared using the MATCONTROL function **freqresh**.

C.2   TESTING THE CONTROLLABILITY

In order to test the controllability of the system, the MATCONTROL function **cntrlhs (see Chapter 6)**, based on the decomposition of the pair (A, B) to a **controller-Hessenberg** form is used.

```
tol = 1e−13;
info = cntrlhs(A, B, tol)
info = 1
```

Conclusion: The system is controllable.

C.3   TESTING THE OBSERVABILITY

The observability is tested by using the MATCONTROL function **obserhs**, based on the decomposition of the pair (A, C) to a **observer-Hessenberg form (See Chapter 6)** with the **same tolerance** as above.

```
info = obserhs(A, C, tol)
info = 1
```

Conclusion: The System is observable.

C.4 TESTING THE STABILITY

In order to test the asymptotic stability of the system, the eigenvalues of the matrix A are computed using the MATLAB function **eig.** They are:

$$\{-147.2000, -153.1189, -56.0425, -37.5446, -15.5478, -4.6610, -3.3013,$$
$$- 3.8592, -0.3047\}.$$

Conclusion: The system is asymptotically stable but it has a small eigenvalue $\lambda = -0.3047$ (relative to the other eigenvalues).

C.5 LYAPUNOV STABILIZATION

The Lyapunov stabilization technique is now used to move the eigenvalues further to the left of the complex plane. The MATCONTROL function **stablyapc** is used for this purpose. This function requires an upper bound β of the spectrum of A, which is taken as the Frobenius norm of A.

```
beta = norm(A, 'fro')
beta = 292.6085
K_lyap = stablyapc(A, B, beta)
```

The feedback matrix $K_{-\text{lyap}}$ is:

$$K_{-\text{lyap}} = 10^2 \begin{bmatrix} -3.3819 & -0.2283 & -56.4126 \\ 5118.1388 & 1207.4106 & 15424.1947 \\ -237858.9775 & -57713.9866 & -997148.5316 \\ -544.7145 & 220.0287 & 15199.8829 \\ 31495.6810 & 7491.5724 & 125946.2030 \\ -4510.0481 & 20403.0258 & -5516.3430 \\ 85.8840 & -495.7330 & 40.1441 \\ -39007.7960 & 182650.2401 & -43969.5212 \\ 38435.8476 & -150078.6710 & 61412.4200 \end{bmatrix}.$$

The eigenvalues of the corresponding closed-loop matrix are:

$$\{-292.6085 \pm 644.6016i, -292.6085 \pm 491.8461i, -292.6085 \pm 145.4054i,$$
$$- 292.6085 \pm 49.3711i, -292.6085\}.$$

Note that these close-loop eigenvalues now are much further to the left of the complex plane than the open-loop ones.

C.6 POLE-PLACEMENT DESIGN

It is now desired to move all the above nine eigenvalues to the negative real-axis with equal spacing in the interval $[-||A||_F/9, -||A||_F)]$. The pole-placement technique is used for this purpose. The MATCONTROL function **polercm**, which implements the recursive multi-input pole-placement algorithm (**Algorithm 11.3.1**) is used to do so.

$$
\begin{aligned}
&\text{eig}_{-\text{rcm}} = -[1 : 9] * \text{beta}/9 \\
&\quad K_{-\text{rcm}} = \text{polercm}(A, B, \text{eig}_{-\text{rcm}})
\end{aligned}
$$

The feedback matrix $K_{-\text{rcm}}$ in this case is:

$K_{-\text{rcm}}$

$$
= 10^5 \begin{bmatrix}
-0.1088 & 14.0002 & -1358.6004 & 17.6295 & 171.8716 & 1.2245 & 0.0034 & 4.8847 & -5.8828 \\
-0.0153 & 2.1371 & -207.6062 & 2.6939 & 26.2618 & 0.1865 & 0.0005 & 0.7408 & -0.8998 \\
-0.0357 & -0.5495 & -74.6029 & 1.3318 & 9.3685 & 0.0670 & 0.0002 & 0.2666 & -0.3206
\end{bmatrix} .
$$

The eigenvalues of the corresponding closed-loop matrix are

$$
\{-292.6085, -260.0965, -227.5844, -195.0724, -162.5603, -130.0482,
$$
$$
-97.5362, -65.0241 \text{ and } -32.5121\}.
$$

C.7 THE LQR AND LQG DESIGNS

Recall (**Chapter 10**) that the LQR design is used to find the optimal control-law

$$
u^0(t) = K_{-\text{lqr}} x(t)
$$

such that the objective functional $J = \int_0^\infty (x^T Q x(t) + u(t) R u(t)) dt$ is minimized subject to $\dot{x} = Ax + Bu$, $x(0) = x_0$. The gain matrix $K_{-\text{lqr}}$ is obtained by solving the CARE: $XA + A^T X - XBR^{-1}B^T X + Q = 0$. The MATLAB function **lqr** with $R = eye(3)$, $N = zeros(9, 3)$, and $Q = eye(9)$ is used for this purpose.

$$
\boxed{K_{-\text{lqr}} = \text{lqr}(A, B, Q, R, N)}
$$

The optimal gain matrix $K_{-\text{lqr}}$ is:

$K_{-\text{lqr}}$

$$
= 10^{-1} \begin{bmatrix}
0.1187 & 0.0728 & 0.0228 & 0.0012 & -0.0007 & 0.0018 & 0.0003 & 0.0042 & 0.0044 \\
0.2443 & -0.3021 & 0.0084 & -0.0465 & -0.0673 & -0.0138 & -0.0023 & -0.0464 & -0.0439 \\
-2.8408 & -0.5942 & -0.4540 & 0.0855 & 0.2102 & 0.0061 & 0.0003 & 0.0496 & 0.0378
\end{bmatrix} .
$$

The eigenvalues of the corresponding closed-loop system are:

$\{-153.1201, -147.1984, -56.0452, -37.5442, -15.5463, -4.6789, -3.3090,$
$-3.8484,$ and $-0.3366\}$.

Note that these closed-loop eigenvalues are quite close to the open-loop ones.
Also, $||K_{-lqr}||$ is much smaller than that of $||K_{-rcm}||$.

To implement the above control law, one needs to have the knowledge of the
state vector $x(t)$; however, in practice only a few of the variables are measured
and the remaining ones need to be estimated. There are several ways to do so (see
Chapter 12 and discussions later here in this section). If the Kalman estimator
K_{-est} is used for this purpose, the design is called LQG (**Linear Quadratic
Gaussian**) design.

The Kalman estimator approximates the state of a stochastic linear system

$$\dot{x}(t) = Ax(t) + Bu(t) + Gw(t) \qquad \text{(state equation)}$$
$$y_m(t) = Cx(t) + Du(t) + Hw(t) + v(t) \qquad \text{(measured equation)}$$

with known inputs $u(t)$, process noise $w(t)$, measurement noise $v(t)$, and noise
covariances

$$Q_n = E[ww^T], \quad R_n = E[vv^T], \quad N_n = E[wv^T]$$

where $E[\cdot]$ denotes the expected value of an stochastic variable. The Kalman
estimator has input $(u; y_m)$ and generates the optimal estimates (y_l, x_l) of (y, x)
given by:

$$\dot{x}_e = Ax_e + Bu + L(y_m - Cx_e - Du)$$
$$\begin{bmatrix} y_l \\ x_l \end{bmatrix} = \begin{bmatrix} C \\ I \end{bmatrix} x_e + \begin{bmatrix} D \\ 0 \end{bmatrix} u$$

where L is the filter gain determined by solving an algebraic Riccati equation(**See
Chapter 12**).

To perform the LQG design, MATLAB functions **kalman** and **lqgreg** are used
as follows:

```
sysA = ss(A, B, C, D);
Qn = 1E−3 * eye(3); Rn = 1E−3 * eye(3);
[K−est, L] = kalman(sysA, Qn, Rn)
```

The filter gain matrix L is:

$$L = 10^{-3} \begin{bmatrix} -0.0007 & 1.0703 & 1.6155 & 1.9729 & 2.8499 & 2.1636 & 1.3816 & 0.7990 & 1.5340 \\ 0.0176 & 0.6284 & 0.9780 & 1.1679 & 1.5765 & 1.2251 & 0.7990 & 0.4962 & 0.9469 \\ 0.0349 & 1.2163 & 1.8921 & 2.2676 & 3.0810 & 2.3701 & 1.5340 & 0.9469 & 1.8112 \end{bmatrix}^T$$

Using the matrices K_{-est} and L, the LQG regulator can now be designed. The MATLAB command for finding an LQG regulator is **lqgreg**.

$$RLQG = \text{lqgreg}(K_{-est}, K_{-lqr})$$

The resulting regulator $RLQG$ has input y_m and the output $u = -K_{-lqr}x_{-e}$ as shown below:

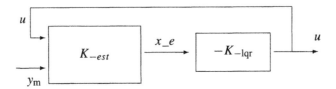

Figure C.1, shows the **impulse response** of thes system in five cases: (i) uncontrolled, (ii) controlled via Lyapunov stabilization, (iii) controlled via pole-placement, (iv) controlled via LQR design, and (v) controlled via LQG design.

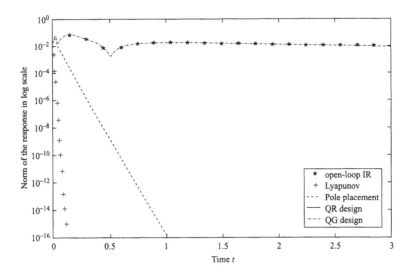

FIGURE C.1:   Comparison of the impulse responses.

Note that the LQR and LQG responses cannot be distinguished from each other. Computations were done in **Simulink 4** with **MATLAB 6**.

C.8   STATE-ESTIMATION (OBSERVER): KALMAN ESTIMATOR VS. SYLVESTER EQUATION ESTIMATOR

Our next goal is to compare two procedures for state estimation of the system: the **Kalman estimator approach** and **the Sylvester-observer approach**.

Recall from **Chapter 12** that the Sylvester-equation approach for state-estimation is based on solving the Sylvester-observer equation: $XA - FX = GC$. The MATCONTROL function **sylvobsmb** which implements Algorithm 12.7.2 (A Recursive Block Triangular Algorithm) for this purpose, is used here. Using the data of the case study and the observer eigenvalues as $ev = [-2, -4 \pm 2i, -5, -6, -7]^T$, we obtain

$$[X, F, G] = \text{sylvobsmb}(A, C, \text{ev});$$

$$X = \begin{bmatrix} -5734.5147 & 5470.8582 & -1106.8206 & 52.2506 & 287.3781 & 0. & 0. & 0. & 727.5156 \\ -8.4146 & 13.4543 & -0.0962 & -0.0139 & -0.5931 & 0. & 0. & 0. & -1.2949 \\ 6.8075 & -9.1950 & 0.8500 & -0.0354 & 0.2121 & 0. & 0. & 0. & 0.5729 \\ 0.5214 & -0.8505 & 0.0685 & -0.0029 & 0.0782 & 0. & 0. & 0. & 0.2327 \\ 0. & 0. & 0. & 0. & 1. & 0. & -0.0213 & 0. & 3.1234 \\ 0. & 0. & 0. & 0. & 0. & 1. & 1.5234 & 0. & -7.1875 \end{bmatrix},$$

$$F = \begin{bmatrix} -2. & 0. & 0. & 0. & 0. & 0. \\ -0.0042 & -5. & 0. & 0. & 0. & 0. \\ 0. & 0.2435 & -6. & 0. & 0. & 0. \\ 0. & 0. & -0.4901 & -7. & 0. & 0. \\ 0. & 0. & 0. & -115.4430 & -4. & -2. \\ 0. & 0. & 0. & 0. & 2. & -4. \end{bmatrix},$$

$$G = 10^1 \begin{bmatrix} 0. & 0. & 0. & 0. & 0.6094 & -21.8109 \\ 1367.7293 & -2.4345 & 1.0771 & 0.4375 & 5.8721 & -8.1925 \\ -1793.4391 & 3.0741 & -1.1006 & -0.4058 & -5.3316 & 19.2128 \end{bmatrix}^T,$$

where F was chosen to be a stable matrix. The error in the solution X: $\|XA - FX - GC\|_F = 1.2246 \cdot 10^{-11}$.

Figure C.2 shows the comparison of relative errors, between actual and estimated states in two cases: **Kalman estimator** and **Sylvester-equation estimator**. The

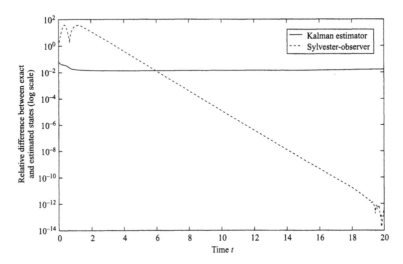

FIGURE C.2: Comparison between Kalman and Sylvester-observer Estimations.

quantity plotted is

$$r(t) = \frac{\|x(t) - \hat{x}(t)\|}{\|x(t)\|}$$

where $\hat{x}(t)$ is the estimate given by the estimator in each case.

The plot shows that error in the Sylvester-observer estimator approaches to zero faster than the Kalman estimator as the time increases.

C.9   SYSTEM IDENTIFICATION AND MODEL REDUCTION

In order to perform system identification tasks, we recall that our system has the transfer function

$$H(s) = C(sI - A)^{-1}B = \sum_{i=1}^{\infty} \frac{CA^i B}{s^i}.$$

The quantities $H_i = CA^i B$, $i = 1, 2, 3, \ldots$ are called the **Markov parameters**; they are usually obtained from input-output experiments. The frequency response is defined by $G(j\omega) = H(j\omega)$ where ω is a nonnegative real number and $j = \sqrt{-1}$.

After directly computing the first 9 Markov parameters, H_i, $i = 1, \ldots, 9$; MAT-CONTROL functions **minresvd** and **minremsvd** are used to perform system identification:

$[A_s, B_s, C_s]$ = minresvd(4,[H1 H2 H3 H4 H5 H6 H7 H8 H9],1e-8)

The resulting model obtained by **minresvd** is oversized. We, therefore, applyl a model reduction technique to this identified oversized model. The MATCONTROL function **modreds** (see **Chapter 14**) is used for this purpose, obtaining a reduced-order model (A_{-r}, B_{-r}, C_{-r}).

$$N = 4; \text{tol} = 1e-13;$$
$$[A_{-s}, B_{-s}, C_{-s}] = \text{minresvd}(N, H_{-i}, \text{tol})$$
$$[A_{-r}, B_{-r}, C_{-r}] = \text{modreds}(A_{-s}, B_{-s}, C_{-s}, 9)$$

The frequency response function **freqresh** from MATCONTROL is then invoked to compute frequency responses in a chosen frequency range for all these models: **the original**, the **model identified by the SVD algorithm(Algorithm 9.3.1)**, the **model identified by the modified SVD algorithm (Algorithm 9.3.2)**, and the **model identified by the SVD algorithm followed by the model reduction technique.**

Let $(A_{-sm}, B_{-sm}, C_{-sm})$ denote the system identified by the function **minremsvd**:

$$\text{omega} = 1:.1:100;$$
$$G = \text{freqresh}(A, B, C, \text{omega})$$
$$G_s = \text{freqresh}(A_s, B_s, C_s, \text{omega})$$
$$G_{sm} = \text{freqresh}(A_{sm}, B_{sm}, C_{sm}, \text{omega})$$
$$G_r = \text{freqresh}(A_r, B_r, C_r, \text{omega})$$

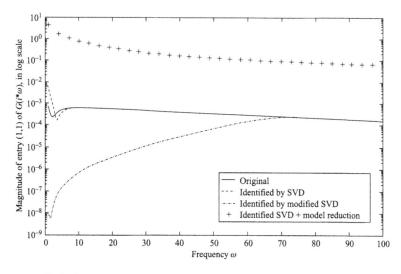

FIGURE C.3: Comparison between Frequency Responses.

Figure C.3 shows a comparison between the magnitude of the entry (1,1) of the original frequency response G, the frequency response G_s of the system identified by the SVD method, the frequency response G_{sm} of the system identified by the modified SVD method, and the frequency response G_{-r} of the system (A_{-r}, B_{-r}, C_{-r}), which is obtained from (A_{-s}, B_{-s}, C_{-s}) followed by model reduction.

References

Benner, P., Laub, A., and Mehrmann, V. *A Collection of benchmark examples for the numerical solution of algebraic Riccati equations I: continuous-time case. Technische Universität Chemnitz-Zwickau*, SPC Report 95–22, 1995.

Patnaik, L., Viswanadham, N., and Sarma, I. *Computer control algorithms for a tubular ammonia reactor*. IEEE Trans. Automat. Control, AC-25, pp. 642–651, 1980.

INDEX

Printed and bound by CPI Group (UK) Ltd, Croydon, CR0 4YY

03/10/2024

01040418-0014